MATHEMATICAL PAPERS.

London: C. J. CLAY and SONS,
CAMBRIDGE UNIVERSITY PRESS WAREHOUSE,
AVE MARIA LANE.
Glasgow: 263, ARGYLE STREET.

Leipzig: F. A. BROCKHAUS.
New York: MACMILLAN AND CO.

THE COLLECTED

MATHEMATICAL PAPERS

OF

ARTHUR CAYLEY, Sc.D., F.R.S.,

LATE SADLERIAN PROFESSOR OF PURE MATHEMATICS IN THE UNIVERSITY OF CAMBRIDGE.

VOL. X.

CAMBRIDGE:
AT THE UNIVERSITY PRESS.
1896

CAMBRIDGE:

PRINTED BY J. AND C. F. CLAY,

AT THE UNIVERSITY PRESS.

ADVERTISEMENT.

THE present volume contains 76 papers, numbered 630 to 705, published for the most part in the years 1876 to 1880.

The Table for the ten volumes is

Vol.	I.	Numbers	1	to	100,
,,	II.	,,	101	,,	158,
,,	III.	,,	159	,,	222,
,,	IV.	,,	223	,,	299,
,,	V.	,,	300	,,	383,
,,	VI.	,,	384	,,	416,
,,	VII.	,,	417	,,	485,
,,	VIII.	,,	486	,,	555,
,,	IX.	,,	556	,,	629,
,,	X.	,,	630	,,	705.

A. R. FORSYTH.

11 *June*, 1896.

CONTENTS.

[An Asterisk means that the paper is not printed in full.]

CLASSIFICATION.

630.

ON AN EXPRESSION FOR $1 \pm \sin(2p+1)u$ IN TERMS OF $\sin u$.

[From the *Messenger of Mathematics*, vol. v. (1876), pp. 7, 8.]

WRITE $\sin u = x$, then we have

$$\begin{array}{ll} \sin u = x, & \cos u = \sqrt{(1-x^2)}, \\ \sin 3u = 3x - 4x^3, & \cos 3u = (1-4x^2)\sqrt{(1-x^2)}, \\ \sin 5u = 5x - 20x^3 + 16x^5, & \cos 5u = (1-12x^2+16x^4)\sqrt{(1-x^2)}, \\ \&c. & \&c. \end{array}$$

It is hence clear, that in general

$$1 - \sin(2p+1)u = (1 \pm x)\{(1,\ \ x)^p\}^2,$$
$$1 + \sin(2p+1)u = (1 \mp x)\{(1,\ -x)^p\}^2,$$

where $(1,\ x)^p$ denotes a rational and integral function of x of the order p, and $(1,\ -x)^p$ the same function of $-x$; for it is only in this manner that we can have

$$\cos^2(2p+1)u = (1-x^2)\{[1,\ x^2]^p\}^2.$$

We, in fact, find

$$\begin{aligned} 1 + \sin u &= 1 + x, \\ 1 - \sin 3u &= (1+x)(1-2x)^2, \\ 1 + \sin 5u &= (1+x)(1+2x-4x^2)^2, \\ 1 - \sin 7u &= (1+x)(1-4x-4x^2+8x^3)^2, \\ &\&c. \end{aligned}$$

and it thus appears that the form is

$$1 + (-)^p \sin(2p+1)u = (1+x)\{(1,\ x)^p\}^2.$$

To find herein the expression of the factor $(1, x)^p$, write $u = \frac{1}{2}\pi - \theta$ and consequently $x = \cos\theta$; we have therefore

$$1 + \cos(2p+1)\theta = (1+\cos\theta)\{(1, x)^p\}^2,$$

where in the second factor on the right-hand side x is retained to stand for its value $\cos\theta$. This gives

$$2\cos^2(p+\tfrac{1}{2})\theta = 2\cos^2\tfrac{1}{2}\theta\{(1, x)^p\}^2,$$

or, what is the same thing,

$$(1, x)^p = \frac{\cos(p+\frac{1}{2})\theta}{\cos\frac{1}{2}\theta},$$

viz. this is

$$= \cos p\theta - \sin p\theta \frac{\sin\frac{1}{2}\theta}{\cos\frac{1}{2}\theta},$$

which is

$$= \cos p\theta - \sin p\theta \frac{1-\cos\theta}{\sin\theta}.$$

We have

$$\begin{aligned}\cos p\theta + i\sin p\theta &= \{x + i\sqrt{(1-x^2)}\}^p \\ &= X + i\sqrt{(1-x^2)}\, Y, \text{ suppose,}\end{aligned}$$

where X, Y are rational and integral functions of x of the orders p and $p-1$ respectively; that is,

$$\cos p\theta = X, \quad \sin p\theta = \sin\theta \,.\, Y,$$

and we have therefore

$$(1, x)^p = X - Y(1-x),$$

which is the required expression for $(1, x)^p$. For instance

$$p = 3, \quad X + i\sqrt{(1-x^2)}\, Y = \{x + i\sqrt{(1-x^2)}\}^3;$$

that is,

$$\begin{aligned} X &= \quad -3x \qquad + 4x^3 \\ Y = -1 + 4x^2, \text{ and } \therefore \quad -(1-x)Y &= 1 - x - 4x^2 + 4x^3 \\ \hline \text{so that} \qquad X - (1-x)Y &= 1 - 4x - 4x^2 + 8x^3, = (1, x)^3,\end{aligned}$$

and hence

$$1 - \sin 7u = (1+x)(1 - 4x - 4x^2 + 8x^3)^2,$$

which agrees with a result already obtained.

The foregoing value of $(1, x)^p$ may also be written

$$(1, x)^p = \frac{1}{\sin\theta}\{\sin(p+1)\theta - \sin p\theta\},$$

which however is not practically so convenient.

The formula corresponds to a like formula in elliptic functions, viz. writing $\text{sinam}\, u = x$, the numerator of $1 + (-)^p \,\text{sinam}\,(2p+1)u$ is

$$= (1+x)\{(1, x)^{2p(p+1)}\}^2,$$

which is $(1+x)$ multiplied by the square of a rational and integral function of x.

631.

SYNOPSIS OF THE THEORY OF EQUATIONS.

[From the *Messenger of Mathematics*, vol. v. (1876), pp. 39—49.]

THE following was proposed as one of the subjects of a Dissertation for the Trinity Fellowships:

Synopsis of the theory of equations; i.e. a statement in a logical order, of the divisions of the subject and the leading questions and theorems, but without demonstrations.

In the subject "Theory of Equations," the term equation is used to denote an equation of the form $x^n - p_1x^{n-1} + \dots \pm p_n = 0$, where $p_1, p_2, \dots, p_n$ are regarded as known, and x as a quantity to be determined; for shortness, the equation is written $f(x) = 0$.

The equation may be *numerical;* that is, the coefficients $p_1, p_2, \dots, p_n$ are then numbers; understanding by number, a quantity of the form $\alpha + \beta i$, where α and β have any positive or negative real values whatever; or say, each of these is regarded as susceptible of continuous variation from an indefinitely large negative to an indefinitely large positive value: and i denotes $\sqrt{(-1)}$.

Or the equation may be *algebraic;* viz. the coefficients are then not restricted to denote, or are not explicitly considered as denoting, numbers.

I. We consider first numerical equations.

A number a (real or imaginary), such that substituted for x it makes the function $x^n - p_1x^{n-1} + \dots \pm p_n$ to be $= 0$, or say, such that it satisfies the equation, is said to be a root of the equation; viz. a being a root, we have

$$a^n - p_1a^{n-1} + \dots \pm p_n = 0, \text{ or say } f(a) = 0;$$

and it is then shown that $x - a$ is a factor of the function $f(x)$, viz. that we have $f(x) = (x - a) f_1(x)$, where $f_1(x)$ is a function $x^{n-1} - q_1x^{n-2} + \dots \pm q_{n-1}$, of the order $n - 1$, with numerical coefficients $q_1, q_2, \dots, q_{n-1}$.

In general, a is not a root of the equation $f_1(x)=0$; but it may be so, viz. $f_1(x)$ may contain the factor $x-a$; when this is so, $f(x)$ will contain the factor $(x-a)^2$; writing then $f(x)=(x-a)^2 f_2(x)$, and assuming that a is not a root of the equation $f_2(x)=0$, $x=a$ is then said to be a double root of the equation. Similarly, $f(x)$ may contain the factor $(x-a)^3$ and no higher power, and then $x=a$ is said to be a triple root; and so on.

Supposing, in general, that $f(x)=(x-a)^\alpha F(x)$, where α is a positive integer which may be $=1$, and Fx is of the order $n-\alpha$, then if b is a root different from a, we shall have $x-b$ a factor (in general a simple one, but it may be a multiple one) of $F(x)$, and $f(x)$ will in this case become $=(x-a)^\alpha(x-b)^\beta \Phi(x)$, where β is a positive integer which may be $=1$, and Φx is of the order $n-\alpha-\beta$. The original equation $fx=0$ is in this case said to have α roots each $=a$, β roots each $=b$, and so on.

We have the *theorem*, a numerical equation of the order n has in every case n roots, viz. there exist n numbers $a, b, \ldots$ (in general, all of them distinct, but they may arrange themselves in groups of equal values) such that

$$f(x)=(x-a)(x-b)(x-c)\ldots \text{ identically.}$$

If an equation has equal roots, these can in general be determined; the case is at any rate a special one, which may be here omitted from consideration. It is therefore, in general, assumed that the equation $f(x)=0$ under consideration has all its roots unequal. If the coefficients $p_1, p_2, \ldots$ are all or any one or more of them imaginary, then the equation $f(x)=0$, separating the real and imaginary parts, may be written $F(x)+i\Phi(x)=0$, where $F(x)$, $\Phi(x)$ are each of them a function with real coefficients; and it thus appears that the equation $f(x)=0$ with imaginary coefficients has not in general any real root; supposing it to have a real root a, this must be at once a root of each of the equations $F(x)=0$ and $\Phi(x)=0$.

But an equation with real coefficients may have as well imaginary as real roots; and we have further the *theorem* that for such an equation the imaginary roots enter in pairs, viz. $\alpha+\beta i$ being a root, then will also $\alpha-\beta i$ be a root.

Considering an equation with real coefficients, the question arises as to the number and situation of its real roots; this is completely resolved by means of *Sturm's theorem*, viz. we form a series of functions $f(x)$, $f'(x)$, $f_2(x), \ldots, f_n(x)$ (a constant) of the degrees $n, n-1, \ldots, 2, 1, 0$ respectively; and substituting therein for x any two real values a and b, we find by means of the resulting signs of these functions how many real roots of $f(x)$ lie between the limits a, b.

The same thing can frequently be effected with greater facility by other means, but the only general method is the one just referred to.

In the general case of an equation with imaginary (it may be real) coefficients, the like question arises as to the situation of the (real or imaginary) roots, viz. if for facility of conception we regard the constituents α, β of a root $\alpha+\beta i$ as the coordinates of a point *in plano*, and accordingly represent the root by such point; then drawing in the plane any closed curve or "contour," the question is how many roots lie within such contour.

This is solved *theoretically* by means of a theorem of Cauchy's, viz. writing in the original equation $x+iy$ in place of x, the function $f(x+iy)$ becomes $=P+iQ$, where P and Q are each of them a rational and integral function (with real coefficients) of (x, y). Imagining the point (x, y) to travel along the contour, and considering the number of changes of sign from $-$ to $+$ and from $+$ to $-$ of the fraction $\frac{P}{Q}$ corresponding to passages of the fraction through zero (that is, to values for which P becomes $=0$, disregarding those for which Q becomes $=0$), the difference of these numbers determines the number of roots within the contour. The investigation leads to a proof of the before-mentioned theorem, that a numerical equation of the order n has precisely n roots.

But, for the actual determination, it is necessary to consider a rectangular contour, and to apply to each of its sides separately a method such as that of Sturm's theorem; and thus the actual determination ultimately depends on a method such as that of Sturm's theorem.

Recurring to the case of an equation with real coefficients, it is important to *separate* the real roots, viz. to determine limits, such that each real root lies alone by itself between two limits l and m. This can be done (with more or less difficulty according to the nearness of the real roots) by repeated applications of Sturm's theorem, or otherwise.

The same thing would be useful, and can theoretically be effected, in regard to the roots of an equation generally, viz. we may, by lines parallel to the axes of x and y respectively, divide the plane into rectangles such that each (real or imaginary) root lies alone by itself in a given rectangle; but the ulterior theory, even as regards the imaginary roots of an equation with real coefficients, has not been developed, and the remarks which immediately follow have reference only to equations with real coefficients, and to the real roots of such equations.

Supposing the roots separated as above, so that a certain root is known to lie alone by itself between two given limits, then it is possible by various processes (Horner's, or Lagrange's method of continued fractions) to obtain to any degree of approximation the numerical value of the real root in question, and thus to obtain (approximately as above) the values of the several real roots.

The real roots can also frequently be obtained, without the necessity of a previous separation of the roots, by other processes of approximation—Newton's, as completed by Fourier, or by a method given by Encke—and the problem of their determination to any degree of approximation may be regarded as completely solved. But this is far from being practically the case even as regards the imaginary roots of such equations, or as regards the roots of an equation with imaginary coefficients.

A class of numerical equations which need to be considered, are the binomial equations $x^n-a=0$, where a, $=\alpha+\beta i$, is a complex number. The foregoing conclusions apply, viz. there are always n roots, which it may be shown are all unequal. Supposing one of these is θ, so that $\theta^n = a$, then, assuming $x = \theta y$, we have $y^n - 1 = 0$, which equation (like the more general one $x^n - a = 0$) has precisely n roots; it is shown that these are $1, \omega, \omega^2, .., \omega^{n-1}$, where ω is a complex number $\alpha + \beta i$ such that $\alpha^2 + \beta^2 = 1$, or, what is the same thing, a complex number of the form $\cos\theta + i\sin\theta$; and it then at once appears that θ may be taken $= \frac{2\pi}{n}$. We have thus the trigonometrical solution of the equation $x^n - 1 = 0$. We may also obtain a like trigonometrical solution of the first-mentioned equation $x^n - a = 0$. We are thus led to the notion (a numerical) of the radical $a^{\frac{1}{n}}$, regarded as an n-valued function, viz. any one of these being denoted by $\sqrt[n]{(a)}$, then the series of values is

$$\sqrt[n]{(a)},\ \omega\sqrt[n]{(a)}, .., \ \omega^{n-1}\sqrt[n]{(a)}.$$

Or we may, if we please, use $\sqrt[n]{(a)}$, instead of $a^{\frac{1}{n}}$, as a symbol to denote the n-valued function.

It is not necessary, as regards the equation $x^n - 1 = 0$, to refer here to the distinctions between the cases n a prime, and a composite, number.

As the coefficients of an algebraical equation may be numerical, all which follows in regard to algebraical equations, is (with, it may be, some few modifications) applicable to numerical equations; and hence, concluding for the present this subject, it will be convenient to pass on to algebraical equations.

II. We consider, secondly, an algebraical equation

$$x^n - p_1 x^{n-1} + \ldots = 0,$$

and we here *assume* the existence of roots, viz. we assume that there are n quantities $a, b, c, \ldots$ (in general, all of them different, but in particular cases they may become equal in sets in any manner), such that

$$x^n - p_1 x^{n-1} + \ldots = (x-a)(x-b)\ldots.$$

Or, looking at the question in a different point of view, and starting with the roots $a, b, c, \ldots$ as given, we express the product of the n factors $x-a$, $x-b, \ldots$ in the foregoing form, and thus arrive at an equation of the order n having the n roots $a, b, c, \ldots$. In either case, we have

$$p_1 = \Sigma a,\ p_2 = \Sigma ab, .., \ p_n = abc\ldots,$$

viz. regarding the coefficients $p_1, p_2, .., p_n$ as given, then we assume the existence of roots $a, b, c, \ldots$ such that $p_1 = \Sigma a$, &c., or regarding the roots as given, then we write p_1, p_2, &c., to denote the functions Σa, Σab, &c.

It is to be noticed that, in virtue of

$$x^n - p_1 x^{n-1} + \ldots = (x-a)(x-b), \text{ \&c.},$$

or of the equivalent equations $p_1 = \Sigma a$, &c., then

$$a^n - p_1 a^{n-1} + \ldots = 0,$$
$$b^n - p_1 b^{n-1} + \ldots = 0,$$
&c.,

(viz. it is for this reason that $a, b, \ldots$ are said to be roots of $x^n - p_1 x^{n-1} + \ldots = 0$); and, moreover, that conversely from the last-mentioned equations, assuming that $a, b, \ldots$ are all different, we deduce

$$p_1 = \Sigma a,\ p_2 = \Sigma ab,\ \&c.,$$

and

$$x^n - p_1 x^{n-1} + \ldots = (x-a)(x-b)\ldots.$$

Observe that, if for instance $a = b$, then the two equations $a^n - p_1 a^{n-1} + \ldots = 0$, $b^n - p_1 b^{n-1} + \ldots = 0$ would reduce themselves to a single equation, which would not of itself express that a was a double root, that is, that $(x-a)^2$ was a factor of $x^n - p_1 x^{n-1} + \&c.$; but by considering b as the limit of $a + h$, h indefinitely small, we obtain a second equation

$$na^{n-1} - (n-1)p_1 a^{n-2} + \ldots = 0,$$

which, with the first, expresses that a is a double root; and then the whole system of equations leads, as before, to the equations $p_1 = \Sigma a$, &c. But this in passing: the general case is when the roots are all unequal.

We have then the *theorem* that every rational symmetrical function of the roots is a rational function of the coefficients; this is an easy consequence from the less general theorem, every rational and integral symmetrical function of the roots is a rational and integral function of the coefficients.

In particular, the sums of powers Σa^2, Σa^3, &c., are rational and integral functions of the coefficients.

An ordinary process, as regards the expression of other functions $\Sigma a^\alpha b^\beta$, &c., in terms of the coefficients, is to make them depend on the functions Σa^α, &c., but this is *very objectionable;* the true theory consists in showing that we have systems of equations

$$p_1 = \Sigma a,$$

$$\begin{cases} p_2 = \Sigma ab, \\ p_1^2 = \Sigma a^2 + 2\Sigma ab, \end{cases}$$

$$\begin{cases} p_3 = \Sigma abc, \\ p_1 p_2 = \Sigma a^2 b + 3\Sigma abc, \\ p_1^3 = \Sigma a^3 + 3\Sigma a^2 b + 6\Sigma abc, \end{cases}$$

&c., &c.

where, in each system, there are precisely as many equations as there are root-functions on the right-hand side, e.g. 3 equations and 3 functions Σabc, $\Sigma a^2 b$, Σa^3. Hence, in each system, the root-functions can be determined linearly in terms of the powers and products of the coefficients.

It follows that it is possible to determine an equation (of an assignable order) having for roots any given (unsymmetrical) functions of the roots of a given equation. For example, in the case of a quartic equation, roots (a, b, c, d), it is possible to find an equation having the roots ab, ac, ad, bc, bd, cd, being therefore a sextic equation; viz. in the product $(y-ab)(y-ac)(y-ad)(y-bc)(y-bd)(y-cd)$, the coefficients of the several powers of y will be symmetrical functions of a, b, c, d, and therefore rational and integral functions of the coefficients of the original quartic equation.

In connexion herewith, the question arises as to the number of values (obtained by permutations of the roots) of given unsymmetrical functions of the roots; for instance, with roots (a, b, c, d) as before, how many values are there of the function $ab+cd$; or, better, how many functions are there of this form; the answer is 3, viz. $ab+cd$, $ac+bd$, $ad+bc$; or, again, we may ask whether it is possible to obtain functions of a given number of values, 3-valued, 4-valued functions, &c.

We have, moreover, the very important *theorem* that, given the value of any unsymmetrical function, e.g. $ab+cd$, it is in general possible to determine rationally the value of any similar function, e.g. $(a+b)^3+(c+d)^3$.

The *à priori* ground of this theorem may be illustrated by means of a numerical equation. Suppose, e.g. that the roots of a quartic equation are 1, 2, 3, 4; then if it is given that $ab+cd=14$, this in effect determines a, b to be 1, 2 (viz. $a=1$, $b=2$, or else $a=2$, $b=1$) and c, d to be 3, 4 (viz. $c=3$, $d=4$, or else $c=4$, $d=3$); and it therefore in effect determines $(a+b)^3+(c+d)^3$ to be $=370$, and not any other value. And we can in the same way account for cases of failure as regards particular equations; thus, the roots being 1, 2, 3, 4, as above, $a^2b=2$ determines a to be $=1$ and b to be $=2$; but if the roots had been 1, 2, 4, 16, then $a^2b=16$ does not uniquely determine a and b, but only makes them to be 1 and 16, or else 2 and 4, respectively.

As to the *à posteriori* proof, assume, for instance, $t_1=ab+cd$, $y_1=(a+b)^3+(c+d)^3$, and so $t_2=ac+db$, $y_2=(a+c)^3+(d+b)^3$, &c.—in the present case there are only the functions t_1, t_2, t_3 and y_1, y_2, y_3—then $y_1+y_2+y_3$, $t_1y_1+t_2y_2+t_3y_3$, $t_1^2y_1+t_2^2y_2+t_3^2y_3$ will be respectively symmetrical functions of the roots of the quartic, and therefore rational and integral functions of its coefficients, that is, they will be known.

Imagine, in the first instance, that t_1, t_2, t_3 are all known; then the equations being linear in y_1, y_2, y_3, these can be expressed rationally in terms of known functions of the coefficients and of t_1, t_2, t_3, that is, y_1, y_2, y_3 will be known. But observe further, that y_1 is obtained as a function of t_1, t_2, t_3 symmetrical as regards t_2, t_3; it can consequently be expressed as a rational function of t_1 and of t_2+t_3, t_2t_3, or, what is the same thing, of t_1 and $t_1+t_2+t_3$, $t_1t_2+t_1t_3+t_2t_3$, $t_1t_2t_3$; but these last will be symmetrical functions of the roots, and as such expressible rationally in terms of the coefficients: that is, y_1 will be expressed as a rational function of t_1 and of the coefficients, or, t_1 being known, y_1 will be rationally determined.

We may consider now the question of the algebraical solution of equations, or, more accurately, that of the *solution of equations by radicals.*

In the case of a quadric equation $x^2+px+q=0$, we can find for x, by the assistance of the sign $\sqrt{(\ )}$ or $(\ )^{\frac{1}{2}}$, an expression for x as a two-valued function of the coefficients p, q, such that, substituting this value in the equation, the equation is thereby identically satisfied, viz. we have

$$x = -\tfrac{1}{2}p \pm \sqrt{(\tfrac{1}{4}p^2 - q)},$$

giving

$$\begin{aligned} x^2 &= \ \ \tfrac{1}{2}p^2 - q \mp p\sqrt{(\tfrac{1}{4}p^2-q)} \\ +px &= -\tfrac{1}{2}p^2 \qquad \pm p\sqrt{(\tfrac{1}{4}p^2-q)} \\ +q &= \qquad +q \\ \hline x^2+px+q &= 0, \end{aligned}$$

and the equation is on this account said to be algebraically solvable, or, more accurately, to be *solvable by radicals.* Or we may, by writing $x=-\frac{1}{2}p+z$, reduce the equation to $z^2=\frac{1}{4}p^2-q$, viz. to an equation of the form $z^2=a$, and, in virtue of its being thus reducible, we may say that the equation is solvable by radicals. And the question for an equation of any higher order is, say of the order n, can we by means of radicals, that is, by aid of the sign $\sqrt[m]{(\ )}$ or $(\ )^{\frac{1}{m}}$, using as many as we please of such signs and with any values of m, find an n-valued function (or any function) of the coefficients, which substituted for x in the equation shall satisfy it identically.

It will be observed that the coefficients $p, q, \ldots$ are not explicitly considered as numbers, but that even if they do denote numbers, the question whether a numerical equation admits of solution by radicals is wholly unconnected with the before-mentioned theorem of the existence of the n roots of such an equation. It does not even follow that, in the case of a numerical equation solvable by radicals, the algebraical expression of x gives the numerical solution; but this requires explanation. Consider, first, a numerical quadric equation with imaginary coefficients; in the formula $x=-\frac{1}{2}p\pm\sqrt{(\frac{1}{4}p^2-q)}$, substituting for p, q their given numerical values we obtain for x an expression of the form $x=\alpha+\beta i\pm\sqrt{(\gamma+\delta i)}$, where α, β, γ, δ are real numbers; this value substituted in the numerical equation would satisfy it identically and it is thus an algebraical solution; but there is no obvious *à priori* reason why the expression $\sqrt{(\gamma+\delta i)}$ should have a value $=c+di$, where c and d are real numbers calculable by the extraction of a root or roots of real numbers; it appears upon investigation that $\sqrt{(\gamma+\delta i)}$ has such a value calculable by means of the radical expression $\sqrt{\{\sqrt{(\gamma^2+\delta^2)}\pm\gamma\}}$; and hence that the algebraical solution of a quadric equation does in every case give the numerical solution of a numerical quadric. The case of a numerical cubic will be considered presently.

A cubic equation can be solved by radicals, viz. taking for greater simplicity the cubic in the reduced form $x^3-qx-r=0$, and writing $x=a+b$, this will be a solution if only $3ab=q$, and $a^3+b^3=r$, or say $\frac{1}{2}(a^3+b^3)=\frac{1}{2}r$; whence

$$\tfrac{1}{2}(a^3-b^3) = \pm\sqrt{(\tfrac{1}{4}r^2 - \tfrac{1}{27}q^3)},$$

and therefore

$$a = \sqrt[3]{\{\tfrac{1}{2}r \pm \sqrt{(\tfrac{1}{4}r^2 - \tfrac{1}{27}q^3)}\}},$$

a six-valued function of q, r. But then writing $b = \dfrac{q}{3a}$, we have, as may be shown, $a+b$ a three-valued function of the coefficients; it would have been wrong to complete the solution by writing $b = \sqrt[3]{\{\tfrac{1}{2}r \pm \sqrt{(\tfrac{1}{4}r^2 - \tfrac{1}{27}q^3)}\}}$, since here $(a+b)$ would be given as a 9-valued function, having only 3 of its values roots, and the other 6 values being irrelevant. An interesting variation of the solution is to write $x = ab(a+b)$, giving $a^3b^3(a^3+b^3) = r$ and $3a^3b^3 = q$, or say $\tfrac{1}{2}(a^3+b^3) = \tfrac{3}{2}\dfrac{r}{q}$, $a^3b^3 = \tfrac{1}{3}q$; whence

$$\{\tfrac{1}{2}(a^3 - b^3)\}^2 = \frac{9}{q^2}(\tfrac{1}{4}r^2 - \tfrac{1}{27}q^3),$$

and therefore

$$a = \sqrt[3]{\left\{\tfrac{3}{2}\frac{r}{q} \pm \frac{3}{q}\sqrt{(\tfrac{1}{4}r^2 - \tfrac{1}{27}q^3)}\right\}},\quad b = \sqrt[3]{\left\{\tfrac{3}{2}\frac{r}{q} \mp \frac{3}{q}\sqrt{(\tfrac{1}{4}r^2 - \tfrac{1}{27}q^3)}\right\}},$$

and here although a, b are each of them a 6-valued function, yet, as may be shown, $ab(a+b)$ is only a 3-valued function.

In the case of a numerical cubic, even when the coefficients are real, substituting their values in the expression

$$x = \sqrt[3]{\{\tfrac{1}{2}r \pm \sqrt{(\tfrac{1}{4}r^2 - \tfrac{1}{27}q^3)}\}} + [\tfrac{1}{3}q \div \sqrt[3]{\{\tfrac{1}{2}r \pm \sqrt{(\tfrac{1}{4}r^2 - \tfrac{1}{27}q^3)}\}}],$$

this *may* depend on an expression of the form $\sqrt[3]{(\gamma + \delta i)}$, where γ and δ are real numbers (viz. it will do so if $\tfrac{1}{4}r^2 - \tfrac{1}{27}q^3$ is a negative number), and here we *cannot* by the extraction of any root or roots of real numbers reduce $\sqrt[3]{(\gamma+\delta i)}$ to the form $c + di$, c and d real numbers; hence, here the algebraical solution does not give the numerical solution. It is to be added that the case in question, called the "irreducible case," is that wherein the three roots of the cubic equation are all real; if the roots are one real and two imaginary, then, contrariwise, the quantity under the cube root is real, and the algebraical solution gives the numerical one.

The irreducible case is solvable by a trigonometrical formula, but this is not a solution by radicals; it consists, in effect, in reducing the given numerical cubic (not to a cubic of the form $z^3 = a$, solvable by the extraction of a cube root, but) to a cubic of the form $4x^3 - 3x = a$, corresponding to the equation $4\cos^3\theta - 3\cos\theta = \cos 3\theta$ which serves to determine $\cos\theta$ when $\cos 3\theta$ is known.

A quartic equation is solvable by radicals; and it may be remarked, that the existence of such a solution depends on the existence of 3-valued functions such as $ab + cd$, of the four roots (a, b, c, d); by what precedes, $ab + cd$ is the root of a cubic equation, which equation is solvable by radicals; hence $ab + cd$ can be found by radicals; and since $abcd$ is a given value, ab and cd can each be found by radicals. But by what precedes, if ab be known, then any similar function, say $a + b$, is obtainable rationally; and, consequently, from the values of $a + b$ and ab we may by radicals obtain the value of a or b, that is, an expression for a root of the given quartic expression; the expression finally obtained is 4-valued, corresponding to the different values of the several radicals which enter therein, and we have therefore the expression by radicals of each of the four roots of the quartic equation. But when the quartic is numerical, the same thing arises as in the cubic: the algebraical expression does not in every case give the numerical one.

It will be understood from the foregoing explanation as to the quartic, how in the next following case, that of a quintic equation, the question of the solvability by radicals depends on the existence or non-existence of k-valued functions of the five roots (a, b, c, d, e); a fundamental theorem on the subject is that a rational function of 5 letters, if it has less than 5, cannot have more than 2 values; viz. that there are no 3-valued, or 4-valued, functions of 5 letters; and by reasoning, depending in part upon this theorem, Abel showed that a general quintic equation is not solvable by radicals: and *à fortiori* the general equation of any order higher than 5 is not solvable by radicals.

The general theory of the solvability of an equation by radicals depends very much on Vandermonde's remark, that supposing an equation is solvable (by radicals) and that we have therefore an algebraical expression of x in terms of the coefficients, then substituting for the coefficients their values in terms of the roots, the resulting value of the expression must reduce itself to any one at pleasure of the róots $a, b, c, \ldots$; thus in the case of the quadric equation where the solution is $x = +\frac{1}{2}p \pm \sqrt{(\frac{1}{4}p^2 - q)}$, writing for p, q their values $a+b$, ab, this is $x = \frac{1}{2}[(a+b) \pm \sqrt{\{(a-b)^2\}}]$, $=a$ or b according to the value of the radical. But it is not considered necessary in the present sketch to go further into the theory of the solvability of an equation by radicals. It may be proper to remark that, for quintic equations, there are solutions analogous to the trigonometrical solution of a cubic equation, viz. the quintic equation is here in effect reduced to some special form of quintic equation; for instance, to Jerrard's form $x^5 + ax + b = 0$ or to some form presenting itself in the theory of elliptic functions; but the solutions in question are not solutions by radicals. And there are various other interesting parts of the theory which have been excluded from consideration.

632.

ON ARONHOLD'S INTEGRATION-FORMULA.

[From the *Messenger of Mathematics*, vol. V. (1876), pp. 88—90.]

THE fundamental theorem in Aronhold's Memoir, "Ueber eine neue algebraische Behandlungsweise der Integrale...$\Pi(x, y)\,dx$, &c.," *Crelle*, t. LXI. (1863), pp. 95—145, is a theorem of *indefinite* integration. The form is

$$\Lambda\int\frac{dx}{(\alpha x+\beta y+\gamma)(hx+by+f)}=\log\frac{(a\xi+h\eta+g)\,x+(h\xi+b\eta+f)\,y+g\xi+f\eta+c}{\alpha x+\beta y+\gamma},$$

where y is a certain irrational function of x, determined by a quadric equation, and the other symbols denote constants connected by certain relations; viz. writing, for shortness,

$$U=(a,\ b,\ c,\ f,\ g,\ h\between x,\ y,\ 1)^2,\quad =(a,\ldots\between x,\ y,\ 1)^2 \text{ for shortness,}$$

that is,

$$=ax^2+2hxy+by^2+2fy+2gx+c\,;$$

$$W=(a,\ b,\ c,\ f,\ g,\ h\between x,\ y,\ 1\between \xi,\ \eta,\ 1),\quad =(a,\ldots\between x,\ y,\ 1\between \xi,\ \eta,\ 1),$$

that is,

$$=(ax+hy+g)\,\xi+(hx+by+f)\,\eta+gx+fy+c,$$

or

$$(a\xi+h\eta+g)\,x+(h\xi+b\eta+f)\,y+g\xi+f\eta+c\,;$$

$$(P,\ Q,\ R)=(ax+hy+g,\quad hx+by+f,\quad gx+fy+c),$$

$$(P_0,\ Q_0,\ R_0)=(a\xi+h\eta+g,\quad h\xi+b\eta+f,\quad g\xi+f\eta+c),$$

$$\Omega=\alpha x+\beta y+\gamma,$$

$$\Omega_0=\alpha\xi+\beta\eta+\gamma,$$

$$(A,\ B,\ C,\ F,\ G,\ H)=(bc-f^2,\ ca-g^2,\ ab-h^2,\ gh-af,\ hf-bg,\ fg-ch),$$

then y is determined as a function of x by the equation $U=0$, that is,

$$(a,\ b,\ c,\ f,\ g,\ h\between x,\ y,\ 1)^2=0;$$

or, what is the same thing,

$$by=-\{hx+f+\sqrt{(-Cx^2+2Gx-A)}\};$$

the constants α, β, ξ, η are such that

$$(a,\ b,\ c,\ f,\ g,\ h\between \xi,\ \eta,\ 1)^2=0,$$

$$\alpha\xi+\beta\eta+\gamma=0,$$

that is,

$$\Omega_0=0;$$

and the value of Λ is given by

$$\Lambda^2=-(A,\ B,\ C,\ F,\ G,\ H\between \alpha,\ \beta,\ \gamma)^2.$$

The theorem may therefore be written

$$\Lambda\int\frac{dx}{\Omega Q}=\log\frac{W}{\Omega},$$

where the several symbols have the significations explained above.

The verification is as follows. We ought to have

$$\frac{\Lambda\,dx}{\Omega Q}=\frac{P_0dx+Q_0dy}{W}-\frac{\alpha\,dx+\beta\,dy}{\Omega},$$

when dx, dy satisfy the relation $P\,dx+Q\,dy=0$, viz. substituting for dy the value $-\dfrac{P\,dx}{Q}$, the equation becomes

$$\frac{\Lambda}{\Omega}=\frac{P_0Q-PQ_0}{W}-\frac{\alpha Q-\beta P}{\Omega},$$

that is, substituting for Ω its value,

$$\Lambda W=(P_0Q-PQ_0)(\alpha x+\beta y+\gamma)-(\alpha Q-\beta P)\,W.$$

On the right-hand side, substituting for W its value,

$$\text{coeff. }\alpha=x(P_0Q-PQ_0)-Q(P_0x+Q_0y+R_0),\quad =Q_0R-QR_0,$$

$$\text{coeff. }\beta=y(P_0Q-PQ_0)+P(P_0x+Q_0y+R_0),\quad =R_0P-RP_0,$$

(as at once appears by aid of the relation $U=Px+Qy+R=0$),

$$\text{coeff. }\gamma\qquad\qquad =P_0Q-PQ_0.$$

The equation to be verified thus is

$$\Lambda W=\begin{vmatrix}\alpha, & \beta, & \gamma\\ P_0, & Q_0, & R_0\\ P, & Q, & R\end{vmatrix};$$

which, substituting therein for P, Q, R, P_0, Q_0, R_0, their values, and writing

$$(\lambda,\ \mu,\ \nu) = (\eta - y,\ x - \xi,\ \xi y - \eta x),$$

is in fact

$$\Lambda W = (A, \ldots \between \lambda,\ \mu,\ \nu \between \alpha,\ \beta,\ \gamma).$$

We have identically

$$(a, \ldots \between x,\ y,\ 1)^2 . (a, \ldots \between \xi,\ \eta,\ 1)^2 - W^2 = (A, \ldots \between \lambda,\ \mu,\ \nu)^2,$$

which, in virtue of $(a, \ldots \between \xi,\ \eta,\ 1)^2 = 0$, gives

$$W^2 = -(A, \ldots \between \lambda,\ \mu,\ \nu)^2;$$

and since $\Lambda^2 = -(A, \ldots \between \alpha,\ \beta,\ \gamma)^2$, the equation is thus

$$\sqrt{\{-(A, \ldots \between \alpha,\ \beta,\ \gamma)^2\}} . \sqrt{\{-(A, \ldots \between \lambda,\ \mu,\ \nu)^2\}} = (A, \ldots \between \lambda,\ \mu,\ \nu \between \alpha,\ \beta,\ \gamma),$$

that is,

$$(A, \ldots \between \alpha,\ \beta,\ \gamma)^2 . (A, \ldots \between \lambda,\ \mu,\ \nu)^2 - [(A, \ldots \between \lambda,\ \mu,\ \nu \between \alpha,\ \beta,\ \gamma)]^2 = 0.$$

The left-hand side is here identically

$$= K (a, \ldots \between \gamma\mu - \beta\nu,\ \alpha\nu - \gamma\lambda,\ \beta\lambda - \alpha\mu)^2:$$

substituting for λ, μ, ν their values, we find

$$(\gamma\mu - \beta\nu,\ \alpha\nu - \gamma\lambda,\ \beta\lambda - \alpha\mu) = (x\Omega_0 - \xi\Omega,\ y\Omega_0 - \eta\Omega,\ z\Omega_0 - \zeta\Omega);$$

viz. in virtue of $\Omega_0 = 0$, these are $= -\xi\Omega,\ -\eta\Omega,\ -\zeta\Omega$, and the quadric function is $= K\Omega^2 (a, \ldots \between \xi,\ \eta,\ 1)^2$, vanishing in virtue of the relation $(a, \ldots \between \xi,\ \eta,\ 1)^2 = 0$.

The equation in question

$$\sqrt{\{-(A \ldots \between \alpha,\ \beta,\ \gamma)^2\}} . \sqrt{\{-(A \ldots \between \lambda,\ \mu,\ \nu)^2\}} = (A \ldots \between \lambda,\ \mu,\ \nu \between \alpha,\ \beta,\ \gamma)$$

is thus verified, and the theorem is proved.

633.

NOTE ON MR MARTIN'S PAPER, "ON THE INTEGRALS OF SOME DIFFERENTIALS."

[From the *Messenger of Mathematics*, vol. v. (1876), p. 163.]

The Note refers to a detail in a process of integration.

634.

THEOREMS IN TRIGONOMETRY AND ON PARTITIONS.

[From the *Messenger of Mathematics*, vol. v. (1876), p. 164, and p. 188.]

If

$$A+B+C+F+G+H=0,$$

then

$$\begin{vmatrix} \sin\overline{A+F}\sin\overline{B+F}\sin\overline{C+F}, & \cos F, & \sin F \\ \sin\overline{A+G}\sin\overline{B+G}\sin\overline{C+G}, & \cos G, & \sin G \\ \sin\overline{A+H}\sin\overline{B+H}\sin\overline{C+H}, & \cos H, & \sin H \end{vmatrix}=0.$$

Let u_n = number of partitions of n, no part less than 2, the order attended to; e.g. if $n=7$, the partitions are 7, 52, 25, 43, 34, 322, 232, 223, $u_7=8$; the series is

$$\begin{aligned} u_2 &= 1, \\ u_3 &= 1, \\ u_4 &= 2, \\ u_5 &= 3, \\ u_6 &= 5, \\ u_7 &= 8, \\ u_8 &= 13, \\ u_9 &= 21, \end{aligned}$$

where each term is the sum of the next preceding two terms.

635.

NOTE ON THE DEMONSTRATION OF CLAIRAUT'S THEOREM.

[From the *Messenger of Mathematics*, vol. v. (1876), pp. 166, 167.]

It seems worth while to indicate what the leading steps of the demonstration are.

The potential of the Earth's mass upon an external or superficial point is taken to be

$$V, \quad = \frac{V_0}{r} + \frac{V_1}{r^2} + \frac{V_2}{r^3} + \&\text{c.},$$

where V_1, V_2, V_3, ... are Laplace's functions of the angular coordinates.

The surface is assumed to be a nearly spherical surface $r = a(1+u)$, where $u = u_1 + u_2 + \&\text{c.}$, and u_1, u_2, ... are Laplace's functions of the angular coordinates. To be a surface of equilibrium, with an equation $V + \frac{1}{2}\omega^2 r^2 \sin^2\theta = C$, the latter must be equivalent to the equation $r = a(1+u)$, and *it follows* that we have

$$\begin{aligned} V_1 &= V_0 a u_1, \\ V_2 &= V_0 a^2 u_2 - \tfrac{1}{2}\omega^2 a^5 (\tfrac{1}{3} - \cos^2\theta), \\ V_3 &= V_0 a^3 u_3, \\ &\&\text{c.}, \end{aligned}$$

which values are to be substituted in the expression for V.

The whole force of gravity (due to the attraction and the centrifugal force) is taken to be $g, = -\frac{d}{dr}(V + \frac{1}{2}\omega^2 r^2 \sin^2\theta)$, and *it follows* that

$$g = \frac{V_0}{a^2}(1 + u_2 + 2u_3 + \ldots) - \tfrac{2}{3}\omega^2 a - \tfrac{5}{2}\omega^2 a(\tfrac{1}{3} - \cos^2\theta),$$

which is of the form

$$g = G\left\{1 + u_2 - \tfrac{5}{2}\frac{\omega^2 a}{G}\left(\tfrac{1}{3} - \cos^2\theta\right) + 2u_3 + \ldots\right\}.$$

Taking the Earth to be the spheroid of revolution

$$r = a\left\{1 + \epsilon\left(\tfrac{1}{3} - \cos^2\theta\right)\right\},$$

then

$$u_2 = \epsilon\left(\tfrac{1}{3} - \cos^2\theta\right), \quad u_3 = 0, \text{ \&c.},$$

and the equation is

$$g = G\left\{1 - \left(\tfrac{5}{2}\frac{\omega^2 a}{G} - \epsilon\right)\left(\tfrac{1}{3} - \cos^2\theta\right)\right\},$$

or say

$$g = G\left\{1 - \left(\tfrac{5}{2}m - \epsilon\right)\left(\tfrac{1}{3} - \cos^2\theta\right)\right\},$$

where $m, = \dfrac{\omega^2 a}{G}$, is the ratio of the centrifugal force at the equator to the force of gravity, which is the theorem in question. The expression "it follows" has been twice used as meaning it follows as a mere analytical consequence, in the proper degree of approximation, the steps of the deduction being purposely omitted.

636.

ON THE THEORY OF THE SINGULAR SOLUTIONS OF DIFFERENTIAL EQUATIONS OF THE FIRST ORDER.

[From the *Messenger of Mathematics*, vol. VI. (1877), pp. 23—27.]

IN continuation of the former paper with this title (*Messenger*, vol. II., 1873, pp. 6—12, [545]), I propose to discuss various particular examples, chiefly of cases in which the differential equation is of the form $(L, M, N \between p, 1)^2 = 0$, where L, M, N are rational and integral functions of (x, y), and whether it admits or does not admit of an integral equation $(P, Q, R \between c, 1)^2 = 0$, where P, Q, R are rational and integral functions of (x, y).

The singular solution of the differential equation

$$(L, M, N \between p, 1)^2 = 0,$$

if there be a singular solution, is $S = 0$, where S is either $= LN - M^2$, or a factor of $LN - M^2$. But in general $LN - M^2$ is an indecomposable function, such that $LN - M^2 = 0$ is not a solution of the differential equation, and this being so, there is no singular solution; viz. a differential equation $(L, M, N \between p, 1)^2 = 0$, where L, M, N are rational and integral functions of (x, y), has not in general any singular solution.

Consider now a system of algebraical curves $U = 0$, where U is as regards (x, y) a rational and integral function of the order m, and depends in any manner on an arbitrary parameter C*. *I say that there is always a proper envelope,* which envelope is the singular solution of the differential equation obtained by the elimination of C from the equation $U = 0$, and the derived equation in regard to (x, y). *It follows that the differential equation* $(L, M, N \between p, 1)^2 = 0$, *which has no singular solution, does not admit of an integral of the form in question* $U = 0$, *viz. an integral representing a system of algebraic curves.*

* The expressions in the text may be understood as extending to the case where U is a function of any number (α) of constants $c_1, c_2, \ldots, c_\alpha$, connected by an $(\alpha - 1)$fold relation, U thus virtually depending on a single arbitrary parameter.

The theorem just referred to, that the system of algebraic curves $U=0$ has always an envelope, is an interesting theorem, which I proceed to prove. Assume that in general, that is, for an arbitrary value of the parameter, the equation $U=0$ represents a curve of the order m, with δ nodes and κ cusps (and therefore of the class n, with i inflexions and τ double tangents, the numbers $m, \delta, \kappa, n, \tau, i$ being connected by Plücker's equations); for particular values of the parameter, the values of δ and κ may be increased, or the curve may break up, but this is immaterial.

The consecutive curve $U+\delta c d_c U=0$ is a curve of the same order m, with δ nodes and κ cusps, consecutive to the nodes and cusps of the original curve U, and the two curves intersect in m^2 points; but of these, there are 2 coinciding with each node, and 3 coinciding with each cusp of the curve $U=0$, as at once appears by drawing a curve with a node or a cusp, and the consecutive curve with a consecutive node or cusp; the number of the remaining intersections is $=m^2-2\delta-3\kappa$, and the envelope is the locus of these $m^2-2\delta-3\kappa$ points. Observe that the two curves have in common n^2 tangents; but of these, 2 coincide with each double tangent and 3 coincide with each stationary tangent of the curve $U=0$, viz. the number of the remaining common tangents is $=n^2-2\tau-3i$ (which is $=m^2-2\delta-3\kappa$): and that these $n^2-2\tau-3i$ common tangents are indefinitely near to the $m^2-2\delta-3\kappa$ common points respectively, and are in fact the tangents of the envelope at the $m^2-2\delta-3\kappa$ points respectively. Now in an algebraic curve we have $m+n=m^2-2\delta-3\kappa$, viz. the number $m^2-2\delta-3\kappa$ cannot be $=0$, and we have therefore always an envelope the locus of the system of the $m^2-2\delta-3\kappa$ points. It might be thought that the conclusion extends to transcendental curves; if this were so, the result would prove too much, viz. it would follow that a differential equation $(L, M, N \between p, 1)^2=0$ without a singular solution had no general integral; but it will appear by an example that the theorem as to the envelope does not extend to transcendental curves.

Ex. 1.

$$p^2-(1-y^2)=0, \text{ that is, } dy^2-(1-y^2)\,dx^2=0.$$

Here there is no algebraical integral, but there is a quasi-algebraical integral of the form $(P, Q, R \between c, 1)^2=0$; viz. starting with the form $y=\sin(x+C)$ and expressing $\sin C$ and $\cos C$ rationally in terms of a new parameter, this is

$$c^2(y+\cos x)-2c\sin x+(y-\cos x)=0,$$

where the coefficients are one-valued functions of (x, y). The discriminant of the differential equation in regard to p and that of the integral equation in regard to c are each $=y^2-1$, and we have a true singular solution $y^2-1=0$.

Ex. 2.

$$(1-x^2)\,p^2-(1-y^2)=0,$$

that is,

$$(1-x^2)\,dy^2-(1-y^2)\,dx^2=0.$$

We have here an algebraic integral of the proper form, which is at once derived from the circular form

$$C=\cos^{-1}x+\cos^{-1}y$$

by changing the constant, viz. this is

$$c^2 - 2cxy - (1 - x^2 - y^2) = 0.$$

The two discriminants are here each $= (x^2 - 1)(y^2 - 1)$, and we have

$$(x^2 - 1)(y^2 - 1) = 0$$

as a true singular solution. The curves are in fact the system of conics (ellipses and hyperbolas) each touching the four lines $x = 1$, $x = -1$, $y = 1$, $y = -1$.

Ex. 3.

$$(1 - y^2)p^2 - 1 = 0, \text{ that is, } (1 - y^2)\,dy^2 - dx^2 = 0.$$

This is an extremely interesting example: the curve is the orthogonal trajectory of the system of sinusoids $y = \sin(x + c)$, which is the integral of Example 1; and we thus at once see that the real portion of the curve is wholly included between the lines $y = -1$, $y = +1$, being an infinite continuous curve, having a series of equidistant cusps alternately at the one and the other line, and obtained by the continued repetition of the finite portion included between two consecutive cusps on the same line. The discriminant of the differential equation equated to zero gives $y^2 - 1 = 0$, the equation of the two lines in question; but this does not satisfy the differential equation, and it is consequently not a singular solution; by what precedes, it appears that it is, in fact, a cusp-locus.

We thus see that the curves which represent the integral equation have no *real* envelope; but it is to be further shown that there is no imaginary envelope, and that the curve obtained by the elimination of the parameter is, in fact, made up of a (imaginary) node-locus and of the foregoing cusp-locus.

The curve is properly represented by taking x, y each of them a one-valued function of the parameter θ, viz. we may write

$$y = \cos\theta,$$

$$x = c + \tfrac{1}{2}\theta - \tfrac{1}{4}\sin 2\theta.$$

In fact, these values give

$$\frac{dy}{d\theta} = -\sin\theta,\ \frac{dx}{d\theta} = \tfrac{1}{2}(1 - \cos 2\theta) = \sin^2\theta,$$

and therefore

$$p = -\frac{1}{\sin\theta} = \frac{-1}{\sqrt{(1 - y^2)}},$$

that is, $(1 - y^2)p^2 - 1 = 0$, the differential equation.

It is obvious that to a given value of the parameter there corresponds a single point of the curve; and it is to be shown that, conversely, to a given point of the curve corresponds in general a single value of the parameter.

Suppose the coordinates of the given point are $y = \cos\alpha$, $x = c + \frac{1}{2}\alpha - \frac{1}{4}\sin 2\alpha$, where α is a determinate quantity; then, to find θ, we have

$$\cos\theta = \cos\alpha, \quad 2\theta - \sin 2\theta = 2\alpha - \sin 2\alpha.$$

The first equation gives $\theta = 2m\pi \pm \alpha$, and the second equation then is

$$4m\pi \pm 2\alpha \mp \sin 2\alpha = 2\alpha - \sin 2\alpha;$$

viz. taking the upper signs, this is $4m\pi = 0$, giving $m = 0$ and $\theta = \alpha$; and, taking the lower signs, it is $m\pi = \alpha - \sin\alpha$, which, α being given, is not in general satisfied; hence to the given point there corresponds *only* the value α of the parameter θ. If, however, α is such that $\alpha - \sin\alpha$ is equal to a multiple of π, say $r\pi$, then the last-mentioned equation is satisfied by the value $m = r$, so that to the given point of the curve correspond the two values α and $2r\pi - \alpha$ of the parameter; these values are in general unequal, and the point is then a node; but they may be equal, viz. this is so if $\alpha = r\pi$ (the point on the curve being then $y = \cos r\pi$, $= \pm 1$, $x = c + \frac{1}{2}r\pi$), and the point is then a cusp; showing what was known, that there are on each of the lines $y = -1$, $y = +1$, an infinite series of equidistant cusps.

More definitely, suppose $\alpha = r\pi \pm \beta$, where β is a root of the equation $2\beta - \sin 2\beta = 0$, then

$$\sin 2\alpha = \pm \sin 2\beta, \quad 2\alpha - \sin 2\alpha = 2r\pi \pm (2\beta - \sin 2\beta) = 2r\pi,$$

and to the given point on the curve correspond the two values α and $2r\pi - \alpha$ of the parameter. If $\beta = 0$, we have, as above, the cusps on the two lines $y = +1$, $y = -1$ respectively; but if β be an imaginary root of the equation $2\beta - \sin 2\beta = 0$, then we have an infinite series of nodes on the imaginary line $y = \cos r\pi \cos\beta$; and there are an infinite number of such lines corresponding to the different imaginary roots of the equation $2\beta - \sin 2\beta = 0$.

From the form in which the equation of the curve is given, we cannot directly form the equation of the envelope by equating to zero the discriminant in regard to the constant c; but we may determine the intersections of the curve by the consecutive curve (corresponding to a value $c + \delta c$ of the constant), and thus determine the locus of these intersections.

Consider for a moment the curves belonging to the constants c, c_1, and let θ, θ_1 be the values of the parameter θ belonging to the points of intersection; we have $\cos\theta = \cos\theta_1$, $4c + 2\theta - \sin 2\theta = 4c_1 + 2\theta_1 - \sin 2\theta_1$; we have $\theta_1 = 2r\pi + \theta$, but we cannot thereby satisfy the second equation; or else $\theta_1 = 2r\pi - \theta$, giving

$$4c + 2\theta - \sin 2\theta = 4c_1 + 4r\pi - 2\theta + \sin 2\theta,$$

that is, $2\theta - \sin 2\theta = 2c_1 - 2c + 2r\pi$; and we have thus corresponding to any given value of r a series of values of θ, viz. these are $\theta = r\pi + \beta$, where β is any root of the equation

$$2\beta - \sin 2\beta = 2c_1 - 2c.$$

In particular, taking $c_1 = c$, the intersections are given by $\theta = r\pi + \beta$, where β is any root of the equation $2\beta - \sin 2\beta = 0$; viz. we have thus an infinite number of intersections lying on each of the lines $y = \cos r\pi \cos \beta$. If $\beta = 0$, the intersections lie on the two lines $y = 1$, $y = -1$ respectively; if β be an imaginary root of the equation $2\beta - \sin 2\beta = 0$, then they lie on the imaginary lines $y = \cos r\pi \cos \beta$. But by what precedes, it is clear that in the former case the intersections are nothing else than the cusps on the lines $y = 1$, $y = -1$; and in the latter case nothing else than the nodes on the lines $y = \cos r\pi \cos \beta$; viz. there is no proper envelope, but instead thereof we have lines of cusps and of nodes.

Ex. 4.

$$(1 - y^2) p^2 - (1 - x^2) = 0,$$

that is,

$$(1 - y^2) dy^2 - (1 - x^2) dx^2 = 0.$$

I have not examined this; the curve is the series of orthogonal trajectories of the conics of Example 2, and the integral equation may be represented by $y = \cos \theta$, $x = \cos \phi$, where $c = (2\theta - \sin 2\theta) - (2\phi - \sin 2\phi)$.

Equating to zero the discriminant of the differential equation, we have $(1 - y^2)(1 - x^2) = 0$, viz. the four lines $x = 1$, $x = -1$, $y = 1$, $y = -1$; this is not an envelope, but a locus of cusps.

637.

ON A DIFFERENTIAL EQUATION IN THE THEORY OF ELLIPTIC FUNCTIONS.

[From the *Messenger of Mathematics*, vol. VI. (1877), p. 29.]

IN the differential equation

$$Q^2 - Q\left(k + \frac{1}{k}\right) - 3 = 3(1 - k^2)\frac{dQ}{dk},$$

considered *Messenger*, t. IV., pp. 69 and 110, [594] and [597], writing $Q = x$ and $k + \frac{1}{k} = y$, the equation becomes

$$dy = \frac{3(y^2 - 4)\,dx}{3 + xy - x^2},$$

and we have, as a particular solution,

$$y = \tfrac{1}{4}\left(x^3 - 6x - \frac{3}{x}\right).$$

To verify this, observe that from the value of y

$$dy = \frac{3}{4x^2}(x^2 - 1)^2\,dx, \quad 3 + xy - x^2 = \tfrac{1}{4}(x^2 - 1)(x^2 - 9),$$

and the equation becomes

$$\frac{3}{4x^2}(x^2 - 1)^2 = \frac{\frac{3}{16x^2}\{(x^4 - 6x^2 - 3)^2 - 64x^2\}}{\frac{1}{4}(x^2 - 1)(x^2 - 9)},$$

viz. this is

$$(x^2 - 1)^3(x^2 - 9) = (x^4 - 6x^2 - 3)^2 - 64x^2,$$

which is right.

638.

ON A q-FORMULA LEADING TO AN EXPRESSION FOR E_1.

[From the *Messenger of Mathematics*, vol. VI. (1877), pp. 63—66.]

It is to be shown that we have identically

$$(1+2q+2q^4+2q^9+\ldots)^4-16\left(\frac{q}{1-q^2}+\frac{2q^2}{1-q^4}+\frac{3q^3}{1-q^6}+\ldots\right)$$
$$=\frac{1-9q-25q^3+49q^6+81q^{10}-\ldots}{1-q-q^3+q^6+q^{10}-\ldots} \quad \ldots\ldots\ldots\ldots (A);$$

or, what is the same thing,

$$(1-2q+2q^4-2q^9+\ldots)^4-16\left(\frac{-q}{1-q^2}+\frac{2q^2}{1-q^4}-\frac{3q^3}{1-q^6}+\ldots\right)$$
$$=\frac{1+9q+25q^3+49q^6+81q^{10}+\ldots}{1+q+q^3+q^6+q^{10}+\ldots} \quad \ldots\ldots\ldots\ldots (B),$$

where the form (A) is that intended to be made use of, but the form (B) is rather more convenient for the demonstration.

We have

$$(1-2q+2q^4-2q^9+\ldots)^4=1+8\left\{\frac{-q}{1+q}+\frac{2q^2}{1+q^2}-\frac{3q^3}{1+q^3}+\ldots\right\},$$

(Jacobi, *Fund. Nova*, p. 188, *Ges. Werke*, t. I., p. 239), taking the formula as there written down, and changing q into $-q$.

Also, if for a moment

$$X=1+q+q^3+q^6+q^{10}+\&c.,$$

and

$$X'=\frac{dX}{dq},$$

so that

$$qX' = q + 3q^3 + 6q^6 + 10q^{10} + \&c.,$$

then

$$X + 8qX' = 1 + 9q + 25q^3 + 49q^6 + 81q^{10} + \&c.,$$

so that the right-hand side of (B) is

$$\frac{X + 8qX'}{X}, \quad = 1 + 8q\frac{X'}{X}.$$

But (*Fund. Nova*, p. 185, *Ges. Werke*, t. I., p. 237),

$$X = \frac{1-q^2 . 1-q^4 . 1-q^6 \ldots}{1-q \ . 1-q^3 . 1-q^5 \ldots},$$

so that

$$\frac{X'}{X} = \frac{-2q}{1-q^2} - \frac{4q^3}{1-q^4} - \frac{6q^5}{1-q^6} - \ldots$$

$$+ \frac{1}{1-q} + \frac{3q^2}{1-q^3} + \frac{5q^4}{1-q^5} + \ldots.$$

And the equation (B) intended to be proved thus becomes

$$1 + \ 8\left\{\frac{-q}{1+q} + \frac{2q^2}{1+q^2} - \frac{3q^3}{1+q^3} + \ldots\right\}$$

$$-16\left\{\frac{-q}{1-q^2} + \frac{2q^2}{1-q^4} - \frac{3q^3}{1-q^6} + \ldots\right\}$$

$$= 1 + 8q\left\{\frac{-2q}{1-q^2} - \frac{4q^3}{1-q^4} - \frac{6q^5}{1-q^6} - \ldots\right.$$

$$\left. + \frac{1}{1-q} + \frac{3q^2}{1-q^3} + \frac{5q^4}{1-q^5} + \ldots\right\};$$

viz. omitting the terms unity, dividing by $8q$, and transposing, this is

$$-\frac{1}{1+q} + \frac{2q}{1+q^2} - \frac{3q^2}{1+q^3} + \ldots$$

$$+\frac{2}{1-q^2} - \frac{4q}{1-q^4} + \frac{6q^2}{1-q^6} - \ldots$$

$$+\frac{2q}{1-q^2} + \frac{4q^3}{1-q^4} + \frac{6q^5}{1-q^6} + \ldots$$

$$-\frac{1}{1-q} - \frac{3q^2}{1-q^3} - \frac{5q^4}{1-q^5} - \ldots = 0.$$

The second and third lines unite together, and the equation becomes

$$-\frac{1}{1+q} + \frac{2q}{1+q^2} - \frac{3q^2}{1+q^3} + \frac{4q^3}{1+q^4} - \ldots$$

$$+\frac{2}{1-q} - \frac{4q}{1+q^2} + \frac{6q^2}{1-q^3} - \frac{8q^3}{1+q^4} + \ldots$$

$$-\frac{1}{1-q} - \frac{3q^2}{1-q^3} - \frac{5q^4}{1-q^5} - \frac{7q^6}{1-q^7} - \ldots = 0;$$

or, collecting and arranging,

$$-\frac{1}{1+q}-\frac{2q}{1+q^2}-\frac{3q^2}{1+q^3}-\frac{4q^3}{1+q^4}-\frac{5q^4}{1+q^5}-\dots$$
$$+\frac{1}{1-q}\qquad+\frac{3q^2}{1-q^3}\qquad+\frac{5q^4}{1-q^5}+\dots=0,$$

an identity which it is easy to verify to any number of terms. But to prove it directly, we have only to add the pairs of terms in the alternate columns; calling the left-hand side Fq, we thus obtain

$$Fq=2q\left\{-\frac{1}{1+q^2}-\frac{2q^2}{1+q^4}-\frac{3q^4}{1+q^6}-\dots\right.$$
$$\left.+\frac{1}{1-q^2}\qquad+\frac{3q^4}{1-q^6}+\dots\right\};$$

viz. this equation is $Fq=2qF(q^2)$; and thence

$$Fq=2^2q^{1+2}F(q^4)=2^3q^{1+2+4}F(q^8)=\text{\&c.};$$

we thus have $Fq=0$.

The equation (B), or, what is the same thing, the equation (A) is thus proved.

Reverting to the equation (A), we have

$$(1+2q+2q^4+\dots)^4=\frac{4K^2}{\pi^2},$$

(Jacobi, *Fund. Nova*, p. 188, *Ges. Werke*, t. I., p. 239),

$$\left(\frac{q}{1-q^2}+\frac{2q^2}{1-q^4}+\dots\right)=\frac{K^2}{2\pi^2}\left(1-\frac{E_1}{K}\right),$$

(*ib.*, p. 135; *ib.*, p. 189),

if $q=e^{-\frac{\pi K'}{K}}$, and K, E_1 are the complete functions F_1k, E_1k.

The left-hand side of the equation is thus

$$\frac{4K^2}{\pi^2}-\frac{8K^2}{\pi^2}\left(1-\frac{E_1}{K}\right),\quad=\frac{4K^2}{\pi^2}\left(-1+\frac{2E_1}{K}\right),$$

and we have

$$\left(-1+\frac{2E_1}{K}\right)=\frac{\pi^2}{4K^2}\cdot\frac{1-9q^1-25q^3+49q^6+81q^{10}-\dots}{1-q^1-q^3+q^6+q^{10}-\dots},$$

which is a new expression for E_1 as a q-function. The expression on the right-hand side presents itself, Clebsch, *Theorie der Elasticität* (Leipzig, 1862), p. 162, and must have been obtained by him as a value for $\left(-1+\dfrac{2E_1}{K}\right)$; but there is no statement that this is so, nor anything to show how this form of q-function was arrived at. Mr Todhunter called my attention to the passage in Clebsch.

639.

AN ELEMENTARY CONSTRUCTION IN OPTICS.

[From the *Messenger of Mathematics*, vol. VI. (1877), pp. 81, 82.]

CONSIDER two lines meeting at a point P, and a point A; through A, draw at right angles to AP, a line meeting the two lines in the points U, V respectively; and through the same point A draw any other line meeting the two lines in the

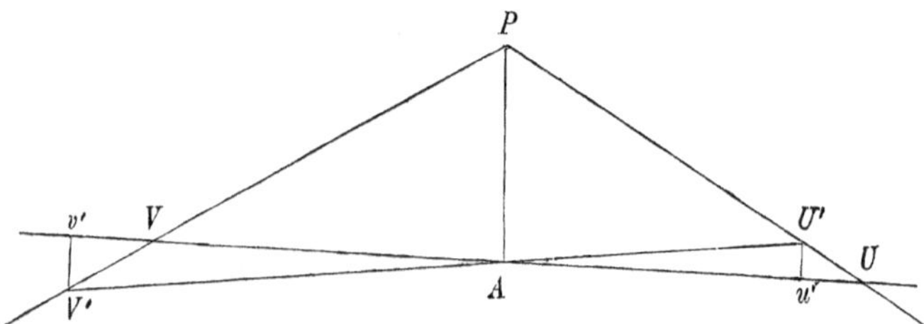

points U', V' respectively; also let the points u', v' be the feet of the perpendiculars let fall from U', V' respectively on the line UV; then we have

$$\frac{1}{Au'}+\frac{1}{Av'}=\frac{1}{AU}+\frac{1}{AV}.$$

The theorem can be proved at once without any difficulty. It answers to the optical construction, according to which, if UPV represents the path of a ray through a convex lens AP, then the thin pencil, axis $U'P$ and centre U', converges after refraction to the point V', where $U'V'$ are *in lineâ* with A the centre of the lens; considering as usual the inclinations to the axis as small, we have approximately $AV'=Av'$, $AU'=Au'$, and the theorem is

$$\frac{1}{AU'}+\frac{1}{AV'}=\frac{1}{AU}+\frac{1}{AV}, \quad =\frac{1}{AF},$$

if AF is the focal length of the lens.

In the original theorem, the line UV need not be at right angles to AP, but may be any line whatever; the projecting lines $U'u'$ and $V'v'$ must then be parallel to AP, and the theorem remains true.

640.

FURTHER NOTE ON MR MARTIN'S PAPER "ON THE INTEGRALS OF SOME DIFFERENTIALS."

[From the *Messenger of Mathematics*, vol. VI. (1877), p. 82.]

See paper, Number 633; this further note relates also to a detail.

641.

ON THE FLEXURE OF A SPHERICAL SURFACE.

[From the *Messenger of Mathematics*, vol. VI. (1877), pp. 88—90.]

It is known that an inextensible spherical surface, or to fix the ideas the spherical quadrilateral included between two arcs of meridian and two arcs of parallel, may be bent in suchwise as to be part of a surface of revolution, the meridians and parallels of the spherical surface being meridians and parallels of the new surface, and, moreover, the radius of each parallel of the spherical surface being in the new surface altered in the constant ratio k to 1. We have, in fact, on the spherical surface, writing p for the latitude and q for the longitude, and the radius being unity,

$$x = \cos p \cos q,$$
$$y = \cos p \sin q,$$
$$z = \sin p,$$

values which give

$$dx^2 + dy^2 + dz^2 = dp^2 + \cos^2 p \, dq^2.$$

This last equation is satisfied by the values

$$x = \cos p \cos \frac{q}{k},$$

$$y = \cos p \sin \frac{q}{k},$$

$$z = E(k,\ p),$$

where $E(k, p), = \int_0 \surd(1 - k^2 \sin^2 p)\, dp$, is the elliptic function of the second kind; or rather, this is so when $k < 1$, but the same notation may be used when $k > 1$. These values give the deformation in question.

The two cases to be considered are $k < 1$, and $k > 1$; we take in each case a spherical quadrilateral $ABCD$ (fig. 1), bounded by AB (an arc of the equator), the arc of parallel CD, and the two arcs of meridian AD and BC. In the first case, there is no limit to the latitude $AD, = BC$, or taking these $= 90°$, we may in place of the quadrilateral $ABCD$ consider the birectangular triangle ABE; the new form of this is $A'B'E'$, where the radius OA', $=k \, . \, OA$, $=k$, is less than the original radius

Fig. 1.

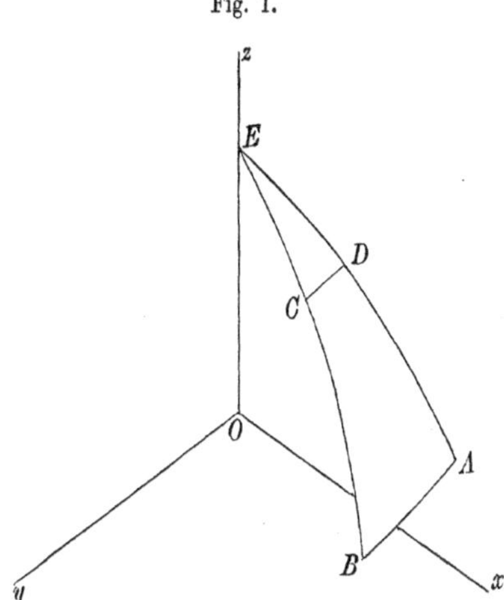

unity, but OE', $=E_1k$, is greater than the same radius unity. The surface has at E' a conical point, the semi-aperture of the cone being $=\tan^{-1}\frac{1}{k'}$, if as usual $k'=\sqrt{(1-k^2)}$; to verify this, writing for convenience $q=0$, we have for the meridian section $x=\cos p$, $z=E(k, p)$, and thence $\frac{dx}{dz}=-\frac{\sqrt{(1-k^2\sin^2 p)}}{\sin p}$, which for $p=90°$ gives $\frac{dz}{dx}=-k'$. Observe also that for $p=0$, $\frac{dz}{dx}=\infty$, viz. the surface of revolution cuts the plane of the equator at right angles.

There is no limit to the arc AB, it may be $=360°$, viz. we must in this case cut the hemisphere along a meridian to allow of the deformation; or it may exceed 360°, the hemisphere spherical surface being in this case conceived of as wrapping indefinitely over itself, and we may instead of the half lune $E'A'B'$, consider the lune included between two meridians extending from pole to pole, and therefore the whole spherical surface, conceived of as wrapping indefinitely over itself; the result is, that this may be deformed into a surface of revolution, which, in its general form, resembles that obtained by the revolution of an arc less than a semi-circle round its chord; the half-chord being greater, and the versed-sine less than the radius of the original sphere.

If $k>1$, there is obviously a limit to the latitude AD, $=BC$, of the spherical quadrilateral; viz. this is equal to $\sin^{-1}\frac{1}{k}$. Supposing that in the quadrilateral $ABCD$ (fig. 1) the latitude has this limiting value, then (see fig. 2) the new form is $A''B''C''D''$, where along the bounding arc $C''D''$ the tangent plane is horizontal; viz. as before $\frac{dz}{dx}=-\frac{\sqrt{(1-k^2\sin^2 p)}}{\sin p}$, $=0$ for $p=\sin^{-1}\frac{1}{k}$. It is to be observed, that the radii for the parallels $A''B''$ and $C''D''$ are k and $k\cos p$ respectively; the difference of these is $k(1-\cos p)$, which, however great k is, must be less than the arc of meridian $A''D''$, $=p$; substituting for k the value $\dfrac{1}{\sin p}$, the condition is $\dfrac{1-\cos p}{\sin p}<p$, viz. this

Fig. 2.

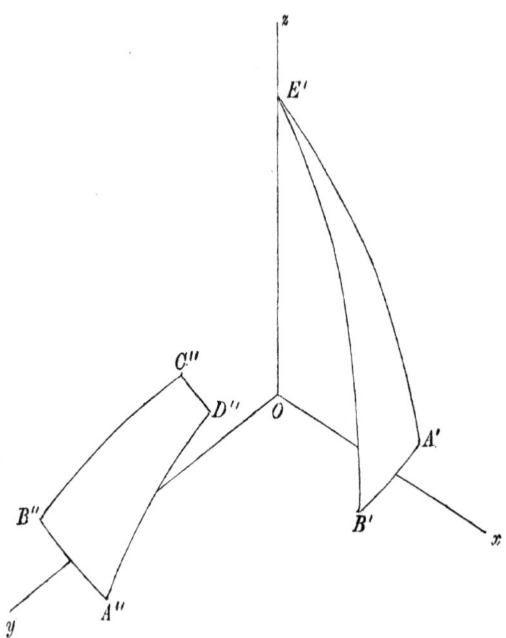

is $\tan\frac{1}{2}p<p$, which is true for every value up to $p=90°$. But, more than this, we should have

$$k^2(1-\cos p)^2+E^2(k,\,p)<p^2,$$

viz. writing as before $k=\dfrac{1}{\sin p}$, this is

$$E^2\left(\frac{1}{\sin p},\,p\right)<p^2-\tan^2\tfrac{1}{2}p\,;$$

this must be true, although (relating as it does to a form of E for which k is greater than 1) there might be some difficulty in verifying it.

There is, as in the first case, no limit to the value of AB, viz. this may be $=360°$, the spherical zone being then cut along a meridian, or it may be greater than $360°$; and, moreover, the spherical quadrilateral may extend south of the equator, but of course so that the limiting south latitude does not extend beyond the foregoing value $\sin^{-1}\dfrac{1}{k}$: viz. we may have a zone between the latitudes $\pm\sin^{-1}\dfrac{1}{k}$, which may be a complete zone from longitude $0°$ to $360°$ or to any greater value than $360°$. The result is, that the zone is deformed into a surface of revolution, which in its general form resembles that obtained by the revolution of a half-circle or half-ellipse about a line parallel to and beyond its bounding diameter, the bounding half-diameter being less, and the greatest radius of rotation greater, than the radius of the original sphere.

642.

ON A DIFFERENTIAL RELATION BETWEEN THE SIDES OF A QUADRANGLE.

[From the *Messenger of Mathematics*, vol. VI. (1877), pp. 99—101.]

LET the sides and diagonals YZ, ZX, XY, OX, OY, OZ of a quadrangle be f, g, h, a, b, c, and let the component triangles be denoted as follows:

$$\begin{aligned} A &= \Delta YZO, &= (b, c, f), \\ B &= \Delta ZXO, &= (c, a, g), \\ C &= \Delta XYO, &= (a, b, h), \\ \Omega &= \Delta XYZ, &= (f, g, h), \end{aligned}$$

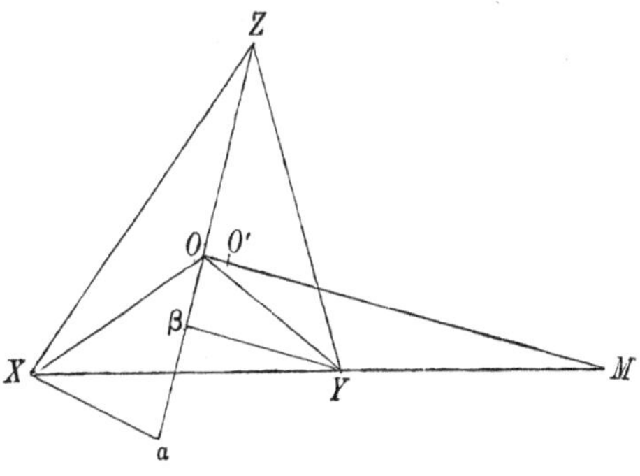

viz. A, B, C, Ω are the triangles whose sides are (b, c, f), (c, a, g), (a, b, h), (f, g, h) respectively, so that $\Omega = A + B + C$. Then we have between (a, b, c, f, g, h) an equation giving rise to a differential relation, which may be written

$$\Omega\,(Aada + Bbdb + Ccdc) - (BCfdf + CAgdg + ABhdh) = 0.$$

This may be proved geometrically and analytically. First, for the geometrical proof, it is enough to prove that, when a and b alone vary, the relation between the increments is $Aada + Bbdb = 0$; for then a and g alone varying, the relation between the increments will be $\Omega ada - Cgdg = 0$ (as to the negative sign it is clear from the figure that a, g will increase or diminish together): and we thence at once infer the general relation.

We have consequently to prove that, considering a and b as alone variable,

$$Aada + Bbdb = 0;$$

or, what is the same thing,

$$ada : -bdb = XOZ : YOZ.$$

The points XYZ remain fixed; but O moves through the infinitesimal arc OO', centre Z, which may be considered as situate in the right line OM drawn from O at right angles to ZO, and meeting XY produced in the point M. And then, writing for a moment $\angle OXY = X$, $\angle OYX = Y$, $\angle OMY = M$, we find at once

$$da = OO' \cos(X+M),$$

$$-db = OO' \cos(Y-M);$$

that is,

$$-\frac{da}{db} = \frac{\cos(X+M)}{\cos(Y-M)}, \quad \text{or} \quad -\frac{ada}{bdb} = \frac{a\cos(X+M)}{b\cos(Y-M)}.$$

But drawing $X\alpha$, $Y\beta$ each of them at right angles to ZO, we have $a\cos(X+M) = X\alpha$, $b\cos(Y-M) = Y\beta$, and evidently $XOZ : YOZ = X\alpha : Y\beta$; whence the equation is $-\dfrac{ada}{bdb} = \dfrac{XOZ}{YOZ}$, which is the required relation.

For the analytical proof, it is to be observed that the relation between a, b, c, f, g, h is a quadric relation in the quantities a^2, b^2, c^2, f^2, g^2, h^2 respectively; this may be written

	1	a^2	a^4	
1	$b^2g^4+b^4g^2+c^2h^4+c^4h^2-(b^2+c^2)g^2h^2-(g^2+h^2)b^2c^2$	$-(b^2-c^2)(g^2-h^2)$		
$u =$ f^2	$+(b^2-h^2)(c^2-g^2)$	$-b^2-c^2-g^2-h^2$	$+1$	$=0:$
f^4		$+1$		

say for a moment this is $A + Ba^2 + Ca^4 = 0$, where

$$\begin{aligned} A = &\ b^2g^4+b^4g^2+c^2h^4+c^4h^2 \qquad +f^2(b^2-h^2)(c^2-g^2) \\ &-(b^2+c^2)g^2h^2-(g^2+h^2)b^2c^2 \\ B = &-(b^2-c^2)(g^2-h^2) \qquad +f^2(-b^2-c^2-g^2-h^2)+f^4, \\ C = &\qquad\qquad f^2; \end{aligned}$$

then we have as usual

$$\frac{du}{d\,.\,a^2}ada + \frac{du}{d\,.\,b^2}bdb + \&c. = 0,$$

where

$$\tfrac{1}{2}\frac{du}{d\,.\,a^2} = Ca^2 + \tfrac{1}{2}B.$$

But in virtue of $u=0$, we have

$$(Ca^2 + \tfrac{1}{2}B)^2 = C(Ca^4 + Ba^2 + A) + \tfrac{1}{4}(B^2 - AC),$$

that is, $\dfrac{du}{d\,.\,a^2} = \surd(B^2 - 4AC)$; and here $B^2 - 4AC$ is a quartic function of f^2, which is easily seen to reduce itself to the form

$$f^2 - (g+h)^2 f^2 - (g-h)^2 f^2 - (b+c)^2 f^2 - (b-c)^2.$$

The coefficients of bdb, cdc, &c., are given as expressions of the like form; substituting their values, the differential relation is

$$\surd\{f^2 - (g+h)^2 f^2 - (g-h)^2 f^2 - (b+c)^2 f^2 - (b-c)^2\}\, ada + \&\text{c.} = 0,$$

which is, in fact, the foregoing result.

It is right to notice that there are in all 16 linear factors,

$$\begin{array}{cccc|cccc} f+g+h, & b+c+f, & c+a+g, & a+b+h & \text{say } d\,, & f\,, & g\,, & h\,, \\ -f+g+h, & -b+c+f, & -c+a+g, & -a+b+h & d'\,, & f'\,, & g'\,, & h'\,, \\ f-g+h, & b-c+f, & c-a+g, & a-b+h & d''\,, & f''\,, & g''\,, & h''\,, \\ f+g-h, & b+c-f, & c+a-g, & a+b-h & d''', & f''', & g''', & h'''; \end{array}$$

and this being so, the coefficients of ada, bdb, cdc, fdf, gdg, hdh, are

$$\begin{array}{ll} \surd(dd'd''d'''\,.\,ff'f''f'''), & -\surd(gg'g''g'''\,.\,hh'h''h'''), \\ \surd(dd'd''d'''\,.\,gg'g''g'''), & -\surd(hh'h''h'''\,.\,ff'f''f'''), \\ \surd(dd'd''d'''\,.\,hh'h''h'''), & -\surd(ff'f''f'''\,.\,gg'g''g'''), \end{array}$$

respectively.

We may imagine the quadrilateral $ZOXY$ composed of the four rods ZO, OX, XY, YZ (lengths c, a, h, f as before) jointed together at the angles, and kept in equilibrium by forces B, G acting along the diagonals $OY\,(=b)$, $ZX\,(=a)$ respectively. We have c, a, h, f given constants, and the relation $\phi(a, b, c, f, g, h) = 0$, which connects the six quantities is the relation between the two variable diagonals (g, b); by what precedes, the differential relation $\phi' g\,.\,dg + \phi' b\,.\,db = 0$ is equivalent to $\Omega Bbdb - CAgdg = 0$. By virtual velocities we have as the condition of equilibrium $Bdb + Gdg = 0$; hence, eliminating db, dg we have

$$\frac{B}{\Omega Bb} = -\frac{G}{CAg};$$

or, say

$$\frac{B}{b} : -\frac{G}{g} = \frac{1}{\Delta XYO\,.\,\Delta ZYO} : \frac{1}{\Delta XYZ\,.\,\Delta ZXO},$$

viz. the forces, divided by the diagonals along which they act, are proportional to the reciprocals of the products of the two pairs of triangles which stand on these diagonals respectively. The negative sign shows, what is obvious, that the forces must be, one of them a pull, the other a push.

643.

ON A QUARTIC CURVE WITH TWO ODD BRANCHES.

[From the *Messenger of Mathematics*, vol. VI. (1877), pp. 107, 108.]

It is a known theorem that the branches of a plane curve are even or odd; viz. two even branches, or an even and an odd branch (whether of the same curve or of different curves) intersect in an even number (it may be 0, and this is to be understood throughout) of real points; but two odd branches (of the same curve or of different curves) intersect in an odd number of real points *.

In particular, a right line is an odd branch, and hence it meets any even branch of a curve in an even number of real points, and an odd branch in an odd number of real points; or (what is the same thing) an even branch is one which is met by any right line whatever in an even number of real points; and an odd branch is one that is met by any right line whatever in an odd number of real points.

It is to be observed, that the simple term branch is used to denote what has been called a *complete* branch, viz. the partial branches which touch an asymptote at its opposite extremities are considered as parts of one and the same branch, and so in other cases. Thus a quadric curve, whether ellipse, parabola, or hyperbola, is one even branch; a cubic curve is either one odd branch, or else it is an odd branch and an even branch; and generally a curve of an odd order has always an odd number of odd branches, and a curve of an even order has always an even number of odd branches.

A curve without nodes has at most one odd branch; for if there were two, these would intersect in a real point, which would be a real node on the curve. In particular, a quartic curve having two odd branches must have a real node; this however may be, as in the instance about to be given, a node at infinity.

A simple instance of a quartic curve with two odd branches is that represented by the equation

$$(x^2-1)(y^2+1)-2mxy=0,$$

* The two branches must be distinct branches; a branch whether odd or even does not of necessity intersect itself (have upon it any real node), but it may intersect itself in an odd, or an even, number of real points.

or, what is the same thing,

$$y=\frac{1}{x^2-1}\left[mx\pm\sqrt{\left\{-(x^2-\alpha^2)\left(x^2-\frac{1}{\alpha^2}\right)\right\}}\right],$$

where

$$\alpha^2+\frac{1}{\alpha^2}=2+m^2,$$

or say

$$\alpha^2=\tfrac{1}{2}\{2+m^2+m\sqrt{(4+m^2)}\},$$

$$\frac{1}{\alpha^2}=\tfrac{1}{2}\{2+m^2-m\sqrt{(4+m^2)}\},$$

so that m being positive $\alpha>1$, and the curve consists of two real branches included between the lines $x=\alpha$, $x=\frac{1}{\alpha}$, and the lines $x=-\alpha$, $x=-\frac{1}{\alpha}$ respectively; each of these lines touches the curve in a real point, viz. x having any one of the last-mentioned values, the value of y at the point of contact is $y=\frac{mx}{x^2-1}$; and between each pair of lines we have the asymptote $x=+1$ or $x=-1$. Hence the curve has the form shown in the figure, and it is thereby evident, that each branch of the curve is met by

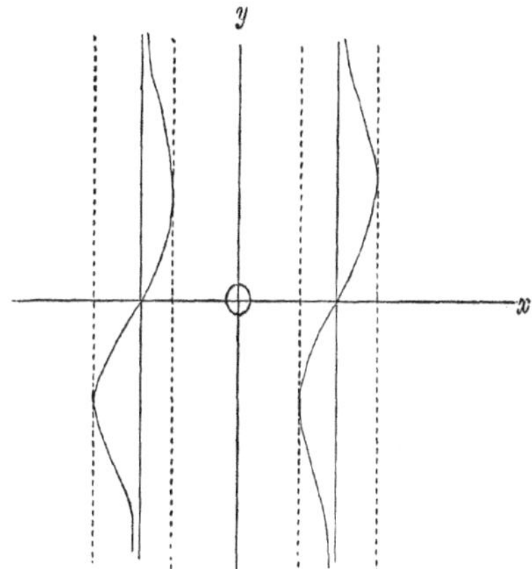

any real right line whatever in one real point, or else in three real points. The numerical values in the figure are $\alpha=\frac{3}{2}$, $m=\frac{5}{6}$, whence also $x=\alpha$ or $-\frac{1}{\alpha}$, $y=1$, and $x=-\alpha$ or $\frac{1}{\alpha}$, $y=-1$.

The curve has two nodes at infinity, viz. writing the equation in the form

$$(x^2-z^2)(y^2+z^2)-mxyz^2=0,$$

that is,

$$x^2y^2+z^2(x^2-y^2-mxy)+z^4=0,$$

it appears that the points $(z=0,\ x=0)$, $(z=0,\ y=0)$ are each of them a node. The first of these $(z=0,\ x=0)$ is the real intersection of the two odd branches: the other of them is a conjugate point.

644.

NOTE ON MAGIC SQUARES.

[From the *Messenger of Mathematics*, vol. VI. (1877), p. 168.]

IN a magic square of any odd order, formed according to the ordinary process, there is a tolerably simple analytical expression for the number which occupies any given compartment; thus taking the square of 21, let the dexter diagonals (N.W. to S.E.) commencing from the N.E. corner compartment, be numbered 1, 2, 3,.., 20, 21, 20′, 19′,.., 2′, 1′, the diagonals of course containing these numbers of compartments respectively; and in any diagonal let the compartments reckoning from the top line be numbered 1, 2, 3,.., respectively; then if $D_{\theta,\phi}$ (or $D'_{\theta,\phi}$ as the case may be) denotes the number in the compartment ϕ of the diagonal θ or θ', we have

$$\begin{aligned}
D_{2\theta+1,\phi} &= 20\theta + 10 + \phi,\\
D_{2\theta\ \ ,\phi} &= 20\theta + 231 + \phi\,(-21),\\
D'_{2\theta+1,\phi} &= -22\theta + 430 + \phi,\\
D'_{2\theta\ \ ,\phi} &= -22\theta + 231 + \phi\,(-21),
\end{aligned}$$

where in the second and fourth expressions the term -21 is to be retained only if $\phi > \theta$; if $\phi \not> \theta$, it is to be omitted. There would be a like formulæ for a square of any odd order, and it would be easy to write down the formulæ for the general value $2n+1$: but I have preferred to give them for a specific case.

645.

A SMITH'S PRIZE PAPER, 1877.

[From the *Messenger of Mathematics*, vol. VI. (1877), pp. 173—182.]

THE paper was as follows:

1. Show (independently of the theory of roots) how, if x satisfies an equation of the order n, a given rational function of x can in general be expressed as a rational and integral function of the order $n-1$. State the theorem in a more precise form, so as to make it true universally.

2. Investigate the form of the factors of $1 \pm \sin(2n+1)x$ considered as a function of $\sin x$; and give the formulæ in the two cases, $2n+1=3$ and 5 respectively.

3. Write down the substitutions which do not alter the function $ab+cd$; and explain the constitution of the group.

4. Find in a form adapted for calculation an approximate value for the sum of the middle $2\alpha+1$ terms of the expansion of $(1+1)^{2n}$, n being a large number, and α small in comparison therewith.

Obtain thence a complete and precise statement of the theorem that in a large number of tosses the numbers of heads and tails will probably be nearly equal.

5. A point in space is represented on a given plane by its projections from two fixed points. Show how a problem relating to points, lines, and planes, is thereby reduced to a problem *in plano;* and apply the method to construct the line of intersection of two planes each passing through three given points.

6. A weight is supported on a tripod of three unequal legs resting on a smooth horizontal plane, their feet connected in pairs by strings of given lengths. Show how to determine the tensions of the several strings.

7. Explain the ordinary configuration of a system of isoparametric lines on a spherical surface; for instance, what is the configuration when there are two points of minimum value, and one point of maximum value, of the parameter?

8. Find the attraction of an infinite circular cylinder, of uniform density, on a given exterior or interior point.

9. Determine the number of arbitrary constants contained in the equation of a surface of the order r which passes through the curve of intersection of two given surfaces of the orders m and n respectively.

10. Find, for the several values of p, the number of the conics passing through p given points and touching $5-p$ given lines; and, in each case, show how to obtain (in point-coordinates or line-coordinates, as may be most simple) the equations of the conics satisfying the conditions in question.

11. Investigate the theory of the linear transformation of a ternary quadric function into itself.

12. Explain the theory of the solution of a partial differential equation, given function of x, y, z, p, q, r = arbitrary constant H; where p, q, r are the differential coefficients of the dependent variable u in regard to the independent variables x, y, z respectively.

I propose, not (as in former years) to give complete solutions, but only to notice in more or less detail the leading points in the several questions.

1. The expression is of course required, not only for a given integral function of x, but for a given fractional function. The case where the given function is integral presents no difficulty; when the given function is fractional, the most simple case is where it is $=\dfrac{1}{x-a}$; supposing the equation to be $f(x)=0$, here dividing $f(x)$ by $x-a$, we have a quotient $R(x)$ which is a rational and integral function of an order not exceeding $n-1$, and a remainder which is $=f(a)$; that is,

$$\frac{f(x)}{x-a}=R(x)+\frac{f(a)}{x-a};$$

or, in virtue of the given equation $\dfrac{f(a)}{x-a}=-R(x)$, viz. we have thus $\dfrac{1}{x-a}$ in the required form. But if $f(a)=0$, then we do not obtain such an expression of $\dfrac{1}{x-a}$. It has to be shown that the like considerations apply to any fractional function, and the precise form of the theorem is, that any rational function of x which does not become infinite for any value of x satisfying the given equation, can be expressed as a rational and integral function of an order not exceeding $n-1$.

2. The function $1-\sin(2n+1)x$ is a rational and integral function of $\sin x$, of the order $2n+1$; which if n is even (or $2n+1=4p+1$) contains, as is at once seen, the factor $1-\sin x$, but if n is odd (or $2n+1=4p-1$) the factor $(1+\sin x)$. Suppose that any other factor is $1-\dfrac{\sin x}{\sin\alpha}$, where $\sin\alpha$ not $=\pm 1$; then this will be a double factor if only $\sin x=\sin\alpha$ satisfies the condition

$$0=\frac{d}{d\,.\sin x}\{1-\sin(2n+1)x\},$$

that is, $0=\frac{\cos(2n+1)x}{\cos x}$; the value in question gives $\sin(2n+1)x=1$, and therefore $\cos(2n+1)x=0$; and it does not give $\cos x=0$; hence every such factor $1-\frac{\sin x}{\sin\alpha}$ is a double factor, or we have

$$1-\sin(2n+1)x=(1\pm\sin x)\,\Pi\left(1-\frac{\sin x}{\sin\alpha}\right)^2.$$

Or the like result might be obtained by considering instead of $1-\sin(2n+1)x$, the more general function

$$\sin(2n+1)\alpha\pm\sin(2n+1)x,$$

and finally assuming $\alpha=\frac{1}{2}\pi$.

3. Relates to a theory which is not, but ought to be, treated of in the text books of the University. See Serret's *Algèbre Supérieure*, t. II., Sect. IV.

The substitutions which leave $ab+cd$ unaltered are

1	1, that is, the letters remain unchanged,
α	(ab), that is, a and b are interchanged,
β	(cd), that is, c and d are interchanged,
γ	$(ab)\,(cd)$, that is, a and b and also c and d are simultaneously interchanged,
δ	$(ac)\,(bd)$, same with a and c, b and d,
ϵ	$(ad)\,(bc)$, same with a and d, b and c,
ζ	$(acbd)$, that is, we cyclically change a into c, c into b, b into d, and d into a,
θ	$(adbc)$, that is, we cyclically change a into d, d into b, b into c, and c into a,

viz. we have eight substitutions $1, \alpha, \beta, \gamma, \delta, \epsilon, \zeta, \theta$ forming a group; that is, the product of any two of them, in either order, is a substitution of the group (or, what is the same thing, the effect of the successive performance of the two upon any arrangement $abcd$ is the same as that of the performance thereon of some other substitution of the group); thus we have $\alpha^2=1$, $\beta^2=1$, $\gamma^2=1$, $\alpha\beta=\beta\alpha=\gamma$, &c.; the system of these equations, which verify that the set of substitutions form a group, defines the constitution of the group—thus to take a more simple instance, a group of 4 may be $1, \alpha, \alpha^2, \alpha^3$ $(\alpha^4=1)$ or $1, \alpha, \beta, \alpha\beta$, $(\alpha^2=1, \beta^2=1, \alpha\beta=\beta\alpha)$.

4. The expression of the general coefficient is

$$=\frac{1.2\ldots 2n}{1.2\ldots n-\alpha.1.2\ldots n+\alpha},$$

which can be transformed by the well-known formula

$$1.2\ldots n=n^{n+\frac{1}{2}}\sqrt{(\pi)}\,e^{-n},$$

viz. the coefficient thus becomes

$$=\frac{2^{2n}}{\sqrt{(n\pi)}}\frac{1}{\left(1-\frac{\alpha}{n}\right)^{n-\alpha+\frac{1}{2}}\left(1+\frac{\alpha}{n}\right)^{n+\alpha+\frac{1}{2}}}.$$

Now α is supposed small in comparison with n, and the factors in the denominator have the logarithms

$$(n-\alpha+\tfrac{1}{2})\log\left(1-\frac{\alpha}{n}\right),\ =(n-\alpha+\tfrac{1}{2})\left(-\frac{\alpha}{n}+\tfrac{1}{2}\frac{\alpha^2}{n^2}\right),\ =-\alpha+\tfrac{1}{2}\frac{\alpha^2}{n},$$

and

$$(n+\alpha+\tfrac{1}{2})\log\left(1+\frac{\alpha}{n}\right),\qquad\qquad =\ \alpha+\tfrac{1}{2}\frac{\alpha^2}{n};$$

hence the denominator is $=e^{\frac{\alpha^2}{n}}$, and the final approximate value of the coefficient is

$$=\frac{2^{2n}}{\sqrt{(n\pi)}}\,e^{-\frac{\alpha^2}{n}}.$$

Hence, converting as usual the sum into a definite integral, we have the sum of the $2\alpha+1$ coefficients

$$=\frac{2^{2n}}{\sqrt{(n\pi)}}\int_{-\alpha}^{\alpha}e^{-\frac{\alpha^2}{n}}\,d\alpha,$$

or, what is the same thing,

$$=\frac{2^{2n}}{\sqrt{(\pi)}}\int_{-\frac{\alpha}{\sqrt{n}}}^{\frac{\alpha}{\sqrt{n}}}e^{-x^2}dx.$$

For the chance that the number of tosses lies between $n+\alpha$ and $n-\alpha$, this has merely to be divided by 2^{2n}; hence writing $\alpha=kn$, the chance that the number may be between $n(1+k)$ and $n(1-k)$ is

$$=\frac{1}{\sqrt{(\pi)}}\int_{-k\sqrt{n}}^{k\sqrt{n}}e^{-x^2}\,dx,$$

where observe that the integral, taken with the limits ∞, $-\infty$ has the value $\sqrt{(\pi)}$.

Considering k as a given fraction however small, by increasing n we make $k\sqrt{(n)}$ as large as we please, and therefore the integral, as nearly as we please $=\sqrt{(\pi)}$, or the chance as nearly as we please $=1$; and hence the complete and precise statement of the theorem, viz. by sufficiently increasing the number of tosses, the probability that the deviation from equality shall be any given percentage (as small as we please) of the whole number of tosses, can be made as nearly as we please equal to certainty.

Further, restoring α in place of kn, the chance of a number between $n+\alpha$ and $n-\alpha$ is

$$=\frac{1}{\sqrt{(\pi)}}\int_{-\frac{\alpha}{\sqrt{n}}}^{\frac{\alpha}{\sqrt{n}}}e^{-x^2}\,dx,$$

which when $\frac{\alpha}{\sqrt{(n)}}$ is small is $=\frac{2\alpha}{\sqrt{(n\pi)}}$, $\left(\text{more accurately } \frac{2\alpha+1}{\sqrt{(n\pi)}}, \text{ when } \alpha \text{ is small}\right)$; hence, however large α is, the chance of a deviation from equality not exceeding $\pm\alpha$, continually *diminishes* with n, and by making n sufficiently large becomes as small as we please.

5. The point is represented in the given plane by two points which lie *in lined* with a fixed point (say O) of that plane, viz. O is the intersection of the given plane by the line which joins the two projecting points.

A line is represented on the given plane by two lines, viz. these are the projections of the line from the two given points; each point of the line is represented by the points of intersection of the two lines by any line through O.

A plane may be represented on the given plane by means of its trace thereon, and of the two points (*in lined* with O) which represent any point of the plane.

Thus any problem relating to points, lines, and planes, in space is converted into a problem of plane geometry. For instance, to find the trace on the given plane of a plane through three given points A, B, C, the three points are represented by means of the pairs of points A_1, A_2; B_1, B_2; C_1, C_2, the points of each pair lying *in lined* with O; the required trace passes through the intersections with the given plane of the lines BC, CA, AB respectively, and we hence find it as the line through the three points which are the intersections of B_1C_1, B_2C_2, of C_1A_1, C_2A_2, and of A_1B_1, A_2B_2 respectively; that these points are in a line is a theorem of plane geometry, which, if not previously known, would have at once been given by the construction.

6. The solution ought obviously to be obtained from the principle of virtual velocities; taking a, b, c for the lengths of the legs, f, g, h for the lengths of the strings, and z for the height of the summit, z is a known function of a, b, c, f, g, h, $\left(z \text{ is in fact } =\frac{3V}{\Delta}\right.$, where V, the volume of the tetrahedron, is a given function of a, b, c, f, g, h; and Δ, the area of the base, is a given function of $\left.f, g, h\right)$. Writing then F, G, H for the tensions, and W for the weight, and regarding z, f, g, h as variable, the principle gives

$$Wdz + Fdf + Gdg + Hdh = 0,$$

that is,

$$F, G, H, = -W\frac{dz}{df}, \; -W\frac{dz}{dg}, \; -W\frac{dz}{dh},$$

respectively.

7. The ordinary case is when an isoparametric line has on one side of it larger values, on the other side of it smaller values of the parameter; the case where the isoparametric line is a line of maximum, or of minimum, parameter is excluded.

The lines in the neighbourhood of a point of maximum, or of minimum, parameter are ovals surrounding the point in question, each oval being itself surrounded by the consecutive oval. Supposing that there are two points of minimum parameter, we have round each of them a series of ovals, until at length an oval belonging to the one of them comes to unite itself with an oval belonging to the other, the two ovals altering themselves into a figure of eight. Surrounding this we have a closed curve (in the first instance a deeply twice-indented oval) which (in the case supposed of there being, besides the two points of minimum parameter, a single point of maximum parameter) is in fact an oval surrounding the point of maximum parameter, and the remaining curves are the series of ovals surrounding that point. If we project stereographically from the point of maximum parameter (so that this point is represented by the points at infinity) we have a figure of eight, each loop containing within it a series of continually diminishing closed curves, and the figure of eight itself surrounded by a series of continually increasing closed curves.

8. The investigation by means of the Potential presents the difficulty that the Potential of the infinite cylinder has no determinate value, as at once appears from the limiting case where the cylinder is reduced to a right line; the difficulty is perhaps rather apparent than real, inasmuch as the partial differential equations contain only differential coefficients $\frac{dV}{dr}$, $\frac{d^2V}{dr^2}$, where $\frac{dV}{dr}$ as representing an attraction, and therefore also $\frac{d^2V}{dr^2}$, are determinate. But it is safer to work directly with the Attraction; the Attraction of an infinite line acts in the perpendicular plane through the attracted point, and is inversely proportional to the distance; the problem is thus reduced to the plane problem of a circle of uniform density, force varying as $(\text{distance})^{-1}$, attracting a point in its own plane. This is precisely similar to the case of a sphere with the ordinary law of attraction; dividing the circle into rings, each ring exerts an attraction $=0$ upon an interior point, and an attraction as if collected at the centre upon an exterior point. Hence, writing a for the radius of the cylinder, and r for the distance of the attracted point, the attraction is $=\pi r$ for an interior point, and $=\frac{\pi a^2}{r}$ for an exterior point.

9. The theory is precisely the same as for curves; taking the surfaces to be $U=0$ of the order m, and $V=0$ of the order n, the general form of the equation of a surface of the order r (r not less than m or n) is $LU+MV=0$, where L is the general function of the order $r-m$, and M the general function of the order $r-n$; and so long as r is less than $m+n$, we obtain the required number of arbitrary constants as the sum of the numbers of the coefficients of L and of M, less unity. But as soon as r is $=m+n$ a modification arises, viz. we obtain here an identity by assuming $L=V$, $M=-U$, and so for any larger value of r, we have an identity by assuming $L=V\phi$, $M=-U\phi$, where ϕ is the general function of the order $r-m-n$.

10. The numbers are known to be 1, 2, 4, 4, 2, 1, which values are obtained most easily (though not in the way which is theoretically most interesting) by finding for

the first three cases the equation of the required conic in point-coordinates; and then, by changing these into line-coordinates, we have the equations for the remaining three cases.

$p=5$: 5 points. The equation of the conic is

$$(a, b, c, f, g, h \between x, y, z)^2 = 0,$$

and we have 5 linear equations to determine the ratios of the coefficients; the number is therefore $=1$.

$p=4$: 4 points and 1 line. Taking $U=0$ and $V=0$, the equations of any two conics each passing through the four points, the equation of the required conic will be $U+\lambda V=0$, and the condition of touching a given line gives a quadric equation for λ; the number is therefore $=2$.

$p=3$: 3 points and 2 lines. In the same manner, by taking $U=0$, $V=0$, $W=0$, for the equations of any three conics through the three points; or if the equations of the lines containing the three points in pairs are $x=0$, $y=0$, $z=0$, then the equations of the three conics are $yz=0$, $zx=0$, $xy=0$, and the equation of any conic through these points is $fyz+gzx+hxy=0$; the conditions of touching two given lines $\xi x+\eta y+\zeta z=0$ and $\xi' x+\eta' y+\zeta' z=0$, are

$$\sqrt{f}\sqrt{\xi}+\sqrt{g}\sqrt{\eta}+\sqrt{h}\sqrt{\zeta}=0, \quad \sqrt{f}\sqrt{\xi'}+\sqrt{g}\sqrt{\eta'}+\sqrt{h}\sqrt{\zeta'}=0;$$

we have thus the ratios $\sqrt{f} : \sqrt{g} : \sqrt{h}$ linearly determined in terms of $\sqrt{\xi}$, $\sqrt{\eta}$, &c.; there is no loss of generality in taking $\sqrt{\xi}$, $\sqrt{\xi'}$ each with a determinate sign, the signs of $\sqrt{\eta}$, &c. being then arbitrary, we have 2^4, $=16$ values of $\sqrt{f} : \sqrt{g} : \sqrt{h}$, and therefore one-fourth of this $=4$, for the number of values of $f : g : h$; that is, the number is $=4$.

11. This is a known theory; taking x_1, y_1, z_1 for the linear functions of x, y, z, which are such that

$$(a, b, c, f, g, h \between x_1, y_1, z_1)^2 = (a, b, c, f, g, h \between x, y, z)^2,$$

then assuming $x_1, y_1, z_1 = 2\xi - x,\ 2\eta - y,\ 2\zeta - z$ respectively, we have

$$(a, \ldots \between 2\xi - x,\ 2\eta - y,\ 2\zeta - z)^2 = (a, \ldots \between x, y, z)^2,$$

or, omitting terms which destroy each other, and throwing out the factor 4, this is

$$(a, \ldots \between \xi, \eta, \zeta)^2 = (a, \ldots \between \xi, \eta, \zeta \between x, y, z),$$

an equation which is satisfied identically by assuming

$$\begin{aligned} a\xi+h\eta+g\zeta &= ax+hy+gz \quad . \quad -\nu\eta+\mu\zeta, \\ h\xi+b\eta+f\zeta &= hx+by+fz+\nu\xi \quad . \quad -\lambda\zeta, \\ g\xi+f\eta+c\zeta &= gx+fy+cz-\mu\xi+\lambda\eta \quad . \quad , \end{aligned}$$

where λ, μ, ν are arbitrary; viz. multiplying by ξ, η, ζ, and adding we have the equation in question. The three equations determine ξ, η, ζ as linear functions of x, y, z; and we have thence x_1, y_1, z_1 as linear functions of x, y, z; viz. this is a solution containing three arbitrary constants λ, μ, ν.

12. The partial differential equation might equally well have been proposed in the form, given function of x, y, z, p, q, $r=0$, viz. the equation then is $\phi(x, y, z, p, q, r)=0$, the general partial differential equation involving the three independent variables x, y, z, and the derived functions p, q, r of the dependent variable u, *but not involving the dependent variable* u. The question is therefore in effect as follows: to find p, q, r functions of x, y, z connected by the foregoing equation, and, moreover, such that $pdx+qdy+rdz$ is an exact differential; for then writing $u=\int(pdx+qdy+rdz)$, we have the solution of the given partial differential equation.

Whatever be the method adopted, it comes out that the solution depends on the integration of the system of ordinary differential equations

$$\frac{dp}{-\frac{d\phi}{dx}}=\frac{dq}{-\frac{d\phi}{dy}}=\frac{dr}{-\frac{d\phi}{dz}}=\frac{dx}{\frac{d\phi}{dp}}=\frac{dy}{\frac{d\phi}{dq}}=\frac{dz}{\frac{d\phi}{dr}},$$

and the answer consists first in showing this, and secondly, in showing how from an integral or integrals of the system we pass to the solution of the partial differential equation.

Considering the partial differential equation in the form actually proposed, we may instead of ϕ write H, where H will stand for that given function of x, y, z, p, q, r which is the value of the arbitrary constant H; making this change, and putting the foregoing equal quantities equal to the differential dt of a new variable, the system of ordinary differential equations is

$$\frac{dp}{dt}=-\frac{dH}{dx},\quad \frac{dq}{dt}=-\frac{dH}{dy},\quad \frac{dr}{dt}=-\frac{dH}{dz},$$

$$\frac{dx}{dt}=\frac{dH}{dp},\quad \frac{dy}{dt}=\frac{dH}{dq},\quad \frac{dz}{dt}=\frac{dH}{dr},$$

where H is a given function of x, y, z, p, q, r. This is, in fact, the Hamiltonian system of equations; and it was in view to the connexion that the partial differential equation was proposed in its actual form.

646.

ON THE GENERAL EQUATION OF DIFFERENCES OF THE SECOND ORDER.

[From the *Quarterly Journal of Pure and Applied Mathematics*, vol. XIV. (1877), pp. 23—25.]

CONSIDER the equation of differences

$$u_x = a_{x-1}\, u_{x-1} + b_{x-2}\, u_{x-2},$$

viz. we have

$$\begin{aligned} u_2 &= a_1u_1 + b_0u_0, \\ u_3 &= a_2u_2 + b_1u_1, \\ u_4 &= a_3u_3 + b_2u_2, \\ u_5 &= a_4u_4 + b_3u_3, \\ u_6 &= a_5u_5 + b_4u_4, \\ &\&c., \end{aligned}$$

and thence

$$u_3 = \left|\begin{array}{l} a_2a_1 \\ +b_1 \end{array}\right| u_1 + \quad a_2 \quad b_0u_0,$$

$$u_4 = \begin{array}{l} a_3a_2a_1 \\ +a_3b_1 \\ +a_1b_2 \end{array}\left|\begin{array}{l} u_1 + \\ \\ \\ \end{array}\right. \begin{array}{l} a_3a_2 \\ +b_2 \\ \\ \end{array}\left|\begin{array}{l} b_0u_0, \\ \\ \\ \end{array}\right.$$

$$u_5 = \begin{array}{l} a_4a_3a_2a_1 \\ +a_4a_3b_1 \\ +a_4a_1b_2 \\ +a_2a_1b_3 \\ +b_1b_3 \end{array}\left|\begin{array}{l} u_1 + \\ \\ \\ \\ \\ \end{array}\right. \begin{array}{l} a_4a_3a_2 \\ +a_4b_2 \\ +a_2b_3 \end{array}\left|\begin{array}{l} b_0u_0, \\ \\ \\ \end{array}\right.$$

$$u_6 = \left.\begin{array}{l} a_5a_4a_3a_2a_1 \\ +\,a_5a_4a_3b_1 \\ +\,a_5a_4a_1b_2 \\ +\,a_5a_2a_1b_3 \\ +\,a_3a_2a_1b_4 \\ +\,a_5b_3b_1 \\ +\,a_3b_4b_1 \\ +\,a_1b_4b_2 \end{array}\right| u_1 + \left.\begin{array}{l} a_5a_4a_3a_2 \\ +\,a_5a_4b_2 \\ +\,a_5a_2b_3 \\ +\,a_3a_2b_4 \\ +\,b_4b_2 \end{array}\right| b_0u_0,$$

&c.

It is now easy to see the law; viz. writing for instance

$$u_6 = 54321\,.\,u_1 + 5432\,.\,b_0u_0,$$

then 54321 has a leading term $a_5a_4a_3a_2a_1$: it has terms derived from this by changing any pair a_2a_1 into b_1, a_3a_2 into b_2, a_4a_3 into b_3, a_5a_4 into b_4: it has terms derived by changing any two pairs a_4a_3, a_2a_1 into b_3b_1; a_5a_4, a_2a_1 into b_4b_1; a_5a_4, a_3a_2 into b_4b_2, and so on; where observe that the expression a pair denotes the product of two consecutive a's.

And, similarly, 5432 has a leading term $a_5a_4a_3a_2$; the other terms being derived from this in the same manner precisely.

The solution of $u_x = lx\,(au_{x-1} - u_{x-2})$ is included in, and might be deduced from the foregoing, but it is convenient to obtain it separately. Supposing for greater simplicity that $u_{-1} = 0$, $u_0 = 1$ (or, what is the same thing, $u_0 = 1$, $u_1 = l_1a$), then we find

$$u_0 = 1,$$

$$u_1 = l_1a,$$

$$u_2 = l_2l_1a^2 - l_2,$$

$$u_3 = l_3l_2l_1a^3 - \left|\begin{array}{l} l_3l_2 \\ +\,l_3l_1 \end{array}\right| a,$$

$$u_4 = l_4l_3l_2l_1a^4 - \left|\begin{array}{l} l_4l_3l_2 \\ +\,l_4l_3l_1 \\ +\,l_4l_2l_1 \end{array}\right| a^2 + l_4l_2,$$

$$u_5 = l_5l_4l_3l_2l_1a^5 - \left|\begin{array}{l} l_5l_4l_3l_2 \\ +\,l_5l_4l_3l_1 \\ +\,l_5l_4l_2l_1 \\ +\,l_5l_3l_2l_1 \end{array}\right| a^3 + \left|\begin{array}{l} l_5l_4l_2 \\ +\,l_5l_3l_2 \\ +\,l_5l_3l_1 \end{array}\right| a,$$

&c.,

viz. we may for example write

$$u_5 = l_5\, 4321\,.\, a^5 - 4321\,(\cdot)\, a^3 + 4321\,(:)\, a\,;$$

where

$$4321 \text{ denotes } l_4 l_3 l_2 l_1:$$

in 4321 ($\cdot$), we omit successively each number, viz. we thus obtain

$$432 + 431 + 421 + 321\,,$$
$$= l_4 l_3 l_2 + l_4 l_3 l_1 + l_4 l_2 l_1 + l_3 l_2 l_1:$$

in 4321 (:), we omit successively each two non-consecutive numbers, viz. the omitted numbers being 1, 3; 1, 4; 2, 4, we obtain

$$42 + 32 + 31,$$
$$= l_4 l_2 + l_3 l_2 + l_3 l_1:$$

and so on, the omissions being each three numbers, each four numbers, &c., no two of them being consecutive; thus in 654321 ($\because$), the omissions are 5, 3, 1, and 6, 4, 2; or the symbol is

$$642 + 531\,,$$
$$= l_6 l_4 l_2 + l_5 l_3 l_1.$$

As an application, a solution of the differential equation $\dfrac{d}{dx}\left(x\dfrac{dy}{dx}\right) + (x - a)\,y = 0$ is $y = u_0 + u_1 x + u_2 x^2 + \text{\&c.}$, where $n^2 u_n = a u_{n-1} - u_{n-2}$, and in particular $1^2 u_1 = a u_0$; the equation of differences is thus of the form in question, and retaining l_n in place of its value, $= n^2$, the solution is $u_0 = 1$, $u_1 = l_1 a$, $u_2 = l_2 l_1 a^2 - l_2$, &c. *ut suprà*. The differential equation was considered by the Rev. H. J. Sharpe, who mentioned it to Prof. Stokes.

647.

ON THE QUARTIC SURFACES REPRESENTED BY THE EQUATION, SYMMETRICAL DETERMINANT = 0.

[From the *Quarterly Journal of Pure and Applied Mathematics*, vol. XIV. (1877), pp. 46—52.]

CONSIDER the equation

$$\nabla = \begin{vmatrix} a, & h, & g, & l \\ h, & b, & f, & m \\ g, & f, & c, & n \\ l, & m, & n, & d \end{vmatrix} = 0,$$

where for the moment $(a, b, \ldots)$ denote linear functions of the coordinates (x, y, z, w). This is a quartic surface having 10 nodes; viz. if we write $(A, B, \ldots)$ for the first minors of the determinant, then the cubic surfaces $A = 0$, $B = 0$, ... have in common 10 points which are nodes of the quartic surface.

Suppose that (a, b, c, f, g, h) are linear functions of the form (x, y); then, observing that every term of ∇ contains as a factor

$$\begin{vmatrix} a, & h, & g \\ h, & b, & f \\ g, & f, & c \end{vmatrix},$$

or one of its first minors, it is clear that the line $x = 0$, $y = 0$ is a double line on the surface. But the number of nodes is now less than 10; in fact, writing $(x = 0, y = 0)$, we make each of the first minors of ∇ to vanish; that is, the cubic surfaces, which by their intersection determine the nodes, have in common the line $(x = 0, y = 0)$, and there is a diminution in the number of their common intersections. I do not pursue the enquiry, but pass to a different question.

I, in fact, take the terms $(a, \ldots)$ of the determinant to be homogeneous functions of (x, y, z, w) of the degrees

$$\begin{vmatrix} 0, & 1, & 1, & 0 \\ 1, & 2, & 2, & 1 \\ 1, & 2, & 2, & 1 \\ 0, & 1, & 1, & 0 \end{vmatrix},$$

respectively, viz. a, d, l are constants, g, h, m, n linear functions, and b, c, f quadric functions of the coordinates; $\nabla = 0$ still represents a quartic surface; and it appears by a general formula that the number of nodes is $= 8$. But we can easily show this directly; and further, that the 8 nodes are the intersections of three quadric surfaces; or say that the quartic surface is *octadic*. For denoting as before the first minors by $A, \ldots,$ then B, C, F are each of them a quadric function of the coordinates, viz. we have

$$\begin{aligned} B &= d\,(ac - g^2\,) - cl^2 \;\; - an^2 \;+ 2gln, \\ C &= d\,(ab - h^2\,) - am^2 - bl^2 \;\; + 2hlm, \\ F &= d\,(gh - af) + l^2f \;\; + mna - nlh - lmg, \end{aligned}$$

and we have identically

$$BC - F^2 = (ad - l^2)\,\nabla,$$

so that throwing out the constant factor $ad - l^2$, the equation of the surface is

$$BC - F^2 = 0,$$

and it has 8 nodes, the intersections of the three quadric surfaces $B = 0$, $C = 0$, $F = 0$. By equating to zero any other minor of the determinant ∇, we have a surface passing through the 8 nodes; we have for instance the quartic surface

$$\begin{vmatrix} a, & h, & g \\ h, & b, & f \\ g, & f, & c \end{vmatrix} = 0.$$

Suppose now (and in all that follows) that, the degrees being as already mentioned, we further assume that b, c, f are quadric functions of the form $(x, y)^2$, g, h linear functions of the form (x, y); then since each term of ∇ contains either

$$\begin{vmatrix} a, & h, & g \\ h, & b, & f \\ g, & f, & c \end{vmatrix}$$

or one of its first minors, it is clear that the line $(x = 0,\ y = 0)$ is a double line on the surface. But in the present case there is not any diminution in the number of the nodes; in fact, writing $x = 0$, $y = 0$, and therefore b, c, f, g, h each $= 0$ (but not $a = 0$), the minors B, C, F none of them vanish; that is, the line $x = 0$, $y = 0$ is not a line on any one of the quadric surfaces, and the quadric surfaces intersect as before in an octad of points.

The equation $\nabla = 0$ thus represents a quartic surface having a double line, and also 8 nodes forming an octad.

We may without loss of generality write $d = 0$; in fact, the determinant is unaltered if we add to the fourth column θ times the first column, and then to the fourth line θ times the first line; the determinant is thus of the original form, but in place of d it has $d + 2\theta l + \theta^2 a$, which by properly determining θ can be made $= 0$. And then changing the original l, m, n, the equation is

$$\nabla = \begin{vmatrix} a, & h, & g, & l \\ h, & b, & f, & m \\ g, & f, & c, & n \\ l, & m, & n, & 0 \end{vmatrix} = 0.$$

Or, writing for shortness,

$$K = \begin{vmatrix} a, & h, & g \\ h, & b, & f \\ g, & f, & c \end{vmatrix},$$

and denoting the minors hereof by (a, b, c, f, g, h), then the equation is

$$\nabla = (\mathrm{a, b, c, f, g, h} \between l, m, n)^2 = 0,$$

where the degree of K is 4, and the degrees of a, b, c, f, g, h are 4, 2, 2, 2, 3, 3 respectively, those of l, m, n being 0, 1, 1 respectively.

The nodes are, as before, the intersections of the quadric surfaces $B = 0$, $C = 0$, $F = 0$, viz. (d being now $= 0$) the values are

$$\begin{aligned} -B &= cl^2 - 2gln + an^2, \\ -C &= bl^2 - 2hlm + am^2, \\ F &= fl^2 - glm - hln + amn. \end{aligned}$$

But, according to a previous remark, the nodes lie also on the quartic surface $K = 0$; viz. this is a set of four planes intersecting in the line $x = 0$, $y = 0$.

Now, in general, any plane through the line $x = 0$, $y = 0$ meets the surface in this line twice and in a conic; if the plane is $y = \theta x$, we have

$$a,\ b,\ c,\ f,\ g,\ h = a',\ b'x^2,\ c'x^2,\ f'x^2,\ g'x,\ h'x,$$

where a', b', c', f', g', h' are functions of θ of the degrees (0, 2, 2, 2, 1, 1) respectively; and thence also

$$\mathrm{a, b, c, f, g, h} = \mathrm{a}'x^4,\ \mathrm{b}'x^2,\ \mathrm{c}'x^2,\ \mathrm{f}'x^2,\ \mathrm{g}'x^3,\ \mathrm{h}'x^3,$$

where a′, b′, c′, f′, g′, h′ are functions of θ of the degrees 4, 2, 2, 2, 3, 3 respectively; the equation of the surface thus becomes $(\mathrm{a}', \mathrm{b}', \mathrm{c}', \mathrm{f}', \mathrm{g}', \mathrm{h}' \between lx, m, n)^2 = 0$; viz. this is a quadric equation which, combined with the equation $y - \theta x = 0$, determines the conic in question. But for each of the planes $K=0$, we have $(\mathrm{a}', \mathrm{b}', \mathrm{c}', \mathrm{f}', \mathrm{g}', \mathrm{h}' \between lx, m, n)^2$ a perfect square, or the conic a two-fold line; we have thus the 8 nodes lying in pairs on four lines, say the four "rays," in the four planes $K=0$ respectively; each of these rays meets the double line $x=0$, $y=0$ in a point; and we have thus on the double line 4 points, which are in fact pinch-points of the surface (as to this presently). It has just been stated that for the plane passing through the nodal line and a ray, the conic is a two-fold line (the ray twice) containing upon it a pair of nodes; more properly, the conic is the point-pair composed of the two nodes.

We can find through the nodes four different plane-pairs; in fact, forming the equation

$$-B+2\lambda F-\lambda^2 C=0,$$

this is

$$l^2(c+2\lambda f+\lambda^2 b)-2l(g+\lambda h)(n+\lambda m)+a(n+\lambda m)^2=0;$$

or, as this may also be written,

$$[a(n+\lambda m)-l(g+\lambda h)]^2+l^2(\mathrm{b}-2\lambda\mathrm{f}+\lambda^2\mathrm{c})=0,$$

where b, c, f and therefore also $\mathrm{b}-2\lambda\mathrm{f}+\lambda^2\mathrm{c}$ are of the form $(x, y)^2$; say that we have $\mathrm{b}-2\lambda\mathrm{f}+\lambda^2\mathrm{c}=(p, q, r \between x, y)^2$, where p, q, r are of course quadric functions of λ; determining λ by the quartic equation $pr-q^2=0$, we have $\mathrm{b}-2\lambda\mathrm{f}+\lambda^2\mathrm{c}$ a perfect square, $=(\alpha x+\beta y)^2$ suppose; and we have thus the plane-pair

$$[a(n+\lambda m)-l(g+\lambda h)]^2-l^2(\alpha x+\beta y)^2=0$$

containing the eight nodes; viz. there are four such plane-pairs. The two planes of a plane-pair intersect in a line called an "axis"; that is, we have four axes each meeting the nodal line; and we have thus also through the nodal line and the four axes respectively four planes, which are "pinch-planes" of the quartic surface (as to this presently).

It has just been seen that the equation $B-2\lambda F+C\lambda^2=0$ (where λ is arbitrary) is expressible in the form

$$[a(n+\lambda m)-l(g+\lambda h)]^2+l^2(p, q, r \between x, y)^2=0,$$

viz. this is the equation of a quadric cone having its vertex on the nodal line at the point $x=0$, $y=0$, $an-lg+\lambda(am-lh)=0$; this is, in fact, a cone *touching* the surface, as at once appears by writing the equation of the cone in the form

$$\frac{1}{C}\{BC-F^2+(\lambda C-F)^2\}=0,$$

that is,

$$\frac{1}{C}\{-l^2\nabla+(\lambda C-F)^2\}=0;$$

we thus see that, taking for vertex any point whatever on the nodal line, there is a circumscribed quadric cone.

For each of the above-mentioned four values of λ, the quadric cone breaks up into a plane-pair; each plane of the plane-pair is thus a "trope" or plane touching the surface along a conic; viz. this is the conic passing through the intersection of the plane (or say of an axis) with the nodal line and through four nodes of the surface. We have thus 8 tropes, intersecting in pairs in the four axes (and the intersection of each axis with the nodal line being a pinch-point). Moreover, joining the nodes in pairs, we have four rays, each meeting the nodal line, the plane through it and the nodal line being a pinch-plane; this is illustrated in the figure.

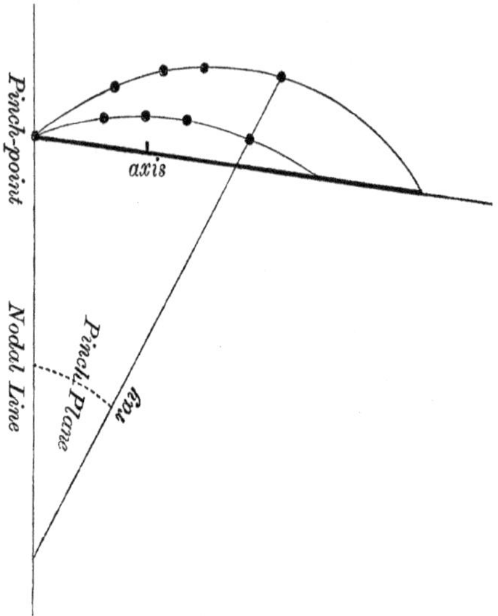

As to the pinch-planes and pinch-points, remark first that a plane through the nodal line is in general a bitangent plane, its two points of contact being the points where the conic in such plane meets the nodal line. When the two points of contact come to coincide, the plane is a pinch-plane; viz. this happens when the plane passes through a ray, the conic being then the ray twice repeated. And secondly, at a point on the nodal line there are in general two tangent planes, viz. these are the tangent planes to the quadric cone belonging to such point; when the two tangent-planes come to coincide the point is a pinch-point, and this happens when the point is the intersection of the nodal line with an axis, for then (the quadric cone breaking up into the two tropes through the axis) the two tangent planes become the plane through the axis taken twice.

Each section through the nodal line is a conic, and the polar of the nodal line in regard to this conic is a point; the locus of this point (for different sections through the nodal line) is a right line which may be called simply the "polar." To prove this, considering the section by the plane $y = \theta x$, we have to find the pole of the line $x=0$ in regard to the conic

$$(\mathrm{a}', \mathrm{b}', \mathrm{c}', \mathrm{f}', \mathrm{g}', \mathrm{h}' \between lx, m, n)^2 = 0;$$

this is $lx : m : n = a' : h' : g'$, viz. if $g = g_0 x + g_1 y$, $h = h_0 x + h_1 y$, this is

$$lx : m : n = a : g_0 + g_1\theta : h_0 + h_1\theta,$$

or joining hereto the equation $y = \theta x$, we have

$$lx : ly : m : n = a : a\theta : g_0 + g_1\theta : h_0 + h_1\theta,$$

where l, a, g_0, g_1, h_0, h_1 are constants; m, n are linear functions of the coordinates (x, y, z, w). The equations represent, it is clear, a right line which is the polar in question; and they may be written

$$\frac{lx}{a} = \frac{h_1 m - g_1 n}{h_1 g_0 - h_0 g_1}, \quad \frac{ly}{a} = -\frac{h_0 m - g_0 n}{h_1 g_0 - h_0 g_1}.$$

When the plane passes through a ray, the conic becomes, as was stated, the point-pair composed of the two nodes in such ray; the harmonic in regard to these two points of the intersection of the ray with the nodal line is thus a point on the polar: that is, the polar meets the ray; and the two nodes are situate harmonically in regard to the intersections of the ray with the nodal line and the polar respectively.

The polar may be arrived at in a different manner, viz. if instead of a plane through the nodal line we consider a point on the nodal line, this is the vertex of a circumscribed quadric cone; and taking the polar plane of the nodal line in regard to this cone, then considering the point as variable, the different polar planes all pass through a line which is the polar in question. And hence, taking for the point the intersection of the nodal line with an axis, it appears that the axis meets the polar; and, moreover, that the two tropes through the axis are harmonics in regard to the planes through the axis, and the polar and nodal line respectively.

Collecting the foregoing results, we have a quartic surface as follows:

We have two lines, a nodal line and a polar; meeting each of these, four lines called "rays" and four other lines called "axes." On each ray, harmonically in regard to its intersections with the nodal line and the polar, two nodes of the surface (in all 8 nodes): through each axis, harmonically in regard to the planes through it and the nodal line and the axis respectively, two tropes of the surface (in all 8 tropes). In each trope (or, what is the same thing, in its conic of contact) are 4 nodes; through each node (or, what is the same thing, touching its tangential quadricone) are 4 tropes; the relation of the nodes and tropes may be thus represented, viz. taking the pairs of nodes to be 1, 2; 3, 4; 5, 6; 7, 8; and those of tropes to be I, II; III, IV; V, VI; VII, VIII; then we have

	I	II	III	IV	V	VI	VII	VIII
1	·		·		·		·	
2		·		·		·		·
3	·		·			·		·
4		·		·	·		·	
5	·			·	·			·
6		·	·			·	·	
7	·			·		·	·	
8		·	·		·			·

viz. reading horizontally or vertically, the dots show the tropes through each node, or the nodes in each trope.

The plane through any ray and the nodal line is a pinch-plane of the surface, its point of contact being the intersection of the ray with the nodal line; and the intersection of each axis with the nodal line is a pinch-point of the surface, the tangent plane being the plane through the axis and the nodal line; the surface has thus 4 pinch-planes and 4 pinch-points.

648.

ALGEBRAICAL THEOREM.

[From the *Quarterly Journal of Pure and Applied Mathematics*, vol. XIV. (1877), p. 53.]

I WISH to put on record the following theorem, given by me as a Senate-House Problem, January, 1851.

If $\{\alpha+\beta+\gamma+\ldots\}^p$ denote the expansion of $(\alpha+\beta+\gamma+\ldots)^p$, retaining those terms $N\alpha^a\beta^b\gamma^c\ldots$ only in which

$$b+c+d+\ldots \not> p-1, \quad c+d+\ldots \not> p-2, \text{ \&c., \&c.,}$$

then

$$x^n = (x+\alpha)^n - n\,\{\alpha\}^1 (x+\alpha+\beta)^{n-1} + \tfrac{1}{2}n\,(n-1)\,\{\alpha+\beta\}^2 (x+\alpha+\beta+\gamma)^{n-2}$$
$$- \tfrac{1}{6}n\,(n-1)\,(n-2)\,\{\alpha+\beta+\gamma\}^3 (x+\alpha+\beta+\gamma+\delta)^{n-3} + \text{\&c.}$$

The theorem, in a somewhat different and imperfectly stated form, is given, Burg, *Crelle*, t. I. (1826), p. 368, as a generalisation of Abel's theorem,

$$(x+\alpha)^n = x^n + n\alpha\,(x+\beta)^{n-1} + \tfrac{1}{2}n\,(n-1)\,\alpha\,(\alpha-2\beta)\,(x+2\beta)^{n-2}$$
$$+ \tfrac{1}{6}\,(n-1)\,(n-2)\,(n-3)\,\alpha\,(\alpha-3\beta)^2\,(x+3\beta)^2 + \text{\&c.}$$

649.

ADDITION TO MR GLAISHER'S NOTE ON SYLVESTER'S PAPER, "DEVELOPMENT OF AN IDEA OF EISENSTEIN."

[From the *Quarterly Journal of Pure and Applied Mathematics*, vol. XIV. (1877), pp. 83, 84.]

THE formula (11) [in the Note], under a slightly different form, is demonstrated by me in an addition [263] to Sir J. F. W. Herschel's paper "On the formulæ investigated by Dr Brinkley, &c.," *Phil. Trans.* t. CL., 1860, pp. 321—323. The demonstration is in effect as follows: let u denote a series of the form $1+bx+cx^2+dx^3+\ldots$, and let u^i (where i is positive or negative, integer or fractional) denote the development of the i-th power of u, continued up to the term which involves x^n, the terms involving higher powers of x being rejected; $u^0, u^1, u^2, \ldots$, and generally u^s will denote in like manner the developments of these powers up to the terms involving x^n, or, what is the same thing, they will be the values of u^i corresponding to $i=0, 1, 2, \ldots, s$. By the formula $u^i = 1+\frac{i}{1}(u-1)+\frac{i\,.\,i-1}{1\,.\,2}(u-1)^2+\ldots$ as far as the term involving $(u-1)^n$, u^i is a rational and integral function of i of the degree n, and can therefore be expressed in terms of the values $u^0, u^1, u^2, \ldots, u^n$ which correspond to $i=0, 1, 2, \ldots, n$. Let s have any one of the last-mentioned values, then the expression

$$\frac{i\,.\,i-1\,.\,i-2\ldots i-n}{i-s}\;\frac{1}{s\,.\,s-1\ldots 2\,.\,1\,.\,-1\,.\,-2\ldots -(n-s)},$$

which as regards i is a rational and integral function of the degree n (the factor $i-s$ which occurs in the numerator and denominator being of course omitted), vanishes for each of the values $i=0, 1, 2, \ldots, n$, except only for the value $i=s$, in which case it becomes equal to unity. The required formula is thus seen to be

$$u^i = \Sigma\left\{\frac{i\,.\,i-1\,.\,i-2\ldots i-n}{i-s}\;\frac{1}{s\,.\,s-1\ldots 2\,.\,1\,.\,-1\,.\,-2\ldots -(n-s)}\,u^s\right\},$$

where the summation extends to the several values $s=0, 1, 2,.., n$; or, what is the same thing, it is

$$u^i = \Sigma \left\{ \frac{i \,.\, i-1 \,.\, i-2 \ldots i-n}{i-s} \; \frac{(-)^{n-s} 1}{1 \,.\, 2 \ldots s \,.\, 1 \,.\, 2 \ldots (n-s)} \, u^s \right\},$$

or, changing the sign of i, it is

$$u^{-i} = \Sigma \left\{ \frac{i \,.\, i+1 \,.\, i+2 \ldots i+n}{i+s} \; \frac{(-)^s 1}{1 \,.\, 2 \ldots s \,.\, 1 \,.\, 2 \ldots n-s} \, u^s \right\},$$

where, as before, s has the values 0, 1, 2,.., n successively. Or, what is the same thing, we have

$$C_{-i,\, n} = \Sigma \left\{ \frac{i \,.\, i+1 \,.\, i+2 \ldots i+n}{i+s} \; \frac{(-)^s 1}{1 \,.\, 2 \ldots s \,.\, 1 \,.\, 2 \ldots n-s} \, C_{s,\, n} \right\},$$

where the term corresponding to $s=0$, as containing the factor $C_{0,\, n}$ vanishes except in the case $n=0$ (for which it is $=1$); and omitting this evanescent term, this is in fact the formula (11).

650.

ON A QUARTIC SURFACE WITH TWELVE NODES.

[From the *Quarterly Journal of Pure and Applied Mathematics*, vol. XIV. (1877), pp. 103—106.]

WRITE for shortness

$$\begin{aligned} a &= \beta-\gamma, & f &= \alpha-\delta, & af &= p, \\ b &= \gamma-\alpha, & g &= \beta-\delta, & bg &= q, \\ c &= \alpha-\beta, & h &= \gamma-\delta, & ch &= r; \end{aligned}$$

then, θ being a variable parameter, the surface in question is the envelope of the quadric surface

$$(\alpha+\theta)^2\, aghX^2 + (\beta+\theta)^2\, bhfY^2 + (\gamma+\theta)^2\, cfgZ^2 + (\delta+\theta)^2\, abcW^2 = 0;$$

viz. this is

$$\Sigma\alpha^2 aghX^2 \,.\, \Sigma aghX^2 - \Sigma\alpha aghX^2 = 0.$$

There are no terms in X^4, &c.; the coefficient of Y^2Z^2 is

$$\gamma^2 cfg \,.\, bfh + \beta^2 bfh \,.\, cfg - 2\beta bfh \,.\, \gamma cfg,$$

which is

$$= bcf^2gh\,(\beta-\gamma)^2, \; = a^2bcf^2gh, \; = abcfgh \,.\, p.$$

Hence the whole equation divides by $abcfgh$, and throwing out this factor, the result is

$$p\,(Y^2Z^2 + X^2W^2) + q\,(Z^2X^2 + Y^2W^2) + r\,(X^2Y^2 + Z^2W^2) = 0,$$

or, observing that $p+q+r=0$, this may also be written

$$p\,(YZ+XW)^2 + q\,(ZX+YW)^2 + r\,(XY+ZW)^2 = 0,$$

and also

$$p(YZ-XW)^2+q(ZX-YW)^2+r(XY-ZW)^2=0.$$

The more general equation

$$(p,\ q,\ r,\ l,\ m,\ n\between YZ+XW,\ ZX+YW,\ XY+ZW)^2=0$$

represents a quartic surface (octadic) having the 8 nodes

$$\begin{array}{ll} (1,\ 0,\ 0,\ 0), & (\bar{1},\ 1,\ 1,\ 1), \\ (0,\ 1,\ 0,\ 0), & (1,\ \bar{1},\ 1,\ 1), \\ (0,\ 0,\ 1,\ 0), & (1,\ 1,\ \bar{1},\ 1), \\ (0,\ 0,\ 0,\ 1), & (1,\ 1,\ 1,\ \bar{1}). \end{array}$$

We have

	$d_X U =$		$d_Y U =$
$p.$	$XW^2\ +YZW$	$p.$	$YZ^2\ +XZW$
$q.$	$YW^2\ +YZW$	$q.$	$YW^2\ +XZW$
$r.$	$ZW^2\ +YZW$	$r.$	$YX^2\ +XZW$
$l.$	$2XYZ+W(Y^2+Z^2)$	$l.$	$2XYW+Z(W^2+X^2)$
$m.$	$2XYW+Z(W^2+Y^2)$	$m.$	$2YZX\ +W(Z^2+X^2)$
$n.$	$2XZW+Y(W^2+Z^2)$,	$n.$	$2YZW\ +X(W^2+Z^2)$,

	$d_Z U =$		$d_W U =$
$p.$	$Y^2Z\ +XYW$	$p.$	$X^2W\ +XYZ$
$q.$	$X^2Z\ +XYW$	$q.$	$Y^2W\ +XYZ$
$r.$	$W^2Z\ +XYW$	$r.$	$Z^2W\ +XYZ$
$l.$	$2ZXW+Y(W^2+X^2)$	$l.$	$2WYZ\ +X(Y^2+Z^2)$
$m.$	$2YZW+X(W^2+Y^2)$	$m.$	$2WZX\ +Y(Z^2\ +X^2)$
$n.$	$2ZXY+W(X^2+Y^2)$,	$n.$	$2WXY+Z(X^2+Y^2)$.

Hence there will be a node

$$\begin{array}{l} 1,\ \bar{1},\ \bar{1},\ 1,\ \text{if } p+q+r+2l-2m-2n=0, \\ \bar{1},\ 1,\ \bar{1},\ 1,\ \ldots\ p+q+r-2l+2m-2n=0, \\ \bar{1},\ \bar{1},\ 1,\ 1,\ \ldots\ p+q+r-2l-2m+2n=0, \\ 1,\ 1,\ 1,\ 1,\ \ldots\ p+q+r+2l+2m+2n=0; \end{array}$$

or say there will be

1 of these nodes if $p+q+r+2l+2m+2n=0$,

2 .... $p+q+r+2l=0,\ m+n=0$,

3 .... $p+q+r=2l=-2m=-2n$,

4 .... $p+q+r=0,\ l=0,\ m=0,\ n=0$;

viz. the surface having the 12 nodes is the original surface

$$p(YZ+XW)^2+q(ZX+YW)^2+r(XY+ZW)^2,$$

where

$$p+q+r=0.$$

The Jacobian of the quadrics

$$YZ+XW=0,\ ZX+YW=0,\ XY+ZW=0,$$

is

$$\left\|\begin{matrix} W, & Z, & Y, & X \\ Z, & W, & X, & Y \\ Y, & X, & W, & Z \end{matrix}\right\| = 0;$$

viz. the equations are

$$\begin{aligned}
X^3 - X(Y^2+Z^2+W^2) + 2YZW &= 0,\\
Y^3 - Y(Z^2+X^2+W^2) + 2ZXW &= 0,\\
Z^3 - Z(X^2+Y^2+W^2) + 2XYW &= 0,\\
W^3 - W(X^2+Y^2+Z^2) + 2XYW &= 0,
\end{aligned}$$

each of which is satisfied in virtue of any one of the pairs of equations

$$\begin{array}{l|l}
(Y-Z=0,\ X-W=0) & (Y+Z=0,\ X+W=0),\\
(Z-X=0,\ Y-W=0) & (Z+X=0,\ Y+W=0),\\
(X-Y=0,\ Z-W=0) & (X+Y=0,\ Z+W=0),
\end{array}$$

so that the Jacobian curve is, in fact, the six lines represented by these equations.

Any two of the three tetrads form an octad, the 8 points of intersection of three quadric surfaces; a figure representing the relation of the 12 points to each other may be constructed without difficulty.

Each tetrad is a sibi-conjugate tetrad *quoad* the quadric $X^2+Y^2+Z^2+W^2=0$. The three tetrads are not on the same quadric surface.

651.

ON A SPECIAL SURFACE OF MINIMUM AREA.

[From the *Quarterly Journal of Pure and Applied Mathematics*, vol. XIV. (1877), pp. 190—196.]

A VERY remarkable form of the surface of minimum area was obtained by Prof. Schwarz in his memoir "Bestimmung einer speciellen Minimal-fläche," Berlin, 1871, [*Ges. Werke*, t. I., pp. 6—125], crowned by the Academy of Sciences at Berlin. The equation of the surface is

$$1+\mu\nu+\nu\lambda+\lambda\mu=0,$$

where λ, μ, ν are functions of x, y, z respectively, viz.

$$x=-\int_\lambda^\infty \frac{d\theta}{\sqrt{(\frac{3}{4}\theta^4+\frac{5}{2}\theta^2+\frac{3}{4})}},$$

and y, z are the same functions of μ, ν respectively. A direct verification of the theorem that this is a surface of minimum area, satisfying, that is, the differential equation

$$r(1+q^2)-2pqs+t(1+p^2)=0,$$

is given in the memoir; but the investigation may be conducted in quite a different manner, so as to be at once symmetrical and somewhat more general, viz. we may enquire whether there exists a surface of minimum area

$$1+\mu\nu+\nu\lambda+\lambda\mu=0,$$

where the determining equations are

$$\lambda'^2=a\lambda^4+b\lambda^2+c,$$

$$\mu'^2=a\mu^4+b\mu^2+c,$$

$$\nu'^2=a\nu^4+b\nu^2+c,$$

$\left(\lambda' = \frac{d\lambda}{dx}, \text{ \&c.}\right)$. I find that the coefficients a, b, c must satisfy four homogeneous quadric equations, which, in fact, admit of simultaneous solution, and that in three distinct ways; viz. assuming $a = 1$, the solutions are

$$\begin{aligned} a = 1,\quad b &= \tfrac{10}{3},\quad c = 1, \\ a = 1,\quad b &= -2,\quad c = 1, \\ a = 1,\quad b &= -\tfrac{3}{2},\quad c = -\tfrac{1}{3}; \end{aligned}$$

that is,

$$\lambda'^2 = \lambda^4 + \tfrac{10}{3}\lambda^2 + 1 \ \{= \tfrac{4}{3}(\tfrac{3}{4}\lambda^4 + \tfrac{5}{2}\lambda^2 + \tfrac{3}{4})\},$$

which gives Schwarz's surface:

$$\lambda'^2 = \lambda^4 - 2\lambda^2 + 1 \ \text{ or } \ \lambda' = \pm(\lambda^2 - 1),$$

which, it is easy to see, gives only $x + y + z = \text{const.}$; and

$$\lambda'^2 = \lambda^4 - \tfrac{2}{3}\lambda^2 - \tfrac{1}{3},\quad = (\lambda^2 - 1)(\lambda^2 + \tfrac{1}{3}),$$

which is a surface similar in its nature to Schwarz's surface.

The investigation is as follows: the condition to be satisfied by a surface of minimum area $U = 0$ is

$$(\mathrm{a} + \mathrm{b} + \mathrm{c})(X^2 + Y^2 + Z^2) - (\mathrm{a}, \mathrm{b}, \mathrm{c}, \mathrm{f}, \mathrm{g}, \mathrm{h})(X, Y, Z)^2 = 0,$$

where (X, Y, Z) are the first derived coefficients and (a, b, c, f, g, h) the second derived coefficients of U in regard to the coordinates. Considering U as a function of λ, μ, ν, which are functions of x, y, z respectively, and writing (L, M, N) and (a, b, c, f, g, h) for the first and second derived functions of U in regard to λ, μ, ν, also λ', λ'' for the first and second derived functions of λ in regard to x, and so for μ', μ'' and ν', ν'': we have

$$(X, Y, Z) = (L\lambda', M\mu', N\nu'),$$

$$(\mathrm{a}, \mathrm{b}, \mathrm{c}, \mathrm{f}, \mathrm{g}, \mathrm{h}) = (a\lambda'^2 + L\lambda'', b\mu'^2 + M\mu'', c\nu'^2 + N\nu'', f\mu'\nu', g\nu'\lambda', h\lambda'\mu'),$$

and for the particular surface $U = 1 + \mu\nu + \nu\lambda + \lambda\mu = 0$, the values are

$$(L, M, N, a, b, c, f, g, h) = (\mu + \nu, \nu + \lambda, \lambda + \mu, 0, 0, 0, 1, 1, 1).$$

Hence the condition is found to be

$$\begin{aligned} & 2\mu'^2\nu'^2(\lambda + \mu)(\lambda + \nu) \\ & + 2\nu'^2\lambda'^2(\mu + \nu)(\mu + \lambda) \\ & + 2\lambda'^2\mu'^2(\nu + \lambda)(\nu + \mu) \\ & - \lambda''(\mu + \nu)\{(\lambda + \nu)^2\mu'^2 + (\lambda + \mu)^2\nu'^2\} \\ & - \mu''(\nu + \lambda)\{(\mu + \lambda)^2\nu'^2 + (\mu + \nu)^2\lambda'^2\} \\ & - \nu''(\lambda + \mu)\{(\nu + \mu)^2\lambda'^2 + (\nu + \lambda)^2\mu'^2\} = 0, \end{aligned}$$

or say this is

$$2\Sigma\mu'^2\nu'^2(\lambda+\mu)(\lambda+\nu)$$
$$-\Sigma\lambda''(\mu+\nu)\{(\lambda+\nu)^2\mu'^2+(\lambda+\mu)^2\nu'^2\}=0.$$

We have to write in this equation $\lambda'^2=a\lambda^4+b\lambda^2+c$, and therefore $\lambda''=2a\lambda^3+b\lambda$, &c.; the left-hand side, call it Ω, is a symmetrical function of λ, μ, ν, and is consequently expressible as a rational function of

$$p, =\lambda+\mu+\nu,$$
$$q, =\mu\nu+\nu\lambda+\lambda\mu,$$
$$r, =\lambda\mu\nu.$$

We ought to have $\Omega=0$, not identically, but in virtue of the equation $1+q=0$, that is, Ω should divide by $1+q$; or, what is the same thing, Ω should vanish on writing therein $q=-1$.

To effect the reduction as easily as possible, observe that we have $(\lambda+\mu)(\lambda+\nu)=\lambda^2+q$; and therefore

$$\Sigma\mu'^2\nu'^2(\lambda+\mu)(\lambda+\nu)=\Sigma\lambda^2\mu'^2\nu'^2+q\Sigma\mu'^2\nu'^2.$$

Similarly, in the second term,

$$(\mu+\nu)(\lambda+\nu)^2=(\nu+\lambda)(\nu^2+q) \text{ and } (\mu+\nu)(\lambda+\mu)^2=(\mu+\lambda)(\mu^2+q).$$

The complete value of Ω thus is

$$\Omega=2(Aq+B)-[(C+D)q+E+F],$$

where

$$A=\Sigma\lambda^2\mu'^2\nu'^2, \qquad B=\Sigma\mu'^2\nu'^2,$$
$$C=\Sigma\lambda\lambda''(\nu^2\mu'^2+\mu^2\nu'^2), \quad D=\Sigma\lambda''(\nu^2\mu'^2+\mu^2\nu'^2),$$
$$E=\Sigma\lambda\lambda''(\mu'^2+\nu'^2), \qquad F=\Sigma\lambda''(\nu\mu'^2+\mu\nu'^2).$$

We find without difficulty

$$\begin{aligned}A= {} & a^2\,(q^4-4q^2pr+4qr^2+2p^2r^2)\\ &+ab\,(-2q^3+q^2p^2+4qpr-3r^2-2p^3r)\\ &+ac\,(4q^2-8qp^2+8pr+2p^4)\\ &+b^2\,(q^2-2pr)\\ &+bc\,(-4q+2p^2)\\ &+c^2\,(3),\end{aligned}$$

$$\begin{aligned}B= {} & a^2\,(q^2r^2+2pr^3)\\ &+ab\,(-4qr^2+2p^2r^2)\\ &+ac\,(-2q^3+q^2p^2+4qpr-3r^2-2p^3r)\\ &+b^2\,(3r^2)\\ &+bc\,(2q^2-4pr)\\ &+c^2\,(-2q+p^2),\end{aligned}$$

$$
\begin{aligned}
C = {} & a^2 \,(\;4q^4 - 16q^2pr + 16qr^2 + 8p^2r^2) \\
& + ab\,(-6q^3 + 3q^2p^2 + 12qpr - 9r^2 - 6p^3r) \\
& + ac\,(\;8q^2 - 16qp^2 + 16pr + 4p^4) \\
& + b^2\,(\;2q^2 - 4pr) \\
& + bc\,(-4q + 2p^2),
\end{aligned}
$$

$$
\begin{aligned}
D = {} & a^2\,(\;2q^2pr - 2qr^2 - 4p^2r^2) \\
& + ab\,(-4qpr + 2p^3r) \\
& + ac\,(-4q^2 + 2qp^2 - 2pr) \\
& + b^2\,(\;2pr) \\
& + bc\,(\;2q),
\end{aligned}
$$

$$
\begin{aligned}
E = {} & + a^2\,(\;4q^2r^2 - 8pr^3) \\
& + ab\,(-12qr^2 + 6p^2r^2) \\
& + ac\,(-\;4q^3 + 2q^2p^2 + 8qpr - 6r^2 - 4p^3r) \\
& + b^2\,(\;6r^2) \\
& + bc\,(\;2q^2 - 4pr),
\end{aligned}
$$

$$
\begin{aligned}
F = {} & a^2\,(\;4pr^3) \\
& + ab\,(\;q^2pr + 3qr^2 - 2p^2r^2) \\
& + ac\,(\;4q^3 - 12qpr + 12r^2) \\
& + b^2\,(\;qpr - 3r^2) \\
& + bc\,(-\;2q^2 + qp^2 - pr),
\end{aligned}
$$

where in each line the terms are arranged according to their order in p, r.

Substituting, we find

$$
\begin{aligned}
\Omega = {} & a^2\,(-2q^5 + 6q^3pr - 8q^2r^2) \\
& + ab\,(\;2q^4 - q^3p^2 - q^2pr + 4qr^2) \\
& + ac\,(\;- 2q^2p^2 + 14qpr - 12r^2) \\
& + b^2\,(\;- 3qpr + 3r^2) \\
& + bc\,(-2q^2 + qp^2 - 3pr) \\
& + c^2\,(\;2q + 2p^2);
\end{aligned}
$$

viz. writing $q = -1$, this is

$$
\begin{aligned}
\Omega = {} & a^2\,(\;2 - 6pr - 8r^2) \\
& + ab\,(\;2 + p^2 - pr - 4r^2) \\
& + ac\,(\;- 2p^2 - 14pr - 12r^2) \\
& + b^2\,(\;3pr + 3r^2) \\
& + bc\,(-2 - p^2 - 3pr) \\
& + c^2\,(-2 - 2p^2);
\end{aligned}
$$

or, what is the same thing, it is

$$\begin{aligned} = & \;(\;2a^2+2ab \qquad\qquad -2bc-2c^2) \\ & +p^2\,(\qquad ab-\;2ac \qquad -\;bc+2c^2) \\ & +pr\,(-6a^2-\;ab-14ac+3b^2-3bc\qquad) \\ & +r^2\,(-8a^2-4ab-12ac+3b^2\qquad\qquad); \end{aligned}$$

so that, writing for convenience $a=1$, the equations to be satisfied are

$$\begin{aligned} 2\;-2c^2+2\,(1-\;c)\;b&=0, \\ -\;2c+2c^2+\;(1-\;c)\;b&=0, \\ -6-14c+3b^2-\;(1+3c)\;b&=0, \\ -8-12c+3b^2-\qquad\quad 4b&=0. \end{aligned}$$

The first and second are $(1-c)(2+2c+2b)=0$ and $(1-c)(-2c+b)=0$; viz. they give $c=1$, or else $b=-\frac{2}{3}$, $c=\frac{1}{3}$. In the former case, the third and fourth equations each become $3b^2-4b-20=0$, that is $(3b-10)(b-2)=0$; in the latter case, they are satisfied identically; hence we have for a, b, c the three systems of values mentioned at the beginning.

This completes the investigation; but it is interesting to find the values assumed by the other factor of Ω on substituting therein for a, b, c the foregoing several systems of values. We have in general

$$\begin{aligned} \Omega = & \qquad -2a^2q^3+2abq^4-\;2bcq^2+2c^2q \\ & +p^2\,(-\;abq^3-2acq^2+\;bcq\;+2c^2\qquad) \\ & +pr\,(\;6a^2q^3-\;abq^2+14acq\;-3b^2q-3bc) \\ & +r^2\,(-8a^2q^2+4abq-12ac\;+3b^2\qquad) \\ = & -2a^2(q^5+1)\qquad +2ab\,(q^4-1)-2bc\,(q^2-1)+2c^2(q+1) \\ & +p^2\;\{-ab\;(q^3+1)-2ac\,(q^2-1)+bc\,(q+1)\} \\ & +pr\;\{\;6a^2(q^3+1)-\;ab\,(q^2-1)+(14ac-3b^2)\,(q+1)\} \\ & +r^2\;\{\qquad\qquad -8a^2\;(q^2-1)+4ab\,(q+1)\} \\ = & (q+1)\left\{\begin{array}{l} -2a^2(q^4-q^3+q^2-q+1)+2ab\,(q^3-q^2+q-1)-2bc\,(q-1)+2c^2 \\ +p^2\;\{-\;ab\,(q^2-q+1)-2ac\,(q-1)+bc\} \\ +pr\;\{\;6a^2\,(q^2+q+1)-\;ab\,(q-1)+(14ac-3b^2)\} \\ +r^2\;\{\qquad\qquad -8a^2\,(q-1)+\;4ab\} \end{array}\right\}. \end{aligned}$$

Hence writing, first, $a=c=1$, $b=\frac{10}{3}$, we obtain, after some reductions,

$$\Omega=(q+1)\{-2q\,(q-1)(q^2-\tfrac{10}{3}q+1)+p^2(q-1)(-\tfrac{10}{3}q-2)+pr(6q^2-\tfrac{28}{3}q-10)+r^2-8q+\tfrac{64}{3}\};$$

secondly, writing $a=c=1$, $b=-2$, we obtain

$$\Omega=(q+1)\{-2\,(q+1)^2(q^2+1)+p^2.\,2\,(q-1)^2+2pr\,(3q^2-2q+6)-8r^2q\};$$

and, thirdly, writing $a=1$, $b=-\frac{1}{3}$, $c=-\frac{2}{3}$, we obtain

$$\Omega=(q+1)\,\{(-2q^4+\tfrac{4}{3}q^3-\tfrac{4}{3}q^2+\tfrac{8}{9}q)+p^2(-\tfrac{1}{3}q^2+\tfrac{5}{3}q-\tfrac{13}{9})+pr\,(6q^2-\tfrac{17}{3}q-\tfrac{10}{3})+r^2(-8q+\tfrac{20}{3})\}.$$

652.

ON A SEXTIC TORSE.

[From the *Quarterly Journal of Pure and Applied Mathematics*, vol. XIV. (1877), pp. 229—235.]

THE torse having for its edge of regression or cuspidal edge the curve defined by the equations $x = \cos\phi$, $y = \sin\phi$, $z = \cos 2\phi$, is an interesting and convenient one for the construction of a model, and it is here considered partly from that point of view.

The edge is a quadriquadric curve, the intersection of the cylinder $x^2 + y^2 = 1$ with the parabolic hyperboloid $z = x^2 - y^2$; the cylinder regarded as a cone having its vertex at infinity on the line $x = 0$, $y = 0$, viz. the vertex is on the hyperboloid, or the curve is a nodal quadriquadric (the node being thus an isolated point at infinity on the line in question), and the torse is consequently of the order $8 - 2, = 6$, viz. it is a sextic torse.

The edge is a bent oval situate on the cylinder $x^2 + y^2 = 1$, such that, regarding ϕ as the azimuth (or angle measured along the circular base from its intersection with the axis of x), the altitude z is given by the equation $z = \cos 2\phi$; viz. there are in the plane xz, or, say in the planes xz, $x'z$, two maxima altitudes $z = 1$, and in the plane yz, or, say in the planes yz and $y'z$, two minima altitudes $z = -1$. The sections by these principal planes are, as is seen at once, nodal curves on the surface; they are, in fact, the cubic curves $z = 3 - \frac{2}{x^2}$, viz. here as x increases from ± 1 to $\pm\infty$, z increases from the before-mentioned value 1 to 3, and $z = -3 + \frac{2}{y^2}$, viz. as y increases from ± 1 to $\pm\infty$, z decreases from the before-mentioned value -1 to -3. The two half-sheets (which meet in the cuspidal edge) intersect each other along these nodal lines, in suchwise that the section of the surface by any axial plane (plane through the line $x = 0$, $y = 0$) is a curve having a cusp on the cuspidal edge, and such that when the axial plane coincides with either of the principal planes $x = 0$, $y = 0$, the two half-branches of the curve coincide together with the portions which lie outside the cylinder $x^2+y^2=1$, in fact, the portions referred to above, of the nodal curve in the plane in question; the portions which lie inside the cylinder are acnodal or isolated curves without any real sheet through them. It may be added, in the way of general description, that the section of the surface by any cylinder $x^2+y^2=c^2$ $(c>1)$ is a curve of the form $z=C\cos(2\theta\pm B)$, θ the angle along the base of the cylinder from the intersection with the axis of x; C, B are functions of c; viz. we have for the two half sheets respectively

$$z=C\cos(2\theta+B) \text{ and } z=C\cos(2\theta-B),$$

each curve having thus the two maxima $+C$, and the two minima $-C$; and the two curves intersect each other at the four points in the two principal planes respectively; viz. the points for which $\theta=0, 90^\circ, 180^\circ, 270^\circ$, and $z=C\cos B, -C\cos B, C\cos B, -C\cos B$ accordingly.

Proceeding to discuss the surface analytically, we have for the equations of a generating line

$$\frac{x-\cos\phi}{-\sin\phi}=\frac{y-\sin\phi}{\cos\phi}=\frac{z-\cos 2\phi}{-2\sin 2\phi}, \quad =\rho \text{ suppose},$$

or say

$$\begin{aligned} x&=\cos\ \phi-\ \rho\sin\ \phi,\\ y&=\sin\ \phi+\ \rho\cos\ \phi,\\ z&=\cos 2\phi-2\rho\sin 2\phi, \end{aligned}$$

which equations, considering therein ρ, ϕ as arbitrary parameters, determine the surface.

Writing $x=0$, we find $y=\dfrac{1}{\sin\phi}$, and then $z=-3+2\sin^2\phi$, viz. we have

$$x=0,\quad z=-3+\frac{2}{y^2}, \text{ for section in plane } yz;$$

and, similarly, writing $y=0$, we find $x=\dfrac{1}{\cos\phi}$, and then $z=3-2\cos^2\phi$, viz.

$$y=0,\quad z=3-\frac{2}{x^2} \text{ for section by plane } xz.$$

By what precedes, these are nodal curves, crunodal for the portions

$$(y=\pm 1 \text{ to } \pm\infty,\ z=-1 \text{ to } -3) \text{ and } (x=\pm 1 \text{ to } \pm\infty,\ z=1 \text{ to } 3)$$

respectively, acnodal for the remaining portions $y<\pm 1$, $x<\pm 1$ respectively.

Writing $x=r\cos\theta$, $y=r\sin\theta$, so that the coordinates of a point on the surface are r, θ, z, where $r=\sqrt{(x^2+y^2)}$ is the projected distance, θ is the azimuth from the axis of x, and z is the altitude, we have

$$\begin{aligned} r\cos\theta&=\cos\ \phi-\ \rho\sin\ \phi,\\ r\sin\theta&=\sin\ \phi+\ \rho\cos\ \phi,\\ z&=\cos 2\phi-2\rho\sin 2\phi. \end{aligned}$$

We have $r^2 = 1+\rho^2$; and thence also, if $\tan\alpha = 2\rho, = \pm 2\sqrt{(r^2-1)}$, that is,

$$\cos\alpha = \frac{1}{\sqrt{(4r^2-3)}},\quad \sin\alpha = \pm\frac{2\sqrt{(r^2-1)}}{\sqrt{(4r^2-3)}},$$

then

$$z = \sqrt{(4r^2-3)}\cos(2\phi+\alpha),$$

showing that for a given value of r (or section by the cylinder $x^2+y^2=r^2$) the maximum and minimum values of z are $z = \pm\sqrt{(4r^2-3)}$.

But proceeding to eliminate ϕ, we find

$$r^2\cos 2\theta = (1-\rho^2)\cos 2\phi - 2\rho\sin 2\phi,$$

$$r^2\sin 2\theta = 2\rho\cos 2\phi + (1-\rho^2)\sin 2\phi;$$

or multiplying these by $1+3\rho^2$ and $2\rho^3$ and adding

$$r^2\{(1+3\rho^2)\cos 2\theta + 2\rho^3\sin 2\theta\} = (1+\rho^2)^2(\cos 2\phi - 2\rho\sin 2\phi),$$

that is,

$$r^2\{(3r^2-2)\cos 2\theta \pm 2(r^2-1)^{\frac{3}{2}}\sin 2\theta\} = r^4 z;$$

or, finally,

$$r^2 z = (3r^2-2)\cos 2\theta \pm 2(r^2-1)^{\frac{3}{2}}\sin 2\theta,$$

which is the equation of the surface in terms of the coordinates r, θ, z.

Observing that $(3r^2-2)^2 + 4(r^2-1)^3 = r^4(4r^2-3)$, we may write

$$r^2\sqrt{(4r^2-3)}\cos\beta = 3r^2-2,$$

$$r^2\sqrt{(4r^2-3)}\sin\beta = 2(r^2-1)^{\frac{3}{2}},$$

and therefore also

$$\tan\beta = \frac{2(r^2-1)^{\frac{3}{2}}}{3r^2-2},$$

and the equation thus becomes

$$z = \sqrt{(4r^2-3)}\cos(2\theta \mp \beta),$$

where z is the altitude belonging to the azimuth θ in the cylindrical section, radius r. The maxima and minima altitudes are $\pm\sqrt{(4r^2-3)}$, and these correspond to the values $\theta = \pm\frac{1}{2}\beta$, $\frac{1}{2}\pi \pm \frac{1}{2}\beta$, $\pi \pm \frac{1}{2}\beta$, $\frac{3}{2}\pi \pm \frac{1}{2}\beta$; it is to be further noticed that when $r=1$, we have $\beta = 0$, but as r increases and becomes ultimately infinite, β increases to $\frac{1}{2}\pi$, that is, $\frac{1}{2}\beta$ increases from 0 to $\frac{1}{4}\pi$.

It may be noticed that the surface is a peculiar kind of deformation, obtained by giving proper rotations to the several cylindrical sections of the surface $z = \sqrt{(4r^2-3)}\cos 2\theta$; viz. in rectangular coordinates this is $r^2 z = \sqrt{(4r^2-3)}(x^2-y^2)$, that is,

$$(x^2+y^2)^2 z^2 - \{4(x^2+y^2)-3\}(x^2-y^2)^2 = 0.$$

To obtain the equation in rectangular coordinates, we have

$$\left\{r^2 z - \frac{3r^2-2}{r^2}(x^2-y^2)\right\}^2 - 16(r^2-1)^3\frac{x^2y^2}{r^4} = 0,$$

viz. this is

$$r^4 z^2 - 2z(3r^2-2)(x^2-y^2) + (3r^2-2)^2\left(1 - \frac{4x^2y^2}{r^4}\right) - 16(r^2-1)^3\frac{x^2y^2}{r^4} = 0,$$

or, what is the same thing, it is

$$r^4z^2 - 2z(3y^2-2)(x^2-y^2) + (3r^2-2)^2 - \frac{4x^2y^2}{r^4}\{4(r^2-1)^3 + (3r^2-2)^2\} = 0,$$

viz. the term in $\{\ \}$ being $r^4(4r^2-3)$, this is

$$r^4z^2 - 2z(3r^2-2)(x^2-y^2) + (3r^2-2)^2 - 4x^2y^2(4r^2-3) = 0,$$

or say

$$z^2(x^2+y^2)^2 - 2z(3x^2+3y^2-2)(x^2-y^2) + (3x^2+3y^2-2)^2 - 4x^2y^2(4x^2+4y^2-3) = 0.$$

This may also be written

$$\{z(x^2-y^2) - 3x^2 - 3y^2 + 2\}^2 + 4x^2y^2(z^2 - 4x^2 - 4y^2 + 3) = 0,$$

a form which puts in evidence the nodal curves

$$x = 0,\ xy^2 = -3y^2 + 2, \text{ and } y = 0,\ zx^2 = 3x^2 - 2.$$

It shows also that the quadric cone $z^2 - 4x^2 - 4y^2 + 3 = 0$ touches the surface along the curve of intersection with the surface $z(x^2-y^2) - 3(x^2+y^2) + 2 = 0$. This is, in fact, the curve of maxima and minima of the cylindrical sections, viz. reverting to the form $z = \sqrt{(4r^2-3)}\cos(2\theta \mp \beta)$, or, if for greater clearness, attending only to one sheet of the surface, we write it $z = \sqrt{(4r^2-3)}\cos(2\theta - \beta)$, we have a maximum, $z = \sqrt{(4r^2-3)}$, for $2\theta = \beta$ (or $2\pi + \beta$), giving

$$\cos 2\theta = \cos\beta, \quad = \frac{3r^2-2}{r^2\sqrt{(4r^2-3)}}, \quad = \frac{3r^2-2}{r^2z}:$$

and a minimum, $z = -\sqrt{(4r^2-3)}$, for $2\theta = \pi + \beta$ (or $3\pi + \beta$), giving

$$\cos 2\theta = -\cos\beta = -\frac{3r^2-2}{r^2\sqrt{(4r^2-3)}}, \quad = \frac{3r^2-2}{r^2z};$$

viz. the locus is $z^2 = 4(r^2-3)$, $z(x^2-y^2) = 3r^2-2$; and for $z = \sqrt{(4r^2-3)}\cos(2\theta+\beta)$ we find the same locus, viz. the equations of the locus are

$$z^2 - 4x^2 - 4y^2 + 3 = 0, \quad z(x^2-y^2) - 3x^2 - 3y^2 + 2 = 0,$$

as above.

To put in evidence the cuspidal edge, write for a moment $\zeta = z - x^2 + y^2$, the equation becomes

$$\{\zeta(x^2-y^2) + (r^2-1)(r^2-2) - 4x^2y^2\}^2 + 4x^2y^2\{\zeta^2 + 2\zeta(x^2-y^2) + (r^2-1)(r^2-3) - 4x^2y^2\} = 0;$$

viz. this is

$$\zeta^2r^4 + 2\zeta(x^2-y^2)(r^2-1)(r^2-2) + (r^2-1)^2(r^2-2)^2 - 4x^2y^2(r^2-1)^2 = 0,$$

or writing the last term thereof in the form

$$-\{r^2 - (x^2-y^2)^2\}(r^2-1)^2,$$

and then putting $r^2 = 1 + U$, the equation is

$$\zeta^2(1 + 2U + U^2) + 2\zeta U(U-1)(x^2-y^2) + U^2(U-1)^2 - U^2\{(U+1)^2 - (x^2-y^2)^2\} = 0;$$

viz. this is

$$\{\zeta - U(x^2-y^2)\}^2 + 2U\{\zeta^2 + \zeta U(x^2-y^2) - 2U^2\} + \zeta^2U^2 = 0,$$

showing the cuspidal edge $\zeta=0$, $U=0$, viz. $z=x^2-y^2$, $x^2+y^2=1$. Moreover, along the cuspidal edge the surface is touched by $\zeta-U(x^2-y^2)=0$, that is, by $z-(x^4-y^4)=0$; and at the points where this tangent surface again meets the surface we have $(x^2-y^2)^2(x^2+y^2+3)-4=0$; viz. the surface contains upon itself the curve represented by this last equation, and $z-(x^4-y^4)=0$.

As a verification, in the form

$$\{z(x^2-y^2)-3x^2-3y^2+2\}^2+4x^2y^2(z^2-4x^2-4y^2+3)=0$$

of the equation of the surface, write $z=x^4-y^4$. If for a moment $x^2+y^2=\lambda$, $x^2-y^2=\mu$, then the value of z is $z=\lambda\mu$, and the equation becomes

$$(\lambda\mu^2-3\lambda+2)^2+(\lambda^2-\mu^2)(\lambda^2\mu^2-4\lambda+3)=0,$$

that is,

$$\mu^2(\lambda^4-6\lambda^2+8\lambda-3)-4\lambda^3+12\lambda^2-12\lambda+4=0;$$

or, what is the same thing,

$$(\lambda-1)^3\{\mu^2(\lambda+3)-4\}=0,$$

so that we have $(\lambda-1)^3=0$, or else $\mu^2(\lambda+3)-4=0$; viz. $(x^2+y^2-1)^3=0$, or else $(x^2-y^2)^2(x^2+y^2+3)-4=0$, agreeing with the former result.

In polar coordinates, the surface is touched along the cuspidal edge by the surface $z=r^4\cos 2\theta$, and where this again meets the surface we have $r^4(r^2+3)\cos^2 2\theta-4=0$.

For the model, taking the unit to be 1 inch, I suppose that for the edge of regression we have

$$x=2\cos\phi,\ y=2\sin\phi,\ z=5+(\cdot45)\cos 2\phi;$$

viz. the curve is situate on a cylinder radius 2 inches. And I construct in zinc-plate the cylindric sections, or say the templets, for one sheet of the surface, for the several radii 2, 3,.., 8 inches; taking the radius as k inches, the circumference of the cylinder, or entire base of the flattened templet, is $=2k\pi$; and the altitude, writing 2θ in place of $2\theta-\beta$ as above, is given by the formula $z=5+(\cdot45)\sqrt{(k^2-3)}\cos 2\theta$, so that the half altitude of the wave is $=(\cdot45)\sqrt{(k^2-3)}$; having this value, the curve is at once constructed geometrically. We have, moreover, $\cos\beta=\dfrac{3k^2-8}{k^2\sqrt{(k^2-3)}}$; the numerical values then are

k	$2k\pi$	$(\cdot45)\sqrt{(k^2-3)}$	$\dfrac{3k^2-8}{k^2\sqrt{(k^2-3)}}$	$\frac{1}{2}\beta$
2	12·57	0·45	1·00	0°
3	18·85	1·10	·86	15
4	25·13	1·62	·69	23
5	31·42	2·11	·57	27½
6	37·70	2·59	·48	30½
7	43·98	3·05	·42	32½
8	50·27	3·51	·36	34

the altitudes in the successive templets being thus included between the limits $5\pm 0\cdot45$, $5\pm 1\cdot10$,.., $5\pm 3\cdot51$.

653.

ON A TORSE DEPENDING ON THE ELLIPTIC FUNCTIONS.

[From the *Quarterly Journal of Pure and Applied Mathematics*, vol. XIV. (1877), pp. 235—241.]

ON attempting to cover with paper one half-sheet of the foregoing sextic torse, [652], I found that the paper assumed approximately the form of a circular annulus of an angle exceeding 360°, and this led me to consider the general theory of the construction of a torse in paper, and, in particular, to consider the torses such that when developed into a plane the edge of regression becomes a circular arc. It is scarcely necessary to remark that, to construct in paper a circular annulus of an angle exceeding 360°, we have only to take a complete annulus, cut it along a radius, and then insert (gumming it on to the two terminal radii) a portion of an equal circular annulus; drawing from each point of the inner circular boundary a half-tangent, and considering these half-tangents as rigid lines, the paper will bend round them so as to form the half-sheet of a torse having for its edge of regression this inner boundary, which will assume the form of a closed curve with two equal and opposite maxima and two equal and opposite minima, described on a cylinder, and being *approximately* such as the curve given by the equations

$$x = \cos\theta,\ y = \sin\theta,\ z = m\cos 2\theta.$$

Considering, in general, an arc PQ (without inflexions) of any curve, and drawing at the consecutive points P, P', P'', &c. the several half-tangents $PT, P'T', P''T'', \ldots$, then, considering these as rigid lines and bending the paper round them, we have the half-sheet of a torse, having for its edge of regression the curve in question now bent into a curve of double curvature. It is, moreover, clear that the edge of regression has at each point thereof the same radius of absolute curvature as the original plane curve; in fact, if in the plane curve $PP' = ds$, and the angle $T'PT$ between the consecutive half-tangents PT and $P'T'$ be $= d\phi$, these quantities ds and

$d\phi$ remain unaltered in the curve of double curvature; and the radius of absolute curvature is given by the equation $\rho d\phi = ds$. In particular when, as above, the arc is a circular one, say of radius $=\alpha$, then, however the paper is bent, the edge of regression has at each point thereof the radius of absolute curvature $=\alpha$.

Consider on any given surface, at a given point P thereof, and in a given direction, an element of length PP', then (under the restrictions presently mentioned) we can determine the consecutive element $P'P''$, such that the curve $PP'P''\ldots$ shall have at P a radius of absolute curvature $=\alpha$; in fact, r being the radius of curvature of the normal section of the surface through the element PP', the radius of curvature of the section inclined at an angle θ to the normal section is $=r\cos\theta$; so that we have only to take the section at the inclination $\theta, =\cos^{-1}\frac{\alpha}{r}$ to the normal section, and we have the consecutive element $P'P''$ such that the radius of absolute curvature of the curve $PP'P''$ is $=\alpha$. The necessary restriction, of course, is that $r>\alpha$; thus, if at the given point P the two principal radii of curvature are of the same sign (to fix the ideas, let the two principal radii and also α be each of them positive), then we may on the surface determine a direction PQ, for which the radius of curvature of the normal section is $=\alpha$; and then the direction of the element PP' may be any direction between PQ and the direction PR, corresponding to the greatest of the two principal radii.

Having obtained the element $P'P''$, we may, if the radius of absolute curvature at P' be given, construct the next element $P'P''$, and so on; that is to say, on a given surface starting from a given point P and given initial direction PP', we can (under a restriction, as above, as to the curvature at the different points of the surface) construct a curve having at the successive points thereof given values of the radius of absolute curvature; viz., the value may be given either as a function of the coordinates of the point on the surface, or as a function of the length of the curve measured say from the initial point P; it is in this last manner that in what follows the value of the radius of absolute curvature is assumed to be given.

We may thus, taking on paper an arc PQ with its half-tangents, apply it to a given surface, the point P to a given point, and the infinitesimal arc PP' to an element PP' in a given direction from the given point; and we thus obtain the half-sheet of a torse having for its edge of regression a determinate curve upon the surface. In particular, the arc PQ may be circular of the radius α, and the surface be a circular cylinder of radius a; and we thus obtain the torse having for edge of regression a curve on the cylinder radius α, and such that the radius of absolute curvature is at each point $=a$. There are three cases according as $a>\alpha$, $a=\alpha$, or $a<\alpha$; it is to be remarked that if $a>\alpha$, then the curve must at each point cut the generating line of the cylinder at an angle not exceeding $\cos^{-1}\frac{\alpha}{a}$, but that in the other two cases the angle may have any value whatever; and, further, that in every case when the angle is $=0$, viz. when the curve touches a generating line of the cylinder, then the osculating plane of the curve coincides with the tangent plane of the cylinder.

The analytical theory is very simple. Taking x, y, z as functions of the length s, we have

$$\left(\frac{dx}{ds}\right)^2 + \left(\frac{dy}{ds}\right)^2 + \left(\frac{dz}{ds}\right)^2 = 1;$$

the condition, which expresses that the radius of absolute curvature is $= a$, then is

$$\left(\frac{d^2x}{ds^2}\right)^2 + \left(\frac{d^2y}{ds^2}\right)^2 + \left(\frac{d^2z}{ds^2}\right)^2 = \frac{1}{a^2}.$$

By what precedes, the point (x, y, z) may be taken to be upon a given surface, say upon the cylinder $x^2 + y^2 = \alpha^2$; and we may then write $x = \alpha\cos\theta$, $y = \alpha\sin\theta$. Taking instead of s any independent variable u whatever, and using accents to denote the derived functions in regard to u, the equations become

$$\begin{aligned} x'^2 + y'^2 + z'^2 \qquad &= s'^2, \\ x''^2 + y''^2 + z''^2 - s''^2 &= \frac{1}{a^2} s'^4, \\ x = \alpha\cos\theta,\ y = \alpha\sin\theta. \end{aligned}$$

From the last two equations we obtain

$$x'^2 + y'^2 = \alpha^2\theta'^2,\ x''^2 + y''^2 = \alpha^2(\theta''^2 + \theta'^4),$$

and the first two equations thus become

$$\begin{aligned} \alpha^2\theta'^2 + \qquad z'^2 \qquad &= s'^2, \\ \alpha^2(\theta''^2 + \theta'^4) + z''^2 - s''^2 &= \frac{1}{a^2} s'^4, \end{aligned}$$

and from the first of these we find

$$s'' = \frac{\alpha^2\theta'\theta'' + z'z''}{(\alpha^2\theta'^2 + z'^2)^{\frac{1}{2}}},$$

whence the second equation is

$$\alpha^2(\theta''^2 + \theta'^4) + z''^2 - \frac{(\alpha^2\theta'\theta'' + z'z'')^2}{(\alpha^2\theta'^2 + z'^2)} = \frac{(\alpha^2\theta'^2 + z'^2)^2}{a^2},$$

or reducing, this is

$$(\alpha^2\theta'^2 + z'^2)(\theta''^2 + \theta'^4) + (\theta'^2 z''^2 - 2\theta'\theta'' z' z'' - \alpha^2\theta'^2\theta''^2) = \frac{1}{a^2\alpha^2}(\alpha^2\theta'^2 + z'^2)^3.$$

Taking here θ as the independent variable, we have $\theta' = 1$, $\theta'' = 0$, and the equation becomes

$$(\alpha^2 + z'^2) + z''^2 = \frac{1}{a^2\alpha^2}(\alpha^2 + z'^2)^3;$$

or, what is the same thing,

$$z''^2 = \frac{1}{a^2\alpha^2}(\alpha^2 + z'^2)^3 - (\alpha^2 + z'^2)^2.$$

Write here

$$\alpha^2 + z'^2 = \Omega^2,$$

then

$$z'' = \frac{\Omega\Omega'}{\sqrt{(\Omega^2 - \alpha^2)}},$$

and the equation becomes

$$\frac{\Omega'^2}{\Omega^2 - \alpha^2} = \frac{\Omega^4}{a^2\alpha^2} - 1,$$

or say

$$\frac{a\alpha d\Omega}{\sqrt{(\Omega^2 - \alpha^2 \,.\, \Omega^4 - a^2\alpha^2)}} = d\theta,$$

viz. this equation determines Ω as a function of θ, and we then have

$$\begin{cases} ds = \Omega d\theta, \\ dz = \sqrt{(\Omega^2 - \alpha^2)}\, d\theta, \\ x = \alpha \cos\theta, \\ y = \alpha \sin\theta, \end{cases}$$

equations which determine x, y, z, s as functions of the parameter θ, and give thus the edge of regression of the torse in question.

It is clear that the formulæ are very much simplified in the case $a = \alpha$, where the radius of absolute curvature a is equal to the radius α of the cylinder; but it is worth while to develope the general case somewhat further.

Considering the elliptic functions $\operatorname{sn} u$, $\operatorname{cn} u$, $\operatorname{dn} u$, to the modulus $k\,(=k') = \frac{1}{\sqrt{(2)}}$, assume

$$\Omega = -\frac{\sqrt{(a\alpha)}}{k}\frac{\operatorname{dn} u}{\operatorname{sn} u},$$

then

$$d\Omega = -\frac{\sqrt{(a\alpha)}}{k}\frac{\operatorname{cn} u\, du}{\operatorname{sn}^2 u},$$

$$\begin{aligned} \Omega^2 - \alpha^2 &= \frac{a\alpha}{k^2 \operatorname{sn}^2 u}\left(\operatorname{dn}^2 u - \frac{\alpha}{a} k^2 \operatorname{sn}^2 u\right), \\ &= \frac{a\alpha}{k^2 \operatorname{sn}^2 u}\left\{1 - \left(1 + \frac{\alpha}{a}\right) k^2 \operatorname{sn}^2 u\right\}, \\ \Omega^4 - a^2\alpha^4 &= \frac{a^2\alpha^2}{k^4 \operatorname{sn}^4 u}(\operatorname{dn}^4 u - k^4 \operatorname{sn}^4 u), \\ &= \frac{a^2\alpha^2}{k^4 \operatorname{sn}^4 u}(1 - 2k^2 \operatorname{sn}^2 u),\ = \frac{a^2\alpha^2}{k^4 \operatorname{sn}^4 u}\operatorname{cn}^2 u, \end{aligned}$$

and hence

$$d\theta = \frac{k^2 \operatorname{sn} u\, du}{\sqrt{\left\{1 - \left(1 + \frac{\alpha}{a}\right) k^2 \operatorname{sn}^2 u\right\}}},$$

$$ds = \frac{k\sqrt{(a\alpha)}\operatorname{dn} u\, du}{\sqrt{\left\{1 - \left(1 + \frac{\alpha}{a}\right) k^2 \operatorname{sn}^2 u\right\}}},$$

$$dz = k\sqrt{(a\alpha)}\, du.$$

We have thus $z = k\sqrt{(a\alpha)}\,u$, no constant of integration being required, viz. u is a mere constant multiple of z: and the first and second equations then give s and θ as functions of u, that is, of z; but it is obviously convenient to retain u instead of expressing it in terms of z. As regards the form of these integrals observe that, writing $\operatorname{sn} u = \lambda$, we have

$$du = \frac{d\lambda}{\sqrt{\{1-\lambda^2 . 1-k^2\lambda^2\}}},$$

and thence

$$d\theta = \frac{k^2\lambda d\lambda}{\sqrt{\left\{1-\lambda^2 . 1-k^2\lambda^2 . 1-\left(1+\frac{a}{\alpha}\right)k^2\lambda^2\right\}}},$$

$$ds = \frac{k\sqrt{(a\alpha)}\,d\lambda}{\sqrt{\left\{1-\lambda^2 . 1-\left(1+\frac{\alpha}{a}\right)k^2\lambda^2\right\}}},$$

each of which is in fact reducible to elliptic integrals, but I do not further pursue this general case.

In the particular case $a = \alpha$, we have

$$1-\left(1+\frac{\alpha}{a}\right)k^2 \operatorname{sn}^2 u = \operatorname{cn}^2 u,$$

and the equations become

$$d\theta = \frac{k^2 \operatorname{sn} u\, du}{\operatorname{cn} u}, \quad ds = \frac{k\alpha \operatorname{dn} u\, du}{\operatorname{cn} u},$$

which admit of immediate integration; viz. we have

$$\theta = \tfrac{1}{2}\frac{k^2}{k'}\log\frac{\operatorname{dn} u + k'}{\operatorname{dn} u - k'},$$

or determining the constant so that θ may vanish for $u=0$, say

$$\theta = \tfrac{1}{2}\frac{k^2}{k'}\log\left(\frac{\operatorname{dn} u + k'}{\operatorname{dn} u - k'} . \frac{1-k'}{1+k'}\right);$$

and

$$s = \tfrac{1}{2}k\alpha\log\left(\frac{1+\operatorname{sn} u}{1-\operatorname{sn} u}\right);$$

viz. to verify these results we have

$$\frac{d\theta}{du} = \tfrac{1}{2}\frac{k^2}{k'} . k^2 \operatorname{sn} u \operatorname{cn} u\left\{\frac{1}{\operatorname{dn} u + k'} - \frac{1}{\operatorname{dn} u - k'}\right\},$$

$$= \frac{k^4 \operatorname{sn} u \operatorname{cn} u}{\operatorname{dn}^2 u - k'^2}, \quad = k^2\frac{\operatorname{sn} u}{\operatorname{cn} u},$$

and

$$\frac{ds}{du} = \tfrac{1}{2}k\alpha . \operatorname{cn} u \operatorname{dn} u\left\{\frac{1}{1+\operatorname{sn} u} + \frac{1}{1-\operatorname{sn} u}\right\},$$

$$= \frac{k\alpha \operatorname{cn} u \operatorname{dn} u}{1-\operatorname{sn}^2 u}, \quad = \frac{k\alpha \operatorname{dn} u}{\operatorname{cn} u}.$$

Hence, recurring to the original equations, and writing for convenience $a = \alpha = 1$, we see that a solution of the simultaneous equations

$$\left(\frac{dx}{ds}\right)^2 + \left(\frac{dy}{ds}\right)^2 + \left(\frac{dz}{ds}\right)^2 = 1,$$

$$\left(\frac{d^2x}{ds^2}\right)^2 + \left(\frac{d^2y}{ds^2}\right)^2 + \left(\frac{d^2z}{ds^2}\right)^2 = 1,$$

is

$$x = \cos\theta,\ y = \sin\theta,\ z = ku,$$

$$\theta = \tfrac{1}{2}\frac{k^2}{k'}\log\left(\frac{\operatorname{dn} u + k'}{\operatorname{dn} u - k'}\cdot\frac{1-k'}{1+k'}\right),\quad s = \tfrac{1}{2}ka\log\left(\frac{1+\operatorname{sn} u}{1-\operatorname{sn} u}\right),$$

where, as before, $k = k' = \dfrac{1}{\sqrt{(2)}}$.

Restoring the radius α, and writing the system in the form

$$x = \alpha\cos\theta,\ y = \alpha\sin\theta,\ z = k\alpha u,$$

$$\theta = \tfrac{1}{2}\frac{k^2}{k'}\log\left(\frac{\operatorname{dn} u + k'}{\operatorname{dn} u - k'}\cdot\frac{1-k'}{1+k'}\right),\quad s = \tfrac{1}{2}k\alpha\log\left(\frac{1+\operatorname{sn} u}{1-\operatorname{sn} u}\right),$$

we see that, as u passes from $u=0$ to $u=K$, and therefore z from $z=0$ to $z=k\alpha K$ $\left(K \text{ the complete function } F_1\left\{\dfrac{1}{\sqrt{(2)}}\right\}\right)$, then θ and s each pass from 0 to ∞; and, similarly, as u passes from $u=0$ to $u=-K$, that is, as z passes from 0 to $-k\alpha K$, then θ passes from 0 to ∞, and s from $s=0$ to $s=-\infty$; viz. the curve makes in each direction an infinity of revolutions about the cylinder. Developing the cylinder, $\alpha\theta$ becomes an x-coordinate; viz. we have thus the plane curve

$$z = k\alpha u,$$

$$x = \tfrac{1}{2}\frac{k^2\alpha}{k'}\log\left(\frac{\operatorname{dn} u + k'}{\operatorname{dn} u - k'}\cdot\frac{1-k'}{1+k'}\right),$$

which is a curve extending from the origin in the direction x positive, to touch at infinity the two parallel asymptotes $z = \pm k\alpha K$; and conversely, when such a plane curve is wound about the cylinder, there will be in each direction an infinity of revolutions round the cylinder.

654.

ON CERTAIN OCTIC SURFACES.

[From the *Quarterly Journal of Pure and Applied Mathematics*, vol. XIV. (1877), pp. 249—264.]

I. CONSIDER the torse generated by the tangents of the quadriquadric curve, the intersection of the two quadric surfaces

$$\mathrm{a}x^2+\mathrm{b}y^2+\mathrm{c}z^2+\mathrm{d}w^2=0,$$

$$\mathrm{a}'x^2+\mathrm{b}'y^2+\mathrm{c}'z^2+\mathrm{d}'w^2=0;$$

then, writing

$$\mathrm{bc}'-\mathrm{b}'\mathrm{c}=a',\quad \mathrm{ad}'-\mathrm{a}'\mathrm{d}=f',$$

$$\mathrm{ca}'-\mathrm{c}'\mathrm{a}=b',\quad \mathrm{bd}'-\mathrm{b}'\mathrm{d}=g',$$

$$\mathrm{ab}'-\mathrm{a}'\mathrm{b}=c',\quad \mathrm{cd}'-\mathrm{c}'\mathrm{d}=h',$$

and therefore

$$a'f'+b'g'+c'h'=0,$$

the equation of the torse, writing for greater convenience (a, b, c, f, g, h) in place of (a', b', c', f', g', h'), but understanding these letters as signifying the accented letters (a', b', c', f', g', h'), is

$$\begin{aligned} & a^4f^2y^4z^4+b^4g^2z^4x^4+c^4h^2x^4y^4 \\ & +a^2f^4x^4w^4+b^2g^4y^4w^4+c^2h^4z^4w^4 \\ & +2b^2c^2ghx^4y^2z^2-2c^2f^2ahx^4y^2w^2+2b^2f^2agx^4z^2w^2 \\ & +2c^2a^2hfy^4z^2x^2-2a^2g^2bfy^4z^2w^2+2c^2g^2bhy^4x^2w^2 \\ & +2a^2b^2fgz^4x^2y^2-2b^2h^2cgz^4x^2w^2+2a^2h^2cfz^4y^2w^2 \\ & -2bcg^2h^2w^4y^2z^2-2cah^2f^2w^4z^2x^2-2abf^2g^2w^4x^2y^2 \\ & +2(bg-ch)(ch-af)(af-bg)x^2y^2z^2w^2=0. \end{aligned}$$

If in this equation we write

$$a'^2f' = a, \quad a'f'^2 = f,$$
$$b'^2g' = b, \quad b'g'^2 = g,$$
$$c'^2h' = c, \quad c'h'^2 = h;$$

and therefore

$$\text{bc}' - \text{b}'\text{c} = \frac{a}{\sqrt[3]{(af)}}, \quad \text{ad}' - \text{a}'\text{d} = \frac{f}{\sqrt[3]{(af)}},$$

$$\text{ca}' - \text{c}'\text{a} = \frac{b}{\sqrt[3]{(bg)}}, \quad \text{bd}' - \text{b}'\text{d} = \frac{g}{\sqrt[3]{(bg)}},$$

$$\text{ab}' - \text{a}'\text{b} = \frac{c}{\sqrt[3]{(ch)}}, \quad \text{cd}' - \text{c}'\text{d} = \frac{h}{\sqrt[3]{(ch)}},$$

and consequently

$$(af)^{\frac{1}{3}} + (bg)^{\frac{1}{3}} + (ch)^{\frac{1}{3}} = 0;$$

then the equation becomes

$$\begin{aligned}
&a^2y^4z^4 + b^2z^4x^4 + c^2x^4y^4 \\
&+ f^2x^4w^4 + g^2y^4w^4 + h^2z^4w^4 \\
&+ 2bcx^4y^2z^2 - 2cfx^4y^2w^2 + 2bfx^4z^2w^2 \\
&+ 2cay^4z^2x^2 - 2agy^4z^2w^2 + 2cgy^4x^2w^2 \\
&+ 2abz^4x^2y^2 - 2bhz^4x^2w^2 + 2ahz^4y^2w^2 \\
&- 2ghw^4y^2z^2 - 2hfw^4z^2x^2 - 2fgw^4x^2y^2 \\
&+ 2\{(bg)^{\frac{1}{3}} - (ch)^{\frac{1}{3}}\}\{(ch)^{\frac{1}{3}} - (af)^{\frac{1}{3}}\}\{(af)^{\frac{1}{3}} - (ch)^{\frac{1}{3}}\}\,x^2y^2z^2w^2 = 0.
\end{aligned}$$

This same equation, without the relation

$$(af)^{\frac{1}{3}} + (bg)^{\frac{1}{3}} + (ch)^{\frac{1}{3}} = 0,$$

and with an arbitrary coefficient for $x^2y^2z^2w^2$; or say, the equation

$$\begin{aligned}
&a^2y^4z^4 + b^2z^4x^4 + c^2x^4y^4 \\
&+ f^2x^4w^4 + g^2y^4w^4 + h^2z^4w^4 \\
&+ 2bcx^4y^2z^2 - 2cfx^4y^2w^2 + 2bfx^4z^2w^2 \\
&+ 2cay^4z^2x^2 - 2agy^4z^2w^2 + 2cgy^4x^2w^2 \\
&+ 2abz^4x^2y^2 - 2bhz^4x^2w^2 + 2ahz^4y^2w^2 \\
&- 2ghw^4y^2z^2 - 2hfw^4z^2x^2 - 2fgw^4x^2y^2 \\
&+ 2kx^2y^2z^2w^2 = 0,
\end{aligned}$$

where a, b, c, f, g, h, k are arbitrary coefficients, is the general equation of an octic surface having the four nodal curves

$$\begin{array}{llllll}
x = 0, & & \cdot & hz^2w^2 & - gw^2y^2 + ay^2z^2 & = 0, \\
y = 0, & -hz^2w^2 & & \cdot & + fw^2x^2 + bz^2x^2 & = 0, \\
z = 0, & gy^2w^2 & - fw^2x^2 & & \cdot \quad + cx^2y^2 & = 0, \\
w = 0, & -ay^2z^2 & - bz^2x^2 & - cx^2y^2 & \cdot & = 0.
\end{array}$$

In fact, the equation of the surface may be written in the form

$$\begin{aligned} & w^4 \{f^2x^4 + g^2y^4 + h^2z^4 - 2ghy^2z^2 - 2hfz^2x^2 - 2fgx^2y^2\} \\ + 2w^2 & \begin{Bmatrix} - cfx^4y^2 - agy^4z^2 - bhz^4x^2 + 2kx^2y^2z^2 \\ + bfx^4z^2 + cgy^4x^2 + ahz^4y^2 \end{Bmatrix} \\ + \quad & \{ay^2z^2 + bz^2x^2 + cx^2y^2\}^2 = 0, \end{aligned}$$

which puts in evidence the nodal curve

$$w = 0, \; - ay^2z^2 - bz^2x^2 - cx^2y^2 = 0:$$

there are three similar forms which put in evidence the other three nodal curves.

The four curves are so related to each other that every line which meets three of them meets also the fourth curve; there is consequently a singly infinite series of lines meeting each of the four curves; these break up into four series of lines each forming an octic scroll, and each scroll has the four curves for nodal curves respectively; that is, each scroll is a surface included under the foregoing general equation, and derived from it by assigning a proper value to the constant k. To determine these values, write

$$\begin{cases} \lambda + \mu + \nu = 0, \\ \dfrac{af}{\lambda^2} + \dfrac{bg}{\mu^2} + \dfrac{ch}{\nu^2} = 0, \end{cases}$$

equations which give four systems of values for the ratios $(\lambda : \mu : \nu)$. We have then

$$k = af\frac{\nu - \mu}{\lambda} + bg\frac{\lambda - \nu}{\mu} + ch\frac{\mu - \nu}{\lambda},$$

viz. k has four values corresponding to the several values of $(\lambda : \mu : \nu)$.

The scroll in question is M. De La Gournerie's scroll Σ_1; the equation of the scroll Σ_1 is consequently obtained from the octic equation by writing therein the last-mentioned value of k.

It is to be noticed that k is, in effect, determined by a quartic equation; and, that, for a certain relation between the coefficients, this equation will have a twofold root. Assuming that this relation is satisfied, and assigning to k its twofold value, the resulting scroll becomes a torse; that is, two of the four scrolls coincide together and degenerate into a torse; corresponding to the remaining two values of k we have two scrolls, *companions* of the torse. In order to a twofold value of k, we must have

$$\frac{af}{\lambda^3} = \frac{bg}{\mu^3} = \frac{ch}{\nu^3};$$

and thence

$$(af)^{\frac{1}{3}} + (bg)^{\frac{1}{3}} + (ch)^{\frac{1}{3}} = 0;$$

or, what is the same thing,

$$(af + bg + ch)^3 - 27abcfgh = 0.$$

If for a moment we write $af=\alpha^3$, $bg=\beta^3$, $ch=\gamma^3$, and, therefore, $\alpha+\beta+\gamma=0$; then for the twofold root, we have $\lambda : \mu : \nu = \alpha : \beta : \gamma$, and consequently

$$\begin{aligned} k &= \alpha^2(\gamma-\beta)+\beta^2(\alpha-\gamma)+\gamma^2(\beta-\alpha) \\ &= (\alpha-\beta)(\beta-\gamma)(\gamma-\alpha), \end{aligned}$$

that is,

$$k = \{(af)^{\frac{1}{3}}-(bg)^{\frac{1}{3}}\}\{(bg)^{\frac{1}{3}}-(ch)^{\frac{1}{3}}\}\{(ch)^{\frac{1}{3}}-(af)^{\frac{1}{3}}\},$$

which agrees with the result in regard to the octic torse.

If in the octic equation we write (x, y, z, w) in place of (x^2, y^2, z^2, w^2), then we have the quartic equation

$$\begin{aligned} & a^2y^2z^2 + b^2z^2x^2 + c^2x^2y^2 \\ & + f^2x^2w^2 + g^2y^2w^2 + h^2z^2w^2 \\ & + 2bcx^2yz - 2cfx^2yw + 2bfx^2zw \\ & + 2cay^2zx - 2agy^2zw + 2cgy^2xw \\ & + 2abz^2xy - 2bhz^2xw + 2ahz^2yw \\ & - 2ghw^2yz - 2hfw^2zx - 2fgw^2xy \\ & + 2kxyzw = 0, \end{aligned}$$

which is the equation of a quartic surface touched by the planes $x=0$, $y=0$, $z=0$, $w=0$, in the four conics

$$\begin{array}{llllll} x=0, & . & hzw & -gwy & +ayz & =0, \\ y=0, & -hzw & . & +fwx & +bzx & =0, \\ z=0, & gyw & -fwx & . & +cxy & =0, \\ w=0, & -ayz & -bzx & -cxy & . & =0, \end{array}$$

respectively.

II. The octic surface

$$\begin{aligned} U = {} & b^2c^2f^2x^8 + c^2a^2g^2y^8 + a^2b^2h^2z^8 + f^2g^2h^2w^8 \\ & - 2a^2cg\,(bg-ch)\,y^6z^2 - 2b^2ah\,(ch-af)\,z^6x^2 - 2c^2bf\,(af-bg)\,x^6y^2 \\ & + 2a^2bh\,(\quad,,\quad)\,y^2z^6 + 2b^2cf\,(\quad,,\quad)\,z^2x^6 + 2c^2ag\,(\quad,,\quad)\,x^2y^6 \\ & - 2f^2bc\,(\quad,,\quad)\,x^6w^2 - 2g^2ca\,(\quad,,\quad)\,y^6w^2 - 2h^2ab\,(\quad,,\quad)\,z^6w^2 \\ & + 2f^2gh\,(\quad,,\quad)\,x^2w^6 + 2g^2hf\,(\quad,,\quad)\,y^2w^6 + 2h^2fg\,(\quad,,\quad)\,z^2w^6 \\ & + f^2(b^2g^2+c^2h^2-4bgch)\,w^4x^4 + g^2(c^2h^2+a^2f^2-4chaf)\,w^4y^4 + h^2(a^2f^2+b^2g^2-4abfg)\,w^4z^4 \\ & + a^2(\qquad,,\qquad)\,y^4z^4 + b^2(\qquad,,\qquad)\,z^4x^4 + c^2(\qquad,,\qquad)\,x^4y^4 \\ & \qquad - 2gh\,(bcgh-a^2f^2-2afbg-2afch)\,w^4y^2z^2 \\ & \qquad - 2bh\,(\qquad\qquad,,\qquad\qquad)\,z^4x^2w^2 \\ & \qquad + 2cg\,(\qquad\qquad,,\qquad\qquad)\,y^4x^2w^2 \\ & \qquad + 2bc\,(\qquad\qquad,,\qquad\qquad)\,x^4y^2z^2 \end{aligned}$$

$$
\begin{array}{lll}
- 2hf\,(cahf - b^2g^2 - 2bgaf & - 2bgch) & w^4z^2x^2 \\
- 2cf\,(& \text{,,} \qquad\qquad) & x^4y^2w^2 \\
+ 2ah\,(& \text{,,} \qquad\qquad) & z^4y^2w^2 \\
+ 2ca\,(& \text{,,} \qquad\qquad) & y^4x^2z^2 \\
- 2fg\,(abfg - c^2h^2 - 2chaf & - 2chbg) & w^4x^2y^2 \\
- 2ag\,(& \text{,,} \qquad\qquad) & y^4z^2w^2 \\
+ 2bf\,(& \text{,,} \qquad\qquad) & x^4z^2w^2 \\
+ 2ab\,(& \text{,,} \qquad\qquad) & z^4x^2y^2 \\
+ 2\Omega x^2y^2z^2w^2 = 0, & &
\end{array}
$$

where the values of the coefficients indicated by (,,) are at once obtained by the proper interchanges of the letters, and where Ω is an arbitrary coefficient, is a surface having the four nodal conics

$$
\begin{array}{lllll}
x = 0, & . & cy^2 & - bz^2 & + fw^2 = 0, \\
y = 0, & - cx^2 & . & + az^2 & + gw^2 = 0, \\
z = 0, & bx^2 & - ay^2 & . & hw^2 = 0, \\
w = 0, & - fx^2 & - gy^2 & - hz^2 & . = 0.
\end{array}
$$

In fact, writing the equation under the form

$$w^2\Theta + (fx^2 + gy^2 + hz^2)^2 \times (b^2c^2x^4 + c^2a^2y^4 + a^2b^2z^4 - 2a^2bcy^2z^2 - 2b^2caz^2x^2 - 2c^2abx^2y^2) = 0,$$

we put in evidence the nodal conic $w=0$, $fx^2 + gy^2 + hz^2 = 0$: and similarly for the other nodal conics.

It is to be observed, that the complete section by the plane $w = 0$ is the conic $fx^2 + gy^2 + hz^2 = 0$, twice repeated, and the quartic

$$b^2c^2x^4 + c^2a^2y^4 + a^2b^2z^4 - 2a^2bcy^2z^2 - 2ab^2cz^2x^2 - 2abc^2x^2y^2 = 0:$$

the latter being the system of four lines

$$\frac{x}{\sqrt{(a)}} + \frac{y}{\sqrt{(b)}} + \frac{z}{\sqrt{(c)}} = 0, \quad \frac{x}{\sqrt{(a)}} + \frac{y}{\sqrt{(b)}} - \frac{z}{\sqrt{(c)}} = 0,$$

$$\frac{x}{\sqrt{(a)}} - \frac{y}{\sqrt{(b)}} + \frac{z}{\sqrt{(c)}} = 0, \quad \frac{x}{\sqrt{(a)}} - \frac{y}{\sqrt{(b)}} - \frac{z}{\sqrt{(c)}} = 0.$$

The plane in question, $w = 0$, meets the other nodal conics in the six points

$$(x=0,\ by^2 - cz^2 = 0), \quad (y=0,\ cz^2 - ax^2 = 0), \quad (z=0,\ ax^2 - by^2 = 0),$$

which six points are the angles of the quadrilateral formed by the above-mentioned four lines.

The four conics are such, that every line meeting three of these conics meets also the fourth conic. The lines in question form a double system: each of these systems has, in reference to any pair of nodal conics, a homographic property as follows; viz. considering for example the two conics in the planes $z=0$ and $w=0$ respectively, if a line meets these conics in the points P and Q respectively, and through these points respectively and the line $x=0$, $y=0$ we draw planes, then the system of the P planes and the system of the Q planes correspond homographically to each other, the coincident planes of the two systems being the planes $x=0$ and $y=0$ respectively.

Conversely, if through the line $(x=0,\ y=0)$ we draw the two homographically related planes meeting the two conics in the points P and Q respectively, then, for a proper value (determined by a quadratic equation) of the constant $k\,(=\Theta \div \theta)$ which determines the homographic relation, the line PQ will be a line meeting each of the four conics, and will belong to one or other of the above-mentioned two systems, as k is equal to one or the other of the two roots of the quadratic equation. The scroll generated by the lines meeting each of the four conics, or what is the same thing, any three of these conics, is *primâ facie* a scroll of the order 16; but by what precedes, it appears that this scroll breaks up into two scrolls, which will be each of the order 8. Moreover, each scroll has the four conics for nodal curves; and since the equation $U=0$ is the general equation of an octic surface having the four conics for nodal curves, it follows, that the equation of the scroll is derived from that of the octic surface $U=0$, by assigning a proper value to the indeterminate coefficient Ω; so that there are in fact two values of Ω, for each of which the surface $U=0$ becomes a scroll.

To sustain the foregoing conclusions, take $x=\theta' y$, $x=\theta y$ for the equations of the two planes through the line $(x=0,\ y=0)$, which meet the z-conic and w-conic in the points P and Q respectively; then the equations of the line PQ are

$$\sqrt{(f\theta^2+g)}\,(x-\theta' y)\ +\sqrt{(-h)}\,(\theta'-\theta)\,z=0,$$
$$-\sqrt{(b\theta'^2-a)}\,(x-\theta y)^2+\sqrt{(-h)}\,(\theta'-\theta)\,w=0,$$

or, writing therein $\theta'=k\theta$, the equations are

$$\sqrt{(f\theta^2\ \ +g)}\,(x-k\theta y)+\sqrt{(-h)}\,(k-1)\,\theta z\ =0,$$
$$-\sqrt{(bk^2\theta^2-a)}\,(x-\ \ \theta y)+\sqrt{(-h)}\,(k-1)\,\theta w=0.$$

To find where the line in question meets the plane $y=0$, we have

$$\sqrt{(f\theta^2\ \ +g)}\,x+\sqrt{(-h)}\,(k-1)\,\theta z=0,$$
$$-\sqrt{(bk^2\theta^2-a)}\,x+\sqrt{(-h)}\,(k-1)\,\theta w=0,$$

and thence

$$(f\theta^2\ \ +g)\,x^2+h\,(k-1)^2\,\theta^2 z^2\ =0,$$
$$(bk^2\theta^2-a)\,x^2+h\,(k-1)^2\,\theta^2 w^2=0,$$

or multiplying a, g and adding

$$(af+bgk^2)\,x^2+h\,(k-1)^2\,(az^2+gw^2)=0,$$

or assuming

$$af+bgk^2+ch\,(k-1)^2=0,$$

the equation is

$$-cx^2+az^2+gw^2=0.$$

That is, k being determined by the quadric equation $af+bgk^2+ch(k-1)^2=0$, the line PQ meets the y-conic $y=0$, $-cx^2+az^2+gw^2=0$; and, in a similar manner, it appears that the line PQ also meets the x-conic $x=0$, $cy^2-bz^2+fw^2=0$.

Writing for greater symmetry $1:-k:k-1=\lambda:\mu:\nu$, we have

$$\lambda+\mu+\nu=0,$$

$$af\lambda^2+bg\mu^2+ch\nu^2=0,$$

so that there are two systems of values of (λ, μ, ν) corresponding to, and which may be used in place of, the two values of k respectively.

Starting now from the equations

$$(f\theta^2+g)(k\theta y-x)^2+h(k-1)^2\theta^2z^2=0,$$

$$(bk^2\theta^2-a)(\theta y-x)^2+h(k-1)^2\theta^2w^2=0,$$

the elimination of θ from these equations leads to an equation $U=0$, of the above mentioned form but with a determinate value of the coefficient.

The process, although a long one, is interesting and I give it in some detail.

Elimination of θ from the foregoing equations.

We have

$$U=M\Pi\,[(f\theta^2+g)(k\theta y-x)^2+h(k-1)^2\theta^2z^2],$$

where Π denotes the product of the expressions corresponding to the four roots of the equation

$$(bk^2\theta^2-a)(\theta y-x)^2+h(k-1)^2\theta^2w^2=0.$$

Observing that this equation does not contain z, and that the expression under the sign Π does not contain w, it is at once seen that the product Π is in regard to (z, w) a rational and integral function of the form $(z^2, w^2)^4$; and since, in regard to (z, w), U is also a rational and integral function of the same form $(z^2, w^2)^4$, it is clear that the factor M does not contain z or w, but is a function of only (x, y). To determine it we may write $z=0$, $w=0$: this gives

$$c^2(bx^2-ay^2)^2(fx^2+gy^2)^2=M\Pi\,(f\theta^2+g)(k\theta y-x)^2,$$

where

$$(bk^2\theta^2-a)(\theta y-x)^2=0,$$

and the values of θ are therefore $+\dfrac{\sqrt{(a)}}{k\sqrt{(b)}}$, $-\dfrac{\sqrt{(a)}}{k\sqrt{(b)}}$, $\dfrac{x}{y}$, $\dfrac{x}{y}$. Hence substituting and observing that

$$c^2h^2(k-1)^4=(af+bgk^2)^2,$$

it is easy to find

$$M = \frac{y^4}{x^4} \frac{b^4k^4}{h^2(k-1)^8},$$

that is, we have

$$\frac{x^4}{y^4} \frac{h^2(k-1)^8}{b^4k^4} U = \Pi\,[(f\theta^2+g)(k\theta y - x)^2 + h(k-1)^2\,\theta^2 z^2],$$

where

$$(bk^2\theta^2 - a)(\theta y - x)^2 + h(k-1)^2\,\theta^2 w^2 = 0.$$

If for greater convenience we write $\theta = \frac{x}{y}\phi$, then this formula becomes

$$\frac{y^4}{x^4} \frac{h^2(k-1)^8}{b^4k^4} U = \Pi\,[(fx^2\phi^2 + gy^2)(k\phi - 1)^2 + h(k-1)^2 z^2\phi^2],$$

where

$$(bk^2x^2\phi^2 - ay^2)(\phi - 1)^2 + h(k-1)^2 w^2\phi^2 = 0,$$

or, what is the same thing,

$$\left(\phi^2 - \frac{ay^2}{bk^2x^2}\right)(\phi-1)^2 + \frac{h(k-1)^2 w^2}{bk^2x^2}\phi^2 = 0.$$

Suppose that the terms in U which contain z^2 are $= \Theta z^2$; then we have

$$\frac{y^4}{x^4} \frac{h(k-1)^6}{b^4k^4} \Theta = \Sigma\phi_1{}^2\Pi'(fx^2\phi^2 + gy^2)(k\phi-1)^2,$$

or, what is the same thing,

$$\Theta = \frac{b^4k^4}{h(k-1)^6} \frac{x^4}{y^4} \Sigma\phi_1{}^2\Pi'(fx^2\phi^2 + gy^2)(k\phi-1)^2,$$

where Π' refers to the remaining three roots ϕ_2, ϕ_3, ϕ_4; this may also be written

$$\Theta = \frac{b^4k^4}{h(k-1)^6} \frac{x^4}{y^4} \Pi\,(fx^2\phi^2 + gy^2)(k\phi-1)^2 . \Sigma \frac{\phi^2}{(fx^2\phi^2 + gy^2)(k\phi-1)^2}.$$

Hence, observing that we have identically

$$\left(\phi^2 - \frac{ay^2}{bk^2x^2}\right)(\phi-1)^2 + \frac{h(k-1)^2w^2}{bk^2x^2}\phi^2 = (\phi-\phi_1)(\phi-\phi_2)(\phi-\phi_3)(\phi-\phi_4),$$

and writing $\phi = \pm\frac{iy\sqrt{(g)}}{x\sqrt{(f)}}$, $\phi = \frac{1}{k}$, $\{i = \sqrt{(-1)}$ as usual$\}$, we find

$$\Pi\,\{\phi x\sqrt{(f)} \pm iy\sqrt{(g)}\} = \frac{h(k-1)^2}{bk^2}[c\,\{x\sqrt{(f)} \pm iy\sqrt{(g)}\}^2 fgw^2]\,y^2,$$

$$\Pi\,(k\phi - 1) \qquad = \frac{(k-1)^2}{bx^2}(bx^2 - ay^2 + hw^2);$$

whence, writing for shortness

$$\begin{aligned}\Delta &= [c\,\{x\sqrt{(f)} + iy\sqrt{(g)}^2\} - fgw^2][c\,\{x\sqrt{(f)} - iy\sqrt{(g)}^2\} - fgw^2],\\ &= c^2f^2x^4 + c^2g^2y^4 + f^2g^2w^4 + 2cfg^2y^2w^2 - 2cf^2gx^2w^2 + 2c^2fgx^2y^2,\end{aligned}$$

we find

$$\Pi\,(fx^2\phi^2+gy^2)=\frac{h^2\,(k-1)^4}{b^2k^4}\,\Delta y^4,$$

$$\Pi\,(k\phi-1)^2=\frac{(k-1)^4}{b^2}\,(bx^2-ay^2+hw^2)\,\frac{1}{x^4},$$

and thence

$$\Pi\,(fx^2\phi^2+gy^2)\,(k\phi-1)^2=\frac{h^2\,(k-1)^8}{b^4k^4}\,\Delta\,(bx^2-ay^2+hw^2)^2\,\frac{y^4}{x^4},$$

and consequently

$$\Theta=h\,(k-1)^2\,\Delta\,(bx^2-ay^2+hw^2)^2.\,\Sigma\,\frac{\phi^2}{(fx^2\phi^2+gy^2)\,(k\phi-1)^2}.$$

Hence, writing

$$\frac{\phi^2}{(fx^2\phi^2+gy^2)\,(k\phi-1)^2}=\frac{A}{(k\phi-1)^2}+\frac{B}{k\phi-1}+\frac{C}{x\phi\,\sqrt{(f)}+iy\,\sqrt{(g)}}+\frac{D}{x\phi\,\sqrt{(f)}-iy\,\sqrt{(g)}},$$

we may calculate separately the terms

$$\Sigma\left\{\frac{A}{(k\phi-1)^2}+\frac{B}{k\phi-1}\right\},$$

and

$$\Sigma\left\{\frac{C}{x\phi\,\sqrt{(f)}+iy\,\sqrt{(g)}}+\frac{D}{x\phi\,\sqrt{(f)}-iy\,\sqrt{(g)}}\right\}.$$

The first of these is

$$=\frac{1}{(k-1)^2\,(fx^2+k^2gy)^2\,(bx^2-ay^2+hw^2)^2}\,\{x,\ y,\ w\}^6,$$

if for shortness

$$\begin{aligned}\{x,\ y,\ w\}^6=(fx^2+k^2gy^2)\,[4\,\{(2-k)\,bx^2-ay^2+h\,(1-k)\ w^2\}^2\\ -2\,(bx^2-ay^2+hw^2)\,\{(6-6k+k^2)\,bx^2-ay^2+h\,(1-k)^2\,w^2\}]\\ +4k^2\,(k-1)\,gy^2\,(bx^2-ay^2+hw^2)\,\{(2-k)\,bx^2-ay^2+h\,(1-k)\ w^2\}:\end{aligned}$$

the second is

$$=\frac{2}{(k-1)^2\,(fx^2+k^2gy^2)^2\,\Delta}\,(x,\ y,\ w)^6,$$

if for shortness

$$\begin{aligned}(x,\ y,\ w)^6=\{(fx^2-k^2gy^2)\,(cfx^2-cgy^2-fgw^2)-4ckfgx^2y^2\}\\ \times\{fgbk^2x^2+[2ch\,(k-1)^2+af\,]\,gy^2+fgh\,(k-1)^2\,w^2\}\\ +2\,\{k^2bg-ch\,(k-1)^2\}\,fgx^2y^2\,\{c\,(k+1)\,(fx^2-kgy^2)-kfgw^2\};\end{aligned}$$

and hence

$$\Theta=\frac{1}{(fx^2+k^2gy^2)^2}\,[h\Delta\,\{x,\ y,\ w\}^6+2\,(bx^2-ay^2+hw^2)^2\,(x,\ y,\ w)^6],$$

which must be a rational and integral function of $(x,\ y,\ w)$.

In partial verification of this, observe that, because U contains the terms

$$2b^2cf\,(ch-af\,)\,x^6z^2+2\Omega x^2y^2z^2w^2,$$

Θ should contain the terms

$$2b^2cf(ch-af)x^6+2\Omega x^2y^2w^2,$$

viz. in Θ the term in x^6 should be $=2b^2cf(ch-af)x^6$.

Now writing $y=0$, $w=0$, we have

$$\begin{aligned}\Delta &= c^2f^2x^4,\\ \{x, y, w\}^6 &= b^2f\{4(2-k)^2-2(6-6k+k^2)\}x^6,\\ &= b^2f(4-4k+2k^2)x^6,\\ (x, y, w)^6 &= bcf^3gk^2x^6;\end{aligned}$$

and hence the required term of Θ is x^6 multiplied by

$$b^2c^2fh(4-4k+2k^2)+2b^3cfgk^2:$$

viz. the coefficient is

$$\begin{aligned}&= 2b^2cf[ch(2-2k+k^2)+bgk^2],\\ &= 2b^2cf[ch+ch(1-k)^2+bgk^2],\end{aligned}$$

which in virtue of the relation $af+bgk^2+ch(1-k)^2$ becomes, as it should do,

$$=2b^2cf(ch-af).$$

The actual division by $(fx^2+k^2gy^2)^2$ would, however, be a very tedious process, and it is to be observed, that we only require to know the term $2\Omega x^2y^2w^2$ of Θ. We may therefore adopt a more simple course as follows: the terms of Θ which contain w^2 are $=(Ax^4+2\Omega x^2y^2+By^4)w^2$, hence writing for a moment

$$\{x, y, w\}^6=P+Qw^2, \quad (x, y, w)^6=R+Sw^2,$$

and observing that we have

$$\begin{aligned}\Delta &= c^2(fx^2+gy^2)^2-2c^2fg(fx^2-gy^2)w^2+\&c.,\\ (bx^2-ay^2+hw^2)^2 &= (bx^2-ay^2)^2+2h(bx^2-ay^2)w^2+\&c.,\end{aligned}$$

we have

$$\begin{aligned}(fx^2+k^2gy^2)^2(Ax^4+2\Omega x^2y^2+By^4) = {}& c^2h(fx^2+gy^2)^2Q-2c^2fgh(fx^2-gy^2)P\\ &+(bx^2-ay^2)^2S+2h(bx^2-ay^2)R.\end{aligned}$$

But in this identical equation we may write $x^2=a$, $y^2=b$, which gives

$$(af+k^2bg)^2(Aa^2+2\Omega ab+Bb^2)=c^2h(af+bg)^2Q-2c^2fgh(af-bg)P;$$

and from the equation

$$\{x, y, w\}^6=P+Qw^2,$$

we have

$$\begin{aligned}P+Qw^2 &= (af+bgk^2)\begin{bmatrix}4\{(1-k)ab+(1-k)hw^2\}^2\\ -2hw^2(5-6k+k^2)ab\end{bmatrix}\\ &\quad +4k^2(k-1)bghw^2(1-k)ab,\\ &= -ch(k-1)^2\{4(k-1)^2(a^2b^2+2w^2abh)-2(5-6k+k^2)hw^2\}\\ &\quad -4k^2(k-1)^2ab^2ghw^2,\end{aligned}$$

that is,

$$P = -4ch(k-1)^4 a^2b^2,$$
$$Q = \quad (k-1)^2 abh\,\{ch(-6k^2+4k+2) - 4k^2b\},$$

whence

$$(k-1)^2(Aa^2 + 2\Omega ab + Bb^2) = ab(af+bg)^2[ch(-6k^2+4k+2) - 4k^2bg] \\ + 8fg(af-bg)(k-1)^2(ab)^2.$$

But we have

$$Aa^2 + Bb^2 = -2ab(af-bg)(-afbg + c^2h^2 + 2chaf + 2chbg),$$

and thence

$$2(k-1)^2\Omega = (af+bg)^2[ch(-6k^2+4k+2) - 4k^2bg] \\ +(k-1)^2(af-bg)\begin{pmatrix}-2afbg + 2c^2h^2 + 4chaf + 4chbg \\ +8afbg\end{pmatrix},$$

or

$$(k-1)^2\Omega = (af+bg)^2[ch(-3k^2+2k+1) - 2k^2bg] \\ +(k-1)^2(af-bg)[3afbg + c^2h^2 + 2chaf + 2chbg].$$

Writing $-3k^2+2k+1 = -3(k-1)^2 - 4(k-1)$, this is

$$\Omega = (af+bg)^2\left[ch\left(-3 - \frac{4}{k-1}\right) - \frac{2k^2}{(k-1)^2}bg\right] + (af-bg)[3afbg + c^2h^2 + 2chaf + 2chbg],$$

or since $1 : -k : k-1 = \lambda : \mu : \nu$; and writing for shortness $(af, bg, ch) = (\alpha, \beta, \gamma)$, this is

$$\Omega = (\alpha+\beta)^2\left\{\gamma\left(-3 - \frac{4\lambda}{\nu}\right) - \frac{2\mu^2}{\nu^2}\beta\right\} + (\alpha-\beta)\{3\alpha\beta + \gamma^2 + 2\gamma\alpha + 2\gamma\beta\},$$

which is the value of Ω: viz. the conclusion arrived at is that, eliminating θ from the equations

$$(f\theta^2 \;\; + g)(k\theta y - x)^2 + h(k-1)^2\theta^2 z^2 = 0,$$
$$(bk^2\theta^2 - a)(\;\theta y - x)^2 + h(k-1)^2\theta^2 w^2 = 0:$$

where k denotes a determinate function of af, bg, ch, viz. writing $af, bg, ch = \alpha, \beta, \gamma$ and $1 : -k : k-1 = \lambda : \mu : \nu$, we have

$$\lambda + \;\; \mu + \;\; \nu = 0,$$
$$\alpha\lambda^2 + \beta\mu^2 + \gamma\nu^2 = 0,$$

equations which serve to determine k: the result of the elimination is the octic equation

$$b^2c^2f^2x^8 + \ldots + 2\Omega x^2y^2z^2w^2 = 0,$$

where Ω has the last-mentioned value.

The value of Ω is unsymmetrical in its form, and there are apparently six values; viz. writing

$$A = (\beta+\gamma)^2 \left\{\alpha\left(-3-\frac{4\mu}{\lambda}\right)-\frac{2\nu^2}{\lambda^2}\gamma\right\}+(\beta-\gamma)(S+\beta\gamma+\alpha^2),$$

$$B = (\gamma+\alpha)^2 \left\{\beta\left(-3-\frac{4\nu}{\mu}\right)-\frac{2\lambda^2}{\mu^2}\alpha\right\}+(\gamma-\alpha)(S+\gamma\alpha+\beta^2),$$

$$C = (\alpha+\beta)^2 \left\{\gamma\left(-3-\frac{4\lambda}{\nu}\right)-\frac{2\mu^2}{\nu^2}\beta\right\}+(\alpha-\beta)(S+\alpha\beta+\gamma^2),$$

$$A_1 = (\beta+\gamma)^2 \left\{\alpha\left(-3-\frac{4\nu}{\lambda}\right)-\frac{2\mu^2}{\lambda^2}\beta\right\}-(\beta-\gamma)(S+\beta\gamma+\alpha^2),$$

$$B_1 = (\gamma+\alpha)^2 \left\{\beta\left(-3-\frac{4\lambda}{\mu}\right)-\frac{2\nu^2}{\mu^2}\gamma\right\}-(\gamma-\alpha)(S+\gamma\alpha+\beta^2),$$

$$C_1 = (\alpha+\beta)^2 \left\{\gamma\left(-3-\frac{4\mu}{\nu}\right)-\frac{2\lambda^2}{\nu^2}\alpha\right\}-(\alpha-\beta)(S+\alpha\beta+\gamma^2),$$

where for shortness $S=2(\beta\gamma+\gamma\alpha+\alpha\beta)$, the six values would be A, B, C, A_1, B_1, C_1. But we have really

$$A=B=C=-A_1=-B_1=-C_1;$$

so that Ω has really only two values, equal and of opposite signs, or, what is the same thing, Ω^2 has a unique value. In fact, writing for shortness

$$\lambda+\mu+\nu=P,\quad \alpha\lambda^2+\beta\mu^2+\gamma\nu^2=X,$$

we find at once the identity

$$\lambda^2(A+A_1)=(\beta+\gamma)^2(-2X-4\lambda\alpha P),$$

so that $A=-A_1$, in value of $P=0$, $X=0$. And similarly $B=-B_1$, $C=-C_1$.

But the demonstration of the equation $A=B$ is more complicated. We have

$$\begin{aligned}A-B= &-3\alpha(\beta+\gamma)^2-4\alpha(\beta+\gamma)^2\frac{\mu}{\lambda}-2\gamma(\beta+\gamma)^2\frac{\nu^2}{\lambda^2}+(\beta-\gamma)(S+\beta\gamma+\alpha^2)\\ &+3\beta(\gamma+\alpha)^2+4\beta(\gamma+\alpha)^2\frac{\nu}{\mu}+2\alpha(\gamma+\alpha)^2\frac{\lambda^2}{\mu^2}-(\gamma-\alpha)(S+\gamma\alpha+\beta^2),\end{aligned}$$

that is,

$$\begin{aligned}\lambda^2\mu^2(A-B)= &\{-3\alpha(\beta+\gamma)^2+3\beta(\gamma+\alpha)^2+(\beta-\gamma)(S+\beta\gamma+\alpha^2)-(\gamma-\alpha)(S+\gamma\alpha+\beta^2)\}\lambda^2\mu^2\\ &-4\alpha(\beta+\gamma)^2\lambda\mu^3\\ &-2\gamma(\beta+\gamma)^2\nu^2\mu^2\\ &+4\beta(\gamma+\alpha)^2\nu\lambda^2\mu\\ &+2\alpha(\gamma+\alpha)^2\lambda^4,\end{aligned}$$

or, denoting for a moment the coefficient of $\lambda^2\mu^2$ by K, and writing also $\gamma\nu^2 = X - \alpha\lambda^2 - \beta\mu^2$, $\nu = P - \lambda - \mu$, this is

$$\begin{aligned} = \ & K\lambda^2\mu^2 \\ & - 4\alpha(\beta+\gamma)^2\lambda\mu^2 \\ & - 2\ (\beta+\gamma)^2\mu^2(X - \alpha\lambda^2 - \beta\mu^2) \\ & + 4\beta(\gamma+\alpha)^2\lambda^2\mu(P-\lambda-\mu) \\ & + 2\alpha(\gamma+\alpha)^2\lambda^4, \end{aligned}$$

$$\begin{aligned} = & - 2\ (\beta+\gamma)^2\mu^2X + 4\beta(\gamma+\alpha)^2\lambda^2\mu P \\ & + 2\alpha(\gamma+\alpha)^2\lambda^4 \\ & - 4\beta(\gamma+\alpha)^2\lambda^3\mu \\ & + \{-4\beta(\gamma+\alpha)^2 + K + 2\alpha(\beta+\gamma)^2\}\lambda^2\mu^2 \\ & \quad - 4\alpha(\beta+\gamma)^2\lambda\mu^3 \\ & \quad + 2\beta(\beta+\gamma)^2\mu^4, \end{aligned}$$

and here the coefficient of $\lambda^2\mu^2$ is found to be

$$= 2\{\alpha\beta(\alpha+\beta) + \gamma(\alpha-\beta)^2 - 3\gamma^2(\alpha+\beta)\}.$$

Hence, the terms without X or P are $= 2\nabla$, where

$$\begin{aligned} \nabla = \ & \quad \alpha(\gamma+\alpha)^2\lambda^4 \\ & - 2\beta(\gamma+\alpha)^2\lambda^3\mu \\ & + \{\alpha\beta(\alpha+\beta) + \gamma(\alpha-\beta)^2 - 3\gamma^2(\alpha+\beta)\}\lambda^2\mu^2 \\ & - 2\alpha(\beta+\gamma)^2\lambda\mu^3 \\ & + \beta\ (\beta+\gamma)^2\mu^3, \end{aligned}$$

and this is identically

$$= + \left.\begin{matrix} (\alpha+\gamma)\lambda^2 \\ 2\gamma\lambda\mu \\ +(\beta+\gamma)\mu^2 \end{matrix}\right\} \times \left\{\begin{matrix} \alpha(\gamma+\alpha)\lambda^2 \\ -2(\beta\gamma+\gamma\alpha+\alpha\beta)\lambda\mu \\ +\beta(\beta+\gamma)\mu^2, \end{matrix}\right.$$

where observing that

$$\alpha\lambda^2 + \beta\mu^2 + \gamma(P-\lambda-\mu)^2 = X,$$

we have the first factor

$$(\alpha+\gamma)\lambda^2 + (\beta+\gamma)\mu^2 + 2\gamma\lambda\mu = X - \gamma P^2 + 2\gamma P(\lambda+\mu),$$

and consequently

$$\begin{aligned} \lambda^2\mu^2(A-B) = & -2(\beta+\gamma)^2\mu^2X + 4\beta(\gamma+\alpha)^2\lambda^2\mu P \\ & + 2\{X - \gamma P^2 + 2\gamma P(\lambda+\mu)\}\{\alpha(\gamma+\alpha)\lambda^2 - 2(\beta\gamma+\gamma\alpha+\alpha\beta)\lambda\mu + \beta(\beta+\gamma)\mu^2\}; \end{aligned}$$

viz. in virtue of $P=0$, $X=0$, we have $A=B$. And thus

$$A = B = C = -A_1 = -B_1 = -C_1:$$

so that the only values of Ω are, say, A and $-A$.

Reverting to the original equations

$$(f\theta^2 \;+g)(k\theta y - x)^2 + h(k-1)^2\,\theta^2 z^2 = 0,$$

$$(bk^2\theta^2 - a)(\;\theta y - x)^2 + h(k-1)^2\,\theta^2 w^2 = 0,$$

say these are

$$(\mathrm{a},\ \mathrm{b},\ \mathrm{c},\ \mathrm{d},\ \mathrm{e}\,\c\,\theta,\ 1)^4 = 0,$$

$$(\mathrm{a}',\ \mathrm{b}',\ \mathrm{c}',\ \mathrm{d}',\ \mathrm{e}'\,\c\,\theta,\ 1)^4 = 0,$$

then the coefficients in the two equations have the values

$$\begin{array}{ll} fk^2y^2, & bk^2y^2, \\ -2kfxy, & -2bk^2xy, \\ fx^2 + gky^2 + h(k-1)^2 z^2, & bk^2x^2 - ay^2 + h(k-1)^2 w^2, \\ -2gkxy, & 2axy, \\ gx^2, & -ax^2, \end{array}$$

where observe that only c contains z^2, and only c' contains w^2. The result of the elimination is

$$\left|\begin{array}{cccccccc} & & & \mathrm{a}, & \mathrm{b}, & \mathrm{c}, & \mathrm{d}, & \mathrm{e} \\ & & \mathrm{a}, & \mathrm{b}, & \mathrm{c}, & \mathrm{d}, & \mathrm{e}, & \\ & \mathrm{a}, & \mathrm{b}, & \mathrm{c}, & \mathrm{d}, & \mathrm{e}, & & \\ \mathrm{a}, & \mathrm{b}, & \mathrm{c}, & \mathrm{d}, & \mathrm{e}, & & & \\ & & & \mathrm{a}', & \mathrm{b}', & \mathrm{c}', & \mathrm{d}', & \mathrm{e}' \\ & & \mathrm{a}', & \mathrm{b}', & \mathrm{c}', & \mathrm{d}', & \mathrm{e}', & \\ & \mathrm{a}', & \mathrm{b}', & \mathrm{c}', & \mathrm{d}', & \mathrm{e}', & & \\ \mathrm{a}', & \mathrm{b}', & \mathrm{c}', & \mathrm{d}', & \mathrm{e}', & & & \end{array}\right| = 0;$$

viz. here the only terms which contain z^8 and w^8 are

$$\mathrm{c}^2\mathrm{a}'^2\mathrm{e}'^2 + \mathrm{c}'^2\mathrm{a}^2\mathrm{e}^2,$$

and hence the terms in z^8 and w^8 are

$$h^4(k-1)^8 z^8 \,.\, a^2b^2k^4x^4y^4 + h^4(k-1)^8 w^8 \,.\, f^2g^2k^4x^4y^4,$$

viz. these are

$$= h^2k^4(k-1)^8 x^4y^4 (a^2b^2h^2z^8 + f^2g^2h^2w^8),$$

or assuming that the determinant contains as a factor the function $b^2c^2f^2x^8 + \ldots + 2\Omega x^2y^2z^2w^2$, with a properly determined value of Ω, we see that the other factor is $= h^2k^4(k-1)^8 x^4y^4$, which agrees with a preceding result.

655.

A MEMOIR ON DIFFERENTIAL EQUATIONS.

[From the *Quarterly Journal of Pure and Applied Mathematics*, vol. XIV. (1877), pp. 292—339.]

WE have to do with a set of variables, which is either unipartite $(x, y, z, \ldots)$, or else bipartite $(x, y, z, \ldots; p, q, r, \ldots)$, the variables in the latter case corresponding in pairs x and p, y and q, &c.

A letter not otherwise explained denotes a function of the variables. Any such letter may be put $=$ const., viz. we thereby establish a relation between the variables; and when this is so, we use the *same* letter to denote the constant value of the function. Thus the set being $(x, y, z; p, q, r)$, H may denote a given function $pqr - xyz$; and then, if $H =$ const., we have $pqr - xyz = H$ (a constant). This notation, when once clearly understood, is I think a very convenient one.

The present memoir relates chiefly to the following subjects:

A. Unipartite set $(x, y, z, \ldots)$. The differential system

$$\frac{dx}{X} = \frac{dy}{Y} = \frac{dz}{Z} = \ldots,$$

and connected therewith the linear partial differential equation

$$X\frac{d\theta}{dx} + Y\frac{d\theta}{dy} + Z\frac{d\theta}{dz} + \ldots = 0:$$

also the lineo-differential

$$Xdx + Ydy + Zdz + \ldots.$$

B. Bipartite set $(x, y, z, \ldots; p, q, r, \ldots)$. The Hamiltonian system

$$\frac{dx}{\frac{dH}{dp}} = \frac{dy}{\frac{dH}{dq}} = \frac{dz}{\frac{dH}{dr}} = \ldots = \frac{dp}{-\frac{dH}{dx}} = \frac{dq}{-\frac{dH}{dy}} = \frac{dr}{-\frac{dH}{dz}} = \ldots,$$

and connected therewith the linear partial differential equation

$$\frac{dH}{dp}\frac{d\theta}{dx} - \frac{dH}{dx}\frac{d\theta}{dp} + \frac{dH}{dq}\frac{d\theta}{dy} - \frac{dH}{dy}\frac{d\theta}{dq} + \dots = 0,$$

otherwise written

$$(H,\ \theta),\ = \frac{d\,(H,\ \theta)}{d\,(p,\ x)} + \frac{d\,(H,\ \theta)}{d\,(q,\ y)} + \dots,\ = 0,$$

where H denotes a given function of the variables: also the Hamiltonian system as augmented by an equality $=dt$, and as augmented by this and another equality

$$= dV \div \left(p\frac{dH}{dp} + q\frac{dH}{dq} + r\frac{dH}{dr}\dots\right).$$

C. Bipartite set $(x,\ y,\ z, \dots;\ p,\ q,\ r, \dots)$. The partial differential equation $H = \text{const.}$, where, as before, H is a given function of the variables, but $p,\ q,\ r, \dots$ are now the differential coefficients in regard to $x,\ y,\ z, \dots$ respectively of a function V of these variables, or, what is the same thing, there exists a function

$$V = \int (pdx + qdy + rdz + \dots),$$

of the variables $x,\ y,\ z, \dots$.

In what precedes, I have written $(x,\ y,\ z, \dots)$ to denote a set of any number n of variables, and $(x,\ y,\ z, \dots;\ p,\ q,\ r, \dots)$ to denote a set of any even number $2n$ of variables, and the investigations are for the most part applicable to these general cases. But for greater clearness and facility of expression, I usually consider the case of a set $(x,\ y,\ z,\ w)$, or $(x,\ y,\ z;\ p,\ q,\ r)$, &c., as the case may be, consisting of a definite number of variables.

The greater part of the theory is not new, but I think that I have presented it in a more compact and intelligible form than has hitherto been done, and I have added some new results.

Introductory Remarks. Art. Nos. 1 to 3.

1. As already noticed, a letter not otherwise explained is considered as denoting a function of the variables of the set; but when necessary we indicate the variables by a notation such as $z = z(x,\ y)$; z is here a function (known or unknown as the case may be) of the variables $x,\ y$, the z on the right-hand side being in fact a functional symbol. And thus also $z = z(x,\ y),\ = \text{const.}$ denotes that the function $z(x,\ y)$ of the variables $x,\ y$ has a constant value, which constant value is $=z$, viz. we thus indicate a relation between the variables $x,\ y$.

2. The variables $x,\ y$, &c., may have infinitesimal increments $dx,\ dy$, &c.; and the equations of connexion between the variables then give rise to linear relations between these increments, the coefficients therein being differential coefficients and,

as such, represented in the usual notation; thus if $z=z(x, y)$, we have $dz=\frac{dz}{dx}dx+\frac{dz}{dy}dy$, where $\frac{dz}{dx}, \frac{dz}{dy}$ are the so-called partial differential coefficients of z in regard to x, y respectively. If we have $y=y(x)$, then also $dy=\frac{dy}{dx}dx$, and the foregoing equation becomes

$$dz=\left(\frac{dz}{dx}+\frac{dz}{dy}\frac{dy}{dx}\right)dx;$$

but considering the two equations $z=z(x, y)$ and $y=y(x)$ as determining z as a function of x, say $z=z(x)$, we have $dz=\frac{d(z)}{dx}dx$; whence comparing the two formulæ

$$\frac{d(z)}{dx}=\frac{dz}{dx}+\frac{dz}{dy}\frac{dy}{dx},$$

where $\frac{d(z)}{dx}$ is the so-called total differential coefficient of z in regard to x. The distinction is best made, not by any difference of notation $\frac{d(z)}{dx}, \frac{dz}{dx}$, but by appending in any case of doubt the equations or equation used in the differentiation. Thus we have $\frac{dz}{dx}$ where $z=z(x, y)$; or, as the case may be, $\frac{dz}{dx}$ where $z=z(x, y)$ and $y=y(x)$.

3. A relation between increments is always really a relation between differential coefficients: but we use the increments for symmetry and conciseness, as in the case of a differential system $\frac{dx}{X}=\frac{dy}{Y}=\frac{dz}{Z}$, or in a question relating to the lineo-differential $Xdx+Ydy+Zdz$, for instance in the question whether this can be put $=du$.

Notations. Art. Nos. 4 to 6.

4. Functional determinants. If $a, b, c, \ldots$ are functions of the variables $x, y, z, w, \ldots$, then the determinants

$$\begin{vmatrix} \frac{da}{dx}, & \frac{da}{dy} \\ \frac{db}{dx}, & \frac{db}{dy} \end{vmatrix}, \quad \begin{vmatrix} \frac{da}{dx}, & \frac{da}{dy}, & \frac{da}{dz} \\ \frac{db}{dx}, & \frac{db}{dy}, & \frac{db}{dz} \\ \frac{dc}{dx}, & \frac{dc}{dy}, & \frac{dc}{dz} \end{vmatrix}, \text{ \&c.,}$$

are for shortness represented by

$$\frac{d(a, b)}{d(x, y)}, \frac{d(a, b, c)}{d(x, y, z)}, \text{ \&c.,}$$

the notation being especially used in the first-mentioned case where the symbol is $\frac{d(a, b)}{d(x, y)}$. It is sometimes convenient to extend this notation, and for instance use $\frac{d(a, b)}{d(x, y, z)}$ to denote the series of determinants

$$\begin{vmatrix} \frac{da}{dx}, & \frac{da}{dy}, & \frac{da}{dz} \\ \frac{db}{dx}, & \frac{db}{dy}, & \frac{db}{dz} \end{vmatrix},$$

which can be formed by selecting in every way two columns to form thereout a determinant; the equation

$$\frac{d(a, b)}{d(x, y, z)} = 0$$

will then denote that each of these determinants is $= 0$.

The analogous notation

$$\frac{d(a, b, c)}{d(x, y)}$$

would denote non-existent determinants, viz. there are here not columns enough to form with them a determinant: and the notation is not required.

5. In the case of a bipartite set $(x, y, z, \ldots; p, q, r, \ldots)$, if a, b are any functions of these variables, we consider the derivative

$$(a, b) = \frac{d(a, b)}{d(p, x)} + \frac{d(a, b)}{d(q, y)} + \frac{d(a, b)}{d(r, z)} + \ldots,$$

viz. (a, b) is used to denote the sum of the functional determinants on the right hand.

6. Taking again $(x, y, z, w, \ldots)$ as the variables, then in the theory of the lineo-differential $Xdx + Ydy + Zdz + Wdw + \ldots$, we use certain derivative functions analogous to Pfaffians. They may be thus defined; viz. considering the numbers 1, 2, 3, 4, ... as corresponding to the variables $x, y, z, w, \ldots$ respectively, we have

$$1 = X,\ 2 = Y,\ 3 = Z,\ 4 = W,\ \&\text{c.},$$

$$12 = \frac{dX}{dy} - \frac{dY}{dx},\ 13 = \frac{dX}{dz} - \frac{dZ}{dx},\ \&\text{c.},$$

$$123 = 1 . 23 + 2 . 31 + 3 . 12$$

$$= X\left(\frac{dY}{dz} - \frac{dZ}{dy}\right) + Y\left(\frac{dZ}{dx} - \frac{dX}{dz}\right) + Z\left(\frac{dX}{dy} - \frac{dY}{dx}\right),$$

$$1234 = 12 . 34 + 13 . 42 + 14 . 23$$

$$= \left(\frac{dX}{dy} - \frac{dY}{dx}\right)\left(\frac{dZ}{dw} - \frac{dW}{dz}\right) + \left(\frac{dX}{dz} - \frac{dZ}{dx}\right)\left(\frac{dW}{dy} - \frac{dY}{dw}\right) + \left(\frac{dX}{dw} - \frac{dW}{dx}\right)\left(\frac{dY}{dz} - \frac{dZ}{dy}\right),$$

and, adding for greater distinctness the next following cases,

$$12345 = 1 . 2345 + 2 . 3451 + 3 . 4512 + 4 . 5123 + 5 . 1234,$$

$$123456 = 12 . 3456 + 13 . 4561 + 14 . 5612 + 15 . 6123 + 16 . 2345,$$

where of course 2345, &c., have the significations mentioned above.

Dependency of Functions. Art. Nos. 7 and 8.

7. Two or more functions of the same variables may be independent, or else dependent or connected; viz. in the latter case any one of the functions is a function of the others $a = a(x)$, $b = b(x)$, the functions a, b are dependent, but if

$$a = a(x, y), \quad b = b(x, y),$$

then the condition of dependency is

$$\frac{d(a, b)}{d(x, y)} = 0,$$

and, similarly, if $a = a(x, y, z)$, $b = b(x, y, z)$, then the conditions of dependency are

$$\frac{d(a, b)}{d(x, y, z)} = 0,$$

viz. if the equations thus represented are all of them satisfied, the functions are dependent, but if not, then they are independent.

Observe that, when $a = a(x, y, z)$, $b = b(x, y, z)$ as above, if we choose to attend only to the variables x, y, treating z as a mere constant, there is then a single condition of dependency $\dfrac{d(a, b)}{d(x, y)} = 0$, and so if we attend only to the variable x, treating y, z as mere constants, then a and b are dependent. Thus when $a = x$, $b = x^2 + y$, the functions a, b are independent if we attend to both the variables x, y; dependent if y be regarded as a constant.

8. Further when $a = a(x, y)$, $b = b(x, y)$, $c = c(x, y)$, the functions a, b, c are dependent; but when $a = a(x, y, z)$, $b = b(x, y, z)$, $c = c(x, y, z)$, the condition of dependency is

$$\frac{d(a, b, c)}{d(x, y, z)} = 0:$$

and so when $a = a(x, y, z, w)$, $b = b(x, y, z, w)$, $c = c(x, y, z, w)$, the conditions of dependency are

$$\frac{d(a, b, c)}{d(x, y, z, w)} = 0;$$

viz. if all the equations thus represented are satisfied, the functions are dependent; but if not, then they are independent. And so in other cases.

The General Differential System. Art. Nos. 9 to 22.

9. Taking the set of variables to be (x, y, z, w), the system is

$$\frac{dx}{X}=\frac{dy}{Y}=\frac{dz}{Z}=\frac{dw}{W},$$

and we associate with this the linear partial differential equation

$$X\frac{d\theta}{dx}+Y\frac{d\theta}{dy}+Z\frac{d\theta}{dz}+W\frac{d\theta}{dw}=0.$$

10. It is tolerably evident that the differential equations establish between x, y, z, w a threefold relation depending upon three arbitrary constants; in fact, regarding (x, y, z, w) as the coordinates of a point in four-dimensional space, and starting from any given point, the differential equations determine the ratios of the increments dx, dy, dz, dw, that is, the direction of passage to a consecutive point; and then again taking for x, y, z, w the coordinates of this point, the same equations give the direction of passage to the next consecutive point, and so on. The locus of the point is therefore a curve, or we have between the coordinates a threefold relation, and (the initial point being arbitrary) we have a curve of the system through each point of the four-dimensional space, viz. the relation must involve three arbitrary constants. But this being so, the constants will be expressible as functions of the coordinates, viz. the threefold relation involving the three constants will be expressible in the form $a=$ const., $b=$ const., $c=$ const., where a, b, c denote respectively functions of the coordinates (x, y, z, w).

11. Supposing that one of the relations is $a=$ const., it is clear that the increment

$$da, =\frac{da}{dx}dx+\frac{da}{dy}dy+\frac{da}{dz}dz+\frac{da}{dw}dw,$$

must become $=0$, on substituting therein for dx, dy, dz, dw, the values X, Y, Z, W to which by virtue of the differential equations they are proportional, viz. that we must have identically

$$X\frac{da}{dx}+Y\frac{da}{dy}+Z\frac{da}{dz}+W\frac{da}{dw}=0.$$

Conversely, when this is so, we have $da=0$, by virtue of the differential equation.

We say that a is a solution of the partial differential equation, and an integral of the differential equations, viz. any solution of the partial differential equation is an integral of the differential equations, and any integral of the differential equations is a solution of the partial differential equation, or, this being so, we may in general without risk of ambiguity, say simply a is an integral*; similarly b and c are integrals, and, by what precedes, there are three integrals a, b, c.

* Viz. we use indifferently, in regard to the differential equations and to the partial differential equation, the term integral, which is appropriate to the differential equations; the appropriate term in regard to the partial differential equation would be solution.

Observe that, in speaking of an integral a, we mean a function of the variables; the differential equations give between the variables the relation $a = \text{const.}$, and when this is so, we use the same letter a to denote the constant value of this function.

12. In speaking of the three integrals a, b, c we mean independent integrals; any function whatever ϕa of an integral a, or any function whatever $\phi(a, b)$ of two integrals a, b, is an integral (viz. it is an integral of the differential equations, and also a solution of the partial differential equation), but such dependent integrals give nothing new, and we require a third independent integral c, viz. we need this to express the threefold relation between the variables, given by the differential equations, and also to express the general solution $\phi(a, b, c)$ of the partial differential equation.

13. By what precedes the analytical condition, in order that the integrals a, b, c may be independent, is that they are such as not to satisfy the relations

$$\frac{d(a, b, c)}{d(x, y, z, w)} = 0.$$

14. We moreover see *à posteriori*, that there cannot be more than three independent integrals; in fact, if a, b, c, d are integrals, then, considering them as solutions of the partial differential equation, we have four equations which by the elimination therefrom of X, Y, Z, W, give

$$\frac{d(a, b, c, d)}{d(x, y, z, w)} = 0,$$

and this is the very equation which expresses that a, b, c, d are not independent.

15. The notion of the integrals may be arrived at somewhat differently thus: take a, b, c, d any functions of the variables, and write

$$A = X\frac{da}{dx} + Y\frac{da}{dy} + Z\frac{da}{dz} + W\frac{da}{dw},$$

and the like for B, C, D; then replacing the original variables x, y, z, w by the new variables a, b, c, d, the differential equations become

$$\frac{da}{A} = \frac{db}{B} = \frac{dc}{C} = \frac{dd}{D},$$

where A, B, C, D are to be (by means of the given values of a, b, c, d as functions of x, y, z, w) expressed as functions of a, b, c, d. If then $A = 0$, $B = 0$, $C = 0$, the differential equations become

$$\frac{da}{0} = \frac{db}{0} = \frac{dc}{0} = \frac{dd}{D};$$

viz. we have $da = 0$, $db = 0$, $dc = 0$, and therefore $a = \text{const.}$, $b = \text{const.}$, $c = \text{const.}$, that is, we have the integrals a, b, c as before.

16. There is no general process for obtaining an integral a of the differential equations. Supposing such integral known, we can introduce it as a variable, in place of one of the original variables, say w, viz. we thus reduce the system to

$$\frac{dx}{X}=\frac{dy}{Y}=\frac{dz}{Z}=\frac{da}{0},$$

where X, Y, Z now denote the values assumed by these functions upon expressing therein w as a function of x, y, z, a, viz. they are now functions of x, y, z, a. The system thus breaks up into $da=0$ and the system

$$\frac{dx}{X}=\frac{dy}{Y}=\frac{dz}{Z},$$

in which last (by virtue of the first equation, or $a=$ const.) a is to be regarded as a constant; the original system of three equations between four variables is thus reduced to a system of two equations between three variables. Supposing b to be an integral of this reduced system, b is given as a function of x, y, z, a, but upon substituting herein for a its value as a function of x, y, z, w, we have b a function of the original variables x, y, z, w, and b is then a second integral of the original system.

17. In like manner supposing a and b to be known, we reduce the system to the single equation

$$\frac{dx}{X}=\frac{dy}{Y},$$

where X, Y are now functions of x, y, a, b; supposing an integral hereof to be c, we have c a function of x, y, a, b; but upon substituting herein for a, b their values as functions of x, y, z, w, we have c a function of x, y, z, w, and as such it is the third integral of the original system.

18. It may be remarked that if, to the original system, we join on an equality $=dt$, viz. if we consider the system

$$\frac{dx}{X}=\frac{dy}{Y}=\frac{dz}{Z}=\frac{dw}{W}\;(=dt),$$

where X, Y, Z, W are as before functions of the variables (x, y, z, w), then the integrals a, b, c of the original system being known, we can by means of them express for instance X as a function of x, a, b, c, and we have then, const. $=t-\int\frac{dx}{X}$, where the integration is to be performed regarding a, b, c as constants; writing $\int\frac{dx}{X}=\tau$, but after the integration replacing a, b, c by their values as functions of x, y, z, w, we have τ a function of x, y, z, w; and we say that $t-\tau$ is an integral; putting it $=$const. we use also τ to denote the constant value of the integral $t-\int\frac{dx}{X}$ in question. Observe that here, the integrals a, b, c being known, the last integral $t-\tau$ is obtained by a quadrature.

19. The result would have been similar, if the adjoined equality had been $=\dfrac{dt}{T}$ (T a function of x, y, z, w), but in reference to subsequent matter, I retain the equality $=dt$, and adjoin a second equality $=\dfrac{dV}{\Omega}$ (Ω a function of x, y, z, w); we have then the integral $t-\tau$ as before, and another integral $V-\int\dfrac{\Omega dx}{X}$, where Ω, X are first expressed as functions of x, a, b, c, but after the integration a, b, c are replaced by their values as functions of (x, y, z, w), say this is the integral $V-\lambda$; this, when the integrals a, b, c are known, is (like $t-\tau$) obtained by a quadrature.

20. Attending only to the adjoined equality $=dt$, we can by means of the four integrals express each of the variables x, y, z, w as a function of a, b, c, $t-\tau$; viz. these four equations, regarding therein $t-\tau$ as a variable parameter, are in fact equivalent to the equations $a=\text{const.}$, $b=\text{const.}$, $c=\text{const.}$, which connect the variables x, y, z, w with the integrals a, b, c regarded as constants.

21. All that precedes is of course applicable to a system of $n-1$ equations between n variables, the number of independent integrals being $=n-1$.

22. I take an example with the three variables x, y, z; the differential equations being

$$\frac{dx}{x(y-z)}=\frac{dy}{y(z-x)}=\frac{dz}{z(x-y)},$$

and therefore the partial differential equation

$$x(y-z)\frac{d\theta}{dx}+y(z-x)\frac{d\theta}{dy}+z(x-y)\frac{d\theta}{dz}=0.$$

The integrals are $a=x+y+z$, $b=xyz$; and it will be shown how either of these integrals being known, the system is reduced to a single equation between two variables, say x, y.

First, a being known, $=x+y+z$ as before, we have

$$x(y-z)=x(x+2y-a),\quad y(z-x)=y(a-2x-y),$$

and the system is

$$\frac{dx}{x(x+2y-a)}=\frac{dy}{y(a-2x-y)},$$

which has the integral $b=xy(a-x-y)$; observe that this is a solution of the partial differential equation

$$x(x+2y-a)\frac{d\theta}{dx}+y(a-2x-y)\frac{d\theta}{dy}=0.$$

For a putting its value we find $b=xyz$.

Secondly, b being known, $=xyz$ as before, we have

$$x(y-z)=xy-\frac{y}{b},\quad y(z-x)=\frac{b}{x}-xy,$$

and the system is

$$\frac{dx}{xy-\frac{b}{y}}=\frac{dy}{\frac{b}{x}-xy},$$

which has the integral $a=x+y+\frac{b}{xy}$; observe that this is a solution of the partial differential equation

$$\left(xy-\frac{b}{y}\right)\frac{d\theta}{dx}+\left(\frac{b}{x}-xy\right)\frac{d\theta}{dy}=0.$$

For b putting its value, we find $a=x+y+z$.

The Multiplier. Art. Nos. 23 to 29.

23. First, if there are only two variables (x, y), the system consists of the single equation

$$\frac{dx}{X}=\frac{dy}{Y},$$

which may be written

$$Ydx-Xdy=0.$$

Hence, if a be an integral, we have

$$\frac{da}{dx}dx+\frac{da}{dy}dy=0;$$

the two will agree if there exists a function M such that

$$\frac{da}{dx}=MY,\quad \frac{da}{dy}=-MX,$$

and thence, in virtue of the identity

$$\frac{d}{dy}\frac{da}{dx}=\frac{d}{dx}\frac{da}{dy},$$

we find

$$\frac{dMX}{dx}+\frac{dMY}{dy}=0;$$

or, as this may also be written,

$$X\frac{dM}{dx}+Y\frac{dM}{dy}+M\left(\frac{dX}{dx}+\frac{dY}{dy}\right)=0,$$

as the condition to determine the multiplier M. Supposing M known, we have $M(Ydx-Xdy)=da$, or say $a=\int M(Ydx-Xdy)$, viz. the integral a is determined by a quadrature.

24. In the case of three variables (x, y, z), the system is

$$\frac{dx}{X}=\frac{dy}{Y}=\frac{dz}{Z},$$

or, writing these in the form

$$Ydz-Zdy=0,\quad Zdx-Xdz=0,\quad Xdy-Ydx=0,$$

the course which immediately suggests itself is to seek for factors L, M, N, such that, a being an integral, we may have

$$L(Ydz-Zdy)+M(Zdx-Xdz)+N(Xdy-Ydx)=da,$$

but this does not lead to any result. The course taken by Jacobi is quite a different one: he, in fact, determines a multiplier M connected with *two* integrals a, b.

25. Supposing that a, b are independent integrals, we have

$$X\frac{da}{dx}+Y\frac{da}{dy}+Z\frac{da}{dz}=0,$$

$$X\frac{db}{dx}+Y\frac{db}{dy}+Z\frac{db}{dz}=0;$$

and determining from these equations the ratio of the quantities X, Y, Z, we may, it is clear, write

$$MX,\ MY,\ MZ=\frac{d(a,\ b)}{d(y,\ z)},\quad \frac{d(a,\ b)}{d(z,\ x)},\quad \frac{d(a,\ b)}{d(x,\ y)}.$$

It may be shown that we have identically

$$\frac{d}{dx}\frac{d(a,\ b)}{d(y,\ z)}+\frac{d}{dy}\frac{d(a,\ b)}{d(z,\ x)}+\frac{d}{dz}\frac{d(a,\ b)}{d(x,\ y)}=0,$$

and we thence deduce

$$\frac{d(MX)}{dx}+\frac{d(MY)}{dy}+\frac{d(MZ)}{dz}=0;$$

or, what is the same thing,

$$X\frac{dM}{dx}+Y\frac{dM}{dy}+Z\frac{dM}{dz}+M\left(\frac{dX}{dx}+\frac{dY}{dy}+\frac{dZ}{dz}\right)=0,$$

as the condition for determining the multiplier M.

26. The use is as follows: supposing that M is known, and supposing also that one integral a of the system is known, we can then by a quadrature determine the other integral b. Thus, supposing that we know the integral a, $=a(x,\ y,\ z)$, we can by means of this integral express z in terms of x, y, a; and hence we may regard the unknown integral b as expressed in the like form, $b=b(x,\ y,\ a)$. The original values of $\frac{db}{dx}$, $\frac{db}{dy}$, $\frac{db}{dz}$ become on this supposition

$$\frac{db}{dx}+\frac{db}{da}\frac{da}{dx},\quad \frac{db}{dy}+\frac{db}{da}\frac{da}{dy},\quad \frac{db}{da}\frac{da}{dz},$$

and we thence find

$$\frac{d(a, b)}{d(y, z)}, \frac{d(a, b)}{d(z, x)}, \frac{d(a, b)}{d(x, y)} = -\frac{da}{dz}\frac{db}{dy}, \frac{da}{dz}\frac{db}{dx}, \frac{d(a, b)}{d(x, y)}.$$

We have therefore

$$MX, MY = -\frac{da}{dz}\frac{db}{dy}, \frac{da}{dz}\frac{db}{dx},$$

and, consequently,

$$db = \frac{db}{dx}dx + \frac{db}{dy}dy, = \frac{M}{\frac{da}{dz}}(Ydx - Xdy);$$

viz. M, $\frac{da}{dz}$, Y, X being all of them expressed as functions of x, y, a, the expression on the right-hand is a complete differential, and we have

$$b = \int \frac{M}{\frac{da}{dz}}(Ydx - Xdy);$$

that is, the integral b is determined by a quadrature.

27. Thus, in the example No. 22,

$$\frac{dX}{dx} + \frac{dY}{dy} + \frac{dZ}{dz} = 0,$$

and a value of the multiplier is $= 1$. Supposing that the given integral is $a = x + y + z$, then $\frac{da}{dz} = 1$, and we have accordingly 1 as the multiplier of the equation

$$y(a - 2x - y)dx + x(a - x - 2y)dy = 0,$$

viz. this equation is integrable *per se*. Supposing the given integral to be $b = xyz$, then $\frac{db}{dz} = xy$, viz. we have $\frac{1}{xy}$ as the multiplier of the equation

$$\left(\frac{b}{x} - xy\right)dx + \left(\frac{b}{y} - xy\right)dy = 0,$$

and we thus in each case obtain the other integral as before.

28. The foregoing result may be presented in a more symmetrical form by taking in place of x, y any two variables $u = u(x, y, z)$, $v = v(x, y, z)$.

Supposing the integral a known as before, the system then is

$$\frac{du}{U} = \frac{dv}{V} = \frac{da}{0},$$

where $U, V = X\frac{du}{dx} + Y\frac{du}{dy} + Z\frac{du}{dz}$, $X\frac{dv}{dx} + Y\frac{dv}{dy} + Z\frac{dv}{dz}$, these being expressed as functions of u, v, a; or, what is the same thing, we have $Vdu - Udv = 0$, a being in this equation regarded as a constant.

From the foregoing values of MX, MY, MZ, we deduce

$$MU,\ MV = \frac{d(u,\ a,\ b)}{d(x,\ y,\ z)},\ \frac{d(v,\ a,\ b)}{d(x,\ y,\ z)}.$$

But forming the values of du, dv, da, db, we have an equation, determinant $= 0$, which equation may be written

$$du\frac{d(v,\ a,\ b)}{d(x,\ y,\ z)} - dv\frac{d(a,\ b,\ u)}{d(x,\ y,\ z)} + da\frac{d(b,\ u,\ v)}{d(x,\ y,\ z)} - db\frac{d(u,\ v,\ a)}{d(x,\ y,\ z)} = 0;$$

or, writing herein $da = 0$, this is

$$du\frac{d(v,\ a,\ b)}{d(x,\ y,\ z)} - dv\frac{d(u,\ a,\ b)}{d(x,\ y,\ z)} - db\frac{d(u,\ v,\ a)}{d(x,\ y,\ z)} = 0,$$

viz. this is

$$M(Vdu - Udv) = db\frac{d(u,\ v,\ a)}{d(x,\ y,\ z)},$$

or say

$$b = \int\left\{M \div \frac{d(u,\ v,\ a)}{d(x,\ y,\ z)}\right\}(Vdu - Udv),$$

where, on the right-hand side, everything must be expressed in terms of u, v, a. It thus appears that on expressing the final equation as a relation $Vdu - Udv = 0$ between the variables u and v, the multiplier hereof is $M \div \dfrac{d(u,\ v,\ a)}{d(x,\ y,\ z)}$. If $u,\ v = x,\ y$, this agrees with a foregoing result.

29. The theory is precisely the same for any number of variables. Thus, if there are four variables x, y, z, w, we have

$$MX,\ MY,\ MZ,\ MW = \frac{d(a,\ b,\ c)}{d(y,\ z,\ w)},\ -\frac{d(a,\ b,\ c)}{d(z,\ w,\ x)},\ \frac{d(a,\ b,\ c)}{d(w,\ x,\ y)},\ -\frac{d(a,\ b,\ c)}{d(x,\ y,\ z)};$$

and, we have between the functions on the right-hand an identical relation, in virtue of which

$$\frac{d(MX)}{dx} + \frac{d(MY)}{dy} + \frac{d(MZ)}{dz} + \frac{d(MW)}{dw} = 0;$$

then, supposing that a value of M is known, and also any two integrals a, b, and that by means of these the equation to be finally integrated is expressed as a relation $Vdu - Udv = 0$ between any two variables u and v, the multiplier of this is

$$= M \div \frac{d(u,\ v,\ a,\ b)}{d(x,\ y,\ z,\ w)},$$

where U, V and this multiplier are to be expressed in terms of u, v, a, b.

The general result is that, given a value of the multiplier, and also all but one of the integrals, the final integral is expressible by a quadrature.

Pfaffian Theorem. Art. No. 30.

30. According as

the variables are	we have	
x,	Xdx	$= du$,
x, y,	$Xdx + Ydy$	$= \lambda du$,
x, y, z,	$Xdx + Ydy + Zdz$	$= \lambda du + \quad dv$,
x, y, z, w,	$Xdx + Ydy + Zdz + Wdw$	$= \lambda du + \mu dv$,

and so on; viz. the theorem is that, taking for instance two variables, a given lineo-differential $Xdx + Ydy$ is $= \lambda du$, that is, there exist λ, u functions of x, y, which verify this identity, or, what is the same thing, such that we have

$$X,\ Y = \lambda \frac{du}{dx},\quad \lambda \frac{du}{dy};$$

and so, in the case of three variables, there exist λ, u, v functions of x, y, z, such that

$$X,\ Y,\ Z = \lambda \frac{du}{dx} + \frac{dv}{dx},\quad \lambda \frac{du}{dy} + \frac{dv}{dy},\quad \lambda \frac{du}{dz} + \frac{dv}{dz}.$$

The problem of determining the functions on the right-hand side is known as the Pfaffian Problem; this I do not at present consider, but only assume that there exist such functions.

The Hamiltonian System, its derivation from the general System. Art. Nos. 31 to 34.

31. Considering a bipartite set $(x, y, z: p, q, r)$, the general system of differential equations may be written

$$\frac{dx}{P} = \frac{dy}{Q} = \frac{dz}{R} = \frac{dp}{-X} = \frac{dq}{-Y} = \frac{dr}{-Z}.$$

But by the Pfaffian theorem we may write

$$Xdx + Ydy + Zdz + Pdp + Qdq + Rdr = \xi d\rho + \eta d\sigma + \zeta d\tau,$$

viz. there exist $\xi, \eta, \zeta, \rho, \sigma, \tau$ functions of the variables x, y, z, p, q, r, such that we have

$$X = \xi \frac{d\rho}{dx} + \eta \frac{d\sigma}{dx} + \zeta \frac{d\tau}{dx}, \ldots,\quad P = \xi \frac{d\rho}{dp} + \eta \frac{d\sigma}{dp} + \zeta \frac{d\tau}{dp},\ \ldots,$$

and we have the foregoing general system expressed by means of these given functions $\xi, \eta, \zeta, \rho, \sigma, \tau$ of the variables.

32. But the lineo-differential

$$Xdx + Ydy + Zdz + Pdp + Qdq + Rdr$$

may be of a more special form; for instance, it may be a sum of two terms $=\xi d\rho+\eta d\sigma$: or, finally, it may be a single term $=\xi d\rho$, and in this case we have the Hamiltonian system, viz. writing H in place of ρ, if we have

$$Xdx+Ydy+Zdz+Pdp+Qdq+Rdr=\xi dH,$$

where H is a given function of the variables, then the system is

$$\frac{dx}{\frac{dH}{dp}}=\frac{dy}{\frac{dH}{dq}}=\frac{dz}{\frac{dH}{dr}}=\frac{dp}{-\frac{dH}{dx}}=\frac{dq}{-\frac{dH}{dy}}=\frac{dr}{-\frac{dH}{dz}},$$

which is the system in question.

33. Any integral a of the system is a solution of

$$\frac{dH}{dp}\frac{d\theta}{dx}+\frac{dH}{dq}\frac{d\theta}{dy}+\frac{dH}{dr}\frac{d\theta}{dz}-\frac{dH}{dx}\frac{d\theta}{dp}-\frac{dH}{dy}\frac{d\theta}{dq}-\frac{dH}{dz}\frac{d\theta}{dr}=0;$$

viz. writing, as above,

$$(H,\ \theta)=\frac{d(H,\ \theta)}{d(p,\ x)}+\frac{d(H,\ \theta)}{d(q,\ y)}+\frac{d(H,\ \theta)}{d(r,\ z)}$$

$$=\text{last-mentioned expression},$$

the partial differential equation is $(H,\ \theta)=0$; and, conversely, any solution of this equation is an integral of the differential equations.

34. It is obvious that a solution of $(H,\ \theta)=0$ is H; hence the entire system of independent solutions may be taken to be $H,\ a,\ b,\ c,\ d$; or, if we choose to consider a set of five independent solutions $a,\ b,\ c,\ d,\ e$, then we have $H=H(a,\ b,\ c,\ d,\ e)$ a function of these solutions.

An Identity in regard to the Functions $(H,\ \theta)$. Art. Nos. 35 and 36.

35. Taking the variables to be $(x,\ y,\ z,\ p,\ q,\ r)$, and $H,\ a,\ b$ to be any functions of these variables, we have the identity

$$(H,\ (a,\ b))+(a,\ (b,\ H))+(b,\ (H,\ a))=0,$$

which is now to be proved. For this purpose we write it in the slightly different form

$$((a,\ b),\ H)=(a,\ (b,\ H))-(b,\ (a,\ H)).$$

The first term on the right-hand side is

$$\left(\frac{da}{dp}\frac{d}{dx}+\frac{da}{dq}\frac{d}{dy}+\frac{da}{dr}\frac{d}{dz}-\frac{da}{dx}\frac{d}{dp}-\frac{da}{dy}\frac{d}{dq}-\frac{da}{dz}\frac{d}{dr}\right)$$

operating upon

$$\left(\frac{db}{dp}\frac{dH}{dx}+\frac{db}{dq}\frac{dH}{dy}+\frac{db}{dr}\frac{dH}{dz}-\frac{db}{dx}\frac{dH}{dp}-\frac{db}{dy}\frac{dH}{dq}-\frac{db}{dz}\frac{dH}{dr}\right);$$

and if we herein attend to the terms which contain the second differential coefficients of H, these are symmetrical functions of a, b. For instance,

$$\frac{d^2H}{dx^2},\ \text{coefficient is}\ \frac{da}{dp}\frac{db}{dp},$$

$$\frac{d^2H}{dx\,dy}\quad ,,\quad ,,\quad \frac{da}{dp}\frac{db}{dq}+\frac{da}{dq}\frac{db}{dp},$$

$$\frac{d^2H}{dx\,dp}\quad ,,\quad ,,\quad -\frac{da}{dp}\frac{db}{dx}-\frac{da}{dx}\frac{db}{dp},$$

$$\frac{d^2H}{dx\,dq}\quad ,,\quad ,,\quad -\frac{da}{dp}\frac{db}{dy}-\frac{da}{dy}\frac{db}{dp}.$$

Hence, forming the like terms of the second terms $(b, (a, H))$ and subtracting, the terms in question all vanish: and we thus see that $(a, (b, H))-(b, (a, H))$ is a linear function of the differential coefficients

$$\frac{dH}{dx},\ \frac{dH}{dy},\ \frac{dH}{dz},\ \frac{dH}{dp},\ \frac{dH}{dq},\ \frac{dH}{dr}.$$

36. Attending to any one of these, suppose $\frac{dH}{dx}$, the coefficient of this

$$\text{in } (a, (b, H)) \text{ is } =\left(a, \frac{db}{dp}\right)$$

$$\text{in } (b, (a, H))\ ,,\quad \left(b, \frac{da}{dp}\right),\quad =-\left(\frac{da}{dp}, b\right),$$

wherefore, in the difference of these, it is

$$\left(a, \frac{db}{dp}\right)+\left(\frac{da}{dp}, b\right),\quad =\frac{d}{dp}(a, b).$$

Hence, for the several terms

$$\frac{dH}{dx},\ \frac{dH}{dy},\ \frac{dH}{dz},\ \frac{dH}{dp},\ \frac{dH}{dq},\ \frac{dH}{dr},$$

the coefficients are

$$\left(\frac{d}{dp},\ \frac{d}{dq},\ \frac{d}{dr},\ -\frac{d}{dx},\ -\frac{d}{dy},\ -\frac{d}{dz}\right)(a, b);$$

or, what is the same thing, we have

$$(a, (b, H))-(b, (a, H))=((a, b), H),$$

the identity in question.

The Poisson-Jacobi Theorem. Art. Nos. 37 to 39.

37. The foregoing identity shows that if $(H, a)=0$, and $(H, b)=0$, then also $(H, (a, b))=0$; or, what is the same thing, if a and b are solutions of the partial differential equation $(H, \theta)=0$, then also (a, b) is a solution; or, say, if a, b are integrals, then also (a, b) is an integral.

Supposing that the set is (x, y, z, p, q, r), so that there are in all five integrals a, b, c, d, e, then the theorem may be otherwise stated, we have (a, b) a function of the integrals a, b, c, d, e.

Observe that, knowing only the integrals a and b, we find (a, b) as a function of x, y, z, p, q, r, this may be $=0$, or a determinate constant, or it may be such a function that by virtue of the given values of a and b it reduces itself to a function of a and b; in any of these cases the theorem does *not* determine a new integral. But if contrariwise the value of (a, b), obtained as above as a function of the variables, is not a function of a, b, then it is a new integral which may be called c.

38. To obtain in this way a new integral, we require two integrals a, b other than H; for knowing only the integrals a, H, the theorem gives only (a, H) an integral, and we have of course $(a, H)=0$, viz. we do not obtain a new integral.

But starting from two integrals a, b other than H, we *may* obtain as above a new integral c; and then again (a, c) and (b, c) will be integrals, one or both of which may be new. And it may therefore happen that in this way we obtain all the independent integrals a, b, c, d, e; or the process may on the other hand terminate, without giving all the independent integrals.

The theory is obviously applicable throughout to the case of a bipartite set $(x, y, z, \ldots, p, q, r, \ldots)$ of $2n$ variables.

39. It may be remarked here that, in the Hamiltonian system, a value of the multiplier is $M=1$; and consequently, if in any way all but one of the integrals, that is, $2n-2$ integrals, be known, the remaining integral can be found by a quadrature.

It is further to be noticed that, if we adjoin a new variable t and a term $=dt$ to the system of equations; then the $2n-1$ integrals of the original system being known, all the original variables can be expressed in terms of the $2n-1$ integrals regarded as constants and of one of the variables say x: we then have

$$dt = dx \div \frac{dH}{dp},$$

or

$$t-\epsilon = \int dx \div \frac{dH}{dp},$$

or say

$$\epsilon = t - \int dx \div \frac{dH}{dp},$$

viz. if after the integration we suppose the $2n-1$ integrals replaced each of them by its value, we have

$$\epsilon = t - \phi(x, y, z, \ldots, p, q, r, \ldots),$$

which is the remaining or $2n$th integral of the original system as augmented by the term $=dt$.

The Poisson-Jacobi theorem peculiar to the Hamiltonian Form. Art. Nos. 40 to 45.

40. Taking for greater simplicity the set (x, y, p, q), and writing

$$Xdx + Ydy + Pdp + Qdq = \xi d\rho + \eta d\sigma,$$

then the general system

$$\frac{dx}{P} = \frac{dy}{Q} = \frac{dp}{-X} = \frac{dq}{-Y},$$

becomes

$$\frac{dx}{\xi\frac{d\rho}{dp} + \eta\frac{d\sigma}{dp}} = \frac{dy}{\xi\frac{d\rho}{dq} + \eta\frac{d\sigma}{dq}} = \frac{dp}{-\left(\xi\frac{d\rho}{dx} + \eta\frac{d\sigma}{dx}\right)} = \frac{dq}{-\left(\xi\frac{d\rho}{dy} + \eta\frac{d\sigma}{dy}\right)};$$

and the corresponding partial differential equation (θ the independent variable) is

$$\xi(\rho, \theta) + \eta(\sigma, \theta) = 0,$$

where

$$(\rho, \theta) = \frac{d(\rho, \theta)}{d(p, x)} + \frac{d(\rho, \theta)}{d(q, y)}, \quad (\sigma, \theta) = \frac{d(\sigma, \theta)}{d(p, x)} + \frac{d(\sigma, \theta)}{d(q, y)}.$$

It is to be shown that if a, b are solutions, viz. if we have

$$\xi(\rho, a) + \eta(\sigma, a) = 0,$$
$$\xi(\rho, b) + \eta(\sigma, b) = 0,$$

implying of course

$$(\rho, a)(\sigma, b) - (\rho, b)(\sigma, a) = 0,$$

then it is *not* in general true that we have (a, b) a solution; that is, *not* in general true that

$$\xi(\rho, (a, b)) + \eta(\sigma, (a, b)) = 0;$$

the condition for the truth of this equation is in fact $\frac{\eta}{\xi}$ = a function of ρ, σ, but when this is so, $\xi d\rho + \eta d\sigma$ is λdH, viz. there exist λ, H functions of ρ, σ (and therefore ultimately of x, y, p, q) satisfying this equation, and the system is really Hamiltonian.

41. We consider whether it is true that

$$\xi(\rho, (a, b)) + \eta(\sigma, (a, b)) = 0.$$

We have identically

$$((a, b), \rho) + ((b, \rho), a) + ((\rho, a), b) = 0,$$
$$((a, b), \sigma) + ((b, \sigma), a) + ((\sigma, a), b) = 0,$$

so that multiplying by ξ, η, and adding, the equation in question is

$$\xi[((b, \rho), a) + ((\rho, a), b)] + [((b, \sigma), a) + ((\sigma, a), b)] = 0.$$

But in virtue of the equations satisfied by a, b, we may write

$$(\rho, a) = \ \ l\eta, \ (b, \rho) = -(\rho, b) = -m\eta,$$
$$(\sigma, a) = -l\xi, \ (b, \sigma) = -(\sigma, b) = \ \ m\xi,$$

where l, m are indeterminate functions of x, y, p, q; and the equation in question now becomes

$$\xi[-(m\eta, a)+(l\eta, b)]+\eta[(m\xi, a)-(l\xi, b)]=0;$$

that is,

$$\begin{aligned}&\xi[-m(\eta, a)-\eta(m, a)+l(\eta, b)+\eta(l, b)]\\&+\eta[\ \ m(\xi, a)+\xi(m, a)-l(\xi, b)-\xi(l, b)]=0;\end{aligned}$$

viz. omitting the terms which destroy each other, this is

$$-m\xi(\eta, a)+l\xi(\eta, b)+m\eta(\xi, a)-l\eta(\xi, b)=0.$$

Substituting for $m\xi$, &c., their values, we have

$$(\sigma, b)(\eta, a)-(\sigma, a)(\eta, b)+(\rho, b)(\xi, a)-(\rho, a)(\xi, b)=0;$$

and the question is whether this is implied in the equations

$$\xi(\rho, a)+\eta(\sigma, a)=0,$$
$$\xi(\rho, b)+\eta(\sigma, b)=0.$$

42. Write $\eta=\kappa\xi$, the equation in question is

$$(\sigma, b)(\kappa\xi, a)-(\sigma, a)(\kappa\xi, b)+(\rho, b)(\xi, a)-(\rho, a)(\xi, b)=0;$$

that is,

$$\left.\begin{matrix}(\sigma, b)\,\kappa(\xi, a)+\xi(\sigma, b)(\kappa, a)\\-(\sigma, a)\,\kappa(\xi, b)-\xi(\sigma, a)(\kappa, b)\end{matrix}\right\}+\{(\rho, b)(\xi, a)-(\rho, a)(\xi, b)\}=0;$$

viz.

$$(\xi, a)[(\rho, b)+\kappa(\sigma, b)]-(\xi, b)[(\rho, a)+\kappa(\sigma, a)]+\xi[(\sigma, b)(\kappa, a)-(\sigma, a)(\kappa, b)]=0;$$

and we wish to see whether this is implied in

$$(\rho, a)+\kappa(\sigma, a)=0,$$
$$(\rho, b)+\kappa(\sigma, b)=0,$$

which give

$$(\sigma, b)(\rho, a)-(\sigma, a)(\rho, b)=0;$$

or, what is the same thing, whether these last equations imply

$$(\sigma, b)(\kappa, a)-(\sigma, a)(\kappa, b)=0.$$

Suppose κ is a function of ρ, σ, then, as is at once seen,

$$(\kappa, a)=\frac{d\kappa}{d\rho}(\rho, a)+\frac{d\kappa}{d\sigma}(\sigma, a),$$

$$(\kappa, b)=\frac{d\kappa}{d\rho}(\rho, b)+\frac{d\kappa}{d\sigma}(\sigma, b),$$

and thence

$$(\sigma, b)(\kappa, a)-(\sigma, a)(\kappa, b)=\frac{d\kappa}{d\rho}[(\sigma, b)(\rho, a)-(\sigma, a)(\rho, b)];$$

viz. κ being a function of ρ and σ, the two equations imply the third.

43. But we wish to prove the converse, viz. that, if the two equations imply the third, then κ is a function of ρ, σ.

Now the equations

$$(\sigma,\ b)(\kappa,\ a)-(\sigma,\ a)(\kappa,\ b)=0,\quad (\sigma,\ b)(\rho,\ a)-(\sigma,\ a)(\rho,\ b)=0,$$

are transformable into

$$\begin{array}{ll}\dfrac{d(\sigma,\ \kappa)}{d(p,\ x)}\dfrac{d(b,\ a)}{d(p,\ x)}=0, & \dfrac{d(\sigma,\ \rho)}{d(p,\ x)}\dfrac{d(b,\ a)}{d(p,\ x)}=0\\ +q,\ y, & +q,\ y,\\ +p,\ q, & +p,\ q,\\ +p,\ y, & +p,\ y,\\ +x,\ q, & +x,\ q,\\ +x,\ y, & +x,\ y,\end{array}$$

the lines after the first being the corresponding terms with q, y, &c. instead of p, x. And if independently of the values of a, b, one of these equations implies the other, we must have

$$\frac{d(\sigma,\ \kappa)}{d(p,\ x)},\ \frac{d(\sigma,\ \kappa)}{d(q,\ y)},\ \frac{d(\sigma,\ \kappa)}{d(p,\ q)},\ \frac{d(\sigma,\ \kappa)}{d(p,\ y)},\ \frac{d(\sigma,\ \kappa)}{d(x,\ q)},\ \frac{d(\sigma,\ \kappa)}{d(x,\ y)},$$

proportional to the like expressions with σ, ρ instead of σ, κ; say these are

$$\frac{d(\sigma,\ \kappa)}{d(p,\ x)}=\Lambda\frac{d(\sigma,\ \rho)}{d(p,\ x)},\ \text{\&c.}$$

44. Assume κ a function of ρ, σ, x, y; we have

$$\frac{d(\sigma,\ \kappa)}{d(p,\ x)}=\frac{d\sigma}{dp}\left(\frac{d\kappa}{d\rho}\frac{d\rho}{dx}+\frac{d\kappa}{d\sigma}\frac{d\sigma}{dx}+\frac{d\kappa}{dx}\right)-\frac{d\sigma}{dx}\left(\frac{d\kappa}{d\rho}\frac{d\rho}{dp}+\frac{d\kappa}{d\sigma}\frac{d\sigma}{dp}\right)=\frac{d\kappa}{d\rho}\frac{d(\sigma,\ \rho)}{d(p,\ x)}+\frac{d\kappa}{dx}\frac{d\sigma}{dp},\ \text{\&c.};$$

the equations thus become

$$\begin{aligned}
\frac{d\kappa}{d\rho}\frac{d(\sigma,\ \rho)}{d(p,\ x)}+\frac{d\kappa}{dx}\frac{d\sigma}{dp}&=\Lambda\frac{d(\sigma,\ \rho)}{d(p,\ x)},\\
\frac{d\kappa}{d\rho}\frac{d(\sigma,\ \rho)}{d(q,\ y)}+\frac{d\kappa}{dy}\frac{d\sigma}{dq}&=\Lambda\frac{d(\sigma,\ \rho)}{d(q,\ y)},\\
\frac{d\kappa}{d\rho}\frac{d(\sigma,\ \rho)}{d(p,\ q)}+\quad 0\quad&=\Lambda\frac{d(\sigma,\ \rho)}{d(p,\ q)},\\
\frac{d\kappa}{d\rho}\frac{d(\sigma,\ \rho)}{d(p,\ y)}+\frac{d\kappa}{dy}\frac{d\sigma}{dp}&=\Lambda\frac{d(\sigma,\ \rho)}{d(p,\ y)},\\
\frac{d\kappa}{d\rho}\frac{d(\sigma,\ \rho)}{d(q,\ x)}+\frac{d\kappa}{dx}\frac{d\sigma}{dq}&=\Lambda\frac{d(\sigma,\ \rho)}{d(q,\ x)},\\
\frac{d\kappa}{d\rho}\frac{d(\sigma,\ \rho)}{d(x,\ y)}-\frac{d(\sigma,\kappa)}{d(x,y)}&=\Lambda\frac{d(\sigma,\ \rho)}{d(x,\ y)}.
\end{aligned}$$

Hence, unless $\dfrac{d(\sigma, \rho)}{d(p, q)}=0$, we have $\Lambda=\dfrac{d\kappa}{d\rho}$. The remaining five equations then are

$$\frac{d\kappa}{dx}\frac{d\sigma}{dp}=0,\quad \frac{d\kappa}{dy}\frac{d\sigma}{dq}=0,$$

$$\frac{d\kappa}{dy}\frac{d\sigma}{dp}=0,\quad \frac{d\kappa}{dx}\frac{d\sigma}{dq}=0,\quad \frac{d\sigma}{dx}\frac{d\kappa}{dy}-\frac{d\sigma}{dy}\frac{d\kappa}{dx}=0,$$

which give, and are satisfied by $\dfrac{d\kappa}{dx}=0$, $\dfrac{d\kappa}{dy}=0$, viz. we then have κ a function of ρ, σ without x, y which is the theorem in question.

45. The proof fails if $\dfrac{d(\sigma, \rho)}{d(p, q)}=0$. But here, unless also $\dfrac{d(\sigma, \rho)}{d(x, y)}=0$, we can, by assuming in the first instance κ a function of ρ, σ, p, q, prove in like manner that κ is a function of only ρ and σ; if however we have as well $\dfrac{d(\sigma, \rho)}{d(p, q)}=0$ and $\dfrac{d(\sigma, \rho)}{d(x, y)}=0$, the last-mentioned process would also fail, but it can be shown the conclusion holds good in this case also; hence the conclusion that the Poisson-Jacobi theorem holds good only for a Hamiltonian system.

Conjugate Integrals of the Hamiltonian System. Art. Nos. 46 to 51.

46. For greater clearness, let $n=4$, or let the variables be x, y, z, w, p, q, r, s; the system of differential equations therefore is

$$\frac{dx}{\dfrac{dH}{dp}}=\frac{dy}{\dfrac{dH}{dq}}=\frac{dz}{\dfrac{dH}{dr}}=\frac{dw}{\dfrac{dH}{ds}}=\frac{dp}{-\dfrac{dH}{dx}}=\frac{dq}{-\dfrac{dH}{dy}}=\frac{dr}{-\dfrac{dH}{dz}}=\frac{ds}{-\dfrac{dH}{dw}},$$

and any integral hereof is as before a solution of $(H, \theta)=0$. Assume that the integrals are H, a, b, c, d, e, f, so that

$$(H, a)=0,\quad (H, b)=0,\quad (H, c)=0,\quad (H, d)=0,\quad (H, e)=0,\quad (H, f)=0.$$

Considering here a as denoting any integral whatever, that is, any solution whatever of the partial differential equation $(H, \theta)=0$, it is to be shown that it is possible to determine θ so as to satisfy as well this equation $(H, \theta)=0$, as also the new equation $(a, \theta)=0$.

47. We, in fact, satisfy the first equation by taking

$$\theta, =\theta(H, a, b, c, d, e, f),$$

any function whatever of the seven integrals. But, θ having this value, we find

$$(a, \theta)=(a, H)\frac{d\theta}{dH}+(a, a)\frac{d\theta}{da}+(a, b)\frac{d\theta}{db}+(a, c)\frac{d\theta}{dc}+(a, d)\frac{d\theta}{dd}+(a, e)\frac{d\theta}{de}+(a, f)\frac{d\theta}{df};$$

or, since the first two terms on the right-hand vanish, the equation $(a, \theta)=0$ thus becomes

$$(a, b)\frac{d\theta}{db}+(a, c)\frac{d\theta}{dc}+(a, d)\frac{d\theta}{dd}+(a, e)\frac{d\theta}{de}+(a, f)\frac{d\theta}{df}=0.$$

But by the Poisson-Jacobi theorem (a, b), &c., are each of them a solution of $(H, \theta)=0$, viz. they are each of them a function of H, a, b, c, d, e, f. This is then a linear partial differential equation wherein the variables are H, a, b, c, d, e, f; or, since there are no terms in $\frac{d\theta}{dH}, \frac{d\theta}{da}$, we may regard a, H as constants, and treat it as a linear partial differential equation in b, c, d, e, f, the solutions of the equation being in fact the integrals, or any functions of the integrals, of

$$\frac{db}{(a, b)}=\frac{dc}{(a, c)}=\frac{dd}{(a, d)}=\frac{de}{(a, e)}=\frac{df}{(a, f)}.$$

48. Suppose any four integrals are b', c', d', e', so that a general integral is $\phi(H, a, b', c', d', e')$, then b', c', d', e' quà functions of H, a, b, c, d, e, f are integrals of the original equation $(H, \theta)=0$; hence *changing the notation* and writing b, c, d, e in place of these accented letters we have (H, a, b, c, d, e) as solutions of the two equations $(H, \theta)=0$, $(a, \theta)=0$; viz. a being any integral of the first of these equations, we see how to find four other integrals (b, c, d, e) which are such that

$$\begin{aligned}&(H, a)=0,\quad (H, b)=0,\quad (H, c)=0,\quad (H, d)=0,\quad (H, e)=0,\\ &\qquad\qquad\quad (a, b)=0,\quad (a, c)=0,\quad (a, d)=0,\quad (a, e)=0.\end{aligned}$$

49. We proceed in the same course and endeavour to find θ, so that not only $(H, \theta)=0$, $(a, \theta)=0$, but also $(b, \theta)=0$. Assuming here $\theta=\theta(H, a, b, c, d, e)$ an arbitrary function of the integrals, the first and second equations are satisfied; for the third equation, we have

$$(b, \theta)=(b, H)\frac{d\theta}{dH}+(b, a)\frac{d\theta}{da}+(b, b)\frac{d\theta}{db}+(b, c)\frac{d\theta}{dc}+(b, d)\frac{d\theta}{dd}+(b, e)\frac{d\theta}{de};$$

viz. the first three terms here vanish, and the equation $(b, \theta)=0$ becomes

$$(b, c)\frac{d\theta}{dc}+(b, d)\frac{d\theta}{dd}+(b, e)\frac{d\theta}{de}=0,$$

where, b, c, d, e being solutions as well of $(H, \theta)=0$ as of $(a, \theta)=0$, we have (b, c) a solution of these two equations, and as such a function of H, a, b, c, d, e; and so (b, d) and (b, e) are each of them a function of the same variables. The above is therefore a linear partial differential equation wherein the variables are H, a, b, c, d, e, but as the equation does not contain $\frac{d\theta}{dH}, \frac{d\theta}{da}$, or $\frac{d\theta}{db}$, we may regard H, a, b as constants; and the solutions of the equation are, in fact, the integrals of

$$\frac{dc}{(b, c)}=\frac{dd}{(b, d)}=\frac{de}{(b, e)}.$$

50. Supposing that any two integrals are c', d', so that a general integral is $\phi(H, a, b, c', d')$, then c', d' quà functions of H, a, b, c, d, e are integrals of the former equations $(H, \theta)=0$, $(a, \theta)=0$, so that again *changing the notation*, and writing c, d instead of the accented letters, we have (H, a, b, c, d) as solutions of the three equations $(H, \theta)=0$, $(a, \theta)=0$, $(b, \theta)=0$, viz. a being any solution of the first equation, and b any solution of the first and second equations, we see how to find two others c, d, of the same two equations, which are such that

$$\begin{aligned} (H, a)=0,\quad (H, b)=0,\quad &(H, c)=0,\quad (H, d)=0,\\ (a\, , b)=0,\quad &(a\, , c)=0,\quad (a\, , d)=0,\\ &(b\, , c)=0,\quad (b\, , d)=0\, ; \end{aligned}$$

or, attending only to the integrals H, a, b, c, these are integrals of the equations $(H, \theta)=0$, $(a, \theta)=0$, $(b, \theta)=0$, such that

$$(H, a)=0,\quad (H, b)=0,\quad (H, c)=0,\quad (a, b)=0,\quad (a, c)=0,\quad (b, c)=0.$$

We here say that H, a, b, c are a system of conjugate solutions. Attempting to continue the process, it would appear that there is not any new independent integral d, such that $(H, d)=0$, $(a, d)=0$, $(b, d)=0$, $(c, d)=0$ (the first three of these are satisfied by the integral d found above, but the last of them is not); we may, however, taking d an arbitrary function of H, a, b, c, replace H by d; viz. we thus have the four integrals a, b, c, d, such that

$$(a, b)=0,\quad (a, c)=0,\quad (a, d)=0,\quad (b, c)=0,\quad (b, d)=0,\quad (c, d)=0,$$

and which are consequently said to form a conjugate system.

51. The process is of course general, and it shows how, in the case of a Hamiltonian system of $2n$ variables, it is possible to find a system $H, a, b, \dots, f$ consisting of H and $n-1$ other integrals, or, if we please, a system of n integrals $a, b, \dots, f, g$, such that the derivative of any two integrals whatever of the system is $=0$; any such system is termed a conjugate system.

Hamiltonian System—the function V. Art. Nos. 52 to 58.

52. Taking a Hamiltonian system with the original variables x, y, z, p, q, r, we adjoin the two new variables t, V, forming the extended system

$$\frac{dx}{\frac{dH}{dp}}=\frac{dy}{\frac{dH}{dq}}=\frac{dz}{\frac{dH}{dr}}=\frac{dp}{-\frac{dH}{dx}}=\frac{dq}{-\frac{dH}{dy}}=\frac{dr}{-\frac{dH}{dz}}=dt=\frac{dV}{p\frac{dH}{dp}+q\frac{dH}{dq}+r\frac{dH}{dr}}.$$

Supposing the integrals of the original system to be a, b, c, d, e, we have $H=H(a, b, c, d, e)$ a determinate function of these integrals; also an integral $\tau=t-\phi(x, y, z, p, q, r)$ and an integral $\lambda=V-\psi(x, y, z, p, q, r)$; these integrals, exclusive of the last of them, serve to express x, y, z, p, q, r as functions of $a, b, c, d, e, t-\tau$; and the last integral then gives $V=\lambda+$ a function of the last-mentioned quantities.

53. We consider the differential expression

$$dV - p\,dx - q\,dy - r\,dz,$$

which, treating the integrals as constants, that is, in the expressions of V, x, y, z, regarding t as the only variable, is at once seen to be $=0$; hence, if we regard all the integrals as variables, the value is

$$= d\lambda + A da + B db + C dc + D dd + E de,$$

without any term in $d\tau$, since this enters originally in the form $dt - d\tau$, and therefore disappears with dt.

The coefficients A, B, C, D, E are of course functions of a, b, c, d, e, $t-\tau$; it is to be shown that they contain $t-\tau$ linearly, viz. that in these coefficients respectively the coefficients of $t-\tau$ are

$$\frac{dH}{da},\ \frac{dH}{db},\ \frac{dH}{dc},\ \frac{dH}{dd},\ \frac{dH}{de},$$

where H is expressed as above in the form $H(a, b, c, d, e)$; this being so, the entire term in $t-\tau$ will be $(t-\tau)\,dH$; each coefficient, for instance A, has besides a part A', which is a function of a, b, c, d, e without $t-\tau$, or *changing the notation* and writing the unaccented letters to denote these parts of the original coefficients, the final result is

$$dV - p\,dx - q\,dy - r\,dz = (t-\tau)\,dH + d\lambda + A da + B db + C dc + D dd + E de,$$

where H stands for its value $H(a, b, c, d, e)$, and A, B, C, D, E are functions of a, b, c, d, e without $t-\tau$.

54. To prove the theorem, we have

$$A = \frac{dV}{da} - p\frac{dx}{da} - q\frac{dy}{da} - r\frac{dz}{da},$$

and thence

$$\frac{dA}{dt} = \frac{d^2V}{da\,dt} - \frac{dp}{dt}\frac{dx}{da} - \frac{dq}{dt}\frac{dy}{da} - \frac{dr}{dt}\frac{dz}{da} - p\frac{d^2x}{da\,dt} - q\frac{d^2y}{da\,dt} - r\frac{d^2z}{da\,dt}$$

$$= \frac{d}{da}\left\{\frac{dV}{dt} - p\frac{dx}{dt} - q\frac{dy}{dt} - r\frac{dz}{dt}\right\}$$

$$+ \frac{dp}{da}\frac{dx}{dt} + \frac{dq}{da}\frac{dy}{dt} + \frac{dr}{da}\frac{dz}{dt} - \frac{dp}{dt}\frac{dx}{da} - \frac{dq}{dt}\frac{dy}{da} - \frac{dr}{dt}\frac{dz}{da};$$

and then substituting for $\dfrac{dV}{dt}$, &c., their values from the system of differential equations, the first line vanishes, and the second line becomes

$$= \frac{dH}{dp}\frac{dp}{da} + \frac{dH}{dq}\frac{dq}{da} + \frac{dH}{dr}\frac{dr}{da} + \frac{dH}{dx}\frac{dx}{da} + \frac{dH}{dy}\frac{dy}{da} + \frac{dH}{dz}\frac{dz}{da},$$

$$= \frac{dH}{da};$$

and hence $A=(t-\tau)\frac{dH}{da}+A'$, and the like for the other coefficients B, C, D, E, which is the theorem in question.

55. We may have between two coefficients of the formula, for instance, D and E, a relation $\frac{dD}{de}=\frac{dE}{dd}$, and I will for the present assume, without proving it, the theorem that if a, b, c are conjugate integrals, then this relation $\frac{dD}{de}-\frac{dE}{dd}=0$, holds good, merely mentioning that the proof depends on the consideration of certain symbols $[a, b]$, which are the converses, so to speak, of the symbols (a, b), viz. considering the variables x, y, z, p, q, r as given functions of $a, b, c, d, e, t-\tau$, then we have

$$[a, b]=\frac{d(p, x)}{d(a, b)}+\frac{d(q, y)}{d(a, b)}+\frac{d(r, z)}{d(a, b)}.$$

The assumption is used only in the two following Nos. 56 and 57.

56. Supposing then that a, b, c are conjugate integrals, we have $\frac{dD}{de}-\frac{dE}{dd}=0$, and there exists therefore ϕ, a function of a, b, c, d, e, such that

$$d\phi=A'\,da+B'\,db+C'\,dc+D\,dd+E\,de,$$

(A', B', C' functions of the same quantities a, b, c, d, e), we have therefore

$$dV-p\,dx-q\,dy-r\,dz=d\lambda+(t-\tau)\,dH+d\phi+(A-A')\,da+(B-B')\,db+(C-C')\,dc.$$

Taking as above a, b conjugate integrals $(a, b)=0$, and c any function whatever of a, b, H, then a, b, c are conjugate integrals, and the formula holds good. Suppose further that a, b, H are absolute constants, then $dH=0$, $da=0$, $db=0$, $dc=0$, and the formula becomes

$$dV-p\,dx-q\,dy-r\,dz=d\lambda+d\phi;$$

or, writing this under the form,

$$p\,dx+q\,dy+r\,dz=dV-d\lambda-d\phi,$$

it follows that $p\,dx+q\,dy+r\,dz$ is an exact differential, a theorem which may be stated as follows: viz. if a, b are conjugate integrals of the Hamiltonian system, and if from the equations $H=\text{const.}$, $a=\text{const.}$, $b=\text{const.}$, we express p, q, r as functions of x, y, z, then $p\,dx+q\,dy+r\,dz$ is an exact differential; or, what is the same thing, p, q, r are the differential coefficients $\frac{dU}{dx}, \frac{dU}{dy}, \frac{dU}{dz}$ of U a function of x, y, z. This is, in fact, a fundamental theorem in regard to the partial differential equation $H=\text{const.}$, and it will presently be proved in a different manner.

57. If, as before, a and b are conjugate integrals, then, writing as we may do λ in place of $\lambda+\phi$, and finding V as a function of x, y, z, a, b, H from the equation

$$V=\lambda+\int(p\,dx+q\,dy+r\,dz),$$

and again treating a, b, H as variable, we have

$$dV - p\,dx - q\,dy - r\,dz = d\lambda + (t-\tau)\,dH + A\,da + B\,db,$$

where A, B are functions of the integrals a, b, c, d, e, that is, they are themselves integrals, which may be taken for the integrals d, e, or we have

$$dV - p\,dx - q\,dy - r\,dz = d\lambda + (t-\tau)\,dH + d\,da + e\,db;$$

we have therefore

$$\frac{dV}{da} = d,\quad \frac{dV}{db} = e,$$

equations which, on substituting therein for a, b, H their values as functions of x, y, z, p, q, r, determine the integrals d, e, which with a, b, H or a, b, c, are the remaining integrals of the Hamiltonian system; and further

$$\frac{dV}{dH} = t - \tau,$$

which, when in like manner, we substitute therein for a, b, H, their values as functions of x, y, z, p, q, r, determines τ, the remaining integral of the system as increased by the equality $= dt$.

58. Reverting to the general theorem No. 52, let x_0, y_0, z_0, p_0, q_0, r_0, t_0 be corresponding values of the variables x, y, z, p, q, r, t; and let a_0, &c., ..., V_0 be the same functions of x_0, y_0, z_0, p_0, q_0, r_0, t_0 that a, &c., ..., V are of the variables; we have $a = a_0, \ldots, e = e_0$, and corresponding to the equation

$$dV - p\,dx - q\,dy - r\,dz = d\lambda + (t-\tau)\,dH + A\,da + \ldots + E\,de,$$

the like equation

$$dV_0 - p_0\,dx_0 - q_0\,dy_0 - r_0\,dz_0 = d\lambda + (t_0-\tau)\,dH + A\,da + \ldots + E\,de.$$

Hence, subtracting

$$dV - dV_0 = (t - t_0)\,dH + p\,dx + q\,dy + r\,dz - p_0\,dx_0 - q_0\,dy_0 - r_0\,dz_0,$$

or, considering only H as an absolute constant,

$$dV - dV_0 = \qquad p\,dx + q\,dy + r\,dz - p_0\,dx_0 - q_0\,dy_0 - r_0\,dz_0;$$

viz. if from the equations $H = \text{const.}$, $a = a_0$, $b = b_0$, $c = c_0$, $d = d_0$, $e = e_0$, we express p, q, r, p_0, q_0, r_0 as functions of x, y, z, x_0, y_0, z_0, H, then

$$p\,dx + q\,dy + r\,dz - p_0\,dx_0 - q_0\,dy_0 - r_0\,dz_0$$

will be an exact differential. And in particular regarding x_0, y_0, z_0 as constants, then $p\,dx + q\,dy + r\,dz$ is an exact differential, viz. there exists a function

$$V = \lambda + \int (p\,dx + q\,dy + r\,dz).$$

We have thus again arrived at a solution of the partial differential equation $H = \text{const.}$

The Partial Differential Equation $H = const.$ Art. Nos. 59 to 70.

59. In what just precedes we have, in fact, brought the theory of the Hamiltonian system into connexion with a partial differential equation, viz. we have determined the variables p, q, r as functions of x, y, z such that $p\,dx + q\,dy + r\,dz$ is an exact differential $= dV$; but we now consider the subject in a more regular manner.

The partial differential equation is $H = \text{const.}$ viz. here H denotes, in the first instance, a given function of p, q, r, x, y, z, where p, q, r are the differential coefficients of a function V of x, y, z, or, what is the same thing, there exists a function V of x, y, z such that $p\,dx + q\,dy + r\,dz = dV$; and then, this function H being constant, we use the same letter H to denote the constant value of the function. The equation $H = \text{const.}$ is the most general form of a partial differential equation of the first order which contains the independent variable only through its differential coefficients p, q, r, and it is for convenience put in a form containing the arbitrary constant H, which constant might without loss of generality be put $= 0$ or $=$ any other determinate value.

60. We seek to determine p, q, r as functions of x, y, z, satisfying the given equation $H = \text{const.}$, and such that we have $p\,dx + q\,dy + r\,dz$ an exact differential $= dV$; this would be done if we can find two other equations $K = \text{const.}$ and $L = \text{const.}$, such that the values of p, q, r obtained from the three equations give p, q, r functions having the property in question. Attending to only two of the equations, say $H = \text{const.}$ and $K = \text{const.}$, we have here p, q, r functions of x, y, z, such that $p\,dx + q\,dy + r\,dz$ is an exact differential, and two of the equations which serve to determine p, q, r as functions of x, y, z are $H = \text{const.}$, $K = \text{const.}$ We have to prove the following fundamental theorem, viz. that $(H, K) = 0$.

61. In fact, from the equations $H = \text{const.}$, $K = \text{const.}$, treating x, y, z as independent variables, we have

$$\frac{dH}{dx} + \frac{dH}{dp}\frac{dp}{dx} + \frac{dH}{dq}\frac{dq}{dx} + \frac{dH}{dr}\frac{dr}{dx} = 0,$$

$$\frac{dK}{dx} + \frac{dK}{dp}\frac{dp}{dx} + \frac{dK}{dq}\frac{dq}{dx} + \frac{dK}{dr}\frac{dr}{dx} = 0;$$

and if from these equations in order to eliminate $\frac{dp}{dx}$ we multiply by $\frac{dK}{dp}$, $-\frac{dH}{dp}$, and add, we find

$$\frac{d(K, H)}{d(p, x)} + \qquad\qquad + \frac{d(K, H)}{d(p, q)}\frac{dq}{dx} + \frac{d(K, H)}{d(p, r)}\frac{dr}{dx} = 0;$$

and, in precisely the same way,

$$\frac{d(K, H)}{d(q, y)} + \frac{d(K, H)}{d(q, p)}\frac{dp}{dy} + \qquad\qquad + \frac{d(K, H)}{d(q, r)}\frac{dr}{dy} = 0,$$

$$\frac{d(K, H)}{d(r, z)} + \frac{d(K, H)}{d(r, p)}\frac{dp}{dz} + \frac{d(K, H)}{d(r, q)}\frac{dq}{dz} + \qquad\qquad = 0.$$

Adding these together, we have

$$(K, H)+\frac{d(K, H)}{d(q, r)}\left(\frac{dr}{dy}-\frac{dq}{dz}\right)+\frac{d(K, H)}{d(r, p)}\left(\frac{dp}{dz}-\frac{dr}{dx}\right)+\frac{d(K, H)}{d(p, q)}\left(\frac{dq}{dx}-\frac{dp}{dy}\right)=0;$$

viz. if $p\,dx+q\,dy+r\,dz$ be an exact differential, then $(H, K)=0$, which is the theorem in question.

62. In the case where the variables are (x, y, p, q), we have simply

$$(K, H)+\frac{d(K, H)}{d(p, q)}\left(\frac{dq}{dx}-\frac{dp}{dy}\right)=0;$$

viz. $p\,dx+q\,dy$ being a complete differential, $(K, H)=0$. Conversely, if $(K, H)=0$, then $\frac{dq}{dx}-\frac{dp}{dy}=0$, and $p\,dx+q\,dy$ is an exact differential; viz. this is so unless $\frac{d(K, H)}{d(p, q)}=0$; this equation would imply that K, H considered as functions of p, q, are functions one of the other: and, supposing it to hold good, we could not from the equations $H=0$, $K=0$ determine p, q as functions of x, y, for, eliminating one of the variables p, q, the other would disappear of itself. We hence obtain the complete statement of the converse theorem, viz. the functions H, K being such that it is possible from the equations $H=0$, $K=0$ to express p, q as functions of x, y, then, if $(H, K)=0$, we have $p\,dx+q\,dy$ an exact differential.

63. Returning to the case of the variables (x, y, z, p, q, r), if p, q, r are determined as functions of x, y, z by the three equations $H=0$, $K=0$, $L=0$, then, by what precedes, in order that $p\,dx+q\,dy+r\,dz$ may be a complete differential, we must have $(H, K)=0$, $(H, L)=0$, $(K, L)=0$; and it further appears that if these equations are satisfied, then we have, conversely,

$$\frac{dr}{dy}-\frac{dq}{dz}=0,\quad \frac{dp}{dz}-\frac{dr}{dx}=0,\quad \frac{dq}{dx}-\frac{dp}{dy}=0,$$

that is, $p\,dx+q\,dy+r\,dz$ is an exact differential; viz. this is the case unless we have between H, K, L the relation

$$\begin{vmatrix} \frac{d(H, K)}{d(q, r)}, & \frac{d(H, K)}{d(r, p)}, & \frac{d(H, K)}{d(p, q)} \\ H, L\ , & H, L\ , & H, L \\ K, L\ , & K, L\ , & K, L \end{vmatrix}=0,$$

where in the determinant the second and third lines are the same functions of H, L and K, L respectively that the first line is of H, L.

The determinant is, in fact, equal to the square of

$$\frac{d(H, K, L)}{d(p, q, r)},$$

and, if it vanish, it is impossible, by means of the equations $H=0$, $K=0$, $L=0$, to determine p, q, r as functions of x, y, z. Hence, if the last-mentioned equations are such that by means of them it is possible to effect the determination, and if, moreover, $(H, K)=0$, $(H, L)=0$, $(K, L)=0$, then $p\,dx+q\,dy+r\,dz$ will be an exact differential.

64. Considering H as given, we have, by what precedes, K, L solutions of the linear partial differential equation $(H, \theta)=0$; and since also K, L must be such that $(K, L)=0$, they are conjugate solutions; or in conformity with what precedes, using the small letters a, b instead of K, L, we have the following theorem for the integration of the partial differential equation $H=\text{const.}$, where as before H is a given function of x, y, z, p, q, r.

Find a and b, such that H, a, b are a system of conjugate solutions of the linear partial differential equation $(H, \theta)=0$: then from the equations $H=\text{const.}$, $a=\text{const.}$, $b=\text{const.}$, determining p, q, r as functions of a, b, H, and in the result treating these quantities as constants, we have $p\,dx+q\,dy+r\,dz$ an exact differential $=dV$, and thence

$$V=\lambda+\int(p\,dx+q\,dy+r\,dz),$$

an expression for V containing the three arbitrary constants λ, a, b, and therefore a complete solution of the given partial differential equation $H=\text{const.}$

The theorem applies to the case where n has any value whatever, viz. if there are n variables x, y, z, ..., then we have to find the $n-1$ integrals a, b, c, ..., constituting with H a system of conjugate integrals; and the theorem holds good.

In particular, if $n=2$, or the independent variables are x and y, then we find any solution a of the partial differential equation $(H, \theta)=0$; the values p, q derived from the equations $H=\text{const.}$, $a=\text{const.}$, give $V=\lambda+\int(p\,dx+q\,dy)$, a complete solution.

65. But there is a different solution depending on the consideration of corresponding values; viz. if the independent variables be as before x, y, z, p, q, r, and if x_0, y_0, z_0, p_0, q_0, r_0 are corresponding values of x, y, z, p, q, r, then, taking a, b, c, d, e to be integrals of $(H, \theta)=0$: so that H is here a given function of a, b, c, d, e, since the number of independent variables is $=5$: and representing by a_0, b_0, c_0, d_0, e_0 the like functions of x_0, y_0, z_0, p_0, q_0, r_0, we form the equations

$$H=\text{const.},\ a=a_0,\ b=b_0,\ c=c_0,\ d=d_0,\ e=e_0.$$

We have the theorem that, expressing by means of these equations p, q, r, as functions of x, y, z, x_0, y_0, z_0, H, and regarding therein x_0, y_0, z_0, H as constants, we have $p\,dx+q\,dy+r\,dz$ an exact differential, and therefore

$$V=\lambda+\int(p\,dx+q\,dy+r\,dz),$$

a solution of the equation $H=\text{const.}$ involving the arbitrary constants λ, x_0, y_0, z_0 (one more than required for a complete solution).

The theorem is here stated in the form proper for the solution of the partial differential equation $H=\text{const.}$; a more general statement will be given further on.

66. I take first $n=2$, or the independent variables to be x, y; here p, q are determined by the equations $a=a_0$, $b=b_0$, $c=c_0$, $H=\text{const.}$, and it is to be shown that $p\,dx+q\,dy=dV$.

Considering p, q, p_0, q_0 as functions of the independent variables x, y, then differentiating in regard to x, and eliminating $\dfrac{dp}{dx}$, $\dfrac{dp_0}{dx}$, $\dfrac{dq_0}{dx}$, we find

$$\begin{vmatrix} \dfrac{da}{dx}+\dfrac{da}{dq}\dfrac{dq}{dx}, & \dfrac{da}{dp}, & \dfrac{da_0}{dp_0}, & \dfrac{da_0}{dq_0} \\ \dfrac{db}{dx}+\dfrac{db}{dq}\dfrac{dq}{dx}, & \dfrac{db}{dp}, & \dfrac{db_0}{dp_0}, & \dfrac{db_0}{dq_0} \\ \dfrac{dc}{dx}+\dfrac{dc}{dq}\dfrac{dq}{dx}, & \dfrac{dc}{dp}, & \dfrac{dc_0}{dp_0}, & \dfrac{dc_0}{dq_0} \\ \dfrac{dH}{dx}+\dfrac{dH}{dq}\dfrac{dq}{dx}, & \dfrac{dH}{dp}, & 0, & 0 \end{vmatrix}=0,$$

viz. this is

$$\frac{d(a_0,\,b_0)}{d(p_0,\,q_0)}\left\{\frac{d(H,\,c)}{d(p,\,x)}+\frac{d(H,\,c)}{d(p,\,q)}\frac{dq}{dx}\right\}+\&c.=0.$$

But in the same way

$$\frac{d(a_0,\,b_0)}{d(p_0,\,q_0)}\left\{\frac{d(H,\,c)}{d(q,\,y)}+\frac{d(H,\,c)}{d(q,\,p)}\frac{dp}{dy}\right\}+\&c.=0;$$

adding these two equations we have

$$\frac{d(a_0,\,b_0)}{d(p_0,\,q_0)}\left\{(H,\,c)+\frac{d(H,\,c)}{d(p,\,q)}\left(\frac{dq}{dx}-\frac{dp}{dy}\right)\right\}+\&c.=0,$$

the terms denoted by the &c. being the like terms with b, c, a and c, a, b in place of a, b, c. We have $(H,\,a)=0$, $(H,\,b)=0$, $(H,\,c)=0$, and the equation, in fact, is

$$\left\{\Sigma\frac{d(a_0,\,b_0)}{d(p_0,\,q_0)}\frac{d(H,\,c)}{d(p,\,q)}\right\}\left(\frac{dq}{dx}-\frac{dp}{dy}\right)=0;$$

viz. we have $\dfrac{dq}{dx}-\dfrac{dp}{dy}=0$, the condition for an exact differential.

67. Coming now to the case where the independent variables are x, y, z, we proceed in the same way with the equations $H=\text{const.}$, $a=a_0$, $b=b_0$, $c=c_0$, $d=d_0$, $e=e_0$. Differentiating in regard to x, and eliminating

$$\frac{dp}{dx},\ \frac{dq}{dx},\ \frac{dp_0}{dx},\ \frac{dq_0}{dx},\ \frac{dr_0}{dx},$$

we find for $\frac{dr}{dx}$ the equation

$$\frac{d(c_0, d_0, e_0)}{d(p_0, q_0, r_0)}\left\{\frac{dr}{dx}\frac{d(a, b, H)}{d(r, p, q)}+\frac{d(a, b, H)}{d(x, p, q)}\right\}+\&c.=0.$$

We have in the same way for $\frac{dp}{dz}$ the equation

$$\frac{d(c_0, d_0, e_0)}{d(p_0, q_0, r_0)}\left\{\frac{dp}{dz}\frac{d(a, b, H)}{d(p, r, q)}+\frac{d(a, b, H)}{d(z, r, q)}\right\}+\&c.=0,$$

whence, adding, we obtain

$$\frac{d(c_0, d_0, e_0)}{d(p_0, q_0, r_0)}\left\{\left(\frac{dr}{dx}-\frac{dp}{dz}\right)\frac{d(a, b, H)}{d(r, p, q)}+\frac{d(a, b, H)}{d(x, p, q)}+\frac{d(a, b, H)}{d(r, z, q)}\right\}+\&c.=0,$$

where the terms denoted by the &c. are the like terms corresponding to the different permutations of the letters a, b, c, d, e.

The equation may be simplified; we have identically

$$-\frac{da}{dq}(b, H)-\frac{db}{dq}(H, a)-\frac{dH}{dq}(a, b)=\frac{d(a, b, H)}{d(x, p, q)}+\frac{d(a, b, H)}{d(z, r, q)};$$

or, since $(H, a)=0$, $(H, b)=0$, the left-hand side is simply $-\frac{dH}{dq}(a, b)$, and the equation becomes

$$\frac{d(c_0, d_0, e_0)}{d(p_0, q_0, r_0)}\left\{\left(\frac{dr}{dx}-\frac{dp}{dz}\right)\frac{d(a, b, H)}{d(r, p, q)}-\frac{dH}{dq}(a, b)\right\}+\&c.=0.$$

68. This ought to give $\frac{dr}{dx}-\frac{dp}{dz}=0$; it will, if only

$$\Sigma\left\{\frac{d(c_0, d_0, e_0)}{d(p_0, q_0, r_0)}(a, b)\right\}=0,$$

which is thus the condition which has to be proved. By the Poisson-Jacobi theorem, (a, b) is a function of a, b, c, d, e: if we write

$$(a_0, b_0)=\frac{d(a_0, b_0)}{d(p_0, x_0)}+\frac{d(a_0, b_0)}{d(q_0, y_0)}+\frac{d(a_0, b_0)}{d(r_0, z_0)},$$

then (a_0, b_0) is the same function of a_0, b_0, c_0, d_0, e_0; but these are equal to a, b, c, d, e respectively, and we then have $(a, b)=(a_0, b_0)$, and the theorem to be proved is

$$\Sigma\left\{\frac{d(c_0, d_0, e_0)}{d(p_0, q_0, r_0)}(a_0, b_0)\right\}=0.$$

But, substituting for (a_0, b_0) its value, the function on the left-hand side is, it is easy to see, the sum of the three functional determinants

$$\frac{d(a_0, b_0, c_0, d_0, e_0)}{d(p_0, q_0, r_0, p_0, x_0)},\ \frac{d(a_0, b_0, c_0, d_0, e_0)}{d(p_0, q_0, r_0, q_0, y_0)},\ \frac{d(a_0, b_0, c_0, d_0, e_0)}{d(p_0, q_0, r_0, r_0, z_0)},$$

and each of these, as containing the same letter twice in the denominator, that is, as having two identical columns, is $=0$; the theorem is thus proved. And in the same way $\frac{dp}{dy}-\frac{dq}{dx}$, $\frac{dq}{dz}-\frac{dr}{dy}$ are each $=0$; that is, $p\,dx+q\,dy+r\,dz=dV$.

69. The proof would fail if the factors multiplying $\frac{dr}{dx}-\frac{dp}{dy}$, &c., or any one of these factors, were $=0$. I have not particularly examined this, but the meaning must be that here the equations $a=a_0$, &c., $H=$ const., fail to give for p, q, r expressions as functions of x, y, z, x_0, y_0, z_0, H; whenever such expressions are obtainable, we have

$$p\,dx+q\,dy+r\,dz=dV.$$

The proof in the case of a greater number of variables, say in the next case where the independent variables are x, y, z, w, would probably present greater difficulty, but I have not examined this.

70. Taking the independent variables to be x and y, we may from the equations $a=a_0$, $b=b_0$, $c=c_0$, $H=$ const. (which last equation may also be written $H=H_0=$ const.) find p, q, p_0, q_0 as functions of x, y, x_0, y_0, H; and we have then the theorem that, considering only H as a constant,

$$p\,dx+q\,dy-p_0\,dx_0-q_0\,dy_0=dV.$$

To show this, we have to prove the further equations $\frac{dp}{dx_0}+\frac{dp_0}{dx}=0$, &c.; we find

$$\frac{dp}{dx_0}\Sigma\left\{\frac{d(b_0,\ c_0)}{d(p_0,\ q_0)}\,\frac{d(a,\ H)}{d(p,\ q)}\right\}-\frac{dH}{dq}\,\frac{d(a_0,\ b_0,\ c_0)}{d(x_0,\ p_0,\ q_0)}=0,$$

$$\frac{dp_0}{dx}\Sigma\left\{\frac{d(b,\ c)}{d(p,\ q)}\,\frac{d(a_0,\ H_0)}{d(p_0,\ q_0)}\right\}-\frac{dH_0}{dq_0}\,\frac{d(a,\ b,\ c)}{d(x,\ p,\ q)}=0,$$

and it is to be shown that the coefficients of $\frac{dp}{dx_0}$, $\frac{dp_0}{dx}$ are equal and of opposite signs, and that the other two terms are equal; viz. this being so, subtracting the two equations, we have the required relation $\frac{dp}{dx_0}+\frac{dp_0}{dx}=0$. Now H, H_0 are the same functions of a, b, c and of a_0, b_0, c_0; and there is no real loss of generality in assuming $c=H$, $c_0=H_0$; but this being so, the first coefficient is

$$\frac{d(b_0,\ H_0)}{d(p_0,\ q_0)}\,\frac{d(a,\ H)}{d(p,\ q)}+\frac{d(H_0,\ a_0)}{d(p_0,\ q_0)}\,\frac{d(b,\ H)}{d(p,\ q)},$$

and the second is

$$\frac{d(b,\ H)}{d(p,\ q)}\,\frac{d(a_0,\ H_0)}{d(p_0,\ q_0)}+\frac{d(H,\ a)}{d(p,\ q)}\,\frac{d(b_0,\ H_0)}{d(p_0,\ q_0)},$$

which only differ by their signs. As regards the other two terms, we have identically

$$\frac{da}{dq}(b,\ H)+\frac{db}{dq}(H,\ a)+\frac{dH}{dq}(a,\ b)=\frac{d(a,\ b,\ H)}{d(x,\ p,\ q)},$$

which, in virtue of $(a, H)=0$, $(b, H)=0$, becomes

$$\frac{dH}{dq}(a\,,\,b\,)=\frac{d(a,\,b,\,H)}{d(x,\,p,\,q)};$$

similarly,

$$\frac{dH_0}{dq_0}(a_0,\,b_0)=\frac{d(a_0,\,b_0,\,H_0)}{d(x_0,\,p_0,\,q_0)}.$$

Hence the terms in question are

$$-\frac{dH}{dq}\frac{dH_0}{dq_0}(a_0,\,b_0),\quad -\frac{dH}{dq}\frac{dH_0}{dq_0}(a,\,b),$$

which are equal in virtue of $(a, b)=(a_0, b_0)$; and, similarly, the other conditions might be proved. But the proof would be more difficult in the case of a greater number of variables.

Examples. Art. Nos. 71 to 79.

71. The variables are taken to be x, y, z, p, q, r. As a first example, which will serve as an illustration of most of the preceding theorems, suppose $pqr-1=H$; the Hamiltonian system, with the adjoined equalities, is here

$$\frac{dx}{qr}=\frac{dy}{rp}=\frac{dz}{pq}=\frac{dp}{0}=\frac{dq}{0}=\frac{dr}{0}=dt=\frac{dV}{3pqr}.$$

The integrals of the original system may be taken to be

$$\begin{aligned} a &= p,\\ b &= q,\\ c &= r,\\ d &= qy-px,\\ e &= rz-px, \end{aligned}$$

and there is of course the integral $H=pqr-1$, which is connected with the foregoing five integrals by the relation $H=abc-1$.

We form at once the equations

$$\begin{array}{llll} (a,\,b)=0, & (a,\,c)=0, & (a,\,d)=-a, & (a,\,e)=-a,\\ & (b,\,c)=0, & (b,\,d)=b, & (b,\,e)=0,\\ & & (c,\,d)=0, & (c,\,e)=c,\\ & & & (d,\,e)=0; \end{array}$$

hence it happens that no two of these integrals a, b, c, d, e give by the Poisson-Jacobi theorem a new integral. To show how the theorem might have given a new integral, suppose that the known integrals had been $\alpha=p+q$, and $e=rz-px$, then $(\alpha, e)=-p$: or the theorem gives the new integral $a=p$.

We have as a conjugate system a, b, c; also the conjugate systems H, a, b; H, a, c; H, b, c; H, b, e; H, c, d; H, d, e; but the first three of these, considering therein H as standing for its value $abc-1$, are substantially equivalent to the first-mentioned system (a, b, c).

72. Postponing the consideration of the augmented system, we now consider the partial differential equation $pqr = 1+H$, where H is a given constant and p, q, r denote the differential coefficients of a function V. The most simple solution is that given by the conjugate system H, a, b, viz. here p, q, r are determined by the equations $p=a$, $q=b$, $pqr=1+H$, that is, $r=\dfrac{1+H}{ab}$; or, introducing for symmetry the constant c, where $abc=1+H$ as before, then $r=c$, and we have

$$V=\lambda+\int(adx+bdy+cdz),\quad =\lambda+ax+by+cz,$$

where a, b, c are connected by the just-mentioned equation $abc=1+H$. This is therefore a solution containing say the arbitrary constants λ, a, b, and, as such, is a complete solution.

But any other conjugate system gives a complete solution, and a very elegant one is obtained from the system H, d, e. Writing for symmetry $\beta-\alpha$, $\gamma-\alpha$ in place of d, e, we have here to find p, q, r from the equations

$$H=pqr-1,\quad qy-px=\beta-\alpha,\quad rz-px=\gamma-\alpha;$$

or, if we assume $\theta=px-\alpha$, then

$$H=pqr-1;\ px,\ qy,\ rz=\theta+\alpha,\ \theta+\beta,\ \theta+\gamma$$

respectively, whence

$$(1+H)\,xyz=(\theta+\alpha)(\theta+\beta)(\theta+\gamma),$$

which equation determines θ as a function of x, y, z (in fact, it is a function of the product xyz), and then

$$p,\ q,\ r=\frac{\theta+\alpha}{x},\quad \frac{\theta+\beta}{y},\quad \frac{\theta+\gamma}{z},$$

and we have

$$V=\lambda+\int\left(\frac{\theta+\alpha}{x}dx+\frac{\theta+\beta}{y}dy+\frac{\theta+\gamma}{z}dz\right).$$

There is no difficulty in effecting the integration directly by introducing θ as a new variable, and we find

$$V=\lambda+3\theta-\alpha\log\frac{\theta+\alpha}{x}-\beta\log\frac{\theta+\beta}{y}-\gamma\log\frac{\theta+\gamma}{z}.$$

Or, starting from this form, we may verify it by differentiation; the value of dV is

$$d\theta\left(3-\frac{\alpha}{\theta+\alpha}-\frac{\beta}{\theta+\beta}-\frac{\gamma}{\theta+\gamma}\right)+\frac{\alpha\,dx}{x}+\frac{\beta\,dy}{y}+\frac{\gamma\,dz}{z},$$

where the term in $d\theta$ is

$$=\theta\, d\theta\left(\frac{1}{\theta+\alpha}+\frac{1}{\theta+\beta}+\frac{1}{\theta+\gamma}\right),$$

which, from the equation which determines θ, is

$$=\theta\left(\frac{dx}{x}+\frac{dy}{y}+\frac{dz}{z}\right),$$

and the value of dV is thus

$$=\frac{\theta+\alpha}{x}\,dx+\frac{\theta+\beta}{y}\,dy+\frac{\theta+\gamma}{z}\,dz.$$

The solution contains apparently the four constants λ, α, β, γ, but there is no loss of generality in writing, for instance, $\alpha=0$, and the number of constants really contained in the solution may be regarded as 3.

73. To show how the equations $H=\text{const.}$, $a=a_0$, $b=b_0$, $c=c_0$, $d=d_0$, $e=e_0$ give a solution; remarking that these equations are $pqr-1=H$, $p=p_0$, $q=q_0$, $r=r_0$, $qy-px=q_0y_0-p_0x_0$, $rz-px=r_0z_0-p_0x_0$, we find

$$p\,(x-x_0)=q\,(y-y_0)=r\,(z-z_0),$$

and consequently p, q, $r=$

$$\sqrt[3]{(1+H)}\,\frac{(y-y_0)^{\frac{1}{3}}(z-z_0)^{\frac{1}{3}}}{(x-x_0)^{\frac{2}{3}}},\quad \sqrt[3]{(1+H)}\,\frac{(z-z_0)^{\frac{1}{3}}(x-x_0)^{\frac{1}{3}}}{(y-y_0)^{\frac{2}{3}}},\quad \sqrt[3]{(1+H)}\,\frac{(x-x_0)^{\frac{1}{3}}(y-y_0)^{\frac{1}{3}}}{(z-z_0)^{\frac{2}{3}}},$$

respectively: whence

$$V=\lambda+\int(p\,dx+q\,dy+r\,dz),$$

$$=\lambda+3\sqrt[3]{(1+H)}\,(x-x_0)^{\frac{1}{3}}(y-y_0)^{\frac{1}{3}}(z-z_0)^{\frac{1}{3}},$$

which is the solution involving the four constants λ, x_0, y_0, z_0.

If in the foregoing value of V we consider x_0, y_0, z_0 as variables, then p, q, r having the values just mentioned, and p_0, q_0, r_0 being equal to these respectively, we obviously have

$$dV=p\,dx+q\,dy+r\,dz-p_0\,dx_0-q_0\,dy_0-r_0\,dz_0.$$

74. Considering now the augmented Hamiltonian system, we join to the foregoing integrals a, b, c, d, e, the new integrals $t-\tau=\dfrac{x}{qr}$ and $V-\lambda=3px$. And then expressing all the quantities in terms of $t-\tau$,

$$x=bc\,(t-\tau),$$

$$y=ca\,(t-\tau)+\frac{d}{b},$$

$$z=ab\,(t-\tau)+\frac{e}{c}.$$

$$p=a,\ q=b,\ r=c,\ H=abc-1,$$

$$V=\lambda+3abc\,(t-\tau).$$

Forming from these the expression for $dV - p\,dx - q\,dy - r\,dz$, the term in $dt - d\tau$ disappears; there is a term in $t - \tau$, the coefficient of which is

$$3d\,.\,abc - a\,d\,.\,bc - b\,d\,.\,ca - c\,d\,.\,ab,$$

which is $= d\,.\,abc$, or the term is $(t-\tau)\,dH$; and we have, finally,

$$dV - p\,dx - q\,dy - r\,dz = d\lambda + (t-\tau)\,dH - bd\frac{d}{b} - cd\frac{e}{c};$$

viz. t enters only in the combination $(t-\tau)\,dH$, which is the fundamental theorem. Considering H as a determinate constant, this term disappears.

We may show how this formula leads to the solution of the partial differential equation $pqr = 1 + H$; treating H as a definite constant, then in order that the formula may give $dV - p\,dx - q\,dy - r\,dz = d\lambda$, or $V = \lambda + \int (p\,dx + q\,dy + r\,dz)$, as before, the last two terms of the formula must disappear; this will be the case if $\frac{d}{b}$ and $\frac{e}{c}$ are constants, or, say, $d = b\beta$, $e = c\gamma$, β and γ being constants. But, this being so, we have $q\beta = qy - px$, $r\gamma = rz - px$, that is, $px = q(y-\beta) = r(z-\gamma)$, $pqr = 1 + H$, giving the values of p, q, r; and then

$$V = \lambda + \int (p\,dx + q\,dy + r\,dz), \quad = \lambda + 3\sqrt[3]{(1+H)}\,x^{\frac{1}{3}}(y-\beta)^{\frac{1}{3}}(z-\gamma)^{\frac{1}{3}},$$

which is substantially the same solution as is obtained above by a different process. Or, again, observing that we have

$$dV - p\,dx - q\,dy - r\,dz = d\lambda + (t-\tau)\,dH - dd - de - \frac{d}{b}\,db - \frac{e}{c}\,dc,$$

then, taking H, b, c constants, we have

$$dV - p\,dx - q\,dy - r\,dz = d\lambda - dd - de,$$

which, changing the value of λ, gives the before-mentioned solution

$$V = \lambda + ax + by + cz, \quad (abc = 1 + H).$$

75. As a second example, suppose

$$H = \tfrac{1}{2}(p^2 + q^2 + r^2 - x^2 - y^2 - z^2);$$

the augmented system is

$$\frac{dx}{p} = \frac{dy}{q} = \frac{dz}{r} = \frac{dp}{x} = \frac{dq}{y} = \frac{dr}{z} = dt = \frac{dV}{p^2 + q^2 + r^2},$$

corresponding to the dynamical problem of the motion of a particle acted upon by a repulsive central force equal to the distance.

The integrals of the original system may be expressed in various forms, viz. the quotient of any two of the expressions $x+p$, $y+q$, $z+r$, or of any two of the expressions $x-p$, $y-q$, $z-r$ is an integral, or again the product of any expression of the first set into any expression of the second set is an integral: we may take as integrals

$$\alpha = x^2 - p^2,\quad \beta = y^2 - q^2,\quad \gamma = z^2 - r^2,\quad \delta = \frac{y+q}{x+p},\quad \epsilon = \frac{z+r}{x+p}.$$

We have then

$$dt = \frac{dx}{\sqrt{(x^2-\alpha)}},\ \text{that is,}\ t - \tau = \log\{x + \sqrt{(x^2-\alpha)}\} = \log(x+p),$$

giving $x+p = e^{t-\tau}$, and thence the other quantities $x-p$, $y+q$, &c. For greater symmetry, I introduce a new set of constants a, b, c, a', b', c', and I write also $e^{t-\tau} = T$, $e^{-t+\tau} = T'$ (where $TT' = 1$). We then have

$$\begin{aligned} x &= aT + a'T', & p &= aT - a'T', \\ y &= bT + b'T', & q &= bT - b'T', \\ z &= cT + c'T', & r &= cT - c'T'; \end{aligned}$$

also, comparing with the values obtained as above,

$$\begin{aligned} a &= \tfrac{1}{2}, & b &= \tfrac{1}{2}\delta, & c &= \tfrac{1}{2}\epsilon, \\ a' &= \tfrac{1}{2}\alpha, & b' &= \tfrac{1}{2}\frac{\beta}{\delta}, & c' &= \tfrac{1}{2}\frac{\gamma}{\epsilon}. \end{aligned}$$

We have, moreover,

$$H = -2(aa' + bb' + cc') = -\tfrac{1}{2}(\alpha + \beta + \gamma).$$

76. We find

$$p^2 + q^2 + r^2 = H + (a^2 + b^2 + c^2)T^2 + (a'^2 + b'^2 + c'^2)T'^2,$$

and thence

$$\begin{aligned} V &= \lambda + \int (p^2 + q^2 + r^2)\,dt \\ &= \lambda + H(t-\tau) + \tfrac{1}{2}(a^2 + b^2 + c^2)T^2 - \tfrac{1}{2}(a'^2 + b'^2 + c'^2)T'^2. \end{aligned}$$

We may from this obtain the expression for

$$dV - p\,dx - q\,dy - r\,dz,$$

when everything is variable. The terms in $(dt - d\tau)$, as is obvious, disappear; omitting these from the beginning, we have

$$dV = d\lambda + (t-\tau)\,dH + (a\,da + b\,db + c\,dc)\,T^2 - (a'da' + b'db' + c'dc')\,T'^2:$$

also

$$\begin{aligned} p\,dx &= (aT - a'T')(T da + T' da'), \\ &= da\,(aT^2 - a') + da'\,(-a'T'^2 + a): \end{aligned}$$

thence forming the analogous expressions for $q\,dy$ and $r\,dz$, we have

$$\begin{aligned} p\,dx + q\,dy + r\,dz = {} & (a\,da + b\,db + c\,dc)\,T^2 - (a'da' + b'db' + c'dc')\,T'^2 \\ & - (a'da + b'db + c'dc) \quad + (a\,da' + b\,db' + c\,dc'), \end{aligned}$$

whence

$$dV - p\,dx - q\,dy - r\,dz = d\lambda + (t-\tau)\,dH + a'da + b'db + c'dc - a\,da' - b\,db' - c\,dc';$$

or, in place of a, b, c, a', b', c', introducing $\alpha, \beta, \gamma, \delta, \epsilon$, and attending to the value of H,

$$dV - p\,dx - q\,dy - r\,dz = d\lambda + (t-\tau)\,dH + \tfrac{1}{2}\,dH + \tfrac{1}{2}\frac{\beta}{\delta}\,d\delta + \tfrac{1}{2}\frac{\gamma}{\epsilon}\,d\epsilon.$$

77. Suppose H, δ, ϵ absolute constants, this becomes

$$d(V-\lambda) = p\,dx + q\,dy + r\,dz,$$

or

$$V = \lambda + \int (p\,dx + q\,dy + r\,dz),$$

and we have thus a solution of the partial differential equation

$$p^2 + q^2 + r^2 = x^2 + y^2 + z^2 + 2H;$$

viz. p, q, r are here to be determined as functions of x, y, z by the equations

$$\begin{aligned} p^2 + q^2 + r^2 &= x^2 + y^2 + z^2 + 2H, \\ y + q &= \delta(x+p), \\ z + r &= \epsilon(x+p). \end{aligned}$$

We have

$$2H + x^2 + y^2 + z^2 = p^2 + \{y - \delta(x+p)\}^2 + \{z - \epsilon(x+p)\}^2;$$

or, on the right-hand side, writing $p^2 = (x+p)^2 - 2x(x+p) + x^2$,

" left " " $x^2 = (x-p)^2 - 2x(x+p) + p^2$,

the equation is

$$(1+\delta^2+\epsilon^2)(x+p)^2 - 2(x+\delta y+\epsilon z)(x+p) - 2H = 0,$$

which gives p as a function of x, y, z. But the result is a complicated one, except in the case $H=0$; we then have

$$x + p = \frac{2(x+\delta y+\epsilon z)}{1+\delta^2+\epsilon^2},$$

$$y + q = \frac{2\delta(x+\delta y+\epsilon z)}{1+\delta^2+\epsilon^2},$$

$$z + r = \frac{2\epsilon(x+\delta y+\epsilon z)}{1+\delta^2+\epsilon^2},$$

and thence

$$V = \lambda - \tfrac{1}{2}(x^2+y^2+z^2) + \frac{(x+\delta y+\epsilon z)^2}{1+\delta^2+\epsilon^2},$$

a complete solution of the partial differential equation

$$p^2 + q^2 + r^2 = x^2 + y^2 + z^2.$$

More symmetrically, we have the solution

$$V = \lambda - \tfrac{1}{2}(x^2 + y^2 + z^2) + \frac{(ax + by + cz)^2}{a^2 + b^2 + c^2},$$

as can be at once verified.

78. In the same particular case $H = 0$, introducing the corresponding values $p_0, q_0, r_0, x_0, y_0, z_0$, we find a very simple expression for $V - V_0$, as a function of x, y, z, x_0, y_0, z_0. We have, writing $T_0 = e^{t_0 - \tau}$, $T_0' = e^{-t_0 + \tau}$, and therefore $T_0 T_0' = 1$,

$$x_0 = aT_0 + a'T_0'; \quad p_0 = aT_0 - a'T_0',$$

$$y_0 = bT_0 + b'T_0', \quad q_0 = bT_0 - b'T_0',$$

$$z_0 = cT_0 + c'T_0', \quad r_0 = cT_0 - c'T_0',$$

and thence

$$x - x_0 = a(T - T_0) + a'\left(\frac{1}{T} - \frac{1}{T_0}\right), = (T - T_0)\left(a - \frac{a'}{TT_0}\right),$$

$$x + x_0 = a(T + T_0) + a'\left(\frac{1}{T} + \frac{1}{T_0}\right), = (T - T_0)\left(a + \frac{a'}{TT_0}\right).$$

Forming the analogous quantities $y - y_0$, &c., we deduce

$$(x - x_0)^2 + (y - y_0)^2 + (z - z_0)^2 = (T - T_0)^2 \left\{a^2 + b^2 + c^2 + (a'^2 + b'^2 + c'^2)\frac{1}{T^2 T_0^2}\right\},$$

$$(x + x_0)^2 + (y + y_0)^2 + (z + z_0)^2 = (T + T_0)^2 \left\{a^2 + b^2 + c^2 + (a'^2 + b'^2 + c'^2)\frac{1}{T^2 T_0^2}\right\}.$$

But we have

$$V - V_0 = \tfrac{1}{2}\left\{(a^2 + b^2 + c^2)(T^2 - T_0^2) - (a'^2 + b'^2 + c'^2)\left(\frac{1}{T^2} - \frac{1}{T_0^2}\right)\right\}$$

$$= \tfrac{1}{2}(T^2 - T_0^2)\left\{a^2 + b^2 + c^2 + (a'^2 + b'^2 + c'^2)\frac{1}{T^2 T_0^2}\right\},$$

and hence the required formula

$$V - V_0 = \tfrac{1}{2}\sqrt{\{(x - x_0)^2 + (y - y_0)^2 + (z - z_0)^2\}}\sqrt{\{(x + x_0)^2 + (y + y_0)^2 + (z + z_0)^2\}},$$

or, say, for shortness,

$$= \tfrac{1}{2}\sqrt{(R)}\sqrt{(S)}.$$

79. We ought, therefore, to have

$$\tfrac{1}{2} d\sqrt{(R)}\sqrt{(S)} = p\,dx + q\,dy + r\,dz - p_0\,dx_0 - q_0\,dy_0 - r_0\,dz_0,$$

where p, q, r, p_0, q_0, r_0 denote as above, and consequently

$$p^2 + q^2 + r^2 = x^2 + y^2 + z^2, \quad p_0^2 + q_0^2 + r_0^2 = x_0^2 + y_0^2 + z_0^2.$$

We have in fact

$$p=\tfrac{1}{2}\left\{\frac{\surd(S)}{\surd(R)}(x-x_0)+\frac{\surd(R)}{\surd(S)}(x+x_0)\right\},\ \text{\&c.},$$

$$p_0=\tfrac{1}{2}\left\{-\frac{\surd(S)}{\surd(R)}(x-x_0)+\frac{\surd(R)}{\surd(S)}(x+x_0)\right\},\ \text{\&c.};$$

and thence

$$p^2+q^2+r^2=\tfrac{1}{4}\{R+S+2(x^2+y^2+z^2-x_0^2-y_0^2-z_0^2)\},\ =x^2+y^2-z^2,$$

$$p_0^2+q_0^2+r_0^2=\tfrac{1}{4}\{R+S-2(x^2+y^2+z^2-x_0^2-y_0^2-z_0^2)\},\ =x_0^2+y_0^2+z_0^2,$$

or the last-mentioned results are thus verified.

Partial Differential Equation containing the Dependent Variable: Reduction to Standard Form. Art. Nos. 80, 81.

80. The equation $H=\text{const.}$ is the most general form of a partial differential equation not containing the dependent variable V; but if a partial differential equation does contain the independent variable, we can, by regarding this as one of the dependent variables, and in place of it introducing a new independent variable, exhibit the equation in the standard form $H=\text{const.}$, H being here a homogeneous function of the order zero in the differential coefficients. Thus, if the independent variables are x, y, the dependent variable z, and its differential coefficients p, q, then the given partial differential equation may be H, $=H(\mathrm{p}, \mathrm{q}, x, y, z)$, $=\text{const.}$ But we may determine z as a function of x, y by an equation $V=\text{const.}$, V being a desired function of x, y, z; and then writing p, q, r for the differential coefficients $\frac{dV}{dx}$, $\frac{dV}{dy}$, $\frac{dV}{dz}$, we have $\mathrm{p}=-\frac{p}{r}$, $\mathrm{q}=-\frac{q}{r}$, and the proposed partial differential equation becomes

$$H\left(-\frac{p}{r},\ -\frac{q}{r},\ x,\ y,\ z\right)=\text{const.}$$

viz. this is an equation containing only the differential coefficients p, q, r of the dependent variable V, a function of x, y, z. And, moreover, H is homogeneous of the order zero in p, q, r; consequently

$$p\frac{dH}{dp}+q\frac{dH}{dq}+r\frac{dH}{dr}=0,$$

or, in the augmented Hamiltonian system, the last equality is $=\frac{dV}{0}$, so that an integral is $V=\text{const.}$; as already stated, this is the equation by which z is determined as a function of x, y.

81. Thus, if the given partial differential equation be $\mathrm{pq}-z=H$, we here consider the equation $\frac{pq}{r^2}-z=H$. The Hamiltonian system is

$$\frac{r^2dx}{q}=\frac{r^2dy}{p}=\frac{-r^3dz}{2pq}=\frac{dp}{0}=\frac{dq}{0}=\frac{dr}{1}\left(=\frac{dV}{0}\right),$$

having the integrals

$$a = p,$$

$$b = q,$$

$$c = px - qy,$$

$$d = \frac{1}{r} + \frac{x}{q},$$

$$e = \frac{z}{pq} - \frac{1}{r^2},$$

(whence $H = -abe$). We have H, a, b, a system of conjugate integrals and, in terms of these,

$$p = a,\ q = b,\ r = \sqrt{\left(\frac{ab}{z+H}\right)};$$

hence, writing λ for the constant value of V, we have

$$\lambda = \int\left\{a\,dx + b\,dy + \sqrt{\left(\frac{ab}{z+H}\right)}\,dz\right\},$$

that is,

$$\lambda = ax + by + 2\sqrt{\{ab\,(z+H)\}},$$

or say,

$$4ab\,(z+H) = (ax + by - \lambda)^2,$$

a solution containing really the two constants λ and $\frac{a}{b}$, and thus a complete solution of the given equation $\mathrm{pq} - z = H$. We have, in fact,

$$2ab\,\mathrm{p} = a\,(ax + by - \lambda),$$

$$2ab\,\mathrm{q} = b\,(ax + by - \lambda);$$

that is,

$$4a^2b^2\mathrm{pq} = ab\,(ax + by - \lambda)^2, \ = 4a^2b^2\,(z+H),$$

or

$$\mathrm{pq} = z + H,$$

as it should be.

656.

ON THE THEORY OF PARTIAL DIFFERENTIAL EQUATIONS.

[From the *Mathematische Annalen*, t. XI. (1877), pp. 194—198.]

IN what follows, any letter not otherwise explained denotes a function of certain variables (x, y, p, q), or (x, y, z, p, q, r), &c., as will be stated in each particular case.

An equation $a = \text{const.}$ denotes that the function a of the variables is, in fact, a constant (viz. by such equation we establish a relation between the variables): and when this is so, we use the same letter a to denote the constant value of the function in question; I find this a very convenient notation.

Thus if the variables are x, y, z, p, q, r and if p, q, r are the differential coefficients in regard to x, y, z respectively of a function V of x, y, z, then H (as a letter not otherwise explained) denotes a function of x, y, z, p, q, r and considering it as a given function,

$$\text{H} = \text{const.}$$

will be a partial differential equation containing the constant H. For instance, if H denote the function $pqr - xyz$, $\text{H} = \text{const.}$ is the partial differential equation, $pqr - xyz = \text{H}$ (a given constant).

The integration of the partial differential equation, $\text{H} = \text{const.}$, depends upon that of the linear partial differential equation

$$(\text{H}, \Theta) = 0,$$

where as usual (H, Θ) signifies

$$\frac{\partial(\text{H}, \Theta)}{\partial(p, x)} + \frac{\partial(\text{H}, \Theta)}{\partial(q, y)} + \frac{\partial(\text{H}, \Theta)}{\partial(r, z)}.$$

It can be effected if we know two conjugate solutions a, b of the equation $(\text{H}, \Theta) = 0$, viz. a, b as solutions are such that $(\text{H}, a) = 0$, $(\text{H}, b) = 0$, and (as conjugate solutions) are also such that $(a, b) = 0$; in this case if from the equations

$$\text{H} = \text{const.}, \quad a = \text{const.}, \quad b = \text{const.}$$

we determine p, q, r as functions of x, y, z, the resulting value of $p\,dx+q\,dy+r\,dz$ is an exact differential, and we have

$$V=\lambda+\int(p\,dx+q\,dy+r\,dz),$$

a solution containing three arbitrary constants, λ, a, b, and therefore a complete solution of the proposed partial differential equation $\mathrm{H}=$ const.

But (as is known) there is a different process of integration, for which the conjugate solutions are not required, and which has reference to a system of initial values x_0, y_0, z_0, p_0, q_0, r_0: viz. if the independent solutions of $(\mathrm{H}, \Theta)=0$, are a, b, c, d, e, and if a_0, b_0, c_0, d_0, e_0 denote respectively the same functions of the initial variables that a, b, c, d, e are of x, y, z, p, q, r, then if from the equations

$$a=a_0,\quad b=b_0,\quad c=c_0,\quad d=d_0,\quad e=e_0,\quad \mathrm{H}=\text{const.}$$

we express p, q, r as functions of x, y, z and of x_0, y_0, z_0, H, these last being regarded as constants, we have $p\,dx+q\,dy+r\,dz$ an exact differential, and

$$V=\lambda+\int(p\,dx+q\,dy+r\,dz),$$

a solution containing the constants λ, x_0, y_0, z_0 (that is, one supernumerary constant), and as such a complete solution.

It is interesting to prove directly that $p\,dx+q\,dy+r\,dz$ is an exact differential.

I consider first the more simple case where the variables are p, q, x, y. Here p, q are to be found from the equations

$$a=a_0,\quad b=b_0,\quad c=c_0,\quad \mathrm{H}=\text{const.}$$

and it is to be shown that $p\,dx+q\,dy$ is an exact differential.

Considering p, q, p_0, q_0 as functions of the independent variables x, y, then differentiating in regard to x, and eliminating $\frac{dp}{dx}$, $\frac{dp_0}{dx}$, $\frac{dq_0}{dx}$, we have

$$\begin{vmatrix} \frac{da}{dx}+\frac{da}{dq}\frac{dq}{dx}, & \frac{da}{dp}, & \frac{da_0}{dp_0}, & \frac{da_0}{dq_0} \\ \frac{db}{dx}+\frac{db}{dq}\frac{dq}{dx}, & \frac{db}{dp}, & \frac{db_0}{dp_0}, & \frac{db_0}{dq_0} \\ \frac{dc}{dx}+\frac{dc}{dq}\frac{dq}{dx}, & \frac{dc}{dp}, & \frac{dc_0}{dp_0}, & \frac{dc_0}{dq_0} \\ \frac{d\mathrm{H}}{dx}+\frac{d\mathrm{H}}{dq}\frac{dq}{dx}, & \frac{d\mathrm{H}}{dp}, & 0, & 0 \end{vmatrix}=0,$$

or introducing a well-known notation for functional determinants, and expanding the determinant, this is

$$\frac{\partial(a_0, b_0)}{\partial(p_0, q_0)}\left\{\frac{\partial(\mathrm{H}, c)}{\partial(p, x)}+\frac{\partial(\mathrm{H}, c)}{\partial(p, q)}\frac{dq}{dx}\right\}+\&\text{c.}=0.$$

But in the same way

$$\frac{\partial(a_0, b_0)}{\partial(p_0, q_0)}\left\{\frac{\partial(\mathrm{H}, c)}{\partial(q, y)}+\frac{\partial(\mathrm{H}, c)}{\partial(q, p)}\frac{dp}{dy}\right\}+\&\text{c.}=0;$$

or adding these, attending to the value of (H, c), and observing that $\dfrac{\partial(\mathrm{H}, c)}{\partial(q, p)}=-\dfrac{\partial(\mathrm{H}, c)}{\partial(p, q)}$ we have

$$\frac{\partial(a_0, b_0)}{\partial(p_0, q_0)}\left\{(\mathrm{H}, c)+\frac{\partial(\mathrm{H}, c)}{\partial(p, q)}\left(\frac{dq}{dx}-\frac{dp}{dy}\right)\right\}+\&\text{c.}=0,$$

the terms denoted by the &c. being the like terms with b, c, a and c, a, b in place of a, b, c. We have (H, a) = 0, (H, b) = 0, (H, c) = 0, and the equation in fact is

$$\left\{\Sigma\frac{\partial(a_0, b_0)}{\partial(p, q)}\frac{\partial(\mathrm{H}, c)}{\partial(p, q)}\right\}\left(\frac{dq}{dx}-\frac{dp}{dy}\right)=0;$$

viz. we have $\dfrac{dq}{dx}-\dfrac{dp}{dy}=0$, the condition for the exact differential.

Coming now to the case where the variables are x, y, z, p, q, r, and in the six equations treating p, q, r, p_0, q_0, r_0 as functions of the independent variables x, y, z,—then differentiating with regard to x and proceeding as before, we find for $\dfrac{dr}{dx}$ the equation

$$\frac{\partial(c_0, d_0, e_0)}{\partial(p_0, q_0, r_0)}\left\{\frac{dr}{dx}\frac{\partial(a, b, \mathrm{H})}{\partial(r, p, q)}+\frac{\partial(a, b, \mathrm{H})}{\partial(x, p, q)}\right\}+\&\text{c.}=0.$$

We have, in the same way, for $\dfrac{dp}{dz}$ the equation

$$\frac{\partial(c_0, d_0, e_0)}{\partial(p_0, q_0, r_0)}\left\{\frac{dp}{dz}\frac{\partial(a, b, \mathrm{H})}{\partial(p, r, q)}+\frac{\partial(a, b, \mathrm{H})}{\partial(z, r, q)}\right\}+\&\text{c.}=0;$$

or, adding the two equations,

$$\frac{\partial(c_0, d_0, e_0)}{\partial(p_0, q_0, r_0)}\left\{\left(\frac{dr}{dx}-\frac{dp}{dz}\right)\frac{\partial(a, b, \mathrm{H})}{\partial(r, p, q)}+\frac{\partial(a, b, \mathrm{H})}{\partial(x, p, q)}+\frac{\partial(a, b, \mathrm{H})}{\partial(z, r, q)}\right\}+\&\text{c.}=0,$$

where the terms denoted by the &c. indicate the like terms corresponding to the different partitions of the letters a, b, c, d, e.

The equation may be simplified; we have identically

$$-\frac{da}{dq}(b, \mathrm{H})-\frac{db}{dq}(\mathrm{H}, a)-\frac{d\mathrm{H}}{dq}(a, b)=\frac{\partial(a, b, \mathrm{H})}{\partial(x, p, q)}+\frac{\partial(a, b, \mathrm{H})}{\partial(z, r, q)},$$

or since (H, a) = 0, (b, H) = 0, the left-hand side is simply $-\frac{dH}{dq}(a, b)$, and the equation becomes

$$\frac{\partial(c_0, d_0, e_0)}{\partial(p_0, q_0, r_0)}\left\{\left(\frac{dr}{dx}-\frac{dp}{dz}\right)\frac{\partial(a, b, H)}{\partial(r, p, q)}-\frac{dH}{dq}(a, b)\right\}+\&c.=0.$$

This ought to give $\frac{dr}{dx}-\frac{dp}{dz}=0$, and it will do so if only

$$\Sigma\left\{\frac{\partial(c_0, d_0, e_0)}{\partial(p_0, q_0, r_0)}(a, b)\right\}=0;$$

this is then the equation which has to be proved. By the Poisson-Jacobi theorem, (a, b) is a function of a, b, c, d, e: if we write

$$(a_0, b_0)=\frac{\partial(a_0, b_0)}{\partial(p_0, x_0)}+\frac{\partial(a_0, b_0)}{\partial(q_0, y_0)}+\frac{\partial(a_0, b_0)}{\partial(r_0, z_0)},$$

then (a_0, b_0) is the same function of a_0, b_0, c_0, d_0, e_0; but these are $=a, b, c, d, e$ respectively, and we thence have $(a, b)=(a_0, b_0)$, and the theorem to be proved is

$$\Sigma\left\{\frac{\partial(c_0, d_0, e_0)}{\partial(p_0, q_0, r_0)}(a_0, b_0)\right\}=0.$$

But substituting for (a_0, b_0) its value, the function on the left-hand is (it is easy to see) the sum of the three functional determinants

$$\frac{\partial(a_0, b_0, c_0, d_0, e_0)}{\partial(p_0, q_0, r_0, p_0, x_0)}+\frac{\partial(a_0, b_0, c_0, d_0, e_0)}{\partial(p_0, q_0, r_0, q_0, y_0)}+\frac{\partial(a_0, b_0, c_0, d_0, e_0)}{\partial(p_0, q_0, r_0, r_0, z_0)},$$

each of which vanishes as containing the same letter twice in the denominator, that is, as having two identical columns; and the theorem in question is thus proved. And in the same way $\frac{dp}{dy}-\frac{dq}{dx}$, $\frac{dq}{dz}-\frac{dr}{dy}$ are each $=0$: or we have $p\,dx+q\,dy+r\,dz$ an exact differential.

The proof would fail if the factors multiplying $\frac{dq}{dx}-\frac{dp}{dy}$, &c., or if any one of these factors, were $=0$; I have not particularly examined this, but the meaning would be, that here the equations in question $a=a_0$, &c., H = const., are such as not to give rise to expressions for p, q, r as functions of x, y, z, x_0, y_0, z_0, H, as assumed in the theorem; whenever such expressions are obtainable, then we have $p\,dx+q\,dy+r\,dz$ an exact differential.

The proof in the case of a greater number of variables, say in the next case where the variables are x, y, z, w, p, q, r, s, would present more difficulty—but I have not proceeded further in the question.

It is worth while to put the two processes into connexion with each other: taking in each case the variables to be x, y, z, p, q, r, and the partial differential equation to be H = const.;

In the one case, a, b being conjugate solutions of $(\text{H}, \Theta) = 0$,

from the equations $\text{H} = \text{const.}$, $a = \text{const.}$, $b = \text{const.}$,

we find p, q, r functions of x, y, z, H, a, b:

and then $p\,dx + q\,dy + r\,dz$ is an exact differential.

In the other case, a, b, c, d, e being the solutions of $(\text{H}, \Theta) = 0$,

from the equations $\text{H} = \text{const.}$, $a = a_0$, $b = b_0$, $c = c_0$, $d = d_0$, $e = e_0$,

we find p, q, r functions of x_0, y_0, z_0, H:

and then $p\,dx + q\,dy + r\,dz$ is an exact differential.

It may be added that, if from the last mentioned equations we determine also p_0, q_0, r_0 as functions of x, y, z, x_0, y_0, z_0, then considering only H as a constant, we ought to have $p\,dx + q\,dy + r\,dz - p_0\,dx_0 - q_0\,dy_0 - r_0\,dz_0$ an exact differential; I have not examined the direct proof.

Cambridge, 28 *Nov.*, 1876.

657.

NOTE ON THE THEORY OF ELLIPTIC INTEGRALS.

[From the *Mathematische Annalen*, t. XII. (1877), pp. 143—146.]

THE equation

$$\frac{Mdy}{\sqrt{1-y^2\,.\,1-k^2y^2}}=\frac{dx}{\sqrt{1-x^2\,.\,1-k^2x^2}}$$

is integrable algebraically when M is rational: and so long as the modulus is arbitrary, then conversely, in order that the equation may be integrable algebraically, M must be rational. For particular values however of the modulus, the equation is integrable algebraically for values of the form M, or (what is the same thing) $\frac{1}{M}$, = a rational quantity $\pm$ square root of a negative rational quantity, say $=\frac{1}{p}(l+m\sqrt{-n})$, where l, m, n, p are integral and n is positive; we may for shortness call this a half-rational numerical value. The theory is considered by Abel in two Memoirs in the *Astr. Nach.* Nos. 138 & 147 (1828), being the Memoirs* XIII & XIV in the *Œuvres Complètes* (Christiania 1839). I here reproduce the investigation in a somewhat altered (and, as it appears to me, improved) form.

Putting the two differentials each $=du$, we have $x=\text{sn}(u+\alpha)$, $y=\text{sn}\left(\frac{u}{M}+\beta\right)$; and the question is whether there exists an algebraical relation between these functions, or, what is the same thing, an algebraical relation between the functions $x=\text{sn}\,u$ and $y=\text{sn}\,\frac{u}{M}$.

Suppose that A and B are independent periods of $\text{sn}\,u$; so that $\text{sn}(u+A)=\text{sn}\,u$, $\text{sn}(u+B)=\text{sn}\,u$, and that every other period is $=mA+nB$, where m and n are integers. Then if u has successively the values u, $u+A$, $u+2A$, etc., the value of x

[* They are the Memoirs XIX. and XX. in the *Œuvres Complètes*, t. I., Christiania, 1881.]

remains always the same, and if x and y are algebraically connected, y can have only a finite number of values: there are consequently integer values p', p'' for which $\operatorname{sn}\frac{1}{M}(u+p'A)=\operatorname{sn}\frac{1}{M}(u+p''A)$: or writing $u-p'A$ for u and putting $p''-p'=p$, there is an integer value p for which $\operatorname{sn}\frac{1}{M}(u+pA)=\operatorname{sn}\frac{1}{M}u$.

Similarly there is an integer value q for which $\operatorname{sn}\frac{1}{M}(u+qB)=\operatorname{sn}\frac{1}{M}u$; and we are at liberty to assume $q=p$; for if the original values are unequal, we have only in the place of each of them to substitute their least common multiple.

We have thus an integer p, for which

$$\operatorname{sn}\frac{1}{M}(u+pA)=\operatorname{sn}\frac{1}{M}u,$$

$$\operatorname{sn}\frac{1}{M}(u+pB)=\operatorname{sn}\frac{1}{M}u.$$

There are consequently integers m, n, r, s such that

$$\frac{pA}{M}=mA+nB,$$

$$\frac{pB}{M}=rA+sB,$$

equations which will constitute a single relation $\frac{p}{M}=m$, if $m=s$, $r=n=0$; but in every other case will be two independent relations. In the case first referred to, the modulus is arbitrary and M is rational.

But excluding this case, the equations give

$$B(mA+nB)=A(rA+sB),$$

or, what is the same thing,

$$rA^2-(m-s)AB-nB^2=0,$$

an equation which implies that the modulus has some one value out of a set of given values. The ratio $A:B$ of the two periods is of necessity imaginary, and hence the integers m, n, r, s must be such that $(m-s)^2+nr$ is negative.

The foregoing equations may be written

$$\left(m-\frac{p}{M}\right)A+\qquad nB=0,$$

$$rA+\left(s-\frac{p}{M}\right)B=0,$$

whence eliminating A and B we have

$$\left(m-\frac{p}{M}\right)\left(s-\frac{p}{M}\right)-nr=0,$$

that is,

$$\left(\frac{p}{M}\right)^2 - (m+s)\frac{p}{M} + ms - nr = 0,$$

and consequently

$$\frac{p}{M} = \tfrac{1}{2}(m+s) \pm \tfrac{1}{2}\sqrt{(m-s)^2 + nr},$$

where, by what precedes, the integer under the radical sign is negative: and we have thus the above mentioned theorem.

As a very general example, consider the two rational transformations

$$z = (x,\ u,\ v);\ \text{mod. eq. } Q(u,\ v) = 0;\ \frac{Ndz}{\sqrt{1-z^2 \,.\, 1-v^8z^2}} = \frac{dx}{\sqrt{1-x^2 \,.\, 1-u^8x^2}},$$

$$y = (z,\ v,\ w);\ \text{mod. eq. } P(v,\ w) = 0;\ \frac{Mdy}{\sqrt{1-y^2 \,.\, 1-w^8y^2}} = \frac{dz}{\sqrt{1-z^2 \,.\, 1-v^8z^2}}:$$

viz. z is taken to be a rational function of x, and of the modular fourth roots u, v; and y to be a rational function of z, and of the modular fourth roots v, w; the transformations being (to fix the ideas) of different orders. We have y a rational function of x, corresponding to the differential relation

$$\frac{MNdy}{\sqrt{1-y^2 \,.\, 1-w^8y^2}} = \frac{dx}{\sqrt{1-x^2 \,.\, 1-u^8x^2}}.$$

Suppose here $w^8 = u^8$, or say $w = \theta u$, θ being an eighth root of unity: we then have $Q(u,\ v) = 0$, $P(v,\ \theta u) = 0$, equations which determine u. The differential equation is then

$$\frac{MNdy}{\sqrt{1-y^2 \,.\, 1-u^8y^2}} = \frac{dx}{\sqrt{1-x^2 \,.\, 1-u^8x^2}},$$

an equation the algebraical integral of which is $y =$ a rational function of x as above: hence, by what precedes, we have

$$\frac{1}{MN} = \frac{1}{2p}\{m + s \pm \sqrt{(m-s)^2 + nr}\},$$

a half-rational numerical value, as above.

To explain what the algebraical theorem implied herein is, observe that the equations $Q(u,\ v) = 0$, $P(v,\ \theta u) = 0$, give for u an algebraical equation. Admitting θ as an adjoint radical, suppose that an irreducible factor is $\phi(u)$, and take u to be determined by the equation $\phi u = 0$; then v, and consequently also any rational function $\frac{1}{MN}$ of u, v, can be expressed as a rational integral function of u, of a degree which is at most equal to the degree of the function ϕu less unity. The theorem is that, in virtue of the equation $\phi u = 0$, this rational function of u becomes equal to a half-rational numerical value as above. Thus in a simple case, which actually presented itself, the equation $\phi u = 0$ was $u^2 - 4u + 1 = 0$; and $\frac{1}{MN}$ had the value $u - 2$, which in virtue of this equation becomes $= \pm\sqrt{-3}$.

Thus if the second transformation be the identity $z=y$, $w=v$, $M=1$: we have $v=\theta u$; and the equations are

$$y=(x,\ u,\ \theta u),\quad Q(u,\ \theta u)=0,\quad \frac{N\,dy}{\sqrt{1-y^2\,.\,1-u^8y^2}}=\frac{dx}{\sqrt{1-x^2\,.\,1-u^8x^2}}.$$

In particular, if the relation between y, x be given by the cubic transformation

$$y=\frac{\dfrac{v+2u^3}{v}x+\dfrac{u^6}{v^2}x^3}{1+vu^2(v+2u^3)x^2},$$

so that the modular equation $Q(u,\ v)=0$ is $u^4-v^4+2uv(1-u^2v^2)=0$; then, writing herein $v=\theta u$, and taking θ a prime eighth root of unity, that is, a root of $\theta^4+1=0$, we have

$$Q(u,\ \theta u)=-2\theta^3u^2(\theta u^2+\theta^2+u^4);$$

viz. disregarding the factor u^2, the equation for u is $u^4+\theta u^2+\theta^2=0$; or, if ω be an imaginary cube root of unity ($\omega^2+\omega+1=0$), this is $(u^2-\omega\theta)(u^2-\omega^2\theta)=0$; so that a value of u^2 is $u^2=-\omega\theta$.

Assuming then $\theta^4+1=0$, $v=\theta u$ and $u^2=-\omega\theta$, we have $(v+2u^3)v=\theta^3\omega(1+2\omega)$, $=\theta^3\omega(\omega-\omega^2)$; $\dfrac{v+2u^3}{v}=\omega-\omega^2$; $\dfrac{u^6}{v^2}=\omega^2$, $(v+2u^3)vu^2=-\omega^2(\omega-\omega^2)$, $u^8=\omega^4\theta^4=-\omega$; and the formula becomes

$$y=\frac{(\omega-\omega^2)x+\omega^2x^3}{1-\omega^2(\omega-\omega^2)x^2},$$

giving

$$\frac{dy}{\sqrt{1-y^2\,.\,1+\omega y^2}}=\frac{(\omega-\omega^2)\,dx}{\sqrt{1-x^2\,.\,1+\omega x^2}},$$

where as before $\omega^2+\omega+1=0$, a result which can be at once verified. We have $(\omega-\omega^2)^2=-3$; or the coefficient $\omega-\omega^2$ in the differential equation is $=\sqrt{-3}$, which is of the form mentioned in the general theorem.

We might, instead of $z=y$, have assumed between y and z the relation corresponding to any other of the six linear transformations of an elliptic integral, and thus have obtained in each case, for a properly determined value of the modulus, a cubic transformation to the same modulus.

Cambridge, 10 *April*, 1877.

658.

ON SOME FORMULÆ IN ELLIPTIC INTEGRALS.

[From the *Mathematische Annalen*, t. XII. (1877), pp. 369—374.]

I REPRODUCE in a modified form an investigation contained in the memoir, Zolotareff, "Sur la méthode d'intégration de M. Tchébychef," *Mathematische Annalen*, t. V. (1872), pp. 560—580.

Starting from the quartic

$$(a,\ b,\ c,\ d,\ e)(x,\ 1)^4,\ \ = a\,.\,x-\alpha\,.\,x-\beta\,.\,x-\gamma\,.\,x-\delta,$$

we derive from it the quartic

$$(a_1,\ b_1,\ c_1,\ d_1,\ e_1)(x_1,\ 1)^4 = a_1\,.\,x_1-\alpha_1\,.\,x_1-\beta_1\,.\,x_1-\gamma_1\,.\,x_1-\delta_1,$$

where, writing for shortness

$$\begin{aligned}\lambda &= -\alpha+\beta+\gamma-\delta,\\ \mu &= \ \ \ \alpha-\beta+\gamma-\delta,\\ \nu &= \ \ \ \alpha+\beta-\gamma-\delta,\end{aligned}$$

the roots of the new quartic are

$$\begin{aligned}\alpha_1 &= \theta+\frac{\mu\nu}{2\lambda},\\ \beta_1 &= \theta+\frac{\nu\lambda}{2\mu},\\ \gamma_1 &= \theta+\frac{\lambda\mu}{2\nu},\\ \delta_1 &= \theta,\end{aligned}$$

θ being arbitrary: the differences of the roots α_1, β_1, γ_1, δ_1 are, it will be observed, functions of the differences of the roots α, β, γ, δ.

We assume $a_1 = a = 1$, nevertheless retaining in the formulæ a_1 or a (each meaning 1), whenever, for the sake of homogeneity, it is convenient to do so. The relations between the remaining coefficients b_1, c_1, d_1, e_1, and b, c, d, e, are of course to be calculated from the formulæ $-4b = \Sigma\alpha$, $6c = \Sigma\alpha\beta$, &c., and the like formulæ $-4b_1 = \Sigma\alpha_1$, $6c_1 = \Sigma\alpha_1\beta_1$, &c. We thus have

$$-4b_1 = 4\theta + \tfrac{1}{2}\ \Sigma\frac{\mu\nu}{\lambda},$$

$$6c_1 = 6\theta^2 + \tfrac{3}{2}\theta\ \Sigma\frac{\mu\nu}{\lambda} + \tfrac{1}{4}\Sigma\lambda^2,$$

$$-4d_1 = 4\theta^3 + \tfrac{3}{2}\theta^2\,\Sigma\frac{\mu\nu}{\lambda} + \tfrac{1}{2}\theta\Sigma\lambda^2 + \tfrac{1}{8}\lambda\mu\nu,$$

$$e_1 = \theta^4 + \tfrac{1}{2}\theta^3\,\Sigma\frac{\mu\nu}{\lambda} + \tfrac{1}{4}\theta^2\,\Sigma\lambda^2 + \tfrac{1}{8}\theta\lambda\mu\nu,$$

where $\Sigma\dfrac{\mu\nu}{\lambda} = \dfrac{1}{\lambda\mu\nu}\Sigma\lambda^2\mu^2$.

Writing, for shortness,

$$\begin{aligned}
C &= ac - b^2,\\
D &= a^2d - 3abc + 2b^3,\\
E &= a^3e - 4a^2bd + 6ab^2c - 3b^4 = a^2I - 3C^2,\\
I &= ae - 4bd + 3c^2,\\
J &= ace - ad^2 - b^2e + 2bcd - c^3,\\
B &= \frac{-a^2I + 12C^2}{4D},
\end{aligned}$$

we have

$$\begin{aligned}
\Sigma\lambda &= -4(b+\delta),\\
\Sigma\lambda^2 &= -48C,\\
\Sigma\lambda\mu &= 24C + 8(b+\delta)^2,\\
\lambda\mu\nu &= 32D,\\
\Sigma\lambda^2\mu^2 &= 64(-a^2I + 12C^2),
\end{aligned}$$

where the last equation may be verified by means of the formula

$$(\Sigma\lambda\mu)^2 = \Sigma\lambda^2\mu^2 + 2\lambda\mu\nu\,\Sigma\lambda.$$

And we hence obtain

$$\begin{aligned}
a_1 &= 1,\\
b_1 &= -\theta - B,\\
c_1 &= \theta^2 + 2B\theta - 2C,\\
d_1 &= -\theta^3 - 3B\theta^2 + 6C\theta - D,\\
e_1 &= \theta^4 + 4B\theta^3 - 12C\theta^2 + 4D\theta.
\end{aligned}$$

And consequently

$$(a_1,\ b_1,\ c_1,\ d_1,\ e_1)(x_1,\ 1)^4 = (1,\ -B,\ -2C,\ -D,\ 0)(x_1-\theta,\ 1)^4.$$

Hence also

$$I_1 = a_1e_1 - 4b_1d_1 + 3c_1^2 = -4BD + 12C^2 = a^2I\,;$$

$$\begin{aligned} J_1 = a_1c_1e_1 - a_1d_1^2 - b_1^2e_1 + 2b_1c_1d_1 - c_1^3 &= -D^2 + 8C^3 - 4BCD \\ &= -D^2 + 8C^3 + C(a^2I - 12C^2) \\ &= a^2CI - 4C^3 - D^2 \\ &= a^3J\,; \end{aligned}$$

where, as regards this last equation $a^2CI - 4C^3 - D^2 = a^3J$, observe that C and D are the leading coefficients of the Hessian H and the cubicovariant Φ of the quartic function U, and hence that the identity $-\Phi^2 = JU^3 - IU^2\mathrm{H} + 4\mathrm{H}^3$, attending only to the term in x^6, becomes $-D^2 = a^3J - a^2CI + 4C^3$, which is the equation in question.

We thus have $I_1 = I$, $J_1 = J$; viz. the functions $(a, b, c, d, e)(x, 1)^4$, $(a_1, b_1, c_1, d_1, e_1)(x_1, 1)^4$, are linearly transformable the one into the other, and that by a unimodular substitution $x_1 = \rho x + \sigma$, $y_1 = \rho' x + \sigma'$, where $\rho\sigma' - \rho'\sigma = 1$. It may be remarked that we have $(a, b, c, d, e)(x, 1)^4 = (1, 0, C, D, E)(x + b, 1)^4$; and hence the theorem may be stated in the form: the quartic functions $(1, 0, C, D, E)(x, 1)^4$, and $(1, -B, -2C, -D, 0)(x_1, 1)^4$, are transformable the one into the other by a unimodular substitution: or again, substituting for E its value $a^2I - 3C^2$, $= -4BD + 9C^2$, the quartic functions

$$(1,\ 0,\ C,\ D,\ -4BD + 9C^2)(x,\ 1)^4, \text{ and } (1,\ -B,\ -2C,\ -D,\ 0)(x_1,\ 1)^4$$

are linearly transformable the one into the other by a unimodular substitution. In this last form B, C, D are arbitrary quantities; it is at once verified that the invariants I, J have the same values for the two functions respectively; and the theorem is thus self-evident.

Reverting to the expressions for $\alpha_1, \beta_1, \gamma_1, \delta_1$ we obtain

$$\alpha_1 - \delta_1 = \frac{\mu\nu}{2\lambda}\,;\quad \beta_1 - \gamma_1 = \frac{\lambda}{2\mu\nu}(\nu^2 - \mu^2),\quad = \frac{\alpha-\delta\,.\,\beta-\gamma}{\alpha_1 - \delta_1},$$

$$\beta_1 - \delta_1 = \frac{\nu\lambda}{2\mu}\,;\quad \gamma_1 - \alpha_1 = \frac{\mu}{2\nu\lambda}(\lambda^2 - \nu^2),\quad = \frac{\beta-\delta\,.\,\gamma-\alpha}{\beta_1 - \delta_1},$$

$$\gamma_1 - \delta_1 = \frac{\lambda\mu}{2\nu}\,;\quad \alpha_1 - \beta_1 = \frac{\nu}{2\lambda\mu}(\mu^2 - \lambda^2),\quad = \frac{\gamma-\delta\,.\,\alpha-\beta}{\gamma_1 - \delta_1}.$$

Hence also

$$\begin{aligned} &\alpha - \delta\,.\,\beta - \gamma,\quad \beta - \delta\,.\,\gamma - \alpha,\quad \gamma - \delta\,.\,\alpha - \beta \\ &= \alpha_1 - \delta_1\,.\,\beta_1 - \gamma_1,\quad \beta_1 - \delta_1\,.\,\gamma_1 - \alpha_1,\quad \gamma_1 - \delta_1\,.\,\alpha_1 - \beta_1, \end{aligned}$$

which agrees with the foregoing equations $I_1 = I$ and $J_1 = J$, since I, J are functions of the first set of quantities and I_1, J_1 the like functions of the second set; in fact, $I = \frac{1}{24}(P^2 + Q^2 + R^2)$, and $J = \frac{1}{432}(Q-R)(R-P)(P-Q)$, if for a moment the quantities are called P, Q, R.

We consider now the differential expression $\dfrac{dx}{\sqrt{x-\alpha\,.\,x-\beta\,.\,x-\gamma\,.\,x-\delta}}$; to transform this into the elliptic form, assume

$$k^2=-\frac{\alpha-\beta\,.\,\gamma-\delta}{\gamma-\alpha\,.\,\beta-\delta},\quad \operatorname{sn}^2 a=\frac{\gamma-\alpha}{\gamma-\delta};$$

(where a is of course not the coefficient, $=1$, heretofore represented by that letter: as a will only occur under the functional signs sn, cn, dn, there is no risk of ambiguity). And then further

$$x=\frac{\alpha\operatorname{sn}^2 u-\delta\operatorname{sn}^2 a}{\operatorname{sn}^2 u-\operatorname{sn}^2 a}.$$

Forming the equations

$$k^2\operatorname{sn}^2 a=-\frac{\alpha-\beta}{\beta-\delta},\quad k^2\operatorname{sn}^4 a=-\frac{\gamma-\alpha\,.\,\alpha-\beta}{\gamma-\delta\,.\,\beta-\delta},$$

we deduce without difficulty

$$\operatorname{sn}^2 a=\frac{\gamma-\alpha}{\gamma-\delta},\quad \frac{\operatorname{sn}^2 u}{\operatorname{sn}^2 a}=\frac{x-\delta}{x-\alpha},$$

$$\operatorname{cn}^2 a=\frac{\alpha-\delta}{\gamma-\delta},\quad \frac{\operatorname{cn}^2 u}{\operatorname{cn}^2 a}=\frac{x-\gamma}{x-\alpha},$$

$$\operatorname{dn}^2 a=\frac{\alpha-\delta}{\beta-\delta},\quad \frac{\operatorname{dn}^2 u}{\operatorname{dn}^2 a}=\frac{x-\beta}{x-\alpha},$$

$$1-k^2\operatorname{sn}^4 a=\frac{(\alpha-\delta)(-\alpha+\beta+\gamma-\delta)}{\beta-\delta\,.\,\gamma-\delta}=\frac{\lambda(\alpha-\delta)}{\beta-\delta\,.\,\gamma-\delta},$$

the use of which last equation will presently appear.

We hence obtain

$$2\operatorname{sn} u\operatorname{cn} u\operatorname{dn} u\,du=-(\alpha-\delta)\operatorname{sn}^2 a\frac{dx}{(x-\alpha)^2},$$

$$\operatorname{sn} u\operatorname{cn} u\operatorname{dn} u\quad=\quad\operatorname{sn} a\operatorname{cn} a\operatorname{dn} a\frac{\sqrt{x-\alpha\,.\,x-\beta\,.\,x-\gamma\,.\,x-\delta}}{(x-\alpha)^2},$$

and consequently

$$2du=-\frac{(\alpha-\delta)\operatorname{sn} a}{\operatorname{cn} a\operatorname{dn} a}\,\frac{dx}{\sqrt{x-\alpha\,.\,x-\beta\,.\,x-\gamma\,.\,x-\delta}},$$

or, reducing the coefficient,

$$\frac{dx}{\sqrt{x-\alpha\,.\,x-\beta\,.\,x-\gamma\,.\,x-\delta}}=\frac{-2}{\sqrt{\gamma-\alpha\,.\,\beta-\delta}}\,du,$$

which is the required formula.

We next have

$$\operatorname{sn}^2 2a=\frac{4\operatorname{sn}^2 a\operatorname{cn}^2 a\operatorname{dn}^2 a}{(1-k^2\operatorname{sn}^4 a)^2}=\frac{4\beta-\delta\,.\,\gamma-\alpha}{\lambda^2}=\frac{\gamma_1-\alpha_1}{\gamma_1-\delta_1},$$

in virtue of the foregoing values

$$\gamma_1-\alpha_1=\frac{2\mu}{\nu\lambda}(\beta-\delta)(\gamma-\alpha) \text{ and } \gamma_1-\delta_1=\frac{\lambda\mu}{2\nu}.$$

Moreover

$$k^2=-\frac{\alpha-\beta\,.\,\gamma-\delta}{\gamma-\alpha\,.\,\beta-\delta}=-\frac{\alpha_1-\beta_1\,.\,\gamma_1-\delta_1}{\gamma_1-\alpha_1\,.\,\beta_1-\delta_1}.$$

Hence the like formulæ with the same value of k^2, and with $2a$ in place of a, will be applicable to the like differential expression in x_1: viz. assuming

$$x_1=\frac{\alpha_1\operatorname{sn}^2 u_1-\delta_1\operatorname{sn}^2 2a}{\operatorname{sn}^2 u_1-\operatorname{sn}^2 2a}$$

we have

$$\frac{dx_1}{\sqrt{x_1-\alpha_1\,.\,x_1-\beta_1\,.\,x_1-\gamma_1\,.\,x_1-\delta_1}}=\frac{-2}{\sqrt{\gamma_1-\alpha_1\,.\,\beta_1-\delta_1}}du_1.$$

We have thus the integral of the differential equation

$$\frac{dx_1}{\sqrt{x_1-\alpha_1\,.\,x_1-\beta_1\,.\,x_1-\gamma_1\,.\,x_1-\delta_1}}=\frac{dx}{\sqrt{x-\alpha\,.\,x-\beta\,.\,x-\gamma\,.\,x-\delta}}$$

(the two quartic functions being of course connected as before); viz. assuming x, x_1 functions of u, u_1 respectively as above and recollecting that $\gamma_1-\alpha_1\,.\,\beta_1-\delta_1=\gamma-\alpha\,.\,\beta-\delta$, we have $du_1=du$; and therefore $u_1=u+f$ (f an arbitrary constant); the required integral is thus given by the equations

$$\frac{\operatorname{sn}^2 u}{\operatorname{sn}^2 a}=\frac{x-\delta}{x-\alpha};\quad \frac{\operatorname{sn}^2(u+f)}{\operatorname{sn}^2 2a}=\frac{x_1-\delta_1}{x_1-\alpha_1};\quad (f \text{ the constant of integration}).$$

Using the formula

$$\operatorname{sn}(u+f)=\frac{\operatorname{sn}^2 u-\operatorname{sn}^2 f}{\operatorname{sn} u\operatorname{cn} f\operatorname{dn} f-\operatorname{sn} f\operatorname{cn} u\operatorname{dn} u},$$

we obtain

$$\frac{x_1-\delta_1}{x_1-\alpha_1}\operatorname{sn}^2 2a=\frac{\{(x-\delta)\operatorname{sn}^2 a-(x-\alpha)\operatorname{sn}^2 f\}^2}{\{\sqrt{x-\alpha\,.\,x-\delta}\operatorname{sn} a\operatorname{cn} f\operatorname{dn} f-\sqrt{x-\beta\,.\,x-\gamma}\operatorname{sn} f\operatorname{cn} a\operatorname{dn} a\}^2},$$

which is the general integral.

We obtain a particular integral of a very simple form by assuming $f=a$, viz. this is

$$\frac{x_1-\delta_1}{x_1-\alpha_1}\operatorname{sn}^2 2a=\frac{\operatorname{sn}^2 a}{\operatorname{cn}^2 a\operatorname{dn}^2 a}\,\frac{(\alpha-\delta)^2}{\{\sqrt{x-\alpha\,.\,x-\delta}-\sqrt{x-\beta\,.\,x-\gamma}\}^2};$$

this is

$$\frac{x_1-\delta_1}{x_1-\alpha_1}\,\frac{\gamma_1-\alpha_1}{\gamma_1-\delta_1}=\frac{\gamma-\alpha\,.\,\beta-\delta}{\{\sqrt{x-\alpha\,.\,x-\delta}-\sqrt{x-\beta\,.\,x-\gamma}\}^2},$$

or writing $\gamma-\alpha\,.\,\beta-\delta=\gamma_1-\alpha_1\,.\,\beta_1-\delta_1$, reducing and inverting, we have

$$\frac{x_1-\alpha_1}{x_1-\delta_1}=\frac{1}{\beta_1-\delta_1\,.\,\gamma_1-\delta_1}\{\sqrt{x-\alpha\,.\,x-\delta}-\sqrt{x-\beta\,.\,x-\gamma}\}^2,$$

which may also be written in the equivalent forms

$$\frac{x_1-\beta_1}{x_1-\delta_1}=\frac{1}{\gamma_1-\delta_1\,.\,\alpha_1-\delta_1}\{\sqrt{x-\beta\,.\,x-\delta}-\sqrt{x-\gamma\,.\,x-\alpha}\}^2,$$

$$\frac{x_1-\gamma_1}{x_1-\delta_1}=\frac{1}{\alpha_1-\delta_1\,.\,\beta_1-\delta_1}\{\sqrt{x-\gamma\,.\,x-\delta}-\sqrt{x-\alpha\,.\,x-\beta}\}^2.$$

In fact, from the first equation we have

$$\frac{\alpha_1-\delta_1\,.\,\beta_1-\delta_1\,.\,\gamma_1-\delta_1}{x_1-\delta_1}=(\beta_1-\delta_1)(\gamma_1-\delta_1)-\{\sqrt{x-\alpha\,.\,x-\delta}-\sqrt{x-\beta\,.\,x-\gamma}\}^2,$$

where the expression on the right-hand side is

$$\delta_1^2-\delta_1(\alpha_1+\beta_1+\gamma_1)+\alpha_1\delta_1+\beta_1\gamma_1-2x^2+x(\alpha+\beta+\gamma+\delta)-\alpha\delta-\beta\gamma+2\sqrt{X},$$

X having here the value

$$X=x-\alpha\,.\,x-\beta\,.\,x-\gamma\,.\,x-\delta.$$

Writing for a moment

$$\begin{aligned} P&=\alpha\delta+\beta\gamma, & P_1&=\alpha_1\delta_1+\beta_1\gamma_1,\\ Q&=\beta\delta+\gamma\alpha, & Q_1&=\beta_1\delta_1+\gamma_1\alpha_1,\\ R&=\gamma\delta+\alpha\beta, & R_1&=\gamma_1\delta_1+\alpha_1\beta_1,\end{aligned}$$

then, by what precedes, Q_1-R_1, R_1-P_1, P_1-Q_1 are equal to $Q-R$, $R-P$, $P-Q$ respectively; that is, $P_1-P=Q_1-Q=R_1-R$, $=$(suppose) Ω, a function symmetrical in regard to $\alpha_1, \beta_1, \gamma_1$; α, β, γ: the equation therefore is

$$\frac{\alpha_1-\delta_1\,.\,\beta_1-\delta_1\,.\,\gamma_1-\delta_1}{x_1-\delta_1}=\delta_1(\delta_1-\alpha_1-\beta_1-\gamma_1)-2x^2+x(\alpha+\beta+\gamma+\delta)+2\sqrt{X}+\Omega,$$

or the relation is symmetrical in regard to $\alpha_1, \beta_1, \gamma_1$; α, β, γ: and the first form implies therefore each of the other two forms.

Cambridge, 8 *May*, 1877.

659.

A THEOREM ON GROUPS.

[From the *Mathematische Annalen*, t. XIII. (1878), pp. 561—565.]

THE following theorem is very simple; but it seems to belong to a class of theorems, the investigation of which is desirable.

I consider a substitution-group of a given order upon a given number of letters; and I seek to *double* the group, that is to derive from it a group of twice the order upon twice the number of letters. This can be effected for *any* group, in a manner which is self-evident and in nowise interesting: but in a different manner for a *commutative* group (or group such that any two of its substitutions satisfy the condition $AB = BA$): it is to be observed that the double group is not in general commutative.

Let the letters of the original group be $abcde\ldots$, we may for shortness write $U = abcde\ldots$; and take U as the primitive arrangement: and let the group then be $1, A, B,\ldots$ where $A, B,\ldots$ represent substitutions: the corresponding arrangements are $U, AU, BU,\ldots$ and these may for shortness be represented by $1, A, B,\ldots$; viz. $1, A, B,\ldots$ represent, properly and in the first instance, substitutions; but when it is explained that they represent arrangements, then they represent the arrangements $U, AU, BU,\ldots$.

For the double group the letters are taken to be $a_1b_1c_1d_1e_1\ldots$ and $a_2b_2c_2d_2e_2\ldots$, $= U_1$ and U_2 suppose, and U_1U_2 is regarded as the primitive arrangement; A_1 and A_2 denote the same substitutions in regard to U_1 and U_2 respectively, that A denotes in regard to U: and so for B_1, B_2, etc.; moreover 12 denotes the substitution $(a_1a_2)(b_1b_2)(c_1c_2)(d_1d_2)(e_1e_2)\ldots$, or interchange of the suffixes 1 and 2. The substitutions A_1, A_2, or any powers of these $A_1{}^\alpha$, $A_2{}^\beta$, are obviously commutative; applying them to the primitive arrangement U_1U_2, we have $A_1{}^\alpha A_2{}^\beta U_1U_2$ and $A_2{}^\beta A_1{}^\alpha U_1U_2$ each $= A_1{}^\alpha U_1 A_2{}^\beta U_2$. But $A_1{}^\alpha$, $A_2{}^\beta$ are not commutative with 12: we have for instance $12A_1{}^\alpha \,.\, U_1U_2 = 12A_1{}^\alpha U_1 \,.\, U_2 = A_1{}^\alpha U_2 \,.\, U_1$, but $A_1{}^\alpha 12U_1U_2 = A_1{}^\alpha \,.\, U_2U_1 = U_2 \,.\, A_1{}^\alpha U_1$. If instead of the substitutions we consider the arrangements obtained by operating upon U_1U_2, then we may for shortness consider for instance A_1A_2 as denoting the arrangement $A_1U_1 . A_2U_2$. But observe that in this use of the symbols the A_1, A_2 are not commutative, A_2A_1 would denote the different arrangement $A_2U_2 . A_1U_1$: in this use of the symbols, 1 would denote U_1U_2, and 12 would denote U_2U_1, but it would be clearer to use 12, 21 as denoting U_1U_2 and U_2U_1 respectively.

These explanations having been given, I remark that in every case the substitution-group 1, A, B,... gives the double group

$$\begin{array}{ccc} 1, & A_1A_2, & B_1B_2, \ldots \\ 12, & 12A_1A_2, & 12B_1B_2, \ldots \end{array}$$

as is at once seen to be true: but further when the original group 1, A, B,... is commutative, then if m be any integer number, such that $m^2 \equiv 1$ (mod. the order of the original group), we have also the double group

$$\begin{array}{ccc} 1, & A_1A_2^m, & B_1B_2^m, \ldots \\ 12, & 12A_1A_2^m, & 12B_1B_2^m, \ldots \end{array}$$

where of course if the order of the original group ($=\mu$ suppose) be prime, we have $m \equiv 1$ or else $m \equiv -1$ (mod. μ), say $m = 1$ or $\mu - 1$; but if the order μ be composite, then the number of solutions may be greater.

The condition in order to the existence of the double group of course is that, in the system of substitutions just written down, the combination of any two substitutions may give a substitution of the system. And this is in fact the case in virtue of the formulæ

$$1^\circ. \quad A_1A_2^m . \quad B_1B_2^m = A_1B_1(A_2B_2)^m,$$

$$2^\circ. \quad A_1A_2^m . 12B_1B_2^m = 12A_1^mB_1(A_2^mB_2)^m,$$

$$3^\circ. \quad 12A_1A_2^m . \quad B_1B_2^m = 12(A_1B_1)(A_2B_2)^m,$$

$$4^\circ. \quad 12A_1A_2^m . 12B_1B_2^m = A_1^mB_1(A_2^mB_2)^m,$$

inasmuch as 1, A, B,... being a group, AB and A^mB are each of them a substitution of the group, $= C$ suppose; we have of course in like manner $A_1B_1 = C_1$, $A_2B_2 = C_2$, etc., and the right-hand sides of the four formulæ are thus of the forms $C_1C_2^m$, $12C_1C_2^m$, $12C_1C_2^m$, $C_1C_2^m$ respectively, viz. these are substitutions of the system.

To prove for instance the formula 2°, considering the arrangements obtained by operating upon U_1U_2, we have

$$B_1B_2^mU_1U_2 = B_1B_2^m, \; 12B_1B_2^mU_1U_2 = B_2B_1^m, \quad A_1A_2^m \, 12B_1B_2^mU_1U_2 = A_2^mB_2 \, A_1B_1^m,$$

where of course the expressions on the right-hand side denote arrangements. By reason that the original group is commutative $(A^mB)^m$ is $= A^{m^2}B^m$ or since $m^2 \equiv 1$ (mod. μ) this is $= AB^m$; hence also $(A_2^mB_2)^m = A_2B_2^m$: hence, considering as before the arrangements obtained by operating on U_1U_2, we have

$$(A_2^m B_2)^m U_1U_2 = 1 \,.\, A_2B_2^m;\quad A_1^m B_1 (A_2^m B_2)^m U_1U_2 = A_1^m B_1 A_2 B_2^m,$$

and

$$12A_1^m B_1 (A_2^m B_2)^m U_1U_2 = A_2^m B_2 A_1 B_1^m,$$

where of course the right-hand sides denote arrangements. Hence in the formula 2°, the two substitutions operating on U_1U_2 give each of them the same arrangement $A_2^m B_2 A_1 B_1^m$, that is, the two substitutions are equal. And similarly the other formulæ 1°, 3°, 4° may be proved.

By interchanging A and B, in the formulæ I obtain

$$1^\circ.\quad A_1A_2^m \,.\, B_1B_2^m = A_1B_1 (A_2B_2)^m;$$
$$B_1B_2^m \,.\, A_1A_2^m = B_1A_1 (B_2A_2)^m = A_1B_1 (A_2B_2)^m,$$

which is

$$= A_1A_2^m \,.\, B_1B_2^m;$$

$$2^\circ \text{ and } 3^\circ.\quad A_1A_2^m \,.\, 12B_1B_2^m = 12A_1^m B_1 (A_2^m B_2)^m;$$
$$12B_1B_2^m \,.\, A_1A_2^m = 12B_1A_1 (B_2A_2)^m = 12A_1B_1 (A_2B_2)^m,$$

which is not

$$= A_1A_2^m \,.\, 12B_1B_2^m;$$

$$3^\circ \text{ and } 2^\circ.\quad 12A_1A_2^m \,.\, B_1B_2^m = 12A_1B_1 (A_2B_2)^m;$$
$$B_1B_2^m \,.\, 12A_1A_2^m = 12A_1B_1^m (A_2B_2^m)^m = 12A_1B_1^m A_2^m B_2,$$

which is not

$$= 12A_1A_2^m \,.\, B_1B_2^m;$$

$$4^\circ.\quad 12A_1A_2^m \,.\, 12B_1B_2^m = A_1^m B_1 (A_2^m B_2)^m;$$
$$12B_1B_2^m \,.\, 12A_1A_2^m = A_1B_1^m (A_2B_2^m)^m = (A_1^m B_1)^m A_2^m B_2,$$

which is not

$$= 12A_1A_2^m \,.\, 12B_1B_2^m.$$

That is, in the double group any two substitutions of the form $A_1A_2^m$ are commutative, but a substitution of this form is not in general commutative with a substitution of the form $12B_1B_2^m$, nor are two substitutions of the last-mentioned form $12A_1A_2^m$ in general commutative with each other; hence the double group is not in general commutative.

In the formula 4°, writing $B = A$, we have

$$(12A_1A_2^m)^2 = A_1^{m+1} A_2^{m^2+m} = A_1^{m+1} \,.\, A_2^{m+1};$$

hence, if λ is the least integer value such that

$$\lambda (m+1) \equiv 0 \pmod{\mu},$$

we have $(12A_1A_2^m)^{2\lambda} = 1$, viz. in the double group the substitutions of the second row are each of them of an order not exceeding 2λ, the substitution 12 being of course of the order 2. In particular, if $m = \mu - 1$, then $\lambda = 1$: and the substitutions of the second row are each of them of the order 2.

As the most simple instance of the theorem, suppose that the original group is the group 1, (abc), (acb), or say 1, Θ, Θ^2, of the cyclical substitutions upon the 3 letters abc. Here $m^2 \equiv 1$ (mod. 3) or except $m=1$ the only solution is $m=2$, and thence $\lambda=1$. The double group is a group of the order 6 on the letters $a_1b_1c_1a_2b_2c_2$: viz. writing $\Theta=(abc)$, and therefore $\Theta_1=(a_1b_1c_1)$, $\Theta_1{}^2=(a_1c_1b_1)$, $\Theta_2=(a_2b_2c_2)$, $\Theta_2{}^2=(a_2c_2b_2)$, also writing $12=\alpha$, the substitutions are

$$1,\quad \Theta_1\Theta_2{}^2,\quad \Theta_1{}^2\Theta_2,$$

$$\alpha,\quad \alpha\Theta_1\Theta_2{}^2,\quad \alpha\Theta_1{}^2\Theta_2,$$

the arrangements corresponding to the second row of substitutions are $a_2b_2c_2a_1b_1c_1$, $b_2c_2a_2c_1a_1b_1$, $c_2a_2b_2b_1c_1a_1$, viz. the substitutions are $(a_1a_2)(b_1b_2)(c_1c_2)$, $(a_1b_2)(b_1c_2)(c_1a_2)$, $(a_1c_2)(b_1a_2)(c_1b_2)$, each of them of the second order as they should be.

I take the opportunity of mentioning a further theorem. Let μ be the order of the group, and a the order of any term A thereof, a being of course a submultiple of μ: and let the term A be called quasi-positive when $\mu\left(1-\frac{1}{a}\right)$ is even, quasi-negative when $\mu\left(1-\frac{1}{a}\right)$ is odd. The theorem is that the product of two quasi-positive terms, or of two quasi-negative terms, is quasi-positive; but the product of a quasi-positive term and a quasi-negative term is quasi-negative. And it follows hence that, either the terms of a group are all quasi-positive, or else one half of them are quasi-positive and the other half of them are quasi-negative.

The proof is very simple: a term A of the group operating on the μ terms $(1, A, B, C, \ldots)$ of the group, gives these same terms in a different order, or it may be regarded as a substitution upon the μ symbols $1, A, B, C, \ldots$; so regarded it is a *regular* substitution (this is a fundamental theorem, which I do not stop to prove), and hence since it must be of the order a it is a substitution composed of $\frac{\mu}{a}$ cycles, each of a letters. But in general a substitution is positive or negative according as it is equivalent to an even or an odd number of inversions; a cyclical substitution upon a letters is positive or negative according as $a-1$ is even or odd; and the substitution composed of the $\frac{\mu}{a}$ cycles is positive or negative according as $\frac{\mu}{a}(a-1)$, that is, $\mu\left(1-\frac{1}{a}\right)$, is even or odd. Hence by the foregoing definition, the term A, according as it is quasi-positive or quasi-negative, corresponds to a positive substitution or to a negative substitution; and such terms combine together in like manner with positive and negative substitutions.

Cambridge, 3rd April, 1878.

660.

ON THE CORRESPONDENCE OF HOMOGRAPHIES AND ROTATIONS.

[From the *Mathematische Annalen*, t. XV. (1879), pp. 238—240.]

IT is a fundamental notion in Prof. Klein's theory of the "Icosahedron" that homographies correspond to rotations (of a solid body about a fixed point): in such wise that, considering the homographies which correspond to two given rotations, the homography compounded of these corresponds to the rotation compounded of the two rotations.

Say the two homographies are $A + Bp + Cq + Dpq = 0$, $A_1 + B_1q + C_1r + D_1qr = 0$, then, eliminating q, the compound homography is $A_2 + B_2p + C_2r + D_2pr = 0$, where

$$A_2,\ B_2,\ C_2,\ D_2 = B_1A - A_1C,\ B_1B - A_1D,\ D_1A - C_1C,\ D_1B - C_1D;$$

and the theorem is that, corresponding to these, we have rotations depending on the parameters (λ, μ, ν), $(\lambda_1, \mu_1, \nu_1)$, $(\lambda_2, \mu_2, \nu_2)$ respectively, such that the third rotation is that compounded of the first and second rotations. The question arises to find the expression for the parameters of the homography in terms of the parameters of the corresponding rotation.

The rotation (λ, μ, ν) is taken to denote a rotation through an angle ϑ about an axis the inclinations of which to the axes of coordinates are f, g, h, the values of λ, μ, ν then being $= \tan\frac{1}{2}\vartheta\cos f$, $\tan\frac{1}{2}\vartheta\cos g$, $\tan\frac{1}{2}\vartheta\cos h$ respectively: $(\lambda_1, \mu_1, \nu_1)$ and $(\lambda_2, \mu_2, \nu_2)$ have of course the like significations; and then, if (λ, μ, ν) refer to the first rotation, and $(\lambda_1, \mu_1, \nu_1)$ to the second rotation, the values of $(\lambda_2, \mu_2, \nu_2)$ for the rotation compounded of these are taken to be*:

$$\lambda_2 = \lambda + \lambda_1 + \mu\nu_1 - \mu_1\nu,$$
$$\mu_2 = \mu + \mu_1 + \nu\lambda_1 - \nu_1\lambda,$$
$$\nu_2 = \nu + \nu_1 + \lambda\mu_1 - \lambda_1\mu,$$

* The numerators might equally well have been $\lambda + \lambda_1 - (\mu\nu_1 - \mu_1\nu)$, etc., but there is no essential difference: we pass from one set of formulæ to the other by reversing the signs of all the symbols: and hence, by properly fixing the sense of the rotations, the signs may be made to be + as in the text. Assuming this to be so, if we then reverse the order of the component rotations, we have for the new compound rotation the like formulæ with the signs − instead of +; but this in passing. The formulæ, virtually due to Rodrigues, are given in my paper "On the motion of rotation of a solid body," *Camb. Math. Journal*, t. III. (1843), [6].

each divided by

$$1-\lambda\lambda_1-\mu\mu_1-\nu\nu_1;$$

and if we then write for λ, μ, ν, the quotients x, y, z each divided by w, and in like manner for λ_1, μ_1, ν_1 and λ_2, μ_2, ν_2, the quotients x_1, y_1, z_1 each divided by w_1, and x_2, y_2, z_2 each divided by w_2, the formulæ for the composition of the rotations are

$$\begin{aligned}
x_2 &= xw_1 + x_1w + yz_1 - y_1z,\\
y_2 &= yw_1 + y_1w + zx_1 - z_1x,\\
z_2 &= zw_1 + z_1w + xy_1 - x_1y,\\
w_2 &= ww_1 - xx_1 - yy_1 - zz_1;
\end{aligned}$$

and the question is to express A, B, C, D as functions of (x, y, z, w), such that A_1, B_1, C_1, D_1 denoting the like functions of (x_1, y_1, z_1, w_1), A_2, B_2, C_2, D_2 shall denote the like functions of (x_2, y_2, z_2, w_2).

It is found that the required conditions are satisfied by assuming

$$A,\ B,\ C,\ D = ix-y,\quad -iz+w,\quad -iz-w,\quad -ix-y,$$

(where $i=\sqrt{-1}$ as usual): in fact, we then have

$$\begin{aligned}
A_2 &= B_1A - A_1C\\
&= (-iz_1+w_1)(ix-y)-(ix_1-y_1)(-iz-w)\\
&= i\,(xw_1+x_1w+yz_1-y_1z)-(yw_1+y_1w+zx_1-z_1x)\\
&= -y_2+ix_2,
\end{aligned}$$

as it should be: and we verify in like manner the values of B_2, C_2 and D_2 respectively.

The result consequently is that we have the homography

$$(ix-y)+(-iz+w)p+(-iz-w)q+(-ix-y)pq=0$$

corresponding to the rotation $\left(\frac{x}{w}, \frac{y}{w}, \frac{z}{w}\right)$: where $\frac{x}{w}, \frac{y}{w}, \frac{z}{w}$ are the parameters of rotation, $\tan\frac{1}{2}\vartheta\cos f$, $\tan\frac{1}{2}\vartheta\cos g$, $\tan\frac{1}{2}\vartheta\cos h$.

I remark as regards the geometrical theory that, if we consider two lines J and K fixed in space, and a third line L fixed in the solid body and moveable with it; then, for any given position of the solid body, the three lines J, K, L are directrices of a hyperboloid, the generatrices whereof meet each of the three lines: and these generatrices determine, say on the fixed lines J and K, two series of points corresponding homographically to each other: that is, corresponding to any given position of the solid body we have a homography. But it is not immediately obvious how we can thence obtain the foregoing analytical formulæ.

Cambridge, 3 *April*, 1879.

661.

ON THE DOUBLE ϑ-FUNCTIONS.

[From the *Proceedings of the London Mathematical Society*, vol. IX. (1878), pp. 29, 30.]

PROF. CAYLEY gave an account of researches* on which he is engaged upon the double ϑ-functions. In regard to these, he establishes in a strictly analogous manner the theory of the single ϑ-functions, the process for the single functions being in fact as follows:—Considering u, x as connected by the differential relation

$$\delta u = \frac{\delta x}{\sqrt{a-x\,.\,b-x\,.\,c-x\,.\,d-x}},$$

then, if A, B, C, D, Ω denote functions of u, viz. for shortness, the single letters are used, instead of writing them as functional symbols, $A(u)$, $B(u)$, &c., then, by way of definition of these functions (called, the first four of them ϑ-functions, and the last an ω-function), we assume

$$A,\ B,\ C,\ D = \Omega\sqrt{a-x},\ \Omega\sqrt{b-x},\ \Omega\sqrt{c-x},\ \Omega\sqrt{d-x}$$

respectively, together with one other equation, as presently mentioned. Without in any wise defining the meaning of Ω, we then obtain a set of equations of the form

$$A\delta B - B\delta A = \Omega^2\sqrt{c-x\,.\,d-x}\,\delta u,$$

(mere constant coefficients are omitted), or, what is the same thing,

$$A\delta B - B\delta A = CD\,\delta u,$$

which are differential equations defining the nature of the ratio-functions $A : B : C : D$. If, proceeding to second differential coefficients, we attempt to form the expressions for $A\delta^2 A - (\delta A)^2$, &c., these involve multiples of $\Omega\delta^2\Omega - (\delta\Omega)^2$; in order to obtain a con-

[* See paper, number 665.]

venient form, we assume $\Omega\delta^2\Omega - (\delta\Omega)^2 = \Omega^2 M (\delta u)^2$, where M is a function of x. We thus obtain an equation $A\delta^2 A - (\delta A)^2 = \Omega^2\mathfrak{A}(\delta u)^2$, where the value of $\mathfrak{A}$ depends upon that of M. The value of M has to be taken so as to simplify as much as may be the expression of $\mathfrak{A}$, but so that M shall be a symmetrical function of the constants a, b, c, d: this last condition is assigned in order that the like simplification may present itself in the analogous relations $B\delta^2 B - (\delta B)^2 = \Omega^2\mathfrak{B}(\delta u)^2$, &c. The proper expression of M is found to be

$$M = -2x^2 + x(a+b+c+d) + a^2 + b^2 + c^2 + d^2 - 2bc - 2ca - 2ab - 2ad - 2bd - 2cd,$$

viz. M having this value, the one other equation above referred to is

$$\Omega\delta^2\Omega - (\delta\Omega)^2 = \Omega^2 M (\delta u)^2;$$

and we then have a system of four equations such as

$$A\delta^2 A - (\delta A)^2 = \Omega^2\mathfrak{A}(\delta u)^2,$$

where $\mathfrak{A}$ is a linear function of x, and where consequently $\Omega^2\mathfrak{A}$ can be expressed as a linear function of any two of the four squares A^2, B^2, C^2, D^2.

To establish the connexion with the Jacobian H and Θ functions, the differential relation between u, x may be taken to be

$$\delta u = \frac{\delta x}{\sqrt{x \,.\, 1-x \,.\, 1-k^2 x}};$$

viz. we have here $d = \infty$, and to adapt the formulæ to this value it is necessary to write $\frac{u}{\sqrt{d}}$ instead of u, and make other easy changes. The result is that Ω differs from D by a constant factor only, so that, instead of the five functions A, B, C, D, Ω, we have only the four functions A, B, C, D. The equation $\Omega\delta^2\Omega - (\delta\Omega)^2 = \Omega^2 M(\delta u)^2$ is replaced by an equation $D\delta^2 D - (\delta D)^2 = D^2\mathfrak{D}(\delta u)^2$, or say $\delta^2(\log D) = \mathfrak{D}(\delta u)^2$, which gives a result of the form

$$D = e^{u + \lambda^2 \int \delta u \int \delta u \frac{A^2}{D^2}},$$

showing that D differs from Jacobi's $\Theta(u)$ only by an exponential factor of the form $Ce^{\lambda u^2}$. And it then further appears that A, B, C differ only by factors of the like form from the three numerator functions $\mathrm{H}(u)$, $\mathrm{H}(u+K)$, $\Theta(u+K)$, so that, neglecting constant factors, the functions

$$\frac{A}{D},\ \frac{B}{D},\ \frac{C}{D} \text{ are equal to } \frac{\mathrm{H}(u)}{\Theta(u)},\ \frac{\mathrm{H}(u+K)}{\Theta(u)},\ \frac{\Theta(u+K)}{\Theta(u)};$$

that is, to the elliptic functions sn u, cn u, dn u.

662.

ON THE DOUBLE Θ-FUNCTIONS IN CONNEXION WITH A 16-NODAL QUARTIC SURFACE.

[From the *Journal für die reine und angewandte Mathematik* (Crelle), t. LXXXIII. (1877), pp. 210—219.]

I HAVE before me Göpel's memoir, "Theoriae transcendentium Abelianarum primi ordinis adumbratio levis," Crelle's Journal, t. XXXV. (1847), pp. 277—312. Writing P_1, P_2, P_3, etc., in place of his P', P'', P''', etc., also α, β, γ, δ, X', Y', Z', W', in place of his t, u, v, w, T, U, V, W, the system of 16 equations (given p. 287) is

$$\begin{array}{rl}
(1) & P^2 = (\alpha,\ -\beta,\ -\gamma,\ \ \delta)(X', Y', Z', W'),\\
(4) & P_1^2 = (\alpha,\ \ \beta,\ -\gamma,\ -\delta)(X', Y', Z', W'),\\
(9) & P_2^2 = (\alpha,\ -\beta,\ \ \gamma,\ -\delta)(X', Y', Z', W'),\\
(12) & P_3^2 = (\alpha,\ \ \beta,\ \ \gamma,\ \ \delta)(X', Y', Z', W'),\\
(3) & Q^2 = (\beta,\ -\alpha,\ -\delta,\ \ \gamma)(X', Y', Z', W'),\\
(2) & Q_1^2 = (\beta,\ \ \alpha,\ -\delta,\ -\gamma)(X', Y', Z', W'),\\
(11) & Q_2^2 = (\beta,\ -\alpha,\ \ \delta,\ -\gamma)(X', Y', Z', W'),\\
(10) & Q_3^2 = (\beta,\ \ \alpha,\ \ \delta,\ \ \gamma)(X', Y', Z', W'),\\
(13) & R^2 = (\gamma,\ -\delta,\ -\alpha,\ \ \beta)(X', Y', Z', W'),\\
(16) & R_1^2 = (\gamma,\ \ \delta,\ -\alpha,\ -\beta)(X', Y', Z', W'),\\
(5) & R_2^2 = (\gamma,\ -\delta,\ \ \alpha,\ -\beta)(X', Y', Z', W'),\\
(8) & R_3^2 = (\gamma,\ \ \delta,\ \ \alpha,\ \ \beta)(X', Y', Z', W'),\\
(15) & S^2 = (\delta,\ -\gamma,\ -\beta,\ \ \alpha)(X', Y', Z', W'),\\
(14) & S_1^2 = (\delta,\ \ \gamma,\ -\beta,\ -\alpha)(X', Y', Z', W'),\\
(7) & S_2^2 = (\delta,\ -\gamma,\ \ \beta,\ -\alpha)(X', Y', Z', W'),\\
(6) & S_3^2 = (\delta,\ \ \gamma,\ \ \beta,\ \ \alpha)(X', Y', Z', W');
\end{array}$$

viz. we have $P^2 = \alpha X' - \beta Y' - \gamma Z' + \delta W'$, etc. The reason for the apparently arbitrary manner in which I have numbered these equations, will appear further on. I recall that the sixteen double Θ-functions, that is, Θ-functions of two arguments u, u', are*

$$\begin{matrix} P, & P_1, & P_2, & P_3, \\ iQ, & Q_1, & iQ_2, & Q_3, \\ iR, & iR_1, & R_2, & R_3, \\ S, & iS_1, & iS_2, & S_3, \end{matrix}$$

the factor i, $=\sqrt{-1}$, being introduced in regard to the six functions which are odd functions of the arguments. But disregarding the sign, I speak of P^2, P_1^2, ..., Q^2, etc., as the squared functions, or simply as the squares; α, β, γ, δ are constants, depending of course on the parameters of the Θ-functions; X', Y', Z', W', which are however to be eliminated, are themselves Θ-functions to a different set of parameters: the 16 equations express that the squared functions P^2, P_1^2, etc., are linear functions of X', Y', Z', W', and they consequently serve to obtain linear relations between the squared functions: viz. by means of them, Göpel expresses the remaining 12 squares, each in terms of the selected four squares P_1^2, P_2^2, S_1^2, S_2^2, which are linearly independent: that is, he obtains linear relations between five squares, and he seems to have assumed that there were not any linear relations between fewer than five squares.

It appears however by Rosenhain's "Mémoire sur les fonctions de deux variables et à quatre périodes etc.", *Mém. Sav. Étrangers*, t. XI. (1851), pp. 364—468, that there are, in fact, linear relations between four squares, viz. that there exist sixes of squares such that, selecting at pleasure any three out of the six, each of the remaining three squares can be expressed as a linear function of these three squares. Knowing this result, it is easy to verify it by means of the sixteen equations, and moreover to show that there are in all 16 such sixes: these are shown by the following scheme which I copy from Kummer's memoir "Ueber die algebraischen Strahlensysteme u. s. w." *Berlin. Abh.* (1866), p. 66: viz. the scheme is

1	2	3	4	5	6	7	8	9	10	11	12	13	14	15	16
1	2	3	4	5	6	7	8	9	10	11	12	13	14	15	16
9	10	11	12	13	14	15	16	1	2	3	4	5	6	7	8
13	14	15	16	9	10	11	12	5	6	7	8	1	2	3	4
8	7	6	5	4	3	2	1	16	15	14	13	12	11	10	9
7	8	5	6	3	4	1	2	15	16	13	14	11	12	9	10
6	5	8	7	2	1	4	3	14	13	16	15	10	9	12	11.

* The same functions in Rosenhain's notation are

$$\begin{matrix} 00, & 02, & 20, & 22, \\ 01, & 03, & 21, & 23, \\ 10, & 12, & 30, & 32, \\ 11, & 13, & 31, & 33\,; \end{matrix}$$

viz. the figures here written down are the suffixes of his ϑ-functions, $00 = \vartheta_{0,0}(v, w)$, etc.

In fact, to show that any four of the squares, for instance 1, 9, 13, 8, that is, P^2, P_2^2, R^2, R_3^2, are linearly connected, it is only necessary to show that the determinant of coefficients

$$\begin{vmatrix} \alpha, & -\beta, & -\gamma, & \delta \\ \alpha, & -\beta, & \gamma, & -\delta \\ \gamma, & -\delta, & -\alpha, & \beta \\ \gamma, & \delta, & \alpha, & \beta \end{vmatrix}$$

is $=0$, or what is the same thing, that there exists a linear function of the new variables (X, Y, Z, W), which will become $=0$ on putting for these variables the values in any line of this determinant: we have such a function, viz. this is

$$\beta X + \alpha Y - \delta Z - \gamma W,$$

or say

$$[1] \quad (\beta, \alpha, -\delta, -\gamma)(X, Y, Z, W).$$

This function also vanishes if for (X, Y, Z, W) we substitute the values

$$\begin{array}{cccc} \delta, & -\gamma, & \beta, & -\alpha, \\ \delta, & \gamma, & \beta, & \alpha, \end{array}$$

which belong to 7, 6, that is, S_2^2 and S_3^2 respectively. It thus appears that 1, 9, 13, 8, 7, 6, that is, P^2, P_2^2, R^2, R_3^2, S_2^2, S_3^2, are a set of six squares having the property in question. I remark that the process of forming the linear functions is a very simple one; we write down six lines, and thence directly obtain the result, thus

$$\begin{array}{rcccc} 1 & \alpha, & -\beta, & -\gamma, & \delta \\ 9 & \alpha, & -\beta, & \gamma, & -\delta \\ 13 & \gamma, & -\delta, & -\alpha, & \beta \\ 8 & \gamma, & \delta, & \alpha, & \beta \\ 7 & \delta, & -\gamma, & \beta, & -\alpha \\ 6 & \delta, & \gamma, & \beta, & \alpha \\ \hline & \beta, & \alpha, & -\delta, & -\gamma: \end{array}$$

viz. β, α, δ, γ are the letters not previously occurring in the four columns respectively: the first letter β is taken to have the sign +, and then the remaining signs are determined by the condition that, combining the last line with any line above it (e.g. with the line next above it $\beta\delta + \alpha\gamma - \delta\beta - \gamma\alpha$), the sum must be zero.

We find in this way, as the conditions for the existence of the 16 sixes respectively,

$$\begin{array}{ll} [1] & (\beta, \ \alpha, \ -\delta, \ -\gamma)(X, Y, Z, W) = 0, \\ [2] & (\alpha, \ -\beta, \ -\gamma, \ \delta)(X, Y, Z, W) = 0, \\ [3] & (\alpha, \ \beta, \ -\gamma, \ -\delta)(X, Y, Z, W) = 0, \\ [4] & (\beta, \ -\alpha, \ -\delta, \ \gamma)(X, Y, Z, W) = 0, \end{array}$$

$$
\begin{array}{ll}
[5] & (\delta, \quad \gamma, \quad \beta, \quad \alpha)(X, Y, Z, W) = 0, \\
[6] & (\gamma, -\delta, \quad \alpha, -\beta)(X, Y, Z, W) = 0, \\
[7] & (\gamma, \quad \delta, \quad \alpha, \quad \beta)(X, Y, Z, W) = 0, \\
[8] & (\delta, -\gamma, \quad \beta, -\alpha)(X, Y, Z, W) = 0, \\
[9] & (\beta, \quad \alpha, \quad \delta, \quad \gamma)(X, Y, Z, W) = 0, \\
[10] & (\alpha, -\beta, \quad \gamma, -\delta)(X, Y, Z, W) = 0, \\
[11] & (\alpha, \quad \beta, \quad \gamma, \quad \delta)(X, Y, Z, W) = 0, \\
[12] & (\beta, -\alpha, \quad \delta, -\gamma)(X, Y, Z, W) = 0, \\
[13] & (\delta, \quad \gamma, -\beta, -\alpha)(X, Y, Z, W) = 0, \\
[14] & (\gamma, -\delta, -\alpha, \quad \beta)(X, Y, Z, W) = 0, \\
[15] & (\gamma, \quad \delta, -\alpha, -\beta)(X, Y, Z, W) = 0, \\
[16] & (\delta, -\gamma, -\beta, \quad \alpha)(X, Y, Z, W) = 0.
\end{array}
$$

I repeat in a new order the sets of coefficients which belong to the several squares, viz. these are

$$
\begin{array}{lll}
(1) & P^2 & (\alpha, -\beta, -\gamma, \quad \delta), \\
(2) & Q_1^2 & (\beta, \quad \alpha, -\delta, -\gamma), \\
(3) & Q^2 & (\beta, -\alpha, -\delta, \quad \gamma), \\
(4) & P_1^2 & (\alpha, \quad \beta, -\gamma, -\delta), \\
(5) & R_2^2 & (\gamma, -\delta, \quad \alpha, -\beta), \\
(6) & S_3^2 & (\delta, \quad \gamma, \quad \beta, \quad \alpha), \\
(7) & S_2^2 & (\delta, -\gamma, \quad \beta, -\alpha), \\
(8) & R_3^2 & (\gamma, \quad \delta, \quad \alpha, \quad \beta), \\
(9) & P_2^2 & (\alpha, -\beta, \quad \gamma, -\delta), \\
(10) & Q_3^2 & (\beta, \quad \alpha, \quad \delta, \quad \gamma), \\
(11) & Q_2^2 & (\beta, -\alpha, \quad \delta, -\gamma), \\
(12) & P_3^2 & (\alpha, \quad \beta, \quad \gamma, \quad \delta), \\
(13) & R^2 & (\gamma, -\delta, -\alpha, \quad \beta), \\
(14) & S_1^2 & (\delta, \quad \gamma, -\beta, -\alpha), \\
(15) & S^2 & (\delta, -\gamma, -\beta, \quad \alpha), \\
(16) & R_1^2 & (\gamma, \quad \delta, -\alpha, -\beta).
\end{array}
$$

And I remark that, if we connect these with the multipliers $(Y, -X, W, -Z)$, we obtain, except that there is sometimes a reversal of all the signs, the *same* linear functions of (X, Y, Z, W) as are written down under the same numbers in square brackets above: thus (1) gives

$$(\alpha, -\beta, -\gamma, \delta)(Y, -X, W, -Z), \text{ which is } (\beta, \alpha, -\delta, -\gamma)(X, Y, Z, W), = [1];$$

and so (2) gives

$$(\beta,\ \alpha,\ -\delta,\ -\gamma)(Y,\ -X,\ W,\ -Z),\ \text{which is}\ (-\alpha,\ \beta,\ \gamma,\ -\delta)(X,\ Y,\ Z,\ W),$$

or, reversing the signs,

$$(\alpha,\ -\beta,\ -\gamma,\ \delta)(X,\ Y,\ Z,\ W),\ =[2].$$

Comparing with the geometrical theory in Kummer's Memoir, it appears that the several systems of values (1), (2), ..., (16) are the coordinates of the nodes of a 16-nodal quartic surface, which nodes lie by sixes in the singular tangent planes, in the manner expressed by the foregoing scheme, wherein each top number may refer to a singular tangent plane, and then the numbers below it show the nodes in this plane: or else the top number may refer to a node, and then the numbers below it show the singular planes through this node.

And, from what precedes, we have the general result: the 16 squared double Θ-functions correspond (one to one) to the nodes of a 16-nodal quartic surface, in such wise that linearly connected squared functions correspond to nodes in the same singular tangent plane.

The question arises, to find the equation of the 16-nodal quartic surface, having the foregoing nodes and singular tangent planes. Starting from one of the irrational forms, say

$$\sqrt{A\,[1]\,[5]}+\sqrt{B\,[2]\,[6]}+\sqrt{C\,[3]\,[7]}=0;$$

the coefficients A, B, C are readily determined; and the result written at full length is

$$\begin{aligned}&\sqrt{2\,(\alpha\beta-\gamma\delta)\,(\alpha\delta+\beta\gamma)\,(\beta X+\alpha Y-\delta Z-\gamma W)\,(\delta X+\gamma Y+\beta Z+\alpha W)}\\ &+\sqrt{(\alpha^2-\beta^2-\gamma^2+\delta^2)\,(\alpha\gamma-\beta\delta)\,(\alpha X-\beta Y-\gamma Z+\delta W)\,(\gamma X-\delta Y+\alpha Z-\beta W)}\\ &+\sqrt{(\alpha^2+\beta^2-\gamma^2-\delta^2)\,(\alpha\gamma+\beta\delta)\,(\alpha X+\beta Y-\gamma Z-\delta W)\,(\gamma X+\delta Y+\alpha Z+\beta W)}=0.\end{aligned}$$

It is a somewhat long, but nevertheless interesting, piece of algebraical work to rationalise the foregoing equation: the result is

$$\begin{aligned}&(\beta^2\gamma^2-\alpha^2\delta^2)\,(\gamma^2\alpha^2-\beta^2\delta^2)\,(\alpha^2\beta^2-\gamma^2\delta^2)\,(X^4+Y^4+Z^4+W^4)\\ &+(\gamma^2\alpha^2-\beta^2\delta^2)\,(\alpha^2\beta^2-\gamma^2\delta^2)\,(\alpha^4+\delta^4-\beta^4-\gamma^4)\,(Y^2Z^2+X^2W^2)\\ &+(\alpha^2\beta^2-\gamma^2\delta^2)\,(\beta^2\gamma^2-\alpha^2\delta^2)\,(\beta^4+\delta^4-\gamma^4-\alpha^4)\,(Z^2X^2+Y^2W^2)\\ &+(\beta^2\gamma^2-\alpha^2\delta^2)\,(\gamma^2\alpha^2-\beta^2\delta^2)\,(\gamma^4+\delta^4-\alpha^4-\beta^4)\,(X^2Y^2+Z^2W^2)\\ &-2\alpha\beta\gamma\delta\,(\alpha^2+\beta^2+\gamma^2+\delta^2)\,(\alpha^2+\delta^2-\beta^2-\gamma^2)\,(\beta^2+\delta^2-\alpha^2-\gamma^2)\,(\gamma^2+\delta^2-\alpha^2-\beta^2)\,XYZW=0;\end{aligned}$$

or, if we write for shortness

$$\begin{aligned}L&=\beta^2\gamma^2-\alpha^2\delta^2, & F&=\alpha^2+\delta^2-\beta^2-\gamma^2,\\ M&=\gamma^2\alpha^2-\beta^2\delta^2, & G&=\beta^2+\delta^2-\gamma^2-\alpha^2,\\ N&=\alpha^2\beta^2-\gamma^2\delta^2, & H&=\gamma^2+\delta^2-\alpha^2-\beta^2,\\ & & \Delta&=\alpha^2+\beta^2+\gamma^2+\delta^2,\end{aligned}$$

then the result is

$$\begin{aligned} &LMN(X^4+Y^4+Z^4+W^4) \\ &+MN(F\Delta+2L)(Y^2Z^2+X^2W^2) \\ &+NL\,(G\Delta+2M)(Z^2X^2+Y^2W^2) \\ &+LM(H\Delta+2N)(X^2Y^2+Z^2W^2) \\ &-2\alpha\beta\gamma\delta FGH\Delta\, XYZW=0. \end{aligned}$$

It may be easily verified that any one of the sixteen points, for instance $(\alpha, \beta, \gamma, \delta)$, is a node of the surface. Thus to show that the derived function in respect to X, vanishes for $X, Y, Z, W = \alpha, \beta, \gamma, \delta$; the derived function here divides by 2α, and omitting this factor, the equation to be verified is

$$LMN.2\alpha^2+MN(F\Delta+2L)\delta^2+NL(G\Delta+2M)\gamma^2+LM(H\Delta+2N)\beta^2-\beta^2\gamma^2\delta^2FGH\Delta=0,$$

viz. the whole coefficient of LMN is $2(\alpha^2+\beta^2+\gamma^2+\delta^2), =2\Delta$; hence throwing out the factor Δ, the equation becomes

$$2LMN+MNF\delta^2+NLG\gamma^2+LMH\beta^2-\beta^2\gamma^2\delta^2FGH=0.$$

Writing this in the form

$$L(2MN+NG\gamma^2+MH\beta^2)=F\delta^2(GH\beta^2\gamma^2-MN),$$

we find without difficulty $GH\beta^2\gamma^2-MN=-(\beta^2-\gamma^2)^2L$; hence throwing out the factor L, the equation becomes

$$N(2M+G\gamma^2)+MH\beta^2+F\delta^2(\beta^2-\gamma^2)^2=0;$$

we find

$$\begin{aligned} MH\beta^2+F\delta^2(\beta^2-\gamma^2)^2 &= (\alpha^2\beta^2-\gamma^2\delta^2)(2\beta^2\delta^2-\gamma^2(\alpha^2+\beta^2+\delta^2)+\gamma^4) \\ &= N(2\beta^2\delta^2-\gamma^2(\alpha^2+\beta^2+\delta^2)+\gamma^4), \end{aligned}$$

or throwing out the factor N, the equation becomes

$$2M+G\gamma^2+2\beta^2\delta^2-\gamma^2(\alpha^2+\beta^2+\delta^2)+\gamma^4=0,$$

which is at once verified: and similarly it can be shown that the other derived functions vanish, and the point $(\alpha, \beta, \gamma, \delta)$ is thus a node.

The surface seems to be the general 16-nodal surface, viz. replacing X, Y, Z, W by any linear functions of four coordinates, we have thus $4.4-1, =15$ constants, and the equation contains besides the three ratios $\alpha:\beta:\gamma:\delta$, that is, in all 18 constants: the general quartic surface has 34 constants, and therefore the general 16-nodal surface $34-16, =18$ constants: but the conclusion requires further examination.

Göpel and Rosenhain each connect the theory with that of the ultra-elliptic functions involving the radical $\sqrt{X}, =\sqrt{x.1-x.1-lx.1-mx.1-nx}$; viz. it appears by their formulæ (more completely by those of Rosenhain) that the ratios of the 16 squares can be expressed rationally in terms of the two variables x, x', and the radicals

$\sqrt{X}$, $\sqrt{X'}$, X' being the same function of x' that X is of x. We may instead of the preceding form take X to be the general quintic function, or what is better take it to be the sextic function $a-x.b-x.c-x.d-x.e-x.f-x$; and we thus obtain a remarkable algebraical theorem: viz. I say that the 16 squares, each divided by a proper constant factor, are proportional to six functions of the form

$$a-x.a-x',$$

and ten functions of the form

$$\frac{1}{(x-x')^2}\{\sqrt{a-x.b-x.c-x.d-x'.e-x'.f-x'}-\sqrt{a-x'.b-x'.c-x'.d-x.e-x.f-x}\}^2,$$

and consequently that these 16 algebraical functions of x, x' are linearly connected in the manner of the 16 squares; viz. there exist 16 sixes such that, in each six, the remaining three functions can be linearly expressed in terms of any three of them.

To further develop the theory, I remark that the six functions may be represented by A, B, C, D, E, F respectively: any one of the ten functions would be properly represented by $ABC.DEF$, but isolating one letter F, and writing DE to denote DEF, this function $ABC.DEF$ may be represented simply as DE; and the ten functions thus are AB, AC, AD, AE, BC, BD, BE, CD, CE, DE.

Writing for shortness a, b, c, d, e, f, to denote $a-x$, $b-x$, etc., and similarly a', b', c', d', e', f', to denote $a-x'$, $b-x'$, etc., we thus have

(13) $A = aa'$,

(9) $B = bb'$,

(7) $C = cc'$,

(8) $D = dd'$,

(6) $E = ee'$, $(=E)$,

(1) $F = ff'$, $(=F)$,

(3) $DE = \dfrac{1}{(x-x')^2}\{\sqrt{abcd'e'f'}-\sqrt{a'b'c'def}\}^2$, $(=\bar{D})$,

(4) $CE = \dfrac{1}{(x-x')^2}\{\sqrt{abdc'e'f'}-\sqrt{a'b'd'cef}\}^2$, $(=\bar{E})$,

(2) $CD = \dfrac{1}{(x-x')^2}\{\sqrt{abec'd'f'}-\sqrt{a'b'e'cdf}\}^2$,

(14) $BE = \dfrac{1}{(x-x')^2}\{\sqrt{acdb'e'f'}-\sqrt{a'c'd'bef}\}^2$, $(=\bar{B})$,

(16) $BD = \dfrac{1}{(x-x')^2}\{\sqrt{aceb'd'f'}-\sqrt{a'c'e'bdf}\}^2$,

(15) $BC = \dfrac{1}{(x-x')^2}\{\sqrt{adeb'c'f'}-\sqrt{a'd'e'bcf}\}^2$,

(10) $AE = \dfrac{1}{(x-x')^2}\{\sqrt{bcda'e'f'}-\sqrt{b'c'd'aef}\}^2$, $(=\bar{A})$,

$$(12)\quad AD = \frac{1}{(x-x')^2}\{\sqrt{bcea'd'f'} - \sqrt{b'c'e'adf}\}^2,$$

$$(11)\quad AC = \frac{1}{(x-x')^2}\{\sqrt{bdea'c'f'} - \sqrt{b'd'e'acf}\}^2,$$

$$(5)\quad AB = \frac{1}{(x-x')^2}\{\sqrt{cdea'b'f'} - \sqrt{c'd'e'abf}\}^2,$$

where the numbers are in accordance with the foregoing scheme; viz. the scheme becomes

(1)	(2)	(3)	(4)	(5)	(6)	(7)	(8)	(9)	(10)	(11)	(12)	(13)	(14)	(15)	(16)
F	CD	DE	CE	AB	E	C	D	B	AE	AC	AD	A	BE	BC	BD
B	AE	AC	AD	A	BE	BC	BD	F	CD	DE	CE	AB	E	C	D
A	BE	BC	BD	B	AE	AC	AD	AB	E	C	D	F	CD	DE	CE
D	C	E	AB	CE	DE	CD	F	BD	BC	BE	A	AD	AC	AE	B
C	D	AB	E	DE	CE	F	CD	BC	BD	A	BE	AC	AD	B	AE
E	AB	D	C	CD	F	CE	DE	DE	A	BD	BC	AE	B	AD	AC.

There is of course the six A, B, C, D, E, F; for each of these is a linear function of $1, x+x', xx'$, and there is thus a linear relation between any four of them. It would at first sight appear that the remaining sixes were of two different forms, A, B, AB, CE, CD, DE, and F, A, AB, AC, AD, AE; but these are really identical, for taking *any* two letters E, F, the six is E, F, AE, BE, CE, DE, or, as this might be written, E, F, AEF, BEF, CEF, DEF, where AEF means $BCD \cdot AEF$, etc.; and we thus obtain each of the remaining fifteen sixes. The six just referred to, viz. E, F, AE, BE, CE, DE, or changing the notation say $E, F, \bar{A}, \bar{B}, \bar{C}, \bar{D}$ as indicated in the table, thus represents any one of the sixes other than the rational six A, B, C, D, E, F; and there is no difficulty in actually finding each of the fifteen relations between four functions of the six in question, $E, F, \bar{A}, \bar{B}, \bar{C}, \bar{D}$. It is to be observed that every such function as $\bar{A}$ contains the same irrational part

$$\frac{2}{(x-x')^2}\sqrt{abcdefa'b'c'd'e'f'},$$

and that the linear relations involve therefore only the differences $\bar{A}-\bar{B}$, $\bar{A}-\bar{C}$, etc., which are rational. Proceeding to calculate these differences, we have for instance

$$\bar{C}-\bar{D} = \frac{1}{(x-x')^2}(cefa'b'd' + c'e'f'abd - defa'b'c' - d'e'f'abc) = \frac{1}{(x-x')^2}(cd'-c'd)(efa'b'-e'f'ab);$$

or, substituting for a, a', etc. their values $a-x$, $a-x'$, etc., we have

$$cd'-c'd = (x-x')(c-d),$$

$$efa'b'-e'f'ab = (x-x')\begin{vmatrix} 1, & x+x', & xx' \\ 1, & a+b, & ab \\ 1, & e+f, & ef \end{vmatrix},$$

or say for shortness

$$= (x-x')\,[xx'abef].$$

We have therefore

$$\bar{C}-\bar{D}=(c-d)\,[xx'abef];$$

and in like manner we obtain the equations

$$\begin{aligned}\bar{B}-\bar{C}&=(b-c)\,[xx'adef], & \bar{A}-\bar{D}&=(a-d)\,[xx'bcef],\\ \bar{C}-\bar{A}&=(c-a)\,[xx'bdef], & \bar{B}-\bar{D}&=(b-d)\,[xx'caef],\\ \bar{A}-\bar{B}&=(a-b)\,[xx'cdef], & \bar{C}-\bar{D}&=(c-d)\,[xx'abef].\end{aligned}$$

It is now easy to form the system of formulæ

E	F	$\bar{A}$	$\bar{B}$	$\bar{C}$	$\bar{D}$	
		$ae\,.\,af\,.\,bcd$	$-be\,.\,bf\,.\,cda$	$+ce\,.\,cf\,.\,dab$	$-de\,.\,df\,.\,abc$	$=0$
$ad\,.\,bf\,.\,cf$	$-ad\,.\,be\,.\,ce$	$+ef$			$-ef$	$=0$
$bd\,.\,cf\,.\,af$	$-bd\,.\,ce\,.\,ae$		$+ef$	$-ef$		$=0$
$cd\,.\,af\,.\,bf$	$-cd\,.\,ae\,.\,be$			$+ef$	$-ef$	$=0$
$bc\,.\,af\,.\,df$	$-bc\,.\,ae\,.\,de$		$+ef$	$-ef$		$=0$
$ca\,.\,bf\,.\,df$	$-ca\,.\,be\,.\,de$	$-ef$		$+ef$		$=0$
$ab\,.\,cf\,.\,df$	$-ab\,.\,ce\,.\,de$	$+ef$	$-ef$			$=0$
$-af\,.\,bcd$			$+be\,.\,cd$	$+ce\,.\,db$	$+de\,.\,bc$	$=0$
$-bf\,.\,cda$		$+ae\,.\,cd$		$+ce\,.\,da$	$+de\,.\,ac$	$=0$
$-cf\,.\,dab$		$+ae\,.\,bd$	$+be\,.\,da$		$+de\,.\,ab$	$=0$
$-df\,.\,abc$		$+ae\,.\,bc$	$+be\,.\,ca$	$+ce\,.\,ab$		$=0$
	$-ae\,.\,bcd$		$+bf\,.\,cd$	$+cf\,.\,db$	$+df\,.\,bc$	$=0$
	$-be\,.\,cda$	$+af\,.\,cd$		$+cf\,.\,da$	$+df\,.\,ac$	$=0$
	$-ce\,.\,dab$	$+af\,.\,bd$	$+bf\,.\,da$		$+df\,.\,ab$	$=0$
	$-df\,.\,abc$	$+af\,.\,bc$	$+bf\,.\,ca$	$+cf\,.\,ab$		$=0,$

where for shortness ab, ac, etc., are written to denote $a-b$, $a-c$, etc.; also abc, etc., to denote $(b-c)(c-a)(a-b)$, etc.: the equations contain all of them only the differences of $\bar{A}$, $\bar{B}$, $\bar{C}$, $\bar{D}$; thus the first equation is equivalent to

$$ae\,.\,af\,.\,bcd\,(\bar{A}-\bar{D})-be\,.\,bf\,.\,cde\,(\bar{B}-\bar{D})+ce\,.\,cf\,.\,dab\,(\bar{C}-\bar{D})=0,$$

and so in other cases.

Cambridge, 14 *March*, 1877.

663.

FURTHER INVESTIGATIONS ON THE DOUBLE ϑ-FUNCTIONS.

[From the *Journal für die reine und angewandte Mathematik* (Crelle), t. LXXXIII. (1877), pp. 220—233.]

I CONSIDER six letters

$$a,\ b,\ c,\ d,\ e,\ f;$$

a duad ab not containing f may be completed into the triad abf, and then into the double triad $abf.cde$; there are in all ten double triads, represented by the duads

$$ab,\ ac,\ ad,\ ae,\ bc,\ bd,\ be,\ cd,\ ce,\ de,$$

and the whole number of letters and of double triads is $=16$.

Taking x, x' as variables, I form sixteen functions; viz. these are

$$[a] = a-x\,.\,a-x',$$

$$[ab] = \frac{1}{(x-x')^2}\left\{\sqrt{\frac{a-x\,.\,b-x\,.\,f-x}{c-x'\,.\,d-x'\,.\,e-x'}} \pm \sqrt{\frac{a-x'\,.\,b-x'\,.\,f-x'}{c-x\,.\,d-x\,.\,e-x}}\right\}^2,$$

where the function under each radical sign is the product of six factors, the arrangement in two lines being for convenience only: the sign $\pm$ has the same value in all the functions, and it will be observed that the irrational part is

$$= \pm\frac{2}{(x-x')^2}\sqrt{\frac{a-x\,.\,b-x\,.\,c-x\,.\,d-x\,.\,e-x\,.\,f-x}{a-x'\,.\,b-x'\,.\,c-x'\,.\,d-x'\,.\,e-x'\,.\,f-x'}},$$

viz. this has the same value in all the functions.

The general property of the double ϑ-functions is that the squares of the sixteen functions are proportional to constant multiples of the sixteen functions $[a]$, $[ab]$; but this theorem may be presented in a much more definite form, viz. we can determine, and that very simply, the actual expressions for the constant factors; and so we can enunciate the theorem as follows; the squares of the sixteen double ϑ-functions are proportional to sixteen functions $-\{a\}$, $+\{ab\}$; where, in a notation about to be explained,

$$\{a\} = \sqrt{a}\,[a], \quad \{ab\} = \sqrt{ab}\,[ab].$$

Here in the radical $\sqrt{a}$, a is to be considered as standing in the first place for the pentad $bcdef$, which is to be interpreted as a product of differences,

$$= bc\,.\,bd\,.\,be\,.\,bf\,.\,cd\,.\,ce\,.\,cf\,.\,de\,.\,df\,.\,ef,$$

(where bc, bd, etc., denote the differences $b-c$, $b-d$, etc.). Similarly, in the radical $\sqrt{ab}$, ab is to be considered as standing in the first instance for the double triad $abf\,.\,cde$, which is to be interpreted as a product of differences, $= ab\,.\,af\,.\,bf\,.\,cd\,.\,ce\,.\,de$, (where ab, af, etc., denote the differences $a-b$, $a-f$, etc.).

It is convenient to consider a, b, c, d, e, f as denoting real magnitudes taken in decreasing order: in all the products $bcdef$, etc., and in each term abf or cde of a product $abf\,.\,cde$, the letters are to be written in alphabetical order; the differences bc, bd, etc., ab, af, etc., which present themselves in the several products, are thus all of them positive; and the radicals, being all of them the roots of positive quantities, may themselves be taken to be positive.

We have to consider the values of the functions $[a]$, $[ab]$, or $\{a\}$, $\{ab\}$, in the case where the variables x, x' become equal to any two of the letters a, b, c, d, e, f; it is clearly the same thing whether we have for instance $x=b$, $x'=c$, or $x=c$, $x'=b$, etc.: we have therefore to consider for x, x' the fifteen values ab, ac, ..., af, ..., ef; there is besides a sixteenth set of values x, x' each infinite, without any relation between the infinite values.

Taking this case first, x, x' each infinite, and in $[ab]$, etc., the sign $\pm$ to be $+$, we have

$$[a] = xx', \quad [ab] = \frac{4x^3x'^3}{(x-x')^2},$$

or, attending only to the ratios of these values,

$$[a] = 1, \qquad [ab] = \frac{4x^2x'^2}{(x-x')^2},$$

where $\dfrac{4x^2x'^2}{(x-x')^2}$ is infinite, and the values may finally be written

$$[a] = 0, \qquad [ab] = 1;$$

whence also, for x, x' infinite,

$$\{a\} = 0, \qquad \{ab\} = \sqrt{ab},$$

the radical $\sqrt{ab}$ being understood as before.

Suppose next that x, x' denote any two of the letters, for instance a, b; then two of the functions $[a]$ vanish, viz. these are $[a]$, $[b]$, but the remaining four functions acquire determinate values; and moreover four of the functions $[ab]$ vanish, viz. these are $[ab]$, $[cd]$, $[ce]$, $[de]$, for each of which the xx' letters a, b occur in the same triad (the double triads for the four functions are, in fact, *abf.cde*, *cdf.abe*, *cef.abd*, *def.abc*); but the other six functions [*ab*], for which the letters *a*, *b* occur in separate triads, acquire determinate values.

It is important to attend to the signs: for example, if x, $x' = b$, e, we have

$$[c] = ce\,.\,cb, \quad = -bc\,.\,ce$$

$$[ce] = \frac{1}{(be)^2}\frac{cb\,.\,eb\,.\,fb}{ae\,.\,be\,.\,de}, \quad = -\frac{cb\,.\,fb}{ae\,.\,de}, \quad = -\frac{bc\,.\,bf}{ae\,.\,de}.$$

TABLE I. OF THE VALUES OF [*a*], [*ab*], ETC.,

x,	$x' = \infty\ \infty$	*ab*	*ac*	*ad*	*ae*	*af*	*bc*	*bd*
[*a*]	0	0	0	0	0	0	+ *ab.ac*	+ *ab.ad*
[*b*]	0	0	− *ab.bc*	− *ab.bd*	− *ab.be*	− *ab.bf*	0	0
[*c*]	0	+ *ac.bc*	0	− *ac.cd*	− *ac.ce*	− *ac.cf*	0	− *bc.cd*
[*d*]	0	+ *ad.bd*	+ *ad.cd*	0	− *ad.de*	− *ad.df*	+ *bd.cd*	0
[*e*]	0	+ *ae.be*	+ *ae.ce*	+ *ae.de*	0	− *ae.ef*	+ *be.ce*	+ *be.de*
[*f*]	0	+ *af.bf*	+ *af.cf*	+ *af.df*	+ *af.ef*	0	+ *bf.cf*	+ *bf.df*
[*ab*]	+ *abf.cde*	0	$+\frac{ad.ae}{bc.cf}$	$+\frac{ac.ae}{bd.df}$	$+\frac{ac.ad}{be.ef}$	0	$+\frac{ac.bd}{be.cf}$	$+\frac{ad.bc}{be.df}$
[*ac*]	+ *acf.bde*	$-\frac{ad.ae}{bc.bf}$	0	$+\frac{ab.ae}{cd.cf}$	$+\frac{ab.ad}{ce.ef}$	0	$+\frac{ab.bf}{cd.ce}$	0
[*ad*]	+ *adf.bce*	$-\frac{ac.ae}{bd.bf}$	$-\frac{ab.ae}{cd.cf}$	0	$+\frac{ab.ac}{de.ef}$	0	0	$-\frac{ab.bf}{cd.de}$
[*ae*]	+ *aef.bcd*	$-\frac{ac.ae}{be.bf}$	$-\frac{ab.ad}{ce.cf}$	$-\frac{ab.ac}{de.df}$	0	0	0	0
[*bc*]	+ *bcf.ade*	$-\frac{ac.af}{bd.be}$	$-\frac{ab.af}{cd.ce}$	0	0	$-\frac{ab.ac}{df.ef}$	0	$-\frac{ab.be}{cd.df}$
[*bd*]	+ *bdf.ace*	$-\frac{ad.af}{bc.be}$	0	$+\frac{ab.af}{cd.de}$	0	$-\frac{ab.ad}{cf.ef}$	$+\frac{ab.be}{cd.cf}$	0
[*be*]	+ *bef.acd*	$-\frac{ae.af}{bc.bd}$	0	0	$-\frac{ab.af}{ce.de}$	$-\frac{ab.ae}{cf.df}$	$+\frac{ab.bd}{ce.cf}$	$+\frac{ab.bc}{de.df}$
[*cd*]	+ *cdf.abe*	0	$+\frac{ad.af}{bc.ce}$	$+\frac{ac.af}{bd.de}$	0	$-\frac{ac.ad}{bf.ef}$	$+\frac{ac.bd}{bf.ce}$	$+\frac{ad.bc}{bf.de}$
[*ce*]	+ *cef.abd*	0	$+\frac{ae.af}{bc.cd}$	0	$-\frac{ac.af}{be.de}$	$-\frac{ac.ae}{bf.df}$	$+\frac{ac.be}{bf.cd}$	0
[*de*]	+ *def.abc*	0	0	$-\frac{ae.af}{bc.cd}$	$-\frac{ad.af}{be.ce}$	$-\frac{ad.ae}{bf.cf}$	0	$-\frac{ad.be}{bf.cd}$

Here the symbols *be*, *ce*, etc., denote differences; [*ce*] is the product of four differences: the arrangement in two lines is for convenience only.

We thus obtain the series of values of [*a*], [*ab*], etc., which although only required as subsidiary to the determination of the corresponding values of {*a*}, {*ab*}, I nevertheless give in a table.

The signs are given as they were actually obtained, but as we are concerned only with the ratios of the functions, it is allowable to change all the signs in any

FOR THE SIXTEEN SPECIAL VALUES OF x, x'.

be	bf	cd	ce	cf	de	df	ef
$+ab.ae$	$+ab.af$	$+ac.ad$	$+ac.ae$	$+ac.af$	$+ad.ae$	$+ad.af$	$+ae.af$
0	0	$+bc.bd$	$+bc.be$	$+bc.bf$	$+bd.be$	$+bd.bf$	$+be.bf$
$-bc.ce$	$-bc.cf$	0	0	0	$+cd.ce$	$+cd.df$	$+ce.cf$
$-bd.de$	$-bd.bf$	0	$-cd.de$	$-cd.df$	0	0	$+de.df$
0	$-be.ef$	$+ce.de$	0	$-ce.ef$	0	$-de.ef$	0
$+bf.ef$	0	$+cf.df$	$+cf.ef$	0	$+df.ef$	0	0
$+\begin{matrix}ac.bc\\bd.ef\end{matrix}$	0	0	0	$-\begin{matrix}ac.bc\\df.ef\end{matrix}$	0	$-\begin{matrix}ad.bd\\cf.ef\end{matrix}$	$-\begin{matrix}ae.be\\cf.df\end{matrix}$
0	$+\begin{matrix}ab.bc\\df.ef\end{matrix}$	$-\begin{matrix}ad.bc\\ce.df\end{matrix}$	$-\begin{matrix}ae.bc\\cd.ef\end{matrix}$	0	0	$-\begin{matrix}ad.bf\\cd.ef\end{matrix}$	$-\begin{matrix}ae.bf\\ce.df\end{matrix}$
0	$+\begin{matrix}ab.bd\\cf.ef\end{matrix}$	$-\begin{matrix}ac.bd\\cf.de\end{matrix}$	0	$+\begin{matrix}ac.bf\\cd.ef\end{matrix}$	$+\begin{matrix}ae.bd\\cd.ef\end{matrix}$	0	$-\begin{matrix}ae.bf\\cf.de\end{matrix}$
$+\begin{matrix}ab.bf\\ce.de\end{matrix}$	$+\begin{matrix}ab.be\\cf.df\end{matrix}$	0	$+\begin{matrix}ac.be\\cf.de\end{matrix}$	$+\begin{matrix}ac.bf\\ce.df\end{matrix}$	$+\begin{matrix}ad.be\\ce.df\end{matrix}$	$+\begin{matrix}ad.bf\\cf.de\end{matrix}$	0
$-\begin{matrix}ab.bd\\ce.ef\end{matrix}$	0	$-\begin{matrix}ac.bd\\ce.df\end{matrix}$	$-\begin{matrix}ac.be\\cd.ef\end{matrix}$	0	0	$-\begin{matrix}af.bd\\cd.ef\end{matrix}$	$-\begin{matrix}af.be\\ce.df\end{matrix}$
$-\begin{matrix}ad.bc\\de.ef\end{matrix}$	0	$-\begin{matrix}ad.bc\\cf.de\end{matrix}$	0	$+\begin{matrix}af.bc\\cd.ef\end{matrix}$	$+\begin{matrix}ad.be\\cd.ef\end{matrix}$	0	$-\begin{matrix}af.be\\cf.de\end{matrix}$
0	0	0	$+\begin{matrix}ae.bc\\cf.de\end{matrix}$	$+\begin{matrix}af.bc\\ce.df\end{matrix}$	$+\begin{matrix}ae.bd\\ce.df\end{matrix}$	$+\begin{matrix}af.bd\\cf.de\end{matrix}$	0
0	$-\begin{matrix}af.bc\\bd.ef\end{matrix}$	0	$+\begin{matrix}ac.bc\\de.ef\end{matrix}$	0	$+\begin{matrix}ad.bd\\ce.ef\end{matrix}$	0	$-\begin{matrix}af.bf\\ce.de\end{matrix}$
$-\begin{matrix}ae.bc\\bf.de\end{matrix}$	$-\begin{matrix}af.bc\\be.df\end{matrix}$	$-\begin{matrix}ac.bc\\de.df\end{matrix}$	0	0	$+\begin{matrix}ae.be\\cd.df\end{matrix}$	$+\begin{matrix}af.bf\\cd.de\end{matrix}$	0
$-\begin{matrix}ae.bd\\bf.ce\end{matrix}$	$-\begin{matrix}af.bd\\be.cf\end{matrix}$	$-\begin{matrix}ad.bd\\ce.cf\end{matrix}$	$-\begin{matrix}ae.be\\cd.cf\end{matrix}$	$-\begin{matrix}af.bf\\cd.ce\end{matrix}$	0	0	0

column: and it appears that there are four columns in each of which the signs are or can be made all +; whereas in each of the remaining twelve columns the signs are or can be made six of them +, the other four −.

Passing to the values of $\{a\}$, $\{ab\}$, etc., we have for example, from the ab column of the foregoing table,

$$\begin{aligned}\{c\} &= + \sqrt{c} \,.\, ac \,.\, bc,\\ \{d\} &= + \sqrt{d} \,.\, ad \,.\, bd,\\ &\vdots\\ \{ac\} &= - \sqrt{ac} \,.\, \frac{ac \,.\, ae}{bc \,.\, bf},\\ &\vdots\end{aligned}$$

where (since the radicals are all positive) the signs are correct: substituting for the quantities under the radical signs their full values, and squaring the rational parts in order to bring them also under the radical signs, this is

$$\begin{aligned}\{c\} &= + \sqrt{ab \,.\, ad \,.\, ae \,.\, af \,.\, bd \,.\, be \,.\, bf \,.\, de \,.\, df \,.\, ef \,.\, ac^2 \,.\, bc^2},\\ \{d\} &= + \sqrt{ab \,.\, ac \,.\, ae \,.\, af \,.\, bc \,.\, be \,.\, bf \,.\, ce \,.\, cf \,.\, ef \,.\, ad^2 \,.\, bd^2},\\ &\vdots\\ \{ac\} &= - \sqrt{ac \,.\, af \,.\, cf \,.\, bd \,.\, be \,.\, de \,.\, ac^2 \,.\, ae^2 \,.\, bc^2 \,.\, bf^2},\end{aligned}$$

where all the expressions of this (the ab-column) have a common factor,

$$ac \,.\, ad \,.\, ae \,.\, af \,.\, bc \,.\, bd \,.\, be \,.\, bf.$$

Omitting this factor, we find

$$\begin{aligned}\{c\} &= + \sqrt{ab \,.\, ac \,.\, bc \,.\, de \,.\, df \,.\, ef},\\ \{d\} &= + \sqrt{ab \,.\, ad \,.\, bd \,.\, ce \,.\, cf \,.\, ef},\\ &\vdots\\ \{ac\} &= - \sqrt{ad \,.\, ae \,.\, de \,.\, bc \,.\, bf \,.\, cf};\end{aligned}$$

viz. recurring to the foregoing condensed notation, this is

$$\begin{aligned}\{c\} &= + \sqrt{de},\\ \{d\} &= + \sqrt{ce},\\ &\vdots\\ \{ac\} &= - \sqrt{bc},\end{aligned}$$

and, in fact, the terms in the several columns have only the ten values $\sqrt{ab}$, $\sqrt{ac}$, etc. each with its proper sign. I repeat the meaning of the notation: ab stands in the first instance for the double triad $abf \,.\, cde$, and then this denotes a product of differences $ab \,.\, af \,.\, bf \,.\, cd \,.\, ce \,.\, de$. We have thus the following table in which I have in several cases changed the signs of entire columns.

TABLE II. OF THE FUNCTIONS $\{a\}$, $\{ab\}$, ETC., FOR THE SIXTEEN SPECIAL VALUES OF x, x'.

$x, x' =$	$\infty\ \infty$	ab	ac	ad	ae	af	bc	bd	be	bf	cd	ce	cf	de	df	ef
$\{a\}$	0	0	0	0	0	0	$+\sqrt{de}$	$+\sqrt{ce}$	$-\sqrt{cd}$	$-\sqrt{ab}$	$-\sqrt{be}$	$+\sqrt{bd}$	$+\sqrt{ac}$	$+\sqrt{bc}$	$+\sqrt{ad}$	$-\sqrt{ae}$
$\{b\}$	0	0	$-\sqrt{de}$	$-\sqrt{ce}$	$+\sqrt{cd}$	$+\sqrt{ab}$	0	0	0	0	$-\sqrt{ae}$	$+\sqrt{ad}$	$+\sqrt{bc}$	$+\sqrt{ac}$	$+\sqrt{bd}$	$-\sqrt{be}$
$\{c\}$	0	$-\sqrt{de}$	0	$-\sqrt{be}$	$+\sqrt{bd}$	$+\sqrt{ac}$	0	$-\sqrt{ae}$	$+\sqrt{ad}$	$+\sqrt{bc}$	0	0	0	$+\sqrt{ab}$	$+\sqrt{cd}$	$-\sqrt{ce}$
$\{d\}$	0	$-\sqrt{ce}$	$+\sqrt{be}$	0	$+\sqrt{bc}$	$+\sqrt{ad}$	$+\sqrt{ae}$	0	$+\sqrt{ac}$	$+\sqrt{bd}$	0	$-\sqrt{ab}$	$-\sqrt{cd}$	0	0	$-\sqrt{de}$
$\{e\}$	0	$-\sqrt{cd}$	$+\sqrt{bd}$	$+\sqrt{bc}$	0	$+\sqrt{ae}$	$+\sqrt{ad}$	$+\sqrt{ac}$	0	$+\sqrt{be}$	$-\sqrt{ab}$	0	$-\sqrt{ce}$	0	$-\sqrt{de}$	0
$\{f\}$	0	$-\sqrt{ab}$	$+\sqrt{ac}$	$+\sqrt{ad}$	$-\sqrt{ae}$	0	$+\sqrt{bc}$	$+\sqrt{bd}$	$-\sqrt{be}$	0	$-\sqrt{cd}$	$+\sqrt{ce}$	0	$+\sqrt{de}$	0	0
$\{ab\}$	$+\sqrt{ab}$	0	$+\sqrt{bc}$	$+\sqrt{bd}$	$-\sqrt{be}$	0	$+\sqrt{ac}$	$+\sqrt{ad}$	$-\sqrt{ae}$	0	0	0	$-\sqrt{de}$	0	$-\sqrt{ce}$	$+\sqrt{cd}$
$\{ac\}$	$+\sqrt{ac}$	$+\sqrt{bc}$	0	$+\sqrt{cd}$	$-\sqrt{ce}$	0	$+\sqrt{ab}$	0	0	$-\sqrt{de}$	$+\sqrt{ad}$	$-\sqrt{ae}$	0	0	$-\sqrt{be}$	$+\sqrt{bd}$
$\{ad\}$	$+\sqrt{ad}$	$+\sqrt{bd}$	$-\sqrt{cd}$	0	$-\sqrt{de}$	0	0	$-\sqrt{ab}$	0	$-\sqrt{ce}$	$+\sqrt{ac}$	0	$+\sqrt{be}$	$+\sqrt{ae}$	0	$+\sqrt{bc}$
$\{ae\}$	$+\sqrt{ae}$	$+\sqrt{be}$	$-\sqrt{ce}$	$-\sqrt{de}$	0	0	0	0	$-\sqrt{ab}$	$-\sqrt{cd}$	0	$+\sqrt{ac}$	$+\sqrt{bd}$	$+\sqrt{ad}$	$+\sqrt{bc}$	0
$\{bc\}$	$+\sqrt{bc}$	$+\sqrt{ac}$	$-\sqrt{ab}$	0	0	$+\sqrt{de}$	0	$-\sqrt{cd}$	$+\sqrt{ce}$	0	$+\sqrt{bd}$	$-\sqrt{be}$	0	0	$-\sqrt{ae}$	$+\sqrt{ad}$
$\{bd\}$	$+\sqrt{bd}$	$+\sqrt{ad}$	0	$+\sqrt{ab}$	0	$+\sqrt{ce}$	$+\sqrt{cd}$	0	$+\sqrt{de}$	0	$+\sqrt{bc}$	0	$+\sqrt{ae}$	$+\sqrt{be}$	0	$+\sqrt{ac}$
$\{be\}$	$+\sqrt{be}$	$+\sqrt{ae}$	0	0	$+\sqrt{ab}$	$+\sqrt{cd}$	$+\sqrt{ce}$	$+\sqrt{de}$	0	0	0	$+\sqrt{bc}$	$+\sqrt{ad}$	$+\sqrt{bd}$	$+\sqrt{ac}$	0
$\{cd\}$	$+\sqrt{cd}$	0	$+\sqrt{ad}$	$+\sqrt{ac}$	0	$+\sqrt{be}$	$+\sqrt{bd}$	$+\sqrt{bc}$	0	$+\sqrt{ae}$	0	$+\sqrt{de}$	0	$+\sqrt{ce}$	0	$+\sqrt{ab}$
$\{ce\}$	$+\sqrt{ce}$	0	$+\sqrt{ae}$	0	$+\sqrt{ac}$	$+\sqrt{bd}$	$+\sqrt{be}$	0	$+\sqrt{bc}$	$+\sqrt{ad}$	$+\sqrt{de}$	0	0	$+\sqrt{cd}$	$+\sqrt{ab}$	0
$\{de\}$	$+\sqrt{de}$	0	0	$-\sqrt{ae}$	$+\sqrt{ad}$	$+\sqrt{bc}$	0	$-\sqrt{be}$	$+\sqrt{bd}$	$+\sqrt{ac}$	$+\sqrt{ce}$	$-\sqrt{cd}$	$-\sqrt{ab}$	0	0	0

Referring now to Göpel's memoir, *Crelle*, t. XXXV. (1847), pp. 277—312, we have the sixteen double ϑ-functions

$$P,\ P_1,\ P_2,\ P_3;\ \ iQ,\ Q_1,\ iQ_2,\ Q_3;\ \ iR,\ iR_1,\ R_2,\ R_3;\ \ S,\ iS_1,\ iS_2,\ S_3,$$

where the six functions affected with the $i\ (=\sqrt{-1})$ are odd functions, vanishing for the values $u=0$, $u'=0$ of the arguments. It is convenient to take ∞, ∞ as the values of x, x' corresponding to these values $u=0$, $u'=0$: the expressions $\{a\}$ will thus correspond to the six squares $-Q^2$, $-Q_2^2$, $-R^2$, $-R_1^2$, $-S_1^2$, $-S_2^2$, and the expressions $\{ab\}$ to the remaining ten squares P^2, P_1^2, ..., S_3^2; and after some *tâtonnement*, I succeed in establishing the correspondence as follows

$$\begin{array}{cccccccccccccccc} S_2^2, & S_1^2, & R_1^2, & R^2, & Q^2, & Q_2^2, & Q_1^2, & P_1^2, & P^2, & S^2, & P_2^2, & P_3^2, & S_3^2, & Q_3^2, & R_3^2, & R_2^2, \\ =\{a\}, & \{b\}, & \{c\}, & \{d\}, & \{e\}, & \{f\}, & \{ab\}, & \{ac\}, & \{ad\}, & \{ae\}, & \{bc\}, & \{bd\}, & \{be\}, & \{cd\}, & \{ce\}, & \{de\}, \end{array}$$

viz. the sixteen squared double ϑ-functions are proportional to the sixteen expressions $-\{a\}$, $+\{ab\}$, as hereby appearing.

TABLE III. OF THE SIXTEEN FORMS OF

	0	A	B	$A+B$	K	$K+A$	$K+B$	$K+A+B$
$-S_2^2$	$-S_2^2=a$	$-S_3^2=-be$	$-S^2=-ae$	$-S_1^2=b$	$+R_2^2=de$	$R_3^2=ce$	$R^2=-d$	$R_1^2=-c$
$-S_1^2$	$-S_1^2=b$	$-S^2=-ae$	$-S_3^2=-be$	$-S_2^2=a$	$-R_1^2=c$	$-R^2=d$	$-R_3^2=-ce$	$-R_2^2=-de$
$-R_1^2$	$-R_1^2=c$	$-R^2=+d$	$-R_3^2=-ce$	$-R_2^2=-de$	$-S_1^2=b$	$-S^2=-ae$	$-S_3^2=-be$	$-S_2^2=a$
$-R^2$	$-R^2=d$	$-R_1^2=+c$	$-R_2^2=-de$	$-R_3^2=-ce$	$S^2=ae$	$S_1^2=-b$	$S_2^2=-a$	$S_3^2=be$
$-Q^2$	$-Q^2=e$	$-Q_1^2=-ab$	$-Q_2^2=+f$	$-Q_3^2=-cd$	$P^2=ad$	$P_1^2=ac$	$P_2^2=bc$	$P_3^2=bd$
$-Q_2^2$	$-Q_2^2=f$	$-Q_3^2=-cd$	$-Q^2=+e$	$-Q_1^2=-ab$	$P_2^2=bc$	$P_3^2=bd$	$P^2=ad$	$P_1^2=ac$
Q_1^2	$Q_1^2=ab$	$Q^2=-e$	$Q_3^2=cd$	$Q_2^2=-f$	$P_1^2=ac$	$P^2=ad$	$P_3^2=bd$	$P_2^2=bc$
P_1^2	$P_1^2=ac$	$P^2=ad$	$P_3^2=bd$	$P_2^2=bc$	$Q_1^2=ab$	$Q^2=-e$	$Q_3^2=cd$	$Q_2^2=-f$
P^2	$P^2=ad$	$P_1^2=ac$	$P_2^2=bc$	$P_3^2=bd$	$-Q^2=e$	$-Q_1^2=-ab$	$-Q_2^2=f$	$-Q_3^2=-cd$
S^2	$S^2=ae$	$S_1^2=-b$	$S_2^2=-a$	$S_3^2=be$	$-R^2=d$	$-R_1^2=c$	$-R_2^2=-de$	$-R_3^2=-ce$
P_2^2	$P_2^2=bc$	$P_3^2=bd$	$P^2=ad$	$P_1^2=ac$	$-Q_2^2=f$	$-Q_3^2=-cd$	$-Q^2=e$	$-Q_1^2=-ab$
P_3^2	$P_3^2=bd$	$P_2^2=bc$	$P_1^2=ac$	$P^2=ad$	$Q_3^2=cd$	$Q_2^2=-f$	$Q_1^2=ab$	$Q^2=-e$
S_3^2	$S_3^2=be$	$S_2^2=-a$	$S_1^2=-b$	$S^2=ae$	$R_3^2=ce$	$R_2^2=de$	$R_1^2=c$	$R^2=-d$
Q_3^2	$Q_3^2=cd$	$Q_2^2=-f$	$Q_1^2=ab$	$Q^2=-e$	$P_3^2=bd$	$P_2^2=bc$	$P_1^2=ac$	$P^2=ad$
R_3^2	$R_3^2=ce$	$R_2^2=de$	$R_1=-c$	$R^2=-d$	$S_3^2=be$	$S_2^2=-a$	$S_1^2=-b$	$S^2=ae$
R_2^2	$R_2^2=de$	$R_3^2=ce$	$R^2=-d$	$R_1^2=-c$	$-S_2^2=a$	$-S_3^2=-be$	$-S^2=-ae$	$-S_1^2=-b$
	$\infty\infty$	cd	ef	ab	bc	bd	ad	ac

We have, after Göpel (*l.c.* p. 283), a table showing how the ratios of the double ϑ-functions are altered, when the arguments are increased by the quarter-periods

$$A,\ B,\ A+B,\ K,\ L,\ K+L,$$

that is, when u, u' are simultaneously changed into $u+A$, $u'+A'$ or into $u+B$, $u'+B'$ etc. If instead, we consider the squared functions, the table is very much simplified, inasmuch as in place of the coefficients ± 1, $\pm i$, it will contain only the coefficients ± 1: and we may complete the table by extending it to all the combinations 0, A, B, $A+B$, K, $K+A$, $K+B$, $K+A+B$, L, $L+A$, $L+B$, $L+A+B$, $K+L$, $K+L+A$, $K+L+B$, $K+L+A+B$ of the quarter-periods: we have thus a table included in the annexed Table III., viz. attending herein only to the capital letters P, Q, R, S, the sixteen columns of the table show how the ratios of the terms $-S_2^2$, $-S_1^2$, etc., of the first column are altered when the arguments are increased by the foregoing combinations of quarter-periods, as indicated by the headings 0, A, B, etc., of the several columns.

THE SQUARED DOUBLE ϑ-FUNCTIONS.

L	$L+A$	$L+B$	$L+A+B$	$K+L$	$K+L+A$	$K+L+B$	$K+L+A+B$
$-Q_2^2=f$	$-Q_3^2=-cd$	$-Q^2=e$	$-Q_1^2=-ab$	$P_2^2=bc$	$P_3^2=bd$	$P^2=ad$	$P_1^2=ac$
$Q_1^2=ab$	$Q^2=-e$	$Q_3^2=cd$	$Q_2^2=-f$	$P_1^2=ac$	$P^2=ad$	$P_3^2=bd$	$P_2^2=bc$
$P_1^2=ac$	$P^2=ad$	$P_3^2=bd$	$P_2^2=bc$	$Q_1^2=ab$	$Q^2=-e$	$Q_3^2=cd$	$Q_2^2=-f$
$P^2=ad$	$P_1^2=ac$	$P_2^2=bc$	$P_3^2=bd$	$-Q^2=e$	$-Q_1^2=-ab$	$-Q_2^2=f$	$-Q_3^2=-cd$
$S^2=ae$	$S_1^2=-b$	$S_2^2=-a$	$S_3^2=be$	$-R^2=d$	$-R_1^2=c$	$-R_2^2=-de$	$-R_3^2=-ce$
$-S_2^2=a$	$-S_3^2=-be$	$-S^2=-ae$	$-S_1^2=b$	$R_2^2=de$	$R_3^2=ce$	$R^2=-d$	$R_1^2=-c$
$-S_1^2=b$	$-S^2=-ae$	$-S_3^2=-be$	$-S_2^2=a$	$-R_1^2=c$	$-R^2=d$	$-R_3^2=-ce$	$-R_2^2=-de$
$-R_1^2=c$	$-R^2=d$	$-R_3^2=-ce$	$-R_2^2=-de$	$-S_1^2=b$	$-S^2=-ae$	$-S_3^2=-be$	$-S_2^2=a$
$-R^2=d$	$-R_1^2=c$	$-R_2^2=-de$	$-R_3^2=-ce$	$S^2=ae$	$S_1^2=b$	$S_2^2=-a$	$S_3^2=be$
$-Q^2=e$	$-Q_1^2=-ab$	$-Q_2^2=f$	$-Q_3^2=-cd$	$P^2=ad$	$P_1^2=ac$	$P_2^2=bc$	$P_3^2=bd$
$R_2^2=de$	$R_3^2=ce$	$R^2=-d$	$R_1^2=-c$	$-S_2^2=a$	$-S_3^2=-be$	$-S^2=-ae$	$-S_1^2=b$
$R_3^2=ce$	$R_2^2=de$	$R_1^2=-c$	$R^2=-d$	$S_3^2=be$	$S_2^2=-a$	$S_1^2=-b$	$S^2=ae$
$Q_3^2=cd$	$Q_2^2=-f$	$Q_1^2=ab$	$Q^2=-e$	$P_3^2=bd$	$P_2^2=bc$	$P_1^2=ac$	$P^2=ad$
$S_3^2=be$	$S_2^2=-a$	$S_1^2=-b$	$S^2=ae$	$R_3^2=ce$	$R_2^2=de$	$R_1^2=-c$	$R^2=-d$
$P_3^2=bd$	$P_2^2=bc$	$P_1^2=ac$	$P^2=ad$	$Q_3^2=cd$	$Q_2^2=-f$	$Q_1^2=ab$	$Q^2=-e$
$P_2^2=bc$	$P_3^2=bd$	$P^2=ad$	$P_1^2=ac$	$-Q_2^2=f$	$-Q_3^2=-cd$	$-Q^2=e$	$-Q_1^2=-ab$
af	be	ae	bf	de	ce	df	cf

But I have also in the table inserted the values to which $-S_2^2$, $-S_1^2$, etc., are respectively proportional, viz. the table runs $-S_2^2 = a$, $-S_1^2 = b$, etc., (read $-S_2^2 = \{a\}$, $-S_1^2 = \{b\}$, etc., the brackets $\{\ \}$ having been for greater brevity omitted throughout the table), and where it is of course to be understood that $-S_2^2$, $-S_1^2$, etc., are proportional only, not absolutely equal to $\{a\}$, $\{b\}$, etc. And I have also at the foot of the several columns inserted suffixes $\infty\infty$, *ab*, *cd*, etc., which refer to the columns of Table II.

Comparing the first with any other column of the table, for instance with the second column, the two columns respectively signify that

$$\begin{array}{c|c} -S_2^2(u) = \{a\}, & -S_2^2(u+A) = -\{be\}, \\ -S_1^2(u) = \{b\}, & -S_1^2(u+A) = -\{ae\}, \\ \vdots & \vdots \\ Q_1^2(u) = \{ab\}, & Q_1^2(u+A) = -\{e\}, \\ \vdots & \vdots \end{array}$$

where, as before, the sign $=$ means only that the terms are proportional; u is written for shortness instead of (u, u'), and so $u+A$ for $(u+A, u'+A')$, etc.: the variables in the functions $\{a\}$, $\{be\}$, etc. are in each case x, x'. But if in the second column we write $u-A$ for A, then the variables x, x' will be changed into new variables y, y', or the meaning will be

$$\begin{array}{c|c} x,\ x' & y,\ y' \\ -S_2^2(u) = \{a\}, & -S_2^2(u) = -\{be\}, \\ -S_1^2(u) = \{b\}, & -S_1^2(u) = -\{ae\}, \\ \vdots & \vdots \\ Q_1^2(u) = \{ab\}, & Q_1^2(u) = -\{e\}, \\ \vdots & \vdots \end{array}$$

so that, omitting from the table the terms which contain the capital letters P, Q, R, S, except only the outside left-hand column $-S_2^2$, $-S_1^2$, etc., the table indicates that these functions $-S_2^2$, $-S_1^2$, etc., are proportional to the functions $\{a\}$, $\{b\}$, etc., of x, x' given in the first column; also to the functions $-\{be\}$, $-\{ae\}$, etc., of y, y' given in the second column; also to the functions $-\{ae\}$, $-\{be\}$, etc., of z, z' given in the third column; and so on, with a different pair of variables in each of the 16 columns.

Thus comparing any two columns, for instance the first and second, it appears that we can have simultaneously

$$\begin{array}{rcl} x,\ x' & & y,\ y' \\ \{a\} & = & -\{be\}, \\ \{b\} & = & -\{ae\}, \\ \vdots & & \\ \{ab\} & = & -\{e\}, \\ \vdots & & \end{array}$$

(fifteen equations, since the meaning is that the terms are only proportional, not absolutely equal), equivalent to two equations serving to determine x and x' in terms of y and y', or conversely y and y' in terms of x and x'. The functions in each column form in fact 16 sixes, such that any four belonging to the same six are linearly connected; and in any such linear relation between four functions in the left-hand column, substituting for these their values as functions in the right-hand column, we have the corresponding relations between four functions out of a set of six belonging to the right-hand column, or we have an identity $0=0$. I will presently verify this in a particular case.

If in any column we give to the variables the values ∞, ∞ we obtain for the terms in the column the values which the terms of the first column assume on giving to x, x' the values shown at the foot of the column in question; thus, in the second column giving to the variables the values ∞, ∞, the column becomes

$$-\sqrt{be},\ -\sqrt{ae},\ 0,\ 0,\ -\sqrt{ab},\ -\sqrt{cd},\ 0,\ \sqrt{ad},\ \sqrt{ac},\ 0,\ \sqrt{bd},\ \sqrt{bc},\ 0,\ 0,\ \sqrt{de},\ \sqrt{ce}$$

which is, in fact, the cd-column of Table II.: this is of course as it should be, for the values in question are those of the functions $-S_2^2$, $-S_1^2$, etc., on writing therein

$$x,\ x'=c,\ d.$$

The formulæ show that

$$\sqrt{ab},\ \sqrt{ac},\ \sqrt{ad},\ \sqrt{ae},\ \sqrt{bc},\ \sqrt{bd},\ \sqrt{be},\ \sqrt{cd},\ \sqrt{ce},\ \sqrt{de},$$

are, in fact, proportional to

$$k_1^2,\quad \varpi_1^2,\quad \varpi^2,\quad \sigma^2,\quad \varpi_2^2,\quad \varpi_3^2,\quad \sigma_3^2,\quad k_3^2,\quad \rho_3^2,\quad \rho_2^2,$$

(k_1, k_2, ... are Göpel's k', k'', ...). This gives rise to a remarkable theorem, for the ten squares are functions of only four quantities α, β, γ, δ (Göpel's t, u, v, w). For greater clearness, I introduce single letters A, B, ..., J and write

$$\begin{aligned}
A &= abc\,.\,def = (\sqrt{de})^2 = \rho_2^4, &&= (\alpha^2-\beta^2+\gamma^2-\delta^2)^2,\\
B &= abd\,.\,cef = (\sqrt{ce})^2 = \rho_3^4, &&= 4(\alpha\gamma+\beta\delta)^2,\\
C &= abe\,.\,cdf = (\sqrt{cd})^2 = k_3^4, &&= 4(\alpha\delta+\beta\gamma)^2,\\
D &= abf\,.\,cde = (\sqrt{ab})^2 = k_1^4, &&= (\alpha^2-\beta^2-\gamma^2+\delta^2)^2,\\
E &= acd\,.\,bef = (\sqrt{be})^2 = \sigma_3^4, &&= 4(\alpha\beta+\gamma\delta)^2,\\
F &= ace\,.\,bdf = (\sqrt{bd})^2 = \varpi_3^4, &&= (\alpha^2+\beta^2+\gamma^2+\delta^2)^2,\\
G &= acf\,.\,bde = (\sqrt{ac})^2 = \varpi_1^4, &&= 4(\alpha\delta-\beta\gamma)^2,\\
H &= ade\,.\,bcf = (\sqrt{bc})^2 = \varpi_2^4, &&= 4(\alpha\gamma-\beta\delta)^2,\\
I &= adf\,.\,bce = (\sqrt{ad})^2 = \varpi^4, &&= 4(\alpha\beta-\gamma\delta)^2,\\
J &= aef\,.\,bdc = (\sqrt{ae})^2 = \sigma^4, &&= (\alpha^2+\beta^2-\gamma^2-\delta^2)^2;
\end{aligned}$$

viz. it has to be shown that A, B, ..., J, considered as given functions of the six letters a, b, c, d, e, f, are really functions of four quantities α, β, γ, δ; or, what is the same thing, that A, B, ..., J, considered as functions of a, b, c, d, e, f satisfy all those relations which they satisfy when considered as given functions of α, β, γ, δ.

Now considering them as given functions of α, β, γ, δ, they ought to satisfy six relations; and inasmuch as, so considered, they are, in fact, linear functions of

$$\alpha^4+\beta^4+\gamma^4+\delta^4,\quad \alpha^2\beta^2+\gamma^2\delta^2,\quad \alpha^2\gamma^2+\beta^2\delta^2,\quad \alpha^2\delta^2+\beta^2\gamma^2,\quad \alpha\beta\gamma\delta,$$

five of these relations will be linear: there is a sixth non-linear relation, expressible in a variety of different forms, one of them, as is easily verified, being

$$\sqrt{AJ}\pm\sqrt{CG}\pm\sqrt{DF}=0.$$

Now considering A, B, ..., J as given functions of a, b, c, d, e, f, there exist between them linear relations which may be obtained by the consideration of identities of the form

$$\begin{vmatrix} abcd \\ abcdef \end{vmatrix}=0,$$

where the left-hand side is used for shortness to denote the determinant

$$\begin{vmatrix} 1, & 1, & 1, & 1 & & \\ a, & b, & c, & d & & \\ a^2, & b^2, & c^2, & d^2 & & \\ 1, & 1, & 1, & 1, & 1, & 1 \\ a, & b, & c, & d, & e, & f \\ a^2, & b^2, & c^2, & d^2, & e^2, & f^2 \end{vmatrix}=0.$$

We thus obtain between them a system of fifteen linear relations, which present themselves in the form

$$\begin{array}{rl}
(1) & A-J+E-B=0,\\
(2) & -A-I+F-C=0,\\
(3) & A-H+G-D=0,\\
(4) & -B-G+H+C=0,\\
(5) & B-F+I+D=0,\\
(6) & C-E+J-D=0,\\
(7) & -E-D-H+E=0,\\
(8) & E-C-I+G=0,\\
(9) & F-B-J-G=0,\\
(10) & H-A+J-I=0,\\
(11) & -J+D-G+I=0,\\
(12) & J+C-F+H=0,\\
(13) & I+B-E-H=0,\\
(14) & G+A+E-F=0,\\
(15) & D-A+B-C=0,
\end{array}$$

and these are all included in the equations (10), (4), (12), (15), (6), which serve to express G, B, E, F, I in terms of D, H, C, A, J, i.e. ac, ce, eb, bd, da in terms of ab, bc, cd, de, ea, if for the moment we write $G = ac$, etc. But the five linear relations in question are, it is at once seen, satisfied by A, B, ..., J considered as given functions of α, β, γ, δ.

The equation $\sqrt{AJ} \pm \sqrt{DF} \pm \sqrt{CG} = 0$, substituting for A, B, ..., J their values in terms of a, b, c, d, e, f, becomes

$$\sqrt{abc \,.\, def \,.\, aef \,.\, bcd} \pm \sqrt{abf \,.\, cde \,.\, ace \,.\, bdf} \pm \sqrt{abe \,.\, cdf \,.\, acf \,.\, bde} = 0,$$

which (omitting common factors) becomes $\sqrt{bc^2 \,.\, ef^2} \pm \sqrt{bf^2 \,.\, ce^2} \pm \sqrt{be^2 \,.\, cf^2} = 0$; or, taking the proper signs, this is the identity $bc \,.\, ef + be \,.\, fc + bf \,.\, ce = 0$.

It is to be noticed that

$$\begin{array}{ccc} \delta^2 + \alpha^2 - \beta^2 - \gamma^2, & 2(\alpha\beta - \gamma\delta), & 2(\gamma\alpha + \beta\delta), \\ 2(\alpha\beta + \gamma\delta), & \delta^2 + \beta^2 - \gamma^2 - \alpha^2, & 2(\beta\gamma - \alpha\delta), \\ 2(\gamma\alpha - \beta\delta), & 2(\beta\gamma + \alpha\delta), & \delta^2 + \gamma^2 - \alpha^2 - \beta^2, \end{array}$$

each divided by $\delta^2 + \alpha^2 + \beta^2 + \gamma^2$, form a system of coefficients in the transformation between two sets of rectangular coordinates. We have therefore

$$\begin{array}{ccc} \sqrt{ab}, & \sqrt{ad}, & \sqrt{ce}, \\ \sqrt{be}, & \sqrt{de}, & \sqrt{ac}, \\ \sqrt{bc}, & \sqrt{cd}, & \sqrt{ae}, \end{array}$$

each divided by $\sqrt{bd}$ and the several terms taken with proper signs, as a system of coefficients in the transformation between two sets of rectangular axes: a result which seems to be the same as that obtained by Hesse in the Memoir, "Transformations-Formeln für rechtwinklige Raum-Coordinaten"; *Crelle*, t. LXIII. (1864), pp. 247—251.

The composition of the last mentioned system of functions is better seen by writing them under the fuller form $\sqrt{abf \,.\, cde}$, etc.; viz. omitting the radical signs, the terms are

$$\begin{array}{ccc} abf \,.\, cde, & adf \,.\, bce, & abd \,.\, cef, \\ bef \,.\, acd, & def \,.\, abc, & acf \,.\, bde, \\ bcf \,.\, ade, & cdf \,.\, abe, & aef \,.\, bcd, \end{array}$$

each divided by $bdf \,.\, ace$; or, in an easily understood algorithm, the terms are

$$\begin{array}{c|ccc} & bf.d & df.b & bd.f \\ \hline a.ce & bf.d & df.b & bd.f \\ e.ac & bf.d & df.b & bd.f \\ c.ae & bf.d & df.b & bd.f, \end{array}$$

each divided by $bdf \,.\, ace$.

Reverting to the before-mentioned comparison of the first and second columns of Table III., four of the equations are

$$\begin{array}{lllll} x, x' & y, y' & & x, x' & y, y' \\ \{c\} = & \{d\}, & \text{that is,} & \sqrt{c}\,[c] = & \sqrt{d}\,[d], \\ \{d\} = & \{c\}, & \text{that is,} & \sqrt{d}\,[d] = & \sqrt{c}\,[c], \\ \{e\} = - & \{ab\}, & \text{that is,} & \sqrt{e}\,[e] = - & \sqrt{ab}\,[ab], \\ \{f\} = - & \{cd\}, & \text{that is,} & \sqrt{f}\,[f] = - & \sqrt{cd}\,[cd]; \end{array}$$

viz. the four terms on the left-hand side are not absolutely equal, but only proportional, to those on the right-hand side. Substituting for $\sqrt{c}$, $\sqrt{d}$, etc., their values, and introducing on the right-hand side the factor

$$\sqrt{ac\,.\,bc\,.\,ce\,.\,cf\,.\,ad\,.\,bd\,.\,de\,.\,df},$$

the equations become

$$\begin{array}{ll} xx' & yy' \\ [c] = & ac\,.\,bc\,.\,ce\,.\,ef\,[d], \\ [d] = & ad\,.\,bd\,.\,de\,.\,df\,[c], \\ [e] = - & ce\,.\,de\,[ab], \\ [f] = - & cf\,.\,df\,[cd]. \end{array}$$

The functions on the left-hand satisfy the identity

$$def\,[c] - efc\,[d] + fcd\,[e] - cde\,[f] = 0,$$

or, as this may also be written,

$$def\,[c] - cef\,[d] + cdf\,[e] - cde\,[f] = 0.$$

Hence substituting the right-hand values, the whole equation divides by $ce\,.\,de\,.\,cf\,.\,df$; omitting this factor, it becomes

$$ef\,.\,ac\,.\,bc\,[d] - ef\,.\,ad\,.\,bd\,[c] - cd\,\{[ab] - [cd]\} = 0,$$

where the variables are y, y': it is to be shown that this is in fact an identity, and (as it is thus immaterial what the variables are) I change them into x, x'.

We have

$$\begin{aligned} ac\,.\,bc\,[d] - ad\,.\,bd\,[c] &= (a-c)(b-c)(d-x)(d-x') \\ &\quad - (a-d)(b-d)(c-x)(c-x') \\ &= (c-d)\begin{vmatrix} 1, & x+x', & xx' \\ 1, & a+b, & ab \\ 1, & c+d, & cd \end{vmatrix} \\ &= cd\,[xx'abcd], \end{aligned}$$

suppose.

We have moreover

$$[ab]-[cd]=\frac{1}{(x-x')^2}\begin{Bmatrix} abf.c'd'e'+a'b'f'.cde \\ -cdf.a'b'e'-c'd'f'.abe\end{Bmatrix}$$

$$=\frac{1}{(x-x')^2}(abc'd'-a'b'cd)(e'f-ef'),$$

where for the moment a, b, a', etc., are written to denote $a-x$, $b-x$, $a-x'$, etc.; we have then

$$e'f-ef'=(e-x')(f-x)-(e-x)(f-x')$$
$$=-(e-f)(x-x')=-ef(x-x'),$$

and

$$abc'd'-a'b'cd=(a-x)(b-x)(c-x')(d-x')=-(x-x')\begin{vmatrix} 1, & x+x', & xx' \\ 1, & a+b, & ab \\ 1, & c+d, & cd\end{vmatrix}$$
$$-(a-x')(b-x')(c-x)(d-x)$$
$$=-(x-x')[xx'abcd].$$

Hence $[ab]-[cd]=ef[xx'abcd]$, and the equation to be verified becomes

$$(ef.cd-cd.ef)[xx'abcd]=0,$$

viz. this is, in fact, an identity.

Cambridge, 14 *March*, 1877.

664.

ON THE 16-NODAL QUARTIC SURFACE.

[From the *Journal für die reine und angewandte Mathematik* (Crelle), t. LXXXIV. (1877), pp. 238—241.]

PROF. BORCHARDT in the Memoir "Ueber die Darstellung u. s. w." *Crelle*, t. LXXXIII. (1877), pp. 234—243, shows that the coordinates x, y, z, w may be taken as proportional to four of the double ϑ-functions, and that the equation of the surface is then Göpel's relation of the fourth order between these four functions: and he remarks at the end of the memoir that it thus appears that the coordinates x, y, z, w of a point on the surface can be expressed as proportional to algebraic functions, involving square roots only, of two arbitrary parameters ξ, ξ'.

It is interesting to develope the theory from this point of view. Writing, as in my paper, "Further investigations on the double ϑ-functions," pp. 220—233, [663],

$$[a] = aa',$$
$$[b] = bb',$$
$$[c] = cc',$$
$$[d] = dd',$$
$$[e] = ee',$$
$$[f] = ff',$$
$$[ab] = \frac{1}{(\xi - \xi')^2} (\sqrt{abfc'd'e'} - \sqrt{a'b'f'cde})^2,$$
etc.,

where on the right-hand sides $a, b, \ldots, a', \ldots$ denote $a - \xi$, $b - \xi, \ldots, a - \xi', \ldots$ (ξ, ξ' being here written in place of the x, x' of my paper), then the sixteen functions are proportional to constant multiples of the square-roots of these expressions; viz. the correspondence is

$$\begin{array}{cccccc} S_2=\vartheta_{13}, & S_1=\vartheta_{24}, & R_1=\vartheta_3, & R=\vartheta_{04}, & Q=\vartheta_1, & Q_2=\vartheta_{02}, \\ i\sqrt[4]{a}\sqrt{[a]}, & i\sqrt[4]{b}\sqrt{[b]}, & i\sqrt[4]{c}\sqrt{[c]}, & i\sqrt[4]{d}\sqrt{[d]}, & i\sqrt[4]{e}\sqrt{[e]}, & i\sqrt[4]{f}\sqrt{[f]}; \\ Q_1=\vartheta_2, & P_1=\vartheta_{34}, & P=\vartheta_{01}, & S=-\vartheta_{14}, & P_2=\vartheta_{12}, & P_3=\vartheta_5, \\ \sqrt[4]{ab}\sqrt{[ab]}, & \sqrt[4]{ac}\sqrt{[ac]}, & \sqrt[4]{ad}\sqrt{[ad]}, & \sqrt[4]{ae}\sqrt{[ae]}, & \sqrt[4]{bc}\sqrt{[bc]}, & \sqrt[4]{bd}\sqrt{[bd]}; \end{array}$$

$$\begin{array}{cccc} S_3=\vartheta_{23}, & Q_3=\vartheta_0, & R_3=\vartheta_4, & R_2=\vartheta_{03}, \\ \sqrt[4]{be}\sqrt{[be]}, & \sqrt[4]{cd}\sqrt{[cd]}, & \sqrt[4]{ce}\sqrt{[ce]}, & \sqrt[4]{de}\sqrt{[de]}; \end{array}$$

where, under the signs $\sqrt[4]{\ }$, a signifies $bcdef$, that is, $bc\,.\,bd\,.\,be\,.\,bf\,.\,cd\,.\,ce\,.\,cf\,.\,de\,.\,df\,.\,ef$, and ab signifies $abf\,.\,cde$, that is, $ab\,.\,af\,.\,bf\,.\,cd\,.\,ce\,.\,de$, in which expressions bc, bd, ..., ab, af, ... signify the differences $b-c$, $b-d$, ..., $a-b$, $a-f$, ... But in what follows, we are not concerned with the values of these constant multipliers.

Prof. Borchardt's coordinates x, y, z, w are

$$x=\vartheta_0=P;\ \ y=\vartheta_{23}=S_3;\ \ z=\vartheta_{14}=-S;\ \ w=\vartheta_5=P_3;$$

viz. P, S, P_3, S_3 are a set connected by Göpel's relation of the fourth order—and this relation can be found (according to Göpel's method) by showing that Q^2 and R^2 are each of them a linear function of the four squares P^2, P_3^2, S^2, S_3^2, and further that QR is a linear function of PS and P_3S_3; for then, squaring the expression of QR, and for Q^2 and R^2 substituting their values, we have the required relation of the fourth order between P, S, P_3, S_3.

Now we have P, S, P_3, S_3, Q, R = constant multiples of $\sqrt{[ac]}$, $\sqrt{[ab]}$, $\sqrt{[cd]}$, $\sqrt{[bd]}$, $\sqrt{[b]}$, $\sqrt{[c]}$ respectively: and it of course follows that we must have the like relations between these six quantities; viz. we must have $[b]$, $[c]$ each of them a linear function of $[ac]$, $[ab]$, $[cd]$, $[bd]$; and moreover $\sqrt{[b]}\sqrt{[c]}$ a linear function of $\sqrt{[ac]}\sqrt{[ab]}$ and $\sqrt{[bd]}\sqrt{[cd]}$.

As regards this last relation, starting from the formulæ

$$\sqrt{[ac]}=\frac{1}{\xi-\xi'}\{\sqrt{acfb'd'e'}+\sqrt{a'c'f'bde}\},$$

$$\sqrt{[bd]}=\frac{1}{\xi-\xi'}\{\sqrt{bdfa'c'e'}+\sqrt{b'd'f'ace}\},$$

$$\sqrt{[ab]}=\frac{1}{\xi-\xi'}\{\sqrt{abfc'd'e'}+\sqrt{a'b'f'cde}\},$$

$$\sqrt{[cd]}=\frac{1}{\xi-\xi'}\{\sqrt{cdfa'b'e'}+\sqrt{c'd'f'abe}\},$$

we have at once

$$\sqrt{[ac]}\sqrt{[ab]}=\frac{1}{(\xi-\xi')^2}\{(afd'e'+a'f'de)\sqrt{bcb'c'}+(bc'+b'c)\sqrt{adea'd'e'}\},$$

$$\sqrt{[bd]}\sqrt{[cd]}=\frac{1}{(\xi-\xi')^2}\{(dfa'e'+d'f'ae)\sqrt{bcb'c'}+(bc'+b'c)\sqrt{adea'd'e'}\};$$

the difference of these two expressions is

$$=\frac{1}{(\xi-\xi')^2}(ad'-a'd)(fe'-f'e)\sqrt{bcb'c'},$$

where substituting for a, d, e, f, a', ... their values $a-\xi$, $d-\xi$, $e-\xi$, $f-\xi$, $a-\xi'$, ... we have $ad'-a'd=(a-d)(\xi-\xi')$, $fe'-f'e=(f-e)(\xi-\xi')$; also $\sqrt{bcb'c'}=\sqrt{[b]}\sqrt{[c]}$; and we have thus the required relation

$$\sqrt{[ac]}\sqrt{[ab]}-\sqrt{[bd]}\sqrt{[cd]}=-(a-d)(e-f)\sqrt{[b]}\sqrt{[c]}.$$

As regards the first mentioned relation, if for greater generality, θ being arbitrary, we write $[\theta]=\theta\theta'$, that is, $=(\theta-\xi)(\theta-\xi')$, then it is easy to see that there exists a relation of the form

$$\nabla[\theta]=A[ab]+B[ac]+C[bd]+D[cd],$$

where $A+B+C+D=0$. The right-hand side is thus a linear function of the differences $[ab]-[ac]$, $[ab]-[bd]$, $[ab]-[cd]$; and each of these, as the irrational terms disappear and the rational terms divide by $(\xi-\xi')^2$, is a mere linear function of 1, $\xi+\xi'$, $\xi\xi'$; whence there is a relation of the form in question. I found without much difficulty the actual formula; viz. this is

$$(a-d)(b-c)(e-f)\begin{vmatrix}1, & e+f, & ef\\ 1, & b+c, & bc\\ 1, & a+d, & ad\end{vmatrix}[\theta]$$

$$=\begin{vmatrix}1, & e, & f, & ef\\ 1, & b, & c, & bc\\ 1, & d, & a, & ad\\ 1, & \theta, & \theta, & \theta^2\end{vmatrix}[ac]-\begin{vmatrix}1, & e, & f, & ef\\ 1, & c, & b, & bc\\ 1, & d, & a, & ad\\ 1, & \theta, & \theta, & \theta^2\end{vmatrix}[ab]-\begin{vmatrix}1, & e, & f, & ef\\ 1, & b, & c, & bc\\ 1, & a, & d, & ad\\ 1, & \theta, & \theta, & \theta^2\end{vmatrix}[cd]+\begin{vmatrix}1, & e, & f, & ef\\ 1, & c, & b, & bc\\ 1, & a, & d, & ad\\ 1, & \theta, & \theta, & \theta^2\end{vmatrix}[bd],$$

where observe that on the right-hand side the last three determinants are obtained from the first one by interchanging b, c: or a, d: or b, c and a, d simultaneously: a single interchange gives the sign $-$, but for two interchanges the sign remains $+$.

Writing successively $\theta = b$ and $\theta = c$, we obtain

$$(a-d)(e-f)\begin{vmatrix} 1, & e+f, & ef \\ 1, & b+c, & bc \\ 1, & a+d, & ad \end{vmatrix}[b]$$

$$= (a-f)(b-d)(b-e)[ac] - (a-b)(b-f)(d-e)[ab]$$
$$+ (a-b)(b-e)(d-f)[cd] - (a-e)(b-d)(b-f)[bd];$$

$$(a-d)(e-f)\begin{vmatrix} 1, & e+f, & ef \\ 1, & b+c, & bc \\ 1, & a+d, & ad \end{vmatrix}[c]$$

$$= -(a-c)(c-f)(d-e)[ac] + (a-f)(c-d)(c-e)[ab]$$
$$-(a-e)(c-d)(c-f)[cd] + (a-c)(c-e)(d-f)[bd];$$

which values of $[b]$ and $[c]$, combined with the foregoing equation

$$(a-d)(e-f)\sqrt{[b]}\sqrt{[c]} = -\sqrt{[ac]}\sqrt{[ab]} + \sqrt{[cd]}\sqrt{[bd]},$$

give the required quartic equation between $\sqrt{[ac]}$, $\sqrt{[ab]}$, $\sqrt{[cd]}$, $\sqrt{[bd]}$.

Cambridge, 2 August, 1877.

665.

A MEMOIR ON THE DOUBLE ϑ-FUNCTIONS.

[From the *Journal für die reine und angewandte Mathematik* (Crelle), t. LXXXV. (1878), pp. 214—245.]

I RESUME my investigations on these functions; see my two papers, *Crelle*, t. LXXXIII. (1877), pp. 210—233; [662] and [663]. But it is proper in the first instance to develope in a corresponding manner, the theories of the circular (or exponential) functions, and of the single ϑ-functions.

Part I. Preliminary investigations.

Starting from the differential relation

$$\partial u = \frac{\partial x}{\sqrt{a-x\,.\,b-x}}$$

between the variables u and x, I write for shortness the single letters A, B, Ω, instead of functional forms $A(u)$, $B(u)$, $\Omega(u)$, to denote functions of u; and I assume as definitions the equations

$$A = \Omega\sqrt{a-x},$$

$$B = \Omega\sqrt{b-x},$$

and another equation to be presently mentioned: these two equations imply between A, B, Ω the algebraical relation

$$A^2 - B^2 = \Omega^2(a-b).$$

Differentiating, we obtain

$$\partial A = \partial\Omega\,.\,\sqrt{a-x} - \frac{\Omega}{2\sqrt{a-x}}\sqrt{a-x\,.\,b-x}\,\partial u,$$

that is,

$$\partial A = \frac{A}{\Omega}\partial\Omega - \tfrac{1}{2}B\,\partial u,$$

and similarly

$$\partial B = \frac{B}{\Omega}\partial\Omega - \tfrac{1}{2}A\,\partial u,$$

whence

$$A\,\partial B - B\,\partial A = -\tfrac{1}{2}(A^2 - B^2)\,\partial u.$$

Proceeding to a second differentiation, we find

$$\partial^2 A = \frac{A}{\Omega}\partial^2\Omega - \frac{B}{\Omega}\partial\Omega\,\partial u + \tfrac{1}{4}A\,(\partial u)^2,$$

$$\partial^2 B = \frac{B}{\Omega}\partial^2\Omega - \frac{A}{\Omega}\partial\Omega\,\partial u + \tfrac{1}{4}B\,(\partial u)^2,$$

and thence

$$A\,\partial^2 A - (\partial A)^2 = \frac{A^2}{\Omega^2}\{\Omega\,\partial^2\Omega - (\partial\Omega)^2\} + \tfrac{1}{4}(A^2 - B^2)\,(\partial u)^2,$$

$$B\,\partial^2 B - (\partial B)^2 = \frac{B^2}{\Omega^2}\{\Omega\,\partial^2\Omega - (\partial\Omega)^2\} + \tfrac{1}{4}(B^2 - A^2)\,(\partial u)^2.$$

To simplify these we assume (as the third equation above referred to)

$$\Omega\,\partial^2\Omega - (\partial\Omega)^2 = 0.$$

The last-mentioned two equations then become

$$A\,\partial^2 A - (\partial A)^2 = \tfrac{1}{4}(A^2 - B^2)\,(\partial u)^2,$$

$$B\,\partial^2 B - (\partial B)^2 = \tfrac{1}{4}(B^2 - A^2)\,(\partial u)^2,$$

which several equations contain the theory of the functions A, B, Ω: we have as their general integrals

$$\begin{aligned} A &= \tfrac{1}{2}\Lambda e^{\lambda u}\sqrt{a-b}\,\{e^{\frac{1}{2}(u+v)} + e^{-\frac{1}{2}(u+v)}\}, \\ B &= -\tfrac{1}{2}\Lambda e^{\lambda u}\sqrt{a-b}\,\{e^{\frac{1}{2}(u+v)} - e^{-\frac{1}{2}(u+v)}\}, \\ \Omega &= \Lambda e^{\lambda u}, \end{aligned}$$

where Λ, λ, v are arbitrary constants. Forming the quotients $A : \Omega$, $B : \Omega$, and introducing the notations cosh, sinh, of the hyperbolic sine and cosine, also writing for simplicity $v = 0$, the equations give

$$\sqrt{a-x} = \sqrt{a-b}\cosh\tfrac{1}{2}u,$$

$$\sqrt{b-x} = -\sqrt{a-b}\sinh\tfrac{1}{2}u,$$

which express the integral of the differential relation

$$\partial u = \frac{\partial x}{\sqrt{a-x\,.\,b-x}}.$$

Instead of considering in like manner the radical $\sqrt{a-x\,.\,b-x\,.\,c-x}$, I pass at once to the radical $\sqrt{a-x\,.\,b-x\,.\,c-x\,.\,d-x}$; and starting from the differential relation

$$\partial u = \frac{\partial x}{\sqrt{a-x\,.\,b-x\,.\,c-x\,.\,d-x}},$$

and using the single letters A, B, C, D, Ω to denote functions of u, I assume as definitions

$$A = \Omega\,\sqrt{a-x},$$

$$B = \Omega\,\sqrt{b-x},$$

$$C = \Omega\,\sqrt{c-x},$$

$$D = \Omega\,\sqrt{d-x},$$

and another equation to be presently mentioned; A, B, C, D are called ϑ-functions, and Ω is called the ω-function.

But before proceeding further I introduce some locutions which will be useful. In reference to a given set of squares or products, I use the expression *a sum of squares* to denote the sum of all or any of the squares each multiplied by an arbitrary coefficient; and in like manner *a sum of products* to denote the sum of all or any of the products each multiplied by an arbitrary coefficient: in particular, the set may consist of a single square or product only, and the sum of squares or products will then denote the single term multiplied by an arbitrary coefficient. In the present case, we have the quantities $\sqrt{a-x}$, $\sqrt{b-x}$, $\sqrt{c-x}$, $\sqrt{d-x}$, and the squares are $a-x$, $b-x$, etc., which belong all to the same set; but the products (meaning thereby products of *two* quantities) $\sqrt{a-x\,.\,b-x}$, etc., are considered as being each of them a set by itself. A sum of squares is thus a linear function $\lambda+\mu x$, and conversely any such function is a sum of squares; a sum of products means a single term $\nu\sqrt{a-x\,.\,b-x}$ (or $\nu\sqrt{a-x\,.\,c-x}$, etc., as the case may be), and conversely any such function is a sum of products: the coefficients λ, μ, ν may depend upon or contain Ω, and in differential expressions (∂u being therein considered constant) the coefficients λ, μ, ν may contain the factor ∂u or $(\partial u)^2$—and if convenient we may of course express such factor by writing the coefficients in the form $\lambda\,\partial u$, or $\lambda\,(\partial u)^2$ etc., as the case may be.

We may now explain very simply the form, as well of the algebraical relations, as of the differential relations of the first and second orders respectively, which connect the functions A, B, C, D.

The functions A^2, B^2, C^2, D^2 are each of them a sum of squares, and hence there exists a linear relation between any three of these squares. But the products AB, AC, etc., are each of them a sum of products (meaning thereby a single term, as already explained); and hence there is not any linear relation between these products.

Considering the first derived functions ∂A, ∂B, etc., these each contain a term in $\partial\Omega$, which however disappears (as is obvious) from the combinations $A\,\partial B - B\,\partial A$, etc.; and, without in any wise fixing the value of Ω, we in fact find that each of these expressions is a sum of products; the form is, as will appear,

$$A\,\partial B - B\,\partial A = \alpha\Omega^2\sqrt{c-x\,.\,d-x} = \nu\,CD, \text{ etc.*}$$

Passing to the second derived functions and forming the combinations $A\,\partial^2 A - (\partial A)^2$, etc., each of these will contain a multiple of $\Omega\,\partial^2\Omega - (\partial\Omega)^2$, but if we assume this expression $\Omega\,\partial^2\Omega - (\partial\Omega)^2 = \Omega^2 M$, where M is $=(\partial u)^2$ multiplied by a properly determined function of x, then it is found that each of the expressions in question $A\,\partial^2 A - (\partial A)^2$, etc., becomes equal to a sum of squares, that is, to a linear function $\Omega^2(\lambda + \mu x)$: viz. it is equal to a sum of squares formed with the squares A^2, B^2, C^2, D^2.

The foregoing equation

$$\Omega\,\partial^2\Omega - (\partial\Omega)^2 = \Omega^2 M,$$

where M has its proper value, is the other equation above referred to, which, with the equations $A = \Omega\sqrt{a-x}$, etc., serves for the definition of the functions A, B, C, D, Ω; it may be mentioned at once that the proper value is

$$M = \tfrac{1}{4}(\partial u)^2\{-2x^2 + x(a+b+c+d) + \kappa\},$$

where κ is a constant, symmetrical as regards a, b, c, d, which may be taken $=0$, but which is better put

$$= a^2 + b^2 + c^2 + d^2 - ab - ac - ad - bc - bd - cd.$$

For the proof of the formula, I introduce and shall in general employ the abbreviations $(\mathrm{a}, \mathrm{b}, \mathrm{c}, \mathrm{d})$ to denote the differences $a-x$, $b-x$, $c-x$, $d-x$: the differential relation between x, u thus becomes $\partial x = \partial u\sqrt{\mathrm{abcd}}$. I use also the abbreviations $\Omega\,\partial^2\Omega - (\partial\Omega)^2 = \Delta\Omega$, etc.

We have

$$A\,\partial B - B\,\partial A = \Omega^2(\sqrt{\mathrm{a}}\,\partial\sqrt{\mathrm{b}} - \sqrt{\mathrm{b}}\,\partial\sqrt{\mathrm{a}}),$$

the terms in $\partial\Omega$ disappearing: viz. observing that $\partial\mathrm{a} = \partial\mathrm{b} = -\partial x$, this is

$$= -\tfrac{1}{2}\Omega^2\left(\frac{\sqrt{\mathrm{a}}}{\sqrt{\mathrm{b}}} - \frac{\sqrt{\mathrm{b}}}{\sqrt{\mathrm{a}}}\right)\partial x,$$

$$= \tfrac{1}{2}\Omega^2\frac{\mathrm{a}-\mathrm{b}}{\sqrt{\mathrm{ab}}}\,\partial x;$$

or observing that $\mathrm{a}-\mathrm{b} = a-b$, and writing for ∂x its value $=\sqrt{\mathrm{abcd}}\,\partial u$, this is

$$A\,\partial B - B\,\partial A = -\tfrac{1}{2}(a-b)\,\Omega^2\sqrt{\mathrm{cd}}\,\partial u,$$

$$= -\tfrac{1}{2}(a-b)\,\Omega^2\sqrt{c-x\,.\,d-x}\,\partial u,$$

which is the equation expressing $A\,\partial B - B\,\partial A$ as a sum of products: it is further obvious that the value is

$$= -\tfrac{1}{2}(a-b)\,CD\,\partial u.$$

* It is hardly necessary to remark that α, ν contain each of them the factor ∂u; and the like in other cases.

Proceeding next to find the value of ΔA, $= A\,\partial^2 A - (\partial A)^2$, $= A^2\partial^2 \log A$, it is to be remarked that we have in general

$$\Delta PQ = P^2\Delta Q + Q^2\Delta P,$$

and therefore also $\Delta P^2 = 2P^2\Delta P$, and consequently $\Delta \sqrt{P} = \frac{\frac{1}{2}}{P}\Delta P$. Hence starting from $A = \Omega\sqrt{\mathrm{a}}$, we have

$$\Delta A = \mathrm{a}\Delta\Omega + \Omega^2 \frac{\frac{1}{2}}{\mathrm{a}}\Delta\mathrm{a},$$

where $\Delta\mathrm{a} = -\mathrm{a}\,\partial^2 x - (\partial x)^2$. I assume that we have $\Delta\Omega = \Omega^2 M = \frac{1}{4}\Omega^2 S(\partial u)^2$, where S denotes a function of x which is to be determined: the equation thus becomes

$$\Delta A = \tfrac{1}{4}\Omega^2\{\mathrm{a}S(\partial u)^2 - 2\partial^2 x - 2(\partial x)^2\};$$

we have $(\partial x)^2 = \mathrm{abcd}\,(\partial u)^2$, and thence, differentiating and omitting on each side the factor ∂x, we obtain

$$2\partial^2 x = -(\mathrm{abc} + \mathrm{abd} + \mathrm{acd} + \mathrm{bcd}),$$

and the equation becomes

$$\Delta A = \tfrac{1}{4}\Omega^2\{\mathrm{a}(S + \mathrm{bc} + \mathrm{bd} + \mathrm{cd}) - \mathrm{bcd}\}(\partial u)^2,$$

which is to be simplified by assuming a proper value for S; in order that the same simplification may apply to the formulæ for ΔB, etc., it is necessary that S be symmetrical in regard to a, b, c, d.

Writing for the moment b', c', d' to denote $b-a$, $c-a$, $d-a$ respectively, we have $b', c', d' = \mathrm{b}-\mathrm{a}$, $\mathrm{c}-\mathrm{a}$, $\mathrm{d}-\mathrm{a}$, and thence

$$b'c'd' = \mathrm{bcd} - \mathrm{a}(\mathrm{bc} + \mathrm{bd} + \mathrm{cd}) + \mathrm{a}^2(\mathrm{b} + \mathrm{c} + \mathrm{d}) - \mathrm{a}^3,$$

and consequently

$$\mathrm{a}(\mathrm{bc} + \mathrm{bd} + \mathrm{cd}) - \mathrm{bcd} = -b'c'd' + \mathrm{a}^2(\mathrm{b} + \mathrm{c} + \mathrm{d} - \mathrm{a}):$$

hence, in the expression of ΔA, the factor which multiplies $\frac{1}{2}\Omega^2(\partial u)^2$ is

$$\mathrm{a}\{S + \mathrm{a}(\mathrm{b} + \mathrm{c} + \mathrm{d} - \mathrm{a})\} - b'c'd',$$

viz. the expression added to S is

$$(a-x)(b+c+d-a-2x),$$

$$= a(b+c+d-a) - x(a+b+c+d) + 2x^2.$$

Hence assuming

$$S = -2x^2 + x(a+b+c+d) + \kappa,$$

κ being a constant symmetrical in regard to a, b, c, d, which may be at once taken to be $= a^2 + b^2 + c^2 + d^2 - ab - ac - ad - bc - bd - cd$; then writing also

$$\lambda = b^2 + c^2 + d^2 - bc - bd - cd, \quad \mu = -b'c'd' = a-b\,.\,a-c\,.\,a-d,$$

the expression $\mathrm{a}\{S + \mathrm{a}(\mathrm{b}+\mathrm{c}+\mathrm{d}-\mathrm{a})\} - b'c'd'$ becomes $= \mathrm{a}\lambda + \mu$; and the sought for equation thus is

$$\Delta A = A\,\partial^2 A - (\partial A)^2 = \tfrac{1}{4}\Omega^2(\mathrm{a}\lambda + \mu)(\partial u)^2,$$

the equation in Ω being of course

$$\Delta\Omega = \Omega\,\partial^2\Omega - (\partial\Omega)^2 = \tfrac{1}{4}\Omega^2\{-2x^2 + x(a+b+c+d) + \kappa\}(\partial u)^2.$$

The theory in regard to the second derivatives is thus completed.

To adapt the formulæ to elliptic integrals, and ordinary H and Θ functions, the radical must be brought to the form $\sqrt{x\,.\,1-x\,.\,1-k^2x}$. Writing for this purpose

$$a,\ b,\ c,\ d = -k^2I^2,\ 0,\ 1,\ \frac{1}{k^2},\quad (I=\infty),$$

substituting also $\dfrac{2u}{I}$ for u, and $ikI\,.\,A$, iB ($i=\sqrt{-1}$ as usual) for A, B respectively, we find $\sqrt{a-x\,.\,b-x\,.\,c-x\,.\,d-x} = I\sqrt{x\,.\,1-x\,.\,1-k^2x}$; and then

$$2\partial u = \frac{\partial x}{\sqrt{x\,.\,1-x\,.\,1-k^2x}},$$

and

$$A=\Omega,\ B=\Omega\sqrt{x},\ C=\Omega\sqrt{1-x},\ D=\frac{1}{k}\Omega\sqrt{1-k^2x}.$$

Ω is in this case $=A$, a ϑ-function: and in the equation for $\Delta\Omega$, writing A in place of Ω, the equation becomes

$$A\,\partial^2 A - (\partial A)^2 = \tfrac{1}{4}A^2\{-2x^2 + x(-k^2I^2) + \kappa\}\frac{4(\partial u)^2}{I^2},$$

viz. replacing $\dfrac{\kappa}{I^2}$ by a new constant, $=\lambda$ suppose and finally putting $I=\infty$, this is

$$A\,\partial^2 A - (\partial A)^2 = A^2(\lambda - k^2x)(\partial u)^2.$$

The differential equation is satisfied by $x=\mathrm{sn}^2 u$, giving $1-x=\mathrm{cn}^2 u$, $1-k^2x=\mathrm{dn}^2 u$; and the equation for A then is

$$\partial^2 \log A = (\lambda - k^2\,\mathrm{sn}^2 u)(\partial u)^2,$$

or say

$$A = Le^{\frac{1}{2}\lambda u^2 - k^2\int_0 du\int_0 du\,\mathrm{sn}^2 u},$$

viz. by properly assuming the constants L, λ, we shall have $A=$ Jacobi's function Θu: and then $\mathrm{sn}\,u=\dfrac{B}{A}$, $\mathrm{cn}\,u=\dfrac{C}{A}$, $\mathrm{dn}\,u=\dfrac{kD}{A}$, which will give the ordinary expressions of sn, cn, dn in terms of H, Θ.

Part II. The double ϑ-functions.

Passing now to the double ϑ-functions, and writing for a moment

$$\sqrt{X} = \sqrt{a-x\,.\,b-x\,.\,c-x\,.\,d-x\,.\,e-x\,.\,f-x},$$

$$\sqrt{Y} = \sqrt{a-y\,.\,b-y\,.\,c-y\,.\,d-y\,.\,e-y\,.\,f-y},$$

the differential equations which connect u, v with x, y are

$$\partial u = \frac{\partial x}{\sqrt{X}} + \frac{\partial y}{\sqrt{Y}},$$

$$\partial v = \frac{x\,\partial x}{\sqrt{X}} + \frac{y\,\partial y}{\sqrt{Y}}.$$

There are here sixteen ϑ-functions A, B, C, D, E, F, AB, AC, AD, AE, BC, BD, BE, CD, CE, DE, and an associated ω-function Ω, where for shortness I use the single and double letters $A, B, \ldots, AB, \ldots, \Omega$, instead of functional expressions $A(u, v)$, $B(u, v), \ldots, AB(u, v), \ldots, \Omega(u, v)$, to denote functions of the two letters u, v. Writing also (a, b, c, d, e, f) for the differences $a-x$, $b-x$, etc., and (a_1, b_1, c_1, d_1, e_1, f_1) for the differences $a-y$, $b-y$, etc., whence $\sqrt{X} = \sqrt{\mathrm{abcdef}}$ and $\sqrt{Y} = \sqrt{\mathrm{a_1 b_1 c_1 d_1 e_1 f_1}}$, and θ for the difference $x-y$, we have sixteen xy-functions which are represented by

$$\sqrt{a},\ \sqrt{b},\ \sqrt{c},\ \sqrt{d},\ \sqrt{e},\ \sqrt{f},\ \sqrt{ab},\ \sqrt{ac},\ \sqrt{ad},\ \sqrt{ae},\ \sqrt{bc},\ \sqrt{bd},\ \sqrt{be},\ \sqrt{cd},\ \sqrt{ce},\ \sqrt{de},$$

the values of which are

$$\sqrt{a} = \sqrt{\mathrm{aa_1}},\ \text{(six equations)},$$
$$\vdots$$
$$\sqrt{ab} = \frac{1}{\theta}\{\sqrt{\mathrm{abfc_1d_1e_1}} - \sqrt{\mathrm{a_1b_1f_1cde}}\},\ \text{(ten equations)},$$
$$\vdots$$

and the definitions of the sixteen ϑ-functions and the ω-function are

$$A = \Omega\sqrt{a},\quad \text{(six equations)},$$
$$\vdots$$
$$AB = \Omega\sqrt{ab},\ \text{(ten equations)},$$
$$\vdots$$

and one other equation to be afterwards mentioned.

I call to mind that, in a binary symbol such as $\sqrt{ab}$, it is always f that accompanies the two expressed letters a, b: the duad ab is, in fact, an abbreviated expression for the double triad $abf.cde$: and I remark also that I have for greater simplicity omitted certain constant factors which, in my second paper above referred to, were introduced as multipliers of the foregoing functions $\sqrt{a}, \ldots, \sqrt{ab}, \ldots$ I remark also that, to avoid confusion, the square of any one of these functions $\sqrt{a}$ or $\sqrt{ab}$ is always written (not a or ab, but) $(\sqrt{a})^2$ or $(\sqrt{ab})^2$.

I use ∂ as a symbol of total differentiation: thus

$$\partial A = \frac{dA}{du}\partial u + \frac{dA}{dv}\partial v,\quad \partial^2 A = \frac{d^2A}{du^2}(\partial u)^2 + 2\frac{d^2A}{du\,dv}(\partial u\,\partial v) + \frac{d^2A}{dv^2}(\partial v)^2,\ \text{etc.}$$

Moreover I consider ∂u and ∂v as constants, and use single letters λ, L, etc., to denote linear functions $\alpha\,\partial u + \beta\,\partial v$, or quadric functions $\alpha(\partial u)^2 + 2\beta\,\partial u\,\partial v + \gamma(\partial v)^2$ (as the case may be) of these differentials; thus, in speaking of $A\,\partial B - B\,\partial A$ as a sum of products, it is implied that the coefficients of the several products are linear functions of ∂u, ∂v; and so in speaking of $A\,\partial^2 A - (\partial A)^2$ as a sum of squares, it is in like manner implied that the coefficients of the several squares are quadric functions of ∂u, ∂v.

An xy-function is simplex, such as $\sqrt{a}$, or complex, such as $\sqrt{ab}$; the square of the former is $\mathrm{aa}_1 = a^2 - a(x+y) + xy$, which is of the form $\lambda + \mu(x+y) + \nu xy$; the square of the latter is

$$= \frac{1}{\theta^2}\{\mathrm{abfc_1d_1e_1} + \mathrm{a_1b_1f_1cde} - 2\sqrt{XY}\},$$

where observe that the irrational part $-\frac{2}{\theta^2}\sqrt{XY}$ is the same for all these squares: so that, taking any two such squares, their difference is $= \frac{1}{\theta^2}$ multiplied by a rational function of xy: this rational function in fact divides by θ^2, the quotient being a rational and integral function of the foregoing form $\lambda + \mu(x+y) + \nu xy$. Hence selecting any one of the complex functions, say $\sqrt{de}$, the square of any other of the complex functions is equal to the square of this *plus* a term $\lambda + \mu(x+y) + \nu xy$; and hence the square of any function simplex or complex is of the form $\lambda + \mu(x+y) + \nu xy + \rho(\sqrt{de})^2$; this being so, the squares of the xy-functions may be regarded as forming a single set; every sum of squares is a function of this form $\lambda + \mu(x+y) + \nu xy + \rho(\sqrt{de})^2$; and conversely every function of this form is a sum of squares. A sum of squares thus depends upon four arbitrary coefficients λ, μ, ν, ρ; and we may, in an infinity of ways, select four out of the 16 squares such that every sum of squares can be represented as a sum of these four squares each multiplied by the proper coefficient; say as a sum of the selected four squares: in particular, each of the remaining squares can be expressed as a sum of the selected four squares. It appears, by the first of my papers above referred to, that there are systems of four squares connected together by a linear equation: we are not here concerned with such systems; only of course the four selected squares must not belong to such a system.

We have the products of the xy-functions, where by product is meant a product of *two* functions. The number of products is of course $= 120$, but distinguishing these according to the radicals which they respectively contain, they form 30 different sets. Thus we have

$$\sqrt{b}\sqrt{ab} = \frac{1}{\theta}\{\mathrm{b}\sqrt{\mathrm{afb_1c_1d_1e_1}} - \mathrm{b_1}\sqrt{\mathrm{a_1f_1bcde}}\},$$

$$\sqrt{c}\sqrt{ac} = \frac{1}{\theta}\{\mathrm{c} \quad ,, \quad - \mathrm{c_1} \quad ,, \quad \},$$

$$\sqrt{d}\sqrt{ad} = \frac{1}{\theta}\{\mathrm{d} \quad ,, \quad - \mathrm{d_1} \quad ,, \quad \},$$

$$\sqrt{e}\sqrt{ae} = \frac{1}{\theta}\{\mathrm{e} \quad ,, \quad - \mathrm{e_1} \quad ,, \quad \},$$

which four expressions form a set, and there are 15 such sets. The set written down may be called the set af: and the fifteen sets are of course ab, ac, etc.

Again, we have

$$\sqrt{a}\sqrt{b} = \sqrt{\mathrm{aba_1b_1}},$$

$$\sqrt{ac}\sqrt{bc} = \frac{1}{\theta^2}\{(\mathrm{cfd_1e_1} + \mathrm{c_1f_1de})\sqrt{\mathrm{aba_1b_1}} - (\mathrm{ab_1} + \mathrm{a_1b})\sqrt{\mathrm{cdefc_1d_1e_1f_1}}\},$$

$$\sqrt{ad}\sqrt{bd} = \frac{1}{\theta^2}\{(\mathrm{dfc_1e_1} + \mathrm{d_1f_1ce}) \quad ,, \quad - \quad ,, \quad ,, \quad\},$$

$$\sqrt{ae}\sqrt{be} = \frac{1}{\theta^2}\{(\mathrm{efc_1d_1} + \mathrm{e_1f_1cd}) \quad ,, \quad - \quad ,, \quad ,, \quad\},$$

which four expressions form a set, and there are 15 such sets. The set written down may be called the set $\mathrm{aba_1b_1}$: and the fifteen sets are of course $\mathrm{aba_1b_1}$, $\mathrm{aca_1c_1}$, etc. The 15 and 15 sets make in all 30 sets as mentioned above.

The expression, a sum of products, means as already explained a sum of products belonging to the same set; and there are thus 30 forms of a sum of products. The products of the same set are connected by two linear relations, so that, selecting at pleasure any two of the products, the other two products can be expressed each of them as a linear function of these; hence a sum of products contains only two arbitrary coefficients.

Reverting now to the equations $A = \Omega\sqrt{a}$, etc., we see at once the form of the algebraical equations which connect the 16 ϑ-functions. Every squared function $A^2, \ldots, (AB)^2, \ldots$ is a sum of squares, whence selecting (as may be done in a great number of ways) four of these squared functions, each of the remaining 12 squares is a sum of these four squares each multiplied by the proper coefficient; or say it is a sum of the four selected squares. And in like manner the 120 products of two of the 16 functions form 30 sets, such that selecting at pleasure two of the set, the remaining two of the set are each of them a linear function of these.

Considering the first derived functions $\partial A, \partial B, \ldots, \partial AB, \ldots$, each of these contains a term in $\partial\Omega$; but $\partial\Omega$ disappears (as is obvious) from the several combinations $I\partial J - J\partial I$ (I write I and similarly J to denote indifferently a single letter A or a double letter AB): and, without in any wise fixing the value of Ω, we in fact find that each of these expressions is a sum of products.

Passing to the second derived functions, and forming the combinations $A\,\partial^2 A - (\partial A)^2$, etc., or to include the two cases of the single and the double letter, say $I\,\partial^2 I - (\partial I)^2$, each of these will contain a multiple of $\Omega\,\partial^2\Omega - (\partial\Omega)^2$; but if we assume this expression $\Omega\,\partial^2\Omega - (\partial\Omega)^2 = \Omega^2 M$, where M is a quadric function of ∂u, ∂v, the coefficients of $(\partial u)^2$, $\partial u\,\partial v$, $(\partial v)^2$ being properly determined functions of xy, then it is found that each of the expressions in question $I\,\partial^2 I - (\partial I)^2$ becomes equal to a sum of squares.

It is to be observed that M is not altogether arbitrary: the equation as containing terms in $(\partial u)^2$, $\partial u\,\partial v$, and $(\partial v)^2$, in fact represents three partial differential equations, which for an arbitrary value of M would be inconsistent with each other: it is therefore necessary to verify that the value assigned to M is such as to render the three equations consistent with each other, and this will accordingly be done.

The foregoing equation

$$\Omega\,\partial^2\Omega - (\partial\Omega)^2 = \Omega^2 M,$$

where M has its proper value, (or say the three partial differential equations into which this breaks up), constitutes the other equation above referred to, which with the original equations $A = \Omega\sqrt{a}$, etc., serve to define the sixteen ϑ-functions and Ω.

The remainder of the present memoir is occupied with the analytical investigation of the foregoing theorems. Although the mere algebraical work is very long, yet it appears to me interesting, and I have thought it best to give it in detail.

The analytical theory: various subheadings.

The equations

$$\partial u = \frac{\partial x}{\sqrt{X}} + \frac{\partial y}{\sqrt{Y}},$$

$$\partial v = \frac{x\,\partial x}{\sqrt{X}} + \frac{y\,\partial y}{\sqrt{Y}},$$

give

$$\frac{\theta\,\partial x}{\sqrt{X}} = \partial v - y\,\partial u, \qquad -\frac{\theta\,\partial y}{\sqrt{Y}} = \partial v - x\,\partial u,$$

which determine ∂x, ∂y in terms of ∂u, ∂v. A different form is sometimes convenient; writing $\partial\varpi = \partial v - a\,\partial u$, and recollecting that a, a_1 denote $a - x$, $a - y$ respectively, the equations become

$$\frac{\theta\,\partial x}{\sqrt{X}} = \partial\varpi + \mathrm{a}_1\,\partial u, \quad -\frac{\theta\,\partial y}{\sqrt{Y}} = \partial\varpi + \mathrm{a}\,\partial u.$$

Expression for $\partial\sqrt{a}$.

We have

$$\partial\sqrt{a} = \partial\sqrt{\mathrm{aa}_1} = \frac{1}{2\sqrt{\mathrm{aa}_1}}(\mathrm{a}\,\partial\mathrm{a}_1 + \mathrm{a}_1\,\partial\mathrm{a}) = -\frac{1}{2\sqrt{\mathrm{aa}_1}}(\mathrm{a}\,\partial y + \mathrm{a}_1\,\partial x)$$

$$= \frac{1}{2\sqrt{\mathrm{aa}_1}}\frac{1}{\theta}\{\mathrm{a}\sqrt{Y}(\partial v - x\,\partial u) - \mathrm{a}_1\sqrt{X}(\partial v - y\partial u)\};$$

substituting for $\sqrt{X}$, $\sqrt{Y}$ their values $\sqrt{\mathrm{abcdef}}$, $\sqrt{\mathrm{a_1b_1c_1d_1e_1f_1}}$, this is

$$\partial\sqrt{a} = \frac{\tfrac{1}{2}}{\theta}\{\sqrt{\mathrm{ab_1c_1d_1e_1f_1}}\,(\partial v - x\,\partial u) - \sqrt{\mathrm{a_1bcdef}}\,(\partial v - y\,\partial u)\},$$

and by the mere interchange of letters we can of course find $\partial\sqrt{b}$, etc.

Expression for $\partial \sqrt{ab}$.

We have next to find

$$\partial \sqrt{ab} = \partial \frac{1}{\theta} \{\sqrt{\mathrm{abfc_1d_1e_1}} - \sqrt{\mathrm{a_1b_1f_1cde}}\};$$

here

$$\partial\theta = \partial x - \partial y, \quad = \frac{1}{\theta} \{\sqrt{\mathrm{abcdef}}\,(\partial v - y\,\partial u) + \sqrt{\mathrm{a_1b_1c_1d_1e_1f_1}}\,(\partial v - x\,\partial u)\},$$

and consequently $\partial \sqrt{ab}$ contains a term

$$-\frac{\partial\theta}{\theta^2} \{\sqrt{\mathrm{abfc_1d_1e_1}} - \sqrt{\mathrm{a_1b_1f_1cde}}\},$$

which is

$$= \frac{1}{\theta^3} \{ \; (\; -\mathrm{abf}\sqrt{\mathrm{cdec_1d_1e_1}} + \mathrm{cde}\;\; \sqrt{\mathrm{abfa_1b_1f_1}})\,(\partial v - y\,\partial u)$$
$$+ (-\mathrm{c_1d_1e_1}\sqrt{\mathrm{abfa_1b_1f_1}} + \mathrm{a_1b_1f_1}\sqrt{\mathrm{cdec_1d_1e_1}})\,(\partial v - x\,\partial u)\},$$

or, what is the same thing,

$$= \frac{1}{\theta^2} \Bigg\{ \quad \sqrt{\mathrm{abfa_1b_1f_1}} \left(\frac{\mathrm{cde} - \mathrm{c_1d_1e_1}}{\theta} \partial v + \frac{-\mathrm{cde}\,y + \mathrm{c_1d_1e_1}\,x}{\theta} \partial u \right)$$
$$+ \sqrt{\mathrm{cdec_1d_1e_1}} \left(\frac{-\mathrm{abf} + \mathrm{a_1b_1f_1}}{\theta} \partial v + \frac{\mathrm{abf}\,y - \mathrm{a_1b_1f_1}\,x}{\theta} \partial u \right) \Bigg\}.$$

Now

$$\frac{\mathrm{cde} - \mathrm{c_1d_1e_1}}{\theta} = -(cd + ce + de) + (c + d + e)(x + y) - x^2 - xy - y^2,$$

$$\frac{-\mathrm{cde}\,y + \mathrm{c_1d_1e_1}\,x}{\theta} = cde - (c + d + e)\,xy + xy\,(x + y);$$

with the like formulæ with a, b, f instead of c, d, e. Hence the foregoing, or say the first, part of $\partial \sqrt{ab}$ is

$$= \frac{1}{\theta^2} [\sqrt{\mathrm{abfa_1b_1f_1}}\,(\{-(cd + ce + de) + (c + d + e)(x + y) - x^2 - xy - y^2\}\,\partial v$$
$$+ \{ \;\; cde - (c + d + e)\,xy + xy\,(x + y)\}\,\partial u)$$
$$+ \sqrt{\mathrm{cdec_1d_1e_1}}\,(\{ \;\; ab + af + bf \; - (a + b + f)(x + y) + x^2 + xy + y^2\}\,\partial v$$
$$+ \{-abf + (a + b + f)\,xy - xy\,(x + y)\}\,\partial u)].$$

The other or second part of $\partial \sqrt{ab}$, using for shortness an accent to denote differentiation in regard to x or to y, according as it is applied to a function of x or of y, is readily found to be

$$= \frac{\frac{1}{2}}{\theta^2} [\sqrt{\mathrm{abfa_1b_1f_1}}\,(\{-(\mathrm{cde})' - (\mathrm{c_1d_1e_1})'\}\,\partial v + \{ \;\; y\,(\mathrm{cde})' + x\,(\mathrm{c_1d_1e_1})'\}\,\partial u)$$
$$+ \sqrt{\mathrm{cdec_1d_1e_1}}\,(\{ \;\; (\mathrm{abf})' + (\mathrm{a_1b_1f_1})'\}\,\partial v + \{-y\,(\mathrm{abf})' - x\,(\mathrm{a_1b_1f_1})'\}\,\partial u)].$$

Hence uniting the two terms so as to form the complete value of $\partial\sqrt{ab}$, we have first, a term $\frac{1}{\theta^2}\sqrt{\text{abfa}_1\text{b}_1\text{f}_1}\,\partial v$, the coefficient of which is

$$=-(cd+ce+de)+(c+d+e)(x+y)-x^2-xy-y^2$$
$$-\tfrac{1}{2}\{(\text{cde})'+(\text{c}_1\text{d}_1\text{e}_1)'\}:$$

this second line is

$$=cd+ce+de-(c+d+e)(x+y)+\tfrac{3}{2}x^2+\tfrac{3}{2}y^2,$$

or the coefficient is $-\frac{1}{2}x^2+xy-\frac{1}{2}y^2, =-\frac{1}{2}\theta^2$; the term is thus

$$=\tfrac{1}{2}\sqrt{\text{abfa}_1\text{b}_1\text{f}_1}\,\partial v.$$

Secondly, a term in $\frac{1}{\theta^2}\sqrt{\text{cdec}_1\text{d}_1\text{e}_1}\,\partial v$ which is in like manner found to be

$$=-\tfrac{1}{2}\sqrt{\text{cdec}_1\text{d}_1\text{e}_1}\,\partial v.$$

Thirdly, a term $\frac{1}{\theta^2}\sqrt{\text{abfa}_1\text{b}_1\text{f}_1}\,\partial u$, the coefficient of which is

$$=cde-(c+d+e)xy+x^2y+xy^2$$
$$+\tfrac{1}{2}\{y(\text{cde})'+x(\text{c}_1\text{d}_1\text{e}_1)'\}:$$

this second line is

$$=-(cd+ce+de)\tfrac{1}{2}(x+y)+(c+d+e)\,2xy-\tfrac{3}{2}x^2y-\tfrac{3}{2}xy^2,$$

and the coefficient is thus

$$=\tfrac{1}{2}\{2\,cde-(cd+ce+de)(x+y)+(c+d+e)(x^2+y^2)-x^3 \qquad -y^3$$
$$-(c+d+e)(x-y)^2+x^3-x^2y-xy^2+y^3\},$$

which is

$$=\tfrac{1}{2}\{\text{cde}+\text{c}_1\text{d}_1\text{e}_1-(c+d+e)\theta^2+(x+y)\theta^2\},$$

or the term is

$$=\tfrac{1}{2}\sqrt{\text{abfa}_1\text{b}_1\text{f}_1}\left\{\frac{\text{cde}+\text{c}_1\text{d}_1\text{e}_1}{\theta^2}-(c+d+e)+x+y\right\}\partial u.$$

And, fourthly, a term in $\frac{1}{\theta^2}\sqrt{\text{cdec}_1\text{d}_1\text{e}_1}\,\partial u$, which is in like manner found to be

$$=-\tfrac{1}{2}\sqrt{\text{cdec}_1\text{d}_1\text{e}_1}\left\{\frac{\text{abf}+\text{a}_1\text{b}_1\text{f}_1}{\theta^2}-(a+b+f)+x+y\right\}\partial u.$$

Hence combining these several terms, we have finally

$$\partial\sqrt{ab}=\tfrac{1}{2}\sqrt{\text{abfa}_1\text{b}_1\text{f}_1}\left[\quad\partial v+\left(\frac{\text{cde}+\text{c}_1\text{d}_1\text{e}_1}{\theta^2}-c-d-e+x+y\right)\partial u\right]$$
$$+\tfrac{1}{2}\sqrt{\text{cdec}_1\text{d}_1\text{e}_1}\left[-\partial v-\left(\frac{\text{abf}+\text{a}_1\text{b}_1\text{f}_1}{\theta^2}-a-b-f+x+y\right)\partial u\right];$$

and by the mere interchange of letters we can of course find $\partial\sqrt{ac}$, etc.

Expression for $A\,\partial B - B\,\partial A$.

Starting now from the equations $A = \Omega\sqrt{a}$, $B = \Omega\sqrt{b}$, we obtain

$$A\partial B - B\partial A = \Omega^2\{\sqrt{a}\,\partial\sqrt{b} - \sqrt{b}\,\partial\sqrt{a}\}$$

$$= \frac{\tfrac{1}{2}\Omega^2}{\theta}(\sqrt{\mathrm{aa}_1}\,\{\sqrt{\mathrm{ba}_1\mathrm{c}_1\mathrm{d}_1\mathrm{e}_1\mathrm{f}_1}\,(\partial v - x\,\partial u) - \sqrt{\mathrm{b}_1\mathrm{acdef}}\,(\partial v - y\,\partial u)\}$$

$$- \sqrt{\mathrm{bb}_1}\,\{\sqrt{\mathrm{ab}_1\mathrm{c}_1\mathrm{d}_1\mathrm{e}_1\mathrm{f}_1}\,(\quad ,, \quad) - \sqrt{\mathrm{a}_1\mathrm{bcdef}}\,(\quad ,, \quad)\}),$$

$$= \frac{\tfrac{1}{2}\Omega^2}{\theta}\{(\mathrm{a}_1 - \mathrm{b}_1)\sqrt{\mathrm{abc}_1\mathrm{d}_1\mathrm{e}_1\mathrm{f}_1}\,(\partial v - x\,\partial u) - (\mathrm{a} - \mathrm{b})\sqrt{\mathrm{a}_1\mathrm{b}_1\mathrm{cdef}}\,(\partial v - y\,\partial u)\};$$

or since $\mathrm{a}_1 - \mathrm{b}_1 = \mathrm{a} - \mathrm{b} = a - b$, this is

$$A\,\partial B - B\,\partial A = \frac{\tfrac{1}{2}\Omega^2}{\theta}(a - b)\,\{\sqrt{\mathrm{abc}_1\mathrm{d}_1\mathrm{e}_1\mathrm{f}_1}\,(\partial v - x\,\partial u) - \sqrt{\mathrm{a}_1\mathrm{b}_1\mathrm{cdef}}\,(\partial v - y\,\partial u)\},$$

which is a sum of products of the set ab: in fact, the four products of this set are

$$\sqrt{f}\,\sqrt{ab} = \frac{1}{\theta}\{\ \mathrm{f}\,\sqrt{\mathrm{abc}_1\mathrm{d}_1\mathrm{e}_1\mathrm{f}_1} - \mathrm{f}_1\,\sqrt{\mathrm{a}_1\mathrm{b}_1\mathrm{cdef}}\},$$

$$\sqrt{c}\,\sqrt{de} = \frac{1}{\theta}\{-\mathrm{c} \quad ,, \quad + \mathrm{c}_1 \quad ,, \quad\},$$

$$\sqrt{d}\,\sqrt{ce} = \frac{1}{\theta}\{-\mathrm{d} \quad ,, \quad + \mathrm{d}_1 \quad ,, \quad\},$$

$$\sqrt{e}\,\sqrt{cd} = \frac{1}{\theta}\{-\mathrm{e} \quad ,, \quad + \mathrm{e}_1 \quad ,, \quad\};$$

choosing any two of these at pleasure, for instance the first and second, multiplying by $\partial v - c\,\partial u$, $\partial v - f\,\partial u$ and adding, we have

$$\left.\begin{array}{r}(\partial v - c\,\partial u)\,\sqrt{f}\,\sqrt{ab}\\ +(\partial v - f\,\partial u)\,\sqrt{c}\,\sqrt{de}\end{array}\right\} = \left\{\begin{array}{l}\frac{1}{\theta}\{\ (\partial v - c\,\partial u)\,\mathrm{f}\sqrt{\mathrm{abc}_1\mathrm{d}_1\mathrm{e}_1\mathrm{f}_1} - (\partial v - c\,\partial u)\,\mathrm{f}_1\sqrt{\mathrm{a}_1\mathrm{b}_1\mathrm{cdef}}\}\\ \frac{1}{\theta}\{-(\partial v - f\partial u)\,\mathrm{c} \quad ,, \quad + (\partial v - f\partial u)\,\mathrm{c}_1 \quad ,, \quad\},\end{array}\right.$$

where the coefficients $\mathrm{f}(\partial v - c\,\partial u) - \mathrm{c}(\partial v - f\,\partial u)$, and $\mathrm{f}_1(\partial v - c\,\partial u) - \mathrm{c}_1(\partial v - f\,\partial u)$, by substituting for f, c, f_1, c_1 their values, become $= (f - c)(\partial v - x\,\partial u)$ and $(f - c)(\partial v - y\,\partial u)$; and the expression is thus

$$= \frac{f - c}{\theta}\{\sqrt{\mathrm{abc}_1\mathrm{d}_1\mathrm{e}_1\mathrm{f}_1}\,(\partial v - x\,\partial u) - \sqrt{\mathrm{a}_1\mathrm{b}_1\mathrm{cdef}}\,(\partial v - y\,\partial u)\}.$$

Reverting to the original expression for $A\,\partial B - B\,\partial A$, it may be remarked that, if we write $\partial v - a\,\partial u = \partial\varpi$, $\partial v - b\,\partial u = \partial\sigma$, then

$$(a - b)(\partial v - x\,\partial u) = \mathrm{a}\,\partial\sigma - \mathrm{b}\,\partial\varpi,\quad (a - b)(\partial v - y\,\partial u) = \mathrm{a}_1\,\partial\sigma - \mathrm{b}_1\,\partial\varpi,$$

and the formula thus becomes

$$A\,\partial B - B\,\partial A = \frac{\tfrac{1}{2}\Omega^2}{\theta}\{\sqrt{\mathrm{abc_1d_1e_1f_1}}\,(\mathrm{a}\,\partial\sigma - \mathrm{b}\,\partial\varpi) - \sqrt{\mathrm{a_1b_1cdef}}\,(\mathrm{a_1}\,\partial\sigma - \mathrm{b_1}\,\partial\varpi)\}:$$

but I shall not in the sequel use this formula, or the notation $\partial v - b\,\partial u = \partial\sigma$ introduced for obtaining it.

Expression for $A\,\partial AB - AB\,\partial A$.

Starting from the equations $A = \Omega\sqrt{a}$ and $AB = \Omega\sqrt{ab}$, we have

$$A\,\partial AB - AB\,\partial A = \Omega^2\{\sqrt{a}\,\partial\sqrt{ab} - \sqrt{ab}\,\partial\sqrt{a}\},$$

where the term in { } is

$$= \sqrt{\mathrm{aa_1}}\Big[\ \tfrac{1}{2}\sqrt{\mathrm{abfa_1b_1f_1}}\ \Big\{\ \partial v + \Big(\frac{\mathrm{cde} + \mathrm{c_1d_1e_1}}{\theta^2} - c - d - e + x + y\Big)\partial u\Big\}$$

$$+ \tfrac{1}{2}\sqrt{\mathrm{cdec_1d_1e_1}}\Big\{-\partial v - \Big(\frac{\mathrm{abf} + \mathrm{a_1b_1f_1}}{\theta^2} - a - b - f + x + y\Big)\partial u\Big\}\Big]$$

$$-\frac{\tfrac{1}{2}}{\theta^2}(\sqrt{\mathrm{abfc_1d_1e_1}} - \sqrt{\mathrm{a_1b_1f_1cde}})\,[\sqrt{\mathrm{ab_1c_1d_1e_1f_1}}\,(\partial v - x\,\partial u) - \sqrt{\mathrm{a_1bcdef}}\,(\partial v - y\,\partial u)].$$

To reduce this, I write $\partial v - a\,\partial u = \partial\varpi$, and therefore

$$\partial v - x\,\partial u = \partial\varpi + \mathrm{a}\,\partial u,\quad \partial v - y\,\partial u = \partial\varpi + \mathrm{a_1}\,\partial u;$$

then for convenience multiplying by $2\theta^2$, the term is

$$= \mathrm{aa_1}\sqrt{\mathrm{bfb_1f_1}}\{\ \theta^2\,\partial\varpi + [(a - c - d - e + x + y)\,\theta^2 + \mathrm{cde} + \mathrm{c_1d_1e_1}]\,\partial u\}$$

$$+ \sqrt{\mathrm{acdea_1c_1d_1e_1}}\,\{-\theta^2\,\partial\varpi + [\qquad (b + f - x - y)\,\theta^2 - \mathrm{abf} - \mathrm{a_1b_1f_1}]\,\partial u\}$$

$$-(\sqrt{\mathrm{abfc_1d_1e_1}} - \sqrt{\mathrm{a_1b_1f_1cde}})\,\{\sqrt{\mathrm{ab_1c_1d_1e_1f_1}}\,(\partial\varpi + \mathrm{a}\,\partial u) - \sqrt{\mathrm{a_1bcdef}}\,(\partial\varpi + \mathrm{a_1}\,\partial u)\}.$$

The last line hereof is

$$= \qquad \sqrt{\mathrm{bfb_1f_1}}\,\{-\mathrm{ac_1d_1e_1}\,(\partial\varpi + \mathrm{a}\,\partial u) - \mathrm{a_1cde}\,(\partial\varpi + \mathrm{a_1}\,\partial u)\}$$

$$+ \sqrt{\mathrm{acdea_1c_1d_1e_1}}\,\{\qquad \mathrm{b_1f_1}\,(\partial\varpi + \mathrm{a}\,\partial u) + \quad \mathrm{bf}\,(\partial\varpi + \mathrm{a_1}\,\partial u)\}.$$

Hence we have first, a term in $\sqrt{\mathrm{acdea_1c_1d_1e_1}}$, the coefficient of which is

$$= -\theta^2\,\partial\varpi + [(b + f - x - y)\,\theta^2 - \mathrm{abf} - \mathrm{a_1b_1f_1}]\,\partial u + \mathrm{b_1f_1}\,(\partial\varpi + \mathrm{a}\,\partial u) + \mathrm{bf}\,(\partial\varpi + \mathrm{a_1}\,\partial u),$$

viz. this is

$$= \partial\varpi\,(-\theta^2 + \mathrm{b_1f_1} + \mathrm{bf}) + \partial u\,[-(\mathrm{a} - \mathrm{a_1})(\mathrm{bf} - \mathrm{b_1f_1}) + (b + f - x - y)\,\theta^2],$$

where $(\mathrm{b} - \mathrm{b_1})(\mathrm{f} - \mathrm{f_1}) = \theta^2$, that is, $\mathrm{bf} + \mathrm{b_1f_1} - \theta^2 = \mathrm{bf_1} + \mathrm{b_1f}$, also

$$(\mathrm{a} - \mathrm{a_1})(\mathrm{bf} - \mathrm{b_1f_1}) = (b + f - x - y)\,\theta^2,$$

or the coefficient is $= (\mathrm{bf_1} + \mathrm{b_1f})\,\partial\varpi$: viz. the term in question is

$$= \sqrt{\mathrm{acdea_1c_1d_1e_1}}\,(\mathrm{bf_1} + \mathrm{b_1f})\,\partial\varpi.$$

We have then, secondly, a term in $\sqrt{\text{bfb}_1\text{f}_1}$, the coefficient of which is

$$= \text{aa}_1\{\theta^2\partial\varpi + [(a-c-d-e+x+y)\,\theta^2 + \text{cde} + \text{c}_1\text{d}_1\text{e}_1]\,\partial u\}$$

$$- \text{ac}_1\text{d}_1\text{e}_1(\partial\varpi + \text{a}\,\partial u) - \text{a}_1\text{cde}(\partial\varpi + \text{a}_1\partial u),$$

viz. this is

$$= (\text{aa}_1\theta^2 - \text{ac}_1\text{d}_1\text{e}_1 - \text{a}_1\text{cde})\,\partial\varpi + [\text{aa}_1(a-c-d-e+x+y)\,\theta^2 + (\text{a}_1\text{cde} - \text{ac}_1\text{d}_1\text{e}_1)(\text{a}-\text{a}_1)]\,\partial u.$$

We have $\text{a} - \text{a}_1 = -\theta$; also $\text{a}_1\text{cde} - \text{ac}_1\text{d}_1\text{e}_1$

$$= \theta\,\{cde - a(cd+ce+de) + (c+d+e)[a(x+y) - xy] - a(x^2+xy+y^2) + xy(x+y)\},$$

where the coefficient of θ is

$$= -(a-c)(a-d)(a-e) - (c+d+e)[a^2 - a(x+y) + xy] + (a+x+y)[a^2 - a(x+y) + xy],$$

viz. it is

$$= -(a-c)(a-d)(a-e) + \text{aa}_1(a-c-d-e+x+y).$$

Hence the coefficient in question is

$$= (\text{aa}_1\theta^2 - \text{ac}_1\text{d}_1\text{e}_1 - \text{a}_1\text{cde})\,\partial\varpi + (a-c)(a-d)(a-e)\,\theta^2\partial u,$$

and the second term is $= \sqrt{\text{bfb}_1\text{f}_1}$, multiplied by this coefficient.

Hence, observing that the whole has to be multiplied by $\frac{\frac{1}{2}}{\theta^2}\Omega^2$, we find

$$A\,\partial AB - AB\,\partial A = \frac{\frac{1}{2}}{\theta^2}\Omega^2\,\{\sqrt{\text{acdea}_1\text{c}_1\text{d}_1\text{e}_1}\,(\text{bf}_1 + \text{b}_1\text{f})\,\partial\varpi$$

$$+ \sqrt{\text{bfb}_1\text{f}_1}\,[(\text{aa}_1\theta^2 - \text{ac}_1\text{d}_1\text{e}_1 - \text{a}_1\text{cde})\,\partial\varpi + (a-c)(a-d)(a-e)\,\theta^2\,\partial u]\},$$

where I retain $\partial\varpi$ in place of its value, $= \partial v - a\,\partial u$.

This is a sum of products of the set bfb_1f_1: we, in fact, have

$$\sqrt{ac}\,\sqrt{de} = \frac{1}{\theta^2}\{(\text{bf}_1 + \text{b}_1\text{f})\sqrt{\text{acdea}_1\text{c}_1\text{d}_1\text{e}_1} - (\text{acd}_1\text{e}_1 + \text{a}_1\text{c}_1\text{de})\sqrt{\text{bfb}_1\text{f}_1}\},$$

$$\sqrt{ad}\,\sqrt{ce} = \text{,,}\ \{\quad \text{,,} \qquad \text{,,} \qquad - (\text{adc}_1\text{e}_1 + \text{a}_1\text{d}_1\text{ce}) \quad \text{,,}\ \},$$

$$\sqrt{ae}\,\sqrt{cd} = \text{,,}\ \{\quad \text{,,} \qquad \text{,,} \qquad - (\text{aec}_1\text{d}_1 + \text{a}_1\text{e}_1\text{cd}) \quad \text{,,}\ \},$$

$$\sqrt{b}\,\sqrt{f} = \text{,,}\ \{\qquad\qquad\qquad + \theta^2 \quad \text{,,}\ \},$$

and selecting any two of these, for instance the first and the fourth, the coefficient of $\frac{\frac{1}{2}}{\theta^2}\Omega^2$ is at once seen to be of the form $\partial\varpi\,\sqrt{ac}\,\sqrt{de} + K\sqrt{b}\,\sqrt{f}$; and for the determination of K, we have

$$(-\text{acd}_1\text{e}_1 - \text{a}_1\text{c}_1\text{de})\,\partial\varpi + K\theta^2 = (\text{aa}_1\theta^2 - \text{ac}_1\text{d}_1\text{e}_1 - \text{a}_1\text{cde})\,\partial\varpi + (a-c)(a-d)(a-e)\,\theta^2\,\partial u,$$

viz. this gives

$$K\theta^2 = \{\text{aa}_1\theta^2 + (\text{c} - \text{c}_1)(\text{ad}_1\text{e}_1 - \text{a}_1\text{de})\}\,\partial\varpi + (a-c)(a-d)(a-e)\,\theta^2\,\partial u.$$

We then have

$$(\mathrm{c}-\mathrm{c}_1)(\mathrm{ad}_1\mathrm{e}_1-\mathrm{a}_1\mathrm{de})=\theta^2\{-\mathrm{aa}_1+(a-d)(a-e)\},$$

and the whole equation divides by θ^2; substituting for $\partial\varpi$ its value, we find

$$K=(a-d)(a-e)(\partial v-c\,\partial u).$$

Expression for $AC\,\partial AB-AB\,\partial AC$.

Starting in like manner from the equations $AB=\Omega\sqrt{ab}$, $AC=\Omega\sqrt{ac}$, we have

$$AC\,\partial AB-AB\,\partial AC=\frac{\frac{1}{2}}{\theta}\Omega^2,$$

multiplied by

$$\begin{aligned}&(\sqrt{\mathrm{acfb_1d_1e_1}}-\sqrt{\mathrm{a_1c_1f_1bde}})\left\{\begin{aligned}&\sqrt{\mathrm{abfa_1b_1f_1}}\left[\partial\varpi+\left(a-c-d-e+x+y+\frac{\mathrm{cde}+\mathrm{c_1d_1e_1}}{\theta^2}\right)\partial u\right]\\&+\sqrt{\mathrm{cdec_1d_1e_1}}\left[-\partial\varpi+\left(b+f-x-y-\frac{\mathrm{abf}+\mathrm{a_1b_1f_1}}{\theta^2}\right)\partial u\right]\end{aligned}\right\}\\&+(-\sqrt{\mathrm{abfc_1d_1e_1}}-\sqrt{\mathrm{a_1b_1f_1cde}})\left\{\begin{aligned}&\sqrt{\mathrm{acfa_1c_1f_1}}\left[\partial\varpi+\left(a-b-d-e+x+y+\frac{\mathrm{bde}+\mathrm{b_1d_1e_1}}{\theta^2}\right)\partial u\right]\\&+\sqrt{\mathrm{bdeb_1d_1e_1}}\left[-\partial\varpi+\left(c+f-x-y-\frac{\mathrm{acf}+\mathrm{a_1c_1f_1}}{\theta^2}\right)\partial u\right]\end{aligned}\right\},\end{aligned}$$

which, omitting the factor $\frac{\frac{1}{2}}{\theta}\Omega^2$, is

$$\begin{aligned}=&\{\mathrm{afb_1}\sqrt{\mathrm{bca_1d_1e_1f_1}}-\mathrm{a_1f_1b}\sqrt{\mathrm{b_1c_1adef}}\}\left[\partial\varpi+\left(a-c-d-e+x+y+\frac{\mathrm{cde}+\mathrm{c_1d_1e_1}}{\theta^2}\right)\partial u\right]\\&+\{\mathrm{cd_1e_1}\sqrt{\mathrm{b_1c_1adef}}-\mathrm{c_1de}\sqrt{\mathrm{bca_1d_1e_1f_1}}\}\left[-\partial\varpi+\left(b+f-x-y-\frac{\mathrm{abf}+\mathrm{a_1b_1f_1}}{\theta^2}\right)\partial u\right]\\&+\{-\mathrm{afc_1}\sqrt{\mathrm{bca_1d_1e_1f_1}}+\mathrm{a_1f_1c}\sqrt{\mathrm{b_1c_1adef}}\}\left[\partial\varpi+\left(a-b-d-e+x+y+\frac{\mathrm{bde}+\mathrm{b_1d_1e_1}}{\theta^2}\right)\partial u\right]\\&+\{-\mathrm{bde_1}\sqrt{\mathrm{b_1c_1adef}}+\mathrm{b_1de}\sqrt{\mathrm{bca_1d_1e_1f_1}}\}\left[-\partial\varpi+\left(c+f-x-y-\frac{\mathrm{acf}+\mathrm{a_1c_1f_1}}{\theta^2}\right)\partial u\right];\end{aligned}$$

and here the whole coefficient of $\partial\varpi$ is

$$=(\mathrm{b_1}-\mathrm{c_1})(\mathrm{af}-\mathrm{de})\sqrt{\mathrm{bca_1d_1e_1f_1}}-(\mathrm{b}-\mathrm{c})(\mathrm{a_1f_1}-\mathrm{d_1e_1})\sqrt{\mathrm{b_1c_1adef}},$$

viz. observing that $\mathrm{b_1}-\mathrm{c_1}=\mathrm{b}-\mathrm{c}=b-c$, this is

$$=(b-c)\{[af-de-(a+f-d-e)x]\sqrt{\mathrm{bca_1d_1e_1f_1}}-[af-de-(a+f-d-e)y]\sqrt{\mathrm{b_1c_1adef}}\},$$

or, what is the same thing, it is

$$\begin{aligned}=(b-c)\{&[-(a-d)(a-e)+(a+f-d-e)\,\mathrm{a}]\sqrt{\mathrm{bca_1d_1e_1f_1}}\\&-[-(a-d)(a-e)+(a+f-d-e)\,\mathrm{a_1}]\sqrt{\mathrm{b_1c_1adef}}\}.\end{aligned}$$

The coefficient of ∂u contains the factor $\sqrt{\mathrm{bca_1d_1e_1f_1}}$, multiplied by

$$\mathrm{afb_1}\left(a-c-d-e+x+y+\frac{\mathrm{cde}+\mathrm{c_1d_1e_1}}{\theta^2}\right)$$

$$-\,\mathrm{c_1de}\left(\qquad b+f-x-y-\frac{\mathrm{abf}+\mathrm{a_1b_1f_1}}{\theta^2}\right)$$

$$-\,\mathrm{afc_1}\left(a-b-d-e+x+y+\frac{\mathrm{bde}+\mathrm{b_1d_1e_1}}{\theta^2}\right)$$

$$+\,\mathrm{b_1de}\left(\qquad c+f-x-y-\frac{\mathrm{acf}+\mathrm{a_1c_1f_1}}{\theta^2}\right);$$

here the terms divided by θ^2 destroy each other, and the expression of the coefficient of $\sqrt{\mathrm{bca_1d_1e_1f_1}}$ becomes

$$=(\mathrm{b_1}-\mathrm{c_1})\left[\mathrm{af}\,(a-d-e+x+y)+\mathrm{de}\,(f-x-y)\right]+(\mathrm{af}-\mathrm{de})\,(b\mathrm{c_1}-c\mathrm{b_1}),$$

or since $\mathrm{b_1}-\mathrm{c_1}=b-c$, $b\mathrm{c_1}-c\mathrm{b_1}=-(b-c)\,y$, this is

$$=(b-c)\left[\mathrm{af}\,(a-d-e+x+y)+\mathrm{de}\,(f-x-y)-(\mathrm{af}-\mathrm{de})\,y\right],$$

which is

$$=(b-c)\left[\mathrm{af}\,(a-d-e)+\mathrm{de}f+(\mathrm{af}-\mathrm{de})\,x\right],$$

and is readily reduced to

$$(b-c)\left[(a-d)(a-e)f-(a-d)(a-e)\,x\right],\quad =(b-c)(a-d)(a-e)\,\mathrm{f}:$$

viz. the coefficient of ∂u contains the term $(b-c)(a-d)(a-e)\,\mathrm{f}\sqrt{\mathrm{bca_1d_1e_1f_1}}$. There is a like term $-(b-c)(a-d)(a-e)\,\mathrm{f_1}\sqrt{\mathrm{b_1c_1adef}}$, and the two terms together form the whole coefficient of ∂u.

Hence, restoring the outside factor $\frac{\frac{1}{2}}{\theta}\Omega^2$, we have

$AC\partial AB-AB\partial AC$

$$=\frac{\frac{1}{2}}{\theta}\Omega^2(b-c)\Big[\{[-(a-d)(a-e)+(a+f-d-e)\,\mathrm{a}]\sqrt{\mathrm{bca_1d_1e_1f_1}}$$

$$-[-(a-d)(a-e)+(a+f-d-e)\,\mathrm{a_1}]\sqrt{\mathrm{b_1c_1adef}}\}\,\partial\varpi$$

$$+(a-d)(a-e)\{\mathrm{f}\sqrt{\mathrm{bca_1d_1e_1f_1}}-\mathrm{f_1}\sqrt{\mathrm{b_1c_1adef}}\}\,\partial u\Big],$$

where, as before, I retain $\partial\varpi$ instead of its value $=\partial v-a\,\partial u$. This is a sum of products of the set bc: the products, in fact, are

$$\sqrt{a}\sqrt{de}=\frac{1}{\theta}\{-\mathrm{a}\sqrt{\mathrm{bca_1d_1e_1f_1}}+\mathrm{a_1}\sqrt{\mathrm{b_1c_1adef}}\},$$

$$\sqrt{d}\sqrt{ae}=\text{,,}\ \{-\mathrm{d}\quad\text{,,}\quad+\mathrm{d_1}\quad\text{,,}\quad\},$$

$$\sqrt{e}\sqrt{ad}=\text{,,}\ \{-\mathrm{e}\quad\text{,,}\quad+\mathrm{e_1}\quad\text{,,}\quad\},$$

$$\sqrt{f}\sqrt{bc}=\text{,,}\ \{+\mathrm{f}\quad\text{,,}\quad-\mathrm{f_1}\quad\text{,,}\quad\},$$

whence, observing that $\mathrm{a} - \mathrm{f} = \mathrm{a}_1 - \mathrm{f}_1 = a - f$, we have

$$\sqrt{a}\sqrt{de} + \sqrt{f}\sqrt{bc} = -\frac{a-f}{\theta}\{\sqrt{\mathrm{bca}_1\mathrm{d}_1\mathrm{e}_1\mathrm{f}_1} - \sqrt{\mathrm{b}_1\mathrm{c}_1\mathrm{adef}}\}:$$

it is clear that the term in question is at once expressible as a sum formed with the products $\sqrt{a}\sqrt{de}$ and $\sqrt{f}\sqrt{bc}$.

It is to be remarked that there are 15 expressions such as $A\,\partial B - B\,\partial A$, and 45 expressions such as $AC\,\partial AB - AB\,\partial AC$; and that each of these $(15+45=)\,60$ expressions is a sum of products of a set such as ab: and that there are also 60 expressions of the form $A\,\partial AB - AB\,\partial A$, and that each of these is a sum of products of a set such as $\mathrm{aba}_1\mathrm{b}_1$.

Expression of $\Omega\,\partial^2\Omega - (\partial\Omega)^2, \;= \tfrac{1}{4}M\Omega^2$.

We assume $\Omega\,\partial^2\Omega - (\partial\Omega)^2 = \tfrac{1}{4}M\Omega^2$, where M is a quadric function of ∂u, ∂v; suppose

$$M = \mathfrak{A}\,(\partial u)^2 + 2\mathfrak{B}\,\partial u\,\partial v + \mathfrak{C}\,(\partial v)^2.$$

It is to be noticed that the $\mathfrak{A}$, $\mathfrak{B}$, $\mathfrak{C}$ are not all of them arbitrary functions of (x, y) or (u, v); we, in fact, have $\tfrac{1}{4}M = \dfrac{\partial^2\Omega}{\Omega} - \dfrac{(\partial\Omega)^2}{\Omega^2} = \partial^2\log\Omega$; and hence $\mathfrak{A}$, $\mathfrak{B}$, $\mathfrak{C}$ satisfy the conditions

$$\frac{d\mathfrak{A}}{dv} = \frac{d\mathfrak{B}}{du},\quad \frac{d\mathfrak{B}}{dv} = \frac{d\mathfrak{C}}{du}.$$

Taking $\mathfrak{A}$, $\mathfrak{B}$, $\mathfrak{C}$ as functions of x, y, these become

$$\left(\frac{d\mathfrak{A}}{dx} + y\frac{d\mathfrak{B}}{dx}\right)\sqrt{X} = \left(\frac{d\mathfrak{A}}{dy} + x\frac{d\mathfrak{B}}{dy}\right)\sqrt{Y},$$

$$\left(\frac{d\mathfrak{B}}{dx} + y\frac{d\mathfrak{C}}{dx}\right)\sqrt{X} = \left(\frac{d\mathfrak{B}}{dy} + x\frac{d\mathfrak{C}}{dy}\right)\sqrt{Y}.$$

Putting for the moment

$$\begin{array}{lll} \lambda = a + b + c, & \rho = d + e + f, & p = \lambda + \rho, \\ \mu = ab + ac + bc, & \sigma = de + df + ef, & q = \mu + \sigma, \\ \nu = abc, & \tau = def, & r = \nu + \tau, \end{array}$$

I found it convenient to assume

$$\mathfrak{C} = -2(x^2 + xy + y^2) + p(x + y),$$

where observe that $p, = a + b + c + d + e + f$, is symmetrical in regard to the constants a, b, c, d, e, f. And then, $\mathfrak{C}$ having this value, there exists (as is seen at once) a value of $\mathfrak{B}, = 2(x^2y + xy^2) - pxy$, for which

$$\frac{d\mathfrak{B}}{dx} + y\frac{d\mathfrak{C}}{dx} = 0,\quad \frac{d\mathfrak{B}}{dy} + x\frac{d\mathfrak{C}}{dy} = 0,$$

and which thus satisfies the second of the above-mentioned conditions.

Assuming now

$$\mathfrak{A} = -2x^2y^2 + qxy - r(x+y) - \mu\sigma + \Theta,$$

where Θ has to be determined so as that the first of the same conditions may be also satisfied, then substituting this value of $\mathfrak{A}$, we have

$$\left(2y^3 - py^2 + qy - r + \frac{d\Theta}{dx}\right)\sqrt{X} = \left(2x^3 - px^2 + qx - r + \frac{d\Theta}{dy}\right)\sqrt{Y};$$

that is,

$$\left(-\mathrm{a_1b_1c_1} - \mathrm{d_1e_1f_1} + \frac{d\Theta}{dx}\right)\sqrt{X} = \left(-\mathrm{abc} - \mathrm{def} + \frac{d\Theta}{dy}\right)\sqrt{Y},$$

viz. in the terms independent of Θ writing for $\sqrt{X}$, $\sqrt{Y}$ their values, this is

$$(\mathrm{abc} + \mathrm{def})\sqrt{\mathrm{a_1b_1c_1d_1e_1f_1}} - (\mathrm{a_1b_1c_1} + \mathrm{d_1e_1f_1})\sqrt{\mathrm{abcdef}} + \frac{d\Theta}{dx}\sqrt{X} - \frac{d\Theta}{dy}\sqrt{Y} = 0,$$

or, what is the same thing,

$$-(\sqrt{\mathrm{abca_1b_1c_1}} - \sqrt{\mathrm{defd_1e_1f_1}})(\sqrt{\mathrm{defa_1b_1c_1}} - \sqrt{\mathrm{d_1e_1f_1abc}}) + \frac{d\Theta}{dx}\sqrt{X} - \frac{d\Theta}{dy}\sqrt{Y} = 0.$$

But treating Θ as a function of u and v, we have

$$\frac{d\Theta}{dv} = \frac{d\Theta}{dx}\frac{dx}{dv} + \frac{d\Theta}{dy}\frac{dy}{dv} = \frac{1}{\theta}\left(\frac{d\Theta}{dx}\sqrt{X} - \frac{d\Theta}{dy}\sqrt{Y}\right):$$

also

$$\sqrt{de} = \frac{1}{\theta}(\sqrt{\mathrm{defa_1b_1c_1}} - \sqrt{\mathrm{d_1e_1f_1abc}});$$

and we thus reduce the equation to

$$-(\sqrt{\mathrm{abca_1b_1c_1}} - \sqrt{\mathrm{defd_1e_1f_1}})\sqrt{de} + \frac{d\Theta}{dv} = 0.$$

But, referring to the expression for $\partial\sqrt{ab}$, we have, by a mere interchange of letters,

$$\frac{d}{dv}\sqrt{de} = -\tfrac{1}{2}(\sqrt{\mathrm{abca_1b_1c_1}} - \sqrt{\mathrm{defd_1e_1f_1}}),$$

and the formula thus becomes

$$2\sqrt{de}\frac{d}{dv}\sqrt{de} + \frac{d\Theta}{dv} = 0;$$

consequently

$$\Theta = -(\sqrt{de})^2 = -\frac{1}{\theta^2}(\mathrm{abcd_1e_1f_1} + \mathrm{a_1b_1c_1def} - 2\sqrt{XY}),$$

and the value of $\mathfrak{A}$ thus is

$$\mathfrak{A} = \frac{1}{\theta^2}\{-\mathrm{abcd_1e_1f_1} - \mathrm{a_1b_1c_1def} + \theta^2(-2x^2y^2 + qxy - r(x+y) - \mu\sigma)\} + \frac{2}{\theta^2}\sqrt{XY},$$

or, as this may be written,

$$\mathfrak{A} = \frac{1}{\theta^2}(\mathrm{abc} - \mathrm{a_1b_1c_1})(\mathrm{def} - \mathrm{d_1e_1f_1}) - 2x^2y^2 + qxy - r(x+y) - \mu\sigma - \frac{1}{\theta^2}(\sqrt{X} - \sqrt{Y})^2.$$

Here

$$\text{abc} - \text{a}_1\text{b}_1\text{c}_1 = (\nu - \mu x + \lambda x^2 - x^3) - (\nu - \mu y + \lambda y^2 - y^3)$$
$$= \theta\,[-\mu + \lambda (x + y) - (x^2 + xy + y^2)],$$

and similarly

$$\text{def} - \text{d}_1\text{e}_1\text{f}_1 = \theta\,[-\sigma + \rho\,(x + y) - (x^2 + xy + y^2)];$$

the expression of $\mathfrak{A}$ contains therefore the terms

$$[\mu - \lambda (x + y) + x^2 + xy + y^2]\,[\sigma - \rho\,(x + y) + x^2 + xy + y^2] - \mu\sigma - r\,(x + y) + q\,xy - 2x^2y^2,$$

viz. for r, q substituting their values $\nu + \tau$, $\mu + \rho$, these terms are

$$= -(\mu\rho + \sigma\lambda + \nu + \tau)(x + y) + (\mu + \sigma + \lambda\rho)(x + y)^2$$
$$-(\lambda + \rho)(x + y)(x^2 + xy + y^2) + (x^2 + xy + y^2)^2 - 2x^2y^2.$$

The coefficients $\mu\rho + \sigma\lambda + \nu + \tau$, $\mu + \sigma + \lambda\rho$, $\lambda + \rho$ are, in fact, symmetrical functions of a, b, c, d, e, f, viz. writing

$$X = a - x\,.\,b - x\,.\,c - x\,.\,d - x\,.\,e - x\,.\,f - x,$$
$$= \text{A} - \text{B}x + \text{C}x^2 - \text{D}x^3 + \text{E}x^4 - \text{F}x^5 + x^6,$$

that is,

$$\text{A} = abcdef,\quad \text{B} = \Sigma\,abcde,\quad \text{C} = \Sigma\,abcd,\quad \text{D} = \Sigma\,abc,\quad \text{E} = \Sigma\,ab,\quad \text{F} = \Sigma\,a,$$

($\text{F} = a + b + c + d + e + f$, which has in fact previously been called p), we have

$$\mu\rho + \sigma\lambda + \nu + \tau = \text{D},\quad \mu + \sigma + \lambda\rho = \text{E},\quad \lambda + \rho = \text{F},$$

and the terms are

$$= -(x + y)\,\{\text{D} - \text{E}\,(x + y) + \text{F}\,(x^2 + xy + y^2)\} + x^4 + 2x^3y + x^2y^2 + 2xy^3 + y^4;$$

viz. we have

$$\mathfrak{A} = -\frac{1}{\theta^2}(\sqrt{X} - \sqrt{Y})^2$$
$$-(x + y)\,\{\text{D} - \text{E}\,(x + y) + \text{F}\,(x^2 + xy + y^2)\} + (x^4 + 2x^3y + x^2y^2 + 2xy^3 + y^4).$$

To this I join the foregoing values of $\mathfrak{B}$, $\mathfrak{C}$; viz. writing F in place of p, these are

$$\mathfrak{B} = -\text{F}\,xy + 2\,(x^2y + xy^2),$$
$$\mathfrak{C} = \text{F}\,(x + y) - 2\,(x^2 + xy + y^2),$$

where it will be noticed that the values of $\mathfrak{A}$, $\mathfrak{B}$, $\mathfrak{C}$ are all of them symmetrical in regard to the constants a, b, c, d, e, f.

I recall the original form of $\mathfrak{A}$, viz. this was

$$\mathfrak{A} = -\mu\sigma - r\,(x + y) + q\,xy - 2x^2y^2 - (\sqrt{de})^2$$
$$= -(ab + ac + bc)(de + df + ef) - (abc + def)(x + y)$$
$$+ (ab + ac + bc + de + df + ef)\,xy - 2x^2y^2 - (\sqrt{de})^2$$
$$= \mathfrak{A}_0 - (\sqrt{de})^2,$$

suppose; and $\mathfrak{A}$, $\mathfrak{B}$, $\mathfrak{C}$ denoting as above, we have

$$M = \mathfrak{A}(\partial u)^2 + 2\mathfrak{B}\,\partial u\,\partial v + \mathfrak{C}(\partial v)^2, \quad \Omega\,\partial^2\Omega - (\partial\Omega)^2 = \tfrac{1}{4}M\Omega^2.$$

For the subsequent calculation of $A\,\partial^2 A - (\partial A)^2$, it is convenient to transform this expression by introducing therein $\partial\varpi$ in place of ∂v, and a, a_1 in place of x, y. We have

$$\begin{aligned} M &= \{\mathfrak{A}_0 - (\sqrt{de})^2\}(\partial u)^2 + 2\mathfrak{B}\,\partial u\,(\partial\varpi + a\,\partial u) + \mathfrak{C}(\partial\varpi + a\,\partial u)^2 \\ &= \{\mathfrak{A}_0' - (\sqrt{de})^2\}(\partial u)^2 + 2\mathfrak{B}'\,\partial u\,\partial\varpi \qquad + \mathfrak{C}'(\partial\varpi)^2, \end{aligned}$$

suppose, where

$$\begin{aligned} \mathfrak{C}' &= \mathfrak{C}, \\ \mathfrak{B}' &= \mathfrak{B} + a\mathfrak{C}, \\ \mathfrak{A}_0' &= \mathfrak{A}_0 + 2a\mathfrak{B} + a^2\mathfrak{C}. \end{aligned}$$

Writing

$$x = a - \mathrm{a}, \quad y = a - \mathrm{a}_1,$$

we find

$$\begin{aligned} \mathfrak{C} &= -6a^2 + 2a\mathrm{F} + (6a - \mathrm{F})(\mathrm{a} + \mathrm{a}_1) - 2(\mathrm{a}^2 + \mathrm{a}\mathrm{a}_1 + \mathrm{a}_1^2), \\ \mathfrak{B} &= 4a^3 - a^2\mathrm{F} + (-6a^2 + a\mathrm{F})(\mathrm{a} + \mathrm{a}_1) + 2a(\mathrm{a}^2 + \mathrm{a}_1^2) + (8a - \mathrm{F})\mathrm{a}\mathrm{a}_1 - 2(\mathrm{a}^2\mathrm{a}_1 + \mathrm{a}\mathrm{a}_1^2), \\ \mathfrak{A}_0 &= -(ab + ac + bc)(de + df + ef) \\ &\quad - (abc + def)(2a - \mathrm{a} - \mathrm{a}_1) \\ &\quad + (ab + ac + bc + de + df + ef)[a^2 - a(\mathrm{a} + \mathrm{a}_1) + \mathrm{a}\mathrm{a}_1] \\ &\quad - 2[a^2 - a(\mathrm{a} + \mathrm{a}_1) + \mathrm{a}\mathrm{a}_1]^2, \end{aligned}$$

the developed value of which is

$$\begin{aligned} &= -2a^4 + a^3(b + c) + a^2(-bc + de + df + ef) \\ &\quad + a\{-2def - (b + c)(de + df + ef)\} - bc(de + df + ef) \\ &\quad + \{4a^3 - a^2(b + c) - a(de + df + ef) + def\}(\mathrm{a} + \mathrm{a}_1) \\ &\quad - 2a^2(\mathrm{a}^2 + \mathrm{a}_1^2) + \{-8a^2 + a(b + c) + bc + de + df + ef\}\,\mathrm{a}\mathrm{a}_1 \\ &\quad + 4a(\mathrm{a}^2\mathrm{a}_1 + \mathrm{a}\mathrm{a}_1^2) \\ &\quad - 2\mathrm{a}^2\mathrm{a}_1^2, \end{aligned}$$

and thence without difficulty

$$\begin{aligned} \mathfrak{C}' &= -6a^2 + 2a\mathrm{F} + (6a - \mathrm{F})(\mathrm{a} + \mathrm{a}_1) - 2(\mathrm{a}^2 + \mathrm{a}\mathrm{a}_1 + \mathrm{a}_1^2), \\ \mathfrak{B}' &= -8a^3 + a^2\mathrm{F} + (6a - \mathrm{F})\mathrm{a}\mathrm{a}_1 - 2(\mathrm{a}^2\mathrm{a}_1 + \mathrm{a}\mathrm{a}_1^2), \\ \mathfrak{A}_0' &= a^3(b + c) + a^2(-bc + de + df + ef) + a\{-2def - (b + c)(de + df + ef)\} - bc(de + df + ef) \\ &\quad + \{-a^3 + a^2(d + e + f) - a(de + df + ef) + def\}(\mathrm{a} + \mathrm{a}_1) \\ &\quad + \{4a^2 + a(-b - c - 2d - 2e - 2f) + bc + de + df + ef\}\,\mathrm{a}\mathrm{a}_1 \\ &\quad - 2\mathrm{a}^2\mathrm{a}_1^2, \end{aligned}$$

which are the required values.

Expression for $A\,\partial^2 A - (\partial A)^2$: several subheadings.

Writing for shortness $A\,\partial^2 A - (\partial A)^2 = \Delta A$, as before, and so in other cases: then in general $\Delta PQ = P^2\Delta Q + Q^2\Delta P$, and thence $\Delta P^2 = 2P^2\Delta P$ or $\Delta\sqrt{P} = \frac{\frac{1}{2}}{P}\Delta P$. Hence starting from $A = \Omega\sqrt{a} = \Omega\sqrt{\mathrm{aa}_1}$, we have

$$\Delta A = \Delta\Omega\sqrt{\mathrm{aa}_1} = \mathrm{aa}_1\,\Delta\Omega + \frac{\frac{1}{2}\Omega^2}{\mathrm{aa}_1}(\mathrm{a}^2\Delta\mathrm{a}_1 + \mathrm{a}_1{}^2\Delta\mathrm{a}),$$

where

$$\Delta\mathrm{a} = \mathrm{a}\,\partial^2\mathrm{a} - (\partial\mathrm{a})^2 = -\,\mathrm{a}\,\partial^2 x - (\partial x)^2,\ \ \Delta\mathrm{a}_1 = -\,\mathrm{a}\,\partial^2 y - (\partial y)^2.$$

Hence writing

$$\Delta\Omega = \tfrac{1}{4}M\Omega^2,$$

we have

$$\frac{1}{\Omega^2}\Delta A = \tfrac{1}{4}\mathrm{aa}_1 M - \tfrac{1}{2}(\mathrm{a}_1\,\partial^2 x + \mathrm{a}\,\partial^2 x_1) - \tfrac{1}{2}\left\{\frac{\mathrm{a}_1}{\mathrm{a}}(\partial x)^2 + \frac{\mathrm{a}}{\mathrm{a}_1}(\partial y)^2\right\}.$$

But we have

$$\partial x = \frac{\sqrt{X}}{\theta}(\partial v - y\,\partial u),\ \ \partial y = -\frac{\sqrt{Y}}{\theta}(\partial v - x\,\partial u);$$

squaring the first of these and differentiating, we find

$$2\partial x\,\partial^2 x = \left[\left(-\frac{2X}{\theta^3} + \frac{X'}{\theta^2}\right)\partial x + \frac{2X}{\theta^3}\partial y\right](\partial v - y\,\partial u)^2 - 2\partial y\,\partial u\frac{X}{\theta^2}(\partial v - y\,\partial u),$$

where as regards X the accent denotes differentiation as to x (and further on, as regards Y, it denotes differentiation to y), viz. this is

$$= \left[\left(-\frac{2X}{\theta^3} + \frac{X'}{\theta^2}\right)\partial x + \frac{2X}{\theta^3}\partial y\right](\partial v - y\,\partial u)^2 - 2\partial y\,\partial u\frac{X}{\theta^2}(\partial v - y\,\partial u),$$

$$= \left(-\frac{2X}{\theta^3} + \frac{X'}{\theta^2}\right)\partial x\,(\partial v - y\,\partial u)^2 + \frac{2X}{\theta^3}(\partial v - y\,\partial u - \theta\,\partial u)(\partial v - y\,\partial u)\,\partial y,$$

where the second term is

$$\frac{2X}{\theta^3}(\partial v - x\,\partial u)(\partial v - y\,\partial u)\,\partial y,$$

which is

$$= -\frac{2\sqrt{XY}}{\theta^3}(\partial v - x\,\partial u)^2\,\partial x:$$

hence dividing by $2\partial x$, the equation is

$$\partial^2 x = \left(-\frac{X}{\theta^3} + \frac{X'}{2\theta^2}\right)(\partial v - y\,\partial u)^2 - \frac{\sqrt{XY}}{\theta^3}(\partial v - x\,\partial u)^2;$$

and similarly

$$\partial^2 y = \frac{\sqrt{XY}}{\theta^3}(\partial v - y\,\partial u)^2 + \left(\frac{Y}{\theta^3} + \frac{Y'}{2\theta^2}\right)(\partial v - x\,\partial u)^2;$$

and we may in these values in place of $\partial v - y\,\partial u$ and $\partial v - x\,\partial u$ write $\partial\varpi + \mathrm{a}_1\,\partial u$ and $\partial\varpi + \mathrm{a}\,\partial u$ respectively.

Hence in $\frac{1}{\Omega^2}\Delta A$ the irrational part is

$$\tfrac{1}{2}\frac{\sqrt{XY}}{\theta^3}\{\mathrm{a}_1(\partial\varpi+\mathrm{a}\,\partial u)^2-\mathrm{a}(\partial\varpi+\mathrm{a}_1\,\partial u)^2\}$$

$$=\frac{\tfrac{1}{2}\sqrt{XY}}{\theta^3}(\mathrm{a}_1-\mathrm{a})\{(\partial\varpi)^2-\mathrm{aa}_1(\partial u)^2\}=\frac{\tfrac{1}{2}\sqrt{XY}}{\theta^2}\{(\partial\varpi)^2-\mathrm{aa}_1(\partial u)^2\}.$$

But we have

$$(\sqrt{de})^2=\frac{1}{\theta^2}\{\mathrm{abcd_1e_1f_1}+\mathrm{a_1b_1c_1def}-2\sqrt{XY}\},$$

whence

$$\frac{\sqrt{XY}}{\theta^2}=\frac{\tfrac{1}{2}}{\theta^2}(\mathrm{abcd_1e_1f_1}+\mathrm{a_1b_1c_1def})-\tfrac{1}{2}(\sqrt{de})^2;$$

and the term thus is

$$\left[\frac{\tfrac{1}{4}}{\theta^2}(\mathrm{abcd_1e_1f_1}+\mathrm{a_1b_1c_1def})-\tfrac{1}{4}(\sqrt{de})^2\right]\{(\partial\varpi)^2-\mathrm{aa}_1(\partial u)^2\}.$$

Joining hereto the rational part of $\frac{1}{\Omega^2}\Delta A$, and multiplying the whole by 4, we have

$$\frac{4}{\Omega^2}\Delta A=\mathrm{aa}_1M+\left[\mathrm{a}_1\left(\frac{2X}{\theta^3}-\frac{X'}{\theta^2}\right)-\frac{2\mathrm{a}_1}{\mathrm{a}}\frac{X}{\theta^2}\right](\partial\varpi+\mathrm{a}_1\partial u)^2$$

$$+\left[\mathrm{a}\left(-\frac{2Y}{\theta^3}-\frac{Y'}{\theta^2}\right)-\frac{2\mathrm{a}}{\mathrm{a}_1}\frac{Y}{\theta^2}\right](\partial\varpi+\mathrm{a}\,\partial u)^2$$

$$+\left[\frac{1}{\theta^2}(\mathrm{abcd_1e_1f_1}+\mathrm{a_1b_1c_1def})-(\sqrt{de})^2\right]\{(\partial\varpi)^2-\mathrm{aa}_1(\partial u)^2\},$$

where M has its foregoing value $=\{\mathfrak{A}_0'-(\sqrt{de})^2\}(\partial u)^2+2\mathfrak{B}'\,\partial u\,\partial\varpi+\mathfrak{C}'(\partial\varpi)^2$.

First step of the reduction.

Writing $\mathrm{bcdef}=U$, $\mathrm{b_1c_1d_1e_1f_1}=U_1$, then $X=\mathrm{a}U$, $Y=\mathrm{a}_1U_1$, and consequently

$$X'=-U+\mathrm{a}U',\quad Y'=-U_1+\mathrm{a}_1U_1',$$

the accents in regard to U, U_1 denoting differentiations as to x, y respectively: then

$$\mathrm{a}_1\left(\frac{2X}{\theta^3}-\frac{X'}{\theta^2}\right)-\frac{2\mathrm{a}_1}{\mathrm{a}}\frac{X}{\theta^2}=\mathrm{a}_1\left(\frac{2\mathrm{a}U}{\theta^3}+\frac{U-\mathrm{a}U'}{\theta^2}\right)-\frac{2\mathrm{a}_1}{\mathrm{a}}\frac{\mathrm{a}U}{\theta^2}=\mathrm{aa}_1\left(\frac{2U}{\theta^3}-\frac{U'}{\theta^2}\right)-\frac{\mathrm{a}_1U}{\theta^2}:$$

and similarly

$$\mathrm{a}\left(-\frac{2Y}{\theta^3}-\frac{Y'}{\theta^2}\right)-\frac{2\mathrm{a}}{\mathrm{a}_1}\frac{Y}{\theta^2}=\ \ldots\ldots\ldots\ =\mathrm{aa}_1\left(-\frac{2U_1}{\theta^3}-\frac{U_1'}{\theta^2}\right)-\frac{\mathrm{a}U_1}{\theta^2}.$$

The formula thus becomes

$$\frac{4}{\Omega^2}\Delta A = \mathrm{aa}_1\left[M + \left(\frac{2U}{\theta^3} - \frac{U'}{\theta^2}\right)(\partial\varpi + \mathrm{a}_1\,\partial u)^2 + \left(-\frac{2U_1}{\theta^3} - \frac{U_1'}{\theta^2}\right)(\partial\varpi + \mathrm{a}\,\partial u)^2\right.$$

$$\left. - \left\{\frac{1}{\theta^2}(\mathrm{abcd_1e_1f_1} + \mathrm{a_1b_1c_1def}) - (\sqrt{de})^2\right\}(\partial u)^2\right]$$

$$- \frac{\mathrm{a}_1 U}{\theta^2}(\partial\varpi + \mathrm{a}_1\,\partial u)^2 - \frac{\mathrm{a}U_1}{\theta^2}(\partial\varpi + \mathrm{a}\,\partial u)^2$$

$$+ \left\{\frac{1}{\theta^2}(\mathrm{abcd_1e_1f_1} + \mathrm{a_1b_1c_1def}) - (\sqrt{de})^2\right\}(\partial\varpi)^2,$$

viz. substituting for M its value, the term in $(\sqrt{de})^2(\partial u)^2$ disappears, and the formula is

$$\frac{4}{\Omega^2}\Delta A = \mathrm{aa}_1\left[\mathfrak{A}_0'(\partial u)^2 + 2\mathfrak{B}'\partial u\partial\varpi + \mathfrak{C}'(\partial\varpi)^2 + \left(\frac{2U}{\theta^3} - \frac{U'}{\theta^2}\right)(\partial\varpi + \mathrm{a}_1\,\partial u)^2\right.$$

$$\left. + \left(-\frac{2U_1}{\theta^3} - \frac{U_1'}{\theta^2}\right)(\partial\varpi + \mathrm{a}\,\partial u)^2 - \frac{1}{\theta^2}(\mathrm{abcd_1e_1f_1} + \mathrm{a_1b_1c_1def})(\partial u)^2\right]$$

$$- \frac{\mathrm{a}_1 U}{\theta^2}(\partial\varpi + \mathrm{a}_1\,\partial u)^2 - \frac{\mathrm{a}U_1}{\theta^2}(\partial\varpi + \mathrm{a}\,\partial u)^2$$

$$+ \left\{\frac{1}{\theta^2}(\mathrm{abcd_1e_1f_1} + \mathrm{a_1b_1c_1def}) - (\sqrt{de})^2\right\}(\partial\varpi)^2:$$

say for shortness this is

$$\frac{4}{\Omega^2}\Delta A$$

$$= \mathrm{aa}_1\Sigma - \frac{\mathrm{a}_1 U}{\theta^2}(\partial\varpi + \mathrm{a}_1\,\partial u)^2 - \frac{\mathrm{a}U_1}{\theta^2}(\partial\varpi + \mathrm{a}\,\partial u)^2 + \frac{1}{\theta^2}(\mathrm{abcd_1e_1f_1} + \mathrm{a_1b_1c_1def}) - (\sqrt{de})^2(\partial\varpi)^2.$$

Second step of the reduction.

In the reductions which follow, we make as many terms as may be to contain the factor aa_1, so as to simplify as much as possible the portion not containing this factor.

We have $\partial\varpi + \mathrm{a}_1\,\partial u = (\partial\varpi + \theta\,\partial u) + \mathrm{a}\,\partial u$, and consequently

$$(\partial\varpi + \mathrm{a}_1\,\partial u)^2 = (\partial\varpi + \theta\,\partial u)^2 + \mathrm{a}P,$$

where $P = 2\,\partial u\,\partial\varpi + (\mathrm{a} + 2\theta)(\partial u)^2$: similarly $\partial\varpi + \mathrm{a}\,\partial u = (\partial\varpi - \theta\,\partial u) + \mathrm{a}_1\,\partial u$, and therefore

$$(\partial\varpi + \mathrm{a}\,\partial u)^2 = (\partial\varpi - \theta\,\partial u)^2 + \mathrm{a}_1 P_1,$$

where $P_1 = 2\,\partial u\,\partial\varpi + (\mathrm{a}_1 - 2\theta)(\partial u)^2$: the values may also be written

$$P = 2\,\partial u\,\partial\varpi + (2\mathrm{a}_1 - \mathrm{a})(\partial u)^2, \quad P_1 = 2\,\partial u\,\partial\varpi + (2\mathrm{a} - \mathrm{a}_1)(\partial u)^2.$$

The formula thus becomes

$$\frac{4}{\Omega^2}\Delta A = \mathrm{aa}_1\left\{\Sigma - \frac{U}{\theta^2}P - \frac{U_1}{\theta^2}P_1\right\}$$

$$-\frac{\mathrm{a}_1 U}{\theta^2}(\partial\varpi + \theta\,\partial u)^2 - \frac{\mathrm{a}U_1}{\theta^2}(\partial\varpi - \theta\,\partial u)^2 + \frac{1}{\theta^2}(\mathrm{abcd_1e_1f_1} + \mathrm{a_1b_1c_1def})(\partial\varpi)^2$$

$$-(\sqrt{de})^2(\partial\varpi)^2.$$

The second line here is

$$-(\mathrm{a}_1 U + \mathrm{a}U_1)(\partial u)^2 - \frac{2}{\theta}(\mathrm{a}_1 U - \mathrm{a}U_1)\,\partial u\,\partial\varpi$$

$$+\frac{1}{\theta^2}\{-\mathrm{a}_1 U - \mathrm{a}U_1 + \mathrm{abcd_1e_1f_1} + \mathrm{a_1b_1c_1def}\}(\partial\varpi)^2,$$

and the coefficient herein of $(\partial\varpi)^2$ is $= \frac{1}{\theta^2}(\mathrm{ad_1e_1f_1} - \mathrm{a_1def})(\mathrm{bc} - \mathrm{b_1c_1})$. Writing for the moment $d-a$, $e-a$, $f-a = d'$, e', f', we have

$$\frac{1}{\theta}(\mathrm{ad_1e_1f_1} - \mathrm{a_1def}) = \frac{1}{\mathrm{a}_1 - \mathrm{a}}\{\mathrm{a}(d' + \mathrm{a}_1 . e' + \mathrm{a}_1 . f' + \mathrm{a}_1) - \mathrm{a}_1(d' + \mathrm{a} . e' + \mathrm{a} . f' + \mathrm{a})\}$$

$$= -d'e'f' + \mathrm{aa}_1(d' + e' + f' + \mathrm{a} + \mathrm{a}_1),$$

$$\frac{1}{\theta}(\mathrm{bc} - \mathrm{b_1c_1}) = -(b' + c' + \mathrm{a} + \mathrm{a}_1).$$

The whole term in $(\partial\varpi)^2$ is thus

$$= \{(b' + c' + \mathrm{a} + \mathrm{a}_1)d'e'f' + \mathrm{aa}_1(b' + c' + \mathrm{a} + \mathrm{a}_1)(d' + e' + f' + \mathrm{a} + \mathrm{a}_1)\}(\partial\varpi)^2.$$

The coefficient of $\partial u\,\partial\varpi$ is $-\frac{2}{\theta}(\mathrm{a}_1 U - \mathrm{a}U_1)$: viz. this is

$$= \frac{2}{\mathrm{a}_1 - \mathrm{a}}\{\mathrm{a}_1(b' + \mathrm{a} . c' + \mathrm{a} . d' + \mathrm{a} . e' + \mathrm{a} . f' + \mathrm{a}) - \mathrm{a}(b' + \mathrm{a}_1 . c' + \mathrm{a}_1 . d' + \mathrm{a}_1 . e' + \mathrm{a}_1 . f' + \mathrm{a}_1)\},$$

and if

$$b' + \mathrm{a} . c' + \mathrm{a} . d' + \mathrm{a} . e' + \mathrm{a} . f' + \mathrm{a} = \mathrm{B}' + \mathrm{C}'\mathrm{a} + \mathrm{D}'\mathrm{a}^2 + \mathrm{E}'\mathrm{a}^3 + \mathrm{F}'\mathrm{a}^4 + \mathrm{a}^5,$$

that is,

$$\mathrm{B}' = b'c'd'e'f',\quad \mathrm{C}' = \Sigma b'c'd'e',\quad \mathrm{D}' = \Sigma b'c'd',\quad \mathrm{E}' = \Sigma b'c',$$

$$\mathrm{F}' = \Sigma b' = b' + c' + d' + e' + f',$$

this is

$$= 2\{\mathrm{B}' - \mathrm{D}'\mathrm{aa}_1 - \mathrm{E}'\mathrm{aa}_1(\mathrm{a} + \mathrm{a}_1) - \mathrm{F}'\mathrm{aa}_1(\mathrm{a}^2 + \mathrm{aa}_1 + \mathrm{a}_1^2) - \mathrm{aa}_1(\mathrm{a}^3 + \mathrm{a}^2\mathrm{a}_1 + \mathrm{aa}_1^2 + \mathrm{a}_1^3)\}:$$

or say for shortness it is $= -2(\mathrm{B}' - \mathrm{aa}_1\Phi)$ where

$$\Phi = \mathrm{D}' + \mathrm{E}'(\mathrm{a} + \mathrm{a}_1) + \mathrm{F}'(\mathrm{a}^2 + \mathrm{aa}_1 + \mathrm{a}_1^2) + \mathrm{a}^3 + \mathrm{a}^2\mathrm{a}_1 + \mathrm{aa}_1^2 + \mathrm{a}_1^3;$$

the term in question thus is $-2(\mathrm{B}' - \mathrm{aa}_1\Phi)\,\partial u\,\partial\varpi$.

The coefficient of $(\partial u)^2$ is $-(\mathrm{a}_1 U + \mathrm{a}U_1)$, viz. this is

$$-\mathrm{a}_1(b' + \mathrm{a} . c' + \mathrm{a} . d' + \mathrm{a} . e' + \mathrm{a} . f' + \mathrm{a}) - \mathrm{a}(b' + \mathrm{a}_1 . c' + \mathrm{a}_1 . d' + \mathrm{a}_1 . e' + \mathrm{a}_1 . f' + \mathrm{a}_1),$$

which is $= -(\mathrm{a} + \mathrm{a}_1)\mathrm{B}' - \mathrm{aa}_1\Psi$, where

$$\Psi = 2\mathrm{C}' + \mathrm{D}'(\mathrm{a} + \mathrm{a}_1) + \mathrm{E}'(\mathrm{a}^2 + \mathrm{a}_1^2) + \mathrm{F}'(\mathrm{a}^3 + \mathrm{a}_1^3) + \mathrm{a}^4 + \mathrm{a}_1^4.$$

The formula thus is

$$\frac{4}{\Omega^2}\Delta A = \mathrm{aa}_1\left\{\Sigma - \frac{U}{\theta^2}P - \frac{U_1}{\theta^2}P_1 - (b'+c'+\mathrm{a}+\mathrm{a}_1)(d'+e'+f'+\mathrm{a}+\mathrm{a}_1)(\partial\varpi)^2 + 2\Phi\,\partial u\,\partial\varpi - \Psi(\partial u)^2\right\}$$
$$-(\mathrm{a}+\mathrm{a}_1)\,\mathrm{B}'(\partial u)^2 - 2\mathrm{B}'\,\partial u\,\partial\varpi + (b'+c'+\mathrm{a}+\mathrm{a}_1)\,d'e'f'(\partial\varpi)^2 - (\sqrt{de})^2(\partial\varpi)^2.$$

The whole coefficient of aa_1, substituting for Σ, P, P_1, Φ, Ψ their values, and arranging according to $\partial\varpi$, ∂u, is

$$=(\partial\varpi)^2\left\{\mathfrak{C}' + \frac{2U}{\theta^3} - \frac{U'}{\theta^2} + \left(-\frac{2U_1}{\theta^3} - \frac{U_1'}{\theta^2}\right) - (b'+c'+\mathrm{a}+\mathrm{a}_1)(d'+e'+f'+\mathrm{a}+\mathrm{a}_1)\right\}$$

$$+2\,\partial\varpi\,\partial u\left\{\mathfrak{B}' + \mathrm{a}_1\left(\frac{2U}{\theta^3} - \frac{U'}{\theta^2}\right) + \mathrm{a}\left(-\frac{2U_1}{\theta^3} - \frac{U_1'}{\theta^2}\right) - \frac{U}{\theta^2} - \frac{U_1}{\theta^2}\right.$$
$$\left. + \mathrm{D}' + \mathrm{E}'(\mathrm{a}+\mathrm{a}_1) + \mathrm{F}'(\mathrm{a}^2+\mathrm{aa}_1+\mathrm{a}_1^2) + \mathrm{a}^3 + \mathrm{a}^2\mathrm{a}_1 + \mathrm{aa}_1^2 + \mathrm{a}_1^3\right\}$$

$$+(\partial u)^2\left\{\mathfrak{A}_0' + \mathrm{a}_1^2\left(\frac{2U}{\theta^3} - \frac{U'}{\theta^2}\right) + \mathrm{a}^2\left(-\frac{2U_1}{\theta^3} - \frac{U_1'}{\theta^2}\right) + (\mathrm{a}-2\mathrm{a}_1)\frac{U_1}{\theta^2}\right.$$
$$\left. - \frac{1}{\theta^2}(\mathrm{abcd_1e_1f_1} + \mathrm{a_1b_1c_1def}) - 2\mathrm{C}' - \mathrm{D}'(\mathrm{a}+\mathrm{a}_1) - \mathrm{E}'(\mathrm{a}^2+\mathrm{a}_1^2) - \mathrm{F}'(\mathrm{a}^3+\mathrm{a}_1^3) - \mathrm{a}^4 - \mathrm{a}_1^4\right\}:$$

and we have to reduce separately the three coefficients of this formula.

Third step of the reduction.

First, for the coefficient of $(\partial\varpi)^2$; recollecting that $\theta = \mathrm{a}_1 - \mathrm{a}$, we have

$$2\,\frac{U-U_1}{\theta} = -2\mathrm{C}' - 2\mathrm{D}'(\mathrm{a}+\mathrm{a}_1) - 2\mathrm{E}'(\mathrm{a}^2+\mathrm{aa}_1+\mathrm{a}_1^2) - 2\mathrm{F}'(\mathrm{a}^3+\mathrm{a}^2\mathrm{a}_1+\mathrm{aa}_1^2+\mathrm{a}_1^3)$$
$$-2(\mathrm{a}^4+\mathrm{a}^3\mathrm{a}_1+\mathrm{a}^2\mathrm{a}_1^2+\mathrm{aa}_1^3+\mathrm{a}_1^4),$$
$$-(U'+U_1') = 2\mathrm{C}' + 2\mathrm{D}'(\mathrm{a}+\mathrm{a}_1) + 3\mathrm{E}'(\mathrm{a}^2+\mathrm{a}_1^2) + 4\mathrm{F}'(\mathrm{a}^3+\mathrm{a}_1^3) + 5(\mathrm{a}^4+\mathrm{a}_1^4).$$

Adding these, the right-hand side divides by $(\mathrm{a}_1-\mathrm{a})^2$, that is, by θ^2; and the resulting value is

$$=\mathrm{E}' + 2\mathrm{F}'(\mathrm{a}+\mathrm{a}_1) + 3\mathrm{a}^2 + 4\mathrm{aa}_1 + 3\mathrm{a}_1^2.$$

The term $-(b'+c'+\mathrm{a}+\mathrm{a}_1)(d'+e'+f'+\mathrm{a}+\mathrm{a}_1)$, attending to the values of E' and F', is

$$=b'c' + d'e' + d'f' + e'f' - \mathrm{E}' - \mathrm{F}'(\mathrm{a}+\mathrm{a}_1) - \mathrm{a}^2 - 2\mathrm{aa}_1 - \mathrm{a}_1^2;$$

hence the whole coefficient of $(\partial\varpi)^2$ is

$$=\mathfrak{C}' + b'c' + d'e' + d'f' + e'f' + \mathrm{F}'(\mathrm{a}+\mathrm{a}_1) - 2(\mathrm{a}^2+\mathrm{aa}_1+\mathrm{a}_1^2),$$

or substituting for b', c', d', e', f' their values, this is

$$=\mathfrak{C}' + 4a^2 - a(b+c+2d+2e+2f) + bc + de + df + ef + (\mathrm{F}-6a)(\mathrm{a}+\mathrm{a}_1) - 2(\mathrm{a}^2+\mathrm{aa}_1+\mathrm{a}_1^2).$$

Proceeding next to reduce the coefficient of $2\,\partial\varpi\,\partial u$, observing as before that $\theta = \mathrm{a}_1 - \mathrm{a}$, we have

$$\frac{2\mathrm{a}_1U - 2\mathrm{a}U_1}{\theta} = 2\mathrm{B}' - 2\mathrm{D}'\mathrm{aa}_1 - 2\mathrm{E}'\mathrm{aa}_1(\mathrm{a}+\mathrm{a}_1) - 2\mathrm{F}'\mathrm{aa}_1(\mathrm{a}^2+\mathrm{aa}_1+\mathrm{a}_1^2) - 2\mathrm{aa}_1(\mathrm{a}^3+\mathrm{a}^2\mathrm{a}_1+\mathrm{aa}_1^2+\mathrm{a}_1^3),$$

also

$$-(a_1U'+U)-(aU_1'+U_1)=$$
$$-2B'+D'(-a^2+4aa_1-a_1^2)+E'(-a^3+3a^2a_1+3aa_1^2-a_1^3)+F'(-a^4+4a^3a_1+4aa_1^3-a_1^4)$$
$$-a^5+5a^4a_1+5aa_1^4-a_1^5;$$

adding these two expressions, the right-hand side divides by $(a_1-a)^2$, that is, by θ^2, and the resulting value is

$$=-\quad D'-E'(a+a_1)-F'(a^2+a_1^2)\qquad -a^3+a^2a_1+aa_1^2-a_1^3.$$

To this is to be added

$$+2D'+E'(a+a_1)+F'(a^2+aa_1+a_1^2)+a^3+a^2a_1+aa_1^2+a_1^3;$$

we thus see that the whole coefficient of $2\partial u\,\partial\varpi$ is

$$=D'+\qquad F'aa_1+2(a^2a_1+aa_1^2),$$

or say it is

$$=D'+(F-6a)\,aa_1+2(a^2a_1+aa_1^2).$$

Lastly, for the coefficient of $(\partial u)^2$; we have

$$\frac{2a_1^2U-2a^2U_1}{\theta}=2B'(a+a_1)+2C'aa_1-2E'a^2a_1^2-2F'a^2a_1^2(a+a_1)-2a^2a_1^2(a^2+aa_1+a_1^2),$$

and also

$$-a_1^2U'+(a-2a_1)\,U-a^2U_1'+(a_1-2a)\,U=$$
$$-B'(a+a_1)+C'(2a^2-4aa_1+2a_1^2)+D'(a^3+a_1^3)+E'(a^4-2a^3a_1+6a^2a_1^2-2aa_1^3+a_1^4)$$
$$+F'(a^5-2a^4a_1+4a^3a_1^2+4a^2a_1^3-2aa_1^4+a_1^5)+(a^6-2a^5a_1+5a^4a_1^2+5a^2a_1^4-2aa_1^5+a_1^6),$$

whence the sum of these two expressions is

$$=B'(a+a_1)+C'(2a^2-2aa_1+2a_1^2)+D'(a^3+a_1^3)+E'(a^4-2a^3a_1+4a^2a_1^2-2aa_1^3+a_1^4)$$
$$+F'(a^5-2a^4a_1+2a^3a_1^2+2a^2a_1^3-2aa_1^4+a_1^5)+a^6-2a^5a_1+3a^4a_1^2-2a^3a_1^3+3a^2a_1^4-2aa_1^5+a_1^6.$$

We must to this add the term $-(abcd_1e_1f_1+a_1b_1c_1def)$, that is,

$$-a(b'+a.c'+a.d'+a_1.e'+a_1.f'+a_1)-a_1(b'+a_1.c'+a_1.d'+a.e'+a.f'+a).$$

Putting for the moment

$$d'+a.e'+a.f'+a=\tau'+\sigma'a+\rho'a^2+a^3,$$

that is,

$$\tau'=d'e'f',\quad \sigma'=d'e'+d'f'+e'f',\quad \rho'=d'+e'+f',$$

the term is

$$-b'c'\tau'(a+a_1)-(b'+c')\,\tau'(a^2+a_1^2)-\tau'(a^3+a_1^3)-(b'c'+\sigma')(a^3a_1+aa_1^3)$$
$$-(b'+c'+\rho')(a^3a_1^2+a^2a_1^3)-2a^3a_1^3$$
$$-2b'c'\sigma'aa_1-[(b'+c')\,\sigma'+b'c'\rho'](a^2a_1+aa_1^2)-2(b'+c')\,\rho'a^2a_1^2.$$

Adding it to the preceding expression, the sum is

$$= (\text{B}' - b'c'\tau')(\text{a} + \text{a}_1) + \{2\text{C}' - (b' + c')\tau'\}(\text{a}^2 + \text{a}_1^2) + (\text{D}' - \tau')(\text{a}^3 + \text{a}_1^3) + \text{E}'(\text{a}^4 + \text{a}_1^4)$$

$$+ \{-2\text{C}' - 2b'c'\sigma'\}\,\text{aa}_1 - \{(b' + c')\sigma' + b'c'\rho'\}(\text{a}^2\text{a}_1 + \text{aa}_1^2) - (2\text{E}' + b'c' + \sigma')(\text{a}^3\text{a}_1 + \text{aa}_1^3)$$

$$+ \{4\text{E}' - 2(b' + c')\rho'\}\,\text{a}^2\text{a}_1^2$$

$$+ \quad \text{F}'(\text{a}^5 + \text{a}_1^5) \qquad\qquad + \quad \text{a}^6 + \text{a}_1^6$$

$$- \quad 2\text{F}'(\text{a}^4\text{a}_1 + \text{aa}_1^4) \qquad\qquad - 2(\text{a}^5\text{a}_1 + \text{aa}_1^5)$$

$$+ (2\text{F}' - b' - c' - \rho')(\text{a}^3\text{a}_1^2 + \text{a}^2\text{a}_1^3) + 3(\text{a}^4\text{a}_1^2 + \text{a}^2\text{a}_1^4) - 4\text{a}^3\text{a}_1^3.$$

This is, in fact, divisible by $(\text{a}_1 - \text{a})^2$, that is, by θ^2: for we have between the symbols the relations

$$\text{F}' = b' + c' + \rho',$$

$$\text{E}' = b'c' + (b' + c')\rho' + \sigma',$$

$$\text{D}' = b'c'\rho' + (b' + c')\sigma' + \tau',$$

$$\text{C}' = b'c'\sigma' + (b' + c')\tau',$$

$$\text{B}' = b'c'\tau',$$

and we thus reduce the expression to

$$\{2\text{C}' - (b' + c')\tau'\}(\text{a}^2 - 2\text{aa}_1 + \text{a}_1^2) + (\text{D}' - \tau')(\text{a}^3 - \text{a}^2\text{a}_1 - \text{aa}_1^2 + \text{a}_1^3)$$

$$+ \text{E}'(\text{a}^4 - 3\text{a}^3\text{a}_1 + 4\text{a}^2\text{a}_1^2 - 3\text{aa}_1^3 + \text{a}_1^4) + (b' + c')\rho'(\text{a}^3\text{a}_1 - 2\text{a}^2\text{a}_1^2 + \text{aa}_1^3)$$

$$+ \text{F}'(\text{a}^5 - 2\text{a}^4\text{a}_1 + \text{a}^3\text{a}_1^2 + \text{a}^2\text{a}_1^3 - 2\text{aa}_1^4 + \text{a}_1^5)$$

$$+ (\text{a}^6 - 2\text{a}^5\text{a}_1 + 3\text{a}^4\text{a}_1^2 - 4\text{a}^3\text{a}_1^3 + 3\text{a}^2\text{a}_1^4 - 2\text{aa}_1^5 + \text{a}_1^6),$$

viz. effecting the division, the quotient is

$$= 2\text{C}' - (b' + c')\tau' + (\text{D}' - \tau')(\text{a} + \text{a}_1) + \text{E}'(\text{a}^2 + \text{a}_1^2) + \text{F}'(\text{a}^3 + \text{a}_1^3) + \text{a}^4 + 2\text{a}^2\text{a}_1^2 + \text{a}_1^4 - (b'c' + \sigma')\,\text{aa}_1.$$

To this must be added

$$-2\text{C}' \qquad\qquad - \text{D}'(\text{a} + \text{a}_1) \qquad\qquad - \text{E}'(\text{a}^2 + \text{a}_1^2) - \text{F}'(\text{a}^3 + \text{a}_1^3) - (\text{a}^4 + \text{a}_1^4);$$

and we thus obtain the coefficient of $(\partial u)^2$ in the form

$$\mathfrak{A}_0' - (b' + c')\tau' - \tau'(\text{a} + \text{a}_1) - (b'c' + \sigma')\,\text{aa}_1 + 2\text{a}^2\text{a}_1^2,$$

viz. this is

$$= \mathfrak{A}_0' + (b + c - 2a)(a - d)(a - e)(a - f) + (a - d)(a - e)(a - f)(\text{a} + \text{a}_1)$$

$$+ \{-(a - b)(a - c) - (a - d)(a - e) - (a - d)(a - f) - (a - e)(a - f)\}\,\text{aa}_1 + 2\text{a}^2\text{a}_1^2,$$

or finally it is

$$\begin{aligned}= \mathfrak{A}_0' - 2a^4 + a^3(b+c+2d+2e+2f) + a^2\{-(b+c)(d+e+f) - 2(de+df+ef)\} \\ + a\{(b+c)(de+df+ef) + 2def\} - (b+c)def \\ + \{a^3 - a^2(d+e+f) + a(de+df+ef) - def\}(\mathrm{a}+\mathrm{a}_1) \\ + \{-4a^2 + a(b+c+2d+2e+2f) - bc - de - df - ef\}\mathrm{a}\mathrm{a}_1 \\ + 2\mathrm{a}^2\mathrm{a}_1^2.\end{aligned}$$

It is to be observed that the investigation thus far has been entirely independent of the values of $\mathfrak{A}_0'$, $\mathfrak{B}'$, $\mathfrak{C}'$: these values are, in fact, such as to make the coefficients of $(\partial\varpi)^2$, $\partial\varpi\,\partial u$, $(\partial u)^2$ each equal to a constant, and it was really by such a condition that the value of $\mathfrak{C}(=\mathfrak{C}')$ was determined; but if we had thus also determined the values of $\mathfrak{A}_0'$ and $\mathfrak{B}'$, it would not have been apparent that the values of $\mathfrak{A}_0'$, $\mathfrak{B}'$ and $\mathfrak{C}'$ thus determined would be consistent with each other: the foregoing investigation of these values was therefore prefixed.

Completion of the reduction and final expression for ΔA.

But now substituting the values of $\mathfrak{A}_0'$, $\mathfrak{B}'$, $\mathfrak{C}'$, we find

$$\begin{aligned}\text{coeff. of } (\partial\varpi)^2 &= ab+ac+bc+de+df+ef, \\ \text{,, ,, } 2\partial\varpi\,\partial u &= -a^2(a-b-c-d-e-f), \\ \text{,, ,, } (\partial u)^2 &= -2a^4 + 2a^3(b+c+d+e+f) \\ &\quad - a^2(bc+bd+be+bf+cd+ce+cf+de+df+ef) \\ &\quad - \;(bcde+bcdf+bcef+bdef+cdef),\end{aligned}$$

viz. these coefficients belong to the portion which contains the factor $\mathrm{a}\mathrm{a}_1$ of the expression for $\frac{4}{\Omega^2}\Delta A$: the other portion was

$$(b'+c'+\mathrm{a}+\mathrm{a}_1)d'e'f'(\partial\varpi)^2 - (\sqrt{de})^2(\partial\varpi)^2 - 2\mathrm{B}'\partial u\,\partial\varpi - (\mathrm{a}+\mathrm{a}_1)\mathrm{B}'(\partial u)^2,$$

where

$$\mathrm{B}' = b'c'd'e'f', \quad b' = b-a, \text{ etc.}$$

We have thus the complete result, viz. this is

$$\begin{aligned}\frac{4}{\Omega^2}\Delta A = \mathrm{a}\mathrm{a}_1 [&(ab+ac+bc+de+df+ef)(\partial\varpi)^2 \\ &- a^2(a-b-c-d-e-f)\,2\partial\varpi\,\partial u \\ &+ \left\{\begin{array}{l}-2a^4 + 2a^3(b+c+d+e+f) \\ -a^2(bc+bd+be+bf+cd+ce+cf+de+df+ef) \\ -(bcde+bcdf+bcef+bdef+cdef)\end{array}\right\}(\partial u)^2] \\ &- (-2a+b+c+\mathrm{a}+\mathrm{a}_1)(a-d)(a-e)(a-f)(\partial\varpi)^2 - (\sqrt{de})^2(\partial\varpi)^2 \\ &+ (a-b)(a-c)(a-d)(a-e)(a-f)\,2\partial u\,\partial\varpi \\ &+ (\mathrm{a}+\mathrm{a}_1)(a-b)(a-c)(a-d)(a-e)(a-f)(\partial u)^2,\end{aligned}$$

which is obviously a sum of squares.

As a partial verification, I remark that ΔA should be symmetrical in regard to the constants b, c, d, e, f; this is obviously the case as regards the terms in $\partial u\,\partial\varpi$ and $(\partial u)^2$, and it must also be so in regard to the term in $(\partial\varpi)^2$. The whole coefficient of $(\partial\varpi)^2$ is

$$= \mathrm{a}\mathrm{a}_1(ab+ac+bc+de+df+ef)$$
$$-(-2a+b+c+\mathrm{a}+\mathrm{a}_1)(a-d)(a-e)(a-f)-(\sqrt{de})^2,$$

and if we interchange for instance b and d, this coefficient becomes

$$= \mathrm{a}\mathrm{a}_1(ad+ac+cd+be+bf+ef)$$
$$-(-2a+d+c+\mathrm{a}+\mathrm{a}_1)(a-b)(a-e)(a-f)-(\sqrt{be})^2.$$

These two expressions must be equal; viz. we must have

$$(\sqrt{be})^2-(\sqrt{de})^2=-\mathrm{a}\mathrm{a}_1(b-d)(a+c-e-f)+(a-e)(a-f)(b-d)(-a+c+\mathrm{a}+\mathrm{a}_1):$$

the left-hand side is

$$=\frac{1}{\theta^2}(\mathrm{b}\mathrm{d}_1-\mathrm{b}_1\mathrm{d})(\mathrm{e}\mathrm{f}\mathrm{a}_1\mathrm{c}_1-\mathrm{e}_1\mathrm{f}_1\mathrm{a}\mathrm{c}),$$

and we have

$$\mathrm{b}\mathrm{d}_1-\mathrm{b}_1\mathrm{d}=(b-d)\theta;$$

hence, throwing out the factor $b-d$, the equation to be verified becomes

$$\frac{1}{\theta}(\mathrm{e}\mathrm{f}\mathrm{a}_1\mathrm{c}_1-\mathrm{e}_1\mathrm{f}_1\mathrm{a}\mathrm{c})=-\mathrm{a}\mathrm{a}_1(a+c-e-f)+(a-e)(a-f)(-a+c+\mathrm{a}+\mathrm{a}_1).$$

Writing

$$\mathrm{e}=e'+\mathrm{a},\ \text{etc.},\quad \theta=\mathrm{a}_1-\mathrm{a},$$

the left-hand side is

$$(\mathrm{a}+\mathrm{a}_1)e'f'+\mathrm{a}\mathrm{a}_1(e'+f')+c'e'f'-c'\mathrm{a}\mathrm{a}_1,$$

and the right-hand side is

$$-\mathrm{a}\mathrm{a}_1(c'-e'-f')+e'f'(c'+\mathrm{a}+\mathrm{a}_1),$$

and these are equal.

There are of course, in all, six expressions such as ΔA, each of them being by what precedes a sum of squares. And there are besides ten expressions such as

$$\Delta AB,\ =AB\,\partial^2 AB-(\partial AB)^2,$$

each of which should be a sum of squares: but I have not as yet effected the calculation of this expression ΔAB.

Cambridge, 7th December, 1877.

666.

SUR UN EXEMPLE DE RÉDUCTION D'INTÉGRALES ABÉLIENNES AUX FONCTIONS ELLIPTIQUES.

[From the *Comptes Rendus de l'Académie des Sciences de Paris*, t. LXXXV. (Juillet—Décembre, 1877), pp. 265—268; 373, 374; 426—429; 472—475.]

JE reprends l'investigation de M. Hermite par rapport aux intégrales réductibles

$$\int \frac{(1, x)\,dx}{\sqrt{x\,.\,1-x\,.\,1+ax\,.\,1+bx\,.\,1-abx}},$$

publiée sous ce même titre: "Sur un exemple, etc.", (*Annales de la Société scientifique de Bruxelles*, 1876).

Nous avons les constantes a, b et les variables x, y, u, v; et en posant

$$X = x\,.\,1-x\,.\,1+ax\,.\,1+bx\,.\,1-abx,$$

$$Y = y\,.\,1-y\,.\,1+ay\,.\,1+by\,.\,1-aby$$

(et $c=\sqrt{1+a\,.\,1+b}$), M. Hermite a effectué l'intégration, par fonctions elliptiques, des équations différentielles

$$\frac{dx}{\sqrt{X}}+\frac{dy}{\sqrt{Y}}=-\ \frac{2}{c}\ \ (du+dv),$$

$$\frac{x\,dx}{\sqrt{X}}+\frac{y\,dy}{\sqrt{Y}}=-\frac{2}{c\sqrt{ab}}(du-dv);$$

il a en effet trouvé les expressions, au moyen des fonctions elliptiques de u, v, des fonctions symétriques $x+y$, xy, et, de là, des cinq fonctions a, b, c, d, e dont je vais parler.

Au cas d'une fonction X du sixième ordre, on a dans la théorie seize fonctions, savoir six fonctions a, b, c, d, e, f, et dix fonctions abf. cde, ..., ou (avec une notation plus simple) ab, ac, ad, ae, bc, bd, be, cd, ce, de: dans le cas d'une fonction du cinquième ordre, et ainsi dans le cas actuel, l'une des six fonctions, disons f, se réduit à l'unité, et l'on a les cinq fonctions a, b, c, d, e, et les dix fonctions ab, ..., de.

Présentement, ces fonctions sont

$$\mathrm{a} = xy,$$

$$\mathrm{b} = 1 - x \,.\, 1 - y,$$

$$\mathrm{c} = 1 + ax \,.\, 1 + ay,$$

$$\mathrm{d} = 1 + bx \,.\, 1 + by,$$

$$\mathrm{e} = 1 - abx \,.\, 1 - aby.$$

$$\mathrm{ab} = (\sqrt{x \,.\, 1 - x \,.\, 1 + ay \,.\, 1 + by \,.\, 1 - aby} - \sqrt{y \,.\, 1 - y \,.\, 1 + ax \,.\, 1 + bx \,.\, 1 - abx})^2 \div (x - y)^2,$$

$$\mathrm{ac} = (\sqrt{x \,.\, 1 + ax \,.\, 1 - y \,.\, 1 + by \,.\, 1 - aby} - \sqrt{y \,.\, 1 + ay \,.\, 1 - x \,.\, 1 + bx \,.\, 1 - abx})^2 \div (x - y)^2,$$

$$\mathrm{ad} = (\sqrt{x \,.\, 1 + bx \,.\, 1 - y \,.\, 1 + ay \,.\, 1 - aby} - \sqrt{y \,.\, 1 + by \,.\, 1 - x \,.\, 1 + ax \,.\, 1 - abx})^2 \div (x - y)^2,$$

$$\mathrm{ae} = (\sqrt{x \,.\, 1 - abx \,.\, 1 - y \,.\, 1 + ay \,.\, 1 + by} - \sqrt{y \,.\, 1 - aby \,.\, 1 - x \,.\, 1 + ax \,.\, 1 + bx})^2 \div (x - y)^2,$$

$$\mathrm{bc} = (\sqrt{1 - x \,.\, 1 + ax \,.\, y \,.\, 1 + by \,.\, 1 - aby} - \sqrt{1 - y \,.\, 1 + ay \,.\, x \,.\, 1 + bx \,.\, 1 - abx})^2 \div (x - y)^2,$$

$$\mathrm{bd} = (\sqrt{1 - x \,.\, 1 + bx \,.\, y \,.\, 1 + ay \,.\, 1 - aby} - \sqrt{1 - y \,.\, 1 + by \,.\, x \,.\, 1 + ax \,.\, 1 - abx})^2 \div (x - y)^2,$$

$$\mathrm{be} = (\sqrt{1 - x \,.\, 1 - abx \,.\, y \,.\, 1 + ay \,.\, 1 + by} - \sqrt{1 - y \,.\, 1 - aby \,.\, x \,.\, 1 + ax \,.\, 1 - bx})^2 \div (x - y)^2,$$

$$\mathrm{cd} = (\sqrt{1 + ax \,.\, 1 + bx \,.\, y \,.\, 1 - y \,.\, 1 - aby} - \sqrt{1 + ay \,.\, 1 + by \,.\, x \,.\, 1 - x \,.\, 1 - abx})^2 \div (x - y)^2,$$

$$\mathrm{ce} = (\sqrt{1 + ax \,.\, 1 - abx \,.\, y \,.\, 1 - y \,.\, 1 + by} - \sqrt{1 + ay \,.\, 1 - aby \,.\, x \,.\, 1 - x \,.\, 1 + bx})^2 \div (x - y)^2,$$

$$\mathrm{de} = (\sqrt{1 + bx \,.\, 1 - abx \,.\, y \,.\, 1 - y \,.\, 1 + ay} - \sqrt{1 + by \,.\, 1 - aby \,.\, x \,.\, 1 - x \,.\, 1 + ax})^2 \div (x - y)^2,$$

et je remarque que la différence de deux quelconques des fonctions ab, ac, ... est une fonction rationnelle et entière de x, y. On a, par exemple:

$$\mathrm{ac} - \mathrm{ad} = a - b \,.\;\; 1 - ab\,xy,$$

$$\mathrm{bc} - \mathrm{bd} = a - b \,.\, -1 + ab\,(x + y) - ab\,xy,$$

$$\mathrm{be} - \mathrm{cd} = 1 + a \,.\;\; 1 + b - 1 + ab\,xy,$$

$$\mathrm{ce} - \mathrm{de} = a - b \,.\, -1 + (x + y) - ab\,xy.$$

En faisant, comme auparavant, $c = \sqrt{1 + a \,.\, 1 + b}$, et puis

$$ck = \sqrt{a} + \sqrt{b}, \qquad cl = \sqrt{a} - \sqrt{b},$$

$$ck' = 1 - \sqrt{ab}\,, \qquad cl' = 1 + \sqrt{ab}\,;$$

$$\sigma = \mathrm{sn}\,(u, k)\,, \qquad \sigma_1 = \mathrm{sn}\,(v, l),$$

$$\gamma = \mathrm{cn}\,(u, k)\,, \qquad \gamma_1 = \mathrm{cn}\,(v, l),$$

$$\delta = \mathrm{dn}\,(u, k)\,, \qquad \delta_1 = \mathrm{dn}\,(v, l),$$

(où j'écris sn, cn, dn pour sin am, cos am, Δ am), et, pour un moment,

$$\xi = \sqrt{ab}\,(\gamma\sigma_1\delta_1 + \gamma_1\sigma\delta),\quad \eta = c\,(-k'\sigma\gamma_1\delta_1 + l'\sigma_1\gamma\delta),\quad \zeta = \gamma\sigma_1\delta_1 - \gamma_1\sigma\delta\,^*,$$

x, y sont donnés au moyen des fonctions elliptiques σ, γ, δ, σ_1, γ_1, δ_1 de u, v par les équations

$$x + y = \frac{\xi^2 + \zeta^2 - \eta^2}{\xi^2},\quad xy = \frac{\zeta^2}{\xi^2},$$

ou, ce qui est la même chose, on a identiquement

$$\xi^2 z^2 - (\xi^2 + \zeta^2 - \eta^2)\,z + \zeta^2 = \xi^2 \,.\, z - x \,.\, z - y,$$

de manière que x, y sont les racines de l'équation quadrique

$$\xi^2 z^2 - (\xi^2 + \zeta^2 - \eta^2)\,z + \zeta^2 = 0.$$

On a l'identité (due à M. Hermite)

$$(Pz^3 + Qz^2 + Rz + S)^2 - c^2\delta^2\delta_1^2\,(\sigma^2 - \sigma_1^2)^2\,Z$$
$$= [\sigma^2(1 + az)(1 + bz) - c^2 z]\,[\sigma_1^2(1 + az)(1 + bz) - c^2 z] \times [\xi^2 z^2 - (\xi^2 + \zeta^2 - \eta^2)\,z + \zeta^2],$$

ou

$$Z = z \,.\, 1 - z \,.\, 1 + az \,.\, 1 + bz \,.\, 1 - abz;$$

et alors les valeurs de P, Q, R, S sont

$$P = -\,ab\sqrt{ab}\,\sigma\sigma_1\,(\gamma\sigma_1\delta_1 + \gamma_1\sigma\delta),$$
$$Q = \sqrt{ab}\,\sigma\sigma_1\,[-(a + b - \sqrt{ab})\,\gamma\sigma_1\delta_1 - (a + b + \sqrt{ab})\,\gamma_1\sigma\delta] + c^2\sqrt{ab}\,(\delta\sigma_1\gamma_1 + \delta_1\sigma\gamma),$$
$$R = \sigma\sigma_1\,[(a + b - \sqrt{ab})\,\gamma\sigma_1\delta_1 - (a + b + \sqrt{ab})\,\gamma_1\sigma\delta] + c^2\,(\delta\sigma_1\gamma_1 - \delta_1\sigma\gamma),$$
$$S = \sigma\sigma_1\,(\gamma\sigma_1\delta_1 - \gamma_1\sigma\delta),$$

lesquelles peuvent aussi s'écrire comme il suit:

$$P = -\,ab\sigma\sigma_1\xi,$$
$$Q = -\,ab\sigma\sigma_1\zeta - c^2\sqrt{ab}\,\sigma\sigma_1\,(l^2\gamma\sigma_1\delta_1 + k^2\gamma_1\sigma\delta) + c^2\sqrt{ab}\,(\delta\sigma_1\gamma_1 + \delta_1\sigma\gamma),$$
$$R = \sigma\sigma_1\xi + c^2\sigma\sigma_1\,(l^2\gamma\sigma_1\delta_1 - k^2\gamma_1\sigma\delta) + c^2\,(\delta\sigma_1\gamma_1 - \delta_1\sigma\gamma),$$
$$S = \sigma\sigma_1\zeta,$$

et je remarque l'équation

$$P + Q + R + S = c^3\gamma\gamma_1\,(-k'\sigma\gamma_1\delta_1 + l'\sigma_1\gamma\delta)$$
$$= c^2\gamma\gamma_1\eta.$$

En écrivant successivement $z = x$, $z = y$, et en choisissant convenablement les signes des radicaux, on obtient

$$Px^3 + Qx^2 + Rx + S = c\delta\delta_1\,(\sigma^2 - \sigma_1^2)\sqrt{X},$$
$$Py^3 + Qy^2 + Ry + S = c\delta\delta_1\,(\sigma^2 - \sigma_1^2)\sqrt{Y};$$

on conçoit sans peine que c'est à cause de ces expressions rationnelles des radicaux que l'intégration des équations différentielles réussit.

* En écrivant

$$\xi' = \gamma\sigma_1\delta_1 + \gamma_1\sigma\delta,\quad \eta' = -\,k'\sigma\gamma_1\delta_1 + l'\sigma_1\gamma\delta,\quad \zeta' = \gamma\sigma_1\delta_1 - \gamma_1\sigma\delta,$$

on a

$$\xi = \sqrt{ab}\,\xi',\quad \eta = c\eta',\quad \zeta = \zeta';$$

je me sers, dans la suite, de ce symbole

$$\xi_1' = \gamma\sigma_1\delta_1 + \gamma_1\sigma\delta.$$

[Pp. 373—376*; 426—429.] Les valeurs de $x+y$, xy donnent sans beaucoup de peine celles de a, b, c, d, e; mais les réductions pour obtenir les valeurs des dix fonctions ab, …, de sont très pénibles; je donne seulement les résultats. Ces valeurs sont

$$\sqrt{\mathrm{a}} = \frac{1}{ab\xi'} \cdot \quad \gamma\sigma_1\delta_1 \quad - \gamma_1\sigma\delta,$$

$$\sqrt{\mathrm{b}} = \frac{c}{\sqrt{ab\xi'}} \cdot - k'\sigma\gamma_1\delta_1 + l'\sigma_1\gamma\delta,$$

$$\sqrt{\mathrm{c}} = \frac{c}{\sqrt{b\xi'}} \cdot \quad l\delta\sigma_1\gamma_1 \quad - k\delta_1\sigma\gamma,$$

$$\sqrt{\mathrm{d}} = \frac{c}{\sqrt{a\xi'}} \cdot \quad l\delta\sigma_1\gamma_1 \quad + k\delta_1\sigma\gamma,$$

$$\sqrt{\mathrm{e}} = \frac{c}{\xi'} \cdot \quad k'\sigma\gamma_1\delta_1 + l'\sigma_1\gamma\delta,$$

où

$$\xi' = \gamma\sigma_1\delta_1 + \gamma_1\sigma\delta;$$

et puis

$$\sqrt{\mathrm{ab}} = \frac{c}{\xi'} \cdot \gamma\gamma_1\delta\delta_1 - k'l'\sigma\sigma_1,$$

$$\sqrt{\mathrm{ac}} = \frac{c}{\xi'(l\delta\sigma_1\gamma_1 - k\delta_1\sigma\gamma)} \cdot k(l'^2 + l^2\gamma_1^4)\,\sigma\gamma\delta - l(k'^2 + k^2\gamma^4)\,\sigma_1\gamma_1\delta_1,$$

$$\sqrt{\mathrm{ad}} = \frac{c}{\xi'(l\delta\sigma_1\gamma_1 + k\delta_1\sigma\gamma)} \cdot k(l'^2 + l^2\gamma_1^4)\,\sigma\gamma\delta + l(k'^2 + k^2\gamma^4)\,\sigma_1\gamma_1\delta_1,$$

$$\sqrt{\mathrm{ae}} = \frac{c}{\xi'(k'\sigma\gamma_1\delta_1 + l'\sigma_1\gamma\delta)} \cdot k'(l'^2 + l^2\gamma_1^4)\,\sigma\gamma\delta + l'(k'^2 + k^2\gamma^4)\,\sigma_1\gamma_1\delta_1,$$

$$\sqrt{\mathrm{bc}} = \frac{\tfrac{1}{2}c^2}{\xi'} \cdot k'\delta_1^2 + l'\delta^2 - kl\,(k'\sigma^2\gamma_1^2 + l'\sigma_1^2\gamma^2),$$

$$\sqrt{\mathrm{bd}} = \frac{\tfrac{1}{2}c^2}{\xi'} \cdot k'\delta_1^2 + l'\delta^2 + kl\,(k'\sigma^2\gamma_1^2 + l'\sigma_1^2\gamma^2),$$

$$\sqrt{\mathrm{be}} = \frac{c}{\xi'} \cdot - \sigma\sigma_1\delta\delta_1 - \gamma\gamma_1,$$

$$\sqrt{\mathrm{cd}} = \frac{c}{\xi'} \cdot - \sigma\sigma_1\delta\delta_1 + \gamma\gamma_1,$$

$$\sqrt{\mathrm{ce}} = \frac{c}{\xi'} \cdot 1 - \frac{1+a}{c\sqrt{a}}\,k\sigma^2 - \frac{1+a}{c\sqrt{a}}\,l\sigma_1^2 + kl\sigma^2\sigma_1^2,$$

$$\sqrt{\mathrm{de}} = \frac{c}{\xi'} \cdot 1 - \frac{1+b}{c\sqrt{b}}\,k\sigma^2 + \frac{1+b}{c\sqrt{b}}\,l\sigma_1^2 - kl\sigma^2\sigma_1^2.$$

* Voir la note, p. 426 du volume.—Dans la seconde Communication (p. 373), une erreur de composition a fait placer, à la suite de la treizième ligne de la page 374, deux pages et demie de texte qui ne devaient trouver place que dans la Communication suivante. Nous rétablissons intégralement cette seconde Communication: la troisième sera insérée dans le prochain numéro.

Les valeurs de a, b, c, d, e donnent

$$\sqrt{X}\sqrt{Y}=\sqrt{\mathrm{a}}\sqrt{\mathrm{b}}\sqrt{\mathrm{c}}\sqrt{\mathrm{d}}\sqrt{\mathrm{e}}$$

$$=\frac{c^4}{(ab)^2\,\xi'^5}(\gamma\sigma_1\delta_1-\gamma_1\sigma\delta)(-k'\sigma\gamma_1\delta_1+l'\sigma_1\gamma\delta)$$

$$\times(l\delta\sigma_1\gamma_1-k\delta_1\sigma\gamma)(l\delta\sigma_1\gamma_1+k\delta_1\sigma\gamma)(k'\sigma\gamma_1\delta_1+l'\sigma_1\gamma\delta),$$

$$=\frac{c^4}{(ab)^2\,\xi'^6}(\gamma^2\sigma_1^{\,2}\delta_1^{\,2}-\gamma_1^{\,2}\sigma^2\delta^2)(-k'^2\sigma^2\gamma_1^{\,2}\delta_1^{\,2}+l'^2\sigma_1^{\,2}\gamma^2\delta^2)(l^2\delta^2\sigma_1^{\,2}\gamma_1^{\,2}-k^2\delta_1^{\,2}\sigma^2\gamma^2);$$

j'ai vérifié que le signe s'accorde avec celui de la valeur obtenue au moyen des expressions rationnelles de $\sqrt{X}$, $\sqrt{Y}$.

On vérifie en partie les valeurs des fonctions $\sqrt{\mathrm{ab}}$, $\sqrt{\mathrm{ac}},\ldots$, en considérant les différences des carrés de ces fonctions; mais ce calcul n'est pas toujours facile. Par exemple, nous avons

$$\mathrm{ac}-\mathrm{ad}=(a-b)(1-ab\,xy)$$

$$=\frac{c^2kl}{\xi'^2}(\xi'^2-\zeta^2)$$

$$=\frac{4c^2kl}{\xi'^2}\sigma\sigma_1\gamma\gamma_1\delta\delta_1;$$

et cette valeur doit ainsi être égale à

$$\frac{c^2}{\xi'^2}\left\{\frac{1}{(l\delta\sigma_1\gamma_1-k\delta_1\sigma\gamma)^2}[k(l'^2+l^2\gamma_1^{\,4})\sigma\gamma\delta-l(k'^2+k^2\gamma^4)\sigma_1\gamma_1\delta_1]^2\right.$$

$$\left.-\frac{1}{(l\delta\sigma_1\gamma_1+k\delta_1\sigma\gamma)^2}[k(l'^2+l^2\gamma_1^{\,4})\sigma\gamma\delta+l(k'^2+k^2\gamma^4)\sigma_1\gamma_1\delta_1]^2\right\}.$$

Pour voir cela, j'écris pour le moment

$$A=k(l'^2+l^2\gamma_1^{\,4})\sigma\gamma\delta,\qquad B=l(k'^2+k^2\gamma^4)\sigma_1\gamma_1\delta_1,$$

$$\alpha=l\delta\sigma_1\gamma_1,\qquad \beta=k\delta_1\sigma\gamma;$$

l'équation devient ainsi

$$4kl\sigma\sigma_1\gamma\gamma_1\delta\delta_1(\alpha^2-\beta^2)^2=(\alpha+\beta)^2(A-B)^2-(\alpha-\beta)^2(A+B)^2,$$

$$=4[\alpha\beta(A^2+B^2)-AB(\alpha^2+\beta^2)];$$

or, en remarquant que AB et $\alpha\beta$ contiennent chacun le facteur $kl\,\sigma\sigma_1\gamma\gamma_1\delta\delta_1$, cette équation devient

$$(\alpha^2-\beta^2)^2=k^2(l'^2+l^2\gamma_1^{\,4})^2\sigma^2\gamma^2\delta^2+l^2(k'^2+k^2\gamma^4)^2\sigma_1^{\,2}\gamma_1^{\,2}\delta_1^{\,2}$$

$$-(l'^2+l^2\gamma_1^{\,4})(k'^2+k^2\gamma_1^{\,4})(l^2\delta^2\sigma_1^{\,2}\gamma_1^{\,2}+k^2\delta_1^{\,2}\sigma^2\gamma^2),$$

c'est-à-dire

$$(\alpha^2-\beta^2)^2=[k^2(l'^2+l^2\gamma_1^{\,4})\sigma^2\gamma^2-l^2(k'^2+k^2\gamma^4)\sigma_1^{\,2}\gamma_1^{\,2}][(l'^2+l^2\gamma_1^{\,4})\delta^2-(k'^2+k^2\gamma^4)\delta_1^{\,2}];$$

or les deux facteurs à droite se réduisant l'un et l'autre à

$$k^2\sigma^2\gamma^2\delta_1^{\,2} - l^2\sigma_1^{\,2}\gamma_1^{\,2}\delta^2,$$

c'est-à-dire à $(\alpha^2 - \beta^2)$, la vérification est ainsi complétée.

La différence be − cd donne un exemple beaucoup plus simple; on a

$$\begin{aligned} \text{be} - \text{cd} &= 1 + a\,.\,1 + b\,(-1 + ab\,xy) \\ &= \frac{c^2}{\xi'^2}(-\xi'^2 + \zeta^2) \\ &= \frac{c^2}{\xi'^2}(-4\sigma\sigma_1\gamma\gamma_1\delta\delta_1)\,; \end{aligned}$$

l'équation à vérifier est ainsi

$$-4\sigma\sigma_1\gamma\gamma_1\delta\delta_1 = (-\sigma\sigma_1\delta\delta_1 - \gamma\gamma_1)^2 - (-\sigma\sigma_1\delta\delta_1 + \gamma\gamma_1)^2,$$

ce qui est juste.

[Pp. 472—475.] Je donne quelques autres formules dont je me suis servi dans le cours de cette recherche. Partant des expressions de ξ, η, ζ, on a

$$\begin{aligned} d\xi = \lambda du + \lambda_1 dv = \sqrt{ab}\,\{ &\;[-\sigma\delta\sigma_1\delta_1 + \gamma\gamma_1(1 - 2k^2\sigma^2)]\,du \\ &+ [\gamma\gamma_1(1 - 2l^2\sigma_1^{\,2}) - \sigma\sigma_1\delta\delta_1\;]\,dv\}, \\ d\eta = \mu du + \mu_1 dv = c\,\{ &\;[-k'\gamma\delta\gamma_1\delta_1 + l'\sigma\sigma_1(-1 - k^2 + 2k^2\sigma^2)]\,du \\ &+ [k'\sigma\sigma_1(1 + l^2 - 2l^2\sigma_1^{\,2}) + l'\gamma\delta\gamma_1\delta_1\;]\,dv\}, \\ d\zeta = \nu du + \nu_1 dv = \;\{ &\;[-\sigma\delta\sigma_1\delta_1 - \gamma\gamma_1(1 - 2k^2\sigma^2)]\,du \\ &+ [\gamma\gamma_1(1 - 2l^2\sigma_1^{\,2}) + \sigma\delta\sigma_1\delta_1\;]\,dv\}\,; \end{aligned}$$

en prenant pour A, B, C des fonctions telles que

$$A\,d\xi + B\,d\eta + C\,d\zeta = du + dv,$$

on a

$$\begin{aligned} A\lambda + B\mu + C\nu &= 1, \\ A\lambda_1 + B\mu_1 + C\nu_1 &= 1. \end{aligned}$$

Je pose aussi

$$A\xi + B\eta + C\zeta = 0,$$

et au moyen de ces équations, j'obtiens pour A, B, C les valeurs

$$\begin{aligned} A\nabla &= \frac{1}{2\sqrt{ab}}(-U + W), \\ B\nabla &= \frac{1}{2c}V, \\ C\nabla &= \tfrac{1}{2}(-U - W), \end{aligned}$$

où

$$U = l'\delta^2(\delta\sigma_1\gamma_1 + \delta_1\sigma\gamma) + k'\sigma^2\gamma_1^2(l^2\delta\sigma_1\gamma_1 + k^2\delta_1\sigma\gamma),$$

$$W = k'\delta_1^2(\delta\sigma_1\gamma_1 + \delta_1\sigma\gamma) + l'\sigma_1^2\gamma^2(l^2\delta\sigma_1\gamma_1 + k^2\delta_1\sigma\gamma),$$

$$V = 2\left[(l'^2 + l^2\gamma_1^4)\sigma\gamma\delta + (k'^2 + k^2\gamma^4)\sigma_1\gamma_1\delta_1\right],$$

$$\nabla = (k'\sigma\gamma_1\delta_1 + l'\sigma_1\gamma\delta)(l\delta\sigma_1\gamma_1 - k\delta_1\sigma\gamma)(l\delta\sigma_1\gamma_1 + k\delta_1\sigma\gamma);$$

et de là aussi

$$-U + W = -\frac{2\sqrt{ab}}{c}(\delta\delta_1\gamma\gamma_1 - k'l'\sigma\sigma_1)(\gamma\sigma_1\delta_1 + \gamma_1\sigma\delta),$$

$$U + W = \frac{2}{c}\left(\{1 + l^2\sigma^2\sigma_1^2 - \sqrt{ab}\left[(1 + k'l')\sigma^2 - l^2\sigma_1^2\right]\}\delta\sigma_1\gamma_1\right.$$

$$\left. + \{1 - k^2\sigma^2\sigma_1^2 + \sqrt{ab}\left[(1 + k'l')\sigma_1^2 - k^2\sigma^2\right]\}\delta_1\sigma\gamma\right).$$

En admettant l'équation

$$\frac{dx}{\sqrt{X}} + \frac{dy}{\sqrt{Y}} = -\frac{2}{c}(du + dv),$$

on obtient sans peine les relations

$$A\xi = \frac{c}{x-y}\left(\frac{x^2 - x}{\sqrt{X}} - \frac{y^2 - y}{\sqrt{Y}}\right),$$

$$B\frac{\xi^2}{\eta} = \frac{c}{x-y}\left(\frac{x}{\sqrt{X}} - \frac{y}{\sqrt{Y}}\right),$$

$$C\frac{\xi^2}{\zeta} = \frac{c}{x-y}\left(\frac{-x+1}{\sqrt{X}} - \frac{-y+1}{\sqrt{Y}}\right),$$

et, en multipliant par

$$c^2\delta\delta_1(\sigma^2 - \sigma_1^2)\sqrt{XY}, \quad = \frac{c^4ab\eta\zeta\delta\delta_1(\sigma^2 - \sigma_1^2)}{\xi^5}\nabla,$$

et dans les seconds membres, au lieu de

$$c^2\delta\delta_1(\sigma^2 - \sigma_1^2)\sqrt{X}, \quad c^2\delta\delta_1(\sigma^2 - \sigma_1^2)\sqrt{Y},$$

substituant les valeurs

$$Px^3 + Qx^2 + Rx + S, \quad Py^3 + Qy^2 + Ry + S,$$

on obtient, après quelques réductions simples, les équations

$$c^4ab\delta\delta_1(\sigma^2 - \sigma_1^2)\nabla A = ab\sigma\sigma_1\xi\eta\zeta - \sigma\sigma_1\xi^3\eta + c^2\gamma\gamma_1\xi^2\zeta,$$

$$\text{,,} \qquad \nabla B = ab\sigma\sigma_1\zeta(\xi^2 + \zeta^2 - \eta^2) + \sigma\sigma_1\xi^3 - Q\xi\zeta,$$

$$\text{,,} \qquad \nabla C = ab\sigma\sigma_1\eta(-2\xi^2 - \zeta^2 + \eta^2) + Q\xi\eta - c^2\gamma\gamma_1\xi^3,$$

lesquelles satisfont, comme cela doit être, à la condition

$$A\xi + B\eta + C\zeta = 0.$$

Réciproquement, en vérifiant ces identités, ce qui est assez pénible, on obtient une démonstration de l'équation différentielle

$$\frac{dx}{\sqrt{X}} + \frac{dy}{\sqrt{Y}} = -\frac{2}{c}(du + dv).$$

En écrivant, pour plus de simplicité,

$$A\nabla = -\frac{1}{c}\mathfrak{A}'\xi', \quad B\nabla = \frac{1}{c}\mathfrak{B}, \quad C\nabla = -\frac{1}{c}\mathfrak{C},$$

les valeurs de $\mathfrak{A}'$, $\mathfrak{B}$, $\mathfrak{C}$ sont

$$\mathfrak{A}' = \gamma\gamma_1\delta\delta_1 - k'l'\sigma\sigma_1,$$

$$\mathfrak{B} = (l'^2 + l^2\gamma^4)\,\sigma\gamma\delta + (k'^2 + k^2\gamma^4)\,\sigma_1\gamma_1\delta_1,$$

$$\mathfrak{C} = \{1 - l^2\sigma^2\sigma_1{}^2 - \sqrt{ab}\,[(1 + k'l')\,\sigma^2 - l^2\sigma_1{}^2]\}\,\delta\sigma_1\gamma_1$$
$$+ \{1 - k^2\sigma^2\sigma_1{}^2 + \sqrt{ab}\,[(1 + k'l')\,\sigma_1{}^2 - k^2\sigma^2]\}\,\delta_1\sigma\gamma\,;$$

et des trois équations pour $A\xi$, $B\frac{\xi^2}{\eta}$, $C\frac{\xi^2}{\zeta}$, on déduit

$$-\frac{\mathfrak{A}'}{\sqrt{ab}} + B\eta\zeta = \Omega\left(\frac{x^2}{\sqrt{X}} - \frac{y^2}{\sqrt{Y}}\right),$$

$$B\zeta = \Omega\left(\frac{x}{\sqrt{X}} - \frac{y}{\sqrt{Y}}\right),$$

$$-\mathfrak{C}\eta + B\zeta = \Omega\left(\frac{1}{\sqrt{X}} - \frac{1}{\sqrt{Y}}\right),$$

où

$$\Omega = \frac{c^2\nabla\,\eta\zeta}{ab\xi'\,(x-y)}\,;$$

et c'est au moyen de ces équations que j'ai trouvé les valeurs ci-dessus données pour $\sqrt{\mathrm{ab}}$, $\sqrt{\mathrm{ac}}, \ldots$; on a, par exemple,

$$\sqrt{\mathrm{ab}} = \frac{\sqrt{X}\,\sqrt{Y}}{\sqrt{\mathrm{a}}\,\sqrt{\mathrm{b}}\,(x-y)}\left(\frac{x-x^2}{\sqrt{X}} - \frac{y-y^2}{\sqrt{Y}}\right) = \frac{\sqrt{X}\,\sqrt{Y}}{\sqrt{\mathrm{a}}\,\sqrt{\mathrm{b}}\,(x-y)\,\Omega}\,\frac{\mathfrak{A}'\eta\zeta}{\sqrt{ab}},$$

ce qui se réduit sans peine à $\sqrt{\mathrm{ab}} = \frac{c}{\xi'}\mathfrak{A}'$. Les dix fonctions contiennent de cette manière les facteurs suivants:

$$\sqrt{\mathrm{ab}}, \quad \mathfrak{A}',$$

$$\sqrt{\mathrm{ac}}, \quad (1+a)\,\mathfrak{B} - \sqrt{\frac{a}{b}}\,\mathfrak{A}'\eta,$$

$$\sqrt{\mathrm{ad}}, \quad (1+b)\,\mathfrak{B} - \sqrt{\frac{a}{b}}\,\mathfrak{A}'\eta,$$

$$\sqrt{\mathrm{ae}}, \quad (1-ab)\,\mathfrak{B} + \sqrt{ab}\,\mathfrak{A}'\eta,$$

$$\sqrt{\mathrm{bc}}, \quad -\mathfrak{C} + \sqrt{\frac{a}{b}}\,\mathfrak{A}'\zeta,$$

$$\sqrt{\mathrm{bd}},\quad -\mathfrak{C}+\sqrt{\frac{b}{a}}\,\mathfrak{A}'\zeta',$$

$$\sqrt{\mathrm{be}},\quad -\mathfrak{C}-\sqrt{ab}\,\mathfrak{A}'\zeta,$$

$$\sqrt{\mathrm{cd}},\quad \frac{1}{c}(-\sqrt{ab}\,\eta\zeta\mathfrak{A}'+c^2\mathfrak{B}\zeta-\mathfrak{C}\eta),$$

$$\sqrt{\mathrm{ce}},\quad \frac{1}{c}[a\sqrt{ab}\,\eta+\mathfrak{A}'\zeta(1+a)(1-ab)\mathfrak{B}\zeta-\mathfrak{C}\eta],$$

$$\sqrt{\mathrm{de}},\quad \frac{1}{c}[b\sqrt{ab}\,\eta\zeta\mathfrak{A}'+(1+b)(1-ab)\mathfrak{B}\zeta-\mathfrak{C}\eta]:$$

mais il y a des dénominateurs variables qui contiennent des facteurs dont quelques-uns divisent les numérateurs, et la réduction aux formes ci-dessus données m'a coûté assez de peine.

667.

ON THE BICIRCULAR QUARTIC: ADDITION TO PROFESSOR CASEY'S MEMOIR "ON A NEW FORM OF TANGENTIAL EQUATION."

[From the *Philosophical Transactions of the Royal Society of London,* vol. CLXVII. Part II. (1877), pp. 441—460. Received January 24,—Read February 22, 1877.]

PROFESSOR CASEY communicated to me the MS. of the Memoir referred to, and he has permitted me to make to it the present Addition, containing further developments on the theory of the bicircular quartic.

Starting from his theory of the fourfold generation of the curve, Prof. Casey shows that there exist series of inscribed quadrilaterals $ABCD$ whereof the sides AB, BC, CD, DA pass through the centres of the four circles of inversion respectively; or (as it is convenient to express it) the pairs of points (A, B), (B, C), (C, D), (D, A) belong to the four modes of generation respectively, and may be regarded as depending upon certain parameters (his θ, θ', θ'', θ''', or say) ω_1, ω_2, ω_3, ω respectively, any three of these being in fact functions of the fourth. Considering a given quadrilateral $ABCD$, and giving to it an infinitesimal variation, we have four infinitesimal arcs AA', BB', CC', DD'; these are differential expressions, AA' and BB' being of the form $M_1d\omega_1$, BB' and CC' of the form $M_2d\omega_2$, CC' and DD' of the form $M_3d\omega_3$, DD' and AA' of the form $Md\omega$; or, what is the same thing, AA' is expressible in the two forms $Md\omega$ and $M_1d\omega_1$, BB' in the two forms $M_1d\omega_1$ and $M_2d\omega_2$, &c., the identity of the two expressions for the same arc of course depending on the relation between the two parameters. But any such monomial expression $Md\omega$ of an arc AA' would be of a complicated form, not obviously reducible to elliptic functions; Casey does not obtain these monomial expressions at all, but he finds geometrically monomial expressions for the differences and sum $BB'-AA'$, $CC'-BB'$, $DD'+CC'$, $DD'-AA'$ (they cannot be all of them differences), and thence a quadrinomial expression $AA'=N_1d\omega_1+N_2d\omega_2+N_3d\omega_3+Nd\omega$ (his $ds'=\rho\, d\theta+\rho'\, d\theta'+\rho''\, d\theta''+\rho'''\, d\theta'''$); and that without any explicit consideration of the relations which connect the parameters.

I propose to complete the analytical theory by establishing the monomial equations $AA' = Md\omega = M_1 d\omega_1$, &c., and the relations between the parameters $\omega, \omega_1, \omega_2, \omega_3$ which belong to an inscribed quadrilateral $ABCD$, so as to show what the process really is by which we pass from the monomial form to a quadrinomial form

$$AA' \text{ (or } dS) = Nd\omega + N_1 d\omega_1 + N_2 d\omega_2 + N_3 d\omega_3,$$

wherein each term is separately expressible as the differential of an elliptic integral; and further to develop the theory of the transformation to elliptic integrals. We require to establish for these purposes the fundamental formulæ in the theory of the bicircular quartic.

I remark that in the various formulæ $f, g, \theta, \theta_1, \theta_2, \theta_3$ are constants which enter only in the combinations $f+\theta$, $f-g$, $\theta_1-\theta$, $\theta_2-\theta$, $\theta_3-\theta$: that X, Y are taken as current coordinates, and these letters, or the same letters with suffixes, are taken as coordinates of a point or points on the bicircular quartic: and that the letters (x, y), (x_1, y_1), (x_2, y_2), (x_3, y_3) are used throughout as variable parameters, viz. we have

$$(f+\theta)\,x^2 + (g+\theta)\,y^2 = 1,$$
$$(f+\theta_1)\,x_1^2 + (g+\theta_1)\,y_1^2 = 1,$$
$$(f+\theta_2)\,x_2^2 + (g+\theta_2)\,y_2^2 = 1,$$
$$(f+\theta_3)\,x_3^2 + (g+\theta_3)\,y_3^2 = 1;$$

so that $x, y = \dfrac{\cos\omega}{\sqrt{f+\theta}}, \dfrac{\sin\omega}{\sqrt{g+\theta}}$, are functions of a single parameter ω, and similarly (x_1, y_1), (x_2, y_2), (x_3, y_3) are functions of the parameters $\omega_1, \omega_2, \omega_3$ respectively. We sometimes use these or similar expressions of (x, y), &c., as trigonometrical functions of a single parameter; but we more frequently retain the pair of quantities, considered as connected by an equation as above and so as equivalent to a single variable parameter.

Formulæ for the fourfold generation of the Bicircular Quartic. Art. Nos. 1 to 5.

1. We have four systems of a dirigent conic and circle of inversion, each giving rise to the same bicircular quartic: viz. the bicircular quartic is the envelope of a generating circle, having its centre on a dirigent conic, and cutting at right angles the corresponding circle of inversion; or, what is the same thing, it is the locus of the extremities of a chord of the generating circle, which chord passes through the centre of the circle of inversion, and cuts at right angles the tangent (at the centre of the generating circle) to the dirigent conic; the two extremities of the chord are thus inverse points in regard to the circle of inversion. The four systems are represented by letters without suffixes, or with the suffixes 1, 2, 3 respectively; and we say that the system, or mode of generation, is 0, 1, 2, or 3 accordingly.

2. The dirigent conics are confocal, and their squared semiaxes may therefore be represented by $f+\theta$, $g+\theta$: $f+\theta_1$, $g+\theta_1$: $f+\theta_2$, $g+\theta_2$: $f+\theta_3$, $g+\theta_3$, (which are, in

fact, functions of the five quantities $f+\theta$, $f-g$, $\theta_1-\theta$, $\theta_2-\theta$, $\theta_3-\theta$); and we can in terms of these data express the equations as well of the dirigent conics as of the circles of inversion; viz. taking X, Y as current coordinates, the equations are

$$\frac{X^2}{f+\theta}+\frac{Y^2}{g+\theta}=1,\ (X-\alpha)^2+(Y-\beta)^2-\gamma^2=0,\ \text{or}\ X^2+Y^2-2\alpha X-2\beta Y+k=0,$$

$$\frac{X^2}{f+\theta_1}+\frac{Y^2}{g+\theta_1}=1,\ (X-\alpha_1)^2+(Y-\beta_1)^2-\gamma_1^2=0,\ \text{or}\ X^2+Y^2-2\alpha_1 X-2\beta_1 Y+k_1=0,$$

$$\frac{X^2}{f+\theta_2}+\frac{Y^2}{g+\theta_2}=1,\ (X-\alpha_2)^2+(Y-\beta_2)^2-\gamma_2^2=0,\ \text{or}\ X^2+Y^2-2\alpha_2 X-2\beta_2 Y+k_2=0,$$

$$\frac{X^2}{f+\theta_3}+\frac{Y^2}{g+\theta_3}=1,\ (X-\alpha_3)^2+(Y-\beta_3)^2-\gamma_3^2=0,\ \text{or}\ X^2+Y^2-2\alpha_3 X-2\beta_3 Y+k_3=0,$$

where

$$\sqrt{\frac{f+\theta\,.\,f+\theta_1\,.\,f+\theta_2\,.\,f+\theta_3}{f-g}}=(f+\theta)\,\alpha=(f+\theta_1)\,\alpha_1=(f+\theta_2)\,\alpha_2=(f+\theta_3)\,\alpha_3,$$

$$\sqrt{\frac{g+\theta\,.\,g+\theta_1\,.\,g+\theta_2\,.\,g+\theta_3}{g-f}}=(g+\theta)\,\beta=(g+\theta_1)\,\beta_1=(g+\theta_2)\,\beta_2=(g+\theta_3)\,\beta_3,$$

$$f+\theta\,.\,g+\theta\,.\,\gamma^2=\theta-\theta_1\,.\,\theta-\theta_2\,.\,\theta-\theta_3,$$

$$f+\theta_1\,.\,g+\theta_1\,.\,\gamma_1^2=\theta_1-\theta\,.\,\theta_1-\theta_2\,.\,\theta_1-\theta_3,$$

$$f+\theta_2\,.\,g+\theta_2\,.\,\gamma_2^2=\theta_2-\theta\,.\,\theta_2-\theta_1\,.\,\theta_2-\theta_3,$$

$$f+\theta_3\,.\,g+\theta_3\,.\,\gamma_3^2=\theta_3-\theta\,.\,\theta_3-\theta_1\,.\,\theta_3-\theta_2,$$

$$f+g+\theta+\theta_1+\theta_2+\theta_3=k+2\theta=k_1+2\theta_1=k_2+2\theta_2=k_3+2\theta_3.$$

3. The geometrical relations between the dirigent conics and circles of inversion are all deducible from the foregoing formulæ; in particular, the conics are confocal, and as such intersect each two of them at right angles; the circles intersect each two of them at right angles. Considering a dirigent conic and the corresponding circle of inversion, the centres of the remaining three circles are conjugate points in regard as well to the first-mentioned conic, as to the first-mentioned circle; or, what is the same thing, they are the centres of the quadrangle formed by the intersections of the conic and circle.

4. The centre of the conics and the centres of the four circles lie on a rectangular hyperbola, having its asymptotes parallel to the axes of the conics. Given the centres of three of the circles (this determines the centre of the fourth circle) and also the centre of the conic, these four points determine a rectangular hyperbola (which passes also through the centre of the fourth circle); and the axes of the conics are then the lines through the centre, parallel to the asymptotes of the hyperbola.

5. The equation of the bicircular quartic may be expressed in the four forms

$$(X^2+Y^2-k)^2-4[(f+\theta)(X-\alpha)^2+(g+\theta)(Y-\beta)^2]=0,$$
$$(X^2+Y^2-k_1)^2-4[(f+\theta_1)(X-\alpha_1)^2+(g+\theta_1)(Y-\beta_1)^2]=0,$$
$$(X^2+Y^2-k_2)^2-4[(f+\theta_2)(X-\alpha_2)^2+(g+\theta_2)(Y-\beta_2)^2]=0,$$
$$(X^2+Y^2-k_3)^2-4[(f+\theta_3)(X-\alpha_3)^2+(g+\theta_3)(Y-\beta_3)^2]=0,$$

the equivalence of which is easily verified by means of the foregoing relations.

Determination as to Reality. Art. Nos. 6 and 7.

6. To fix the ideas, suppose that $f-g$ is positive; then in order that the centres of the four circles of inversion may be real, we must have $f+\theta.f+\theta_1.f+\theta_2.f+\theta_3$ positive, but $g+\theta.g+\theta_1.g+\theta_2.g+\theta_3$ negative; and this will be the case if $f+\theta$, $f+\theta_1$, $f+\theta_2$, $f+\theta_3$ are all positive, but $g+\theta$, $g+\theta_1$, $g+\theta_2$, $g+\theta_3$ one of them negative, and the other three positive. In reference to a figure which I constructed, I found it convenient to take θ_3, θ_1, θ_0, θ_2 to be in order of increasing magnitude: this being so, we have $f+\theta_3$ positive, $g+\theta_3$ negative; and the other like quantities $f+\theta_1$, $f+\theta_0$, $f+\theta_2$, $g+\theta_1$, $g+\theta_0$, $g+\theta_2$ all positive: we then have γ_3^2 and γ_1^2 each positive, γ_0^2 negative, γ_2^2 positive: viz. the conics and circles are

Hyperbola H_3, corresponding to real circle C_3,
Ellipse E_1, " real circle C_1,
" E_0, " imaginary circle C_0,
(viz. the radius is a pure imaginary),
" E_2, " real circle C_2,

and the confocal ellipses E_1, E_0, E_2 are in order of increasing magnitude. The centre C_0 is here a point within the triangle formed by the remaining three centres C_1, C_2, C_3. It will be convenient to adopt throughout the foregoing determination as to reality.

7. It may be remarked that a circle of a pure imaginary radius γ, $=i\lambda$, where λ is real, may be indicated by means of the concentric circle radius λ, which is the concentric orthotomic circle; and that a circle which cuts at right angles the original circle cuts diametrally (that is, at the extremities of a diameter) the substituted circle radius λ; we have thus a real construction in relation to a circle of inversion of pure imaginary radius.

Investigation of dS. Art. Nos. 8 to 17.

8. The coordinates of a point on the dirigent conic $\dfrac{X^2}{f+\theta}+\dfrac{Y^2}{g+\theta}=1$ may be taken to be $(f+\theta)x$ $(g+\theta)y$: and we hence prove as follows the fundamental

theorem for the generation of the bicircular quartic. Consider the generating circle, centre $(f+\theta)x$, $(g+\theta)y$, which cuts at right angles the circle of inversion

$$(X-\alpha)^2+(Y-\beta)^2=\gamma^2.$$

If for a moment the radius is called δ, then the equation of the generating circle is

$$(X-\overline{f+\theta x})^2+(Y-\overline{g+\theta y})^2=\delta^2;$$

the condition for the intersection at right angles is

$$(\alpha-\overline{f+\theta x})^2+(\beta-\overline{g+\theta y})^2=\gamma^2+\delta^2,$$

and hence eliminating δ^2, the equation of the generating circle is

$$X^2+Y^2-k-2(X-\alpha)(f+\theta)x-2(Y-\beta)(g+\theta)y=0;$$

and considering herein x, y as variable parameters connected by the foregoing equation $(f+\theta)x^2+(g+\theta)y^2=1$, we have as the envelope of this circle the required bicircular quartic.

9. It is convenient to write $R=\frac{1}{2}(X^2+Y^2-k)$. The equation then is

$$R-(X-\alpha)(f+\theta)x-(Y-\beta)(g+\theta)y=0;$$

the derived equation is

$$(X-\alpha)(f+\theta)\,dx+(Y-\beta)(g+\theta)\,dy=0;$$

and from these two equations, together with the equation in (x, y) and its derivative, we find $X-\alpha=Rx$, $Y-\beta=Ry$; from these last equations, and the equations $R=\frac{1}{2}(X^2+Y^2-k)$, $(f+\theta)x^2+(g+\theta)y^2=1$, eliminating x, y, R, we have

$$(f+\theta)(X-\alpha)^2+(g+\theta)(Y-\beta)^2=R^2,$$

that is,

$$(X^2+Y^2-k)^2-4[(f+\theta)(X-\alpha)^2+(g+\theta)(Y-\beta)^2]=0,$$

the required equation of the bicircular quartic.

10. We have thus $X-\alpha=Rx$, $Y-\beta=Ry$, as the equations which serve to determine the bicircular quartic: if from these equations, together with $R=\frac{1}{2}(X^2+Y^2-k)$, we eliminate X and Y, we have R expressed as a function of x, y; and thence also X, Y expressed in terms of x, y; that is, in effect the coordinates X, Y of a point of the bicircular quartic expressed as functions of a single variable parameter. The process gives $2R+k=(\alpha+Rx)^2+(\beta+Ry)^2$, viz. this is

$$R^2(x^2+y^2)-2(1-\alpha x-\beta y)R+\gamma^2=0,$$

or putting for shortness

$$\Omega=(1-\alpha x-\beta y)^2-\gamma^2(x^2+y^2),$$

this is

$$R=\frac{1-\alpha x-\beta y+\sqrt{\Omega}}{x^2+y^2},$$

or say the two values are

$$R = \frac{1 - \alpha x - \beta y + \sqrt{\Omega}}{x^2 + y^2}, \quad R' = \frac{1 - \alpha x - \beta y - \sqrt{\Omega}}{x^2 + y^2};$$

to preserve the generality it is proper to consider $\sqrt{\Omega}$ as denoting a determinate value (the positive or the negative one, as the case may be) of the radical.

11. Considering the root R', we have $X = \alpha + R'x$, $Y = \beta + R'y$; from these equations we obtain

$$dX = R'dx + x\, dR',$$

$$dY = R'dy + y\, dR'.$$

But from the equation for R' we have

$$[R'(x^2 + y^2) - (1 - \alpha x - \beta y)]\, dR' + R'^2 (x\, dx + y\, dy) + R'(\alpha\, dx + \beta\, dy) = 0,$$

that is,

$$-\sqrt{\Omega}\, dR' + R'(X dx + Y dy) = 0;$$

whence

$$dX = R'dx + \frac{R'x}{\sqrt{\Omega}}(X dx + Y dy),$$

$$dY = R'dy + \frac{R'y}{\sqrt{\Omega}}(X dx + Y dy).$$

12. The differentials dx, dy can be expressed in terms of a single differential $d\omega$, viz. writing

$$x = \frac{\cos\omega}{\sqrt{f + \theta}}, \quad y = \frac{\sin\omega}{\sqrt{g + \theta}},$$

and

$$\Theta = (f + \theta)(g + \theta),$$

then we have

$$dx = -\frac{g + \theta}{\sqrt{\Theta}}\, y\, d\omega, \quad dy = \frac{f + \theta}{\sqrt{\Theta}}\, x\, d\omega.$$

It is to be observed that, when the dirigent conic is an ellipse, ω is a real angle, and Θ is positive (whence also $\sqrt{\Theta}$ is real and positive); but when the dirigent conic is a hyperbola, ω is imaginary, and Θ is negative; we have, however, in either case

$$dx^2 + dy^2 = \frac{(f + \theta)^2 x^2 + (g + \theta)^2 y^2}{\Theta}\, d\omega^2,$$

and we may therefore write

$$\frac{d\omega}{\sqrt{\Theta}} = \frac{ds}{\sqrt{(f + \theta)^2 x^2 + (g + \theta)^2 y^2}},$$

where $\sqrt{(f + \theta)^2 x^2 + (g + \theta)^2 y^2}$ is positive; ds is the increment of arc on the conic $(f + \theta) x^2 + (g + \theta) y^2 = 1$, this arc being measured in a determinate sense, and therefore ds being positive or negative as the case may be: $\dfrac{d\omega}{\sqrt{\Theta}}$ has thus a real positive or negative value, even when ω is imaginary, and it is convenient to retain it in the formulæ.

13. It may further be noticed that, if υ denote the inclination to the axis of x of the tangent to the dirigent conic at the point $\sqrt{f+\theta}\cos\omega$, $\sqrt{g+\theta}\sin\omega$, where υ is Casey's θ, then

$$x=\frac{\cos\upsilon}{\sqrt{U}},\quad y=\frac{\sin\upsilon}{\sqrt{U}},\quad \text{where } U=(f+\theta)\cos^2\upsilon+(g+\theta)\sin^2\upsilon,$$

viz. we have

$$\frac{\cos\omega}{\sqrt{f+\theta}}=\frac{\cos\upsilon}{U},\quad \frac{\sin\omega}{\sqrt{g+\theta}}=\frac{\sin\upsilon}{U},$$

giving, as is easily verified, $\dfrac{d\omega}{\sqrt{\Theta}}=\dfrac{d\upsilon}{U}$; we have therefore

$$\frac{d\omega}{(x^2+y^2)\sqrt{\Theta}}=\frac{d\upsilon}{\upsilon(x^2+y^2)}=d\upsilon,$$

or

$$\frac{d\omega}{\sqrt{\Theta}}=(x^2+y^2)\,d\upsilon,$$

which is another interpretation of $\dfrac{d\omega}{\sqrt{\Theta}}$.

14. Substituting for dx, dy their values, the formulæ become

$$dX=\frac{R'}{\sqrt{\Theta}}\left\{-(g+\theta)\,y+\frac{x}{\sqrt{\Omega}}\left(-(g+\theta)\,yX+(f+\theta)\,xY\right)\right\}d\omega,$$

$$dY=\frac{R'}{\sqrt{\Theta}}\left\{\ (f+\theta)\,x+\frac{y}{\sqrt{\Omega}}\left(-(g+\theta)\,yX+(f+\theta)\,xY\right)\right\}d\omega.$$

We have

$$\begin{aligned}xX+yY&=\alpha x+\beta y+(x^2+y^2)\,R'\\&=1-\sqrt{\Omega},\end{aligned}$$

that is,

$$1=\frac{1-xX-yY}{\sqrt{\Omega}};$$

and consequently the foregoing expressions of dX, dY become

$$\begin{aligned}dX&=\frac{R'd\omega}{\sqrt{\Theta}\sqrt{\Omega}}\{(g+\theta)\,y\,(xX+yY-1)+x\left(-(g+\theta)\,yX+(f+\theta)\,xY\right)\}\\&=\frac{R'd\omega}{\sqrt{\Theta}\sqrt{\Omega}}\{(\overline{g+\theta}\,y^2+\overline{f+\theta}\,x^2)\,Y-(g+\theta)\,y\},\end{aligned}$$

$$\begin{aligned}dY&=\frac{R'd\omega}{\sqrt{\Theta}\sqrt{\Omega}}\{(f+\theta)\,x\,(1-xX-yY)+y\left(-(g+\theta)\,yX+(f+\theta)\,xY\right)\}\\&=\frac{R'd\omega}{\sqrt{\Theta}\sqrt{\Omega}}\{(f+\theta)\,x-\left((f+\theta)\,x^2+(g+\theta)\,y^2\right)X\},\end{aligned}$$

or finally

$$dX = \frac{R' d\omega}{\sqrt{\Theta}\sqrt{\Omega}}\{Y - (g+\theta)y\} = \frac{R' d\omega}{\sqrt{\Theta}\sqrt{\Omega}}\{R'y + \beta - (g+\theta)y\},$$

$$dY = \frac{-R' d\omega}{\sqrt{\Theta}\sqrt{\Omega}}\{X - (f+\theta)x\} = \frac{-R' d\omega}{\sqrt{\Theta}\sqrt{\Omega}}\{R'x + \alpha - (f+\theta)x\}.$$

15. We have

$$(R'x + \alpha - \overline{f+\theta}\,x)^2 + (R'y + \beta - \overline{g+\theta}\,y)^2$$

$$= R'^2(x^2+y^2) - 2R'(1 - \alpha x - \beta y)$$

$$+ (\alpha - \overline{f+\theta}\,x)^2 + (\beta - \overline{g+\theta}\,y)^2;$$

viz. this is

$$= (\alpha - \overline{f+\theta}\,x)^2 + (\beta - \overline{g+\theta}\,y)^2 - \gamma^2$$

$= \delta^2$, the radius of the generating circle.

Hence if $dS, = \sqrt{dX^2 + dY^2}$, be the element of arc of the bicircular quartic, this element being taken to be positive, we have

$$dS = \frac{\epsilon' R' \delta\, d\omega}{\sqrt{\Omega}\sqrt{\Theta}},$$

where ϵ' denotes a determinate sign, + or −, as the case may be.

16. I stop to consider the geometrical interpretation; introducing dv, the formula may be written

$$dS = \frac{\epsilon' . R'(x^2+y^2)\,\delta\, dv}{\sqrt{\Omega}},$$

and we have $(x^2+y^2)R' = 1 - \alpha x - \beta y - \sqrt{\Omega}$, or

$$\frac{(x^2+y^2)R'}{\sqrt{\Omega}} = \frac{1 - \alpha x - \beta y}{\sqrt{\Omega}} - 1.$$

Here $\dfrac{1 - \alpha x - \beta y}{\sqrt{x^2+y^2}}$ is the perpendicular from the centre of the circle of inversion upon the tangent to the dirigent conic, and $\dfrac{\sqrt{\Omega}}{\sqrt{x^2+y^2}}$ is the half-chord which this perpendicular forms with the generating circle. Hence $\dfrac{1 - \alpha x - \beta y}{\sqrt{\Omega}} - 1 =$ (perpendicular − half-chord) ÷ half-chord, the numerator being in fact the distance of the element dS (or point X, Y) from the centre of inversion: the formula thus is

$$dS = \pm \frac{\rho . \delta}{\tfrac{1}{2}c}\, dv,$$

where δ is the radius of the generating circle, ρ the distance of the element from the centre of the circle of inversion, and c the chord which this distance forms with the generating circle. If we consider the two points on the generating circle, and write dS' for the element at the other point, then we have

$$(dS \pm dS') = \pm \frac{(\rho - \rho')\,\delta\, dv}{\tfrac{1}{2}c} = 2\delta\, dv,$$

which is Casey's formula $ds' - ds = 2\rho\, d\phi$ (273).

17. The foregoing forms of dX, dY are those which give most directly the required value of dS: but I had previously obtained them in a different form. Writing

$$\Delta = \beta x - \alpha y + (f - g)\, xy,$$

then

$$x\Delta = \beta x^2 - \alpha xy + [(f+\theta)\, x^2 - (g+\theta)\, y^2];$$

or since

$$(f+\theta)\, x^2 = 1 - (g+\theta)\, y^2,$$

this is

$$x\Delta = \beta x^2 - \alpha xy + [1 - (g+\theta)(x^2+y^2)] = y\,(1 - \alpha x - \beta y) + (x^2+y^2)(\beta - (g+\theta)\, y)$$
$$= (x^2+y^2)\,\{yR' + \beta - (g+\theta)\, y\} + y\sqrt{\Omega},$$

that is,

$$x\Delta - y\sqrt{\Omega} = (x^2+y^2)\,\{yR' + \beta - (g+\theta)\, y\};$$

and similarly

$$-y\Delta - x\sqrt{\Omega} = (x^2+y^2)\,\{xR' + \alpha - (f+\theta)\, x\}.$$

We have therefore

$$dX = \frac{R'd\omega}{(x^2+y^2)\sqrt{\Theta}\sqrt{\Omega}}\,(x\Delta - y\sqrt{\Omega}),$$

$$dY = \frac{R'd\omega}{(x^2+y^2)\sqrt{\Theta}\sqrt{\Omega}}\,(y\Delta + x\sqrt{\Omega}),$$

and thence a value of dS which, compared with the former value, gives

$$\Omega + \Delta^2 = (x^2+y^2)\,\delta^2,$$

an equation which may be verified directly.

Formulæ for the Inscribed Quadrilateral. Art. Nos. 18 to 22.

18. We consider on the curve four points, A, B, C, D, forming a quadrilateral, $ABCD$. The coordinates are taken to be (X, Y), (X_1, Y_1), (X_2, Y_2), (X_3, Y_3) respectively. It is assumed that (A, B), (B, C), (C, D), (D, A) belong to the generations 1, 2, 3, 0, and depend on the parameters (x_1, y_1), (x_2, y_2), (x_3, y_3), (x, y) respectively.

We write

$$\Omega = (1 - \alpha x - \beta y)^2 - \gamma^2 (x^2 + y^2),$$
$$\Omega_1 = (1 - \alpha_1 x_1 - \beta_1 y_1)^2 - \gamma_1^2 (x_1^2 + y_1^2),$$
$$\Omega_2 = (1 - \alpha_2 x_2 - \beta_2 y_2)^2 - \gamma_2^2 (x_2^2 + y_2^2),$$
$$\Omega_3 = (1 - \alpha_3 x_3 - \beta_3 y_3)^2 - \gamma_3^2 (x_3^2 + y_3^2);$$

and then, $\sqrt{\Omega}$ denoting as above a determinate value, positive or negative as the case may be, of the radical, and similarly $\sqrt{\Omega_1}$, $\sqrt{\Omega_2}$, $\sqrt{\Omega_3}$ denoting determinate values of these radicals respectively, each radical having its own sign at pleasure, we further write

$$\begin{aligned}
(x^2+y^2)R' &= 1-\alpha x-\beta y-\sqrt{\Omega}, & (x_1^2+y_1^2)R_1 &= 1-\alpha_1x_1-\beta_1y_1+\sqrt{\Omega_1},\\
(x_1^2+y_1^2)R_1' &= 1-\alpha_1x_1-\beta_1y_1-\sqrt{\Omega_1}, & (x_2^2+y_2^2)R_2 &= 1-\alpha_2x_2-\beta_2y_2+\sqrt{\Omega_2},\\
(x_2^2+y_2^2)R_2' &= 1-\alpha_2x_2-\beta_2y_2-\sqrt{\Omega_2}, & (x_3^2+y_3^2)R_3 &= 1-\alpha_3x_3-\beta_3y_3+\sqrt{\Omega_3},\\
(x_3^2+y_3^2)R_3' &= 1-\alpha_3x_3-\beta_3y_3-\sqrt{\Omega_3}, & (x^2+y^2)R &= 1-\alpha x-\beta y+\sqrt{\Omega};
\end{aligned}$$

and this being so, we must have

$$\begin{aligned}
X &= \alpha+R'x=\alpha_1+R_1x_1, & Y &= \beta+R'y=\beta_1+R_1y_1, & R' &= \tfrac{1}{2}(X^2+Y^2-k), & R_1 &= \tfrac{1}{2}(X^2+Y^2-k_1),\\
X_1 &= \alpha_1+R_1'x_1=\alpha_2+R_2x_2, & Y_1 &= \beta_1+R_1'y_1=\beta_2+R_2y_2, & R_1' &= \tfrac{1}{2}(X_1^2+Y_1^2-k_1), & R_2 &= \tfrac{1}{2}(X_1^2+Y_1^2-k_2),\\
X_2 &= \alpha_2+R_2'x_2=\alpha_3+R_3x_3, & Y_2 &= \beta_2+R_2'y_2=\beta_3+R_3y_3, & R_2' &= \tfrac{1}{2}(X_2^2+Y_2^2-k_2), & R_3 &= \tfrac{1}{2}(X_2^2+Y_2^2-k_3),\\
X_3 &= \alpha_3+R_3'x_3=\alpha+Rx, & Y_3 &= \beta_3+R_3'y_3=\beta+Ry, & R_3' &= \tfrac{1}{2}(X_3^2+Y_3^2-k_3), & R &= \tfrac{1}{2}(X_3^2+Y_3^2-k);
\end{aligned}$$

and then from the values of X, Y, R', R, we have

$$\begin{aligned}
\alpha-\alpha_1+R'x-R_1x_1 &= 0,\\
\beta-\beta_1+R'y-R_1y_1 &= 0,\\
(\theta-\theta_1)+R'-R_1 &= 0,
\end{aligned}$$

giving

$$(\beta-\beta_1)(x-x_1)-(\alpha-\alpha_1)(y-y_1)+(\theta-\theta_1)(xy_1-x_1y)=0;$$

and similarly

$$\begin{aligned}
(\beta_1-\beta_2)(x_1-x_2)-(\alpha_1-\alpha_2)(y_1-y_2)+(\theta_1-\theta_2)(x_1y_2-x_2y_1) &= 0,\\
(\beta_2-\beta_3)(x_2-x_3)-(\alpha_2-\alpha_3)(y_2-y_3)+(\theta_2-\theta_3)(x_2y_3-x_3y_2) &= 0,\\
(\beta_3-\beta)(x_3-x)-(\alpha_3-\alpha)(y_3-y)+(\theta_3-\theta)(x_3y-xy_3) &= 0,
\end{aligned}$$

which are the relations connecting the parameters (x, y), (x_1, y_1), (x_2, y_2), (x_3, y_3) of the quadrilateral.

19. We have thus apparently four equations for the determination of four quantities, or the number of quadrilaterals would be finite; but if from the first and second equations we eliminate (x_1, y_1), and if from the third and fourth equations we eliminate (x_3, y_3), we find in each case the same relation between (x, y), (x_2, y_2), viz. this is found to be

$$\Omega\Omega_2=(1-\alpha x_2-\beta y_2)^2(1-\alpha_2x-\beta_2y)^2;$$

and we have thus the singly infinite series of quadrilaterals. We have, of course, between (x_1, y_1), (x_3, y_3) the like relation,

$$\Omega_1\Omega_3=(1-\alpha_1x_3-\beta_1y_3)^2(1-\alpha_3x_1-\beta_3y_1)^2.$$

20. The relation between (x, y), (x_1, y_1) may be expressed also in the two forms:

$$1-\alpha\,(x+x_1)-\beta\,(y+y_1)+(f+\theta_1)\,xx_1+(g+\theta_1)\,yy_1+\frac{x^2+y^2}{xy_1-x_1y}(\overline{\alpha-\alpha_1}y_1-\overline{\beta-\beta_1}x_1)=0,$$

$$1-\alpha_1\,(x+x_1)-\beta_1\,(y+y_1)+(f+\theta)\,xx_1+(g+\theta)\,yy_1+\frac{x_1^2+y_1^2}{x_1y-xy_1}(\overline{\alpha_1-\alpha}y-\overline{\beta_1-\beta}x)=0.$$

In fact, the first of these equations is

$$\{1+(f+\theta_1)\,xx_1+(g+\theta_1)\,yy_1\}\,(xy_1-x_1y)-\{\alpha\,(x+x_1)+\beta\,(y+y_1)\}\,(xy_1-x_1y)$$
$$+\{(\alpha-\alpha_1)\,y_1-(\beta-\beta_1)\,x_1\}\,(x^2+y^2)=0,$$

which, by virtue of the original form of relation, is

$$-\{1+(f+\theta_1)\,xx_1+(g+\theta_1)\,yy_1\}\,\frac{(\beta-\beta_1)\,(x-x_1)-(\alpha-\alpha_1)\,(y-y_1)}{\theta-\theta_1}$$
$$-\{\alpha\,(x+x_1)+\beta\,(y+y_1)\}\,(xy_1-x_1y)+\{(\alpha-\alpha_1)\,y_1-(\beta-\beta_1)\,x_1\}\,(x^2+y^2)=0;$$

or, in the first term, writing

$$-\frac{\beta-\beta_1}{\theta-\theta_1}=\frac{\beta}{g+\theta_1},\quad \frac{\alpha-\alpha_1}{\theta-\theta_1}=\frac{-\alpha}{f+\theta_1},$$

and in the third term

$$\alpha-\alpha_1=-\frac{(\theta-\theta_1)\,\alpha}{f+\theta_1},\quad -(\beta-\beta_1)=\frac{(\theta-\theta_1)\,\beta}{g+\theta_1},$$

this is

$$(1+(f+\theta_1)\,xx_1+(g+\theta_1)\,yy_1)\left(\frac{\beta\,(x-x_1)}{g+\theta_1}-\frac{\alpha\,(y-y_1)}{f+\theta_1}\right)$$
$$-\{\alpha\,(x+x_1)+\beta\,(y+y_1)\}\,(xy_1-x_1y)-\left\{\frac{\alpha\,(\theta-\theta_1)}{f+\theta_1}y_1-\frac{\beta\,(\theta-\theta_1)}{g+\theta_1}x_1\right\}(x^2+y^2)=0.$$

In this equation the coefficients of α and of β are separately $=0$: in fact, the coefficient of β is

$$\frac{x-x_1}{g+\theta_1}+\frac{f+\theta_1}{g+\theta_1}xx_1\,(x-x_1)+(x-x_1)\,yy_1-(y+y_1)\,(xy_1-x_1y)+\frac{\theta-\theta_1}{g+\theta_1}x_1\,(x^2+y^2)$$
$$=\frac{x}{g+\theta_1}\{1-(f+\theta_1)\,x_1^2-(g+\theta_1)\,y_1^2\}-\frac{x_1}{g+\theta_1}\{1-(f+\theta)\,x^2-(g+\theta)\,y^2\}=0;$$

and similarly the coefficient of α is $=0$.

And in like manner the second equation may be verified.

21. The two equations are:

$$1-\alpha x-\beta y-(x^2+y^2)\,R'=\alpha x_1+\beta y_1-(f+\theta_1)\,xx_1-(g+\theta_1)\,yy_1,$$
$$1-\alpha_1x_1-\beta_1y_1-(x_1^2+y_1^2)\,R_1=\alpha_1x+\beta_1y-(f+\theta)\,xx_1-(g+\theta)\,yy_1;$$

or, substituting for R' and R_1 their values, these are

$$\sqrt{\Omega}=\alpha x_1+\beta y_1-(f+\theta_1)\,xx_1-(g+\theta_1)\,yy_1,\quad \sqrt{\Omega_1}=-\alpha_1x-\beta_1y+(f+\theta)\,xx_1+(g+\theta)\,yy_1;$$

and similarly

$$\sqrt{\Omega_1} = \alpha_1 x_2 + \beta_1 y_2 - (f+\theta_2)\, x_1 x_2 - (g+\theta_2)\, y_1 y_2, \quad \sqrt{\Omega_2} = -\alpha_2 x_1 - \beta_2 y_1 + (f+\theta_1)\, x_1 x_2 + (g+\theta_1)\, y_1 y_2$$

$$\sqrt{\Omega_2} = \alpha_2 x_3 + \beta_2 y_3 - (f+\theta_3)\, x_2 x_3 - (g+\theta_3)\, y_2 y_3, \quad \sqrt{\Omega_3} = -\alpha_3 x_2 - \beta_3 y_2 + (f+\theta_2)\, x_2 x_3 + (g+\theta_2)\, y_2 y_3,$$

$$\sqrt{\Omega_3} = \alpha_3 x + \beta_3 y - (f+\theta)\, x_3 x - (g+\theta)\, y_3 y, \quad \sqrt{\Omega} = -\alpha x_3 - \beta y_3 + (f+\theta_3)\, x_3 x + (g+\theta_3)\, y_3 y.$$

Differentiating the equation

$$(\beta-\beta_1)(x-x_1) - (\alpha-\alpha_1)(y-y_1) + (\theta-\theta_1)(xy_1 - x_1 y) = 0,$$

we have

$$[(\beta-\beta_1) + (\theta-\theta_1)\, y_1]\, dx - [(\alpha-\alpha_1) + (\theta-\theta_1)\, x_1]\, dy$$
$$-[(\beta-\beta_1) + (\theta-\theta_1) y]\, dx_1 + [(\alpha-\alpha_1) + (\theta-\theta_1)\, x]\, dy_1 = 0;$$

and writing herein

$$dx = -\frac{(g+\theta)}{\sqrt{\Theta}}\, y\, d\omega, \quad dx_1 = \frac{-(g+\theta_1)}{\sqrt{\Theta_1}}\, y_1\, d\omega_1,$$

$$dy = \frac{(f+\theta)}{\sqrt{\Theta}}\, x\, d\omega, \quad dy_1 = \frac{(f+\theta_1)}{\sqrt{\Theta_1}}\, x_1\, d\omega_1,$$

we find

$$-\frac{d\omega}{\sqrt{\Theta}}\{(g+\theta)(\beta-\beta_1)\, y + (f+\theta)(\alpha-\alpha_1)\, x + (\theta-\theta_1)((f+\theta)\, xx_1 + (g+\theta)\, yy_1)\}$$

$$+\frac{d\omega_1}{\sqrt{\Theta_1}}\{(g+\theta_1)(\beta-\beta_1)\, y_1 + (f+\theta_1)(\alpha-\alpha_1)\, x_1 + (\theta-\theta_1)((f+\theta_1)\, xx_1 + (g+\theta_1)\, yy_1)\} = 0;$$

viz., dividing by $\theta-\theta_1$, this becomes

$$-\sqrt{\Omega_1}\frac{d\omega}{\sqrt{\Theta}} - \sqrt{\Omega}\frac{d\omega_1}{\sqrt{\Theta_1}} = 0, \text{ that is, } \frac{d\omega}{\sqrt{\Theta}\sqrt{\Omega}} + \frac{d\omega_1}{\sqrt{\Theta_1}\sqrt{\Omega_1}} = 0;$$

or, completing the system, we have

$$\frac{d\omega}{\sqrt{\Theta}\sqrt{\Omega}} = \frac{-d\omega_1}{\sqrt{\Theta_1}\sqrt{\Omega_1}} = \frac{d\omega_2}{\sqrt{\Theta_2}\sqrt{\Omega_2}} = \frac{-d\omega_3}{\sqrt{\Theta_3}\sqrt{\Omega_3}},$$

which are the differential relations between the parameters ω, ω_1, ω_2, ω_3, or (x, y), (x_1, y_1), (x_2, y_2), (x_3, y_3).

22. From the equations $X = \alpha + R'x$, $Y = \beta + R'y$, we found

$$dX = \frac{R'd\omega}{\sqrt{\Omega}\sqrt{\Theta}}\{Y - (g+\theta)\, y\},$$

$$dY = \frac{R'd\omega}{\sqrt{\Omega}\sqrt{\Theta}}\{X - (f+\theta)\, x\};$$

the new values, $X = \alpha_1 + R_1 x_1$ and $Y = \beta_1 + R_1 y_1$, give in like manner

$$dX = -\frac{R_1 d\omega_1}{\sqrt{\Omega_1}\sqrt{\Theta_1}}\{Y - (g+\theta_1)\, y_1\},$$

$$dY = -\frac{R_1 d\omega_1}{\sqrt{\Omega_1}\sqrt{\Theta_1}}\{X - (f+\theta_1)\, x_1\};$$

in virtue of the relation just found between $d\omega$ and $d\omega_1$, these two sets of values will agree together if only

$$R'\{Y-(g+\theta)y\}=R_1\{Y-(g+\theta_1)y_1\},$$
$$R'\{X-(f+\theta)x\}=R_1\{X-(f+\theta_1)x_1\}.$$

These are easily verified: the first is

$$R'Y-(g+\theta)(Y-\beta)=(R'-\theta+\theta_1)Y-(g+\theta_1)(Y-\beta_1),$$

viz. this is $(g+\theta)\beta-(g+\theta_1)\beta_1=0$, which is right; and similarly the second equation gives $(f+\theta)\alpha-(f+\theta_1)\alpha_1=0$, which is right.

From the first values of dX, dY, we have, as above,

$$dS=\frac{\epsilon' R'\delta\, d\omega}{\sqrt{\Omega}\sqrt{\Theta}};$$

and the second values give in like manner

$$dS=\frac{\epsilon_1 R_1\delta_1\, d\omega_1}{\sqrt{\Omega_1}\sqrt{\Theta_1}},$$

where ϵ_1 is $=\pm 1$. It will be observed that we have in effect, by means of the relation $(\beta-\beta_1)(x-x_1)-(\alpha-\alpha_1)(y-y_1)+(\theta-\theta_1)(xy_1-x_1y)=0$, proved the identity of the two values of dS.

Considering the quadrilateral $ABCD$, and giving it an infinitesimal variation, so as to change it into $A'B'C'D'$, then dS is the element of arc AA'; and writing in like manner dS_1, dS_2, dS_3 for the elements of arc BB', CC', DD', we have, of course, a like pair of values for each of the elements dS_1, dS_2, dS_3.

Formulæ for the elements of Arc dS, dS_1, dS_2, dS_3. Art. Nos. 23 to 27.

23. The formulæ are

$$dS=\epsilon' R'\delta\,\frac{d\omega}{\sqrt{\Omega}\sqrt{\Theta}}=\epsilon_1 R_1\delta_1\frac{d\omega_1}{\sqrt{\Omega_1}\sqrt{\Theta_1}},$$

$$dS_1=\epsilon_1' R_1'\delta_1\frac{d\omega_1}{\sqrt{\Omega_1}\sqrt{\Theta_1}}=\epsilon_2 R_2\delta_2\frac{d\omega_2}{\sqrt{\Omega_2}\sqrt{\Theta_2}},$$

$$dS_2=\epsilon_2' R_2'\delta_2\frac{d\omega_2}{\sqrt{\Omega_2}\sqrt{\Theta_2}}=\epsilon_3 R_3\delta_3\frac{d\omega_3}{\sqrt{\Omega_3}\sqrt{\Theta_3}},$$

$$dS_3=\epsilon_3' R_3'\delta_3\frac{d\omega_3}{\sqrt{\Omega_3}\sqrt{\Theta_3}}=\epsilon R\,\delta\,\frac{d\omega}{\sqrt{\Omega}\sqrt{\Theta}},$$

where the ϵ's each denote ± 1. Supposing as above that γ^2 is negative, but that γ_1^2, γ_2^2, γ_3^2 are positive; then R', R have opposite signs: but R_1', R_1 have the same sign, as have also R_2' and R_2, and R_3' and R_3. We may take δ, δ_1, δ_2, and δ_3 as each of them positive: the signs of

$$\frac{d\omega}{\sqrt{\Omega}\sqrt{\Theta}},\ \frac{d\omega_1}{\sqrt{\Omega_1}\sqrt{\Theta_1}},\ \frac{d\omega_2}{\sqrt{\Omega_2}\sqrt{\Theta_2}},\ \frac{d\omega_3}{\sqrt{\Omega_3}\sqrt{\Theta_3}} \text{ are } +,\ -,\ +,\ -, \text{ or } -,\ +,\ -,\ +:$$

hence to make dS, dS_1, dS_2, dS_3 all positive,

$$\epsilon',\quad \epsilon_1',\quad \epsilon_2',\quad \epsilon_3',\quad \epsilon_1,\quad \epsilon_2,\quad \epsilon_3,\quad \epsilon,$$

must have either the signs of

$$R',\ -R_1',\ R_2',\ -R_3',\ -R_1,\ R_2,\ -R_3,\ R,$$

or else the reverse signs: hence in either case $\epsilon' = -\epsilon$, $\epsilon_1' = \epsilon_1$, $\epsilon_2' = \epsilon_2$, $\epsilon_3' = \epsilon_3$; or the equations are

$$dS = -\epsilon R'\delta \frac{d\omega}{\sqrt{\Omega}\sqrt{\Theta}} = \epsilon_1 R_1 \delta_1 \frac{d\omega_1}{\sqrt{\Omega_1}\sqrt{\Theta_1}},$$

$$dS_1 = \epsilon_1 R_1'\delta_1 \frac{d\omega_1}{\sqrt{\Omega_1}\sqrt{\Theta_1}} = \epsilon_2 R_2 \delta_2 \frac{d\omega_2}{\sqrt{\Omega_2}\sqrt{\Theta_2}},$$

$$dS_2 = \epsilon_2 R_2'\delta_2 \frac{d\omega_2}{\sqrt{\Omega_2}\sqrt{\Theta_2}} = \epsilon_3 R_3 \delta_3 \frac{d\omega_3}{\sqrt{\Omega_3}\sqrt{\Theta_3}},$$

$$dS_3 = \epsilon_3 R_3'\delta_3 \frac{d\omega_3}{\sqrt{\Omega_3}\sqrt{\Theta_3}} = \epsilon R \delta \frac{d\omega}{\sqrt{\Omega}\sqrt{\Theta}}.$$

24. But we have $R' - R = \frac{-2\sqrt{\Omega}}{x^2+y^2}$, &c.; and hence, putting for shortness

$$\frac{\delta}{(x^2+y^2)\sqrt{\Theta}},\ \frac{\delta_1}{(x_1^2+y_1^2)\sqrt{\Theta_1}},\ \frac{\delta_2}{(x_2^2+y_2^2)\sqrt{\Theta_2}},\ \frac{\delta_3}{(x_3^2+y_3^2)\sqrt{\Theta_3}} = P,\ P_1,\ P_2,\ P_3,$$

$$dS + dS_3 = +2\epsilon P\, d\omega,$$
$$dS_1 - dS = -2\epsilon_1 P_1 d\omega_1,$$
$$dS_2 - dS_1 = -2\epsilon_2 P_2 d\omega_2,$$
$$dS_3 - dS = -2\epsilon_3 P_3 d\omega_3,$$

and consequently

$$dS = \epsilon P d\omega + \epsilon_1 P_1 d\omega_1 + \epsilon_2 P_2 d\omega_2 + \epsilon_3 P_3 d\omega_3,$$
$$dS_1 = \epsilon P d\omega - \epsilon_1 P_1 d\omega_1 + \epsilon_2 P_2 d\omega_2 + \epsilon_3 P_3 d\omega_3,$$
$$dS_2 = \epsilon P d\omega - \epsilon_1 P_1 d\omega_1 - \epsilon_2 P_2 d\omega_2 + \epsilon_3 P_3 d\omega_3,$$
$$dS_3 = \epsilon P d\omega - \epsilon_1 P_1 d\omega_1 - \epsilon_2 P_2 d\omega_2 - \epsilon_3 P_3 d\omega_3,$$

which are the required formulæ for the elements of arc.

25. The determination of the signs has been made by means of the particular figure; but it is easy to see that the pairs of terms could not for instance be $dS - dS_3$, $dS_1 - dS$, $dS_2 - dS_1$, $dS_3 - dS$, or any other pairs such that it would be possible to eliminate dS, dS_1, dS_2, dS_3, and thus obtain an equation such as

$$\epsilon P d\omega + \epsilon_1 P_1 d\omega_1 + \epsilon_2 P_2 d\omega_2 + \epsilon_3 P_3 d\omega_3 = 0;$$

this would, by virtue of the relations between $d\omega$, $d\omega_1$, $d\omega_2$, $d\omega_3$, become

$$\epsilon\frac{\delta\sqrt{\Omega}}{x^2+y^2}-\epsilon_1\frac{\delta_1\sqrt{\Omega_1}}{x_1^2+y_1^2}+\epsilon_2\frac{\delta_2\sqrt{\Omega_2}}{x_2^2+y_2^2}-\epsilon_3\frac{\delta_3\sqrt{\Omega_3}}{x_3^2+y_3^2}=0,$$

an equation not deducible from the relations which connect ω, ω_1, ω_2, ω_3, and which therefore cannot be satisfied by the variable quadrilateral.

26. The differentials of the formulæ are, it will be observed, of the form $Pd\omega$

$$=\frac{\delta\, d\omega}{(x^2+y^2)\sqrt{\Theta}},$$

where $\sqrt{\Theta}, =\sqrt{f+\theta\,.\,g+\theta}$, is a mere constant,

$$x,\ y=\frac{\cos\omega}{\sqrt{f+\theta}},\ \frac{\sin\omega}{\sqrt{g+\theta}},$$

and

$$\delta^2=\{(f+\theta)x-\alpha\}^2+\{(g+\theta)y-\beta\}^2-\gamma^2;$$

viz. the form is

$$\frac{\sqrt{(\cos\omega\sqrt{f+\theta}-\alpha)^2+(\sin\omega\sqrt{g+\theta}-\beta)^2-\gamma^2}}{\sqrt{\Theta}\,.\left(\dfrac{\cos^2\omega}{f+\theta}+\dfrac{\sin^2\omega}{g+\theta}\right)}d\omega,$$

which is, in fact, the same as Casey's form in ϕ, equation (300), his ϕ being $=90^\circ-\omega$.

Writing as before v in place of his θ, the differential expression becomes simply $=\delta\, dv$: but δ^2 expressed as a function of v is an irrational function $M+N\sqrt{U}$, and δ would be the root of such a function; so that, if the form originally obtained had been this form δdv, it would have been necessary to transform it into the first-mentioned form $\dfrac{\delta\, d\omega}{(x^2+y^2)\sqrt{\Theta}}$, in which δ is expressed as a function of (x, y), that is, of ω.

27. The system of course is

$$\begin{aligned}
dS\ &=\epsilon\delta dv+\epsilon_1\delta_1 dv_1+\epsilon_2\delta_2 dv_2+\epsilon_3\delta_3 dv_3,\\
dS_1&=\epsilon\delta dv-\epsilon_1\delta_1 dv_1+\epsilon_2\delta_2 dv_2+\epsilon_3\delta_3 dv_3,\\
dS_2&=\epsilon\delta dv-\epsilon_1\delta_1 dv_1-\epsilon_2\delta_2 dv_2+\epsilon_3\delta_3 dv_3,\\
dS_3&=\epsilon\delta dv-\epsilon_1\delta_1 dv_1-\epsilon_2\delta_2 dv_2-\epsilon_3\delta_3 dv_3,
\end{aligned}$$

where $dv=\dfrac{d\omega}{(x^2+y^2)\sqrt{\Theta}}$, &c.; and this is the most convenient way of writing it.

Reference to Figure. Art. No. 28.

28. I constructed a bicircular quartic consisting of an exterior and interior oval with the following numerical data: ($f+\theta_3=48$, $f+\theta_1=56$, $f+\theta_0=60$, $f+\theta_2=80$; $g+\theta_3=-6$, $g+\theta_1=2$, $g+\theta_0=6$, $g+\theta_2=26$),—not very convenient ones, inasmuch as

the exterior oval came out too large. The annexed figure shows 0, 1, 2, 3, the centres of the circles of inversion, the interior oval, and a portion of the exterior

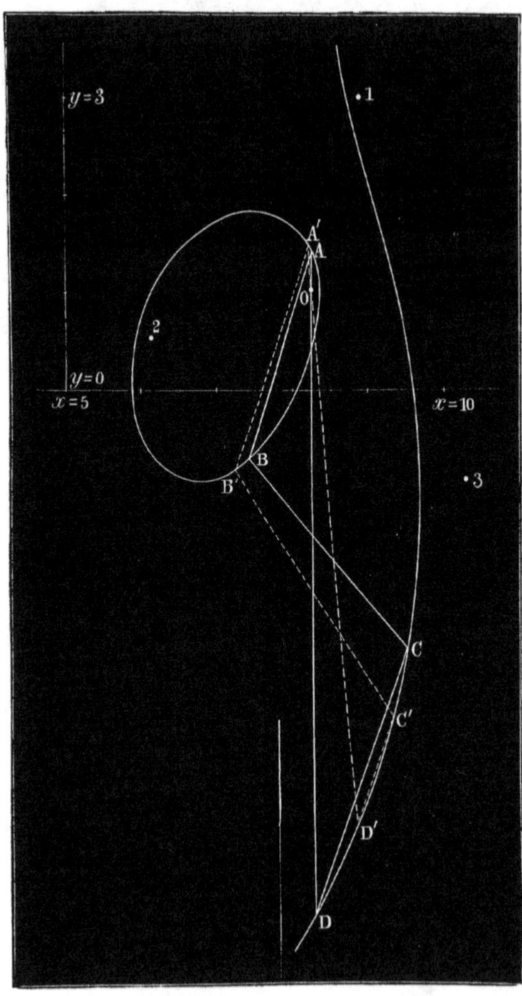

oval, also the origin and axes; it will be seen that the centres 0, 2 lie inside the interior oval, the centres 1, 3 outside the exterior oval: I add further the values

$$\sqrt{f+\theta_3}=6{\cdot}93,\quad \sqrt{-(g+\theta_3)}=2{\cdot}45,\quad \alpha_3=10{\cdot}18,\quad \beta_3=-\ {\cdot}98,$$

$$\sqrt{f+\theta_1}=7{\cdot}48,\quad \sqrt{g+\theta_1}\quad =1{\cdot}41,\quad \alpha_1=\ 8{\cdot}73,\quad \beta_1=+2{\cdot}94,$$

$$\sqrt{f+\theta_0}=7{\cdot}75,\quad \sqrt{g+\theta_0}\quad =2{\cdot}45,\quad \alpha_0=\ 8{\cdot}15,\quad \beta_0=+\ {\cdot}98,$$

$$\sqrt{f+\theta_2}=8{\cdot}94,\quad \sqrt{g+\theta_2}\quad =5{\cdot}09,\quad \alpha_2=\ 6{\cdot}10,\quad \beta_2=+\ {\cdot}23.$$

We thus see how there exists a series of quadrilaterals $ABCD$, where A, B are situate on the interior oval, C, D on the exterior oval. Considering the sides as drawn in the senses A to B, B to C, C to D, D to A: and representing the inclinations, measured from the positive infinity on the axis of x in the sense x to y, by v_1, v_2, v_3, v respectively: then, in passing to the consecutive quadrilateral $A'B'C'D'$, we have v_1 and v_2 decreasing, v_3 and v increasing, that is, dv_1 and dv_2 negative, dv_3 and dv positive; so that, reckoning the elements AA', BB', CC', DD', that is, dS_1, dS_2, dS_3, dS, as each of them positive, we have

$$dS_2 - dS_1 = -2\delta_1 dv_1,$$

$$dS_3 - dS_2 = -2\delta_2 dv_2,$$

$$dS - dS_3 = +2\delta_3 dv_3,$$

$$dS_1 + dS = +2\delta\, dv,$$

and thence

$$dS = \delta dv - \delta_1 dv_1 - \delta_2 dv_2 + \delta_3 dv_3,$$

$$dS_1 = \delta dv + \delta_1 dv_1 + \delta_2 dv_2 - \delta_3 dv_3,$$

$$dS_2 = \delta dv - \delta_1 dv_1 + \delta_2 dv_2 - \delta_3 dv_3,$$

$$dS_3 = \delta dv - \delta_1 dv_1 - \delta_2 dv_2 - \delta_3 dv_3,$$

which are the correct signs in regard to the particular figure.

Reduction of $\int \frac{\delta\, d\omega}{(x^2+y^2)\sqrt{\Theta}}$ *to Elliptic Integrals.* Art. No. 29.

29. The expression in question is

$$\int d\omega \,.\, \frac{\sqrt{(\cos\omega\sqrt{f+\theta}-\alpha)^2+(\sin\omega\sqrt{g+\theta}-\beta)^2-\gamma^2}}{\left\{\frac{\cos^2\omega}{f+\theta}+\frac{\sin^2\omega}{g+\theta}\right\}\sqrt{\Theta}},$$

where $\sqrt{\Theta}$ is a mere constant; and we may apply it to the Gaussian transformation,

$$\cos\omega = \frac{a+a'\cos T+a''\sin T}{c+c'\cos T+c''\sin T},$$

$$\sin\omega = \frac{b+b'\cos T+b''\sin T}{c+c'\cos T+c''\sin T},$$

where the coefficients a, b, c, a', b', c', a'', b'', c'' are such that identically

$$\cos^2\omega+\sin^2\omega-1 = \frac{1}{(c+c'\cos T+c''\sin T)^2}\{\cos^2 T+\sin^2 T-1\}:$$

and also

$$(\cos\omega\sqrt{f+\theta}-\alpha)^2+(\sin\omega\sqrt{g+\theta}-\beta)^2-\gamma^2,$$

that is,

$$\cos^2\omega\,(f+\theta)+\sin^2\omega\,(g+\theta)-2\alpha\sqrt{f+\theta}\cos\omega-2\beta\sqrt{g+\theta}\sin\omega+k,$$

$$=\frac{1}{(c+c'\cos T+c''\sin T)^2}(G_1-G_2\cos^2 T-G_3\sin^2 T).$$

30. It is found that G_1, G_2, G_3 are the roots of a cubic equation

$$(G+\theta-\theta_1)(G+\theta-\theta_2)(G+\theta-\theta_3),$$

which being so, we may assume $G_1=\theta_1-\theta$, $G_2=\theta_2-\theta$, $G_3=\theta_3-\theta$; the second condition, in fact, then is

$$(f+\theta)\cos^2\omega+(g+\theta)\sin^2\omega-2\alpha\sqrt{f+\theta}\cos\omega-2\beta\sqrt{g+\theta}\sin\omega+k$$

$$=\frac{1}{(c+c'\cos T+c''\sin T)^2}\{\theta_1-\theta-(\theta_2-\theta)\cos^2 T-(\theta_3-\theta)\sin^2 T\};$$

and this being so, we find without difficulty the values

$$a^2=\frac{g+\theta_1\,.\,f+\theta_2\,.\,f+\theta_3}{f-g\,.\,\theta_1-\theta_2\,.\,\theta_1-\theta_3},\quad b^2=\frac{f+\theta_1\,.\,g+\theta_2\,.\,g+\theta_3}{g-f\,.\,\theta_1-\theta_2\,.\,\theta_1-\theta_3},\quad c^2=\frac{f+\theta_1\,.\,g+\theta_1}{\theta_1-\theta_2\,.\,\theta_1-\theta_3},$$

$$a'^2=-\frac{g+\theta_2\,.\,f+\theta_1\,.\,f+\theta_3}{f-g\,.\,\theta_2-\theta_1\,.\,\theta_2-\theta_3},\quad b'^2=-\frac{f+\theta_2\,.\,g+\theta_1\,.\,g+\theta_3}{g-f\,.\,\theta_2-\theta_1\,.\,\theta_2-\theta_3},\quad c'^2=-\frac{f+\theta_2\,.\,g+\theta_2}{\theta_2-\theta_1\,.\,\theta_2-\theta_3},$$

$$a''^2=-\frac{g+\theta_3\,.\,f+\theta_1\,.\,f+\theta_2}{f-g\,.\,\theta_3-\theta_1\,.\,\theta_3-\theta_2},\quad b''^2=-\frac{f+\theta_3\,.\,g+\theta_1\,.\,g+\theta_2}{g-f\,.\,\theta_3-\theta_1\,.\,\theta_3-\theta_2},\quad c''^2=-\frac{f+\theta_3\,.\,g+\theta_3}{\theta_3-\theta_1\,.\,\theta_3-\theta_2}.$$

To make these positive, the order of ascending magnitude must, however, be not as heretofore θ_3, θ_1, θ_2, but θ_3, θ_2, θ_1, viz. we must have $f+\theta_1$, $f+\theta_2$, $f+\theta_3$, $g+\theta_1$, $g+\theta_2$, $-(g+\theta_3)$, $\theta_1-\theta_3$, $\theta_1-\theta_2$, $\theta_2-\theta_3$ all positive.

31. The above are the values of the squares of the coefficients; we must have definite relations between the signs of the products aa', bb', ab, &c., viz. we may have

$$a'a''=\frac{f+\theta_1}{f-g\,.\,\theta_2-\theta_3}\sqrt{\frac{\Theta_2\Theta_3}{\theta_3-\theta_1\,.\,\theta_1-\theta_2}},\quad a''a=\frac{f+\theta_2}{f-g\,.\,\theta_3-\theta_1}\sqrt{\frac{-\Theta_3\Theta_1}{\theta_1-\theta_2\,.\,\theta_2-\theta_3}},$$

$$b'b''=\frac{g+\theta_1}{g-f\,.\,\theta_2-\theta_3}\sqrt{\quad ''\quad},\quad b''b=\frac{g+\theta_2}{g-f\,.\,\theta_3-\theta_1}\sqrt{\quad ''\quad},$$

$$c'c''=\frac{1}{\theta_2-\theta_3}\sqrt{\quad ''\quad},\quad c''c=\frac{1}{\theta_3-\theta_1}\sqrt{\quad ''\quad},$$

$$aa'=\frac{f+\theta_2}{f-g\,.\,\theta_1-\theta_2}\sqrt{\frac{-\Theta_1\Theta_2}{\theta_2-\theta_3\,.\,\theta_3-\theta_1}},$$

$$bb'=\frac{g+\theta_2}{g-f\,.\,\theta_1-\theta_2}\sqrt{\quad ''\quad},$$

$$cc'=\frac{1}{\theta_1-\theta_2}\sqrt{\quad ''\quad},$$

and further

$$ab = \frac{1}{f-g\,.\,\theta_3-\theta_1\,.\,\theta_1-\theta_2}\sqrt{-\Theta_1\Theta_2\Theta_3},\quad bc = -\frac{f+\theta_1}{\theta_3-\theta_1\,.\,\theta_1-\theta_2}\sqrt{\frac{g+\theta_1\,.\,g+\theta_2\,.\,g+\theta_3}{g-f}},$$

$$a'b' = \frac{-1}{f-g\,.\,\theta_1-\theta_2\,.\,\theta_2-\theta_3}\sqrt{\quad ,,\quad},\quad b'c' = \frac{f+\theta_2}{\theta_1-\theta_2\,.\,\theta_2-\theta_3}\sqrt{\qquad ,,\qquad},$$

$$a''b'' = \frac{-1}{f-g\,.\,\theta_2-\theta_3\,.\,\theta_3-\theta_1}\sqrt{\quad ,,\quad},\quad b''c'' = \frac{f+\theta_3}{\theta_2-\theta_3\,.\,\theta_3-\theta_1}\sqrt{\qquad ,,\qquad},$$

$$ca = -\frac{g+\theta_1}{\theta_3-\theta_1\,.\,\theta_1-\theta_2}\sqrt{\frac{f+\theta_1\,.\,f+\theta_2\,.\,f+\theta_3}{f-g}},$$

$$c'a' = \frac{g+\theta_2}{\theta_1-\theta_2\,.\,\theta_2-\theta_3}\sqrt{\qquad ,,\qquad},$$

$$c''a'' = \frac{g+\theta_3}{\theta_2-\theta_3\,.\,\theta_3-\theta_1}\sqrt{\qquad ,,\qquad};$$

and also

$$b'c''+b''c' = \frac{2g+\theta_2+\theta_3}{\theta_2-\theta_3}\sqrt{\frac{g+\theta_1.f+\theta_2.f+\theta_3}{g-f.\theta_3-\theta_1.\theta_1-\theta_2}},\quad c'a''+c''a' = \frac{2f+\theta_2+\theta_3}{\theta_2-\theta_3}\sqrt{\frac{f+\theta_1.g+\theta_2.g+\theta_3}{f-g.\theta_3-\theta_1.\theta_1-\theta_2}},$$

$$b''c+bc'' = \frac{2g+\theta_3+\theta_1}{\theta_3-\theta_1}\sqrt{-\frac{g+\theta_2.f+\theta_3.f+\theta_1}{g-f.\theta_1-\theta_2.\theta_2-\theta_3}},\quad c''a+ca'' = \frac{2f+\theta_3+\theta_1}{\theta_3-\theta_1}\sqrt{-\frac{f+\theta_2.g+\theta_3.g+\theta_1}{f-g.\theta_1-\theta_2.\theta_2-\theta_3}},$$

$$bc'+b'c = \frac{2g+\theta_1+\theta_2}{\theta_1-\theta_2}\sqrt{-\frac{g+\theta_3.f+\theta_1.f+\theta_2}{g-f.\theta_2-\theta_3.\theta_3-\theta_1}},\quad ca'+c'a = \frac{2f+\theta_1+\theta_2}{\theta_1-\theta_2}\sqrt{-\frac{f+\theta_3.g+\theta_1.g+\theta_2}{f-g.\theta_2-\theta_3.\theta_3-\theta_1}}.$$

32. These values, in fact, satisfy the several relations which exist between the nine coefficients; viz. the original expressions of $\cos\omega$, $\sin\omega$, in terms of $\cos T$, $\sin T$ give conversely expressions of $\cos T$, $\sin T$ in terms of $\cos\omega$, $\sin\omega$, the two sets being

$$\cos\omega = \frac{a+a'\cos T+a''\sin T}{c+c'\cos T+c''\sin T},\quad \cos T = -\frac{a'\cos\omega+b'\sin\omega-c'}{a\cos\omega+b\sin\omega-c},$$

$$\sin\omega = \frac{b+b'\cos T+b''\sin T}{c+c'\cos T+c''\sin T},\quad \sin T = -\frac{a''\cos\omega+b''\sin\omega-c''}{a\cos\omega+b\sin\omega-c}:$$

and we have then the relations

$$\cos^2\omega+\sin^2\omega-1 = \frac{1}{(c+c'\cos T+c''\sin T)^2}(\cos^2 T+\sin^2 T-1),$$

$$\cos^2 T+\sin^2 T-1 = \frac{1}{(a\cos\omega+b\sin\omega-c)^2}(\cos^2\omega+\sin^2\omega-1),$$

$$(\theta+f)\cos^2\omega+(\theta+g)\sin^2\omega-2\alpha\sqrt{\theta+f}\cos\omega-2\beta\sqrt{\theta+g}\sin\omega+k$$

$$= \frac{1}{(c+c'\cos T+c''\sin T)^2}\{(\theta_1-\theta)-(\theta_2-\theta)\cos^2 T-(\theta_3-\theta)\sin^2 T\},$$

$$(\theta_1-\theta)-(\theta_2-\theta)\cos^2 T-(\theta_3-\theta)\sin^2 T$$

$$= \frac{1}{(a\cos\omega+b\sin\omega-c)^2}\{(\theta+f)\cos^2\omega+(\theta+g)\sin^2\omega-2\alpha\sqrt{\theta+f}\cos\omega-2\beta\sqrt{\theta+g}\sin\omega+k\},$$

giving the four sets each of six equations

$$\begin{aligned}
a^2 + b^2 - c^2 &= -1, & a'a'' + b'b'' - c'c'' &= 0,\\
a'^2 + b'^2 - c'^2 &= +1, & a''a + b''b - c''c &= 0,\\
a''^2 + b''^2 - c''^2 &= +1, & aa' + bb' - cc' &= 0,
\end{aligned}$$

$$\begin{aligned}
-a^2 + a'^2 + a''^2 &= +1, & -bc + b'c' + b''c'' &= 0,\\
-b^2 + b'^2 + b''^2 &= +1, & -ca + c'a' + c''a'' &= 0,\\
-c^2 + c'^2 + c''^2 &= -1, & -ab + a'b' + a''b'' &= 0,
\end{aligned}$$

$$\begin{aligned}
(\theta+f)\,a^2 + (\theta+g)\,b^2 - 2\alpha\sqrt{\theta+f}\,ac - 2\beta\sqrt{\theta+g}\,bc + kc^2 &= \theta_1+\theta,\\
(\theta+f)\,a'^2 + (\theta+g)\,b'^2 - 2\alpha\sqrt{\theta+f}\,a'c' - 2\beta\sqrt{\theta+g}\,b'c' + kc'^2 &= -\theta_2+\theta,\\
(\theta+f)\,a''^2 + (\theta+g)\,b''^2 - 2\alpha\sqrt{\theta+f}\,a''c'' - 2\beta\sqrt{\theta+g}\,b''c'' + kc''^2 &= -\theta_3+\theta,
\end{aligned}$$

$$\begin{aligned}
(\theta+f)\,a'a'' + (\theta+g)\,b'b'' - \alpha\sqrt{\theta+f}\,(a'c''+a''c') - \beta\sqrt{\theta+g}\,(b'c''+b''c') + kc'c'' &= 0,\\
(\theta+f)\,a''a + (\theta+g)\,b''b - \alpha\sqrt{\theta+f}\,(a''c+ac'') - \beta\sqrt{\theta+g}\,(b''c+bc'') + kc''c &= 0,\\
(\theta+f)\,aa' + (\theta+g)\,bb' - \alpha\sqrt{\theta+f}\,(ac'+a'c) - \beta\sqrt{\theta+g}\,(bc'+b'c) + kcc' &= 0,
\end{aligned}$$

$$\begin{aligned}
&(\theta_1-\theta)\,a^2 - (\theta_2-\theta)\,a'^2 - (\theta_3-\theta)\,a''^2 = \theta+f, \text{ or say } (\theta_1+f)\,a^2 - (\theta_2+f)\,a'^2 - (\theta_3+f)\,a''^2 = 0,\\
&(\theta_1-\theta)\,b^2 - (\theta_2-\theta)\,b'^2 - (\theta_3-\theta)\,b''^2 = \theta+g, \quad ,, \quad (\theta_1+g)\,b^2 - (\theta_2+g)\,b'^2 - (\theta_3+g)\,b''^2 = 0,\\
&(\theta_1-\theta)\,c^2 - (\theta_2-\theta)\,c'^2 - (\theta_3-\theta)\,c''^2 = k, \quad ,, \quad \theta_1c^2 - \theta_2c'^2 - \theta_3c''^2 = k+\theta,
\end{aligned}$$

$$\begin{aligned}
-(\theta_1-\theta)\,bc + (\theta_2-\theta)\,b'c' + (\theta_3-\theta)\,b''c'' &= -\beta\sqrt{\theta+g},\\
-(\theta_1-\theta)\,ca + (\theta_2-\theta)\,c'a' + (\theta_3-\theta)\,c''a'' &= -\alpha\sqrt{\theta+f},\\
-(\theta_1-\theta)\,ab + (\theta_2-\theta)\,a'b' + (\theta_3-\theta)\,a''b'' &= 0;
\end{aligned}$$

all which formulæ are in fact satisfied by the foregoing values of the expressions a^2, b^2, a'^2, &c.

33. We then have

$$d\omega = \frac{dT}{c + c'\cos T + c''\sin T};$$

the radical which multiplies $d\omega$ being

$$= \frac{1}{c + c'\cos T + c''\sin T}\sqrt{\theta_1 - \theta_2\cos^2 T - \theta_3\sin^2 T},$$

the differential becomes

$$= \frac{dT\sqrt{\theta_1 - \theta_2\cos^2 T - \theta_3\sin^2 T}}{\left(\dfrac{\cos^2\omega}{f+\theta} + \dfrac{\sin^2\omega}{g+\theta}\right)(c + c'\cos T + c''\sin T)^2\sqrt{\Theta}},$$

that is,

$$= \frac{dT\sqrt{\theta_1 - \theta_2\cos^2 T - \theta_3\sin^2 T}}{\left\{\dfrac{1}{f+\theta}(a + a'\cos T + a''\sin T)^2 + \dfrac{1}{g+\theta}(b + b'\cos T + b''\sin T)^2\right\}\sqrt{\Theta}}.$$

The denominator could, of course, be reduced to the form $(*\between 1, \cos T, \sin T)^2$; but the actual form seems preferable, inasmuch as it puts in evidence the linear factors

$$\frac{1}{\sqrt{f+\theta}}(a + a'\cos T + a''\sin T) \pm \frac{i}{\sqrt{g+\theta}}(b + b'\cos T + b''\sin T),$$

and there seems to be no advantage in further reducing the integral.

668.

ON COMPOUND COMBINATIONS.

[From the *Proceedings of the Lit. Phil. Soc. Manchester*, t. XVI. (1877), pp. 113, 114; *Memoirs*, *ib.*, Ser. III., t. VI. (1879), pp. 99, 100.]

PROF. CLIFFORD'S paper, "On the Types of Compound Statement involving Four Classes," [volume of *Proceedings* quoted, pp. 88—101; *Mathematical Papers*, pp. 1—13], relates mathematically to a question of compound combinations; and it is worth while to consider its connexion with another question of compound combinations, the application of which is a very different one.

Starting with four symbols, A, B, C, D, we have sixteen combinations of the *five types* 1, A, AB, ABC, $ABCD$, $(1+4+6+4+1=16$ as before). But in Prof. Clifford's question 1 means $A'B'C'D'$, A means $AB'C'D'$, &c.; viz. each of the symbols means an aggregate of four assertions; and the 16 symbols are thus *all of the same type*. Considering them in this point of view, the question is as to the number of types of the binary, ternary, &c., combinations of the sixteen combinations; for, according as these are combined,

$$\text{No. of types} = \frac{1,\ 2,\ 3,\ \ 4,\ \ 5,\ \ 6,\ \ 7,\ \ 8,\ \ 9,\ 10,\ 11,\ 12,\ 13,\ 14,\ 15}{1,\ 4,\ 6,\ 19,\ 27,\ 47,\ 55,\ 78,\ 55,\ 47,\ 27,\ 19,\ \ 6,\ \ 4,\ \ 1}$$

together.

In the first mentioned point of view the like question arises, in regard to the sets belonging to the five different types separately or in combination with each other; for instance, taking only the six symbols of the type AB, these may be taken 1, 2, 3, 4, or 5 together, and we have in these cases respectively

$$\text{No. of types} = \frac{1,\ 2,\ 3,\ 4,\ 5}{1,\ 2,\ 2,\ 2,\ 1},$$

as is very easily verified; but if the number of letters $A, B, \ldots$ be greater (say this $=8$), or, instead of letters, writing the numbers 1, 2, 3, 4, 5, 6, 7, 8, then the question is that of the number of types of combination of the 28 duads 12, 13, ..., 78, taken 1, 2, 3, ..., 27 together, a question presenting itself in geometry in regard to the bitangents of a quartic curve (see Salmon's *Higher Plane Curves*, Ed. 2 (1873), pp. 222 *et seq.*): the numbers, so far as they have been obtained, are

$$\text{No. of types} = \frac{1,\ 2,\ 3,\ \ 4, \ldots,\ 24,\ 25,\ 26,\ 27}{1,\ 2,\ 5,\ 11, \ldots,\ 11,\ \ 5,\ \ 2,\ \ 1}.$$

It might be interesting to complete the series, and, more generally, to determine the number of the types of combination of the $\frac{1}{2}n(n-1)$ duads of n letters.

669.

ON A PROBLEM OF ARRANGEMENTS.

[From the *Proceedings of the Royal Society of Edinburgh*, t. IX. (1878), pp. 338—342.]

IT is a well-known problem to find for n letters the number of the arrangements in which no letter occupies its original place; and the solution of it is given by the following general theorem:—viz., the number of the arrangements which satisfy any r conditions is

$$(1-1)(1-2)\ldots\ldots(1-r),$$
$$=1-\Sigma(1)+\Sigma(12)-\ldots\ldots\pm(12\ldots r),$$

where 1 denotes the whole number of arrangements; (1) the number of them which fail in regard to the first condition; (2) the number which fail in regard to the second condition; (12) the number which fail in regard to the first condition, and also in regard to the second condition; and so on: $\Sigma(1)$ means $(1)+(2)+\ldots+(r)$: $\Sigma(12)$ means $(12)+(13)+(2r)+\ldots+(r-1,\ r)$; and so on, up to $(12\ldots r)$, which denotes the number failing in regard to each of the r conditions.

Thus, in the special problem, the first condition is that the letter in the first place shall not be a; the second condition is that the letter in the second place shall not be b; and so on; taking $r=n$, we have the known result,

$$\text{No.}=\Pi n-\frac{n}{1}\Pi(n-1)+\frac{n.n-1}{1.2}\Pi(n-2).+\ldots\pm\frac{n.n-1\ldots 2.1}{1.2\ldots n},$$
$$=1.2.3\ldots n\left\{1-\frac{1}{1}+\frac{1}{1.2}-\frac{1}{1.2.3}+\ldots\pm\frac{1}{1.2.3\ldots n}\right\},$$

giving for the several cases

$$n=2,\ 3,\ 4,\ \ 5,\ \ \ 6,\ \ \ \ 7,\ldots$$
$$\text{No.}=1,\ 2,\ 9,\ 44,\ 265,\ 1854,\ldots$$

I proceed to consider the following problem, suggested to me by Professor Tait, in connexion with his theory of knots: to find the number of the arrangements of n letters $abc\ldots jk$, when the letter in the first place is not a or b, the letter in the second place not b or $c,\ldots$, the letter in the last place not k or a.

Numbering the conditions 1, 2, 3, ..., n, according to the places to which they relate, a single condition is called [1]; two conditions are called [2] or [1, 1], according as the numbers are consecutive or non-consecutive: three conditions are called [3], [2, 1], or [1, 1, 1], according as the numbers are all three consecutive, two consecutive and one not consecutive, or all non-consecutive; and so on: the numbers which refer to the conditions being always written in their natural order, and it being understood that they follow each other cyclically, so that 1 is consecutive to n. Thus, $n=6$, the set 126 of conditions is [3], as consisting of 3 consecutive conditions; and similarly 1346 is [2, 2].

Consider a single condition [1], say this is 1; the arrangements which fail in regard to this condition are those which contain in the first place a or b; whichever it be, the other $n-1$ letters may be arranged in any form whatever; and there are thus $2\Pi(n-1)$ failing arrangements.

Next for two conditions; these may be [2], say the conditions are 1 and 2: or else [1, 1], say they are 1 and 3. In the former case, the arrangements which fail are those which contain in the first and second places ab, ac, or bc: and for each of these, the other $n-2$ letters may be arranged in any order whatever; there are thus $3\Pi(n-2)$ failing arrangements. In the latter case, the failing arrangements have in the first place a or b, and in the third place c or d,—viz. the letters in these two places are $a.c$, $a.d$, $b.c$, or $b.d$, and in each case the other $n-2$ letters may be arranged in any order whatever: the number of failing arrangements is thus $=2.2.\Pi(n-2)$. And so, in general, when the conditions are $[\alpha, \beta, \gamma, \ldots]$, the number of failing arrangements is

$$=(\alpha+1)(\beta+1)(\gamma+1)\ldots\Pi(n-\alpha-\beta-\gamma\ldots).$$

But for $[n]$, that is, for the entire system of the n conditions, the number of failing arrangements is (not as by the rule it should be $=n+1$, but) $=2$,—viz. the only arrangements which fail in regard to each of the n conditions are (as is at once seen), $abc\ldots jk$, and $bc\ldots jka$.

Changing now the notation so that [1], [2], [1, 1], &c., shall denote the *number* of the conditions [1], [2], [1, 1], &c., respectively, it is easy to see the form of the general result. If, for greater clearness, we write $n=6$, we have

$$\begin{array}{llll} 1 & -\Sigma(1) & +\Sigma(12) & -\Sigma(123) \\ \text{No.}=720 & -\{([1]=6)\,2\}\,120 & +\left\{\begin{array}{l}([2]\;\;=6)\,3 \\ +([1,1]=9)\,2.2\end{array}\right\}24 & -\left\{\begin{array}{l}([3]\;\;\;\;=\;\;6)\,4 \\ +([2,1]\;\;=12)\;\;3.2 \\ +([1,1,1]=\;\;2)\,2.2.2\end{array}\right\}6 \end{array}$$

$$\begin{array}{lll} +\Sigma(1234) & -\Sigma(12345) & +(123456) \\ +\left\{\begin{array}{l}([4]\;\;=6)\,5 \\ +([3,1]=6)\,4.2 \\ +([2,2]=3)\,3.3\end{array}\right\}2 & -\{([5]=6)\,6\}\,1 & +\{([6]=1)\,2\}; \end{array}$$

or, reducing into numbers, this is

$$\text{No.} = 720 - 1440 + 1296 - 672 + 210 - 36 + 2, \qquad = 80.$$

The formula for the next succeeding case, $n = 7$, gives

$$\text{No.} = 5040 - 10080 + 9240 - 5040 + 1764 - 392 + 49 - 2, \quad = 579.$$

Those for the preceding cases, $n = 3, 4, 5$, respectively are

$$\text{No.} = 6 - 12 + 9 - 2 \qquad = 1,$$

$$\text{No.} = 24 - 48 + 40 - 16 + 2 \qquad = 2,$$

$$\text{No.} = 120 - 240 + 210 - 100 + 25 - 2 \qquad = 13.$$

We have in general $[1] = n$, $[2] = n$, $[1, 1] = \frac{1}{2}n(n-3)$; and in the several columns of the formulæ the sums of the numbers thus represented are equal to the coefficients of $(1+1)^2$: thus, when $n = 6$ as above, the sums are 6, 15, 20, 15, 6, 1. As regards the calculation of the numbers in question, any symbol $[\alpha, \beta, \gamma]$ is a sum of symbols $[\alpha - \alpha' + \beta - \beta' + \gamma - \gamma' + \ldots]$, where $\alpha' + \beta' + \gamma' + \ldots$ is any partition of $n - (\alpha + \beta + \gamma + \ldots)$; read, of the series of numbers $1, 2, 3, \ldots, n$, taken in cyclical order beginning with any number, retain α, omit α', retain β, omit β', retain γ, omit $\gamma', \ldots$. Thus in particular, $n = 6$, $[1, 1]$ is a sum of symbols $[1 - 3 + 1 - 1]$ and $[1 - 2 + 1 - 2]$; it is clear that any such symbol $[\alpha - \alpha' + \beta - \beta' + \ldots]$ is $= n$ or a submultiple of n (in particular, if n be prime, the symbol is always $= n$): and we thus in every case obtain the value of $[\alpha, \beta, \gamma, \ldots]$ by taking for the negative numbers the several partitions of

$$n - (\alpha + \beta + \gamma + \ldots),$$

and for each symbol

$$[\alpha - \alpha' + \beta - \beta' + \gamma - \gamma' + \ldots],$$

writing its value, $= n$ or a given submultiple of n, as just mentioned. There would, I think, be no use in pursuing the matter further, by seeking to obtain an analytical expression for the symbols $[\alpha, \beta, \gamma, \ldots]$.

For the actual formation of the required arrangements, it is of course easy, when all the arrangements are written down, to strike out those which do not satisfy the prescribed conditions, and so obtain the system in question. Or introducing the notion of substitutions*, and accordingly considering each arrangement as derived by a substitution from the primitive arrangement *abcd...jk*, we can write down the substitutions which give the system of arrangements in which no letter occupies its original place: viz. we must for this purpose partition the n letters into parts, no part less than 2, and then in each set taking one letter (say the first in alphabetical order) as fixed, permute in every possible way the other letters of the set; we thus obtain

* In explanation of the notation of substitutions, observe that (*abcde*) means that *a* is to be changed into *b*, *b* into *c*, *c* into *d*, *d* into *e*, *e* into *a*; and similarly (*ab*) (*cde*) means that *a* is to be changed into *b*, *b* into *a*, *c* into *d*, *d* into *e*, *e* into *c*.

all the substitutions which move every letter. Thus when $n=5$, we obtain the 44 substitutions for the letters *abcde*, viz. these are

(*abcde*), &c., 24 symbols obtained by permuting in every way the four letters *b*, *c*, *d*, *e*;

(*ab*) (*cde*), &c., 20 symbols corresponding to the 10 partitions *ab*, *cde*, and for each of them 2 arrangements such as *cde*, *ced*.

And then if we reject those symbols which contain in any ( ) two consecutive letters, we have the substitutions which give the arrangements wherein the letter in the first place is not *a* or *b*, that in the second place not *b* or *c*, and so on. In particular, when $n=5$, rejecting the substitutions which contain in any ( ), *ab*, *bc*, *cd*, *de*, or *ea*, we have 13 substitutions, which may be thus arranged:—

(*acbed*), (*ac*) (*bed*), (*acebd*), (*adbec*), (*aedbc*),

(*aedbc*), (*bd*) (*aec*),

(*acedb*), (*ce*) (*adb*),

(*aecbd*), (*ad*) (*bec*),

(*adceb*), (*be*) (*adc*).

Here in the first column, performing on the symbol (*acbed*) the substitution (*abcde*), we obtain (*bdcae*), = (*aebdc*), the second symbol; and so again and again operating with (*abcde*), we obtain the remaining symbols of the column; these are for this reason said to be of the same type. In like manner, symbols of the second column are of the same type; but the symbols in the remaining three columns are each of them a type by itself; viz. operating with (*abcde*) upon (*acebd*), we obtain (*bdace*), = (*acebd*); and the like as regards (*adbec*) and (*aedbc*) respectively. The 13 substitutions are thus of 5 different types, or say the arrangements to which they belong, viz.

cebad, *ceabd*, *cdeab*, *deabc*, *eabcd*,

edacb, *edabc*,

caebd, *daebc*,

edbac, *debac*,

daecb, *deacb*,

are of 5 different types. The question to determine for any value of n, the number of the different types, is, it would appear, a difficult one, and I do not at present enter upon it.

670.

[NOTE ON MR MUIR'S SOLUTION OF A "PROBLEM OF ARRANGEMENT."]

[From the *Proceedings of the Royal Society of Edinburgh*, t. IX. (1878), pp. 388—391.]

THE investigation may be carried further: writing for shortness u_3, u_4, &c., in place of $\Psi(3)$, $\Psi(4)$, &c., the equations are

$$\begin{aligned}
u_3 &= 1,\\
u_4 &= 2u_3,\\
u_5 &= 3u_4 + 6u_3 + 1,\\
u_6 &= 4u_5 + 8u_4 + 12u_3,\\
u_7 &= 5u_6 + 10u_5 + 15u_4 + 18u_3 + 1.
\end{aligned}$$

Hence assuming

$$u = u_3 + u_4 x + u_5 x^2 + u_6 x^3 + u_7 x^4 + \ldots,$$

we have

$$\begin{aligned}
u = \frac{1}{1-x^2} &+ u_3(2x + 6x^2 + 12x^3 + 18x^4 + \ldots)\\
&+ u_4(3x^2 + 8x^3 + 15x^4 + 22x^5 + \ldots)\\
&+ u_5(4x^3 + 10x^4 + 18x^5 + 26x^6 + \ldots)\\
&+ u_6(5x^4 + 12x^5 + 21x^6 + 30x^7 + \ldots);
\end{aligned}$$

so that, forming the equation

$$\begin{aligned}
u' \frac{x^2}{(1-x)^2} = \;& u_4(x^2 + 2x^3 + 3x^4 + 4x^5 + \ldots)\\
&+ u_5(2x^3 + 4x^4 + 6x^5 + 8x^6 + \ldots)\\
&+ u_6(3x^4 + 6x^5 + 9x^6 + 12x^7 + \ldots),
\end{aligned}$$

where u' denotes $\dfrac{du}{dx}$, we have

$$u - u'\frac{x^3}{(1-x)^2} = \frac{1}{1-x^2} + (u_3 + u_4x + u_5x^2 + \dots)(2x + 6x^2 + 12x^3 + 18x^4 + \dots)$$
$$= \frac{1}{1-x^2} + u(2x + 6x^2 + 12x^3 + 18x^4 + \dots);$$

or, what is the same thing,

$$u - u'\frac{x^2}{(1-x)^2} = \frac{1}{1-x^2} + u\left\{\frac{2x}{(1-x)^3} - \frac{2x^4}{(1-x)^3(1+x)}\right\};$$

that is,

$$\left\{1 - \frac{2x}{(1-x)^3} + \frac{2x^4}{(1-x)^3(1+x)}\right\} u - \frac{x^2}{(1-x)^2} u' = \frac{1}{1-x^2}.$$

This equation may be simplified: write

$$u = -\frac{1-x^2}{x^4} Q, \quad = \left(-\frac{1}{x^4} + \frac{1}{x^2}\right) Q,$$

then

$$u' = \left(\frac{4}{x^5} - \frac{2}{x^3}\right) Q + \frac{1-x^2}{x^4} Q',$$

and the equation is

$$\left\{-\frac{1-x^2}{x^4} + \frac{2}{x^3}\frac{1+x}{(1+x)^2} - \frac{2}{(1-x)^2} - \frac{4}{x^3}\frac{1}{(1-x)^2} + \frac{2}{x(1-x)^2}\right\} Q + \frac{1+x}{(1+x)x^2} Q' = \frac{1}{1-x^2};$$

that is,

$$\left\{-\frac{1}{x^4} + \frac{1}{x^2} - \frac{2}{x^3(1-x)^2} + \frac{2}{x^2(1-x)^2} + \frac{2}{x(1-x)^2} - \frac{2}{(1-x)^2}\right\} Q + \frac{1+x}{(1-x)x^2} Q' = \frac{1}{1-x^2},$$

viz. this is

$$\left\{-\frac{(1-x)^2}{x^4} + \frac{(1-x)^2}{x^2} - \frac{2}{x^3} + \frac{2}{x^2} + \frac{2}{x} - 2\right\} Q + \frac{1-x^2}{x^2} Q' = \frac{1-x}{1+x},$$

that is,

$$\left\{-\frac{1}{x^4} + \frac{2}{x^2} - 1\right\} Q + \frac{1-x^2}{x^2} Q' = \frac{1-x}{1+x},$$

or

$$-\frac{(1-x^2)^2}{x^4} Q + \frac{1-x^2}{x^2} Q' = \frac{1-x}{1+x};$$

or finally,

$$Q\left(1 - \frac{1}{x^2}\right) + Q' = \frac{x^2}{(1+x)^2},$$

giving

$$Q = e^{-\left(x+\frac{1}{x}\right)} \int \frac{x^2}{(x+1)^2} e^{x+\frac{1}{x}} dx,$$

and thence

$$u = \frac{x^2-1}{x^4} e^{-\left(x+\frac{1}{x}\right)} \int \frac{x^2}{(x+1)^2} e^{\left(x+\frac{1}{x}\right)} dx,$$

which is the value of the generating function

$$u = u_3 + u_4x + u_5x^2 + \&c.$$

But for the purpose of calculation it is best to integrate by a series the differential equation for Q: assuming

$$Q = -q_3x^4 - q_4x^5 - q_5x^6 - \dots,$$

we find

$$\begin{aligned} q_4 &= 4q_3 && - 2, \\ q_5 &= 5q_4 + q_3 && + 3, \\ q_6 &= 6q_5 + q_4 && - 4, \\ q_7 &= 7q_6 + q_5 && + 5, \\ &\vdots \\ q_n &= nq_{n-1} + q_{n-2} + (-)^{n-1}(n-2). \end{aligned}$$

We have thus for $q_3, q_4, q_5, \dots$ the values 1, 2, 14, 82, 593, 4820, ..., and thence

$$u = (1 - x^2)(1 + 2x + 14x^2 + 82x^3 + 593x^4 + 4820x^5 + \dots),$$

viz. writing

	1	2	14	82	593	4820...
			−1	−2	−14	−82
the values of $u_3, u_4, \dots$ are	1,	2,	13,	80,	579,	4738, ...,

agreeing with the results found above.

In the more simple problem, where the arrangements of the n things are such that no one of them occupies its original place, if u_n be the number of arrangements, we have

$$\begin{aligned} u_2 &= 1 &&= 1, \\ u_3 &= 2\,u_2 &&= 2, \\ u_4 &= 3\,(u_3 + u_2) &&= 9, \\ u_5 &= 4\,(u_4 + u_3) &&= 44, \\ &\vdots \\ u_{n+1} &= n\,(u_n + u_{n-1}), \end{aligned}$$

and writing

$$u = u_2 + u_3x + u_4x^2 + \dots,$$

we find

$$u = 1 + (2x + 3x^2)\,u + (x^2 + x^3)\,u';$$

that is,

$$(-1 + 2x + 3x^2)\,u + (x^2 + x^3)\,u' = -1,$$

or, what is the same thing,

$$u' + \left(\frac{3}{x} - \frac{1}{x^2}\right)u = -\frac{1}{x^2(1+x)},$$

whence

$$u = x^{-3}e^{-\frac{1}{x}}\int \frac{-x}{1+x}e^{\frac{1}{x}}\,dx.$$

But the calculation is most easily performed by means of the foregoing equation of differences, itself obtained from the differential equation written in the foregoing form,

$$(-1 + 2x + 3x^2)\,u + (x^2 + x^3)\,u' = -1.$$

671.

ON A SIBI-RECIPROCAL SURFACE.

[From the *Berlin. Akad. Monatsber.*, (1878), pp. 309—313.]

THE question of the generation of a sibi-reciprocal surface—that is, a surface the reciprocal of which is of the same order and has the same singularities as the original surface—was considered by me in the year 1868, see *Proc. London Math. Soc.* t. II. pp. 61—63, [part of 387], where it is remarked that if a surface be considered as the envelope of a quadric surface varying according to given conditions, then the reciprocal surface is given as the envelope of a quadric surface varying according to the reciprocal conditions; whence, if the conditions be sibi-reciprocal, it follows that the surface is a sibi-reciprocal surface. And I gave as instances the surface which is the envelope of a quadric surface touching each of 8 given lines; and also the surface called the "tetrahedroid," which is a homographic transformation of Fresnel's Wave Surface and a particular case of the quartic surface with 16 nodes.

The interesting surface of the order 8, recently considered by Herr Kummer, *Berl. Monatsber.*, Jan. 1878, pp. 25—36, is included under the theory. In fact, if we consider a line L, whereof the six coordinates

$$a,\ b,\ c,\ f,\ g,\ h,$$

satisfy each of the three linear relations

$$f_1a + g_1b + h_1c + a_1f + b_1g + c_1h = 0,$$
$$f_2a + g_2b + h_2c + a_2f + b_2g + c_2h = 0,$$
$$f_3a + g_3b + h_3c + a_3f + b_3g + c_3h = 0,$$

the locus of this line is a quadric surface the equation of which is

$$\begin{aligned} T = (agh)\,x^2 + (bhf)\,y^2 + (cfg)\,z^2 + (abc)\,w^2 \\ + [(abg) - (cah)]\,xw + [(bfg) + (chf)]\,yz \\ + [(bch) - (abf)]\,yw + [(cgh) + (afg)]\,zx \\ + [(caf) - (bcg)]\,zw + [(ahf) + (bgh)]\,xy = 0, \end{aligned}$$

where (agh) is used to denote the determinant $\begin{vmatrix} a_1, & g_1, & h_1 \\ a_2, & g_2, & h_2 \\ a_3, & g_3, & h_3 \end{vmatrix}$, and so for the other symbols. Considering the reciprocal of the line L in regard to the quadric surface $X^2+Y^2+Z^2+W^2=0$, the six coordinates of the reciprocal line are

$$f,\ g,\ h,\ a,\ b,\ c,$$

and it is hence at once seen that the locus of the reciprocal line is the quadric surface obtained from the equation $T=0$ by interchanging therein the symbolical quantities a, b, c and f, g, h: viz. writing also $(\xi, \eta, \zeta, \omega)$ in place of (x, y, z, w), the new equation is

$$\begin{aligned} T' = (fbc)\,\xi^2 + (gca)\,\eta^2 + (hab)\,\zeta^2 + (fgh)\,\omega^2 & \\ + [(fgb)-(hfc)]\,\xi\omega + [(fab)+(hca)]\,\eta\zeta & \\ + [(ghc)-(fga)]\,\eta\omega + [(gbc)+(fab)]\,\zeta\xi & \\ + [(hfa)-(ghb)]\,\zeta\omega + [(hca)+(gbc)\,]\,\xi\eta = 0; & \end{aligned}$$

or, what is the same thing, this equation $T'=0$ is the equation of the original quadric surface (the locus of L) expressed in terms of the plane-coordinates ξ, η, ζ, ω.

Now considering each of the quantities $a_1, b_1, c_1, f_1, g_1, h_1, a_2, b_2$, etc., a_3, b_3, etc., as a given linear function of a variable parameter λ, say $a_1 = a_1' + a_1''\lambda$, $b_1 = b_1' + b_1''\lambda$, etc., the equation $T=0$ takes the form

$$A\lambda^3 + 3B\lambda^2 + 3C\lambda + D = 0,$$

where A, B, C, D are given quadric functions of the coordinates x, y, z, w; and the envelope of the quadric surface $T=0$ is Herr Kummer's surface of the eighth order

$$(AD-BC)^2 - 4(AC-B^2)(BD-C^2) = 0.$$

In like manner the equation $T'=0$ takes the form

$$A'\lambda^3 + 3B'\lambda^2 + 3C'\lambda + D' = 0,$$

where A', B', C', D' are given functions of the coordinates ξ, η, ζ, ω; and we have

$$(A'D'-B'C')^2 - 4(A'C'-B'^2)(B'D'-C'^2) = 0,$$

as the equation of the reciprocal surface; or (what is the same thing) as that of the original surface, regarding ξ, η, ζ, ω as plane-coordinates.

In regard to the foregoing equation $T=0$, it is to be noticed that, if $a_1, b_1, c_1, f_1, g_1, h_1$; a_2, b_2, etc., a_3, b_3, etc., instead of being arbitrary coefficients, were the coordinates of three given lines L_1, L_2, L_3 respectively; that is, if we had

$$a_1f_1 + b_1g_1 + c_1h_1 = 0,$$

$$a_2f_2 + b_2g_2 + c_2h_2 = 0,$$

$$a_3f_3 + b_3g_3 + c_3h_3 = 0,$$

then the three linear relations satisfied by (a, b, c, f, g, h) would express that the line L was a line meeting each of the three given lines L_1, L_2, L_3: the locus is therefore the quadric surface which passes through these three lines; and I have in my paper "On the six coordinates of a Line," *Camb. Phil. Trans.*, t. XI. (1869), pp. 290—323, [435], found the equation to be the foregoing equation $T=0$. But it is easy to see that the same equation subsists in the case where the three equations $a_1f_1+b_1g_1+c_1h_1=0$, etc., are not satisfied. For the several coefficients being perfectly general, any one of the three linear relations may be replaced by a linear combination of these equations; that is, in place of a_1, b_1, c_1, f_1, g_1, h_1, we may write a_1', b_1', c_1', f_1', g_1', h_1', where $a_1'=\theta_1a_1+\theta_2a_2+\theta_3a_3$, $b_1'=\theta_1b_1+\theta_2b_2+\theta_3b_3$, etc.; and these factors θ_1, θ_2, θ_3 may be conceived to be such that the condition in question $a_1'f_1'+b_1'g_1'+c_1'h_1'=0$ is satisfied. Similarly the second set of coefficients may be replaced by a_2', b_2', c_2', f_2', g_2', h_2', where $a_2'=\phi_1a_1+\phi_2a_2+\phi_3a_3$, etc., and the condition $a_2'f_2'+b_2'g_2'+c_2'h_2'=0$ is satisfied: and the third set by a_3', b_3', c_3', f_3', g_3', h_3', where $a_3'=\psi_1a_1+\psi_2a_2+\psi_3a_3$, etc., and the condition $a_3'f_3'+b_3'g_3'+c_3'h_3'=0$ is satisfied. We have therefore an equation $0=(a'g'h')\,x^2+\text{etc.}$, which only differs from the equation $T=0$ by having therein the accented letters in place of the unaccented ones: and, substituting for the accented letters their values, the whole divides by the determinant $(\theta\phi\psi)$, and throwing this out we obtain the required equation $T=0$.

But it is easier to obtain the equation $T=0$ directly. We have

$$\begin{array}{rrrr} . & hy & -gz & +aw=0, \\ -hx & . & +fz & +bw=0, \\ gx & -fy & . & +cw=0, \\ -ax & -by & -cz & .\;=0; \end{array}$$

viz. in virtue of the equation $af+bg+ch=0$ which connects the six coordinates, these four equations are equivalent to two independent equations which are the equations of the line (a, b, c, f, g, h): or, what is the same thing, any three of these equations imply the fourth equation and also the relation $af+bg+ch=0$.

We might, from the three linear relations and any three of the last-mentioned four equations, eliminate a, b, c, f, g, h and so obtain the required equation $T=0$; but it is better, introducing the arbitrary coefficients α, β, γ, δ, to employ all the four equations. The result of the elimination is thus given in the form

$$\left|\begin{array}{ccccccc} \alpha, & w, & & & & -z, & y \\ \beta, & & w, & & z, & & -x \\ \gamma, & & & w, & -y, & x, & \\ \delta, & x, & y, & z, & & & \\ & f_1, & g_1, & h_1, & a_1, & b_1, & c_1 \\ & f_2, & g_2, & h_2, & a_2, & b_2, & c_2 \\ & f_3, & g_3, & h_3, & a_3, & b_3, & c_3 \end{array}\right| = 0,$$

viz. the left-hand side here contains the factor $-(\alpha x+\beta y+\gamma z+\delta w)$; throwing this out, we obtain the required quadric equation $T=0$. If for the calculation of T we compare the terms containing δ, we have

$$Tw = \begin{vmatrix} w, & & & & -z, & y \\ w, & w, & & z, & & -x \\ & & w, & -y, & x, & \\ f_1, & g_1, & h_1, & a_1, & b_1, & c_1 \\ f_2, & g_2, & h_2, & a_2, & b_2, & c_2 \\ f_3, & g_3, & h_3, & a_3, & b_3, & c_3 \end{vmatrix},$$

where observe that, writing $w=0$, the right-hand side vanishes as containing the factor

$$\begin{vmatrix} & -z, & y \\ z, & & -x \\ -y, & x, & \end{vmatrix}.$$

Hence the right-hand side divides by w; and one of its terms being evidently $w^3(abc)$, T contains as it should do the term $(abc)w^2$: the remaining terms can be found without any difficulty, and the foregoing expression for T is thus verified.

672.

ON THE GAME OF MOUSETRAP.

[From the *Quarterly Journal of Pure and Applied Mathematics,* vol. XV. (1878), pp. 8—10.]

IN the note "A Problem in Permutations," *Quarterly Mathematical Journal,* t. I. (1857), p. 79, [161], I have spoken of the problem of permutations presented by this game.

A set of cards—ace, two, three, &c., say up to thirteen—are arranged (in any order) in a circle with their faces upwards; you begin at any card, and count one, two, three, &c., and if upon counting, suppose the number five, you arrive at the card five, the card is thrown out; and beginning again with the next card, you count one, two, three, &c., throwing out (if the case happen) a new card as before, and so on until you have counted up to thirteen, without coming to a card which has to be thrown out. The original question proposed was: for any given number of cards to find the arrangement (if any) which would throw out all the cards in a given order; but (instead of this) we may consider *all* the different arrangements of the cards, and inquire how many of these there are in which all or any given smaller number of the cards will be thrown out; and (in the several cases) in what orders the cards are thrown out. Thus to take the simple case of four cards, the different arrangements, with the cards thrown out in each, are

1, 2, 3, 4	1,
1, 2, 4, 3	1, 3, 4, 2,
1, 3, 2, 4	1,
1, 3, 4, 2	1,
1, 4, 2, 3	1, 2, 3, 4,
1, 4, 3, 2	1,
2, 1, 3, 4	3, 4,
2, 1, 4, 3	—
2, 3, 4, 1	—
2, 3, 1, 4	4,
2, 4, 1, 3	—
2, 4, 3, 1	3, 2,
3, 1, 2, 4	4,
3, 1, 4, 2	—
3, 2, 1, 4	4, 2, 1, 3,
3, 2, 4, 1	2, 3,
3, 4, 1, 2	—
3, 4, 2, 1	—
4, 1, 2, 3	—
4, 1, 3, 2	3,
4, 2, 1, 3	2, 1, 3, 4,
4, 2, 3, 1	3, 1, 2, 4,
4, 3, 1, 2	—
4, 3, 2, 1	—.

Classifying these so as to show in how many arrangements a given card or permutation of cards is thrown out, we have the table

No.	Thrown out.
9	none
4	1
1	3
2	4
1	3, 2
1	2, 3
1	3, 4
1	1, 3, 4, 2
1	1, 2, 3, 4
1	4, 2, 1, 3
1	2, 1, 3, 4
1	3, 1, 2, 4,
24,	

viz. there are nine arrangements in which no card is thrown out, four arrangements in which only the card 1 is thrown out, one arrangement in which only the card 3 is thrown out, and so on.

It will be observed that there are five arrangements in which all the cards are thrown out, each throwing them out in a different order; there are thus only five orders in which all the cards are thrown out.

The general question is of course to form a like table for the numbers 5, 6, ..., or any greater number of cards.

673.

NOTE ON THE THEORY OF CORRESPONDENCE.

[From the *Quarterly Journal of Pure and Applied Mathematics*, vol. xv. (1878), pp. 32, 33.]

If the point P on a given curve U of the order m, and the point Q on a given curve V of the order m', have a (1, 1) correspondence, this implying that the two curves have the same deficiency; then if PQ intersects the consecutive line $P'Q'$ in a point R, the locus of R is a curve W of the class $m+m'$, and the point R on this curve has, in general (but not universally), a (1, 1) correspondence with the point P on U or with the point Q on V. For, considering the correspondence of the points P and R, to a given position of P there corresponds, it is clear, a single position of R; on the other hand, starting from R, the tangent at this point to the curve W meets the curve U in m points and the curve V in m' points, but it is in general only one of the m points and only one of the m' points which are corresponding points on the curves U and V; that is, it is only one of the m points which is a point P; and the correspondence of (P, R) is thus a (1, 1) correspondence.

But the curves U, V may be such that the correspondence of (P, R) is not a (1, 1) but a $(k, 1)$ correspondence; viz., that to a given position of P there corresponds a single position of R, but to a given position of R, k positions of P. To show that this is so, imagine through P a line Π having therewith a $(k, 1)$ correspondence; P being, as above, a point on the curve U, the line in question envelopes a curve W; and the correspondence is such that, for any given position of P on the curve U, we have through it a single position of the line: but, for a given tangent of the curve W, we have upon it k positions of the point P, viz. k of the m intersections of the line with the curve U are points corresponding to the line; this, of course, implies that the curve U is not any curve whatever of the order m, but a curve of a peculiar nature.

Imagine now that we have on the line Π a point Q, having with P a (1, 1) correspondence of a given nature: to fix the ideas, suppose P, Q are harmonics in regard to a given conic: since on each of the lines Π there are k positions of P, there are also on the line k positions of Q, and the locus of these k points Q is a curve V, say of the order m'.

The point P on the curve U and the point Q on the curve V have a (1, 1) correspondence. For, consider P as given: there is a single position of the line Π intersecting V in m' points, but obviously only one of these is the point Q. And consider Q as given: then through Q we have say μ tangents of the curve W; each of these tangents intersects the curve U in m points, k of which are points P, but for a tangent taken at random no one of these is the correspondent of Q; it is, in general, only one of the μ tangents which has upon it k points P, one of them being the point corresponding to Q; that is, to a given position of Q there corresponds a single position of P; and the correspondence of the points (P, Q) is thus a (1, 1) correspondence.

We have thus the point P on the curve U and the point Q on the curve V, which points have with each other a (1, 1) correspondence; and the line Π is the line PQ joining these points; this intersects the consecutive line in a point R; and the locus of R is the curve W. To a given position of P there corresponds a single line Π, and therefore a single position of R; but to a given position of R there correspond k positions of P, viz. drawing at R the tangent to the curve W, this is a line Π having upon it k points P, or the correspondence of (P, Q) is, as stated, a (k, 1) correspondence.

The foregoing considerations were suggested to me by the theory of parallel curves. Take a curve parallel to a given curve, for example, the ellipse; this is a curve of the order δ, such that every normal thereto is a normal at two distinct points; and the curve has as its evolute the evolute of the ellipse, *or, more accurately, the evolute of the ellipse taken twice;* but, attending only to the evolute taken once, each tangent of the evolute is a normal of the parallel curve at two distinct points thereof, and the points of the parallel curve have with those of the evolute not a (1, 1) but a (2, 1) correspondence.

674.

NOTE ON THE CONSTRUCTION OF CARTESIANS.

[From the *Quarterly Journal of Pure and Applied Mathematics*, vol. XV. (1878), p. 34.]

IF $\rho = a + b\cos\theta$, and $r = \frac{1}{2}\{\rho \pm \sqrt{(\rho^2 - c^2)}\}$, then obviously $r^2 - r\rho + \frac{1}{4}c^2 = 0$, that is,

$$r^2 - r(a + b\cos\theta) + \tfrac{1}{4}c^2 = 0,$$

which is the equation of a Cartesian. Here $\rho = a + b\cos\theta$ is the equation of a limaçon or nodal Cartesian, having the origin for the node; and for any given value of θ, deducing from the radius vector of the limaçon the new radius vector r by the above formula $r = \frac{1}{2}\{\rho \pm \sqrt{(\rho^2 - c^2)}\}$, we obtain a Cartesian, or by giving different values to c, a series of Cartesians having the origin for a common focus. The construction is a very convenient one.

675.

ON THE FLEFLECNODAL PLANES OF A SURFACE.

[From the *Quarterly Journal of Pure and Applied Mathematics*, vol. xv. (1878), pp. 49—51.]

If at a node (or double point) of a plane curve there is on one of the branches an inflexion, (that is, if the tangent has a 3-pointic intersection with the branch), the node is said to be a flecnode; and if there is on each of the branches an inflexion, then the node is said to be a fleflecnode. The tangent plane of a surface intersects the surface in a plane curve having at the point of contact a node; if this is a flecnode or a fleflecnode, the tangent plane is said to be a flecnodal or a fleflecnodal plane accordingly. For a quadric surface each tangent plane is fleflecnodal; this is obvious geometrically (since the section is a pair of lines), and it will presently appear that the analytical condition for such a plane is satisfied. In fact, if the origin be taken at a point of a surface, so that $z=0$ shall be the equation of the tangent plane, then in the neighbourhood of the point we have

$$z=(x,\ y)^2+(x,\ y)^3+\&c.;$$

and the condition for a fleflecnodal plane is that the term $(x,\ y)^2$ shall be a factor of the succeeding term $(x,\ y)^3$. Now for a quadric surface the equation is

$$z=\tfrac{1}{2}\{ax^2+2hxy+by^2+2(fy+gx)z+cz^2\};$$

that is,

$$z(1-fy-gx-\tfrac{1}{2}cz)=\tfrac{1}{2}(ax^2+2hxy+by^2),$$

or developing as far as the third order in $(x,\ y)$, we have

$$z=\tfrac{1}{2}(ax^2+2hxy+by^2)(1+fy+gx),$$

so that the condition in question is satisfied.

In what follows, I take for greater simplicity $h=0$, (viz. $x=0$, $y=0$ are here the tangents to the two curves of curvature at the point in question), and to avoid fractions write $2f$, $2g$ in place of f, g respectively; the developed equation of the quadric surface is thus

$$z=\tfrac{1}{2}(ax^2+by^2)+(ax^2+by^2)(gx+fy).$$

I consider the parallel surface, obtained by measuring off on the normal a constant length k. If, as usual, p, q denote $\dfrac{dz}{dx}$ and $\dfrac{dz}{dy}$ respectively, then, in general, (X, Y, Z) being the coordinates of the point on the parallel surface,

$$Z=z+\frac{k}{\sqrt{(1+p^2+q^2)}},$$

$$X=x-\frac{kp}{\sqrt{(1+p^2+q^2)}},$$

$$Y=y-\frac{kq}{\sqrt{(1+p^2+q^2)}}.$$

But in the present case

$$p=ax+3agx^2+2afxy+\ bgy^2,$$

$$q=by+\ afx^2+2bgxy+3bfy^2,$$

whence

$$X=x-k(ax+3agx^2+2afxy+\ bgy^2),$$

$$Y=y-k(by+\ afx^2+2bgxy+3bfy^2);$$

or, putting for convenience,

$$X=(1-ka)\xi,\quad Y=(1-kb)\eta,$$

then, for a first approximation $x=\xi$, $y=\eta$; whence, writing

$$P=3ag\xi^2+2af\xi\eta+\ bg\eta^2,$$

$$Q=\ af\xi^2+2bg\xi\eta+3bf\eta^2,$$

we find

$$x=\xi+\frac{kP}{1-ka},\quad y=\eta+\frac{kQ}{1-kb},$$

and thence

$$p=a\left(\xi+\frac{k}{1+ka}P\right)+P=a\xi+\frac{P}{1-ka},$$

$$q \qquad\qquad =b\eta+\frac{Q}{1-kb}.$$

Hence

$$Z=\tfrac{1}{2}(a\xi^2+b\eta^2)+\frac{ka}{1-ka}\xi P+\frac{kb}{1-kb}\eta Q+(a\xi^2+b\eta^2)(g\xi+f\eta)$$
$$+k\left\{1-\tfrac{1}{2}(a^2\xi^2+b^2\eta^2)-\frac{a\xi P}{1-ka}-\frac{b\eta Q}{1-kb}\right\};$$

or, finally,

$$Z-k=\tfrac{1}{2}\{a(1-ka)\xi^2+b(1-kb)\eta^2\}+(a\xi^2+b\eta^2)(g\xi+f\eta),$$

where, changing the origin to the point $x=0$, $y=0$, $z=k$ on the parallel surface, the coordinates of the consecutive point are $Z-k$, X, $=(1-ka)\,\xi$, and Y, $=(1-kb)\,\eta$.

We cannot, by any determination of the value of k, make the plane $Z-k=0$ a fleflecnodal plane of the parallel surface; but if

$$k=\frac{af^2+bg^2}{a^2f^2+b^2g^2},$$

then

$$1-ka=\frac{bg^2(b-a)}{a^2f^2+b^2g^2},\quad 1-kb=\frac{af^2(a-b)}{a^2f^2+b^2g^2},$$

and the equation becomes

$$Z-k=\tfrac{1}{2}\frac{ab\,(b-a)}{a^2f^2+b^2g^2}(g^2\xi^2-f^2\eta^2)+(a\xi^2+b\eta^2)(g\xi+f\eta);$$

viz. the term of the second has here a factor $g\xi+f\eta$ which divides the term of the third order, and the plane $Z-k=0$ is a flecnodal plane of the parallel surface.

676.

NOTE ON A THEOREM IN DETERMINANTS.

[From the *Quarterly Journal of Pure and Applied Mathematics*, vol. XV. (1878), pp. 55—57.]

It is well known that if 12, &c., denote the determinants formed with the matrix

$$\left\| \begin{matrix} \alpha, & \beta, & \gamma, & \delta \\ \alpha', & \beta', & \gamma', & \delta' \end{matrix} \right\|,$$

then, identically,

$$12 \,.\, 34 + 13 \,.\, 42 + 14 \,.\, 23 = 0.$$

The proper proof of the theorem is obtained by remarking that we have

$$0 = \left| \begin{matrix} \alpha, & \beta, & \gamma, & . \\ \alpha', & \beta', & \gamma', & . \\ \alpha, & \beta, & \gamma, & \delta \\ \alpha', & \beta', & \gamma', & \delta' \end{matrix} \right|,$$

as at once appears by subtracting the first and second lines from the third and fourth lines respectively; and, this being so, the development of the determinant gives the theorem. The theorem might, it is clear, have been obtained in four different forms according as in the determinant the missing terms were taken to be as above (δ, δ'), or to be (α, α'), (β, β'), or (γ, γ'); but the four results are equivalent to each other.

There is obviously a like theorem for the sums of products of determinants formed with the matrix

$$\left\| \begin{matrix} \alpha, & \beta, & \gamma, & \delta, & \epsilon, & \zeta \\ \alpha', & \beta', & \gamma', & \delta', & \epsilon', & \zeta' \\ \alpha'', & \beta'', & \gamma'', & \delta'', & \epsilon'', & \zeta'' \end{matrix} \right\|,$$

viz. the theorem is obtained by development of the determinant in an identical equation, such as

$$0=\begin{vmatrix} \alpha\,, & \beta\,, & \gamma\,, & \delta\,, & .\,, & . \\ \alpha', & \beta', & \gamma', & \delta', & .\,, & . \\ \alpha'', & \beta'', & \gamma'', & \delta'', & .\,, & . \\ \alpha\,, & \beta\,, & \gamma\,, & \delta\,, & \epsilon\,, & \zeta \\ \alpha', & \beta', & \gamma', & \delta', & \epsilon', & \zeta' \\ \alpha'', & \beta'', & \gamma'', & \delta'', & \epsilon'', & \zeta'' \end{vmatrix};$$

but we thus obtain 15 results which are not all equivalent.

If, for shortness, we write

$$\begin{aligned} A&=123\,.\,456,\\ -B&=124\,.\,356,\\ -C&=125\,.\,346,\\ D&=126\,.\,345,\\ -E&=134\,.\,256,\\ -F&=135\,.\,246,\\ G&=136\,.\,245,\\ -H&=145\,.\,236,\\ I&=146\,.\,235,\\ J&=156\,.\,234, \end{aligned}$$

then the fifteen results are

$$\begin{aligned} A+B-C-D&=0,\\ A+B-E-J&=0,\\ A-C+F-I&=0,\\ A-D+G-H&=0,\\ A-E+F+G&=0,\\ A-H-I-J&=0,\\ B-C-G+H&=0,\\ B-D-F+I&=0,\\ B-E+H+I&=0,\\ B-F-G-J&=0,\\ C+D-E-J&=0,\\ C-E+G+I&=0,\\ C-F-H-J&=0,\\ D-E+F+H&=0,\\ D-G-I-J&=0, \end{aligned}$$

which are all satisfied if only

$$\begin{aligned} A&=\;.\quad\;\;.+H+I+J,\\ B&=F+G\quad.\quad.+J,\\ C&=F\quad.+H\quad.+J,\\ D&=\;.\quad G\quad.+I+J,\\ E&=F+G+H+I+J; \end{aligned}$$

and we thus have these five relations between the ten products of determinants $A, B, C, D, E, F, G, H, I, J$.

677.

[ADDITION TO MR GLAISHER'S PAPER "PROOF OF STIRLING'S THEOREM."]

[From the *Quarterly Journal of Pure and Applied Mathematics*, vol. XV. (1878), pp. 63, 64.]

It is easy to extend Mr Glaisher's investigation so as to obtain from it the more approximate value

$$\Pi n = \sqrt{(2\pi)}\, n^{n+\frac{1}{2}}\, e^{-n+\frac{1}{12n}}.$$

We, in fact, have

$$\psi x = e^{2nx + ax^3 + bx^5 + \dots},$$

where $a, b, \dots$ are given functions of n, viz.

$$a = \tfrac{2}{3}\left\{\frac{1}{3^2} + \frac{1}{5^2} + \dots + \frac{1}{(2n+1)^2}\right\},$$

$$b = \tfrac{3}{5}\left\{\frac{1}{3^4} + \frac{1}{5^4} + \dots + \frac{1}{(2n+1)^4}\right\},$$

&c.

And hence writing $x = 1$, we have

$$\psi(1) = \tfrac{1}{2}\frac{1}{2^{2n}\,\Pi^2(n)}(2n+2)^{2n+1} = e^{2n+a+b+\dots},$$

that is,

$$\begin{aligned}\Pi n &= \left(\frac{2n+2}{2}\right)^{\frac{2n+1}{2}} e^{-n-\frac{1}{2}(a+b+\dots)} \\ &= (n+1)^{n+\frac{1}{2}}\, e^{-n-\frac{1}{2}(a+b+\dots)} \\ &= n^{n+\frac{1}{2}}\left(1+\frac{1}{n}\right)^{n+\frac{1}{2}} e^{-n-\frac{1}{2}(a+b+\dots)}.\end{aligned}$$

Hence for $\left(1+\frac{1}{n}\right)^{n+\frac{1}{2}}$ writing $e^{-(n+\frac{1}{2})\log\left(1+\frac{1}{n}\right)}$, the whole exponent of e is

$$(n+\tfrac{1}{2})\log\left(1+\frac{1}{n}\right)-n-\tfrac{1}{2}(a+b+\ldots)$$

$$=(n+\tfrac{1}{2})\left(\frac{1}{n}-\tfrac{1}{2}\frac{1}{n^2}+\tfrac{1}{3}\frac{1}{n^3}-\ldots\right)-n-\tfrac{1}{2}(a+b+\ldots)$$

$$=-n+1+\frac{1}{3.4}\frac{1}{n^2}-\frac{2}{4.6}\frac{1}{n^3}+\frac{3}{5.8}\frac{1}{n^4}-\ldots$$

$$-\tfrac{1}{2}(a+b+\ldots).$$

We have

$$\frac{1}{1^2}+\frac{1}{3^2}+\ldots+\frac{1}{(2n+1)^2}=\text{const.}-\frac{1}{4n}+\text{terms in }\frac{1}{n^2},\ \frac{1}{n^3},\ \&\text{c.}$$

(the constant is in fact $=\frac{1}{8}\pi^2$, but the value is not required), hence $a=\text{const.}-\frac{1}{6n}$ + terms in $\frac{1}{n^2}$, $\frac{1}{n^3}$, &c.; as regards b, c, &c., there are no terms in $\frac{1}{n}$, but we have $b=\text{const.}+\text{terms in }\frac{1}{n^2}$, &c., $c=\text{const.}+\text{terms in }\frac{1}{n^3}$, &c. Hence the whole exponent of e is

$$=-n+C+\frac{1}{12n}+\text{terms in }\frac{1}{n^2},\ \&\text{c.}$$

As in Mr Glaisher's investigation, it is shown that $e^{-C}=\sqrt{(2\pi)}$, and hence neglecting the terms in $\frac{1}{n^2}$, &c., the final result is

$$\Pi n=\sqrt{(2\pi)}\,n^{n+\frac{1}{2}}\,e^{-n+\frac{1}{12n}}.$$

678.

ON A SYSTEM OF QUADRIC SURFACES.

[From the *Quarterly Journal of Pure and Applied Mathematics*, vol. xv. (1878), pp. 124, 125.]

THE following theorem was communicated to me by Dr Klein; "given in regard to a quadric surface two sibi-reciprocal line-pairs, the two tractors (or lines meeting each of the four lines) form a sibi-reciprocal line-pair." This may be presented under a more general form as a theorem relating to the tractors of any two line-pairs. In fact, if a given line-pair is taken to be sibi-reciprocal in regard to a quadric surface, we thereby establish only a four-fold relation between the coefficients of the surface, and the surface will still depend on five arbitrary parameters. Whence if two given line-pairs are taken to be each of them sibi-reciprocal in regard to one and the same quadric surface, we thereby establish only an eight-fold relation and the surface will still depend upon one arbitrary parameter. The theorem thus is: given any two line-pairs, then each of these, and also the pair of tractors, are sibi-reciprocal in regard to a singly infinite system of quadric surfaces.

The question arises, what is this system of quadric surfaces? It is, in fact, the system of surfaces having in common a skew quadrilateral constructed as follows: starting from the two given line-pairs, construct the two tractors, each of them intersected by the given line-pairs in two point-pairs; and on each tractor construct the double or sibi-reciprocal points of the involution thus determined; these double points are the vertices (those on the same tractor being opposite vertices) of the skew quadrilateral; which is consequently at once obtained by joining the two double points on the one tractor with the two double points on the other tractor. The construction is an immediate consequence of the following theorem: consider a skew quadrilateral, and drawing its two diagonals, take a pair of lines cutting each diagonal harmonically; these will be sibi-reciprocal in regard to any quadric surface through the skew quadrilateral.

The condition of passing through a skew quadrilateral is that of passing through a certain system of eight points; in fact, the eight points may be taken to be the four vertices and any four points on the four sides respectively. But observe that the system of the quadric surfaces through *any* eight points has the characteristics (1, 2, 3); viz. there are in the system 1 surface passing through a given point, 2 touching a given line, 3 touching a given plane; the system of surfaces through the same skew quadrilateral has the characteristics (1, 2, 1).

679.

ON THE REGULAR SOLIDS.

[From the *Quarterly Journal of Pure and Applied Mathematics*, vol. xv. (1878), pp. 127—131.]

In a regular solid, or say in the spherical figure obtained by projecting such solid, by lines from the centre, on the surface of a concentric sphere, we naturally consider 1° the summits, 2° the centres of the faces, 3° the mid-points of the sides. But, imagining the five regular figures drawn in proper relation to each other on the same spherical surface, the only points which have thus to be considered are 12 points A, 20 points B, 30 points Θ, and 60 points Φ. These may be, in the first instance, described by reference to the dodecahedron; viz. the points A are the centres of the faces, the points B are the summits, the points Θ are the mid-points of the sides, and the points Φ are the mid-points of the diagonals of the faces (viz. there are thus 5 points Φ in each face of the dodecahedron, or in all 60 points Φ). But reciprocally we may describe them in reference to the icosahedron; viz. the points A are the summits, the points B the centres of the faces, the points Θ the mid-points of the sides, (viz. each point Θ is the common mid-point of a side of the dodecahedron and a side of the icosahedron, which sides there intersect at right angles), and the points Φ are points lying by 3's on the faces of the icosahedron, each point Φ of the face being given as the intersection of a perpendicular $A\Theta$ of the face by a line BB, joining the centres of two adjacent faces and intersecting $A\Theta$ at right angles.

The points A lie opposite to each other in pairs in such wise that, taking any two opposite points as poles, the relative situation is as follows:

A	Longitudes.
1	— ,
5	0°, 72°, 144°, 216°, 288°,
5	36°, 108°, 180°, 252°, 324°,
1	— ,

where the points A in the same horizontal line form a zone of points equidistant from the point taken as the North Pole. And the points B lie also opposite to each other in such wise that, taking two opposite points as poles, the relative situation is as follows:

B	Longitudes.
1	— ,
3	0°, 120°, 240° ,
6	(0°, 120°, 240°) ± 22° 14′,
6	(60°, 180°, 300°) ± 22° 14′,
3	60°, 180°, 300° ,
1	— ,

where the points B in the same horizontal line form a zone of points equidistant from the point taken as the North Pole. Neglecting the $3+3$ points B which lie adjacent to the poles, the remaining 14 points B may be arranged as follows ($\beta = 22°\ 14'$ as above):

B	Longitudes.	
1	—	
6	β, $120° + \beta$, $240° + \beta$	$-\beta$, $120° - \beta$, $240° - \beta$,
6	$60° + \beta$, $180° + \beta$, $300° + \beta$	$60° - \beta$, $180° - \beta$, $300° - \beta$.
1	—	

And taking the two poles separately with each system of the remaining poles, we have 2 systems each of 8 points B, which are, in fact, the summits of a cube (hexahedron); each point B taken as North Pole thus belongs to two cubes; but inasmuch as the cube has 8 summits, the number of the cubes thus obtained is $20 \times 2 \div 8, = 5$; viz. the 20 points B form the summits of 5 cubes, each point B of course belonging to 2 cubes.

It is to be added that, considering the 5 points B which form a face of the dodecahedron, any diagonal BB of this dodecahedron is a side of a cube. We have thus $12 \times 5, = 60$, the number of the sides of the 5 cubes.

It is at once seen that the centres of the faces of a cube are points Θ, and that the mid-points of the sides of the cube are points Φ.

To each cube there corresponds of course an octahedron, the summits being points Θ, the centres of the faces points B, and the mid-points of the sides points Φ; thus, for the five octahedra the summits are the $5 \times 6, = 30$, points Θ; the centres of the faces are $5 \times 8, = 40$, points B (each point B being thus a centre of face for two octahedra), and the mid-points of the sides being the $5 \times 12, = 60$, points Φ.

Finally, considering the 8 points B which belong to a cube, we can, in four different ways, select thereout 4 points B which are the summits of a tetrahedron;

the remaining 4 points B are then the centres of the faces, and the mid-points of the sides are points Θ: there are thus 5×4, $=20$, tetrahedra having 20×4 summits which are the 20 points B each 4 times; 20×4 centres of faces which are the 20 points B each 4 times; and 20×6 mid-points of sides which are the 30 points Θ each 4 times.

It thus appears that, as mentioned above, the five regular figures depend only on the points A, B, Θ, and Φ.

We might take as poles two opposite points A, B, Θ, or Φ; and in each case determine in reference to these the positions of the other points; but for brevity I consider only the case in which we take as poles two opposite points A. We have the following table:

Poles two opposite points A.

	N. P. D.	Longitudes.
A_0	0°	—
$5A_1$	63° 26′	0°, 72°, 144°, 216°, 288°
$5A_2$	116° 34′	36°, 108°, 180°, 252°, 324°
A_3	180°	—
$5B_1$	37° 22′	36°, 108°, ..., 324°
$5B_2$	79° 12′	36°, 108°, ..., 324°
$5B_3$	100° 48′	0°, 72°, ..., 288°
$5B_4$	142° 38′	0°, 72°, ..., 288°
$5\Theta_1$	31° 43′	0°, 72°, ..., 288°
$5\Theta_2$	58° 77′	36°, 108°, ..., 324°
$10\Theta_3$	90°	(0°, 72°, ..., 288°) ± 18°
$5\Theta_4$	121° 43′	0°, 72°, ..., 288°
$5\Theta_5$	148° 17′	36°, 108°, ..., 324°
$5\Phi_1$	13° 16′	36°, 108°, ..., 324°
$10\Phi_2$	52° 52′	(0°, 72°, ..., 288°) ± 9° 44′
$10\Phi_3$	68° 10′	(0°, 72°, ..., 288°) ± 13° 35′
$5\Phi_4$	76° 42′	0°, 72°, ..., 288°
$5\Phi_5$	103° 18′	36°, 108°, ..., 324°
$10\Phi_6$	111° 50′	(36°, 108°, ..., 324°) ± 13° 35′
$10\Phi_7$	127° 8′	(36°, 108°, ..., 324°) ± 9° 44′
$5\Phi_8$	166° 44′	0°, 72°, ..., 108°.

I add for greater completeness the following results, some of which were used in the calculation of the foregoing table. Considering successively (1) the tetrahedral triangle, summits 3 points B, centre a point B; (2) the hexahedral square, summits 4 points B, centre a point Θ; (3) the octahedral triangle, summits 3 points Θ, centre a point B; (4) the icosahedral triangle, summits 3 points A, centre a point B; (5) the dodecahedral pentagon, summits 5 points, centre a point B; and (6), what may be called the small pentagon, summits 5 points Φ lying within a dodecahedral pentagon, and having therewith the common centre B; we may in each case write s the side, r the radius or distance of the centre from a summit, p the perpendicular or distance of the centre from a side. And the values then are

	s	r	p
Tet. Δ	109° 30′	70° 30′	54° 45′
Hex. square	70 30	54 45	45
Oct. Δ	90	54 45	35 15
Icos. Δ	63 26	37 22	20 55
Dod. pentagon	41 50	37 22	31 43
Small pentagon	15 30	13 16	10 48

680.

ON THE HESSIAN OF A QUARTIC SURFACE.

[From the *Quarterly Journal of Pure and Applied Mathematics*, vol. xv. (1878), pp. 141—144.]

THE surface considered is

$$U = k^2w^2\left(\frac{x^2}{a^2}+\frac{y^2}{b^2}+\frac{z^2}{c^2}\right) - (x^2+y^2+z^2)^2 = 0,$$

or say

$$U = k^2w^2P - Q^2 = 0,$$

viz. this may be considered as the central inverse of the ellipsoid

$$\frac{x^2}{a^2}+\frac{y^2}{b^2}+\frac{z^2}{c^2}-1=0.$$

The values of the second derived functions, or terms of the Hessian determinant

$$\left|\begin{array}{cccc} a, & h, & g, & l \\ h, & b, & f, & m \\ g, & f, & c, & n \\ l, & m, & n, & d \end{array}\right|,$$

are

$$\left|\begin{array}{cccc} \frac{k^2}{a^2}w^2-2Q-4x^2, & -4xy\quad, & -4xz\quad, & \frac{2k^2}{a^2}wx \\ -4xy\quad, & \frac{k^2}{b^2}w^2-2Q-4y^2, & -4yz\quad, & \frac{2k^2}{b^2}wy \\ -4xz\quad, & -4yz\quad, & \frac{k^2}{c^2}w^2-2Q-4z^2, & \frac{2k^2}{c^2}wz \\ \frac{2k^2}{a^2}wx\quad, & \frac{2k^2}{b^2}wy\quad, & \frac{2k^2}{c^2}wz\quad, & k^2P \end{array}\right|$$

and we thence have

$$bc - f^2 = \frac{k^4}{b^2c^2} - \frac{k^2w^2}{b^2}(2Q + 4z^2) - \frac{k^2w^2}{c^2}(2Q + 4y^2) + 4Q^2 + 8Q(y^2 + z^2),$$

$$gh - af = 4yz\left(\frac{k^2w^2}{a^2} - 2Q\right),$$

whence, forming the analogous quantities $ca - g^2$, &c., it is easy to obtain

$$\begin{aligned} abc - af^2 - bg^2 - ch^2 + 2fgh \\ &= \frac{k^6w^6}{a^2b^2c^2} \\ &\quad - k^4w^4\left\{2Q\left(\frac{1}{b^2c^2} + \frac{1}{c^2a^2} + \frac{1}{a^2b^2}\right) + 4\left(\frac{x^2}{b^2c^2} + \frac{y^2}{c^2a^2} + \frac{z^2}{a^2b^2}\right)\right\} \\ &\quad + k^2w^2\left\{12Q^2\left(\frac{1}{a^2} + \frac{1}{b^2} + \frac{1}{c^2}\right) - 8QP\right\} \\ &\quad - 24Q^3, \end{aligned}$$

which is to be multiplied by $d, = k^2P$. And

$$\begin{aligned} &-[l^2(bc - f^2) + m^2(ca - g^2) + n^2(ab - h^2) \\ &+ 2mn(gh - af) + 2nl(hf - bg) + 2lm(fg - ch)] \\ &= -\frac{4k^8w^6P}{a^2b^2c^2} \\ &\quad - 4k^6w^4\left[2Q\left\{\frac{x^2}{a^4}\left(\frac{1}{b^2} + \frac{1}{c^2}\right) + \frac{y^2}{b^4}\left(\frac{1}{c^2} + \frac{1}{a^2}\right) + \frac{z^2}{c^4}\left(\frac{1}{a^2} + \frac{1}{b^2}\right)\right\}\right. \\ &\quad\quad \left. + \frac{4y^2z^2}{a^2}\left(\frac{1}{b^2} - \frac{1}{c^2}\right)^2 + \frac{4z^2x^2}{b^2}\left(\frac{1}{c^2} - \frac{1}{a^2}\right)^2 + \frac{4x^2y^2}{c^2}\left(\frac{1}{a^2} - \frac{1}{b^2}\right)^2\right] \\ &\quad + 4k^4w^2\left[4Q^2\left(\frac{x^2}{a^4} + \frac{y^2}{b^4} + \frac{z^2}{c^4}\right)\right. \\ &\quad\quad \left. + 8Q\left\{y^2z^2\left(\frac{1}{b^2} - \frac{1}{c^2}\right)^2 + z^2x^2\left(\frac{1}{c^2} - \frac{1}{a^2}\right)^2 + x^2y^2\left(\frac{1}{a^2} - \frac{1}{b^2}\right)^2\right\}\right], \end{aligned}$$

which is

$$\begin{aligned} &= -\frac{4k^8w^6P}{a^2b^2c^2} \\ &\quad + k^6w^4\left\{8\left(\frac{1}{b^2c^2} + \frac{1}{c^2a^2} + \frac{1}{a^2b^2}\right)PQ - \frac{24}{a^2b^2c^2}Q^2 + 16\left(\frac{x^2}{b^2c^2} + \frac{y^2}{c^2a^2} + \frac{z^2}{a^2b^2}\right)P\right\} \\ &\quad + k^4w^2\left\{-48\left(\frac{x^2}{a^4} + \frac{y^2}{b^4} + \frac{z^2}{c^4}\right)Q^2 + 32P^2Q\right\}. \end{aligned}$$

Hence, uniting the two parts, we have

$$H = k^8w^6\left(-\frac{3}{a^2b^2c^2}P\right)$$

$$+ k^6w^4\left\{\begin{array}{l} 6\left(\dfrac{1}{b^2c^2}+\dfrac{1}{c^2a^2}+\dfrac{1}{a^2b^2}\right)PQ \\ -\dfrac{24}{a^2b^2c^2}Q^2 \\ +12\left(\dfrac{x^2}{b^2c^2}+\dfrac{y^2}{c^2a^2}+\dfrac{z^2}{a^2b^2}\right)P \end{array}\right\}$$

$$+ k^4w^2\left\{\begin{array}{l} 12\left(\dfrac{1}{a^2}+\dfrac{1}{b^2}+\dfrac{1}{c^2}\right)PQ^2 \\ -48\left(\dfrac{x^2}{b^2c^2}+\dfrac{y^2}{c^2a^2}+\dfrac{z^2}{a^2b^2}\right)Q^2 \\ +24P^2Q \end{array}\right\}$$

$$+ k^2\ \{-24PQ^3\}.$$

Writing herein $Q^2 = k^2w^2P - U$, and transposing all the terms which contain U, we have

$$H + k^2U\left\{-\frac{24k^4w^4}{a^2b^2c^2}+12k^2w^2\left(\frac{1}{a^2}+\frac{1}{b^2}+\frac{1}{c^2}\right)P - 48k^2w^2\left(\frac{x^2}{a^4}+\frac{y^2}{b^4}+\frac{z^2}{c^4}\right) - 24PQ\right\}$$

$$= k^6w^4P\left\{\begin{array}{l} -\dfrac{27k^2}{a^2b^2c^2}w^2 \\ +16\left(\dfrac{1}{b^2c^2}+\dfrac{1}{c^2a^2}+\dfrac{1}{a^2b^2}\right)Q \\ +12\left(\dfrac{1}{a^2}+\dfrac{1}{b^2}+\dfrac{1}{c^2}\right)P \\ +12\left(\dfrac{x^2}{b^2c^2}+\dfrac{y^2}{c^2a^2}+\dfrac{z^2}{a^2b^2}\right) \\ -48\left(\dfrac{x^2}{a^4}+\dfrac{y^2}{b^4}+\dfrac{z^2}{c^4}\right) \end{array}\right\}$$

where, in the term in $\{\ \}$, the last four lines are

$$= 18\left(\frac{1}{b^2c^2}+\frac{1}{c^2a^2}+\frac{1}{a^2b^2}\right)Q$$

$$-36\left(\frac{x^2}{a^4}+\frac{y^2}{b^4}+\frac{z^2}{c^4}\right).$$

Hence, writing for shortness

$$\Theta = -\frac{2k^4}{a^2b^2c^2}w^4 + k^2w^2\left(\frac{1}{a^2}+\frac{1}{b^2}+\frac{1}{c^2}\right)P - 4k^2w^2\left(\frac{x^2}{a^4}+\frac{y^2}{b^4}+\frac{z^2}{c^4}\right) - 2PQ,$$

we have

$$H + 12k^2\Theta U = 9k^6w^4P\left\{-\frac{3k^2}{a^2b^2c^2}w^2 + 2\left(\frac{1}{b^2c^2}+\frac{1}{c^2a^2}+\frac{1}{a^2b^2}\right)Q - 4\left(\frac{x^2}{a^4}+\frac{y^2}{b^4}+\frac{z^2}{c^4}\right)\right\}.$$

Hence, recollecting that $U = k^2w^2P - Q^2$, the Hessian curve of the order 32 breaks up into

$$U=0,\ w^4=0, \text{ that is, } Q^2=0,\ w^4=0, \text{ or the nodal conic,}$$

$$w=0,\ Q=0,\ 8 \text{ times (order 16),}$$

$$U=0,\ P=0, \text{ that is, } Q^2=0,\ P=0, \text{ or the quadriquadric,}$$

$$P=0,\ Q=0,\ 2 \text{ times (order 8),}$$

and into a curve (order 8) which is

$$k^2w^2P - Q^2 = 0,$$

$$-\frac{3k^2}{a^2b^2c^2}w^2 + 2\left(\frac{1}{b^2c^2}+\frac{1}{c^2a^2}+\frac{1}{a^2b^2}\right)Q - 4\left(\frac{x^2}{a^4}+\frac{y^2}{b^4}+\frac{z^2}{c^4}\right) = 0,$$

viz. this, the intersection of the surface with a quadric surface, is the proper Hessian curve.

681.

ON THE DERIVATIVES OF THREE BINARY QUANTICS.

[From the *Quarterly Journal of Pure and Applied Mathematics*, vol. XV. (1878), pp. 157—168.]

FOR a reason which will appear, instead of the ordinary factorial notation, I write $\{\alpha 012\}$ to denote the factorial $\alpha\,.\,\alpha+1\,.\,\alpha+2$, and so in other cases; and I consider the series of equations

$$(1) = X,$$

$$(2) = (\{\alpha 0\},\ \{\beta 0\} \between Y,\ -Y'),$$

$$(3) = (\{\alpha 01\},\ 2\{\alpha 1\}\{\beta 1\},\ \{\beta 01\} \between Z,\ -Z',\ Z''),$$

$$(4) = (\{\alpha 012\},\ 3\{\alpha 12\}\{\beta 2\},\ 3\{\alpha 2\}\{\beta 12\},\ \{\beta 012\} \between W,\ -W',\ -W'',\ -W'''),$$

&c.

where

$$X = Y + Y',$$

$$Y = Z + Z',\quad Y' = Z' + Z'',$$

$$Z = W + W',\quad Z' = W' + W'',\quad Z'' = W'' + W''',$$

&c.

We have thus a series of linear equations serving to determine X; Y, Y'; Z, Z', Z''; W, W', W'', W'''; &c. We require in particular the values of X; Y, Y'; Z, Z''; W, W'''; &c., and I write down the results as follow:

$$X = (1),$$

$$\begin{array}{rl} & \overline{\;(1)\quad (2)\;} \\ \{\alpha+\beta 0\}\, Y & = \{\beta 0\},\ +1, \\ \{\quad ,, \quad\}\, Y' & = \{\alpha 0\},\ -1; \end{array}$$

$$\begin{array}{rcllll}
 & & \{\alpha+\beta 2\}(1), & \{\alpha+\beta 1\}(2), & \{\alpha+\beta 0\}(3), & \\ \cline{3-5}
\{\alpha+\beta 012\}Z & = & \{\beta 01\} & , \ +2\{\beta 1\} & , \ +1 & , \\
\{\ \ ,,\ \ \}Z'' & = & \{\alpha 01\} & , \ -2\{\alpha 1\} & , \ +1 & ; \\
 & & \{\alpha+\beta 34\}(1), & \{\alpha+\beta 14\}(2), & \{\alpha+\beta 03\}(3), & \{\alpha+\beta 01\}(4); \\ \cline{3-6}
\{\alpha+\beta 01\ldots 4\}W & = & \{\beta 012\} & , \ +3\{\beta 12\} & , \ +3\{\beta 2\} & , \ +1 \ , \\
\{\ \ ,,\ \ \}W''' & = & \{\alpha 012\} & , \ -3\{\alpha 12\} & , \ +3\{\alpha 2\} & , \ -1 \ ;
\end{array}$$

$$\begin{array}{rclllll}
 & & \{\alpha+\beta 456\}(1), & \{\alpha+\beta 156\}(2), & \{\alpha+\beta 036\}(3), & \{\alpha+\beta 015\}(4), & \{\alpha+\beta 012\}(5); \\ \cline{3-7}
\{\alpha+\beta 01\ldots 6\}U & = & \{\beta 0123\} & , \ +4\{\beta 123\} & , \ +6\{\beta 23\} & , \ +4\{\beta 3\} & , \ +1 \ , \\
\{\ \ ,,\ \ \}U'''' & = & \{\alpha 0123\} & , \ -4\{\alpha 123\} & , \ +6\{\alpha 23\} & , \ -4\{\alpha 3\} & , \ +1 \ ;
\end{array}$$

&c.

read

$$\alpha+\beta\,.\,Y = \beta\,(1)+(2),$$
$$,,\ .\,Y' = \alpha\,(1)-(2),$$

$$\alpha+\beta.\alpha+\beta+1.\alpha+\beta+2.Z = \beta.\beta+1.\alpha+\beta+2.(1)+2.\beta+1.\alpha+\beta+1.(2)+\alpha+\beta.(3),$$
$$,,\qquad ,,\qquad ,,\qquad .Z'' = \alpha.\alpha+1.\alpha+\beta+2.(1)+2.\alpha+1.\alpha+\beta+1.(2)+\alpha+\beta.(3),$$

&c.,

the law being obvious, except as regards the numbers which in the top lines occur in connexion with $\alpha+\beta$ in the $\{\ \ \}$ symbols. As regards these, we form them by successive subtractions as shown by the diagrams

$$\begin{array}{r|l@{\qquad}r|l@{\qquad}r|l}
\underline{34} & 34 & \underline{456} & 456 & \underline{5678} & 5678 \ \&c.; \\
2 & 14 & 3 & 156 & 4 & 1678 \\
11 & 03 & 12 & 036 & 13 & 0378 \\
2 & 01 & 21 & 015 & 22 & 0158 \\
 & & 3 & 012 & 31 & 0127 \\
 & & & & 4 & 0123
\end{array}$$

and the statement of the result is now complete.

In part verification, starting from the Y-formulæ (which are obtained at once), assume

$$\begin{array}{rclll}
 & & \{\alpha+\beta 2\}(1), & \{\alpha+\beta 1\}(2), & \{\alpha+\beta 0\}(3), \\ \cline{3-5}
\{\alpha+\beta 012\}Z & = & \lambda & , \quad \mu & , \quad \nu \\
\{\ \ ,,\ \ \}Z' & = & \lambda' & , \quad \mu' & , \quad \nu' \\
\{\ \ ,,\ \ \}Z'' & = & \lambda'' & , \quad \mu'' & , \quad \nu''
\end{array}$$

we must have

$$\begin{array}{l}
\qquad\qquad\qquad\qquad\qquad\qquad\qquad\qquad\qquad\qquad\quad \overline{(1)\quad (2)} \\
\{\alpha+\beta 012\}.Z+Z' = \{\alpha+\beta 012\}Y, = \{\alpha+\beta 12\}\,(\{\beta 0\},\ +1) \\
\{\ \ ,,\ \ \}.Z'+Z'' = \{\ \ ,,\ \ \}Y', = \{\ \ ,,\ \ \}\,(\{\alpha 0\},\ -1)
\end{array}$$

that is,

$$\{\alpha+\beta 2\}.\lambda+\lambda' = \{\alpha+\beta 12\}\{\beta 0\},$$
$$\{\ \ ,,\ \ \}.\lambda'+\lambda'' = \{\ \ ,,\ \ \}\{\alpha 0\},$$

and further

$$\{\alpha+\beta 2\}\,(\{\alpha 01\},\ -2\,\{\alpha 1\}\,\{\beta 1\},\ \{\beta 01\}\between \lambda,\ \lambda',\ \lambda'')=0,$$

or, what is the same thing,

$$\begin{aligned}\lambda+\lambda' &= \{\alpha+\beta 1\}\,\{\beta 0\},\\ \lambda'+\lambda'' &= \{\ \ ,,\ \ \}\,\{\alpha 0\},\end{aligned}$$

$$(\{\alpha 01\},\ -2\,\{\alpha 1\}\,\{\beta 1\},\ \{\beta 01\}\between \lambda,\ \lambda',\ \lambda'')=0.$$

And in like manner we have

$$\begin{aligned}\mu+\mu' &= \{\alpha+\beta 2\}\,.\ \ 1,\\ \mu'+\mu'' &= \{\ \ ,,\ \ \}\,.-1,\end{aligned}$$

$$(\{\alpha 01\},\ -2\,\{\alpha 1\}\,\{\beta 1\},\ \{\beta 01\}\between \mu,\ \mu',\ \mu'')=0;$$

and

$$\begin{aligned}\nu+\nu' &= 0,\\ \nu'+\nu'' &= 0,\end{aligned}$$

$$(\{\alpha 01\},\ -2\,\{\alpha 1\}\,\{\beta 1\},\ \{\beta 01\}\between \nu,\ \nu',\ \nu'')=0.$$

We hence find without difficulty

$$\begin{aligned}\lambda,\ \mu,\ \nu &= \beta.\beta+1,\quad 2.\beta+1,\ +1, = \{\beta 01\},\ 2\,\{\beta 1\},\ +1,\\ \lambda',\ \mu',\ \nu' &= \alpha.\beta,\quad \alpha-\beta,\ -1, = \{\alpha 0\}\,\{\beta 0\},\ \alpha-\beta,\ -1,\\ \lambda'',\ \mu'',\ \nu'' &= \alpha.\alpha+1,\ -2.\alpha+1,\ +1, = \{\alpha 01\},\ 2\,\{\alpha 1\},\ +1;\end{aligned}$$

viz. for verification of the λ-equations we have

$$\begin{aligned}&\beta.\beta+1.+\ \alpha.\beta\ ,\text{ that is, } \alpha+\beta+1.\beta, = \{\alpha+\beta 1\}\,\{\beta 0\},\\ &\alpha.\beta.\ +\alpha.\alpha+1,\quad ,,\quad \alpha+1+\beta.\alpha, = \{\ \ ,,\ \ \}\,\{\alpha 0\},\end{aligned}$$

and

$$(\alpha.\alpha+1,\ -2.\alpha+1.\beta+1,\ \beta.\beta+1\between \beta.\beta+1,\ \alpha.\beta,\ \alpha.\alpha+1)=0,$$

that is,

$$\alpha.\alpha+1.\beta.\beta+1.-2.\alpha+1.\beta+1.\alpha.\beta.+\beta.\beta+1.\alpha.\alpha+1=0;$$

and similarly the μ- and ν-equations may be verified.

We have thus for the Z's the equations

$$\begin{array}{lcccc} & \{\alpha+\beta 2\}\,(1), & \{\alpha+\beta 1\}\,(2), & \{\alpha+\beta 0\}\,(3), \\ \hline \{\alpha+\beta 012\}\,Z = & \{\beta 01\}, & 2\,\{\beta 1\}, & +1, \\ \{\ \ ,,\ \ \}\,Z' = & \{\alpha 0\}\,\{\beta 0\}, & \alpha-\beta, & -1, \\ \{\ \ ,,\ \ \}\,Z'' = & \{\alpha 01\}, & -2\,\{\alpha 1\}, & +1, \end{array}$$

which include the foregoing expressions for Z and Z''.

We may then take the expressions for the W's to be

$$\begin{array}{lccccc} & \{\alpha+\beta 34\}\,(1), & \{\alpha+\beta 14\}\,(2), & \{\alpha+\beta 03\}\,(3), & \{\alpha+\beta 01\}\,(4), \\ \hline \{\alpha+\beta 0123\}\,W = & \lambda, & \mu, & \nu, & \rho, \\ \{\ \ ,,\ \ \}\,W' = & \lambda', & \mu', & \nu', & \rho', \\ \{\ \ ,,\ \ \}\,W'' = & \lambda'', & \mu'', & \nu'', & \rho'', \\ \{\ \ ,,\ \ \}\,W''' = & \lambda''', & \mu''', & \nu''', & \rho'''; \end{array}$$

and we obtain in like manner the equations

$$\lambda\ +\lambda'\ =\{\alpha+\beta 234\}\,\{\beta 01\},$$

$$\lambda'\ +\lambda''=\{\quad ,, \quad\}\,\{\ \alpha 0\ \}\ \{\beta 0\},$$

$$\lambda''+\lambda'''=\{\quad ,, \quad\}\,\{\alpha 01\},$$

$$(\{\alpha 012\},\ -3\,\{\alpha 12\}\,\{\beta 2\},\ +3\,\{\alpha 2\}\,\{\beta 12\},\ -\{\beta 012\}\between \lambda,\ \lambda',\ \lambda'',\ \lambda''')=0;$$

$$\mu\ +\mu'\ =\{\alpha+\beta 134\}\,.\quad 2\,\{\beta 1\},$$

$$\mu'\ +\mu''=\{\quad ,, \quad\}\,.\quad \alpha-\beta,$$

$$\mu''+\mu'''=\{\quad ,, \quad\}\,.-2\,\{\alpha 1\},$$

$$(\{\alpha 012\},\ -3\,\{\alpha 12\}\,\{\beta 2\},\ +3\,\{\alpha 2\}\,\{\beta 12\},\ -\{\beta 012\}\between \mu,\ \mu',\ \mu'',\ \mu''')=0;$$

$$\nu\ +\nu'\ =\{\alpha+\beta 034\}\,.\quad 1,$$

$$\nu'\ +\nu''=\{\quad ,, \quad\}\,.-1,$$

$$\nu''+\nu'''=\{\quad ,, \quad\}\,.\quad 1,$$

$$(\{\alpha 012\},\ -3\,\{\alpha 12\}\,\{\beta 2\},\ +3\,\{\alpha 2\}\,\{\beta 12\},\ -\{\beta 012\}\between \nu,\ \nu',\ \nu'',\ \nu''')=0;$$

$$\rho\ +\rho'\ =0,$$

$$\rho'\ +\rho''\ =0,$$

$$\rho''+\rho'''=0,$$

$$(\{\alpha 012\},\ -3\,\{\alpha 12\}\,\{\beta 2\},\ +3\,\{\alpha 2\}\,\{\beta 12\},\ -\{\beta 012\}\between \rho,\ \rho',\ \rho'',\ \rho''')=\{\alpha+\beta 01234\}.$$

These give for the $\lambda\rho'''$ square the values

$$\begin{array}{llll}
\{\beta 012\}\ , & 3\,\{\beta 12\}\ , & 3\,\{\beta 2\}\ , & +1, \\
\{\alpha 0\}\,\{\beta 01\}, & 2\alpha-\beta\,.\,\{\beta 1\}, & \alpha-2\beta-2, & -1, \\
\{\alpha 01\}\,\{\beta 0\}, & \alpha-2\beta\,.\,\{\alpha 1\}, & -2\alpha+\beta-2, & +1, \\
\{\alpha 012\}\ , & -3\,\{\alpha 12\}\ , & +3\,\{\alpha 2\}\ , & -1,
\end{array}$$

and so on; the law however of the terms in the intermediate lines is not by any means obvious.

Consider now the binary quantics P, Q, R, of the forms $(*\between x,\ y)^p$, $(*\between x,\ y)^q$, $(*\between x,\ y)^r$; we have for any, for instance for the fourth, order, the derivates

$$P(Q,\ R)^4,\ \ (P,\ (Q,\ R)^3)^1,\ \ (P,\ (Q,\ R)^2)^2,\ \ (P,\ (Q,\ R)^1)^3,\ \ (P,\ QR)^4;$$

and it is required to express

$$Q(P,\ R)^4 \text{ and } R(P,\ Q)^4,$$

each of them as a linear function of these.

I recall that we have $(P, Q)^0 = PQ$, so that the first and the last terms of the series might have been written $(P, (Q, R)^4)^0$ and $(P, (Q, R)^0)^4$ respectively; and, further, that $(P, Q)^1$ denotes $d_xP \,.\, d_yQ - d_yP \,.\, d_xQ$; $(P, Q)^2$ denotes

$$d_x^2P \,.\, d_y^2Q - 2d_xd_yP \,.\, d_xd_yQ + d_y^2P \,.\, d_x^2Q;$$

and so on.

I write (a, b, c, d, e) for the fourth derived functions of any quantic $U, = (*\between x, y)^m$; we have, in a notation which will be at once understood,

$$\begin{aligned}
U &= (a, b, c, d, e\between x, y)^4 \div [m]^4 ,\\
(d_x, d_y)\, U &= (a, b, c, d), (b, c, d, e)(x, y)^3 \div [m-1]^3,\\
(d_x, d_y)^2\, U &= (a, b, c), (b, c, d), (c, d, e)(x, y)^2 \div [m-2]^2,\\
(d_x, d_y)^3\, U &= (a, b), (b, c), (c, d), (d, e)(x, y)^1 \div [m-3]^1,\\
(d_x, d_y)^4\, U &= (a, b, c, d, e);
\end{aligned}$$

and then, taking

$$(a_1, b_1, c_1, d_1, e_1),\quad (a_2, b_2, c_2, d_2, e_2),\quad (a_3, b_3, c_3, d_3, e_3),$$

to belong to P, Q, R, respectively, we must, instead of m, write p, q, r for the three functions respectively.

If we attend only to the highest terms in x, we have

$$\begin{aligned}
U &= ax^4 \div [m]^4 ,\\
(d_x, d_y)\, U &= (a, b)\, x^3 \div [m-1]^3,\\
(d_x, d_y)^2\, U &= (a, b, c)\, x^2 \div [m-2]^2,\\
(d_x, d_y)^3\, U &= (a, b, c, d)\, x \div [m-3]^1,\\
(d_x, d_y)^4\, U &= (a, b, c, d, e).
\end{aligned}$$

Consider now $P(Q, R)^4$, $(P, (Q, R)^3)^1$, &c.; in each case attending only to the term in a_1, and in this term to the highest term in x, we have

$$(1)\quad [p]^4 P(Q, R)^4 = a_2e_3 - 4b_2d_3 + 6c_2c_3 - 4d_2b_3 + e_2a_3 \quad (X),$$

$$\begin{aligned}
(2)\quad [p-1]^3[q-3]^1[r-3]^1(P, (Q, R)^3)^1 = \; & [q-3]^1 \,.\, b_2d_3 - 3c_2c_3 + 3d_2b_3 - e_2a_3 \; (-Y'),\\
+ \; & [r-3]^1 \,.\, a_2e_3 - 3b_2d_3 + 3c_2c_3 - d_2b_3 \; (Y),
\end{aligned}$$

$$\begin{aligned}
(3)\quad [p-2]^2[q-2]^2[r-2]^2(P, (Q, R)^2)^2 = \; & [q-2]^2 \,.\, c_2c_3 - 2d_2b_3 + e_2a_3 \; (Z''),\\
+ \; & 2[q-2]^1[r-2]^1 \,.\, b_2d_3 - 2c_2c_3 + d_2b_3 \; (-Z'),\\
+ \; & [r-2]^2 \,.\, a_2e_3 - 2b_2d_3 + c_2c_3 \; (Z),
\end{aligned}$$

$$\begin{aligned}
(4)\quad [p-3]^1[q-1]^3[r-1]^3(P, (Q, R)^1)^3 = \; & [q-1]^3 \,.\, d_2b_3 - e_2a_3 \quad (-W'''),\\
+ \; & 3[q-1]^2[r-1]^1 \,.\, c_2c_3 - d_2b_3 \quad (W''),\\
+ \; & 3[q-1]^1[r-1]^2 \,.\, b_2d_3 - c_2c_3 \quad (-W'),\\
+ \; & [r-1]^3 \,.\, a_2e_3 - b_2d_3 \quad (W),
\end{aligned}$$

$$(5)\qquad [p-4]^0[q]^4[r]^4(P,\ QR)^4 = \begin{array}{lr} [q]^4 \ .\ e_2a_3 & (U''''), \\ +4[q]^3[r]^1 .\ d_2b_3 & (-U'''), \\ +6[q]^2[r]^2 .\ c_2c_3 & (U''), \\ +4[q]^1[r]^3 .\ b_2d_3 & (-U'), \\ + \quad [r]^4 .\ a_2e_3 & (U). \end{array}$$

Thus, for the second of these equations,

$$(P,\ (Q,\ R)^3)^1 = d_xP \,.\, d_y(Q,\ R)^3 - \&c.;$$

the term in a_1 is $d_y(Q,\ R)^3, = (d_xQ,\ R)^3 + (Q,\ d_yR)^3$, the whole being divided by $[p-1]^3$; where attending only to the highest terms in x, the two terms are respectively

$$(b_2d_3 - 3c_2c_3 + 3d_2b_3 - e_2a_3) \div [r-3]^1,$$

and

$$(a_2e_3 - 3b_2d_3 + 3c_2c_3 - d_2b_3) \div [q-3]^1,$$

which are each divided by $[p-1]^3$ as above; whence, multiplying by

$$[p-1]^3[q-1]^2[r-1]^1,$$

we have the formula in question; and so for the other cases.

Writing now (1), (2), (3), (4), (5) for the left-hand sides of the five equations respectively; and

$$\begin{array}{rrrrr} &&&& X: \\ &&& -Y', & Y: \\ && Z'', & Z', & Z: \\ & -W''', & W'', & -W', & W: \\ U'''', & -U''', & U'', & -U', & U: \end{array}$$

for the literal parts on the right-hand sides of the same equations respectively; then we have

$$\begin{aligned} &X = Y + Y', \\ &Y = Z + Z', \quad Y' = Z' + Z'', \\ &\&c., \end{aligned}$$

and the equations become

$$\begin{aligned} (1) &= X, \\ (2) &= [r-1]^1\, Y - 1\, [q-3]^1\, Y', \\ (3) &= [r-2]^2\, Z - 2[r-2]^1[q-2]^1\, Z' + 1\, [q-2]^2\, Z'', \\ (4) &= [r-1]^3\, W - 3[r-1]^2[q-1]^1\, W' + 3[r-1]^1[q-1]^2\, W'' - 1[q-1]^3\, W''', \\ (5) &= [r]^4\, U - 4\, [r]^3\, [q]^1\, U' + 6\, [r]^2\, [q]^2\, U'' - 4[r]^1[q]^3\, U''' + [q]^4\, U'''', \end{aligned}$$

which are, in fact, the equations considered at the beginning of the present paper, putting therein $\alpha = r-3$ and $\beta = q-3$, they consequently give

$$\begin{array}{lllllll} & & \{q+r-6,\ 456\}(1), & \{q+r-6,\ 156\}(2), & \{q+r-6,\ 036\}(3), & \{q+r-6,\ 015\}(4), & \{q+r-6,\ 012\}(5), \\ \hline +r-6,\ 01\ldots6\}U & = & \{q-3,\ 0123\}, & +4\{q-3,\ 123\}, & +6\{q-3,\ 23\}, & +4\{q-3,\ 3\}, & +1, \\ \text{,,} & \}U'''' = & \{r-3,\ 0123\}, & -4\{r-3,\ 123\}, & +6\{r-3,\ 23\}, & -4\{r-3,\ 3\}, & +1. \end{array}$$

Also, attending as before only to the terms in a, and therein to the highest power of x, we have

$$Q(R,\ P)^4 = a_2 e_3 \div [q]^4,$$
$$R(P,\ Q)^4 = a_3 e_2 \div [r]^4;$$

that is,

$$[q]^4\, Q(R,\ P)^4 = U,\quad [r]^4\, R(P,\ Q)^4 = U'''';$$

and, observing that $\{q+r-6,\ 01\ldots 6\}$ is $=[q+r]^7$, and that $\{q+r-6,\ 456\}$, &c., may be written $\{q-r,\ \bar{2}\bar{1}0\}$, &c., where the superscript bars are the signs $-$, the formulæ become

$$\begin{array}{llllll} & \{q+r,\ \bar{2}\bar{1}0\}(1), & \{q+r,\ \bar{5}\bar{1}0(2), & \{q+r,\ \bar{6}\bar{3}0\}(3), & \{q+r,\ \bar{6}\bar{5}\bar{1}\}(4), & \{q+r,\ \bar{6}\bar{5}\bar{4}\}(5), \\ \hline [q+r]^7[q]^4Q(P,R)^4 = & [q]^4, & +4[q]^3, & +6[q]^2, & +4[q]^1, & +1, \\ [q+r]^7[r]^4R(P,Q)^4 = & [r]^4, & -4[r]^3, & +6[r]^2, & -4[r]^1, & +1. \end{array}$$

Written at full length, the first of these equations (which, as being the fourth in a series, I mark 4th equation) is

$$\begin{array}{lllll} q+r]^7[q]^4Q(P,R)^4 = & 1.q+r & .q+r-1.q+r-2. & [p]^4\ [q]^4 & .P,(Q,R)^4 \quad \text{(4th equat.} \\ & +4.q+r & .q+r-1.q+r-5. & [p-1]^3[q]^3[q-3]^1[r-1]^1 & .(P,(Q,R)^3)^1 \\ & +6.q+r & .q+r-3.q+r-6. & [p-2]^2[q]^2[q-2]^2[r-2]^2 & .(P,(Q,R)^2)^2 \\ & +4.q+r-1 & .q+r-5.q+r-6. & [p-3]^1[q]^1[q-1]^3[r-1]^3 & .(P,(Q,R)^1)^3 \\ & +1.q+r-1 & .q+r-5.q+r-6. & [q]^4\ \ [r]^4 & .P,(Q,R)^4, \end{array}$$

and the other is, in fact, the same equation with q, Q, r, R interchanged with r, R, q, Q; the alternate $+$ and $-$ signs arise evidently from the terms

$$(R,\ Q)^4,\ = (Q,\ R)^4;\ (R,\ Q)^3,\ = -(Q,\ R)^3;\ \&c.,$$

which present themselves on the right-hand side.

It will be observed that the identity has been derived from the comparison of the terms in a, which are the highest terms in x, the other terms not having been written down or considered; but it is easy to see that an identity of the form in question exists, and, this being admitted, the process is a legitimate one.

The preceding equations of the series are

$$\begin{array}{llll} [q+r]^1[q]^1Q(P,R)^1 = & 1.\ [p]^1\ [q]^1 & P(Q,\ R)^1 & \text{(1st equation)} \\ & +1.\ \quad [q]^1\ \ [r]^1 & (P,\ QR)^1; & \\ [q+r]^3[q]^2Q(P,R)^2 = & 1.q+r\ .\ [p]^2\ [q]^2 & P,(Q,\ R)^2 & \text{(2nd equation)} \\ & +2.q+r-1.[p-1]^1[q]^1[q-1]^1[r-1]^1 & (P,(Q,R)^1)^1 & \\ & +1.q+r-2.\ \quad [q]^2\ \ [r]^2 & (P,\ QR)^2; & \\ [q+r]^5[q]^3Q(P,R)^2 = & 1.q+r\ .q+r-1.\ [p]^3\ [q]^3 & P,(Q,\ R)^3 & \text{(3rd equation)} \\ & +3.q+r\ .q+r-3.[p-1]^2[q]^2[q-2]^1[r-2]^1 & (P,(Q,R)^2)^1 & \\ & +3.q+r-1.q+r-4.[p-2]^1[q]^1[q-1]^2[r-1]^2 & (P,\ QR)^3 & \\ & +1.q+r-3.q+r-4.\ \quad [q]^3\ \ [r]^3 & (P,\ QR)^3. & \end{array}$$

From these four equations the law is evident, except as to the numbers subtracted from $q+r$. These are obtained, as explained above, in regard to the numbers added to $\alpha+\beta$ in the { } symbols; transforming the diagrams so as to be directly applicable to the case now in question, they become

$\underline{0}$	0	$\underline{01}$	01	$\underline{012}$	012	$\underline{0123}$	0123
1	1	2	03	3	015	4	0127
1	2	11	14	21	036	31	0158
		2	34	12	156	22	0378
				3	456	13	1678
						4	5678,

showing how the numbers are obtained for the equations 2, 3, 4, 5 respectively. The first equation is

$$(q^2+qr)\,Q\,(P,\,R)=pq\,P\,(Q,\,R)+qr\,[Q\,(P,\,R)+R\,(P,\,Q)],$$

viz. this is

$$\begin{aligned}0=pq\,P\,(Q,\,R)-qr\,Q\,(RP)+qr\,R\,(P,\,Q)\\ +(q^2+qr)\,Q\,(R,\,P);\end{aligned}$$

or, dividing by q, this is

$$0=pP\,(Q,\,R)+qQ\,(R,\,P)+rR\,(P,\,Q),$$

which is a well-known identity.

We may verify any of the equations, though the process is rather laborious, for the particular values

$$P=x^{\frac{1}{2}(p+\alpha)}\,y^{\frac{1}{2}(p-\alpha)},\quad Q=x^{\frac{1}{2}(q+\beta)}\,y^{\frac{1}{2}(q-\beta)},\quad R=x^{\frac{1}{2}(r+\gamma)}\,y^{\frac{1}{2}(r-\gamma)};$$

thus, taking the second equation, we have, omitting common factors,

$$\begin{aligned}(Q,\,R)^2= &\quad q+\beta\,.\,q+\beta-2\,.\,r-\gamma\,.\,r-\gamma-2\\ &-2\quad .\,q+\beta\quad .\,q-\beta\,.\,r+\gamma\,.\,r-\gamma\\ &+\,.\,q-\beta\,.\,q-\beta-2\,.\,r+\gamma\,.\,r+\gamma-2\\ =&\quad \beta^2\,(r^2-r)+\gamma^2\,(q^2-q)-2\beta\gamma\,(q-1)\,(r-1)-qr\,(q+r-2),\end{aligned}$$

$$\begin{aligned}(P,\,(Q,\,R)^1)^1&=(q+\beta\,.\,r-\gamma\,.-.\,q-\beta\,.\,r+\gamma)\,(p+\alpha\,.\,q+r-\beta-\gamma-2\,.-.\,p-\alpha\,.\,q+r+\beta+\gamma-2)\\ &=(\beta r-q\gamma)\,(\alpha\,.\,q+r-2\,.-p\,.\,\beta+\gamma)\\ &=\alpha\beta r\,(r+q-2)-\alpha\gamma q\,(q+r-2)-pr\beta^2+p\,(q-r)\,\beta\gamma+pq\gamma^2,\end{aligned}$$

and from the first of these the expressions of $Q\,(P,\,R)^2$ and $(P,\,QR)^2$ are at once obtained. The identity to be verified then becomes

$$\begin{aligned}&[q+r]^3\,[q]^2\,\{\alpha^2\,(r^2-r)+\gamma^2\,(p^2-p)-2\alpha\gamma\,(p-1)\,(r-1)-pr\,(p+r-2)\}\\ &\quad=\ (q+r)\,[q]^2\,[p]^2\,\{\beta^2\,(r^2-r)+\gamma^2\,(q^2-q)-2\beta\gamma\,(q-1)\,(r-1)-qr\,(q+r-2)\}\\ &\quad\ +2\,(q+r-1)\,[q]^2\,(p-1)\,(r-1)\,\{\alpha\beta r\,(q+r-2)-\alpha\gamma q\,(q+r-2)\\ &\qquad\qquad -pr\beta^2+p\,(q-r)\,\beta\gamma+pq\gamma^2\}\\ &\quad\ +(q+r-2)\,[q]^2\,[r]^2\,\{\alpha^2\,(q+r)\,(q+r-1)+(\beta+\gamma)^2\,(p^2-p)\\ &\qquad\qquad -2\alpha\,(\beta+\gamma)\,(p-1)\,(q+r-1)-p\,(q+r)\,(p+q+r-2)\},\end{aligned}$$

which is easily verified, term by term; for instance, the terms with α, β, or γ, give

$$[q+r]^3 [q]^2 pr (p+r-2) = (q+r) [q]^2 [p]^2 qr (q+r-2) + (q+r-2) [q]^2 [r]^2 p (q+r) (p+q+r-2),$$

which, omitting the factor $(q+r)(q+r-2)[q]^2 pr$, is

$$(q+r-1)(p+r-2) = (p-1) q + (r-1)(p+q+r+2);$$

viz. the right-hand side is

$$(p-1) q + (r-1) q + (r-1)(p+r-2), \quad = (q+r-1)(p+r-2),$$

as it should be.

The equations are useful for the demonstration of a subsidiary theorem employed in Gordan's demonstration of the finite number of the covariants of any binary form U. Suppose that a system of covariants (including the quantic itself) is

$$P, Q, R, S, \ldots;$$

this may be the complete system of covariants; and if it is so, then, T and V being any functions of the form $P^\alpha Q^\beta R^\gamma \ldots$, every derivative $(T, V)^\theta$ must be a term or sum of terms of the like form $P^\alpha Q^\beta R^\gamma \ldots$; the subsidiary theorem is that in order to prove that the case is so, it is sufficient to prove that every derivative $(P, Q)^\theta$, where P and Q are any two terms of the proposed system, is a term or sum of terms of the form in question $P^\alpha Q^\beta R^\gamma \ldots$.

In fact, supposing it shown that every derivative $(T, V)^\theta$ up to a given value θ_0 of θ is of the form $P^\alpha Q^\beta R^\gamma \ldots$, we can by successive application of the equation for $Q(P, R)^{\theta+1}$, regarded as an equation for the reduction of the last term on the right-hand side $(P, QR)^{\theta+1}$, bring first $(P, QR)^{\theta+1}$, and then $(P, QRS)^{\theta+1}, \ldots$, and so ultimately any function $(P, V)^{\theta+1}$, and then again any functions $(PQ, V)^{\theta+1}$, $(PQR, V)^{\theta+1}, \ldots$, and so ultimately any function $(T, V)^{\theta+1}$, into the required form $P^\alpha Q^\beta R^\gamma \ldots$: or the theorem, being true for θ, will be true for $\theta+1$; whence it is true generally.

682.

FORMULÆ RELATING TO THE RIGHT LINE.

[From the *Quarterly Journal of Pure and Applied Mathematics*, vol. XV. (1878), pp. 169—171.]

1. LET λ, μ, ν be the direction-angles of a line; α, β, γ the coordinates of a point on the line; and write

$$a = \cos\lambda,\quad f = \beta\cos\nu - \gamma\cos\mu,$$
$$b = \cos\mu,\quad g = \gamma\cos\lambda - \alpha\cos\nu,$$
$$c = \cos\nu,\quad h = \alpha\cos\mu - \beta\cos\lambda,$$

whence

$$a^2 + b^2 + c^2 = 1,$$
$$af + bg + ch = 0,$$

or the six quantities (a, b, c, f, g, h), termed the coordinates of the line, depend upon four arbitrary parameters.

2. It is at once shown that the condition for the intersection of any two lines (a, b, c, f, g, h), (a', b', c', f', g', h'), is $af' + bg' + ch' + a'f + b'g + c'h = 0$.

3. Given two lines (a, b, c, f, g, h), (a', b', c', f', g', h'), it is required to find their shortest distance, and the coordinates of their line of shortest distance.

Let

$$Ax + By + Cz + D = 0,$$
$$Ax + By + Cz + D' = 0,$$

be parallel planes containing the two lines respectively; then the first plane contains the point $\alpha + r\cos\lambda$, $\beta + r\cos\mu$, $\gamma + r\cos\nu$, and the second contains the point $\alpha' + r'\cos\lambda'$, $\beta' + r'\cos\mu'$, $\gamma' + r'\cos\nu'$; that is, we have

$$A\alpha + B\beta + C\gamma + D = 0,$$
$$A\alpha' + B\beta' + C\gamma' + D' = 0,$$
$$A\cos\lambda + B\cos\mu + C\cos\nu = 0,$$
$$A\cos\lambda' + B\cos\mu' + C\cos\nu' = 0,$$

which last equations may be written

$$Aa + Bb + Cc = 0,$$
$$Aa' + Bb' + Cc' = 0,$$

giving

$$A : B : C = bc' - b'c : ca' - c'a : ab' - a'b,$$

or, if we write

$$\theta = aa' + bb' + cc',$$

and assume, as is convenient,

$$A^2 + B^2 + C^2 = 1,$$

then

$$A,\ B,\ C = \frac{bc' - b'c}{\sqrt{(1-\theta^2)}},\ \frac{ca' - c'a}{\sqrt{(1-\theta^2)}},\ \frac{ab' - a'b}{\sqrt{(1-\theta^2)}},$$

where θ, = cosine-inclination, $= aa' + bb' + cc'$.

Hence, shortest distance $= D - D'$

$$= A(\alpha - \alpha') + B(\beta - \beta') + C(\gamma - \gamma')$$
$$= \frac{1}{\sqrt{(1-\theta^2)}}\{(bc' - b'c)(\alpha - \alpha') + (ca' - c'a)(\beta - \beta') + (ab' - a'b)\}$$
$$= \frac{1}{\sqrt{(1-\theta^2)}}\{a'(c\beta - b\gamma) + b'(a\gamma - c\alpha) + c'(b\alpha - a\beta)$$
$$+ a(c'\beta' - b'\gamma') + b(a'\gamma' - c'\alpha') + c(b'\alpha' - a'\beta')\}$$
$$= \frac{1}{\sqrt{(1-\theta^2)}}(af' + bg' + ch' + a'f + b'g + c'h),\ = \delta \text{ suppose.}$$

The six coordinates of the line of shortest distance are A, B, C, F, G, H, where A, B, C denote as before, and F, G, H are to be determined.

Since the line meets each of the given lines, we have

$$Af + Bg + Ch + Fa + Gb + Hc = 0,$$
$$Af' + Bg' + Ch' + Fa' + Gb' + Hc' = 0,$$

and we have also

$$FA + GB + HC = 0,$$

which equations give F, G, H. Multiplying the first equation by $b'C - c'B$, the second by $Bc - Cb$, and the third by $bc' - b'c$, we find

$$(b'C - c'B)(Af + Bg + Ch) + (Bc - Cb)(Af' + Bg' + Ch') + F\begin{vmatrix} a, & b, & c \\ a', & b', & c' \\ A, & B, & C \end{vmatrix} = 0.$$

Here

$$b'C - c'B = \frac{1}{\sqrt{(1-\theta^2)}}\{b'(ab' - a'b) - c'(ca' - c'a)\}$$
$$= \frac{1}{\sqrt{(1-\theta^2)}}\{a(a'^2 + b'^2 + c'^2) - a'(aa' + bb' + cc')\}$$
$$= \frac{1}{\sqrt{(1-\theta^2)}}(a - a'\theta),$$

and similarly

$$cB - bC = \frac{1}{\sqrt{(1-\theta^2)}}(a' - a\theta).$$

Also, putting for shortness

$$\Omega = \begin{vmatrix} a, & b, & c \\ a', & b', & c' \\ f, & g, & h \end{vmatrix}, \quad \Omega' = \begin{vmatrix} a, & b, & c \\ a', & b', & c' \\ f', & g', & h' \end{vmatrix},$$

we have

$$Af + Bg + Ch = \frac{1}{\sqrt{(1-\theta^2)}}\Omega, \quad Af' + Bg' + Ch' = \frac{1}{\sqrt{(1-\theta^2)}}\Omega',$$

and finally, the determinant which multiplies F is

$$\frac{1}{\sqrt{(1-\theta^2)}}\{(bc' - b'c)^2 + (ca' - c'a)^2 + (ab' - a'b)^2\} = \frac{1}{\sqrt{(1-\theta^2)}}(1-\theta^2), = \sqrt{(1-\theta^2)}.$$

We have thus the value of F; forming in the same way those of G and H, we find

$$F = \frac{-1}{(1-\theta^2)^{\frac{3}{2}}}\{(a - a'\theta)\,\Omega + (a' - a\theta)\,\Omega'\},$$

$$G = \frac{-1}{(1-\theta^2)^{\frac{3}{2}}}\{(b - b'\theta)\,\Omega + (b' - b\theta)\,\Omega'\},$$

$$H = \frac{-1}{(1-\theta^2)^{\frac{3}{2}}}\{(c - c'\theta)\,\Omega + (c' - c\theta)\,\Omega'\},$$

which, with the foregoing equations for A, B, C, give the six coordinates of the line of shortest distance.

683.

ON THE FUNCTION arc sin $(x+iy)$.

[From the *Quarterly Journal of Pure and Applied Mathematics*, vol. XV. (1878), pp. 171—174.]

THE determination of the function in question, the arc to a given imaginary sine, is considered in Cauchy's *Exercises d'Analyse, &c.*, t. III. (1844), p. 382; but it appears, by two hydrodynamical papers by Mr Ferrers and Mr Lamb, *Quarterly Mathematical Journal*, t. XIII. (1874), p. 115, and t. XIV. (1875), p. 40, that the question is connected with the theory of confocal conics.

Taking $c=\sqrt{(a^2-b^2)}$ a positive real quantity which may ultimately be put $=1$, the question is to find the real quantities ξ, η, such that

$$\xi+i\eta=\text{arc sin}\,\frac{1}{c}(x+iy),$$

or say

$$x+iy=c\sin(\xi+i\eta),$$

so that

$$x=c\sin\xi\cos i\eta,\quad iy=c\cos\xi\sin i\eta.$$

It is convenient to remark that if a value of $\xi+i\eta$ be $\xi'+i\eta'$, then the general value is $2m\pi+\xi'+i\eta'$ or $(2m+1)\pi-(\xi'+i\eta')$; hence, η may be made positive or negative at pleasure; $\cos i\eta$ is in each case positive, but $\frac{1}{i}\sin i\eta$ has the same sign as η; hence $\cos\xi$ has the same sign as x, but $\sin\xi$ has the same sign as y or the reverse sign, according as η is positive or negative; for any given values of x and y, we obtain, as will appear, determinate positive values of $\sin^2\xi$ and $\cos^2\xi$; and the square roots of these must therefore be taken so as to give to $\sin\xi$, $\cos\xi$ their proper signs respectively.

Suppose that λ, μ are the elliptic coordinates of the point (x, y); viz. that we have

$$\frac{x^2}{a^2+\lambda}+\frac{y^2}{b^2+\lambda}=1,$$

$$\frac{x^2}{a^2+\mu}+\frac{y^2}{b^2+\mu}=1,$$

where $a^2+\lambda$, $b^2+\lambda$, and $a^2+\mu$ are positive, but $b^2+\mu$ is negative. Calling ρ, σ the distances of the point x, y from the points $(c, 0)$ and $(-c, 0)$, that is, assuming

$$\rho=\sqrt{\{(x-c)^2+y^2\}},$$
$$\sigma=\sqrt{\{(x+c)^2+y^2\}},$$

then we have

$$\sqrt{(a^2+\lambda)}=\tfrac{1}{2}(\sigma+\rho), \text{ whence also } \sqrt{(b^2+\lambda)}=\tfrac{1}{2}\sqrt{\{(\sigma+\rho)^2-4c^2\}},$$
$$\sqrt{(a^2+\mu)}=\tfrac{1}{2}(\sigma\sim\rho), \quad\quad ,, \quad\quad \sqrt{(b^2+\mu)}=\tfrac{1}{2}\sqrt{\{(\sigma\sim\rho)^2-4c^2\}},$$

which equations determine λ, μ as functions of x, y.

Now we have

$$\rho\sigma=\sqrt{\{(x^2+y^2-c^2)^2-4c^2x^2\}}=\sqrt{\{(x^2-y^2-c^2)^2+4x^2y^2\}},$$
$$\rho^2+\sigma^2=2(x^2+y^2+c^2);$$

substituting herein for x, y their values

$$c\sin\xi\cos i\eta,\ -ci\cos\xi\sin i\eta,$$

we find

$$x^2-y^2-c^2=c^2\{\sin^2\xi\cos^2 i\eta+\cos^2\xi\sin^2 i\eta-(\sin^2\xi+\cos^2\xi)(\sin^2 i\eta+\cos^2 i\eta)\}$$
$$=-c^2(\sin^2\xi\sin^2 i\eta+\cos^2\xi\cos^2 i\eta),$$

whence

$$(x^2-y^2-c^2)^2= \quad c^4(\cos^2\xi\cos^2 i\eta+\sin^2\xi\sin^2 i\eta)^2$$
$$+4x^2y^2 \quad\quad -4c^4\sin^2\xi\cos^2\xi\sin^2 i\eta\cos^2 i\eta$$
$$= \quad c^4(\cos^2\xi\cos^2 i\eta-\sin^2\xi\sin^2 i\eta)^2.$$

Hence

$$2\rho\sigma=2c^2(\cos^2\xi\cos^2 i\eta-\sin^2\xi\sin^2 i\eta),$$

and

$$\rho^2+\sigma^2=2c^2(\sin^2\xi\cos^2 i\eta-\cos^2\xi\sin^2 i\eta+1);$$

hence

$$(\rho+\sigma)^2=2c^2(\cos^2 i\eta-\sin^2 i\eta+1),\ =4c^2\cos^2 i\eta,$$
$$(\rho-\sigma)^2=2c^2(\sin^2\xi-\cos^2\xi+1),\ =4c^2\sin^2\xi.$$

Consequently

$$a^2+\lambda=c^2\cos^2 i\eta, \text{ and thence } b^2+\lambda=-c^2\sin^2 i\eta,$$
$$a^2+\mu=c^2\sin^2\xi, \quad\quad ,, \quad\quad b^2+\mu=-c^2\cos^2\xi,$$

values which verify as they should do the equations

$$\frac{x^2}{a^2+\lambda}+\frac{y^2}{b^2+\lambda}=1,$$

$$\frac{x^2}{a^2+\mu}+\frac{y^2}{b^2+\mu}=1,$$

viz. these become

$$\frac{x^2}{c^2\cos^2 i\eta}+\frac{y^2}{-c^2\sin^2 i\eta}=\sin^2\xi+\cos^2\xi,\ =1,$$

$$\frac{x^2}{c^2\sin^2\xi}+\frac{y^2}{-c^2\cos^2\xi}=\cos^2 i\eta+\sin^2 i\eta,\ =1.$$

The same equations, or as we may also write them,

$$\lambda=-a^2\sin^2 i\eta-b^2\cos^2 i\eta,$$

$$\mu=-a^2\cos^2\xi-b^2\sin^2\xi,$$

determine η as a function of λ, and ξ as a function of μ; λ, μ being by what precedes, given functions of x, y.

Or more simply, starting from the last-mentioned values of λ, μ, and substituting these in the expressions

$$x^2=\frac{a^2+\lambda\,.\,a^2+\mu}{a^2-b^2},\quad y^2=\frac{b^2+\lambda\,.\,b^2+\mu}{b^2-a^2},$$

we find

$$x^2=c^2\sin^2\xi\cos^2 i\eta,\quad y^2=-c^2\cos^2\xi\sin^2 i\eta,$$

or say

$$x=c\sin\xi\cos i\eta,\quad iy=c\cos\xi\sin i\eta,$$

whence

$$x+iy=c\sin(\xi+i\eta),$$

the original relation between x, y and ξ, η.

684.

ON A RELATION BETWEEN CERTAIN PRODUCTS OF DIFFERENCES.

[From the *Quarterly Journal of Pure and Applied Mathematics*, vol. XV. (1878), pp. 174, 175.]

CONSIDER the function

$$3\left\{\begin{array}{l} abc\,.\,de \\ +\,bcd\,.\,ea \\ +\,cde\,.\,ab \\ +\,dea\,.\,bc \\ +\,eab\,.\,cd \end{array}\right\} - \left\{\begin{array}{l} abd\,.\,ce \\ +\,bce\,.\,da \\ +\,cda\,.\,eb \\ +\,deb\,.\,ac \\ +\,eac\,.\,bd \end{array}\right\},$$

where

$$\begin{aligned} abc &= (a-b)(b-c)(c-a), \\ ab &= (a-b)(b-a), = -(a-b)^2, \\ &\text{\&c.}; \end{aligned}$$

therefore

$$\begin{aligned} abc &= bca = cab = -bac, \text{ \&c.}; \\ ab &= ba. \end{aligned}$$

It is to be shown that the function vanishes if $e=d$. Writing $e=d$, the value is

$$\begin{aligned} 3\,(bcd\,.\,da + dab\,.\,cd) &- abd\,.\,cd \\ &- bcd\,.\,da \\ &- cda\,.\,db \\ &- dac\,.\,bd, \end{aligned}$$

viz. this is

$$\begin{aligned} &\quad 3\ bcd\,.\,ad - \ abd\,.\,cd \\ &+3\ abd\,.\,cd - \ bcd\,.\,ad \\ &\qquad\qquad - 2acd\,.\,bd \\ &= 2\ bcd\,.\,ad - 2acd\,.\,bd + 2abd\,.\,cd \\ &= 2\,(bcd\,.\,ad + \ cad\,.\,bd + \ abd\,.\,cd), \end{aligned}$$

which is easily seen to vanish; the value is

$$\begin{aligned} &(b-c)(c-d)(d-b)(a-d)^2 = -(b-c)(a-d)^2(b-d)(c-d) \\ &+(c-a)(a-d)(d-c)(b-d)^2 \quad -(c-a)(a-d)(b-d)^2(c-d) \\ &+(a-b)(b-d)(d-a)(c-d)^2 \quad -(a-b)(a-d)(b-d)(c-d)^2: \end{aligned}$$

viz. omitting the factor $(a-d)(b-d)(c-d)$, this is

$$\begin{aligned} &= -(b-c)(a-d) \\ &\quad -(c-a)(b-d) \\ &\quad -(a-b)(c-d), \end{aligned}$$

which vanishes. Hence the function also vanishes if $e=a$, or $a=b$ or $b=c$, or $c=d$; and it is thus a mere numerical multiple of $(a-b)(b-c)(c-d)(d-e)(e-a)$, or say it is $=Mabcde$.

To find M write $e=c$, the equation becomes

$$\begin{aligned} &3abc\,.\,dc - cda\,.\,cb = Mabcdc, \ = Mabc\,.\,dc, \\ &+3bcd\,.\,ca - ac \\ &+3dca\,.\,bc \\ &+3cab\,.\,cd, \end{aligned}$$

viz. this is

$$6abc\,.\,dc + 4dbc\,.\,ac + 4adc\,.\,bc = M\,.\,abc\,.\,dc,$$

giving $M=10$. In fact, we then have

$$-4abc\,.\,dc + 4dbc\,.\,ac + 4adc\,.\,bc = 0,$$

that is,

$$-\ abc\,.\,dc - \ bdc\,.\,ac - \ dac\,.\,bc = 0,$$

which is right. And we have thus the identity

$$3\begin{Bmatrix} abc\,.\,de \\ +bcd\,.\,ea \\ +cde\,.\,ab \\ +dea\,.\,bc \\ +eab\,.\,cd \end{Bmatrix} - \begin{Bmatrix} abd\,.\,ce \\ +bce\,.\,da \\ +cda\,.\,eb \\ +deb\,.\,ac \\ +eac\,.\,bd \end{Bmatrix} = 10\,.\,abcde,$$

or say

$$3\,[abcde] - [acebd] = 10\,\{abcde\}.$$

685.

ON MR COTTERILL'S GONIOMETRICAL PROBLEM.

[From the *Quarterly Journal of Pure and Applied Mathematics*, vol. XV. (1878), pp. 196—198.]

THE very remarkable formulæ contained in Mr Cotterill's paper, "A goniometrical problem, to be solved analytically in one move, or more simply synthetically in two moves," *Quarterly Mathematical Journal*, t. VII. (1866), pp. 259—272, are presented in a form which, to say the least, is not as easily intelligible as might be; and they have not, I think, attracted the attention which they well deserve.

Using his notation, except that I write for angles small roman letters, in order to be able to have the corresponding italic small letters and capitals for the sines and cosines respectively of the same angles, we consider nine angles

$$\begin{matrix} \mathrm{a}, & \mathrm{b}, & \mathrm{c}, \\ \mathrm{d}, & \mathrm{e}, & \mathrm{f}, \\ \mathrm{x}, & \mathrm{y}, & \mathrm{z}, \end{matrix}$$

which are such that the sum of three angles in the same line, or in the same column, is an odd multiple of π. Of course, any four angles such as a, b, d, e are

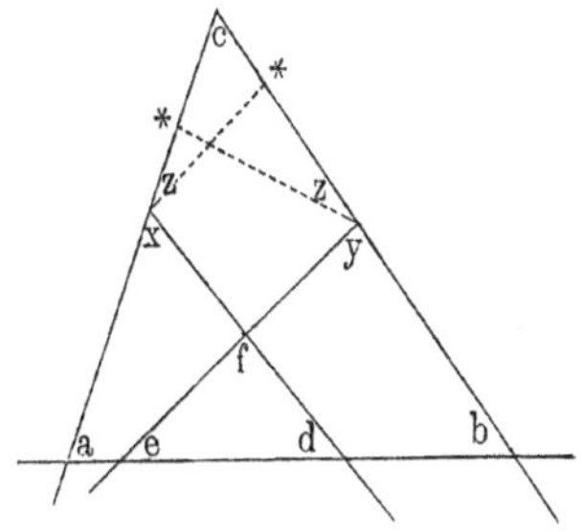

arbitrary, and each of the remaining angles is then determinate save as to an even multiple of π. And it may be remarked that these angles a, b, d, e may represent the inclinations of any four lines to a fifth line, and that the remaining angles are then at once obtained, as in the figure. The small roman letters are here used to denote as well angles as points, being so placed as to show what the angles are which they respectively denote; the points *, * are constructed as the intersections of the lines ac, bc by the circle circumscribed about fxy, and the angle z is the angle which the points *, * subtend at x or y. It will be observed that the sum of the three angles in a line or column is in each case $=\pi$.

But this in passing: the analytical theorem is, *first*, we can form with the sines and cosines of the angles in any two lines or columns a function S presenting itself under two distinct forms, which are in fact equal in value, or say S is a symmetrical function of the two lines or columns, viz. for the first and second lines this is

$$S\begin{pmatrix}\mathrm{a}, & \mathrm{b}, & \mathrm{c}\\ \mathrm{d}, & \mathrm{e}, & \mathrm{f}\end{pmatrix} = d^2 Abc + e^2 Bca + f^2 Cab$$

$$= a^2 Def + b^2 Efd + c^2 Fde,$$

where, as already mentioned, a, A denote sin a, cos a, and so for the other letters.

Secondly, if to the S of any two lines or columns we add twice the product of the six sines, we obtain a sum M which has the same value from whichever two lines or columns we obtain it; or, say M is a symmetrical function of the matrix of the nine angles. Thus

$$M = S\begin{pmatrix}\mathrm{a}, & \mathrm{b}, & \mathrm{c}\\ \mathrm{d}, & \mathrm{e}, & \mathrm{f}\end{pmatrix} + 2abcdef,$$

which is one of a system of six forms each of which (on account of the two forms of the S contained in it) may be regarded as a double form, and the twelve values are all of them equal. There are, moreover, 15 other forms, of M, viz. 3 line-forms, such as

$$bcdx + caey + abfz \text{ (belongs to line a, b, c)},$$

3 column-forms, such as

$$dxbc + xaef + adyz \text{ (belongs to column a, d, x)},$$

and 9 term-forms, such as

$$e^2z^2 + f^2y^2 + 2efyzA \text{ (belongs to term a)},$$

and the $12 + 15$, $= 27$ values are all equal.

The several identities can of course be verified by means of the relations between the nine angles, or rather the derived sine- and cosine-relations

$$C = ab - AB,$$

$$c = aB + bA, \text{ \&c.}$$

Thus, as regards the two forms of $S\begin{pmatrix}\mathrm{a}, & \mathrm{b}, & \mathrm{c}\\ \mathrm{d}, & \mathrm{e}, & \mathrm{f}\end{pmatrix}$, the identity to be verified may be written

$$c\,(d^2Ab + e^2Ba - cFde) = f\,(a^2De + b^2Ed - fCab).$$

Proceeding to reduce the factor $a^2De + b^2Ed - fCab$, if we first write herein $f = eD + dE$, it becomes

$$a^2De + b^2Ed - (eD + dE)\,Cab,$$

which is

$$= aDe\,(a - bC) + bEd\,(b - aC),$$

and then writing $C = ab - AB$, we have $a - bC = a\,(1 - b^2) + bAB$, $= B\,(aB + bA)$, $= Bc$; and, similarly, $b - aC = Ac$; whence the term is $= c\,(aeBD + bdAE)$; or, in the equation to be verified, the right-hand side is $= cf\,(aeBD + bdAE)$, and by a similar reduction, the left-hand side is found to have the same value.

The paper contains various other interesting results.

686.

ON A FUNCTIONAL EQUATION.

[From the *Quarterly Journal of Pure and Applied Mathematics*, vol. XV. (1878), pp. 315—325; *Proceedings of the London Mathematical Society*, vol. IX. (1878), p. 29.]

I WAS led by a hydrodynamical problem to consider a certain functional equation; viz. writing for shortness $x_1 = \dfrac{ax+b}{cx+d}$, this is

$$\phi x - \phi x_1 = (x - x_1)\frac{Ax+B}{Cx+D}.$$

I find by a direct process, which I will afterwards explain, the solution

$$\phi x = \frac{A}{C}x + \frac{\sqrt{\{(a-d)^2+4bc\}}\,(AD-BC)}{C\,(dC-cD)}\int_0^\infty \frac{\sin\xi t \sin\eta t\, dt}{\sin\zeta t \sinh\pi t};$$

where ζ is a constant, but ξ, η are complicated logarithmic functions of x (ξ, η, ζ depend also on the quantities a, b, c, d, C, D); $\sinh\pi t$ denotes as usual the hyperbolic sine, $\frac{1}{2}(e^{\pi t} - e^{-\pi t})$.

The values of ξ, η, ζ are given by the formulæ

$$\lambda + \frac{1}{\lambda} = \frac{a^2+d^2+2bc}{ad-bc},$$

$$\mathrm{a} = ax+b,\quad \mathrm{b} = -dx+b,$$

$$\mathrm{c} = cx+d,\quad \mathrm{d} = cx-a,$$

$$W = C\mathrm{a} + D\mathrm{c},$$

$$Z = C\mathrm{b} + D\mathrm{d},$$

$$R = \lambda\mathrm{c} + \lambda\mathrm{d},$$

$$S = -\mathrm{c} - \mathrm{d},$$

$$R' = W + \frac{1}{\lambda}Z,$$

$$S' = -W - \lambda Z,$$

which determine λ, R, S, R', S' and then

$$\xi = \tfrac{1}{2}\log\frac{RS'}{R'S},\quad \eta = \tfrac{1}{2}\left(\log\lambda + \log\frac{RR'}{SS'}\right),\quad \zeta = \tfrac{1}{2}\log\lambda.$$

There is some difficulty as to the definite integral, on account of the denominator factor $\sin\zeta t$, which becomes $=0$ for the series of values $t = \frac{m\pi}{\zeta}$, but this is a point which I do not enter into.

I will in the first instance verify the result. Writing x_1 in place of x, and taking ξ_1, η_1 to denote the corresponding values of ξ, η, it will be shown that

$$\xi_1 = \xi,\quad \eta_1 = \eta + 2\zeta,$$ see *post*, (1).

Hence in the difference $\phi x - \phi x_1$ we have the integral

$$\int \frac{\sin\xi t\,\{\sin\eta t - \sin(\eta + 2\zeta)\,t\}\,dt}{\sin\zeta t\sinh\pi t},$$

(where and in all that follows the limits are ∞, 0 as before); here, since

$$\sin\eta t - \sin(\eta + 2\zeta)\,t = -2\sin\zeta t\cos(\eta+\zeta)\,t,$$

the factor $\sin\zeta t$ divides out, and the numerator is

$$= -2\sin\xi t\cos(\eta+\zeta)\,t,$$

which is

$$= \sin(\eta+\zeta-\xi)\,t - \sin(\eta+\zeta+\xi)\,t.$$

Hence the integral in question is

$$= \int\frac{\sin(\eta+\zeta-\xi)\,t\,dt}{\sinh\pi t} - \int\frac{\sin(\eta+\zeta+\xi)\,t\,dt}{\sinh\pi t}.$$

Now we have in general

$$\frac{1}{1+\exp.\,\alpha} = \tfrac{1}{2} - \int\frac{\sin\alpha t\,dt}{\sinh\pi t};$$

(this is, in fact, Poisson's formula

$$-\frac{1}{1+k\beta^{2n}} = \tfrac{1}{2} - 2\int\frac{\sin(2n\log\beta + \log k)\,t\,.\,dt}{e^{\pi t} - e^{-\pi t}},$$

in the second Memoir on the distribution of Electricity, &c., *Mém. de l'Inst.*, 1811, p. 223); and hence the value is

$$-\frac{1}{1+\exp.(\eta+\zeta-\xi)} + \frac{1}{1+\exp.(\eta+\zeta+\xi)},$$

or since

$$\eta+\zeta = \log\lambda + \tfrac{1}{2}\log\frac{RR'}{SS'},\quad \xi = \tfrac{1}{2}\log\frac{RS'}{R'S},$$

we have

$$\eta+\zeta-\xi = \log\lambda + \tfrac{1}{2}\log\frac{R'^2}{S'^2} = \log\lambda\frac{R'}{S'},$$

$$\eta+\zeta+\xi = \log\lambda + \tfrac{1}{2}\log\frac{R^2}{S^2} = \log\lambda\frac{R}{S},$$

and the value is thus

$$=-\frac{1}{1+\lambda\dfrac{R'}{S'}}+\frac{1}{1+\lambda\dfrac{R}{S}},\quad =-\frac{(RS'-R'S)\lambda}{(\lambda R'+S')(\lambda R+S)}.$$

Hence, from the assumed value of ϕx, we obtain

$$\phi x-\phi x_1=\frac{A}{C}(x-x_1)-\frac{\sqrt{\{(a-d)^2+4bc\}}\,(AD-BC)(RS'-R'S)\lambda}{C(dC-cD)(\lambda R'+S')(\lambda R+S)}.$$

We have

$$RS'-R'S=\frac{(\lambda-1)(a+d)^2}{ad-bc}(dC-cD)\{cx^2+(d-a)x-b\},$$

$$R\lambda+\ \ S=(\lambda^2-1)(cx+d),\qquad\qquad \text{see } post, (2),$$

$$R'\lambda+\ \ S'=(\lambda-1)(a+d)(Cx+D),$$

or since

$$\frac{cx^2+(d-a)x-b}{cx+d}=x-x_1,$$

this is

$$\phi x-\phi x_1=\frac{A}{C}(x-x_1)-\frac{\sqrt{\{(a-d)^2+4bc\}}\,(AD-BC)}{C}\,\frac{(a+d)\lambda}{(ad-bc)(\lambda^2-1)(Cx+D)}(x-x_1).$$

But from the value of λ,

$$\frac{\lambda}{\lambda^2-1}=\frac{ad-bc}{(a+d)\sqrt{\{(a-d)^2+4bc\}}},$$

and the equation thus is

$$\phi x-\phi x_1=(x-x_1)\left\{\frac{A}{C}-\frac{AD-BC}{C(Cx+D)}\right\},\quad =(x-x_1)\frac{Ax+B}{Cx+D},$$

as it should be.

(1) For the foregoing values of ξ_1, η_1, we require R_1, S_1, R_1', S_1', the values which R, S, R', S' assume on writing therein x_1 for x. We have

$$R_1=\ \ \lambda(cx_1+d)+\ \ (cx_1-a),$$

$$S_1=-\ \ (cx_1+d)-\lambda(cx_1-a):$$

substituting for x_1 its value, we find

$$R_1(cx+d)=(a+d)\lambda(cx+d)-(ad-bc)(\lambda+1),$$

or writing herein

$$ad-bc=\frac{(a+d)^2\lambda}{(\lambda+1)^2},$$

this is

$$R_1(cx+d)=\frac{(a+d)\lambda}{\lambda+1}R;$$

and similarly

$$S_1(cx+d)=\frac{a+d}{\lambda+1}S.$$

We have in like manner

$$R_1' = \quad W_1 + \frac{1}{\lambda} Z_1, \text{ where } W_1 = C(\ \ ax_1 + b) + D(cx_1 + d),$$

$$S_1' = - W_1 - \lambda Z_1, \text{ where } Z_1 = C(-dx_1 + b) + D(cx_1 - a).$$

Substituting for x_1 its value, we find

$$W_1(cx+d) = C[(a+d)(ax+b) - (ad-bc)x] + D[(a+d)(cx+d) - (ad-bc)],$$
$$Z_1(cx+d) = C[\qquad\qquad\qquad -(ad-bc)x] + D[\qquad\qquad\qquad -(ad-bc)]:$$

hence, substituting for $ad-bc$ as before,

$$W_1(cx+d) = \frac{a+d}{(\lambda+1)^2}\{(\lambda+1)^2 W - (a+d)\lambda(Cx+D)\},$$

$$Z_1(cx+d) = \frac{a+d}{(\lambda+1)^2}\{\qquad\qquad -(a+d)\lambda(Cx+D)\},$$

whence without difficulty

$$R_1'(cx+d) = \frac{(a+d)\lambda}{\lambda+1} R',$$

$$S_1'(cx+d) = \quad \frac{a+d}{\lambda+1} S':$$

consequently

$$\frac{R_1 S_1'}{R_1' S_1} = \frac{RS'}{R'S}, \text{ that is, } \xi_1 = \xi,$$

$$\frac{R_1 R_1'}{S_1 S_1'} = \lambda^2 \frac{RR'}{SS'}, \quad \text{,,} \quad , \eta_1 = \log\lambda + \eta, = 2\zeta + \eta,$$

which are the formulæ in question.

(2) For the value of $RS' - R'S$, we have

$$RS' - R'S = (\lambda\mathrm{c} + \mathrm{d})(-W - \lambda Z) - (-\lambda\mathrm{d} - \mathrm{c})\left(W + \frac{Z}{\lambda}\right)$$

$$= \left(-\lambda^2 + \frac{1}{\lambda}\right)\mathrm{c}Z + (\lambda+1)\{(\mathrm{d}-\mathrm{c})W - \mathrm{d}R\}$$

$$= -(\lambda-1)\left\{\left(1 + \lambda + \frac{1}{\lambda}\right)\mathrm{c}Z + (\mathrm{c}-\mathrm{d})W + \mathrm{d}Z\right\};$$

or substituting for $\lambda + \frac{1}{\lambda}$, Z and W their values, this is

$$= \frac{-(\lambda-1)}{ad-bc}\{(a^2+d^2+ad+bc)\,\mathrm{c}\,(\mathrm{b}C + \mathrm{d}D)$$
$$+ (ad-bc)[(\mathrm{c}-\mathrm{d})(\mathrm{a}C + \mathrm{c}D) + \mathrm{d}(\mathrm{b}C + \mathrm{d}D)]\}.$$

In the term in $\{\ \}$, the coefficient of C is

$$[(a^2+d^2+ad+bc)\,\mathrm{b} + (ad-bc)\,\mathrm{a}]\,\mathrm{c} - \mathrm{d}(\mathrm{a}-\mathrm{b})(ad-bc)$$
$$= (a+d)(d\mathrm{b} - b\mathrm{d})\,\mathrm{c} - (a+d)\,\mathrm{d}x\,(ad-bc),$$

and similarly the coefficient of D is

$$[(a^2+d^2+ad+bc)\,\mathrm{d}+(ad-bc)\,\mathrm{c}]\,\mathrm{c}-\mathrm{d}\,(\mathrm{c}-\mathrm{d})\,(ad-bc)$$
$$=(a+d)\,(\mathrm{ad}-\mathrm{cb})\,\mathrm{c}-(a+d)\,\mathrm{d}\,(ad-bc).$$

Hence the whole term in $\{\ \ \}$ is

$$=(a+d)\,\{[(d\mathrm{b}-b\mathrm{d})\,\mathrm{c}-\mathrm{d}\,(ad-bc)\,x]\,C+[(\mathrm{ad}-\mathrm{cb})\,\mathrm{c}-\mathrm{d}\,(ad-bc)]\,D\},$$

which is readily reduced to

$$(a+d)\,(\mathrm{ad}-\mathrm{bc})\,(-dC+cD)\,;$$

also

$$\mathrm{ad}-\mathrm{bc}=(a+d)\,\{cx^2+(d-a)\,x-b\}\,;$$

so that we have

$$RS'-R'S=\frac{(\lambda-1)\,(a+d)^2}{ad-bc}\,(dC-cD)\,[cx^2+(d-a)\,x-b],$$

which is the required value of $RS'-R'S$; and there is no difficulty in obtaining the other two formulæ,

$$R\lambda\;+S\;=(\lambda^2-1)\,(cx+d),$$
$$R'\lambda+S'=(\lambda\;-1)\,(a+d)\,(Cx+D)\,;$$

the verification is thus completed.

To show how the formula was directly obtained, we have

$$\frac{Ax+B}{Cx+D}=\frac{A}{C}-\frac{AD-BC}{C}\,\frac{1}{Cx+D}$$
$$=\frac{A}{C}+\beta x \text{ suppose}\,;$$

the equation then is

$$\phi x-\phi x_1=\frac{A}{C}\,(x-x_1)+(x-x_1)\,\beta x.$$

Hence, if x_1, x_2, x_3, ... denote the successive functions ϑx, $\vartheta^2 x$, $\vartheta^3 x$, &c., we have

$$\phi x_1-\phi x_2=\frac{A}{C}\,(x_1-x_2)+(x_1-x_2)\,\beta x_1,$$

$$\phi x_2-\phi x_3=\frac{A}{C}\,(x_2-x_3)+(x_2-x_3)\,\beta x_2,$$

whence adding, and neglecting ϕx_∞ and x_∞, we have

$$\phi x=\frac{A}{C}\,x+[(x-x_1)\,\beta x+(x_1-x_2)\,\beta x_1+(x_2-x_3)\,\beta x_2+\ldots],$$

where the term in [ ], regarding therein x_1, x_2, x_3, ... as given functions of x, is itself a given function of x; and it only remains to sum the series.

Starting from

$$x_1=\vartheta x=\frac{ax+b}{cx+d},$$

and writing

$$\lambda+\frac{1}{\lambda}=\frac{a^2+d^2+2bc}{ad-bc},$$

then the nth function is given by the formula

$$x_n = \vartheta_n x = \frac{(\lambda^{n+1}-1)(ax+b)+(\lambda^n-\lambda)(-dx+b)}{(\lambda^{n+1}-1)(cx+d)+(\lambda^n-\lambda)(cx-a)}$$

$$= \frac{(\lambda^{n+1}-1)\,\mathrm{a}+(\lambda^n-\lambda)\,\mathrm{b}}{(\lambda^{n+1}-1)\,\mathrm{c}+(\lambda^n-\lambda)\,\mathrm{d}}$$

$$= \frac{\lambda^n P+Q}{\lambda^n R+S},$$

if $P=\lambda\mathrm{a}+\mathrm{b}$, $Q=-\mathrm{a}-\lambda\mathrm{b}$, and as before $R=\lambda\mathrm{c}+\mathrm{d}$, $S=-\mathrm{c}-\lambda\mathrm{d}$.

I stop to remark that λ being real, then if $\lambda>1$ we have λ^n very large for n very large, and $x^n=\frac{P}{R}$ which is independent of n; the value in question is

$$x_n = \frac{\lambda(ax+b)+(-dx+b)}{\lambda(cx+d)+(\ \ cx-a)},$$

which, observing that the equation in λ may be written

$$\frac{\lambda a-d}{c(\lambda+1)} = \frac{b(\lambda+1)}{\lambda d-a},$$

is, in fact, independent of x, and is $=\frac{\lambda a-d}{c(\lambda+1)}$ or $\frac{b(\lambda+1)}{\lambda d-a}$; we have $x_{n-1}=x_n$, or calling each of these two equal values x, we have

$$x = \frac{ax+b}{cx+d},$$

which is the same equation as is obtainable by the elimination of λ from the equations

$$x = \frac{\lambda a-d}{c(\lambda+1)} = \frac{b(\lambda+1)}{\lambda d-a}.$$

The same result is obtained by taking $\lambda<1$ and consequently $x_n=\frac{Q}{S}$.

We find

$$x_{n-1}-x_n = \frac{\lambda^{n-1}P+Q}{\lambda^{n-1}R+S} - \frac{\lambda^n P+Q}{\lambda^n R+S},$$

$$= \frac{-\lambda^{n-1}(\lambda-1)(PS-QR)}{(\lambda^{n-1}R+S)(\lambda^n R+S)},$$

where

$$PS-QR = -(\lambda^2-1)(\mathrm{ad}-\mathrm{bc}) = -(\lambda^2-1)(a+d)\{cx^2+(d-a)x-b\};$$

and therefore

$$x_{n-1}-x_n = \frac{(\lambda-1)(\lambda^2-1)(a+d)\{cx^2+(d-a)x-b\}\lambda^n}{\lambda(\lambda^{n-1}R+S)(\lambda^n R+S)}.$$

Also

$$\beta x_{n-1} = -\frac{AD-BC}{C}\,\frac{1}{Cx_{n-1}+D},$$

where

$$Cx_{n-1}+D=\frac{C(\lambda^{n-1}P+Q)+D(\lambda^{n-1}R+S)}{\lambda^{n-1}R+S}=\frac{R'\lambda^n+S'}{R\lambda^{n-1}+S},$$

where

$$R'=\frac{CP}{\lambda}+\frac{DR}{\lambda},\ =C\left(\text{a}+\frac{\text{b}}{\lambda}\right)+D\left(\text{c}+\frac{\text{d}}{\lambda}\right),$$

$$S'=CQ+DS,\ =C(-\text{a}-\text{b}\lambda)+D(-\text{c}-\text{d}\lambda);$$

viz.

$$R'=W+\frac{1}{\lambda}Z,\ \ S'=-W-\lambda Z,$$

where Z and W denote $\text{a}C+\text{c}D$ and $\text{b}C+\text{d}D$ as before.

We hence obtain

$$(x_{n-1}-x_n)\,\beta x_n=\frac{-(AD-BC)}{C}$$

$$\times\frac{(\lambda-1)(\lambda^2-1)(a+d)\{cx^2+(d-a)x-b\}}{\lambda}\;\frac{\lambda^n}{(R\lambda^n+S)(R'\lambda^n+S')}$$

$$=\frac{-(AD-BC)}{C}$$

$$\times\frac{(\lambda-1)(\lambda^2-1)(a+d)\{cx^2+(d-a)x-b\}}{\lambda(RS'-R'S)}\;\frac{(RS'-R'S)\lambda^n}{(R\lambda^n+S)(R'\lambda^n+S')},$$

or, substituting for $RS'-R'S$ its value in the denominator, this is

$$(x_{n-1}-x_n)\,\beta x_n=-\frac{AD-BC}{C}\;\frac{(ad-bc)(\lambda^2-1)}{(a+d)\lambda(cD-dC)}\;\frac{(RS'-R'S)\lambda^n}{(R\lambda^n+S)(R'\lambda^n+S')}$$

$$=-\frac{\sqrt{\{(a-d)^2+4bc\}}\,(AD-BC)}{C(cD-dC)}\;\frac{(RS'-R'S)\lambda^n}{(R\lambda^n+S)(R'\lambda^n+S')},$$

and thence

$$\phi x=\frac{A}{C}x-\frac{\sqrt{\{(a-d)^2+4bc\}}\,(AD-BC)}{C(cD-dC)}\,\Sigma\,\frac{(RS'-R'S)\lambda^n}{(R\lambda^n+S)(R'\lambda^n+S')},$$

the summation extending from 1 to ∞.

Now the before-mentioned integral formula gives

$$\frac{1}{1+k\lambda^n}=\tfrac{1}{2}-\int\frac{\sin(n\log\lambda+\log k)\,t\,dt}{\sinh\pi t},$$

$$\frac{1}{1+k'\lambda^n}=\tfrac{1}{2}-\int\frac{\sin(n\log\lambda+\log k')\,t\,dt}{\sinh\pi t}.$$

Taking the difference, and then writing $k=\frac{R}{S}$, $k'=\frac{R'}{S'}$, we have under the integral sign

$$\sin\left(n\log\lambda+\log\frac{R}{S}\right)t-\sin\left(n\log\lambda+\log\frac{R'}{S'}\right)t,$$

which is

$$=2\sin\tfrac{1}{2}\left(\log\frac{RS'}{R'S}\right)t\cos\left(n\log\lambda+\tfrac{1}{2}\log\frac{RR'}{SS'}\right)t,$$

which attending to the before-mentioned values of ξ, η, ζ is

$$=2\sin\xi t\cos(2n\zeta-\zeta+\eta)\,t,$$

and the formula thus is

$$\frac{S}{R\lambda^n+S}-\frac{S'}{R'\lambda^n+S'},\ =-\frac{(RS'-R'S)\,\lambda^n}{(R\lambda^n+S)(R'\lambda^n+S')}=-\int\frac{2\sin\xi t\cos(2n\zeta-\zeta+\eta)\,t\,dt}{\sinh\pi t}.$$

We have here

$$\cos(2n\zeta-\zeta+\eta)\,t=\cos 2n\zeta t\cos(\eta-\zeta)\,t-\sin 2n\zeta t\sin(\eta-\zeta)\,t,$$

whence summing from 1 to ∞ by means of the formulæ

$$\cos 2\zeta t+\cos 4\zeta t+\ldots=-\tfrac{1}{2},$$

$$\sin 2\zeta t+\sin 4\zeta t+\ldots=\ \tfrac{1}{2}\cot\zeta t,$$

(which series however are not convergent), the numerator under the integral sign becomes

$$\sin\xi t\,\{-\cos(\eta-\zeta)\,t-\cot\zeta t\sin(\eta-\zeta)\,t\},$$

which is

$$=-\frac{\sin\xi t\sin\eta t}{\sin\zeta t},$$

and the formula thus is

$$\Sigma\frac{(RS'-R'S)\,\lambda^n}{(R\lambda^n+S)(R'\lambda^n+S')}=-\int\frac{\sin\xi t\sin\eta t\,dt}{\sin\zeta t\sinh\pi t};$$

and we therefore find

$$\phi x=\frac{A}{C}x+\frac{\sqrt{\{(a-d)^2+4bc\}}\,(AD-BC)}{C\,(cD-dC)}\int\frac{\sin\xi t\sin\eta t\,dt}{\sin\zeta t\sinh\pi t},$$

which is the result in question.

The solution is a particular one; calling it for a moment (ϕx), then, if the general solution be $\phi x=\Phi x+(\phi x)$, it at once appears that we must have $\Phi x-\Phi x_1=0$; and as it has been shown that $\dfrac{RS'}{R'S}$ is a function of x which remains unaltered by the change of x into x_1, this is satisfied by assuming $\Phi x=f\left(\dfrac{RS'}{R'S}\right)$, an arbitrary function of $\dfrac{RS'}{R'S}$. Hence we may to the foregoing expression of ϕx add this term $f\left(\dfrac{RS'}{R'S}\right)$.

Postscript. The new formula

$$\vartheta^n x=\frac{(\lambda^{n+1}-1)(ax+b)+(\lambda^n-\lambda)(-dx+b)}{(\lambda^{n+1}-1)(cx+d)+(\lambda^n-\lambda)(\ \ cx-a)},$$

where

$$\lambda+\frac{1}{\lambda}=\frac{a^2+d^2+2bc}{ad-bc},$$

for the nth repetition of ϑx, $=\frac{ax+b}{cx+d}$, is a very interesting one. It is to be remembered that, when n is even the numerator and denominator each divide by $\lambda-1$, but when n is odd they each divide by λ^2-1; after such division, then further dividing by a power of λ, they each consist of terms of the form $\lambda^a+\frac{1}{\lambda^a}$, that is, they are each of them a rational function of $\lambda+\frac{1}{\lambda}$. Substituting and multiplying by the proper power of $ad-bc$, the numerator and denominator become each of them a rational and integral function of a, b, c, d of the order $n+1$ when n is even, but of the order n when n is odd; in the former case, however, the numerator and denominator each divide by $a+d$, so that ultimately, whether n be even or odd, the order is $=n$ as it should be.

For example, when $n=2$, the value is

$$\frac{(\lambda^3-1)\mathrm{a}+(\lambda^2-\lambda)\mathrm{b}}{(\lambda^3-1)\mathrm{c}+(\lambda^2-\lambda)\mathrm{d}},\ =\frac{(\lambda^2+\lambda+1)\mathrm{a}+\lambda\mathrm{b}}{(\lambda^2+\lambda+1)\mathrm{c}+\lambda\mathrm{d}},\ =\frac{\left(\lambda+\frac{1}{\lambda}+1\right)\mathrm{a}+\mathrm{b}}{\left(\lambda+\frac{1}{\lambda}+1\right)\mathrm{c}+\mathrm{d}},$$

or, as this may be written,

$$=\frac{\left(\lambda+\frac{1}{\lambda}+2\right)\mathrm{a}-\mathrm{a}+\mathrm{b}}{\left(\lambda+\frac{1}{\lambda}+2\right)\mathrm{c}-\mathrm{c}+\mathrm{d}},$$

where, observing that

$$\lambda+\frac{1}{\lambda}+2=\frac{(a+d)^2}{ad-bc},\quad -\mathrm{a}+\mathrm{b}=-(\mathrm{a}+\mathrm{d})\,x,\quad -\mathrm{c}+\mathrm{d}=-(a+d),$$

the numerator and denominator each divide by $a+d$, and the final value is

$$=\frac{(a+d)(ax+b)-(ad-bc)\,x}{(a+d)(cx+d)-(ad-bc)},\ =\frac{(a^2+bc)\,x+b\,(a+d)}{c\,(a+d)\,x+bc+d^2},$$

which is the proper value of ϑ^2x. But, when $n=3$, the value is

$$\frac{(\lambda^4-1)\mathrm{a}+(\lambda^3-\lambda)\mathrm{b}}{(\lambda^4-1)\mathrm{c}+(\lambda^3-\lambda)\mathrm{d}},\ =\frac{(\lambda^2+1)\mathrm{a}+\lambda\mathrm{b}}{(\lambda^2+1)\mathrm{c}+\lambda\mathrm{d}},\ =\frac{\left(\lambda+\frac{1}{\lambda}\right)\mathrm{a}+\mathrm{b}}{\left(\lambda+\frac{1}{\lambda}\right)\mathrm{c}+\mathrm{d}};$$

and this is

$$=\frac{(a^2+d^2+2bc)(ax+b)+(ad-bc)(-dx+b)}{(a^2+d^2+2bc)(cx+d)+(ad-bc)(\ \ cx-a)},$$

or finally

$$=\frac{(a^3+2abc+bcd)\,x+b\,(a^2+ad+bc+d^2)}{c\,(a^2+ad+bc+d^2)\,x+(abc+2bcd+d^3)},$$

which is the proper value of ϑ^3x.

687.

NOTE ON THE FUNCTION $\vartheta x = a^2(c-x) \div \{c(c-x)-b^2\}$.

[From the *Quarterly Journal of Pure and Applied Mathematics*, vol. XV. (1878), pp. 338—340.]

STARTING from the general form

$$\vartheta x = \frac{\alpha x + \beta}{\gamma x + \delta},$$

we have

$$\vartheta^n x = \frac{(\lambda^{n+1}-1)(\alpha x + \beta) + (\lambda^n - \lambda)(-\delta x + \beta)}{(\lambda^{n+1}-1)(\gamma x + \delta) + (\lambda^n - \lambda)(\gamma x - \alpha)},$$

where

$$\lambda + \frac{1}{\lambda} = \frac{\alpha^2 + \delta^2 + 2\beta\gamma}{\alpha\delta - \beta\gamma}.$$

For the function in question

$$\vartheta x = \frac{a^2(c-x)}{c(c-x)-b^2},$$

(a form which presents itself in the problem of the distribution of electricity upon two spheres), the values of $\alpha, \beta, \gamma, \delta$ are

$$\alpha = -a^2, \quad \beta = a^2c, \quad \gamma = -c, \quad \delta = c^2 - b^2;$$

the equation for λ therefore is

$$\lambda + \frac{1}{\lambda} = \frac{a^4 + (c^2-b^2)^2 - 2a^2c^2}{a^2b^2};$$

or, what is the same thing,

$$\frac{(\lambda+1)^2}{\lambda} = \frac{(a^2+b^2-c^2)^2}{a^2b^2}.$$

Suppose that a, b, c are the sides of a triangle the angles whereof are A, B, C; then $c^2 = a^2 + b^2 - 2ab\cos C$, or we have

$$\frac{(\lambda+1)^2}{\lambda} = 4\cos^2 C;$$

or, writing this under the form

$$\sqrt{(\lambda)} + \frac{1}{\sqrt{(\lambda)}} = 2\cos C,$$

the value of λ is at once seen to be $= e^{2iC}$; and it is interesting to obtain the expression of the nth function in terms of the sides and angles of the triangle.

The numerator and the denominator are

$$\lambda^n P + Q,$$

$$\lambda^n R + S,$$

where

$$P = \lambda(\alpha x + \beta) + (-\delta x + \beta),\quad R = \lambda(\gamma x + \delta) + \gamma x - \alpha,$$

$$Q = -(\alpha x + \beta) - \lambda(-\delta x + \beta),\quad S = -(\gamma x + \delta) - \lambda(\gamma x - \alpha).$$

Hence, writing the numerator and the denominator in the forms

$$\lambda^{\frac{1}{2}n} P + \lambda^{-\frac{1}{2}n} Q,$$

$$\lambda^{\frac{1}{2}n} R + \lambda^{-\frac{1}{2}n} S,$$

these are

$$(P+Q)\cos nC + (P-Q)\, i\sin nC,$$

$$(R+S)\cos nC + (R-S)\, i\sin nC;$$

viz. they are

$$(\lambda-1)(\alpha+\delta)\, x\cos nC + (\lambda+1)\{(\alpha-\delta)\, x + 2\beta\}\, i\sin nC,$$

$$(\lambda-1)(\alpha+\delta)\,.\cos nC + (\lambda+1)\{2\gamma x - (\alpha-\delta)\}\, i\sin nC,$$

or, observing that $\dfrac{\lambda-1}{\lambda+1} = i\tan C$ and removing the common factor $i(\lambda+1)$, they may be written

$$\tan C(\alpha+\delta)\, x\cos nC + \{(\alpha-\delta)\, x + 2\beta\}\sin nC,$$

$$\tan C(\alpha+\delta)\,.\cos nC + \{2\gamma x - (\alpha-\delta)\}\sin nC.$$

Substituting for α, β, γ, δ their values, these are

$$\tan C\{(c^2-a^2-b^2)\, x\cos nC\} + \{(b^2-a^2-c^2)\, x + 2a^2c\}\sin nC,$$

$$\tan C\{(c^2-a^2-b^2)\,.\cos nC\} + \{-2cx - (b^2-a^2-c^2)\}\sin nC,$$

$$= \tan C\{-ab\cos Cx\cos nC\} + \{-ac\cos B\,.\,x + a^2c\}\sin nC,$$

$$\tan C\{-ab\cos Cx\cos nC\} + \{-cx + ac\cos B\}\sin nC,$$

$$= x\{-ab\sin C\cos nC - ac\cos B\sin nC\} + a^2c\sin nC,$$

$$-cx\sin nC + \{ac\cos B\sin nC - ab\sin C\cos nC\};$$

or, writing herein $b\sin C = c\sin B$, these are

$$\begin{array}{ll} -acx\{\sin B\cos nC + \cos B\sin nC\} & +a^2c\sin nC, \\ -cx\sin nC & +ac\{\cos B\sin nC - \sin B\cos nC\}, \end{array}$$

whence finally

$$\vartheta^n x = \frac{a^2\sin nC - ax\sin(nC+B)}{a\sin(nC-B) - x\sin nC}.$$

As a verification, writing $n=1$, we have

$$\vartheta x = \frac{a^2\sin C - ax\sin A}{a\sin(C-B) - x\sin C}$$

$$= \frac{a^2c - acx\dfrac{\sin A}{\sin C}}{ac\dfrac{\sin(C-B)}{\sin C} - cx},$$

or observing that

$$ac\frac{\sin(C-B)}{\sin C} = c^2 - b^2,$$

(for this is $\sin A\sin(C-B) = \sin^2 C - \sin^2 B$), we have

$$\vartheta x = \frac{a^2(c-x)}{c^2-b^2-cx}$$

as it should be. If in the formula for $\vartheta^n x$ we write $x=0$, we have a formula given in the Senate-House Problems, January 14, 1878: it was thus that I was led to investigate the general expression.

688.

GEOMETRICAL CONSIDERATIONS ON A SOLAR ECLIPSE.

[From the *Quarterly Journal of Pure and Applied Mathematics*, vol. XV. (1878), pp. 340—347.]

I CONSIDER, from a geometrical point of view, the phenomena of a solar eclipse over the earth generally; attending at present only to the penumbral cone, the vertex of which I denote by V. It is convenient to regard the earth as fixed, and the sun and moon as moving each of them with its proper motion, and also with the diurnal motion. The penumbral cone meets the earth's surface in a curve which may be called the penumbral curve; viz. when the cone is not completely traversed by the earth's surface, (that is, when only some of the generating lines of the cone meet the earth's surface), the penumbral curve is a single (convex or hour-glass-shaped) oval; separated, as afterwards mentioned, into two parts, one of them lying away from the sun, and having no astronomical significance; but when the cone is completely traversed by the earth's surface, then the penumbral curve consists of two separate (convex) ovals; one of them lying away from the sun and having no astronomical significance, the other lying towards the sun. The intermediate case is when the cone just traverses the earth's surface, or is touched internally by the earth's surface; the penumbral curve is then a figure of eight, one portion of which lies away from the sun, and has no astronomical significance: there is another limiting case when the cone is touched externally by the earth's surface, the penumbral curve being then a mere point.

It is necessary to consider on the earth's surface a curve which may for shortness be termed the horizon; viz. this is the curve of contact of the cone, vertex V, circumscribed about the earth; it is a small circle nearly coincident with the great circle, which is the intersection by a plane through the centre of the earth at right angles to the line from this point to the centre of the sun.

Regarding V as a point in the heavens, capable of being viewed notwithstanding the interposition of the moon; the horizon, as above defined, is the curve separating the portions of the earth's surface for which V is visible and invisible respectively. The horizon does or does not meet the penumbral curve, according as this last consists of a single oval or of two distinct ovals; viz. in the latter case the horizon lies between the two ovals, in the former case the horizon traverses the area of the oval (separating this area into two parts), thus meeting the oval, or penumbral curve, in two points, or say these points separate the oval into two parts; from any point of the one part V is visible, from any point of the other part V is invisible; and from each of the two points themselves V is visible as a point on the horizon in the ordinary sense of the word; that is, there is an exterior contact of the sun and moon visible on the horizon. It is to be observed that, in the limiting cases where the penumbral curve is a mere point and a figure of eight respectively, the horizon passes through the mere point and through the node of the figure of eight respectively.

The two points of intersection of the penumbral curve with the horizon may for shortness be termed critic points. The lines which present themselves in a diagram of a solar eclipse, (see *Nautical Almanac*) are the "northern and south lines of simple contact," say for shortness the "limits"; viz. these are the envelope or, geometrically, a portion of the envelope of the penumbral curve; and the lines of "eclipse begins or ends at sunrise or sunset," say for shortness the critic lines; viz. these are the locus of the critic points.

The point V considered as a point in the heavens is a point occupying a position intermediate between those of the centres of the sun and moon; hence referring it to the surface of the earth by means of a line drawn from the centre, its position on the earth's surface is nearly coincident with that point to which the sun is then vertical; and its motion on the earth's surface is from east to west approximately along the parallel of latitude = sun's declination, and with a velocity of approximately 15° per hour. For any given position of V on the earth's surface, describing with a given angular radius nearly = 90° a small circle (nearly a great circle), this is the horizon; as V moves upon the surface of the earth, the horizon envelopes a curve which is very nearly a parallel, angular radius = sun's declination (there are two such curves in the northern and southern hemispheres respectively, but I attend only to one of them in the proper hemisphere, as will be explained), say this is the horizon-envelope; the horizon in each of its successive positions is thus a curvilinear tangent (nearly a great circle) to this horizon-envelope. If for a given position of V, and also for the consecutive position we consider the corresponding horizons, these intersect in a point K on the horizon-envelope, and the horizon for V is the circle centre V and angular radius VK; K is a point which is very nearly upon, and which may be taken to be upon, the meridian through V; the horizon may be regarded as a tangent which sweeps round the horizon-envelope; to each position thereof there corresponds a position of V, and consequently also a penumbral curve; and (when this is a single oval) the horizon meets it in two points, which are the critic points. It is to be added that, if for a given position of the horizon we consider as well K as the opposite point K_1, (viz., K_1 lies on the great circle KV), then the points K and K_1 divide the horizon into two portions; for any point on one of these portions

V (considered as a point in the heavens) is rising, for a point on the other of them it is setting; and for the points K and K_1 respectively it is moving horizontally; that is, first rising and then setting, or *vice versâ*.

A solar eclipse is of one of two classes; viz. either the penumbral cone completely traverses the earth, so that towards the middle of the eclipse the penumbral curve consists of two separate ovals: or the penumbral cone does not completely traverse the earth, so that throughout the eclipse the penumbral curve consists of a single oval only. In the former case, we have to consider the commencement, during which the penumbral curve passes from a mere point to a figure of eight: the middle, during which it passes from a figure of eight through two ovals to a figure of eight: and the termination, during which it passes from a figure of eight to a mere point. In the latter case, we consider the whole eclipse during which the penumbral curve passes from a mere point through a single oval to a mere point.

In an eclipse of the first class: for the commencement, the penumbral curve is at first a mere point (point of first contact); it then becomes a convex oval, each oval in the first instance inclosing the preceding ones, so that there is not any intersection of two consecutive ovals. We come at last to an oval which is touched north by its consecutive oval, and to an oval which is touched south by its consecutive oval (I presume that the contacts north and south do not take place on the same oval, but I am not sure); and after this, the ovals assume the hour-glass form, each oval intersecting the consecutive oval in two points north and two points south; the ovals thus beginning to form an envelope or limit. There are on each of the ovals two critic points, and we have thus a critic curve commencing at the mere point (point of first contact) and extending in each direction from this point. The point, where an oval is touched by the consecutive oval, is not so far as appears a critic point; that is, the critic curve does not at this point unite itself with the envelope or limit. But the critic curve comes subsequently to unite itself each way with the limit; and, since clearly it cannot intersect the limit, it will at each of these points touch the limit; that is, we have a critic curve extending each way from the point of first contact until it touches the northern limit and until it touches the southern limit. Observe that the penumbral curve, as being at first a mere point or an indefinitely small oval, does not at first contain within itself the point K or K_1: it can only come to do this by passing through a position where the curve passes through K or K_1; viz. K or K_1 would then be a critic point; and I assume for the present that this does not take place. The critic curve at the point of first contact is a curve "eclipse begins at sunrise," and as not coming to pass through a point or K_1, it cannot alter its character; that is, the critic curve, as extending each way from the point of first contact until it comes to touch the northern and southern limits respectively, is a curve "eclipse begins at sunrise"; at the terminal points in question, there is a mere contact of the sun and moon, so that they are points, where the eclipse begins and simultaneously ends at sunrise. Continuing the series of ovals until we arrive at the figure of eight, there are on each of them two critic points, which ultimately unite in the node of the figure of eight; these constitute a critic curve, extending each way from the node of the figure of eight to the contacts with the northern and southern limits respectively. There is, as before, no passage through a point K or K_1, the curve in question thus retains throughout the same character; and by consideration of the two terminal points it at once appears that it is a curve "eclipse ends at sunrise." The above-mentioned critic curves form together an oval touching the northern and southern limits respectively; say this is the sunrise oval.

The termination of the eclipse is similar to this, only the events happen in the reverse order; we have a critic line starting from the node of the figure of eight and extending each way until it comes to touch the northern and southern limits respectively, viz. this is the line "eclipse begins at sunset"; and then, extending each way from the points of contact to reunite itself at the point of last contact, this being the line "eclipse ends at sunset," and the two portions together form an oval touching the northern and southern limits respectively; say this is the sunset oval. It is to be noticed that certain portions of the two limits are generated as the envelope of the penumbral curve during the commencement and during the termination of the eclipse.

For the middle of the eclipse; the penumbral curve, in the first instance a figure of eight, breaks up into two ovals, but only one of these is attended to; and ultimately the oval unites itself with another oval so as to give rise to a new figure of eight. There is thus throughout the middle of the eclipse a single oval; this has, north and south, an envelope which joins itself on to the portions enveloped during the latter part of the commencement and the former part of the termination of the eclipse, and constitutes therewith the northern and southern limits respectively, viz. each of these is considered as extending from a point of contact with the sunrise oval to a point of contact with the sunset oval.

The line K_1VK, or say the meridian line through V, travels westwardly, while the penumbral curve travels eastwardly; the two come to touch each other, and there are then two intersections which ultimately come to the northern and southern limits respectively: the locus of these is a line of "eclipse commences at midday"; as the motion continues, the points of intersection move away from the two limits respectively and ultimately unite at the point where the line KVK_1 again touches the penumbral curve; the locus is the line of "eclipse terminates at midday," the two lines together forming an oval which touches the northern and southern limits respectively and which may be termed the midday oval. In all that precedes, no distinction has been made between the two portions of the horizon-envelope, or the points K and K_1, and either curve and point indifferently may be alone attended to.

Considering now an eclipse of the second kind, the penumbral curve is at first a mere point (the point of first contact) and it then becomes an oval, the successive ovals not at first intersecting each other, but each oval inclosing within itself the preceding ones. Any oval is met by the corresponding horizon in two points P and P', at first coinciding with each other at the point of first contact, and then separating from each other, one of them, say P, moving down towards and ultimately arriving at one of the horizon-envelopes, say to fix the ideas the southern one (which curve is henceforth selected as being, and is called, the horizon-envelope, and the points on this curve are taken to be the points K), viz. P is then a point K on the penumbral curve, I call it K_1. The successive ovals will in the meantime have begun to intersect each other so as to give rise to a northern limit; this will touch the critic line (locus of P, P'), and we have a portion of the critic line extending from the point of first contact, in one direction to the point of contact with the northern limit, and in the other direction to the point K_1 on the horizon-envelope; this is the line "eclipse begins at sunrise." As the horizon continues to sweep on, the other point P', which has not yet reached the horizon-envelope, will gradually approach and ultimately arrive at the horizon-envelope, say at the point K_2; we have thus a second portion of the critic line extending from the contact with the northern envelope to the point K_2; this is the line "eclipse ends at sunrise." The horizon continuing to sweep on, the point P beginning with the position K_1, which is now on the other side of the point of contact of the horizon with the horizon-envelope, will trace out a portion of the critic curve extending from K_1 to a second point of contact with the northern limit; this will be the line of "eclipse begins at sunset." And, finally, the point P from the last-mentioned point of contact, and the point P' from its position K_2, which is now on the other side of the point of contact of the horizon with the horizon-envelope, (that is, P, P' have now each passed through the point of contact of the horizon with the horizon-envelope, and are both of them on the same side thereof, viz. the side opposite to their original side), will come to unite at the point of the last contact; we have thus a fourth portion of the critic curve extending from K_2 to the second point of contact with the northern limit, viz. this is the line "eclipse ends at sunset." The description will be more intelligible by means of the figure, in which 1, 1′, 2, 2′, …, 8, 8′ represent successive corresponding

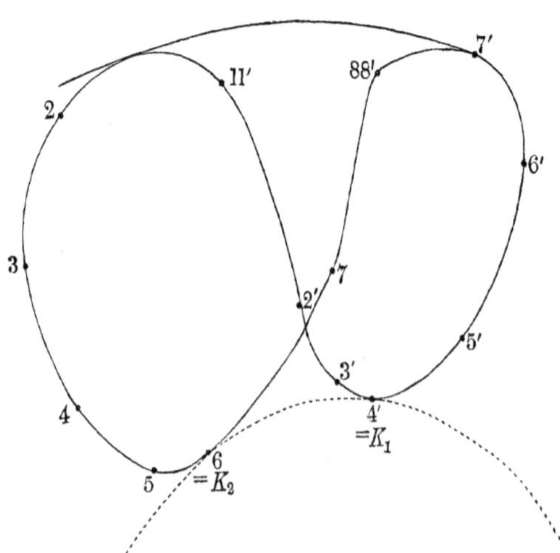

positions of the points P, P', the successive positions of the horizon being given by the right lines 11′, 22′, &c., all of them tangents to the dotted circle or horizon-envelope.

The entire critic line is thus a figure of eight, twice touching the horizon-envelope and also twice touching the limit. If we consider, as before, the intersections of KV with the corresponding penumbral curve, this will be a curve extending from K_1 so as to touch the limit, and thence onward to K_2, the portion from K_1 to the contact with the limit being the line "eclipse begins at transit," and the portion from the limit to K_2 the line "eclipse ends at transit." I say "transit" instead of midday, since for a circumpolar place the phenomenon may happen at one or the other transit of the sun over the meridian. It is to be remarked, that the node of the figure of eight is a point, such that the eclipse there begins at sunrise and ends at sunset; this point does not appear to be an important one in the geometrical theory.

The two loops of the critic line may be of very unequal magnitudes, and in particular one of them may actually vanish; viz. the points K_1 and K_2 then coincide together, and the critic curve is a closed cuspidal curve touching the horizon-envelope at the cusp; moreover, instead of two contacts with the limit there is one proper contact, and an improper contact at the cusp, that is, the limit simply passes through the cusp. And through this special separating case, we pass to the case where, instead of the figure of eight, we have a single oval, not touching the horizon-envelope (viz. the points K_1, K_2 have become imaginary), but still touching the limit twice; this is a distinct type for an eclipse of the second class.

And, similarly, in an eclipse of the first class, where the points K_1, K_2 do not in general exist (viz. geometrically they are imaginary), these points may present themselves in the first instance as two coincident points, viz. instead of the sunrise oval or the sunset oval (as the case may be), we have then a cuspidal curve; or they may be two real points, viz. instead of the same oval, we have then a figure of eight touching the horizon-envelope twice, and also touching each of the two limits. These are thus the several cases.

When the Earth traverses the penumbral cone, the critic curve is

1. A pair of ovals:
2. An oval and a cuspidate oval:
3. An oval and a figure of eight.

And when the Earth does not traverse the penumbral cone, the critic curve is

4. A figure of eight:
5. A cuspidate oval:
6. An oval.

To which may be added the transition case which separates 1 and 4, viz. here the Earth just has an internal contact with the penumbral cone, and the critic curve is

7. Two ovals touching each other.

But of course 2, 5, and 7 are so special that they may be disregarded altogether; and 3 and 6 are of rare occurrence. I have not sufficiently examined the conditions for the occurrence of these forms 3 and 6; my attention was called to them, and indeed to the whole theory, by a question proposed by Prof. Adams in the Cambridge Smith's Prize Examination for 1869.

689.

ON THE GEOMETRICAL REPRESENTATION OF IMAGINARY VARIABLES BY A REAL CORRESPONDENCE OF TWO PLANES.

[From the *Proceedings of the London Mathematical Society*, vol. IX. (1878), pp. 31—39. Read December 13, 1877.]

IN my recently published paper, "Geometrical Illustration of a Theorem relating to an Irrational Function of an Imaginary Variable," *Proceedings of the London Mathematical Society*, t. VIII. (1877), pp. 212—214, [627], I remark as follows:—"If we have v a function of u determined by an equation $f(u, v)=0$, then to any given imaginary value $x+iy$ of u there belong two or more values, in general imaginary, of v; and for the complete understanding of the relation between the two imaginary variables we require to know the series of values $x'+iy'$ which correspond to a given series of values $x+iy$ of v, u respectively. We must, for this purpose, take x, y as the coordinates of a point P in a plane Π, and x', y' as the coordinates of a corresponding point P' in another plane Π'";—and I then proceed to consider the particular case where the equation between u, v is $u^2+v^2=a^2$, that is, where

$$(x+iy)^2+(x'+iy')^2=a^2.$$

The general case is that of an equation $(*)(u,\ 1)^m(v,\ 1)^n=0$, where to each given value, real or imaginary, of u, there correspond n real or imaginary values of v; and to each given value, real or imaginary, of v, there correspond m real or imaginary values of u. And then, writing $u=x+iy$ and $v=x'+iy'$, and regarding (x, y), (x', y') as the coordinates of the points P, P' in the two planes Π, Π' respectively, we have a *real* (m, n) correspondence between the two planes; viz. to each real point P in the first plane there correspond n real points P' in the second plane, and to each real point P' in the second plane there correspond m real points P in the first plane. But such real correspondence of two planes does not of necessity arise from an equation between the two imaginary variables u, v; and the question of the real correspondence of two planes may be considered in itself, without any reference to such origin.

I was under the impression that the theory was a known one; but I have not found it anywhere set out in detail. It is to be noticed that, although intimately connected with, it is quite distinct from (and seems to me to go beyond) that of a Riemann's surface. Riemann represents the value $u, = x + iy$, by a point P whose coordinates are x, y; but he considers $u', = x' + iy'$, as a given imaginary value attached to the point P, without representing this value by a point P', coordinates x', y'.

I proceed to consider the general theory of the real (m, n) correspondence. Points in the first plane are denoted by the unaccented letters $P, Q, ..$; and the corresponding points in the second plane are in general denoted by the *same* letters accented; but there are, as will be explained, special points V, W where the letters are interchanged; viz. to the points V or W in the first plane correspond points W' or V' in the second plane.

1. To a point P there correspond in general n distinct points P'; and as P varies continuously, each of the points P' also varies continuously.

2. There are certain points V called branch-points (Verzweigungspunkte), such that to each point V there correspond two united points, represented by (W'), and $n-2$ other distinct points W'. The points (W') are called cross-points, and the number of them is of course equal to that of the branch-points V.

It is throughout assumed that a point denoted by a letter other than V is not a point V.

3. If the point P, moving continuously, describe a closed curve so as to return to its original position, then, if this curve includes within it no point V (or all the points V)*, each of the corresponding points P' will describe continuously a closed curve returning into its original position. Supposing that the curve described by P is an oval (non-autotomic closed curve), and taking this to be in the first instance an indefinitely small oval, then the curves described by the points P' will in the first instance be each of them an indefinitely small oval; but it is worth while to notice how, as the oval described by P increases, any one of the ovals described by a point P' may become autotomic; viz. if the oval described by P passes through two points Q, Q of the m points Q which correspond in the first plane to the same Q' in the second plane, then Q' will be a node in the closed curve described by that point P' which in the course of its motion comes to pass through Q'. This curve is in general an inloop curve composed of two loops, one wholly within the other (united at the point Q'), and such that they each include one and the same point V' (viz. V' is included within the inner loop): as to this, see *post*, Nos. 9 and 10. It will be observed that this node Q' is not a point (W') nor any other special point of the second plane.

* The two cases of the closed curve including no point V, and including all the points V, are really identical, as the discontinuity at infinity may be disregarded. It is to be observed that, this being so, it follows that the number of the points V must be even.

4. Consider, as before, P as describing a closed curve which does not include within it any point V, and the corresponding points P' as describing each of them a closed curve. As the curve described by P approaches a point V, the curves described by two of the points P' will approach the corresponding point (W'); and when the curve described by P passes through V, the curves described by the two points P' will unite together at this point (W') as a node; viz. they will form a figure of eight*, the crossing being at the cross-point (W'), which corresponds to the branch-point V. And, corresponding to the closed curve described by P, we have this figure of eight (replacing two of the original n closed curves), and $n-2$ closed curves described by the other points P'.

5. Supposing, next, that the closed curve described by P (instead of passing through the point V) includes within it the point V, then the figure of eight transforms itself into a twice-indented oval*. There are on this curve two of the points P' which correspond to the given point P; and as P, moving continuously in its closed curve, returns to its original position, the first of these points P', moving continuously along a portion of the curve, comes to coincide with the original position of the second point P'; while the second point P', moving continuously along the remaining portion of the curve, comes to coincide with the original position of the first point P'; viz. the two portions of the curve are described by the two points P' respectively. The curve may thus be regarded as a bifid curve, belonging to these two points P'. And, corresponding to the closed curve described by P, we have this bifid curve belonging to the two points P', and $n-2$ single closed curves belonging to the other $n-2$ points P' respectively.

6. If the closed curve described by P (including within it a point V) comes to pass through a second point V, the effect will be a new node at the corresponding point (W'); viz. at this point (W') either the bifid curve unites itself with one of the single curves, or two of the single curves unite together, or the bifid curve there cuts itself. And, if the curve described by P comes to include within it this second point V, then in the three cases respectively:—the bifid curve takes to itself the single curve, so that the system then is a trifid curve and $n-3$ single curves; or the two single curves give rise to a bifid curve, so that the system is two bifid curves and $n-4$ single curves; or, lastly, the bifid curve breaks up into two single curves, so that the system resumes its original form of n single curves.

7. We thus see how the closed curve described by P, including within it certain of the points V, may be such as to have corresponding to it an α-fid curve, a β-fid curve, &c., $(\alpha+\beta+\ldots=n)$; viz. an α-fid curve contains upon it α of the points P' which correspond to the original position of P; and then, as P describes

* The name *figure of eight* refers to the case where the two curves which come to unite at (W') are proper ovals (non-autotomic closed curves). They might have one or both of them a node or nodes, as explained in No. 3; and the term would then be inappropriate. And so, lower down, the name *twice-indented oval* is used to express the form into which a proper figure of eight is changed by the disappearance of the node.

continuously its closed curve, returning to its original position, each of these points P' describes a portion of the α-fid curve, passing from its original position to the original position of a point P' next to it upon the α-fid curve; and the like as to a β-fid curve, &c. The numbers $\alpha, \beta, \ldots$ are not of necessity unequal, and we may have sets of equal numbers in any manner. It is hardly necessary to remark that, if the curve described by P passes through any point or points V, then two of the curves described by the points P' will unite together, or it may be that one of these will cut itself at the corresponding point or points (W'); and further that, as in No. 3, if the curve described by P passes through two or more of the points Q which correspond to the same point Q', then any such point Q' will present itself as a node upon the curve belonging to some point, or set of points, P'. But the order of succession in which the original n single curves unite themselves together into multifid curves, or again break up into single curves, cannot, it would appear, be explained in any general manner, and would in each case depend on the nature of the particular correspondence.

8. We may consider the case where the closed curve described by P cuts itself. The curve may here be considered as made up of two or more ovals, or, to use a more appropriate term, say loops, each such loop being a curve not cutting itself; and the case is thus reducible to that before considered, where the curve does not cut itself. Thus, to fix the ideas, let the curve be a figure of eight, the initial position of P being at the crossing, and let neither of the loops contain within it a point V. Then, as P passes continuously along one of the loops, returning to its original position, each of the corresponding points P' describes a closed curve, which will be in the nature of a loop, viz. the initial and final directions of the motion of P not being continuous with each other, the initial and final directions of the motions of each point P' will not be continuous with each other, or there will be at the point P' an abrupt change in the direction of the curve. Similarly, as P describes the other loop of the figure of eight, each of the points P' will describe another loop; and the two loops belonging to the same point P' will unite together so as to form a figure of eight; viz. to the figure of eight described by P there will correspond figures of eight described by the n points P' respectively.

9. But consider next the case where the two loops of the curve described by P include each of them one and the same point V. This implies that one of the two loops lies inside the other, or that the curve is what has been called an inloop curve. As P, which is in the first instance taken to be at the node, passes continuously along one of the loops and returns to its original position, there are two of the points P' such that the first of these passes from its original position to the original position of the second, and the second of them passes from its original position to the original position of the first of them. We have thus two arcs between these two points P'; but inasmuch as the initial and the final directions of motion of the point P are not continuous with each other, these two arcs are not continuous in direction at the two points P', but at each of these points P' the two arcs meet at an angle. As P describes the other loop, we have in like manner two arcs between the same two points P', these arcs at each of the points P' meeting at an

angle; but they join on to the first-mentioned two arcs in such manner as to form two ovals intersecting each other in the two points P'. Corresponding to the inloop curve described by P, we have this pair of intersecting ovals described by two of the points P', and $n-2$ other curves described by the other points P', and being each of them (I assume) an inloop curve.

10. If we attend only to one of the two intersecting ovals, we have in the first plane an inloop curve, and corresponding thereto in the second plane an oval passing through two of the points P' which correspond to the node P of the inloop curve. Interchanging the two planes, and writing Q instead of P, we have in the first plane an oval passing through two of the points Q which correspond to a point Q'; and corresponding to this oval we have in the second plane an inloop curve having this point Q for its node, viz. these are the corresponding figures mentioned in No. 3.

11. Consider a given point Q; and let the corresponding points Q' be called (selecting the suffixes at pleasure) Q_1', Q_2', .., Q_n'. Taking then a point O indefinitely near to Q, the corresponding points O' will be indefinitely near to Q_1', Q_2', .., Q_n' respectively, and they will be called O_1', O_2', .., O_n' accordingly. It is to be observed that by the indefinitely near point O is meant a point such that the distance from O to Q is indefinitely small in comparison with the distance of either of these points from any point V; so that we cannot have from Q to O two indefinitely short paths including between them a point V; or say so that the indefinitely short path from Q to O is determinate.

Proceeding in this manner from Q to O, and so through a succession of indefinitely near points to a distant point S, we seem to determine the suffixes of the corresponding points S'; but, by what precedes, it appears that such determination for a point S is dependent on the path from Q to S; and consequently that we do not thus obtain a proper determination of the suffixes of the points S'. In fact, if we were to pass from Q by a path including one or more of the points V back to Q, we should obtain for the several points Q' respectively suffixes which are in general different from the suffixes originally given to these points respectively.

12. The difficulty is got over as follows:—Considering as before the given point Q, and calling the corresponding points Q_1', Q_2', .., Q_n' at pleasure, we pass from Q to the indefinitely near point O, and thence, by so many paths chosen at pleasure, to the several branch-points V; these paths from O to the several points V are called *barriers*. To fix the ideas, we may consider these as non-autotomic non-intersecting lines drawn from O to the several points V. Consider the barrier from O to one of these points V; as P passes along this barrier from O to V, two of the corresponding points P' will pass from two of the corresponding points O' to the corresponding cross-point (W'); the paths of these two points are called the *counter-barrier* corresponding to the barrier in question; and we have thus in the second plane a system of counter-barriers, each drawn from two points O' to meet in a point (W'). By what precedes, the points O' have each of them a determinate suffix; a counter-barrier is thus drawn from two points with given suffixes, suppose O_1' and O_2', to a

point (W'), and this may be distinguished accordingly as a counter-barrier 12; and in like manner the cross-point (W') through which it passes will be called a cross-point (W_{12}'); and the barrier corresponding hereto, and the branch-point V at which it terminates, will in like manner be called a barrier 12, and a branch-point V_{12}. Each barrier and branch-point will thus have a pair of suffixes; and the corresponding counter-barrier and cross-point will have the same pair of suffixes. It is to be observed that two or more of these corresponding figures may very well have the same pair of suffixes; but that such corresponding figures must be distinguished from each other; thus, if there are two branch-points V_{12}, these may be distinguished as the branch-points αV_{12} and βV_{12}, and the barriers, counter-barriers, and cross-points by means of these same letters α and β, (or otherwise), as may be convenient. It would seem that not only the number of the points V must be even, but the number of each set of points V_{12} must also be even (see *post*, No. 15).

13. It is also to be noticed that the determination of the suffixes of the several points V, &c., depends *first* upon the arbitrary choice of the suffixes of the points Q', and *next* on the choice of the system of barriers; but that, *these being assumed*, the suffixes of the several points V, &c., are completely determinate.

14. Taking now any point S whatever, and supposing that P moves from Q continuously to S by a path which does not meet a barrier, the points P' will move from the several points Q' to the several points S' by paths not meeting the counter-barriers; viz. to each point S' there will be a path from some point Q'; and giving to such point S' the suffix of the point Q', the suffixes of the several points S' which correspond to any point whatever, S, will be completely determined. The determination depends of course on the assumptions referred to No. 13, but not in anywise on the position of the point S.

It will be noticed that, as all the points V are connected together by the barriers, the only closed paths from a point to itself are paths not including any, or including all, of the points V; and that between such paths there is no real distinction.

15. Consider a point P moving continuously in any manner. The several corresponding points $P_1', P_2', .., P_n'$ will each of them move continuously, but the suffixes interchange; viz. when P arrives at and then passes over a barrier $\alpha\beta$, the corresponding points P_α' and P_β' will each arrive at the corresponding counter-barrier $\alpha\beta$, and, on passing over this, P_α' will be changed into P_β' and P_β' into P_α', the other points P' remaining unchanged; and the like in other cases. This in fact includes the whole or the greater part of the foregoing theory. Thus, if P describe a closed curve not cutting any barrier, there will be no change of suffix; and when P returns to its original position each of the corresponding points $P_1', P_2', .., P_n'$ will describe a closed curve, returning to its original position. But suppose that P describes a closed curve, cutting once only a barrier 12; suppose that the path is from P to Q, and then crossing the barrier to R, and thence again to P; P_1' passes to Q_1', and then crossing the counter-barrier it passes from R_2' to P_2'; while at the same time P_2' passes to Q_2', and then crossing the counter-barrier it passes from R_1' to P_1';

viz. we have P_1', P_2' describing the two portions of a bifid curve. If there were only a single branch-point V_{12}, and therefore only a single barrier OV_{12}, then we might have through P a closed curve cutting OV_{12} once only, and including within it the point O, but not including within it the point V_{12}; and here there ought not to be a bifid curve, but the points P_1', P_2' ought to describe each of them a single curve. But suppose there are two points V_{12}, and consequently two barriers OV_{12} (meeting in O); then the closed curve, meeting once only a barrier 12, (viz. it meets only one such barrier, and that once only), must include within one and only one of the two points V_{12}; and in this case there ought to be a bifid curve. It is by such reasoning as this that I infer the foregoing theorem (No. 12), that the number of each set of points V_{12} is even.

16. We may consider how the suffixes are affected when, instead of the original system of barriers, we have a new system of barriers. I suppose that we have in the two cases respectively the same point Q, and the same suffixes for the points Q_1', Q_2',.., Q_n' which correspond thereto. In the first case, passing from Q to an indefinitely near point O, say the red O, we draw from this point to the several points V a set of barriers, say the red barriers; while in the second case, passing from Q to an indefinitely near point O, say the blue O, we draw from this point to the several points V a set of barriers, say the blue barriers; and we then proceed as before, viz. in the first case, drawing from Q to the point S a curve which does not meet any of the red barriers, we determine accordingly the suffixes (say the red suffixes) of the several corresponding points S'; and in the second case, drawing in like manner from Q to S a curve which does not meet any of the blue barriers, we determine accordingly the suffixes (say the blue suffixes) of the same points S'. Now the curve drawn from Q to S so as not to cut any of the red barriers, and which is used for the determination of the red suffixes of the several points S', will in general cut certain of the blue barriers; and, by examining the suffixes of the blue barriers which are thus cut, we determine the blue suffixes of the same points S'; the result of course depending only on the situation of S in one or other of the regions formed by the red barriers and the blue barriers conjointly. In particular, the point S may be so situate that we can from Q to S draw a curve not meeting any red barrier or any blue barrier; and in this case the red suffixes and the blue suffixes are identical.

17. We may imagine the first plane as consisting of n superimposed planes or sheets, say the sheets 1, 2,.., n. Each barrier 12 is considered as a line drawn in the two sheets 1 and 2; and so on in other cases. The point P is considered as a set of superimposed points P_1, P_2,.., P_n moving in the several sheets respectively; under the convention that P_1 moving in the sheet 1, and coming to cross a barrier 12, passes into the sheet 2 and becomes P_2; and the like in other cases. And this being so, we say that to a point P, considered as a point P_α in the sheet α, there corresponds in the second plane one and only one point P_α'; and that P moving continuously in any manner (subject to the change of sheet as just explained), each of the n corresponding points P' will also move continuously, and so that each such point P_α' will return

to its original position, upon the corresponding point P_a returning to its original position and sheet. This is, in fact, Riemann's theory, only instead of the points P' we must speak of the values $x'+iy'$ of the irrational function of $x+iy$.

18. Everything is of course symmetrical as regards the two planes; we have therefore, in the second plane, a system of points V' and of barriers, and in the first plane a system of points (W) and of counter-barriers. To a given point P' in the second plane there correspond m points P in the first plane; and we can (the determination depending on the system of barriers in the second plane) assign to the m points suffixes, thereby distinguishing them as the corresponding points $P_1, P_2,.., P_m$. And we may imagine the second plane as consisting of m superimposed planes or sheets, say the sheets $1, 2, 3,.., m$; the general theorem then is that to a point P or P' in either plane, considered as a point P_a or P_a' in the sheet a or a', there corresponds in the other plane one and only one point P_a' or $P_{a'}$; and that the first-mentioned point in either plane moving continuously in any manner (subject to the proper change of sheet), the corresponding point in the other plane will also move continuously, and will return to its original position and sheet, upon the first-mentioned point returning to its original position and sheet.

19. In all that precedes it has been assumed that, to a branch-point V, there correspond two united points represented by (W') and $n-2$ distinct points W'; the cases of a point (W') composed of three or more united points, or of the points W' uniting themselves in sets in any other manner, would give rise to further specialities.

690.

ON THE THEORY OF GROUPS.

[From the *Proceedings of the London Mathematical Society*, t. IX. (1878), pp. 126—133. Read May 9, 1878.]

I RECAPITULATE the general theory so far as is necessary in order to render intelligible the quasi-geometrical representation of it which will be given.

Let $\alpha, \beta, ..$ be functional symbols each operating upon one and the same number of letters, and producing as its result the same number of functions of these letters. For instance, $\alpha(x, y, z) = (X, Y, Z)$, where the capitals denote each of them a given function of (x, y, z).

Such symbols are susceptible of repetition and combination;

$$\alpha^2(x, y, z) = \alpha(X, Y, Z),$$

or

$$\beta\alpha(x, y, z) = \beta(X, Y, Z),$$

in each case equal to three given functions of (x, y, z); and similarly for α^3, $\alpha^2\beta$, etc.

The symbols are not in general commutative, $\alpha\beta$ not $=\beta\alpha$; but they are associative, $\alpha\beta.\gamma = \alpha.\beta\gamma$, each $=\alpha\beta\gamma$, which has thus a determinate meaning.

Unity as a functional symbol denotes that the letters are unaltered,

$$1(x, y, z) = (x, y, z);$$

whence

$$1\alpha = \alpha 1 = \alpha.$$

The functional symbols *may* be substitutions; $\alpha(x, y, z) = (y, z, x)$, the same letters in a different order. Substitutions can be represented by the notation $\dfrac{yzx}{zxy}$, the substitution which changes xyz into yzx, or, as products of cyclical substitutions, $\alpha = \dfrac{yzx\ wvu}{xyz\ uvw}$, $=(xyz)(uw)$, the product of the cyclical substitutions x into y, y into z, z into x, and u into w, w into u, the letter v being unaltered.

A set of symbols $\alpha, \beta, \gamma, \ldots$, such that the product $\alpha\beta$ of each two of them (in each order, $\alpha\beta$ and $\beta\alpha$) is a symbol of the set, is a group. It is easily seen that 1 is a symbol of every group, and we may therefore give the definition in the form that a set of symbols $1, \alpha, \beta, \gamma, \ldots$ satisfying the foregoing condition is a group. When the number of symbols (or terms) is $=n$, then the group is of the order n; and each symbol α is such that $\alpha^n = 1$, so that a group of the order n is in fact a group of symbolical nth roots of unity.

A group is defined by means of the laws of combinations of its symbols. For the statement of these we may either (by the introduction of powers and products) diminish as much as may be the number of distinct functional symbols; or else, using distinct letters for the several terms of the group, employ a square diagram, as presently mentioned.

Thus, in the first mode, a group is $1, \beta, \beta^2, \alpha, \alpha\beta, \alpha\beta^2$, $(\alpha^2 = 1,\ \beta^3 = 1,\ \alpha\beta = \beta^2\alpha)$, where observe that these conditions imply also $\alpha\beta^2 = \beta\alpha$.

Or in the second mode, calling the symbols $(1, \alpha, \beta, \alpha\beta, \beta^2, \alpha\beta^2)$ of the same group $(1, \alpha, \beta, \gamma, \delta, \epsilon)$, or, if we please, (a, b, c, d, e, f), the laws of combination are given by one or other of the square diagrams:

	1	α	β	γ	δ	ϵ
1	1	α	β	γ	δ	ϵ
α	α	1	γ	β	ϵ	δ
β	β	ϵ	δ	α	1	γ
γ	γ	δ	ϵ	1	α	β
δ	δ	γ	1	ϵ	β	α
ϵ	ϵ	β	α	δ	γ	1

,

a	b	c	d	e	f
b	a	d	c	f	e
c	f	e	b	a	d
d	e	f	a	b	c
e	d	a	f	c	b
f	c	b	e	d	a

where, taking for greater symmetry the second form of the square, observe that the square is such that no letter occurs twice in the same line, or in the same column (or what is the same thing, each of the lines and of the columns contains all the letters). But this is not sufficient in order that the square may represent a group; the square must be such that the substitutions by means of which its several lines are derived from any line thereof are (in a different order) the same substitutions by which the lines are derived from a particular line, or say from the top line. These, in fact, are:

1
$ab \,.\, cd \,.\, ef$,
$ace \;.\; bfd$,
$ad \,.\, be \,.\, cf$,
$aec \;.\; bdf$,
$af \,.\, bc \,.\, de$,

where, for shortness, *ab*, *ace*, &c., are written instead of (*ab*), (*ace*), &c., to denote the cyclical substitutions *a* into *b*, *b* into *a*; and *a* into *c*, *c* into *e*, *e* into *a*, &c.; and it is at once seen that by the same substitutions the lines may be derived from any other line.

It will be noticed that in the foregoing substitution-group each substitution is *regular*, that is, composed of cyclical substitutions each of the same number of letters; and it is easy to see that this property is a general one; each substitution of the substitution-group must be regular.

By what precedes, the group of any order composed of the functional symbols is replaced by a substitution-group upon a set of letters the number of which is equal to the order of the group, and wherein all the substitutions are regular.

The general theory being thus explained, I endeavour to form a substitution-group with the twelve letters *abcdefghijkl*; and I assume that there is one substitution, such as *abc.def.ghi.jkl*, and another substitution, such as *agj.bfi.cek.dhl*. Observe that, if the twelve letters are to be thus arranged in two different ways as a set of four triads, without repetition of any duad, all the ways in which this can be done are essentially similar, and there is no loss of generality in taking the two sets of triads to be those just written down. But the substitution to be formed with either set of triads will be different according as any triad thereof, for instance *agj*, is written in this form or in the reversed form *ajg*. There are thus in all sixteen substitutions which can be formed with the first set of triads, and sixteen substitutions which can be formed with the second set of triads; and the relation of a triad of the first set to a triad of the second set is by no means independent of the selection of the triads out of the two sets respectively. To show this, take the two substitutions quite at random; suppose they are those written down above, say

$$\alpha = abc\,.\,def\,.\,ghi\,.\,jkl, \quad \beta = agj\,.\,bfi\,.\,cek\,.\,dhl\,;$$

and perform these in succession on the primitive arrangement $\Omega = abcdefghijkl$. The operation stands thus:

$$\beta\alpha\Omega = fegkihlbjcda,$$

$$\alpha\Omega = bcaefdhigklj,$$

$$\Omega = abcdefghijkl,$$

whence

$$\beta\alpha, = afhbeijcgl\,.\,dk,$$

is not a regular substitution; and, by what precedes, α, β cannot belong to a group.

But take the substitutions to be

$$\alpha \text{ (as before)} = abc\,.\,def\,.\,ghi\,.\,jkl, \quad \beta = ajg\,.\,bif\,.\,cek\,.\,dhl,$$

then we have

$$\beta\alpha\Omega = iejkbhlfacdg,$$

$$\alpha\Omega = bcaefdhigklj,$$

$$\Omega = abcdefghijkl,$$

whence

$$\beta\alpha = ai\,.\,be\,.\,cj\,.\,dk\,.\,fh\,.\,gl,$$

a regular substitution; and, for anything that appears to the contrary, α, β may belong to a group. It is convenient to mention at once that these two substitutions do, in fact, give rise to a group; viz. the square diagram is

a	*b*	*c*	*d*	*e*	*f*	*g*	*h*	*i*	*j*	*k*	*l*
b	*c*	*a*	*e*	*f*	*d*	*h*	*i*	*g*	*k*	*l*	*j*
c	*a*	*b*	*f*	*d*	*e*	*i*	*g*	*h*	*l*	*j*	*k*
d	*l*	*h*	*a*	*g*	*j*	*e*	*c*	*k*	*f*	*i*	*b*
e	*j*	*i*	*b*	*h*	*k*	*f*	*a*	*l*	*d*	*g*	*c*
f	*k*	*g*	*c*	*i*	*l*	*d*	*b*	*j*	*e*	*h*	*a*
g	*f*	*k*	*l*	*c*	*i*	*j*	*d*	*b*	*a*	*e*	*h*
h	*d*	*l*	*j*	*a*	*g*	*k*	*e*	*c*	*b*	*f*	*i*
i	*e*	*j*	*k*	*b*	*h*	*l*	*f*	*a*	*c*	*d*	*g*
j	*i*	*e*	*h*	*k*	*b*	*a*	*l*	*f*	*g*	*c*	*d*
k	*g*	*f*	*i*	*l*	*c*	*b*	*j*	*d*	*h*	*a*	*e*
l	*h*	*d*	*g*	*j*	*a*	*c*	*k*	*e*	*i*	*b*	*f*

and the substitutions, obtained therefrom by writing successively each line over the top line, are

$$\begin{array}{ll}
1 & = 1, \\
abc\,.\,def\,.\,ghi\,.\,jkl & \alpha, \\
acb\,.\,dfe\,.\,gih\,.\,jlk & \alpha^2, \\
ad\,.\,bl\,.\,ch\,.\,eg\,.\,fj\,.\,ik & \beta^2\alpha\beta^2, \\
aeh\,.\,bjd\,.\,cil\,.\,fkg & \beta\alpha^2, \\
afl\,.\,bkh\,.\,cgd\,.\,eij & \beta^2\alpha, \\
agj\,.\,bfi\,.\,cke\,.\,dlh & \beta^2, \\
ahe\,.\,bdj\,.\,cli\,.\,fgk & \beta\alpha^2\beta\alpha^2, \\
ai\,.\,be\,.\,cj\,.\,dk\,.\,fh\,.\,gl & \beta\alpha, \\
ajg\,.\,bif\,.\,cek\,.\,dhl & \beta, \\
ak\,.\,bg\,.\,cf\,.\,di\,.\,el\,.\,hj & \beta^2\alpha^2, \\
alf\,.\,bkh\,.\,cdg\,.\,eji & \beta^2\alpha\beta^2\alpha.
\end{array}$$

To explain the theory, I introduce the notion of a hemipolyhedron, or say a hemihedron, viz. this is a figure obtained from a polyhedron by the removal of certain faces. In a polyhedron each edge occurs twice (more properly it occurs in the two forms *ab* and *ba*), as belonging to two faces; but in a hemihedron one of these faces must always be removed, so that the edge may occur once only; and again (what is apparently, although not really, a different thing), we may remove two intersecting faces, leaving their edge of intersection; this edge is, in fact, then considered as a bilateral face $ab = ab.ba$, just as *abc* is a trilateral face $abc = ab.bc.ca$. Thus, if in a prism we remove the lateral faces, leaving the lateral edges, and leaving also the terminal faces, we have a hemihedron: thus, the prism being trilateral, say the faces of the hemihedron are *abc*, *def*, *ad*, *be*, *cf*, where *ad*, *be*, *cf* are the edges regarded as bilateral faces. And, for the present purpose, *abc* denotes the cyclical substitution *a* into *b*, *b* into *c*, *c* into *a*; and *ad* denotes in like manner the cyclical substitution (or interchange) *a* into *d*, *d* into *a*.

But the hemihedron about to be considered has no bilateral faces; it is, in fact, the figure composed of the 8 triangular faces of the octo-hexahedron or figure obtained by truncating the summits of a hexahedron (or of an octahedron) so as to obtain a polyhedron of 8 triangular faces and 6 square faces, representing the faces of the octahedron and the hexahedron respectively. The faces of the octo-hexahedron may be taken to be

abc, *def*, *ghi*, *jkl*,

ajg, *bif*, *cek*, *dhl*,

cbfe, *fihd*, *hgjl*, *jack*, *agib*, *klde*,

(where I observe in passing that the symbols are written in such manner that each edge *ab* occurs under the two opposite forms *ab* in *abc* and *ba* in *agib*). And then, omitting the square faces, represented by the third line, we have the hemihedron, wherein as before *abc* denotes the cyclical substitution *a* into *b*, *b* into *c*, *c* into *a*; and so for the other faces.

I represent this by a diagram, the lines of which were red and black, and they

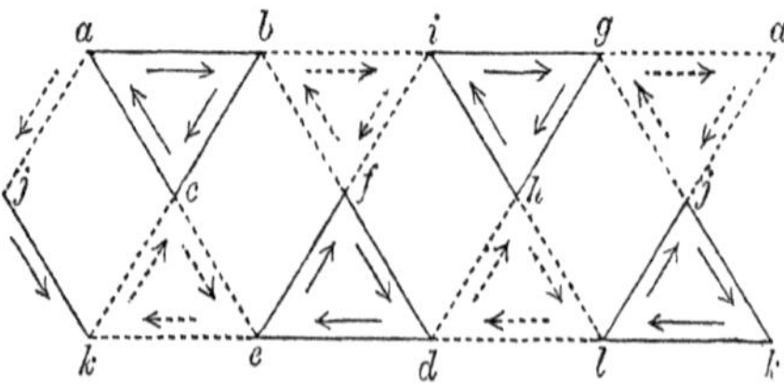

will be thus spoken of, but the black lines are in the woodcut continuous lines, and the red lines broken lines: each face indicates a cyclical substitution, as shown by the arrows. The figure should be in the first instance drawn with the arrows, but without the letters, and these may then be affixed to the several points in a perfectly arbitrary manner; but I have in fact affixed them in such wise that the group given by the diagram, as presently appearing, may (instead of being any other equivalent group) be that group which contains the before-mentioned substitution

$$\alpha = abc \,.\, def \,.\, ghi \,.\, jkl, \text{ and } \beta = ajg \,.\, bif \,.\, cek \,.\, dhl.$$

Observe that in the diagram, considering the lines to be drawn as shown by the arrows, there is *from* any given point whatever only one black line, and only one red line. Let B denote motion along a black line, R motion along a red line (always from a point to the next point); then B^2 will denote motion along two black lines successively, BR (any such symbol being read always from right to left) will denote motion first along a red line, and then along a black line, and so in other cases; a symbol or "route" $\ldots R^\beta B^\alpha$ has thus a perfectly definite signification, determining the path when the initial point is given.

The diagram has the property that every route, leading from any one letter to itself, leads also from every other letter to itself; or say a route leading from a to a, leads also from b to b, from c to $c, \ldots,$ from l to l; and we can thus in the diagram speak absolutely (that is, without restriction as to the initial point) of a route as leading from a point to itself, or say as being equal to unity; *it is in virtue of this property* that the diagram gives a group.

For, assuming the property, it at once follows (1) that two routes, each leading say from the point a to the same point f, lead also from any other point b to one and the same point g. Such routes are said to be equivalent, or equal to each other; and the number of distinct routes (including the route unity) is thus equal to the numbers of the letters, viz. we have only the routes from a to a, to $b, \ldots,$ to l, respectively; (2) a route, leading from a point a to a point f, leads from any other point b to a different point g; and (3) two routes, leading from the same point a to different points b and c, lead also from any other point f to different points k and l. Hence a given route leads from the several points $abc \ldots l$ successively to the same series of points taken in a different order, or we thus obtain a new arrangement of the points; and dealing in this manner successively with the routes from a to a, to $b, \ldots,$ to l, we obtain so many distinct arrangements, beginning with the letters $a, b, c, \ldots, l$ respectively, such that in no two of them does the same letter occupy the same place; we thus obtain a square of 12 such as that already written down, and which is, in fact, the same square, the several routes of course corresponding to the substitutions of the square. The hemihedron thus gives the foregoing group of 12.

Observe that the diagram is composed of the four black triangles representing the substitution $abc \,.\, def \,.\, ghi \,.\, jkl$, and of the four red triangles representing the substitution $ajg \,.\, bif \,.\, cek \,.\, dhl$; viz. these are independent substitutions which by their powers and products serve to express all the substitutions of the group; that they are sufficient appears by the diagram itself, in that every point thereof is (by black and red lines) connected with every other point thereof. The group might have contained three or more independent substitutions, and the diagram would then have contained the like number of differently coloured sets of lines. The essential characters are that the lines of any given colour shall form polygons of the same number of sides (but for different colours the polygons may have different numbers of sides; in particular, for any given colour or colours, the polygons may be bilaterals, represented each by a line with a double arrow pointing opposite ways); that there shall be *from* each point only one line of the same colour; that every point shall be connected with every other point; and finally, that every route leading from one point to itself shall lead also from every other point to itself. When these conditions are satisfied the foregoing investigation in fact shows that the diagram, or say the hemihedron, gives rise to a group.

It may be remarked that we can, if we please, introduce into the diagram a set of lines of a new colour to represent any dependent substitution of the group; thus, in the example considered, a substitution is $aeh.bjd.cil.fkg$, and if we draw these triangles in green (the arrows being from a to e, e to h, h to a, &c.), then there will be *from* each point one black line, one red line, and one green line; any route $\ldots G^\gamma R^\beta B^\alpha$ will thus be perfectly definite, and will have the same properties as a route composed of black and red lines only; and the theory thus subsists without alteration.

I remark, in conclusion, that the group of 12 considered above is, in fact, the group of 12 positive substitutions upon 4 letters $abcd$; viz. the substitutions are 1, abc, acb, abd, adb, acd, adc, bcd, bdc, $ab.cd$, $ac.bd$, $ad.bc$; the groups each contain unity, three substitutions of the order (or index) 2, and 8 substitutions of the order (or index) 3, and their identity can be easily verified.

691.

NOTE ON MR MONRO'S PAPER "ON FLEXURE OF SPACES."

[From the *Proceedings of the London Mathematical Society*, vol. IX. (1878), pp. 171, 172. Read June 13, 1878.]

CONSIDER an element of surface, surrounding a point P; the flexure of the element may be interfered with by the continuity round P, and it is on this account proper to regard the element as cut or slit along a radius drawn from P to the periphery of the element. This being understood, we have the well-known theorem that, considering in the neighbourhood of the origin elements of the surfaces

$$z=\tfrac{1}{2}(ax^2+2hxy+by^2),\quad \text{and}\quad z'=\tfrac{1}{2}(a'x'^2+2h'x'y'+b'y'^2),$$

these will be applicable the one on the other, provided only $ab-h^2=a'b'-h'^2$. But in connexion with Mr Monro's paper it is worth while to give the proof in detail.

It is to be shown that z, z' denoting the above-mentioned functions of (x, y) and (x', y') respectively, it is possible to find (for small values) x', y' functions of x, y such that identically

$$dx'^2+dy'^2+dz'^2=dx^2+dy^2+dz^2.$$

The solution is taken to be $x'=x+\xi$, $y'=y+\eta$, where ξ, η denote cubic functions of x, y. We have then, attending only to the terms of an order not exceeding 3 in x, y,

$$\begin{aligned}dx^2+dy^2+2(dx\,d\xi+dy\,d\eta)+\{(a'x+h'y)\,dx+(h'x+b'y)\,dy\}^2&\\=dx^2+dy^2+\{(ax+hy)\,dx+(hx+by)\,dy\}^2,&\end{aligned}$$

so that the terms dx^2+dy^2 disappear; and then writing

$$d\xi=\frac{d\xi}{dx}dx+\frac{d\xi}{dy}dy,\quad d\eta=\frac{d\eta}{dx}dx+\frac{d\eta}{dy}dy,$$

the equation will be satisfied identically as regards dx, dy if only

$$2\frac{d\xi}{dx} = (ax+hy)^2 - (a'x+h'y)^2,$$

$$\frac{d\xi}{dy} + \frac{d\eta}{dx} = (ax+hy)(hx+by) - (a'x+h'y)(h'x+b'y),$$

$$2\frac{d\eta}{dy} = (hx+by)^2 - (h'x+b'y)^2.$$

Calling the terms on the right-hand side $2\mathfrak{A}$, $\mathfrak{B}$, $2\mathfrak{C}$ respectively, we have

$$\frac{d^2\mathfrak{A}}{dy^2} - \frac{d^2\mathfrak{B}}{dx\,dy} + \frac{d^2\mathfrak{C}}{dx^2} = 0,$$

that is,

$$(h^2 - h'^2) + (h^2 - h'^2) - \{(ab+h^2) - (a'b'+h'^2)\} = 0,$$

or, what is the same thing,

$$a'b' - h'^2 = ab - h^2,$$

a relation which must exist between the constants (a, b, h) and (a', b', h').

It is easy to find the actual values of ξ, η; viz. these are

$$\xi = \tfrac{1}{6}(a^2 - a'^2)\,x^3 + \tfrac{1}{2}(ah - a'h')\,x^2y + \tfrac{1}{2}(h^2 - h'^2)\,x^2y + \tfrac{1}{6}(bh - b'h')\,y^3,$$

$$\eta = \tfrac{1}{6}(ah - a'h')\,x^3 + \tfrac{1}{2}(h^2 - h'^2)\,x^2y + \tfrac{1}{2}(bh - b'h')\,x^2y + \tfrac{1}{6}(b^2 - b'^2)\,y^3,$$

or, what is the same thing, we have

$$\xi = \tfrac{1}{24}\frac{d\Omega}{dx}, \quad \eta = \tfrac{1}{24}\frac{d\Omega}{dy},$$

where

$$\Omega = (a^2 - a'^2)\,x^4 + 4(ah - a'h')\,x^3y + 6(h^2 - h'^2)\,x^2y^2 + 4(bh - b'h')\,xy^3 + (b^2 - b'^2)\,y^4,$$

$$= (ax^2 + 2hxy + by^2)^2 - (a'x^2 + 2h'xy + b'y^2)^2 = 4(z^2 - z'^2),$$

in virtue of the relation $ab - h^2 = a'b' - h'^2$. The resulting values $x' = x + \xi$, $y' = y + \eta$ are obviously the first terms of two series which, if continued, would contain higher powers of (x, y).

692.

ADDITION TO THE MEMOIR ON THE TRANSFORMATION OF ELLIPTIC FUNCTIONS.

[From the *Philosophical Transactions of the Royal Society of London,* vol. CLXIX. Part II. (1878), pp. 419—424. Received February 6,—Read March 7, 1878.]

I HAVE recently succeeded in completing a theory considered in my "Memoir on the Transformation of Elliptic Functions," *Phil. Trans.*, vol. CLXIV. (1874), pp. 397—456, [578],—that of the septic transformation, $n=7$. We have here

$$\frac{1-y}{1+y}=\frac{1-x}{1+x}\left(\frac{\alpha-\beta x+\gamma x^2-\delta x^3}{\alpha+\beta x+\gamma x^2+\delta x^3}\right)^2,$$

a solution of

$$\frac{Mdy}{\sqrt{1-y^2\,.\,1-v^8y^2}}=\frac{dx}{\sqrt{1-x^2\,.\,1-u^8x^2}},$$

where $\frac{1}{M}=1+\frac{2\beta}{\alpha}$; and the ratios $\alpha:\beta:\gamma:\delta$, and the uv-modular equation are determined by the equations

$$u^{14}\alpha^2=v^2\delta^2,$$

$$u^6(2\alpha\gamma+2\alpha\beta+\beta^2)=v^2(\gamma^2+2\gamma\delta+2\beta\delta),$$

$$\gamma^2+2\beta\gamma+2\alpha\delta+2\beta\delta=v^2u^2(2\alpha\gamma+2\beta\gamma+2\alpha\delta+\beta^2),$$

$$\delta^2+2\gamma\delta=v^2u^{10}(\alpha^2+2\alpha\beta);$$

or, what is the same thing, writing $\alpha=1$, the first equation may be replaced by $\delta=\frac{u^7}{v}$, and then, α, δ having these values, the last three equations determine β, γ and the modular equation. If instead of β we introduce M, by means of the relation

$\frac{1}{M}=1+2\beta$, that is, $2\beta=\frac{1}{M}-1$, then the last equation gives $2\gamma=u^3v^3\left(\frac{1}{M}-\frac{u^4}{v^4}\right)$; and $\alpha, \beta, \gamma, \delta$ having these values, we have the residual two equations

$$u^6(2\alpha\gamma+2\alpha\beta+\beta^2)=\quad v^2(\gamma^2+2\gamma\delta+2\beta\delta),$$
$$\gamma^2+2\beta\gamma+2\alpha\delta+\beta\delta=v^2u^2(2\alpha\gamma+2\beta\gamma+2\alpha\delta+\beta^2),$$

viz. each of these is a quadric equation in $\frac{1}{M}$; hence eliminating $\frac{1}{M}$, we have the modular equation; and also (linearly) the value of $\frac{1}{M}$, and thence the values of $\alpha, \beta, \gamma, \delta$ in terms of u, v.

Before going further it is proper to remark that, writing as above $\alpha=1$, then if $\delta=\beta\gamma$, we have

$$1-\beta x+\gamma x^2-\delta x^3=(1-\beta x)(1+\gamma x^2),$$
$$1+\beta x+\gamma x^2+\delta x^3=(1+\beta x)(1+\gamma x^2),$$

and the equation of transformation becomes

$$\frac{1-y}{1+y}=\frac{1-x}{1+x}\left(\frac{1-\beta x}{1+\beta x}\right)^2,$$

viz. this belongs to the cubic transformation. The value of β in the cubic transformation was taken to be $\beta=\frac{u^3}{v}$, but for the present purpose it is necessary to pay attention to an omitted double sign, and write $\beta=\pm\frac{u^3}{v}$; this being so, $\delta=\beta\gamma$, and giving to γ the value $\mp u^4$, δ will have its foregoing value $=\frac{u^7}{v}$. And from the theory of the cubic equation, according as $\beta=\frac{u^3}{v}$ or $=-\frac{u^3}{v}$, the modular equation must be

$$u^4-v^4+2uv(1-u^2v^2)=0, \text{ or } u^4-v^4-2uv(1-u^2v^2)=0.$$

We thus see *à priori*, and it is easy to verify that the equations of the septic transformation are satisfied by the values

$$\alpha=1,\ \beta=\ \frac{u^3}{v},\ \gamma=\ u^4,\ \delta=\frac{u^7}{v}, \text{ and } u^4-v^4+2uv(1-u^2v^2)=0;$$
$$\alpha=1,\ \beta=-\frac{u^3}{v},\ \gamma=-u^4,\ \delta=\frac{u^7}{v}, \text{ and } u^4-v^4-2uv(1-u^2v^2)=0;$$

and it hence follows that in obtaining the modular equation for the septic transformation, we shall meet with the factors $u^4-v^4\pm2uv(1-u^2v^2)$. Writing for shortness $uv=\theta$, these factors are $u^4-v^4\pm2\theta(1-\theta^2)$; the factor for the proper modular equation is $u^8+v^8-\Theta$, where

$$\Theta=8\theta-28\theta^2+56\theta^3-70\theta^4+56\theta^5-28\theta^6+8\theta^7,$$

viz. the equation $(1-u^8)(1-v^8)-(1-uv)^8=0$ is $u^8+v^8-\Theta=0$; and the modular equation, as obtained by the elimination from the two quadric equations, presents itself in the form

$$(u^4-v^4+2\theta-2\theta^3)^2(u^4-v^4-2\theta+2\theta^3)^2(u^8+v^8-\Theta)=0.$$

Proceeding to the investigation, we substitute the values

$$\alpha=1,\ \beta=\tfrac{1}{2}\left(\frac{1}{M}-1\right),\ \gamma=\tfrac{1}{2}u^3v^2\left(\frac{1}{M}-\frac{u^4}{v^4}\right),\ \delta=\frac{u^7}{v},$$

in the residual two equations, which thus become

$$\begin{aligned}&\frac{1}{M^2}(1-v^8) + \frac{2}{M}(1-uv)^3(1+uv)\\ &\qquad + \left\{(1-u^8)-4(1-uv)\left(1+\frac{u^7}{v}\right)\right\}=0,\\ &\frac{1}{M^2}\left\{-u^2v^2(1-uv)^3(1+uv)\right\}+\frac{2}{M}\left\{u^2v^2(1-u^8)+\frac{u^3}{v}(1+u^2v^2)(u^4-v^4)\right\}\\ &\qquad+\left\{\frac{u^{14}}{v^2}+6\frac{u^7}{v}(1-u^2v^2)-u^2v^2\right\}=0,\end{aligned}$$

the first of which is given p. 432 of the "Memoir," [*Coll. Math. Papers*, vol. IX., p. 150]. Calling them

$$\left(\mathrm{a},\ \mathrm{b},\ \mathrm{c}\right)\!\left(\frac{1}{M},\ 1\right)^2=0,\quad \left(\mathrm{a}',\ \mathrm{b}',\ \mathrm{c}'\right)\!\left(\frac{1}{M},\ 1\right)^2=0,$$

we have

$$\frac{1}{M^2}:\frac{2}{M}:1=\mathrm{bc}'-\mathrm{b}'\mathrm{c}:\mathrm{ca}'-\mathrm{c}'\mathrm{a}:\mathrm{ab}'-\mathrm{a}'\mathrm{b},$$

and the result of the elimination therefore is

$$(\mathrm{ca}'-\mathrm{c}'\mathrm{a})^2-4(\mathrm{bc}'-\mathrm{b}'\mathrm{c})(\mathrm{ab}'-\mathrm{a}'\mathrm{b})=0.$$

Write as before $uv=\theta$. In forming the expressions $\mathrm{ca}'-\mathrm{c}'\mathrm{a}$, &c., to avoid fractions we must in the first instance introduce the factor v^2: thus

$$\begin{aligned}v^2(\mathrm{ca}'-\mathrm{c}'\mathrm{a})&=v\{v(1-u^8)-4(1-\theta)(v+u^7)\}\{-\theta^2(1+\theta)(1-\theta)^3\}\\ &\quad-\{u^{14}+6u^6\theta(1-\theta^2)-v^2\theta^2\}\{1-v^8\},\\ &=-\theta^2(1+\theta)(1-\theta)^3\{v^2(-3+4\theta)+u^6(-4\theta+3\theta^2)\}\\ &\quad-\{u^{14}+6u^6(\theta-\theta^3)-v^2\theta^2\}(1-v^8);\end{aligned}$$

but instead of θ^2v^2 writing u^2v^4, the expression on the right-hand side becomes divisible by u^2; and we find

$$\begin{aligned}\frac{v^2}{u^2}(\mathrm{ca}'-\mathrm{c}'\mathrm{a})&=-(1+\theta)(1-\theta)^3\{v^4(-3+4\theta)+u^4(-4\theta^3+3\theta^4)\}\\ &\quad-\{u^{12}+6u^4(\theta-\theta^3)-v^4\}(1-v^8),\end{aligned}$$

and thence

$$-\frac{v^2}{u^2}(\text{ca}'-\text{c}'\text{a}) = u^{12} + u^4(6\theta - 10\theta^3 + 11\theta^4 - 6\theta^5 - 8\theta^6 + 10\theta^7 - 4\theta^8)$$
$$+ v^4(-4 + 10\theta - 8\theta^2 - 6\theta^3 + 11\theta^4 - 10\theta^5 + 6\theta^7) + v^{12}.$$

Similarly we have

$$\frac{v^2}{u^2}(\text{bc}'-\text{b}'\text{c}) = u^{12}(5 - 5\theta + 4\theta^2 - 5\theta^3 + 2\theta^4) + u^4(9\theta - 16\theta^2 + \theta^3 + 10\theta^4 + \theta^5 - 16\theta^6 + 9\theta^7)$$
$$+ v^4(2 - 5\theta + 4\theta^2 - 5\theta^3 + 5\theta^4),$$
$$\frac{v^2}{u^2}(\text{ab}'-\text{a}'\text{b}) = u^4(\theta + \theta^3 - \theta^4) + v^4(2 - 5\theta + 4\theta^2 + 3\theta^3 - 10\theta^4 + 3\theta^5 + 4\theta^6 - 5\theta^7 + 2\theta^8)$$
$$+ v^{12}(-1 + \theta + \theta^3);$$

say these values are

$$u^{12} + pu^4 + qv^4 + v^{12}, \quad \lambda u^{12} + \mu u^4 + \nu v^4, \quad \rho u^4 + \sigma v^4 + \tau v^{12}.$$

The required equation is thus

$$0 = (u^{12} + pu^4 + qv^4 + v^{12})^2 - 4(\lambda u^{12} + \mu u^4 + \nu v^4)(\rho u^4 + \sigma v^4 + \tau v^{12}),$$

viz. the function is

$$\begin{aligned}
&u^{24}\\
&+ u^{16}(2p - 4\lambda\rho)\\
&+ u^8\ (2q\theta^4 + p^2 - 4\lambda\sigma\theta^4 - 4\mu\rho)\\
&+ \quad (2\theta^{12} + 2pq\theta^4 - 4\lambda\tau\theta^{12} - 4\mu\sigma\theta^4 - 4\nu\rho\theta^4)\\
&+ v^8\ (2p\theta^4 + q^2 - 4\mu\tau\theta^4 - 4\nu\sigma)\\
&+ v^{16}(2q - 4\nu\tau)\\
&+ v^{24},
\end{aligned}$$

or say it is

$$= (1,\ b,\ c,\ d,\ e,\ f,\ 1 \between u^{24},\ u^{16},\ u^8,\ 1,\ v^8,\ v^{16},\ v^{24}).$$

Supposing that this has a factor $u^8 - \Theta + v^8$, the form is

$$(u^{16} + Bu^8 + C + Dv^8 + v^{16})(u^8 - \Theta + v^8);$$

and comparing coefficients we have

$$\begin{aligned}
B - \Theta \quad &= b,\\
C - \Theta B + \theta^8 \quad &= c,\\
D\theta^8 - \Theta C + B\theta^8 &= d,\\
\theta^8 - \Theta D + C \quad &= e,\\
-\Theta \quad + D \quad &= f,
\end{aligned}$$

where Θ has the before-mentioned value

$$= (8,\ -28,\ +56,\ -70,\ +56,\ -28,\ +8 \between \theta,\ \theta^2,\ \theta^3,\ \theta^4,\ \theta^5,\ \theta^6,\ \theta^7).$$

From the first, second, and fifth equations, $B = b + \Theta$, $C = c + \Theta B - \theta^8$, $D = f + \Theta$; and the third and fourth equations should then be verified identically. Writing down the coefficients of the different powers of θ, we find

$$
\begin{array}{rl}
2p = & 0+12 \quad 0-20+22-12-16+20-8 \; (\theta^0, .., \theta^8) \\
-4\lambda\rho = & 0-20+20-36+60-44+36-28+8 \quad ,, \\
\hline
b = & 0-\ 8+20-56+82-56+20-\ 8 \quad 0 \quad ,, \\
\Theta = & 0+\ 8-28+56-70+56-28+\ 8 \quad 0 \quad ,, \\
\hline
\therefore\ B = & 0 \quad 0-\ 8 \quad 0+12 \quad 0-\ 8 \quad 0 \quad 0 \quad ,,
\end{array}
$$

that is,

$$B = -8\theta^2 + 12\theta^4 - 8\theta^6;$$

and in precisely the same way the fifth equation gives

$$D = -8\theta^2 + 12\theta^4 - 8\theta^6.$$

We find similarly C from the second equation: writing down first the coefficients of p^2, $2q\theta^4$, $-4\lambda\sigma\theta^4$, and $-4\mu\rho$, the sum of these gives the coefficients of c; and then writing underneath these the coefficients of $B\Theta$ and of $-\theta^8$, the final sum gives the coefficients of C: the coefficients of each line belong to $(\theta^0, \theta^1, .., \theta^{16})$.

$$
\begin{array}{l}
0 \quad 0 \quad 36 \quad 0-120+132+\ 28-316+361-\ 20-340+396-144-112+164-80+16 \\
\qquad\qquad -\ \ 8+\ 20-\ 16-\ 12+\ 22-\ 20 \quad 0+\ 12 \\
\qquad\qquad -\ 40+140-212+140+\ 80-188+168-\ 92-\ 64+176-164+80-16 \\
\quad -36+64-\ 40+\ 60-\ 72+\ 28 \quad 0+\ 68-100+\ 36 \\
\hline
0 \quad 0 \quad 0+64-208+352-272-160+463-160-272+352-208+\ 64 \quad 0 \quad 0 \quad 0 \\
0 \quad 0 \quad 0-64+224-352+224+160-392+160+224-352+224-\ 64 \quad 0 \quad 0 \quad 0 \\
\qquad\qquad\qquad\qquad\qquad -\ 1 \\
\hline
0 \quad 0 \quad 0 \quad 0+\ 16 \quad 0-\ 48 \quad 0+\ 70 \quad 0-\ 48 \quad 0+\ 16 \quad 0 \quad 0 \quad 0 \quad 0,
\end{array}
$$

that is,

$$C = 16\theta^4 - 48\theta^6 + 70\theta^8 - 48\theta^{10} + 16\theta^{12};$$

and in precisely the same way this value of C would be found from the fourth equation. There remains to be verified only the fourth equation $(D+B)\theta^8 - \Theta C = d$, that is,

$$2\theta^8(-8\theta^2 + 12\theta^4 - 8\theta^6) - \Theta C = (2-4\lambda\tau)\theta^{12} + (2pq - 4\mu\sigma - 4\nu\rho)\theta^4,$$

and this can be effected without difficulty.

The factor of the modular equation thus is

$$u^{16} + v^{16} + (-8\theta^2 + 12\theta^4 - 8\theta^6)(u^8 + v^8) + 16\theta^4 - 48\theta^6 + 70\theta^8 - 48\theta^{10} + 16\theta^{12},$$

viz. this is

$$(u^8+v^8)^2+(-4\theta^2+6\theta^4-4\theta^6)\,2\,(u^8+v^8)+16\theta^4-48\theta^6+68\theta^8-48\theta^{10}+16\theta^{12}$$
$$=(u^8+v^8-4\theta^2+6\theta^4-4\theta^6)^2$$
$$=\{(u^4-v^4)^2-4\theta^2(1-\theta^2)^2\}^2,$$

that is,

$$\{u^4-v^4-2\theta(1-\theta^2)\}^2\{u^4-v^4+2\theta(1-\theta^2)\}^2;$$

or the modular equation is

$$\{u^4-v^4-2\theta(1-\theta^2)\}^2\{u^4-v^4+2\theta(1-\theta^2)\}^2(u^8+v^8-\Theta)=0;$$

viz. the first and second factors belong to the cubic transformation; and we have for the proper modular equation in the septic transformation $u^8+v^8-\Theta=0$, or what is the same thing $(1-u^8)(1-v^8)-(1-\theta)^8=0$, that is, $(1-u^8)(1-v^8)-(1-uv)^8=0$, the known result; or, as it may also be written,

$$(\theta-u^8)(\theta-v^8)+7\theta^2(1-\theta)^2(1-\theta+\theta^2)^2=0.$$

The value of M is given by the foregoing relations

$$\frac{1}{M^2}:\frac{2}{M}:1=\lambda u^{12}+\mu u^4+\nu v^4\;:\;-(u^{12}+pu^4+qv^4+v^{12})\;:\;\rho u^4+\sigma v^4+\tau v^{12};$$

but these can be, by virtue of the proper modular equation $u^8+v^8-\Theta=0$, reduced into the form

$$\frac{1}{M^2}:\frac{2}{M}:1=7(\theta-u^8)\;:\;14(\theta-2\theta^2+2\theta^3-\theta^4)\;:\;-\theta+v^8,$$

viz. the equality of these two sets of ratios depends upon the following identities,

$$(-\theta+v^8)(u^{12}+pu^4+qv^4+v^{12})+14(\theta-2\theta^2+2\theta^3-\theta^4)(\rho u^4+\sigma v^4+\tau v^{12})$$
$$=\{-\theta u^4+(1-\theta)(-4-\theta+5\theta^2-\theta^3-4\theta^4)v^4+v^{12}\}(u^8-\Theta+v^8),$$
$$-7(\theta-u^8)(\rho u^4+\sigma v^4+\tau v^{12})-(\theta-v^8)(\lambda u^{12}+\mu u^4+\nu v^4)$$
$$=\{(2\theta+5\theta^2+3\theta^3-2\theta^4-2\theta^5)u^4+(2+2\theta-3\theta^2-5\theta^3-2\theta^4)v^4\}(u^8-\Theta+v^8),$$
$$-2(\theta-2\theta^2+2\theta^3-\theta^4)(\lambda u^{12}+\mu u^4+\nu v^4)+(u^8-\theta)(u^{12}+pu^4+qv^4+v^{12})$$
$$=\{u^{12}+\theta(1-\theta)(3+5\theta+3\theta^2)u^4-\theta v^4\}(u^8-\Theta+v^8),$$

which can be verified without difficulty: from the last-mentioned system of values, replacing θ by its value uv, we then have

$$\frac{1}{M^2}:\frac{2}{M}:1=7u(v-u^7)\;:\;14uv(1-uv)(1-uv+u^2v^2)\;:\;-v(u-v^7),$$

which agree with the values given p. 482 of the "Memoir"; and the analytical theory is thus completed.

693.

A TENTH MEMOIR ON QUANTICS.

[From the *Philosophical Transactions of the Royal Society of London*, vol. CLXIX., Part II. (1878), pp. 603—661. Received June 12,—Read June 20, 1878.]

THE present Memoir, which relates to the binary quintic $(*\between x, y)^5$, has been in hand for a considerable time: the chief subject-matter was intended to be the theory of a canonical form which was discovered by myself and is briefly noticed in Salmon's *Higher Algebra*, 3rd Ed. (1876), pp. 217, 218; writing $a, b, c, d, e, f, g, .., u, v, w$ to denote the 23 *covariants* of the quintic, then a, b, c, d, f are connected by the relation

$$f^2 = -a^3d + a^2bc - 4c^3;$$

and the form contains these covariants thus connected together, and also e; it, in fact, is

$$(1, 0, c, f, a^2b - 3c^2, a^2e - 2cf \between x, y)^5.$$

But the whole plan of the Memoir was changed by Sylvester's discovery of what I term the Numerical Generating Function (N.G.F.) of the covariants of the quintic, and my own subsequent establishment of the Real Generating Function (R.G.F.) of the same covariants. The effect of this was to enable me to establish for any given degree in the coefficients and order in the variables, or as it is convenient to express it, for any given deg-order whatever, a selected system of powers and products of the covariants, say a system of "segregates": these are asyzygetic, that is, not connected together by any linear equation with numerical coefficients; and they are also such that every other combination of covariants of the same deg-order, say every "congregate" of the same deg-order, can be expressed (and that, obviously, in one way only) as a linear function, with numerical coefficients, of the segregates of that deg-order. The number of congregates of a given deg-order is precisely equal to the number of the independent syzygies of the same deg-order, so that these syzygies give in effect the congregates in terms of the segregates: and the proper form in which to exhibit the syzygies is thus to make each of them give a single congregate in terms of the segregates: viz. the left-hand side can always be taken to be a monomial congregate $a^\alpha b^\beta\ldots$ or, to avoid fractions, a numerical multiple of such form; and the right-hand side will then be a linear function, with numerical coefficients, of the segregates of the same deg-order. Supposing such a system of syzygies obtained for a given deg-order, any covariant function (rational and integral function of covariants) is at once expressible as a linear function of the segregates of that deg-order: it is, in fact, only necessary to substitute therein for every monomial congregate its value as a linear function of the segregates. Using the word covariant in its most general sense, the conclusion thus is that every covariant can be expressed, and that in one way only, as a linear function of segregates, or say in the segregate form.

Reverting to the theory of the canonical form, and attending to the relation

$$f^2 = -a^3d + a^2bc - 4c^3,$$

it thereby appears that every covariant multiplied by a power of the quintic itself a, can be expressed, and that in one way only, as a rational and integral function of the covariants a, b, c, d, e, f, linear as regards f: say every covariant multiplied by a power of a can be expressed, and that in one way only, in the "standard" form: as an illustration, take

$$a^2h = 6acd + 4bc^2 + ef.$$

Conversely, an expression of the standard form, that is, a rational and integral function of a, b, c, d, e, f, linear as regards f, not explicitly divisible by a, may very well be really divisible by a power of a (the expression of the quotient of course containing one or more of the higher covariants g, h, &c.), and we say that in this case the expression is divisible, and has for its divided form the quotient expressed as a rational and integral function of covariants. Observe that in general the divided form is not perfectly definite, only becoming so when expressed in the before-mentioned segregate form, and that this further reduction ought to be made. There is occasion, however, to consider these divided forms, whether or not thus further reduced; and moreover it sometimes happens that the non-segregate form presents itself, or can be expressed, with integer numerical coefficients, while the coefficients of the corresponding segregate form are fractional.

The canonical form is peculiarly convenient for obtaining the expressions of the several derivatives (Gordan's *Uebereinanderschiebungen*) $(a,\ b)^1$, $(a,\ b)^2$, &c., (or as I propose to write them $ab1$, $ab2$, &c.), which can be formed with two covariants, the same or different, as rational and integral functions of the several covariants. It will be recollected that in Gordan's theory these derivatives are used in order to establish the system of the 23 covariants: but it seems preferable to have the system of covariants, and by means of them to obtain the theory of the derivatives.

I mention at the end of the Memoir two expressions (one or both of them due to Sylvester) for the N.G.F. of a binary sextic.

The several points above adverted to are considered in the Memoir; the paragraphs are numbered consecutively with those of the former Memoirs upon Quantics.

The Numerical and Real Generating Functions. Art. Nos. 366 to 374, and Table No. 96.

366. I have, in my Ninth Memoir (1871) [462], given what may be called the Numerical Generating Function (N.G.F.) of the covariants of a quartic; this was

$$A(x) = \frac{1 - a^6x^{12}}{1 - ax^4 . 1 - a^2x^4 . 1 - a^2 . 1 - a^3 . 1 - a^3x^6},$$

the meaning being that the number of asyzygetic covariants $a^\theta x^\mu$, of the degree θ in the coefficients and order μ in the variables, or say of the deg-order $\theta . \mu$, is equal to the coefficient of $a^\theta x^\mu$ in the development of this function. And I remarked that the formula indicated that the covariants were made up of (ax^4, a^2x^4, a^2, a^3, a^3x^6), the quartic itself, the Hessian, the quadrinvariant, the cubinvariant, and the cubicovariant, these being connected by a syzygy a^6x^{12} of the degree 6 and order 12. Calling these covariants a, b, c, d, e, so that these italic small letters stand for covariants,

Deg-order.	
1.4	a,
2.0	b,
2.4	c,
3.0	d,
3.6	e,

then it is natural to consider what may be called the Real Generating Function (R.G.F.): this is

$$\frac{1 - e^2}{1 - a . 1 - b . 1 - c . 1 - d . 1 - e};$$

the development of this contains, as it is easy to see, only terms of the form $a^\alpha b^\beta c^\gamma d^\delta$ and $a^\alpha b^\beta c^\gamma d^\delta e$, each with the coefficient $+1$, so that the number of terms of a given deg-order $\theta . \mu$ is equal to the coefficient of $a^\theta x^\mu$ in the first-mentioned function: and these terms of a given deg-order represent the asyzygetic covariants of that deg-order: any other covariant of the same deg-order is expressible as a linear function of them. For instance, deg-order 6.12, the terms of the R.G.F. are a^3d, a^2bc, c^3: there is one more term e^2 of the same deg-order; hence e^2 must be a linear function of these: and in fact

$$e^2 = - a^3d + a^2bc - 4c^3,$$

viz. this is the equation

$$\Phi^2 = - U^3J + U^2IH - 4H^3.$$

367. Sylvester obtained an expression for the N.G.F. of the quintic: this is

$$
\frac{\begin{array}{ll}
a^0\,. & 1 \\
+a^3\,. & x^3+x^5+x^9 \\
+a^4\,. & x^4+x^6 \\
+a^5\,. & x\;+x^3+x^7-x^{11} \\
+a^6\,. & x^2+x^4 \\
+a^7\,. & x\;+x^5-x^9 \\
+a^8\,. & x^2+x^4 \\
+a^9\,. & x^3+x^5-x^7 \\
+a^{10}. & x^2+x^4-x^{10} \\
+a^{11}. & x\;+x^3-x^9 \\
+a^{12}. & x^2-x^8-x^{10} \\
+a^{13}. & x\;-x^7-x^9 \\
+a^{14}. & x^4-x^6-x^8 \\
+a^{15}. & -x^7-x^9 \\
+a^{16}. & x^2-x^6-x^{10} \\
+a^{17}. & -x^7-x^9 \\
+a^{18}. & 1\;-x^4-x^8-x^{10} \\
+a^{19}. & -x^5-x^7 \\
+a^{20}. & -x^2-x^6-x^8 \\
+a^{23}. & -x^{11}
\end{array}}{1-ax^5\,.\,1-a^2x^2\,.\,1-a^2x^6\,.\,1-a^4\,.\,1-a^8\,.\,1-a^{12}};
$$

viz. expanding this function in ascending powers of a, x, then, if a term is $Na^\theta x^\mu$, this means that there are precisely N asyzygetic covariants of the deg-order $\theta\,.\,\mu$.

368. It is known that the number of the irreducible covariants of the binary quintic is $=23$; representing these by the letters $a, b, c, d, e, f, g, h, i, j, k, l, m, n, o, p, q, r, s, t, u, v, w$, ($a$ the quintic itself), the deg-orders of these, and the references* to the tables which give them are

[* See also the paper, 143, in the second volume of this collection.]

Deg-order.		Tab.	Mem.	
1.5	a	13	2	
2.2	b	14	„	
„.6	c	15	„	
3.3	d	16	„	
„.5	e	17	„	
„.9	f	18	„	
4.0	g	19	„	
„.4	h	20	„	
„.6	i	21	„	
5.1	j	22	„	
„.3	k	23	„	
„.7	l	24	„	
6.2	m	83	8	
„.4	n	84	„	
7.1	o	90*	9	
„.5	p	91	„	
8.0	q	25	2	[See also paper 143]
„.2	r	92	9	
9.3	s†			
11.1	t	94	9	
12.0	u	29	3	
13.1	v	95	9	
18.0	w	29A	5.	

Starting from the foregoing expression of the N.G.F. of the quintic, we can, instead of each term $a^\theta x^\mu$, introduce a covariant or product of covariants of the proper deg-order $\theta.\mu$: the mode of doing this depends of course on the different admissible partitions of θ, μ, and it is for some of the terms very indeterminate: for instance, a^5x^{11} is ai, bf, or ce. I found it possible to perform the whole process so as to satisfy a condition which will be presently referred to; and I found

[* See vol. VII. of this collection, p. 348.]

† See end of Memoir. The S of Table 93 has the value $-96(D, M)+16BO-7GK$, but it is better to use the simple value $-(D, M)$; and the S of the present Memoir has this value, say $S=-(d, m)$.

R.G.F. of quintic =	Deg-orders.
$1\ .1-b^5$	0.0 – 10.10
$+d\ .1-ag^2$	3.3 – 12. 8
$+e\ .1-b^2$	3.5 – 7. 9
$+f\ .1-b$	3.9 – 5.11
$+h\ .1-ag^2$	4.4 – 13. 9
$+i\ .1-b^2g$	4.6 – 12.10
$+j\ .1-ag^2$	5.1 – 14. 6
$+k\ .1-b^2$	5.3 – 9. 7
$+l\ .1-bg$	5.7 – 11. 9
$+m\ .1-ag^2$	6.2 – 15. 7
$+n\ .1-b^2g$	6.4 – 14. 8
$+o\ .1-b^3$	7.1 – 13. 7
$+p\ .1-b^2g$	7.5 – 15. 9
$+r\ .1-b^2g$	8.2 – 16. 6
$+dj\ .1-ag^2$	8.4 – 17. 9
$+s\ .1-abg$	9.3 – 16.10
$+hj\ .1-ag^2$	9.5 – 18.10
$+j^2\ .1-ag^2$	10.2 – 19. 7
$+jk\ .1-b^2g$	10.4 – 18. 8
$+t\ .1-b^3$	11.1 – 17. 7
$+jm\ .1-ag^2$	11.3 – 20. 8
$+jo\ .1-bg$	12.2 – 18. 4
$+v\ .1-b^5$	13.1 – 23.11
$+js\ .1-bg$	14.4 – 20. 6
$+jt\ .1-g$	16.2 – 20. 2
$+w\ .1-a$	18.0 – 19. 5

$$\overline{1-a\,.\,1-b\,.\,1-c\,.\,1-g\,.\,1-q\,.\,1-u},$$

where observe that each negative term of the numerator is equal to a positive term multiplied by a power or product of terms a, b, g, contained in the denominator: this is the condition above referred to. The expansion thus consists only of terms each with the coefficient $+1$; for instance, a part of the function is

$$\frac{s(1-abg)}{1-a\,.\,1-b\,.\,1-c\,.\,1-g\,.\,1-q\,.\,1-u},\quad =\frac{s}{1-c\,.\,1-q\,.\,1-u}\cdot\frac{1-abg}{1-a\,.\,1-b\,.\,1-g},$$

where the first factor is the entire series of terms $sc^{\delta}q^{\epsilon}u^{\zeta}$, and the second factor is the series of terms $a^{\alpha}b^{\beta}g^{\gamma}$ omitting only those terms which are divisible by abg: and in the product of the two factors the terms are all distinct, so that the coefficients are still each $=1$. The same thing is true for every other pair of numerator terms: and since the terms arising from each such pair are distinct from each other, in the expansion of the entire function the coefficients are each $=+1$. Hence (as in the case of the quartic) for any given deg-order, the terms in the expansion of the R.G.F. may be taken for the asyzygetic covariants of that deg-order; and if there are any other terms of the same deg-order, each of these must be a linear function, with numerical coefficients, of these asyzygetic covariants: thus deg-order 6.14, the expansion contains only the terms a^2h, acd, bc^2; there is besides a term of the same deg-order, ef, which is not a term of the expansion, and hence ef must be a linear function of a^2h, acd, bc^2; we in fact have $ef=a^2h-6acd-4bc^2$.

The terms in the expansion of the R.G.F. may be called "segregates," and the terms not in the expansion "congregates"; the theorem thus is: every congregate is a linear function, with determinate numerical coefficients, of the segregates of the same deg-order.

369. I stop to remark that the numerator of the R.G.F. may be written in the more compendious form

$$\begin{aligned}&(1-b^5)(1-v)+(1-b^3)(o+t)+(1-b^2)(e+k)+(1-b)f\\&+(1-ag^2)(d+h+j+m+dj+hj+j^2+jm)\\&+(1-bg)(l+jo+js)\\&+(1-b^2g)(i+n+p+jk)\\&+(1-abg)s\\&+(1-g)jt\\&+(1-a)w;\end{aligned}$$

but the first-mentioned form is, I think, the more convenient one.

370. It is to be noticed that the positive terms of the numerator are unity, the seventeen covariants $d, e, f, h, i, j, k, l, m, n, o, p, r, s, t, v, w$, and the products of j by (d, h, j, k, m, o, s, t), where j^2 is reckoned as a product; in all, 26 terms. Disregarding the negative terms of the numerator the expansion would consist of these 26 terms, each multiplied by every combination whatever $a^{\alpha}b^{\beta}c^{\gamma}g^{\delta}q^{\epsilon}u^{\zeta}$ of the denominator terms a, b, c, g, q, u (which for this reason might be called "reiterative"): the effect of the negative terms of the numerator is to remove from the expansion certain of the terms in question, thereby diminishing the number of the segregates: thus as regards the terms belonging to unity, any one of these which contains the factor b^5 is not a segregate but a congregate: and so as regards the terms belonging to d, any one of these which contains the factor ag^2 is a congregate: and the like in other cases.

For a given deg-order we have a certain number of segregates and a certain number of congregates: and the number of independent syzygies of that deg-order is precisely equal to the number of congregates: viz. each such syzygy may be regarded as giving a congregate in terms of the segregates: we have on the left-hand side a congregate, or, to avoid fractions, a numerical multiple of the congregate, and on the right-hand side a linear function, with numerical coefficients, of the segregates.

371. The syzygy is irreducible or reducible; and in the latter case it is, or is not, simply divisible: viz. if the congregate on the left-hand side contains any congregate factor (the other factor being literal), then the syzygy is reducible: it is, in fact, obtainable from the syzygy (of a lower deg-order) which gives the value of such congregate factor. But there are here two cases; multiplying the lower syzygy by the proper factor, the right-hand side may still contain segregates only, and then no further step is required: the original syzygy is nothing else than this lower syzygy, each side multiplied by the factor in question, and it is accordingly said to be simply divisible (S.D.). But contrariwise, the right-hand side, as multiplied, may contain congregates which have to be replaced by their values in terms of the segregates of the same deg-order: the resulting expression is then no longer explicitly divisible by the introduced factor: and the original syzygy, although arising as above from a lower syzygy, is not this lower syzygy each side multiplied by a factor: viz. it is in this case not simply divisible.

For example (see the subsequent Table No. 96, under the indicated deg-orders) (6.6), from the syzygy

$$9d^2 = aj - b^3 + 2bh - cg,$$

we deduce (7.11) the syzygy

$$9ad^2 = a^2j - ab^3 + 2abh - acg,$$

which (all the terms on the right-hand being segregates) requires no further reduction: it is a reducible and simply divisible syzygy. But we have (6.8) a syzygy giving de, and also (6.10) a syzygy giving e^2; multiplying the former of these by e or the latter of them by d, we obtain values of de^2, but in each case the right-hand sides contain terms which are not segregates, and have thus to be further reduced; the final formula (9.13) is

$$3de^2 = -4a^2bj + 3a^2dg + 4ab^4 - 8ab^2h + 4abcg - 12b^2cd,$$

which is not divisible by any factor: the syzygy is thus reducible, but not simply divisible.

A syzygy, which is not in the sense explained reducible, is said to be irreducible.

372. The number of irreducible syzygies is obviously finite: it has, however, the large value 179 as appears from the annexed diagram, showing the congregates determined by these several syzygies, and the deg-orders of the syzygies:—

	1	d	e	f	h	i	j	k	l	m	n	o	p	r	s	t	v	w	j^2
1	b^5 10.10	ag^2 12.8	b^2 7.9	b 5.11	ag^2 13.9	b^2g 12.10	ag^2 14.6	b^2 9.7	bg 11.9	ag^2 15.7	b^2g 14.8	b^3 13.7	b^2g 15.9	b^2g 16.6	abg 16.10	b^3 17.7	b^5 23.11	a 19.5	
d		6.6	6.8	6.12	7.7	7.9	ag^2 17.9	8.6	8.10	9.5	9.7	10.4	10.8	11.5	12.6	14.4	16.4	21.3	13.5
e			6.10	6.14	7.9	7.11	8.6	8.8	8.12	9.7	9.9	10.6	10.10	11.7	12.8	14.6	16.6	21.5	
f				6.18	7.13	7.15	8.10	8.12	8.16	9.11	9.13	10.10	10.14	11.11	12.12	14.10	16.10	21.9	
h					8.8	8.10	ag^2 18.10	9.7	9.11	10.6	10.8	11.5	11.9	12.6	13.7	15.5	17.5	22.4	14.6
i						8.12	9.7	9.9	9.13	10.8	10.10	11.7	11.11	12.8	13.9	15.7	17.7	22.6	
j							ag^2 19.7	b^2g 18.8	10.8	ag^2 20.8	11.5	bg 18.4	12.6	13.3	bg 20.6	g 20.2	18.2	23.1	15.3
k								10.6	10.10	11.5	11.7	12.4	12.8	13.5	14.6	16.4	18.4	23.3	15.5
l									10.14	11.9	11.11	12.8	12.12	13.9	14.10	16.8	18.8	23.7	
m										12.4	12.6	13.3	13.7	14.4	15.5	17.3	19.3	24.2	16.4
n											12.8	13.5	13.9	14.6	15.7	17.5	19.5	24.4	
o												14.2	14.6	15.3	16.4	18.2	20.2	25.1	17.3
p													14.10	15.7	16.8	18.6	20.6	25.5	
r														16.4	17.5	19.3	21.3	26.2	
s															18.6	20.4	22.4	27.3	19.5
t																22.2	24.2	29.1	21.3
v																	26.2	31.1	
w																		36.0	

Each term inside this diagram is a deg-order indicating the congregate determined by an irreducible syzygy: viz. the congregate is the product of the outside covariants in the line and column containing the deg-order, and of the literal factor (if any) placed immediately above the deg-order. Thus, line d and column i, 7.9 indicates the congregate di, but, same line and column j, 17.9 indicates the congregate $dj \cdot ag^2$, $= adg^2j$.

Observe as regards the foregoing diagram, that dj^2 is irreducible (since neither dj nor j^2 is segregate), and similarly j^2h, j^3, &c., are irreducible: we have thus the last or j^2 column of the diagram.

The simply divisible syzygies are infinite in number, as are also the reducible syzygies not simply divisible. There is obviously no use in writing down a simply divisible syzygy; but as regards the reducible syzygies not simply divisible, these require a calculation, and it is proper to give them as far as they have been obtained.

373. The following Table, No. 96, replaces Tables 88 and 89 of my Ninth Memoir. The arrangement is according to deg-orders, and the table is complete up to the deg-order 8.40: it shows for each deg-order the segregate covariants, and also the congregate covariants (if any), and the syzygies which are the expressions of these in terms of the segregates. When there are only segregates these are given in the same horizontal line with the deg-order; for instance, | 5.9 | ab^2, ah, cd, shows that for the deg-order 5.9 the only covariants are the segregates ab^2, ah, cd; but when there are also congregates, the segregates are arranged in the same horizontal line with the deg-order, and the congregates, each in its own horizontal line together with its expression as a linear function of the segregates: thus

5.11	*	ai	ce
	bf	-1	$+1$

, the segregates are ai, ce, and there is a congregate bf which is a linear function of these, $=-ai+ce$. The table gives the irreducible syzygies and also the reducible syzygies which are not simply divisible, but the simply divisible syzygies are indicated each by a reference to the divided syzygy which occurs previously in the table.

374. Any syzygy might of course be directly verified by substituting for the several covariants contained therein their expressions in terms of the coefficients and facients of the quintic. But it is to be remarked that among the syzygies, or easily deducible from them, we have (6.18) the before-mentioned equation $f^2=-a^3d+a^2bc-4c^3$, and also a set of 17 syzygies, the left-hand sides of which are the covariants $g, h, .., u, v, w$, each multiplied by a or a^2, and which lead ultimately to the standard expressions of these covariants respectively, viz. each covariant multiplied by a proper power of a can be expressed as a rational and integral function of a, b, c, d, e, f, linear as regards f. Supposing them thus expressed, a far more simple verification of any syzygy would consist in substituting therein for the several covariants their expressions in the standard form, reducing if necessary by the equation $f^2=-a^3d+a^2bc-4c^3$: but of course, as to the syzygies used for obtaining the standard forms, this is only a verification if the standard forms have been otherwise obtained, or are assumed to be correct.

The 17 syzygies above referred to are

Deg-ord.		
6.10	$a^2g =$	$12abd + 4b^2c + e^2,$
6.14	$a^2h =$	$6acd + 4bc^2 + ef,$
5.11	$ai = -$	$bf + ce,$
6.6	$aj =$	$b^3 - 2bh + cg + 9d^2,$
6.8	$ak = -$	$2bi + 3de,$
6.12	$al =$	$2ci - 3df,$
7.7	$am = -$	$2b^2d - cj + 3dh,$
7.9	$an =$	$b^2e - 6bl - 2ck - fg,$
8.6	$ao =$	$2bn + ej,$
8.10	$ap = -$	$2cn - fj,$
9.5	$aq = -$	$2b^2j + bdg - 12dm + hj,$
9.7	$ar =$	$b^2k + bp - co + hk,$
10.8	$as =$	$3bdk + 3dp + 2im,$
12.6	$at =$	$bjk + jp - 2mn,$
13.5	$18au =$	$2agq + b^2gj + 6bmj - 6dj^2 - ghj + no,$
14.6	$3av =$	$2b^3q - 8b^2j^2 - 2b^2gm + 6bdgj - 12bm^2 + 3et,$
19.5	$18aw =$	$3b^2gt + b^2qo - 4bj^2o - bgmo + 18bmt + 3dgjo - 18djt - 3ght - 6m^2o,$

the last four of these being, however, beyond the limits of the table: the expressions of g, h, i are here in the standard form: the standard forms of the other covariants $j, k, .., u, v, w$, will be given further on.

TABLE No. 96 (Segregates, Congregates, and Syzygies).

Deg-ord.	Congs.	Segregates.
1. 1		1
3		
5		a
2. 0		
2		b
4		
6		c
8		
10		a^2
3. 1		
3		d
5		e
7		ab
9		f
11		ac
13		
15		a^3

TABLE No. 96 (*continued*).

Deg-ord.	Congs.	Segregates.			
4. 0		g			
2					
4		b^2,	h		
6		i			
8		ad,	bc		
10		ae			
12		a^2b,	c^2		
14		af			
16		a^2c			
18					
20		a^4			
5. 1		j			
3		k			
5		ag,	bd		
7		be,	l		
9		ab^2,	ah,	cd	
11	*	ai,	ce		
	bf	-1	$+1$		
13		a^2d,	abc		
15		a^2e,	cf		
17		a^3b,	ac^2		
19		a^2f			
21		a^3c			
23					
25		a^5			
6. 0					
2		bg,	m		
4		n			
6	*	aj,	b^3,	bh,	cg
	$d^2 . 9$	$+1$	-1	$+2$	-1
8	*	ak,	bi		
	$de . 3$	$+1$	$+2$		
10	*	a^2g,	abd,	b^2c,	ch
	e^2	$+1$	-12	-4	.
12	*	abe,	al,	ci	
	$df . 3$	.	-1	$+2$	
14	*	a^2b^2,	a^2h,	acd,	bc^2
	ef	.	$+1$	-6	-4
16	*	a^2i,	ace		
	abf	. . . . S.D. 5.11, bf			

TABLE No. 96 (*continued*).

Deg-ord.	Congs.	Segregates.
6. 18	*	a^3d, a^2bc, c^3
	f^2	-1 $+1$ -4
20		a^3e, acf
22		a^4b, a^2c^2
24		a^3f
26		a^4c
28		
30		a^6
7. 1		o
3		bj, dg
5		bk, eg, p
7	*	aby, am, b^2d, cj
	$dh.3$	0 $+1$ $+2$ $+1$
9	*	an, bl, ck, fg
	b^2e	$+1$ $+6$ $+2$ $+1$
	$di.3$	0 $+1$ $+1$
	eh	0 $+4$ $+2$ $+1$
11	*	a^2j, ab^3, abh, acg, bcd
	ad^2	. . . . S.D. 6.6, d^2
	ei	. . -1 $+6$ -6
13	*	a^2k, abi, bce, cl
	ade	. . . . S.D. 6.8 , de
	b^2f	. . . . S.D. 5.11, bf
	$fh.3$	-1 -2 $+3$ -6
15	*	a^3g, a^2bd, ab^2c, ach, c^2d
	ae^2	. . . . S.D. 6.10, e^2
	fi	. $+1$ -1 $+1$ -6
17	*	a^2be, a^2l, aci, c^2e
	adf	. . . . S.D. 6.12, df
	bcf	. . . . S.D. 5.11, bf
19	*	a^3b^2, a^3h, a^2cd, abc^2
	aef	. . . . S.D. 6.14, ef
21	*	a^3i, a^2ce, c^3f
	a^2bf	. . . . S.D. 5.11, bf
23	*	a^4d, a^3bc, ac^3
	af^2	. . . . S.D. 6.18, f^2

TABLE No. 96 (*continued*).

Deg-ord.	Congs.	Segregates.							
7. 25		a^4e,	a^2cf						
27		a^5b,	a^3c^2						
29		a^4f							
31		a^5c							
33									
35		a^7							
8. 0		g^2,	q						
2		r							
4		b^2g,	bm,	dj,	gh				
6	*	ao,	bn,	gi					
	$dk\,.\,3$	+1	−3	−1					
	ej	+1	−2						
8	*	abj,	adj,	b^4,	b^2h,	bcg,	cm		
	bd^2	. . . .							S.D. 6.6, d^2
	ek	−4	+3	+4	−6	+2			
	$h^2\,.\,3$	+4	−3	−4	+8	−1	+12		
10	*	abk,	aeg,	ap,	b^2i,	cn			
	bde	. . . .							S.D. 6.8, de
	$dl\,.\,9$	+2	.	+3	+1	+3			
	fj	.	.	−1	.	−2			
	$hi\,.\,3$	+1	.	.	+2	−3			
12	*	a^2bg,	a^2m,	ab^2d,	acj,	b^3c,	bch,	c^2g	
	adh	. . . .							S.D. 7.7, dh
	be^2	. . . .							S.D. 6.10, e^2
	cd^2	. . . .							S.D. 6.6, d^2
	el	.	−1	−2	+1	−2	+2		
	fk	.	+1	.	−1	+4	−6	+2	
	i^2	.	.	−1	.	+1	−2	+1	
14	*	a^2n,	abl,	ack,	afg,	bci			
	ab^2e	. . . .							S.D. 7.9, b^2e
	adi	. . . .							S.D. 7.9, di
	aeh	. . . .							S.D. 7.9, eh
	bdf	. . . .							S.D. 6.12, df
	cde	. . . .							S.D. 6.8, de
16	*	a^3j,	a^2b^3,	a^2bh,	a^2cg,	$abcd$,	b^2c^2,	c^2h	
	a^2d^2	. . . .							S.D. 6.6, d^2
	aei	. . . .							S.D. 7.11, ei
	bef	. . . .							S.D. 6.14, ef
	ce^2	. . . .							S.D. 6.10, e^2
	$fl\,.\,3$	+1	−1	+2	−1	−3	−6	+6	

TABLE No. 96 (*continued*).

Deg-ord.	Congs.	Segregates.						
8. 18	*	a^3k,	a^2bi,	$abce$,	acl,	c^2i		
	a^2de	. . . . . . S.D. 6.8, de						
	ab^2f	. . . . . . S.D. 5.11, bf						
	afh	. . . . . . S.D. 7.13, fh						
	cdf	. . . . . . S.D. 6.12, df						
20	*	a^4g,	a^3bd,	a^2b^2c,	a^2ch,	ac^2d,	bc^3	
	a^2e^2	. . . . . . S.D. 6.10, e^2						
	afi	. . . . . . S.D. 7.15, fi						
	bf^2	. . . . . . S.D. 6.18, f^2						
	cef	. . . . . . S.D. 6.14, ef						
22	*	a^3be,	a^3l,	a^2ci,	$abcf$,	ac^2e		
	a^2df	. . . . . . S.D. 6.12, df						
24	*	a^4b^2,	a^4h,	a^3cd,	$a^2b^2c^2$,	c^4		
	a^2ef	. . . . . . S.D. 6.14, ef						
	cf^2	. . . . . . S.D. 6.18, f^2						
26	*	a^4i,	a^3ce,	ac^2f				
	a^3bf	. . . . . . S.D. 5.11, bf						
28	*	a^5d,	a^4bc,	a^2c^3				
	a^2f^2	. . . . . . S.D. 6.18, f^2						
30		a^5e,	a^3cf					
32		a^6b,	a^4c^2					
34		a^5f						
36		a^6c						
38								
40		a^8						
9. 1		gj						
3		bo,	gk,	s				
5	*	ag^2,	aq,	b^2j,	bdg,	hj		
	dm . 12	.	−1	−2	+1	+1		
7	*	ar,	beg,	bp,	co,	gl		
	b^2k . 3	+1	.	−5	−1	+1		
	dn . 3	−1	.	−1	−1			
	em . 3	+2	.	+2	+1	−1		
	hk . 3	+2	.	+2	+4	−1		
	ij	.	.	+1	+1			
9	*	ab^2g,	abm,	adj,	agh,	b^3d,	bcj,	cdg
	bdh . 3	.	+1	.	.	+2	+1	
	d^3 . 27	.	+2	+3	.	+1	+2	−3
	en	+1	+6	.	−1	.	+2	
	ik	.	−1	.	.	.	−3	+3

TABLE No. 96 (*continued*).

Deg-ord.	Congs.	Segregates.										
9. 11	*	a^2o,	abn,	agi,	b^2l,	bck,	ceg,	cp				
	adk											S.D. 8.6, dk
	aej											S.D. 8.6, ej
	b^3e	.	+1	−1	+6	+2	+1					
	bdi											S.D. 7.9, bi
	beh	.	.	−1	+4	+2	+1					
	bfg											S.D. 5.11, bf
	d^2e . 9	+1	−3	−1	+2	+2						
	fm . 3	+1	+3	+1	.	−3	.	−3				
	hl . 3	+1	−3	−1	+2	+5	.	−6				
13	*	a^2bj,	a^2dg,	ab^4,	ab^2h,	$abcg$,	acm,	b^2cd,	c^2j			
	abd^2											S.D. 6.6, d^2
	aek											S.D. 8.8, ek
	ah^2											S.D. 8.8, h^2
	bei											S.D. 7.11, ei
	cdh											S.D. 7.7, dh
	de^2 . 3	−4	+3	+4	−8	+4	.	−12				
	fn	−2	+1	+2	−3	+1	−2	.	+2			
	il	−1	.	+1	−2	+1	−3	−3	+3			
10. 0												
2		bg^2,	bq,	gm,	j^2							
4	*	br,	gn,	jk								
	do . 3	+2		−1								
6	*	agj,	b^3g,	b^2m,	bdj,	bgh,	cg^2,	cq				
	d^2g											S.D. 6.6, d^2
	eo	+1	+2	+12	−12	−2						
	hm . 3	−1	+1	+ 6	+12	−2	+1	−3				
	k^2	+1	−1	− 4	−12	+2	−1					
8	*	abo,	agk,	as,	b^2n,	bgi,	cr					
	bdk											S.D. 8.6, dk
	bej											S.D. 8.6, ej
	deg											S.D. 6.8, de
	dp . 9	−5	.	+3	+15	+5	− 6					
	hn . 3	−6	+1	.	+18	+5	−12					
	im . 3	+1	.	.	− 3	−1	+ 3					
	jl . 3	−5	.	+3	+15	+5	−12					
10	*	a^2g^2,	a^2q,	ab^2j,	$abdg$,	ahj,	b^3h,	b^2cg,	bcm,	cdj,	cgh	
	adm											S.D. 9.5, dm
	b^5 . 8	.	+1	+10	−9	+3	+12	.	+32	−12	−4	
	b^2d^2 . 72	.	−1	− 2	+9	−3	+ 4	−8	−32	+12	+4	
	bek . 2	.	+1	+ 2	−3	+3	.	+4	+32	−12	−4	
	bh^2 . 6	.	−1	− 2	+3	−3	+ 4	−2	− 8	+12	−4	
	d^2h . 27	.	−1	− 2	+3	.	+ 1	−2	− 8	+12	+1	
	e^2g											S.D. 6.10, e^2
	ep	.	.	.	.	−1	.	−2	−12	+ 6	+2	
	fo . 2	.	+1	+ 2	−3	+7	.	+4	+24	−24	−4	
	in . 4	.	−1	− 2	+3	−5	.	.	.	+12		
	kl . 4	.	+1	+ 2	−1	−1	.	.	+16	+12		

TABLE No. 96 (*concluded*).

Deg-ord.	Congs.	Segregates.						
11. 1		go,	t					
3		bgj,	dg^2,	dq,	jm			
5	*	b^2o,	bgk,	bs,	eg^2,	eq,	gp	
	dr . 18	-2	$+5$	-6	.	-3	$+3$	
	ho . 3	-2	$+11$	-24	.	-3	$+6$	
	jn	$+1$	-3	$+6$	.	$+1$	-2	
	km . 6	-2	$+5$	-12	.	-3	$+3$	
12. 0		g^3,	g^2q,	u				
2		gr,	jo					
4	*	b^2g^2,	b^2q,	bgm,	bj^2,	dgj,	g^2h,	hg
	ko	.	-2	-2	-4	$+3$		
	m^2 . 12	.	$+2$	$+1$	$+4$	-3	.	-3
13. 1		g^2j,	jq,	v				
3	*	bgo,	bt,	g^2k,	gs,	kg		
	jr . 2	.	-2	.	$+1$	-1		
	mo . 2	.	-4	.	$+1$	-1		
14. 0								
2		bg^3,	bgq,	bu,	g^2m,	gj^2,	mq,	o^2
4	*	bgr,	bjo,	g^2n,	gjk,	js,	nq	
	dgo	. . . . S.D. 10.4, *do*						
	dt . 18	$+1$	$+2$	.	-1	$+6$	$+3$	
	mr . 12	$+1$	$+2$	.	-1	.	$+3$	

Theory of the Canonical Form. Art. Nos. 375 to 381, and Tables Nos. 97 and 98.

375. As the small italic letters have been used to represent the covariants, different letters are required for the coefficients of the quintic: using also new letters for the facients, I take the quintic to be $(\mathrm{a, b, c, d, e, f} \between \xi, \eta)^5$. Considering a linear transformation of $\frac{1}{\mathrm{a}}(\mathrm{a, b, c, d, e, f} \between \xi, \eta)^5$, viz.

$$\frac{1}{\mathrm{a}}(\mathrm{a, b, c, d, e, f} \between \xi - \mathrm{b}\eta, \mathrm{a}\eta)^5,$$

this is

	ξ^5	$5\xi^4\eta$	$10\xi^3\eta^2$	$10\xi^2\eta^3$	$5\xi\eta^4$	η^5
=	1	$-\mathrm{b}$	$+\mathrm{b}^2$	$-\mathrm{b}^3$	$+\mathrm{b}^4$	$-\mathrm{b}^5$
$+\mathrm{b}$ (		1	$-2\mathrm{b}$	$+3\mathrm{b}^2$	$-4\mathrm{b}^3$	$+5\mathrm{b}^4$)
$+\mathrm{ac}$ (			1	$-3\mathrm{b}$	$+6\mathrm{b}^2$	$-10\mathrm{b}^3$)
$+\mathrm{a}^2\mathrm{d}$ (				1	$-4\mathrm{b}$	$+10\mathrm{b}^2$)
$+\mathrm{a}^3\mathrm{e}$ (					1	$-5\mathrm{b}$)
$+\mathrm{a}^4\mathrm{f}$ (						1),

which is

=(						
	1	0	$\mathrm{ac}+1$	$\mathrm{a}^2\mathrm{d}+1$	$\mathrm{a}^3\mathrm{e}+1$	$\mathrm{a}^4\mathrm{f}+1$
			b^2-1	$\mathrm{abc}-3$	$\mathrm{a}^2\mathrm{bd}-4$	$\mathrm{a}^3\mathrm{be}-5$
				b^3+2	$\mathrm{ab}^2\mathrm{c}+6$	$\mathrm{a}^2\mathrm{b}^2\mathrm{d}+10$
					b^4-3	$\mathrm{ab}^3\mathrm{c}-10$
						b^5+4

$)(\xi, \eta)^5$.

The values of a, b, c, d, e, f, considered for a moment as denoting the leading coefficients of the several covariants ultimately represented by these letters respectively, are

a	b	c	d	e	f
$\mathrm{a}+1$	$\mathrm{ae}+1$	$\mathrm{ac}+1$	$\mathrm{ace}+1$	$\mathrm{a}^2\mathrm{f}+1$	$\mathrm{a}^2\mathrm{d}+1$
	$\mathrm{bd}-4$	b^2-1	ad^2-1	$\mathrm{abe}+5$	$\mathrm{abc}-3$
	c^2+3		$\mathrm{b}^2\mathrm{e}-1$	$\mathrm{acd}+2$	b^3-2
			$\mathrm{bcd}+2$	$\mathrm{b}^2\mathrm{d}+8$	
			c^3-1	bc^2-10	

satisfying, as they should do, the relation

$$f^2 = -a^3d + a^2bc - 4c^3.$$

Hence forming the values of a^2b-3c^2 and a^2e-2cf, it appears that the value of the last-mentioned quintic function is

$$(1,\ 0,\ c,\ f,\ a^2b-3c^2,\ a^2e-2cf)(\xi,\ \eta)^5.$$

Writing herein x, y in place of ξ, η, and now using a, b, c, d, e, f to denote, not the leading coefficients but the covariants themselves (a denoting the original quintic, with ξ, η as facients), we have the form

$$A = (1,\ 0,\ c,\ f,\ a^2b-3c^2,\ a^2e-2cf)(x,\ y)^5,$$

a new quintic, which is the canonical form in question: the covariants hereof (reckoning the quintic itself as a covariant) will be written $A, B, C, .., V, W$, and will be spoken of as capital covariants.

376. The fundamental property is: Every capital covariant, say I, has for its leading coefficient the corresponding covariant i multiplied by a power of a: and this follows as an immediate consequence of the foregoing genesis of A. The covariant i of the form

$$\frac{1}{\text{a}}(\text{a, b, c, d, e, f}\between \xi, \eta)^5$$

has a leading coefficient

$$=\frac{1}{\text{a}^4}(\text{a}^2\text{cf}-\text{a}^2\text{de}+\&\text{c.}),$$

which, when $a, b, c, d, e, f, .., i$ denote leading coefficients, is $=i$ multiplied by a power of a: and upon substituting for the quintic the linear transformation thereof

$$(1,\ 0,\ c,\ f,\ a^2b-3c^2,\ a^2e-2cf\between \xi, \eta)^5,$$

(observing that, in the transformation ξ, η into $\xi-\text{b}\eta$, $\text{a}\eta$, the determinant of substitution is $=\text{a}$), the value is still $=i$ multiplied by a power of a; or using the relation $a=\text{a}$, say the value is $=i$ multiplied by a power of a. Now the covariant i is the same function of the covariants a, b, c, d, e, f that the leading coefficient i is of the leading coefficients a, b, c, d, e, f; hence, the italic letters now denoting covariants, the leading coefficient still is $=i$ multiplied by a power of a: which is the above-mentioned theorem.

377. To show how the transformation is carried out, consider, for example, the covariant B. This is obtained from the corresponding covariant of $(\text{a, b, c, d, e, f}\between \xi, \eta)^5$, that is,

(

ae	1	af	1	bf	1
bd	−4	be	−3	ce	−4
c²	+1	cd	+1	d²	+3

$\between \xi, \eta)^2$,

by changing the variables, and for the coefficients

$$\text{a, b, c, d, e, f}$$

writing

$$1,\ 0,\ c,\ f,\ a^2b-3c^2,\ a^2e-2cf;$$

thus the coefficients are

First.	Second.	Third.
$1\,(a^2b-3c^2)$	$1\,(a^2e-2cf)$	$-4c\,(a^2b-3c^2)$
$+3c^2$	$+2cf$	$+3f^2$
$=\ a^2b$	$=\ a^2e$	$=-4a^2bc+12c^3$
		$+3\,(-a^3d+a^2bc-4c^3)$
		$=\ a^2(-3ad-bc)$;

and we have thus the expression of B (see the Table No. 97); and similarly for the other capital covariants $C, D, .., V, W$: in every case the coefficients are obtained in the standard form, that is, as rational and integral functions of a, b, c, d, e, f, linear as regards f.

378. It will be observed that there is in each case a certain power of a which explicitly divides all the coefficients and is consequently written as an exterior factor: disregarding these exterior factors, the leading coefficients for B, C, D, E, F are b, c, ad, e, f respectively; that for G is $12abd + 4b^2c + e^2$, which must be $= g$ multiplied by a power of a, and (in Table 97) is given as $= a^2g$; similarly, that for H is $6acd + 4bc^2 + ef$, which must be $= h$ multiplied by a power of a, and is given as $= a^2h$: and so in the other cases. The index of a is at once obtained by means of the deg-order, which is in each case inserted at the foot of the coefficient.

For A, B, C, E, F there is no power of a as an interior factor: and for the invariants G, Q, U we may imagine the interior factor thrown together with the exterior factor, ($G = a^6g$, &c.): whence disregarding the exterior factors, we may say that for A, B, C, E, F, G, Q, U the standard forms are also "divided" forms. But take any other covariant—for instance, D: the leading coefficient is ad, having the interior factor a; and this being so it is found that all the following coefficients will divide by a (the quotients being of course expressible only in terms of the covariants subsequent to f): thus the second coefficient of D is $-bf + ce$, and (5.11) we have $-bf + ce = ai$, or the coefficient divided by a is $= i$; and so for the other coefficients of D; or throwing out the factor a, we obtain for D an expression of the form $(d, i, ... \between x, y)^3$, see the Table 98: this is the "divided" form of D: and we have similarly a divided form for every other capital covariant. All that has been required is that each coefficient of the divided form shall be expressed as a rational and integral function of the covariants $a, b, c, .., v, w$: and the form is not hereby made definite: to render it so, the coefficient must be expressed in the segregate form. But there is frequently the disadvantage that we thus introduce fractions; for instance, the last coefficient of D is $= -ci + df$, where to get rid of the congregate term df we have (6.12), $3df = -al + 2ci$, and the segregate form of the coefficient is $= -\frac{1}{3}al + \frac{2}{3}ci$.

379. We have in regard to the canonical form, a differential operator which is analogous to the two differential operators $xd_y - \{xd_y\}$, $yd_x - \{yd_x\}$ considered in the Introductory Memoir (1854), [139]. Let δ denote a differentiation in regard to the constants under the conditions

$$\begin{aligned}
\delta a &= \ 0, \\
\delta b &= \ e, \\
\delta c &= \ 3f, \\
\delta d &= \ \frac{1}{a}(-bf + ce),\ (= i), \\
\delta e &= -6ad - 10bc, \\
\delta f &= \ 2a^2b - 18c^2,
\end{aligned}$$

which (as is at once verified) are consistent with the fundamental relation

$$f^2 = -a^3d + a^2bc - 4c^3;$$

then it is easy to verify that

$$\left(x\frac{d}{dy} - 4cy\frac{d}{dx} - \delta\right)A = 0;$$

and this being so, any other covariant whatever, expressed in the like standard form, is reduced to zero by the operator

$$x\frac{d}{dy} - 4cy\frac{d}{dx} - \delta;$$

and we have thus the means of calculating the covariant when the leading coefficient is known.

Thus, considering the covariant B, the expression of which has just been obtained, $=(B_0, B_1, B_2 \between x, y)^2$, suppose: the equation to be satisfied is

$$\begin{aligned} &x\,(B_1x + 2B_2y \qquad\quad) \\ -&4cy\,(\qquad\quad 2B_0x + B_1y) \\ -&x^2\delta B_0 \;\; - xy\delta B_1 - y^2\delta B_2 \;\; = 0, \end{aligned}$$

viz. we have

$$\begin{aligned} B_1 \qquad\quad &- \delta B_0 = 0, \\ 2B_2 - 8cB_0 &- \delta B_1 = 0, \\ -4cB_1 &- \delta B_2 = 0; \end{aligned}$$

which (omitting, as we may do, the outside factor a^2) are satisfied by the foregoing values B_0, B_1, B_2, $= b$, e, $-3ad - bc$. And if we assume only $B_0 = b$, then the first equation gives at once the value $B_1 = e$, the second equation then gives $B_2 = -3ad - 3bc$; and the third equation is satisfied identically, viz. the equation is

$$-4ce + \delta\,(3ad + bc) = 0,$$

that is,

$$\begin{aligned} &-4ce &&= -4ce &&= 0, \\ &+c\delta b &&+c\,.\,e && \\ &+b\delta e &&+b\,.\,3f && \\ &+3a\delta d &&+3\,(-bf + ce) && \end{aligned}$$

which is right.

Of course every invariant must be reduced to zero by the operation δ: thus we have, see the Table No. 97,

$$\begin{aligned} a^2g = \;& 12abd \\ &+ \;4b^2c \\ &+ \;1e^2, \end{aligned}$$

and thence

				ade	b²f	bce
$a^2\delta g =$	$(12ad+8bc)\,\delta b$	$=$	$(12ad+8bc)\,e$	$=+12$		$+\ 8$
	$+4b^2 \quad .\,\delta c$		$+4b^2 \quad .\,3f$		$+12$	
	$+12ab \quad .\,\delta d$		$+12b\,(-bf+ce)$		-12	$+12$
	$+2e \quad .\,\delta e$		$+2e\,(-6ad-10bc)$	-12		$-20,$

which is $=0$, as it should be.

380. As already remarked, the leading coefficients of H, I, J, &c., are each of them equal to a power of a multiplied by the corresponding covariant $h, i, j, \ldots$; hence, supposing these leading coefficients, or, what is the same thing, the standard expressions of the covariants $h, i, j, \ldots, v, w$ to be known, we can calculate the values of $\delta h, \delta i, \delta j, \ldots, \delta v, \delta w$ $(=0$, since w is an invariant): and the operation δ, instead of being applicable only to the forms containing a, b, c, d, e, f, becomes applicable to forms containing any of the covariants. The values of $\delta a, \delta b, \ldots, \delta v, \delta w$ can, it is clear, be expressed in terms of segregates; and this is obviously the proper form: but for δr, δt, and δv, for which the segregate forms are fractional, I have given also forms with integer coefficients. The entire series is

Deg-order.		
2.8	$\delta a = 0,$	
3.5	$\delta b = e,$	
3.9	$\delta c = 3f,$	
4.6	$\delta d = i,$	
4.8	$\delta e = -6ad-10bc,$	
4.12	$\delta f = 2a^2b-18c^2,$	
5.3	$\delta g = 0,$	
5.7	$\delta h = 2be-4l,$	
5.9	$\delta i = -2ab^2+2ah-18cd,$	
6.4	$\delta j = -n,$	
6.6	$\delta k = -2aj+6b^3-9bh+3cg,$	
6.10	$\delta l = -3abd-7b^2c+7ch,$	
7.5	$\delta m = -bk-p,$	
7.7	$\delta n = 4cj,$	
8.4	$\delta o = b^2g+6bm-6dj-gh,$	
8.8	$\delta p = 8abj-5adg-10b^4+15b^2h-5bcg+10cm,$	
9.3	$\delta q = 0,$	
9.5	$\delta r = \frac{1}{2}(aq+6b^2j-5bdg-jh),$	$=2b^2j-2bdg-6dm,$
10.6	$\delta s = -2agj+2b^3g+3b^2m+21bdj-4bgh+2cg^2-3cq,$	
12.4	$\delta t = \frac{1}{2}(bgm+4bj^2-3dgj-hq),$	$=-b^2q+hq+6m^2,$
13.3	$\delta u = 0,$	
14.4	$\delta v = \frac{1}{6}(-5bgr-10bjo+5gjk-12js-9nq),$	$=-6dt-6mr+nq,$
19.3	$\delta w = 0.$	

It is obvious that for every covariant whatever written in the denumerate form $(I_0, I_1, \ldots \between x, y)^a$, the second coefficient is equal to the first coefficient operated upon by δ; so that the foregoing formulæ give, in fact, the second coefficients of the several covariants.

381. It is worth noticing how very much the formulæ of Table No. 97 simplify themselves, if one of the covariants b, c, d, e vanishes, in particular, if b vanishes. Suppose $b=0$; writing also (although this makes but little difference) $a=1$, we have

$$
\begin{aligned}
a &= 1,\\
b &= 0,\\
c &= c,\\
d &= d,\\
e &= e,\\
f^2 &= -d-4c^3,\\
g &= e^2,\\
h &= 6cd+ef,\\
i &= ce,\\
j &= 9d^2+ce^2,\\
k &= 3de,\\
l &= -3df+2c^2e,\\
m &= 9cd^2+3def-c^2e^2,\\
n &= -6cde-e^2f,\\
o &= 9d^2e+ce^3,\\
p &= -9d^2f+12c^2de+ce^2f,\\
q &= -54cd^3-27d^2ef+18c^2de^2+ce^3f,\\
r &= 9cd^2e+3de^2f-c^2e^3,\\
s &= -27d^3f+54c^2d^2e+9cde^2f-2c^3e^3,\\
t &= -81d^4f-6d^2e^3+216c^2d^3e+54cd^2e^2f-24c^3de^3-c^2e^4f,\\
u &= -27d^5-18cd^3e^2-4d^2e^3f+c^3de^4,\\
v &= -81d^4ef-6d^2e^4+216c^2d^3e^2+54cd^2e^3f-24c^3de^4-1c^2e^5f,
\end{aligned}
$$

w (not calculated).

These values are very convenient for the verification of syzygies, &c. Take, for instance, the before-mentioned relation $\delta v=-6dt-6mr+nq$, that is, if $V=(V_0, V_1 \between x, y)$, then $V_1=-6dt-6mr+nq$: calculating the three products on the right-hand side, observing that f^2 when it occurs is to be replaced by its value $-d-4c^3$, and taking their sum, the figures are as follows:

	$-6dt$	$-6mr$	$+nq$	Sum
d^5f	$+486$			$+486$
d^3e^3	$+36$	$+54$	-27	$+63$
c^2d^4e	-1296	-486	$+324$	-1458
cd^3e^2f	-324	-324	$+216$	-432
cde^5			$+1$	$+1$
$c^3d^2e^3$	$+144$	$+324$	-216	$+252$
c^2de^4f	$+6$	$+36$	-24	$+18$
c^4e^5		-6	$+4$	-2

where the last column is, in fact, what V_1 becomes on writing therein $a=1$, $b=0$. The verification would not of course apply to terms which contain b; thus, (13.3), a derived syzygy is $jr=bt+mo$; and the foregoing values give, as they should do, $jr=mo$: we might for the verification of most of the terms in b use values $a, b, c, d, e, f^2=1, b, 0, d, e, -d$: the only failure would be for terms containing bc.

TABLE No. 97 (Covariants of A, in the af- or standard forms: W is not given).

The several covariants are—

$A=($

1	0	$c+10$	$f+10$	a^2b+5 c^2-15	a^2e+1 $cf-2$
0.0	1.3	2.6	3.9	4.12	5.15

$\between x, y)^5$

$B=a^2($

$b+1$	$e+1$	$ad-3$ $bc-1$
2.2	3.5	4.8

$\between x, y)^2$

$C=($

$c+1$	$f+1$	a^2b+3 c^2-15	a^2e+1 $cf-10$	a^3d+6 a^2bc-3 c^3+15	a^2bf-3 „ $ce+3$ a^0cf+3	a^4b^2-1 a^3cd+2 a^2bc^2+4 „ $ef+1$ a^0c^4-1
2.6	3.9	4.12	5.15	6.18	7.21	8.24

$\between x, y)^6$

TABLE No. 97 (*continued*).

$D = a^2\,($	$ad\ +1$	$bf\ -1$	$a^2b^2\ -1$	$adf\ +1$	$\between(x, y)^3$
		$ce\ +1$	$acd\ +3$	$a^0bcf\ +1$	
			$a^0bc^2\ +4$	,, $c^2e\ -1$	
			,, $ef\ +1$		
	4.8	5.11	6.14	7.17	

$E = a^2\,($	$e+1$	$ad\ -6$	$bf\ -12$	$a^2b^2\ -8$	$a^2be\ -5$	$a^3bd\ -6$	$\between(x, y)^5$
		$a^0bc\ -10$	$ce\ +2$	$acd\ -36$	$adf\ -24$	$a^2b^2c\ -2$	
				$a^0bc^2\ +12$	$a^0bcf\ -4$	,, $e^2\ -1$	
				,, $ef\ -2$	,, $c^2e\ +2$	$ac^2d\ +18$	
						$a^0bc^3\ +6$	
						,, $cef\ +2$	
	3.5	4.8	5.11	6.14	7.17	8.20	

$F = ($	$f+1$	$a^2b\ +2$	$a^2e\ +1$	$a^3d\ +34$	$a^2bf\ -40$	$a^4b^2\ -16$	$a^4be\ -7$	$a^5bd\ +6$	$a^5de\ -12$	$a^6b^3\ -2$	$\between(x, y)^9$
		$a^0c^2\ -18$	$a^0cf\ -36$	$a^2bc\ -42$	,, $ce\ +5$	$a^3cd\ +6$	$a^3df\ +8$	$a^4b^2c\ -22$	$a^4bce\ +11$	$a^5bcd\ +6$	
				$a^0c^3\ +168$	$a^0c^2f\ -126$	$a^2bc^2\ +134$	$a^2bcf\ +8$	,, $e^2\ -1$	,, $b^2f\ -9$	$a^4b^2c^2\ +12$	
						,, $ef\ -5$	,, $c^2e\ +55$	$a^3c^2d\ +54$	$a^3cdf\ +24$	,, $bef\ +3$	
						$a^0c^4\ -252$	$a^0c^3f\ -84$	$a^2bc^3\ +66$	$a^2bc^2f\ +32$	,, $ce^2\ -1$	
								,, $cef\ +38$	,, $c^3e\ -45$	$a^3c^3d\ -14$	
								$a^0c^5\ +72$	$a^0c^4f\ +9$	$a^2bc^4\ -16$	
										,, $c^2ef\ -5$	
										$a^0c^6\ -2$	
	3.9	4.12	5.15	6.18	7.21	8.24	9.27	10.30	11.33	12.36	

$G = a^4$	$abd\ +12$
	$a^0b^2c\ +4$
	,, $e^2\ +1$
	$= a^2g$
	6.10

TABLE No. 97 (*continued*).

$H = a^2($ … $)(x, y)^4$

acd + 6	a^2be + 2	a^2b^2c + 4	a^3de + 2	a^4b^3 + 2
a^0bc^2 + 4	adf + 12	,, e^2 + 1	a^2b^2f + 4	,, d^2 + 6
,, ef + 1	a^0bcf − 8	ac^2d − 36	,, bce − 6	a^3bcd − 2
	,, c^2e − 8	a^0bc^3 − 24	$acdf$ − 12	$a^2b^2c^2$ − 8
		,, cef − 6	a^0bc^2f − 8	,, bef − 3
			,, c^3e + 8	,, ce^2 + 1
				ac^3d + 6
				a^0bc^4 + 4
$= a^2h$				,, c^2ef + 1
6.14	7.17	8.20	9.23	10.26

$I = a^2($ … $)(x, y)^6$

bf − 1	a^2b^2 − 2	adf − 15	a^3bd − 20	a^3de − 5	a^4b^3 + 2	a^4b^2e + 1
ce + 1	acd − 6	a^0bcf + 5	ac^2d + 60	a^2b^2f + 5	,, d^2 − 12	a^3bdf + 3
	a^0bc^2 + 8	,, c^2e − 5		,, bce − 5	a^3bcd − 2	,, cde − 5
	,, ef − 2			$acdf$ + 30	$a^2b^2c^2$ − 6	a^2b^2cf + 1
				a^0bc^2f + 5	,, ce^2 − 2	,, bc^2e − 5
				,, c^3e − 5	ac^3d − 30	,, e^2f − 1
					a^0bc^4 − 8	ac^2df − 3
					,, c^2ef − 2	a^0bc^3f − 1
$= ai$						,, c^4e + 1
5.11	6.14	7.17	8.20	9.23	10.26	11.29

$J = a^4($ … $)(x, y)^1$

a^2b^3 + 1	a^2b^2e − 1
,, d^2 + 9	$a\,bdf$ − 6
$a^0b^2c^2$ − 4	,, cde + 6
,, bef − 2	a^0b^2cf − 4
,, ce^2 + 1	,, bc^2e + 8
	,, e^2f + 1
$= a^3j$	
8.16	9.19

TABLE No. 97 (*continued*).

$K = a^4($ … $)(x, y)^3$

$ade + 3$	$a^2b^3 + 4$	$a^2b^2e + 1$	$a^3b^2d + 6$
$a^0b^2f + 2$	,, $d^2 - 18$	$abdf + 6$	$a^2b^3c + 2$
,, $bce - 2$	$abcd - 18$	$acde - 15$	,, $cd^2 - 18$
	$a^0b^2c^2 - 16$	$a^0b^2cf - 2$	$abc^2d - 30$
	,, $bef - 5$	,, $bc^2e - 2$	$adef - 9$
	,, $ce^2 + 1$	,, $e^2f - 1$	$a^0b^2e^3 - 8$
			,, $bcef - 5$
$= a^2k$			,, $c^2e^2 + 3$
7.13	8.16	9.19	10.22

$L = a^2($ … $)(x, y)^7$

$adf - 3$	$a^3bd - 3$	$a^3de - 12$	$a^4b^3 - 6$	$a^4b^2e - 1$	$a^5b^2d + 15$	$a^5bde - 7$	$a^6b^4 - 2$
$a^0bcf - 2$	$a^2b^2c - 7$	$a^2b^2f - 9$	,, $d^2 - 39$	$a^3bdf + 39$	$a^4b^3c - 9$	$a^4b^3f - 7$	,, $bd^2 + 3$
,, $c^2e + 2$	$ac^2d + 42$	,, $bce + 9$	$a^3bcd + 40$	,, $cde - 14$	,, $cd^2 + 18$	,, $b^2ce + 14$	$a^5b^2cd + 10$
	$a^0bc^3 + 28$	$acdf + 63$	$a^2b^2c^2 + 59$	$a^2b^2cf + 16$	$a^3bc^2d - 33$	,, $d^2f + 12$	,, $de^2 + 2$
	,, $cef + 7$	$a^0bc^2f + 42$	,, $bef + 7$	,, $bc^2e - 12$	,, $def - 3$	$a^3bcdf + 23$	$a^4b^3c^2 + 13$
		,, $c^3e - 42$	,, $ce^2 - 1$	,, $e^2f + 1$	$a^2b^2c^3 + 15$	,, $c^2de - 26$	,, $b^2ef + 4$
			$ac^3d - 210$	$ac^2df - 105$	,, $bcef + 21$	$a^2b^2c^2f + 25$	,, $bce^2 - 2$
			$a^0bc^4 - 140$	$a^0bc^3f - 70$	,, $c^2e^2 - 12$	,, $bc^3e - 53$	,, $c^2d^2 - 15$
			,, $c^2ef - 35$	,, $c^4e + 70$	$ac^4d + 126$	,, $ce^2f - 7$	$a^3bc^3d - 28$
					$a^0bc^5 + 84$	$ac^3df + 21$	,, $cdef - 7$
					,, $c^3ef + 21$	$a^0bc^4f + 14$	$a^2b^2c^4 - 19$
						,, $c^5e - 14$	,, $bc^2ef - 10$
							,, $c^3e^2 + 5$
							$ac^5d - 6$
							$a^0bc^6 - 4$
$= a^2l$							,, $c^4ef - 1$
7.17	8.20	9.23	10.26	11.29	12.32	13.35	14.38

$M = a^4($ … $)(x, y)^2$

$a^3b^2d - 2$	$a^3bde - 1$	$a^4b^4 - 1$
$a^2b^3c - 1$	$a^2b^3f - 1$	,, $bd^2 + 3$
,, $cd^2 + 9$	,, $b^2ce + 2$	$a^3b^2cd + 6$
$a\,bc^2d + 12$	,, $d^2f + 9$	,, $de^2 + 1$
,, $def + 3$	$a\,bcdf + 12$	$a^2b^3c^2 + 5$
$a^0b^2c^3 + 4$	,, $c^2de - 12$	,, $b^2ef + 2$
,, $bcef + 2$	$a^0b^2c^2f + 4$	,, $bce^2 - 1$
,, $c^2e^2 - 1$	,, $bc^3e - 8$	,, $c^2d^2 - 9$
	,, $ce^2f - 1$	$a\,bc^3d - 12$
		,, $cdef - 3$
		$a^0b^2c^4 - 4$
		,, $bc^2ef - 2$
$= a^4m$		,, $c^3e^2 + 1$
10.22	11.25	12.28

TABLE No. 97 (*continued*).

$N = a^4($

$a^2b^2e\ +1$	$a^2b^3c\ +4$	$a^3bde\ +12$	$a^4b^4\ +4$	$a^4b^3e\ +2$
$a\,bdf\ +6$	$,,\ cd^2\ +36$	$a^2b^3f\ +6$	$,,\ bd^2\ +12$	$,,\ d^2e\ +9$
$,,\ cde\ -6$	$a^0b^2c^3\ -16$	$,,\ b^2ce\ -6$	$a^3b^2cd\ +8$	$a^3b^2df\ +6$
$a^0b^2cf\ +4$	$,,\ bcef\ -8$	$,,\ d^2f\ +54$	$,,\ de^2\ +4$	$,,\ bcde\ -10$
$,,\ bc^2e\ -8$	$,,\ c^2e^2\ +4$	$a\,bcdf\ +36$	$a^2b^3c^2\ -12$	$a^2b^3cf\ -2$
$,,\ e^2f\ -1$		$,,\ c^2de\ -36$	$,,\ b^2ef\ -4$	$,,\ b^2c^2e\ -11$
			$,,\ c^2d^2\ -108$	$,,\ be^2f\ -3$
			$a\,bc^3d\ -96$	$,,\ cd^2f\ -18$
			$,,\ cdef\ -24$	$,,\ ce^3\ +1$
			$a^0b^2c^4\ -16$	$a\,bc^2df\ -18$
			$,,\ bc^2ef\ -8$	$,,\ c^3de\ +18$
			$,,\ c^3e^2\ +4$	$a^0b^2c^3f\ -4$
				$,,\ bc^4e\ +8$
$=a^3n$				$,,\ ce^2f\ +1$
9.19	10.22	11.25	12.28	13.31

$)(x, y)^4$

$O = a^6($

$a^2b^3e\ +3$	$a^3b^3d\ -6$
$,,\ d^2e\ +9$	$,,\ d^3\ -54$
$a\,b^2df\ +12$	$a^2b^4c\ -2$
$,,\ bcde\ -12$	$,,\ b^2e^2\ +1$
$a^0b^3cf\ +8$	$,,\ bcd^2\ -18$
$,,\ b^2c^2e\ -20$	$a\,b^2c^2d\ +24$
$,,\ be^2f\ -4$	$,,\ bdef\ +18$
$,,\ ce^3\ +1$	$,,\ cde^2\ -12$
	$a^0b^3c^3\ +8$
	$,,\ b^2cef\ -8$
	$,,\ bc^2e^2\ -10$
$=a^4o$	$,,\ e^3f\ -1$
11.21	12.24

$)(x, y)^1$

$P = a^4($

$a^3bde\ -2$	$a^4b^4\ -2$	$a^4b^3e\ -5$	$a^5b^3d\ -8$	$a^5b^2de\ +6$	$a^6b^5\ +2$
$a^2b^3f\ -1$	$,,\ bd^2\ +12$	$,,\ d^2e\ -9$	$,,\ d^3\ -72$	$a^4b^4f\ +5$	$,,\ b^2d^2\ +12$
$,,\ d^2f\ -9$	$a^3b^2cd\ -10$	$a^3b^2df\ -24$	$a^4b^4c\ +4$	$,,\ b^3ce\ -13$	$a^5b^3cd\ -2$
$a\,bcdf\ -12$	$,,\ de^2\ -5$	$,,\ bcde\ +44$	$,,\ b^2e^2\ -1$	$,,\ bd^2f\ +21$	$,,\ bde^2\ -3$
$,,\ c^2de\ +12$	$a^2b^3c^2\ -2$	$a^2b^3cf\ -6$	$,,\ bcd^2\ -24$	$,,\ cd^2e\ -21$	$,,\ cd^3\ -36$
$a^0b^2c^2f\ -4$	$,,\ b^2e^2f\ -1$	$,,\ b^2c^2e\ +36$	$a^3b^2cd\ +52$	$a^3b^2cdf\ -4$	$a^4b^4c^2\ -12$
$,,\ bc^3e\ +8$	$,,\ bce^2\ +3$	$,,\ be^2f\ +6$	$,,\ bdef\ +10$	$,,\ bc^2de\ +10$	$,,\ b^3ef\ -5$
$,,\ ce^2f\ +1$	$,,\ c^2d^2\ +90$	$,,\ cd^2f\ +90$	$,,\ cde^2\ +8$	$,,\ de^2f\ +4$	$,,\ b^2ce^2\ +4$
	$a\,bc^3d\ +120$	$,,\ ce^3\ -1$	$a^2b^3c^3\ +4$	$a^2b^3c^2f\ -17$	$,,\ bc^2d^2\ -66$
	$,,\ cdef\ +30$	$a\,bc^2df\ +120$	$,,\ b^2cef\ -2$	$,,\ b^2c^3e\ +44$	$,,\ d^2ef\ -18$
	$a^0b^2c^4\ +40$	$,,\ c^3de\ -120$	$,,\ bc^2e^2\ +2$	$,,\ bce^2f\ +13$	$a^3b^2c^3d\ -10$
	$,,\ bc^2ef\ +20$	$a^0b^2c^3f\ +40$	$,,\ c^3d^2\ -180$	$,,\ c^2d^2f\ -45$	$,,\ bcdef\ -4$
	$,,\ c^3e^2\ -10$	$,,\ bc^4e\ -80$	$,,\ e^3f\ +1$	$,,\ c^2e^3\ -5$	$,,\ c^2de^2\ +1$
		$,,\ c^2e^2f\ -10$	$a\,bc^4d\ -240$	$a\,bc^3df\ -60$	$a^2b^3c^4\ +14$
			$,,\ c^2def\ -60$	$,,\ c^4de\ +60$	$,,\ b^2c^2ef\ +11$
			$a^0b^2c^5\ -80$	$a^0b^2c^4f\ -20$	$,,\ bc^3e^2\ -11$
			$,,\ bc^3ef\ -40$	$,,\ bc^5e\ +40$	$,,\ c^4d^2\ +18$
			$,,\ c^4e^2\ +20$	$,,\ c^3e^2f\ -5$	$,,\ ce^3f\ -1$
					$a\,bc^5d\ +26$
					$,,\ c^3def\ +6$
					$a^0b^2c^6\ +8$
					$,,\ bc^4ef\ +4$
$=a^4p$					$,,\ c^5e^2\ -2$
11.25	12.28	13.31	14.34	15.37	16.40

$)(x, y)^5$

TABLE No. 97 (*continued*).

$Q = a^6$

a^4b^5	$-\ 2$
,, b^2d^2	$+\ 18$
a^3b^3cd	$+\ 22$
,, bde^2	$+\ 3$
,, cd^3	$-\ 54$
$a^2b^4c^2$	$+\ 12$
,, b^3ef	$+\ 5$
,, b^2ce^2	$-\ 4$
,, bc^2d^2	$-\ 108$
,, d^2ef	$-\ 27$
$a\,b^2c^3d$	$-\ 72$
,, $bcdef$	$-\ 36$
,, c^2de^2	$+\ 18$
$a^0b^3c^4$	$-\ 16$
,, b^2c^2ef	$-\ 12$
,, bc^3e^2	$+\ 12$
,, ce^3f	$+\ 1$
$= a^6q$	
14 . 30	

$R = a^6\,($ ... $\between x, y)^2$

a^3bde	$-\ 1$	a^4b^5	$+\ 2$	a^4b^4e	$+\ 1$
a^2b^4f	$+\ 1$	,, b^2d^2	$+\ 6$	a^3b^3df	$+\ 3$
,, b^3ce	$-\ 3$	a^3b^3cd	$-\ 2$	,, b^2cde	$-\ 11$
,, bd^2f	$-\ 9$	,, bde^2	$-\ 2$	,, d^3f	$-\ 27$
,, cd^2e	$+\ 9$	,, cd^3	$-\ 54$	,, de^3	$-\ 1$
$a\,b^2cdf$	$-\ 12$	$a^2b^4c^2$	$-\ 8$	a^2b^4cf	$+\ 1$
,, bc^2de	$+\ 24$	,, b^3ef	$-\ 4$	,, b^3c^2e	$-\ 7$
,, de^2f	$+\ 3$	,, b^2ce^2	$+\ 2$	,, b^2e^2f	$-\ 2$
$a^0b^3c^2f$	$-\ 4$	,, bc^2d^2	$-\ 72$	,, bcd^2f	$-\ 45$
,, b^2c^3e	$+\ 12$	,, d^2ef	$-\ 18$	,, bce^3	$+\ 1$
,, bce^2f	$+\ 3$	$a\,b^3c^3d$	$-\ 24$	,, c^2d^2e	$+\ 45$
,, c^2e^3	$-\ 1$	,, $bcdef$	$-\ 12$	$a\,b^2c^2df$	$-\ 24$
		,, c^2de^2	$+\ 6$	,, bc^3de	$+\ 48$
				a^0b^3cf	$-\ 4$
				,, b^2c^4e	$+\ 12$
				,, bc^2e^2f	$+\ 3$
				,, c^3e^3	$-\ 1$
$= a^5r$					
13 . 27		14 . 30		15 . 33	

$S = a^6\,($ ... $\between x, y)^3$

a^4bd^2e	$+\ 9$	a^5b^4d	$+\ 15$	a^5b^3de	$-\ 6$	a^6b^6	$-\ 2$
a^3b^3df	$+\ 7$	,, bd^3	$-\ 27$	,, d^3e	$-\ 27$	,, b^3d^2	$+\ 9$
,, b^2cde	$-\ 12$	a^4b^5c	$+\ 3$	a^4b^5f	$-\ 3$	,, d^4	$-\ 27$
,, d^3f	$-\ 27$	,, b^2cd^2	$-\ 99$	,, b^4ce	$+\ 9$	a^5b^4cd	$+\ 18$
a^2b^4cf	$+\ 2$	,, d^2e^2	$-\ 18$	,, b^2d^2f	$-\ 9$	,, b^2de^2	$+\ 6$
,, b^3c^2e	$-\ 6$	$a^3b^3c^2d$	$-\ 114$	,, bcd^2e	$-\ 18$	,, bcd^3	$-\ 54$
,, bcd^2f	$-\ 54$	,, b^2def	$-\ 33$	a^3b^3cdf	$-\ 9$	$a^4b^5c^2$	$+\ 15$
,, c^2d^2e	$+\ 54$	,, $bcde^2$	$+\ 12$	,, b^2c^2de	$+\ 24$	,, b^4ef	$+\ 6$
$a\,b^2c^2df$	$-\ 36$	,, c^2d^3	$+\ 162$	,, bde^2f	$+\ 3$	,, $b^3c^2d^2$	$-\ 36$
,, bc^3de	$+\ 72$	$a^2b^4c^2$	$-\ 24$	,, cd^3f	$+\ 81$	,, b^3ce^2	$-\ 6$
,, cde^2f	$+\ 9$	,, b^3cef	$-\ 9$	,, cde^3	$-\ 3$	,, $b^2c^2d^2$	$-\ 27$
$a^0b^3c^3f$	$-\ 8$	,, $b^2c^2e^2$	$+\ 9$	$a^2b^4c^2f$	$+\ 6$	,, bd^2ef	$-\ 9$
,, b^2c^4e	$+\ 24$	,, bc^3d^2	$+\ 324$	,, b^3c^3e	$-\ 18$	,, cd^2e^2	$-\ 9$
,, bc^2e^2f	$+\ 6$	,, cd^2ef	$+\ 69$	,, b^2ce^2f	$-\ 9$	$a^3b^3c^3d$	$-\ 54$
,, c^3e^3	$-\ 2$	$a\,b^2c^4d$	$+\ 216$	,, bc^2d^2f	$+\ 162$	,, b^2cdef	$-\ 27$
		,, bc^2def	$+\ 120$	,, bc^2e^3	$+\ 3$	,, bc^2de^2	$+\ 3$
		,, c^3de^2	$-\ 54$	,, c^2d^2e	$-\ 162$	,, c^3d^3	$-\ 54$
		$a^0b^3c^5$	$+\ 48$	$a\,b^2c^3df$	$+\ 108$	,, de^3f	$-\ 2$
		,, b^2c^3ef	$+\ 36$	,, bc^4de	$-\ 216$	$a^2b^4c^4$	$-\ 24$
		,, bc^4e^2	$-\ 36$	,, c^2de^2f	$-\ 27$	,, b^3ce^2f	$-\ 21$
		,, c^2e^3f	$-\ 3$	$a^0b^3c^4f$	$+\ 24$	,, $b^2c^3e^2$	$+\ 21$
				,, b^2c^5e	$-\ 72$	,, bc^4d^2	$-\ 108$
				,, bc^3e^2f	$-\ 18$	,, bce^3f	$+\ 2$
				,, c^4e^3	$+\ 6$	,, c^2d^2ef	$-\ 27$
						$a\,b^2c^5d$	$-\ 72$
						,, bc^3def	$-\ 36$
						,, c^4de^2	$+\ 18$
						$a^0b^3c^6$	$-\ 16$
						,, b^2c^4ef	$-\ 12$
						,, bc^5e^2	$+\ 12$
$= a^6s$						,, c^3e^3f	$+\ 1$
15 . 33		16 . 36		17 . 39		18 . 42	

TABLE No. 97 (*continued*).

$T = ($ … $)(x, y)^1$

a^5b^4de	$+\ 7$	a^6b^7	$+\ 2$
,, bd^3e	$+\ 27$	,, b^4d^2	$+\ 6$
a^4b^6f	$+\ 1$	a^5b^5cd	$-\ 10$
,, b^5ce	$-\ 2$	,, b^3de^2	$-\ 8$
,, b^3d^2f	$+\ 24$	,, b^2cd^2	$-\ 54$
,, b^2cd^2e	$-\ 54$	,, d^3e^2	$-\ 27$
,, d^4f	$-\ 81$	$a^4b^6c^2$	$-\ 14$
,, d^2e^3	$-\ 6$	,, b^5ef	$-\ 7$
a^3b^4cdf	$+\ 16$	,, b^4ce^2	$+\ 9$
,, b^3c^2de	$-\ 76$	,, $b^3c^2d^2$	$-\ 84$
,, b^2de^2f	$-\ 12$	,, b^2d^2ef	$-\ 27$
,, bcd^3f	-216	,, bcd^2e^2	$+\ 9$
,, $bcde^3$	$+\ 5$	,, c^2d^4	$+162$
,, c^2d^3e	$+216$	$a^3b^4c^3d$	$-\ 8$
$a^2b^4c^3e$	$-\ 8$	,, b^3cdef	$+\ 4$
,, $b^2c^2d^2f$	-216	,, $b^2c^2de^2$	$+\ 18$
,, $b^2c^2e^3$	$+\ 2$	,, bc^3d^3	$+432$
,, bc^3d^2e	$+432$	,, bde^3f	$+\ 3$
,, cd^2e^2f	$+\ 54$	,, cd^3ef	$+108$
$a\,b^3c^3df$	$-\ 96$	,, cde^4	$-\ 1$
,, b^2c^4de	$+288$	$a^2b^5c^4$	$+\ 16$
,, bc^2de^2f	$+\ 72$	,, b^4c^2ef	$+\ 20$
,, c^3de^3	$-\ 24$	,, $b^3c^3e^2$	$-\ 24$
$a^0b^4c^4f$	$-\ 16$	,, $b^2c^4d^2$	$+432$
,, b^3c^5e	$+\ 64$	,, b^2ce^3f	$-\ 5$
,, $b^2c^3e^2f$	$+\ 24$	,, bc^2d^2ef	$+216$
,, bc^4e^3	$-\ 16$	,, bc^2e^4	$+\ 1$
,, c^2e^4f	$-\ 1$	,, $c^3d^2e^2$	-108
		$a\,b^3c^5d$	$+192$
		,, b^2c^3def	$+144$
		,, bc^4de^2	-144
		,, c^2de^3f	$-\ 12$
		$a^0b^4c^6$	$+\ 32$
		,, b^3c^4ef	$+\ 32$
		,, $b^2c^5e^2$	$-\ 48$
		,, bc^3e^3f	$-\ 8$
$=a^8t$		,, c^4e^4	$+\ 2$
19.41		20.44	

$U =$

a^5b^6d	$-\ 3$
,, b^3d^3	$+\ 14$
,, d^5	$-\ 27$
a^4b^7c	$-\ 1$
,, b^4cd^2	$+\ 34$
,, $b^2d^2e^2$	$+\ 11$
,, bcd^4	$-\ 81$
$a^3b^5c^2d$	$+\ 32$
,, b^4def	$+\ 10$
,, b^3cde^2	$-\ 6$
,, $b^2c^2d^3$	-144
,, bd^3ef	$-\ 18$
,, cd^3e^2	$-\ 18$
$a^2b^6c^3$	$+\ 8$
,, b^5cef	$+\ 4$
,, $b^4c^2e^2$	$-\ 6$
,, $b^3c^3d^2$	-152
,, b^2cd^2ef	$-\ 60$
,, $bc^2d^2e^2$	$+\ 6$
,, d^2e^3f	$-\ 4$
$a\,b^4c^4d$	$-\ 80$
,, b^3c^2def	$-\ 56$
,, $b^2c^3de^2$	$+\ 48$
,, $bcde^3f$	$+\ 2$
,, c^2de^4	$+\ 1$
$a^0b^5c^5$	$-\ 16$
,, b^4c^3ef	$+\ 16$
,, $b^3c^4e^2$	$+\ 24$
,, $b^2c^2e^3f$	$+\ 4$
,, bc^3e^4	$+\ 1$
$=a^9u$	
21.45	

TABLE No. 97 (*concluded*).

$V = a^{10}\,($

a^6b^8	$-\ 4$	a^6b^7e	$-\ 2$
,, b^5d^2	$-\ 12$	,, b^4d^2e	$-\ 48$
a^5b^6cd	$+\ 20$	,, bd^4e	$-\ 162$
,, b^4de^2	$+\ 23$	a^5b^6df	$-\ 6$
,, b^3cd^3	$+\ 108$	,, b^5cde	$+\ 8$
,, bd^3e^2	$+\ 81$	,, b^3d^3f	$-\ 144$
$a^4b^7c^2$	$+\ 28$	,, b^3de^3	$+\ 8$
,, b^6ef	$+\ 15$	,, b^2cd^3e	$+\ 324$
,, b^5ce^2	$-\ 20$	,, d^5f	$+\ 486$
,, $b^4c^2d^2$	$+\ 168$	,, d^3e^3	$+\ 63$
,, b^3d^2ef	$+\ 78$	a^4b^7cf	$-\ 2$
,, $b^2cd^2e^2$	$-\ 72$	,, b^6c^2e	$+\ 18$
,, bc^2d^4	$-\ 324$	,, b^5e^2f	$+\ 7$
,, d^4ef	$-\ 81$	,, b^4cd^2f	$-\ 144$
,, d^2e^4	$-\ 6$	,, b^4ce^3	$-\ 9$
$a^3b^5c^3d$	$+\ 16$	,, $b^3c^2d^2e$	$+\ 648$
,, b^4cdef	$+\ 8$	,, $b^2d^2e^2f$	$+\ 99$
,, $b^3c^2de^2$	$-\ 112$	,, bcd^4f	$+\ 1458$
,, $b^2c^3d^3$	$-\ 864$	,, bcd^2e^3	$-\ 27$
,, b^2de^3f	$-\ 18$	,, c^2d^4e	$-\ 1458$
,, bcd^3ef	$-\ 432$	$a^3b^5c^2df$	$-\ 32$
,, $bcde^4$	$+\ 7$	,, b^4c^3de	$+\ 208$
,, $c^2d^3e^2$	$+\ 216$	,, b^3cde^2f	$+\ 20$
$a^2b^6c^4$	$-\ 32$	,, $b^2c^2d^3f$	$+\ 1728$
,, b^5c^2ef	$-\ 40$	,, $b^2c^2de^3$	$-\ 40$
,, $b^4c^3e^2$	$+\ 40$	,, bc^3d^3e	$-\ 3456$
,, $b^3c^4d^2$	$-\ 864$	,, bde^4f	$-\ 3$
,, b^3ce^3f	$+\ 10$	,, cd^3e^2f	$-\ 432$
,, $b^2c^2d^2ef$	$-\ 648$	,, cde^5	$+\ 1$
,, $bc^3d^2e^2$	$+\ 648$	$a^2b^4c^2e^2f$	$-\ 20$
,, cd^2e^3f	$+\ 54$	,, $b^3c^3d^2f$	$+\ 1008$
$a\,b^4c^5d$	$-\ 384$	,, $b^3c^3e^3$	$+\ 20$
,, b^3c^3def	$-\ 384$	,, $b^2c^4d^2e$	$-\ 3024$
,, $b^2c^4de^2$	$+\ 576$	,, b^2ce^4f	$+\ 5$
,, bc^2de^3f	$+\ 96$	,, $bc^2d^2e^2f$	$-\ 756$
,, c^3de^4	$-\ 24$	,, bc^2e^5	$-\ 1$
$a^0b^5c^6$	$-\ 64$	,, $c^3d^2e^3$	$+\ 252$
,, b^4c^4ef	$-\ 80$	$a\,b^4c^4df$	$+\ 288$
,, $b^3c^5e^2$	$+\ 160$	,, b^3c^5de	$-\ 1152$
,, $b^2c^3e^3f$	$+\ 40$	,, $b^2c^3de^2f$	$-\ 432$
,, bc^4e^4	$-\ 20$	,, bc^4de^3	$+\ 288$
,, c^2e^5f	$-\ 1$	,, c^2de^4f	$+\ 18$
		a^0b^5cf	$+\ 32$
		,, b^4c^6e	$-\ 160$
		,, $b^3c^4e^2f$	$-\ 80$
		,, $b^2c^5e^3$	$+\ 80$
		,, bc^3e^4f	$+\ 10$
$= a^9v$		,, c^4e^5	$-\ 2$
22 . 46		23 . 49	

$\between x, y)^2$

TABLE No. 98. Covariants of A, divided and (except as to a few coefficients) segregate.

A and B as given in Table 97 were divided and segregate.

C was divided but not segregate: the divided and segregate form is

$C = ($

$c+1$	$f+3$	$a^2b\ +\ 3$	$a^2e\ +\ 1$	$a^3d\ +\ 6$	$a^3i\ +3$	$a^4b^2\ -1$
		a^0c^2-15	a^0cf-10	$a^0bc-\ 3$	a^0c^2f+3	,, $h\ +1$
				,, $c^3\ +15$		$a^3cd\ -4$
						$a^0c^4\ -1$
2.6	3.9	4.12	5.15	6.18	7.21	8.24

$)(x, y)^6.$

D divided and segregate is

$D = a^3($

$d+1$	$i+1$	$ab^2\ -1$	$\div 3$ $al\ -1$
		,, $h\ +1$	$a^0ci\ -1$
		a^0cd-3	
3.3	4.6	5.9	6.12

$)(x, y)^3,$

an integer non-segregate form of the fractional coefficient is

$ci\ -1$
$df\ +1$

E was divided but not segregate: the divided and segregate form is

$E = ($

$e+1$	$ad\ -\ 6$	$ai\ +12$	$a^2b^2\ -\ 8$	$a^2be\ -\ 5$	$a^4g\ -1$
	a^0bc-10	a^0ce-10	,, $h\ -\ 2$	,, $l\ +\ 8$	a^3bd+6
			$acd\ -24$	$aci\ -12$	a^2b^2c+2
			a^0bc^2+20	$a^0c^2e+\ 5$	,, $ch\ +2$
					$ac^2d\ +6$
					a^0bc^3-2
3.5	4.8	5.11	6.14	7.17	8.20

$)(x, y)^5.$

TABLE No. 98 (*continued*).

F was divided but not segregate: the divided and segregate form is

						÷ 3				
$f+1$	a^2b+2	a^2e+1	a^3d+34	a^3i+40	a^4b^2-16	a^4be-21	a^6g-1	a^6k-4	a^6b^3-2	$)(x,y)^9$,
	a^0c^2-18	a^0cf-36	a^2bc-42	a^2ce-35	$,,\ h-5$	a^3l-8	a^5bd+18	a^5bi+1	$,,\ bh+3$	
			a^0c^3+168	a^0c^2f+126	a^3cd+46	$,,\ ci-16$	a^4b^2c-18	a^4bce+2	$,,\ cg-1$	
					a^2bc^2+155	a^2c^2e+189	$,,\ ch+38$	$,,\ cl-8$	$a^4b^2c^2+4$	
					a^0c^4+252	a^0c^3f-252	a^3c^2d-174	a^3c^2i-16	$,,\ c^2h-5$	
							a^2bc^2-86	a^2c^3e-13	a^3c^3d+16	
							a^0c^5+72	a^0c^4f+9	a^2bc^4+4	
									a^0c^6-2	
3.9	4.12	5.15	6.18	7.21	8.24	9.27	10.30	11.33	12.36	

where for an integer non-segregate value of the fractional coefficient, see the original form of F.

G as an invariant was divided and segregate, $G = a^6 \underset{4.0}{g}$.

H divided and segregate is

				÷ 3	÷ 3	
$H = a^4($	$h+1$	$be+2$	a^2g+1	a^2k+2	a^3j+2	$)(x,y)^4$,
		$l-4$	$abd-12$	$abi-8$	a^2b^3+4	
			a^0ch-6	a^0bce-6	$,,\ bh-5$	
				$,,\ cl+12$	$,,\ cg+1$	
					$abcd+12$	
					a^0c^2h+3	
	4.4	5.7	6.10	7.13	8.16	

where the fractional coefficients are =

$ade+2$	a^2b^3+2
a^0b^2f+4	$,,\ d^2+6$
$,,\ bce-6$	$abcd-2$
$,,\ cl+4$	$a^0b^2c^2-8$
	$,,\ bef-3$
	$,,\ c^2h+1$
	$,,\ ce^2+1$

TABLE No. 98 (*continued*).

I divided and segregate is

					$\div 3$	$\div 3$	$\div 3$	
$I = a^3($	$i + 1$	$ab^2 \;\; - \;\; 2$	$al \;\; + \;\; 5$	$a^2bd - 20$	$a^3k \;\; - \;\; 5$	$a^4j \;\; - \;\; 4$	$a^3b^2e \;\; + \;\; 3$	$\between x, y)^6$,
		$,, \; h + \;\; 2$	$a^0ci - 15$	$a^0c^2d + 60$	$a^2bi - 25$	$a^3b^3 \;\; + 10$	$,, \; bl \;\; + \;\; 9$	
		$a^0cd - 18$			$acl \;\; - 30$	$,, \; bh \;\; - \;\; 8$	$,, \; ck \;\; - \;\; 5$	
					$a^0c^2i + 45$	$,, \; cg \;\; + \;\; 4$	$,, fg \;\; - \;\; 3$	
						$a^2bcd - \;\; 6$	$a^2bci \;\; - \;\; 8$	
						$a \, bc^2 - 18$	$a \, b^2cf - \;\; 9$	
						$,, \; c^2h - \;\; 6$	$,, \; bc^2e - 15$	
						$a^0c^3d - 54$	$,, \; c^2l \;\; + \;\; 3$	
							$a^0c^3i \;\; - \;\; 3$	
	4.6	5.9	6.12	7.15	8.18	9.21	10.24	

where the fractional coefficients are =

$a^2de \;\; - \;\; 5$	$a^3b^3 \;\; + \;\; 2$	$a^3b^2e \;\; + 1$
$a \, b^2f + \;\; 5$	$,, \; d^2 \;\; - 12$	$a^2bdf + 3$
$,, \; bce - \;\; 5$	$a^2bcd - \;\; 2$	$,, \; cde \;\; - 5$
$a^0c^2i \;\; - \;\; 5$	$a \, b^2c^2 - \;\; 6$	$a \, b^2cf + 1$
$,, \; cdf + 30$	$,, \; c^2h \;\; - \;\; 2$	$,, \; bc^2e - 5$
	$,, \; ce^2 \;\; - \;\; 2$	$,, \; e^2f \;\; - 1$
	$a^0c^3d - 18$	$a^0c^3i \;\; + 1$
		$,, \; c^2df - 3$

J divided and segregate is

$$J = a^7 \, (j, \; - n \between x, \; y)^1.$$
$$\begin{matrix} 5.6 & 6.4 \end{matrix}$$

K divided and segregate is

$K = a^6($	$k + 1$	$aj \;\; - \;\; 2$	$an \;\; + 1$	$a^2m \;\; - 3$	$\between x, \; y)^3$.
		$a^0b^3 \; + \;\; 6$	$a^0ck \;\; - 3$	$acj \;\; + 1$	
		$,, \; bh - \;\; 9$		$a^0b^3c - 2$	
		$,, \; cg \; + \;\; 3$		$,, \; bch + 3$	
				$,, \; c^2g - 1$	
	5.3	6.6	7.9	8.12	

TABLE No. 98 (*continued*).

L divided and (as to first six coefficients) segregate is

			÷ 3	÷ 3				
$L=a^4($	$l+1$	$abd\ -3$	$a^2k\ -4$	$a^3j\ -13$	$a^3n\ -3$	$a^4m\ -1$	$a^3bde\ -7$	$a^4b^4\ -2$ $)(x, y)^7$,
		$a^0b^2c\ -7$	$abi\ +1$	$a^2b^3\ -5$	$a^2bl\ -45$	$a^3b^2d\ +13$	$a^2b^3f\ -7$	,, $bd^2\ +3$
		,, $ch\ +7$	$a^0cl\ -21$	,, $bh\ -5$	,, $ck\ -20$	,, $cj\ +3$	,, $b^2ce\ +14$	$a^3b^2cd\ +10$
				,, $cg\ +10$	$abci\ -10$	$a^2b^3c\ -13$	,, $d^2f\ +12$	,, $de^2\ +2$
				$abcd\ +30$	$a^0c^2l\ +105$	,, $bch\ +29$	$abcdf\ +23$	$a^2b^3c^2\ +13$
				$a^0b^2c^2\ +105$		,, $c^2g\ -16$	,, $c^2de\ -24$	,, $b^2ef\ +4$
				,, $ch\ -105$		$abc^2d\ -3$	$a^0b^2cf\ +25$	,, $bce^2\ -2$
						$a^0b^2c^3\ -21$	,, $bc^3e\ -55$	,, $c^2d^2\ -15$
						,, $c^3h\ +21$	,, $c^3l\ -3$	$abc^3d\ -28$
							,, $c^2fh\ +2$	,, $cdef\ -7$
							,, $ce^2f\ -7$	$a^0b^2c^4\ -19$
								,, $bc^2ef\ -10$
								,, $c^4h\ -1$
								,, $c^3e^2\ +5$
	5.7	6.10	7.13	8.16	9.19	10.22	11.25	12.28

where the fractional coefficients are =

$a^2b^3\ -6$	$a^2b^2e\ -1$
,, $d^2\ +3$	$abdf\ +39$
$abcd\ +26$	,, $cde\ -22$
$a^0b^2c^2\ +31$	$a^0b^2cf\ +16$
,, $bef\ +7$	,, $bc^2e\ -4$
,, $c^2h\ -7$	,, $c^2l\ +19$
,, $ce^2\ -1$	,, $cfh\ -8$
,, $fl\ -14$	,, $e^2f\ +1$

the last two coefficients have not been reduced to the segregate form.

M divided and segregate is

$M=a^8($	$m+1$	$bk-1$	$abj\ -1$ $)(x, y)^2$.
		$p\ -1$	$adg\ +1$
			,, ${}^0cm-1$
	6.2	7.5	8.8

TABLE No. 98 (*continued*).

N divided and segregate is

$N = a^7($						
	$n+1$	$cj+4$	$ap-6$ $cn-6$	$a^2bj \; -4$,, $dg \; +4$ $a\,b^4 \; +8$,, $b^2h \; -12$,, $bcg \; +4$,, $cm \; -8$ $a^0c^2j \; -4$	$a^3o \; +1$ $a^2bn \; -1$ $acp \; +2$ $a^0c^2n \; +1$	$)(x, y)^4$.
	6.4	7.7	8.10	9.13	10.16	

O divided and segregate is

$O = a^{10}($			
	$o+1$	b^2g+1 $bm+6$ $df-6$ $gh-1$	$)(x, y)^1$.
	7.1	8.4	

P divided and (as to first three coefficients) segregate is

$P = a^8($							
	$p+1$	$a\,bj \; +8$,, $dg \; -5$ $a^0b^4 \; -14$,, $b^2h \; +15$,, $bcg \; -5$,, $cm \; +10$	$a^2o \; +7$ $abn \; -2$ $a^0cp \; -14$	$a^2bm \; +8$,, $dj \; +3$ $a\,b^3d \; +9$,, $bcj \; +13$,, $bdh \; -9$,, $cdg \; -12$,, $d^3 \; -81$ $a^0b^4c \; -5$,, $b^2ch \; -8$,, $b^2e^2 \; -1$,, $bc^2g \; -1$,, $bcd^2 \; -9$,, $bel \; -1$,, $bfk \; +1$,, $c^2m \; -12$,, $dei \; +9$,, $e^2h \; +1$,, $fp \; -2$	$a^2b^2k \; -3$,, $bp \; -3$,, $em \; +2$ $a\,b^2de \; +3$,, $bdl \; +3$,, $bfj \; +8$,, $d^2i \; -27$,, $deh \; -3$,, $dfg \; -8$ $a^0b^3ce \; -7$,, $b^2cl \; +1$,, $bc^2k \; +9$,, $bcdi \; -9$,, $bceh \; +10$,, $c^2eg \; -3$,, $c^2p \; +9$,, $cd^2e \; +18$,, $cfm \; +4$,, $efk \; +3$	$a^3b^2j \; +2$,, $bdg \; -2$ $a^2bcm \; +2$,, $bek \; -1$,, $ep \; -1$ $a\,b^3cd \; +24$,, $bc^2j \; -6$,, $bcdh \; -33$,, $c^2dg \; +15$,, $cd^3 \; -54$,, $dfk \; -9$ $a^0b^4c^3 \; +8$,, $b^2c^2h \; -11$,, $bc^3g \; +3$,, $bc^2d^2 \; -18$,, $bcfk \; -1$,, $c^3m \; -6$,, $cfp \; +2$	$)(x, y)^5$,
	7.5	8.8	9.11	10.14	11.17	12.20	

the last three coefficients have not been reduced to the segregate form.

TABLE No. 98 (*continued*).

Q as an invariant was divided and segregate, $Q = a^{12}\, q.$
8.0

R divided and segregate is

		÷ 2	÷ 3	
$R = a^{11}($	$r + 1$	$aq\ \ + 1$	$abo\ - 1$	$)(x, y)^2,$
		$a^0b^2j\ + 6$	,, $gk\ - 1$	
		,, $bdg\ - 5$	,, $s\ + 3$	
		,, $hj\ - 1$	$a^0cr\ - 3$	
	8.2	9.5	10.8	

where the fractional coefficients are =

$b^2j\ + 2$	$bdk + 3$
$bdg - 2$	$bej\ + 1$
$dm\ - 6$	$cr\ + 1$
	$deg - 1$
	$dp\ + 3$

S divided and (as to the first three coefficients) segregate is

				÷ 2	
$S = a^{12}($	$s + 1$	$agj\ - 2$	$a\,br - 1$	$a^2bq\ + 4$	$)(x, y)^3,$
		$a^0b^3g\ + 2$	,, $do - 1$	$a\,b^3j\ + 4$	
		,, $b^2m\ + 3$	$a^0cs\ - 3$	,, $b^2dg\ - 4$	
		,, $bdj\ + 21$		,, $bdm\ - 31$	
		,, $bgh\ - 4$		,, $d^2j\ - 3$	
		,, $cg^2\ + 2$		$a^0b^6\ + 4$	
		,, $cq\ - 3$		,, $b^3d^2\ + 16$	
				,, $bd^2h\ - 24$	
				,, $den\ + 4$	
				,, $fs\ - 1$	
	9.3	10.6	11.9	12.12	

but the last coefficient is neither segregate nor integer.

TABLE No. 98 (*concluded*).

T divided and segregate is

$$T = a^{16} \left(\begin{array}{c|c} & \div 2 \\ t + 1 & bgm + 1 \\ & bj^2 \ \ + 4 \\ & dgj \ \ - 3 \\ & hq \ \ \ - 1 \\ 11.1 & 12.4 \end{array} \right\Updownarrow x, y)^1,$$

where the fractional coefficient is =

b^2q	-1
hq	$+1$
m^2	$+6$

U as an invariant was divided and segregate, $U = a^{18} \underset{12.0}{u}$.

V divided and segregate is

$$V = a^{19} \left(\begin{array}{c|c} & \div 6 \\ v + 1 & bgr - \ 5 \\ & bjo - 10 \\ & gjk + \ 5 \\ & js \ \ - 12 \\ & nq \ \ - \ 9 \\ 13.1 & 14.4 \end{array} \right\Updownarrow x, y)^1,$$

where the fractional coefficient is =

dt	-6
mr	-6
nq	$+1$

W as an invariant was divided and segregate, $W = a^{27} \underset{18.0}{w}$.

Derivatives. Art. Nos. 382 to 384, and Tables Nos. 99 and 100.

382. I call to mind that any two covariants a, b, the same or different, give rise to a set of derivatives $(a, b)^1$, $(a, b)^2$, $(a, b)^3$, &c., or, as I propose to write them, $ab1$, $ab2$, $ab3$, &c., viz.:

$$\begin{aligned} ab1 &= d_x a \,.\, d_y b - d_y a \,.\, d_x b, \\ ab2 &= d_x^2 a \,.\, d_y^2 b - 2d_x d_y a \,.\, d_x d_y b + d_y^2 a \,.\, d_x^2 b, \\ ab3 &= d_x^3 a \,.\, d_y^3 b - 3d_x^2 d_y a \,.\, d_x d_y^2 b + 3d_x d_y^2 a \,.\, d_x^2 d_y b - d_y^3 a \,.\, d_x^3 b, \\ &\text{\&c.}; \end{aligned}$$

or, as these are symbolically written,

$$ab1 = \overline{12} a_1 b_2, \quad ab2 = \overline{12}^2 a_1 b_2, \quad ab3 = \overline{12}^3 a_1 b_2, \text{ \&c.};$$

where

$$12 = \xi_1\eta_2 - \xi_2\eta_1, \quad = \frac{d}{dx_1}\frac{d}{dy_2} - \frac{d}{dx_2}\frac{d}{dy_1},$$

the differentiations $\frac{d}{dx_1}$, $\frac{d}{dy_1}$ applying to the a_1 and the $\frac{d}{dx_2}$, $\frac{d}{dy_2}$ applying to the b_2, but the suffixes being ultimately omitted: hence if θ be the index of derivation, the derivative is thus a linear function of the differential coefficients of the order θ of the two covariants a and b respectively: and we have the general property that any such derivative, if not identically vanishing, is a covariant. If the a and the b are one and the same covariant, then obviously every odd derivative is $=0$; so that in this case the only derivatives to be considered are the even derivatives $aa2$, $aa4$, &c.: moreover, if the index of derivation θ exceeds the order of either of the component covariants, then also the derivative is $=0$: in particular, neither of the covariants must be an invariant. The degree of the derivative is evidently equal to the sum of the degrees of the component covariants; the order is equal to the sum of the orders less twice the index of derivation.

383. It was by means of the theory of derivatives that Gordan proved (for a binary quantic of any order) that the number of covariants was finite, and, in the particular case of the quintic, established the system of the 23 covariants. Starting from the quantic itself a, then the system of derivatives $aa2$, $aa4$, &c., must include among itself all the covariants of the second degree, and if the entire system of these is, suppose, b, c, &c., then the derivatives $ab1$, $ab2$, &c., $ac1$, $ac2$, &c., must include among them all the covariants of the third degree, and so on for the higher degrees; and in this way, limiting by general reasoning the number of the independent covariants of each degree obtained by the successive steps, the foregoing conclusion is arrived at. But returning to the quintic, and supposing the system of the 23 covariants established, then knowing the deg-order of a derivative we know that it must be a linear function of the segregates of that deg-order; and we thus confirm, *à posteriori*, the results of the derivation theory. I annex the following Table No. 99, showing all the derivatives which present themselves, and for each of them the

covariants as well congregate as segregate of the same deg-order: the congregates are distinguished each by two prefixed dots, ..*bf*, &c. No further explanation of the arrangement is, I think, required. We see from the table in what manner the different covariants present themselves in connexion with the derivation-theory. Thus starting with the quintic itself a, we have the two derivatives $aa4$, $aa2$, which are in fact the covariants of the second degree (deg-orders 2.2 and 2.6 respectively) b and c. For the third degree we have the derivatives $ab2$, $ab1$, $ac5$, $ac4$, $ac3$, $ac2$, $ac1$: the deg-order of $ac5$ is 3.1, and there being no covariants of this deg-order, $ac5$ must, it is clear, vanish identically: $ab2$ and $ac4$ are each of them of the deg-order 3.3, but for this deg-order we have only the covariant d, and hence $ab2$ and $ac4$ must be each of them a numerical multiple of d; similarly, deg-order 3.5, $ab1$ and $ac3$ must be each of them a numerical multiple of e; deg-order 3.7, $ac2$ must be a numerical multiple of ab; and deg-order 3.9, $ac1$ must be a numerical multiple of f: the 7 derivatives, which *primâ facie* might give, each of them, a covariant of the third degree, thus give in fact only the 3 covariants d, e, f; and in order to show according to the theory of derivations that this is so, it is necessary to prove—1°, that $ac5 = 0$; 2°, that $ac4$ and $ab2$ differ only by a numerical factor; 3°, that $ab1$ and $ac3$ differ only by a numerical factor; 4°, that $ac2$ is a numerical multiple of ab: which being so, we have the 3 new covariants. The table shows that

for degrees	2,	3,	4,	5,	6,	7,	8,	9,	10,	11,	12,	13,	14,	15,	16,	17,	18,	19,	20,	21,	22,	23,	24
No. of derivatives =	2,	7,	19,	29,	41,	46,	52,	46,	44,	35,	26,	19,	17,	12,	13,	6,	6,	3,	3,	1,	1,	0,	1

so that the whole number of derivatives is 429, giving the 22 covariants $b, c, \ldots, w$. While it is very remarkable that (by general reasoning, as already mentioned, and with a very small amount of calculation) Gordan should have been able in effect to show this, the great excess of the number of derivatives over that of the covariants seems a reason why the derivations ought not to be made a basis of the theory.

It is to be remarked that we may consider derivatives $pq1$, $pq2$, &c., where p, q instead of being simple covariants are powers or products of covariants, but that these may be made to depend upon the derivatives formed with the simple covariants. (As to this see my paper "On the Derivatives of Three Binary Quantics," *Quart. Math. Journal*, t. xv. (1877), pp. 157—168, [681].)

TABLE No. 99 (Index Table of Derivatives).

Deg.	2					3					
Ord.		0	2	4	6		1	3	5	7	9
			b		c			d	e	ab	f
	aa		4		2	ab		2	1		
						ac	5	4	3	2	1
	2 derivs.					7 derivs.					

TABLE No. 99 (*continued*).

Deg.	4						
Ord.	0	2	4	6	8	10	12
	g		b^2 h	i	ad bc	ae	a^2b c^2
ad		3	2	1			
ae	5	4	3	2	1		
af			5	4	3	2	1
bb	2						
bc			2	1			
cc	6		4		2		

19 derivs.

Deg.	5						
Ord.	1	3	5	7	9	11	13
	j	k	ag bd	be l	ab^2 ah cd	ai $..\,bf$ ce	a^2d abc
ah	4	3	2	1			
ai	5	4	3	2	1		
bd	2	1					
be		2	1				
bf				2	1		
cd		3	2	1			
ce	5	4	3	2	1		
cf		6	5	4	3	2	1

29 derivs.

Deg.	6							
Ord.	0	2	4	6	8	10	12	14
		bg m	n	aj b^3 bh cg $..\,d^2$	ak bi $..\,de$	a^2g abd b^2c ch $..\,e^2$	abe al ci $..\,df$	a^2b^2 a^2h acd bc^2 $..\,ef$
aj			1					
ak		3	2	1				
al		5	4	3	2	1		
bh		2	1					
bi			2	1				
ch		4	3	2	1			
ci	6	5	4	3	2	1		
dd		2						
de		3	2	1				
df				3	2	1		
ee		4		2				
ef			5	4	3	2	1	
ff		8		6		4		2

41 derivs.

TABLE No. 99 (*continued*).

Deg.	7						
Ord.	1	3	5	7	9	11	13
	o	*bj*	*bk*	*abg*	*an*	a^2j	a^2k
		dg	*eg*	*am*	$..\,b^2e$	ab^3	*abi*
			p	b^2d	*bl*	*abh*	$..\,ade$
				cj	*ck*	*acg*	$..\,b^2f$
				..*dh*	..*di*	$..\,ad^2$	*bce*
					..*eh*	*bcd*	*cl*
					fg	..*ei*	..*fh*
am		2	1				
an	4	3	2	1			
bj	1						
bk	2	1					
bl			2	1			
cj			1				
ck		3	2	1			
cl	6	5	4	3	2	1	
dh	3	2	1				
di		3	2	1			
eh	4	3	2	1			
ei	5	4	3	2	1		
fh			4	3	2	1	
fi		6	5	4	3	2	1

46 derivs.

TABLE No. 99 (*continued*).

Deg.	8							
Ord.	0	2	4	6	8	10	12	14
	g^2	r	b^2g	as	abj	abk	a^2bg	a^2n
	q		bm	bn	adg	aeg	a^2m	$..\ ab^2e$
			dj	$..\ dk$	b^4	ap	ab^2d	abl
			gh	$..\ ej$	b^2h	b^2i	acj	ack
				gi	bcg	$..\ bde$	$..\ adh$	$..\ adi$
					$..\ bd^2$	cn	b^3c	$..\ aeh$
					cm	$..\ dl$	bch	afg
					$..\ ek$	$..\ fj$	$..\ be^2$	bci
					$..\ h^2$	$..\ hi$	c^2g	$..\ bdf$
							$..\ cd^2$	$..\ cde$
							$..\ el$	
							$..\ fk$	
							$..\ i^2$	
ao			1					
ap	5	4	3	2	1			
bm	2	1						
bn		2	1					
cm			2	1				
cn		4	3	2	1			
dj		1						
dk	3	2	1					
dl			3	2	1			
ej			1					
ek		3	2	1				
el		5	4	3	2	1		
fj					1			
fk				3	2	1		
fl		7	6	5	4	3	2	1
hh	4		2					
hi		4	3	2	1			
ii	6		4		2			

52 derivs.

TABLE No. 99 (*continued*).

Deg.	9					
Ord.	1	3	5	7	9	11
	gj	bo	ag^2	ar	ab^2g	a^2o
		gk	aq	$..\ b^2k$	abm	abn
		s	b^2j	beg	adj	$..\ adk$
			bdg	bp	agh	$..\ aej$
			$..\ dm$	co	b^3d	agi
			hj	$..\ dn$	bcj	$..\ b^3e$
				$..\ em$	$..\ bdh$	b^2l
				gl	cdg	bck
				$..\ hk$	$..\ d^3$	$..\ bdi$
				$..\ ij$	$..\ en$	$..\ beh$
					$..\ ik$	$..\ bfg$
						ceg
						cp
						$..\ d^2e$
						$..\ fm$
						$..\ hl$
ar		2	1			
bo	1					
bp		2	1			
co			1			
cp	5	4	3	2	1	
dm	2	1				
dn	3	2	1			
em		2	1			
en	4	3	2	1		
fm				2	1	
fn			4	3	2	1
hj		1				
hk	3	2	1			
hl		4	3	2	1	
ij			1			
ik		3	2	1		
il	6	5	4	3	2	1

46 derivs.

TABLE No. 99 (*continued*).

Deg.	10						
Ord.	0	2	4	6	8	10	12
		bg^2	br	agj	abo	a^2g^2	a^2r
		bq	$..do$	b^3g	agk	a^2q	$..ab^2k$
		gm	gn	b^2m	as	ab^2j	$abeg$
		j^2	jk	bdj	b^2n	$abdg$	abp
				bgh	$..bdk$	$..adm$	aco
				cg^2	$..bej$	ahj	$..adn$
				cq	bgi	$..b^5$	$..aem$
				$..d^2g$	cr	b^3h	agl
				$..eo$	$..deg$	b^2cg	$..ahk$
				$..hm$	$..dp$	$..b^2d^2$	$..aij$
				$..k^2$	$..hn$	bcm	b^3i
					$..im$	$..bek$	$..b^2de$
					$..jl$	$..bh^2$	bcn
						cdj	$..bdl$
						cgh	$..bfj$
						$..d^2h$	$..bhi$
						$..e^2g$	$..cdk$
						$..ep$	$..cej$
						$..fo$	cgi
						$..in$	$..d^2i$
						$..kl$	$..deh$
							$..dfg$
as		3	2	1			
br	2	1					
cr			2	1			
do		1					
dp		3	2	1			
eo			1				
ep	5	4	3	2	1		
fo					1		
fp			5	4	3	2	1
hm		2	1				
hn	4	3	2	1			
im			2	1			
in		4	3	2	1		
jk		1					
jl				1			
kk		2					
kl			3	2	1		
ll		6		4		2	

44 derivs.

TABLE No. 99 (*continued*).

Deg.	11				
Ord.	1	3	5	7	9
	go	bgj	b^2o	abg^2	abr
	t	dg^2	bgk	abq	. . ado
		dq	bs	agm	agn
		jm	. . dr	aj^2	ajk
			eg^2	b^3j	. . b^3k
			eq	b^2dg	. . b^2eg
			gp	. . bdm	b^2p
			. . ho	bhj	bco
			. . jn	cgj	. . bdn
			. . km	. . d^2j	. . bem
				. . dgh	bgl
				. . er	. . bhk
				. . io	. . bij
				. . kn	cjk
					cs
					. . d^2k
					. . dej
					. . dgi
					. . egh
					fg^2
					fq
					. . hp
					. . lm
bs	2	1			
cr		3	2	1	
dr	2	1			
er		2	1		
fr				2	1
ho		1			
hp	4	3	2	1	
io			1		
ip	5	4	3	2	1
jm	1				
jn		1			
km	2	1			
kn	3	2	1		
lm	2	1			
ln		4	3	2	1

35 derivs.

TABLE No. 99 (*continued*).

Deg.	12					
Ord.	0	2	4	6	8	10
	g^3	gr	b^2g^2	ago	$abgj$	ab^2o
	g^9	jo	b^2q	at	. . adg^2	$abgk$
	u		bgm	b^2r	adq	abs
			bj^2	. . bdo	ajm	. . adr
			dgj	bgn	b^4g	aeg^2
			g^2h	bjk	b^3m	aeq
			hq	. . dgk	b^2dj	agp
			. . ko	. . ds	b^2gh	. . aho
			. . m^2	. . egj	bcg^2	. . ajn
				g^2i	bcq	. . akm
				. . hr	. . bd^2g	b^3n
				iq	. . beo	. . b^2dk
				. . jp	. . bhm	. . b^2ej
				. . mn	. . bk^2	. . b^2gi
					cgm	bcr
					cj^2	. . $bdeg$
					. . d^2m	. . bdp
					. . dhj	. . bhn
					. . egk	. . bim
					. . es	. . bjl
					. . gh^2	. . cdo
					. . ir	cgn
					. . kp	cjk
					. . lo	. . d^2n
					. . n^2	. . dem
						. . dgl
						. . dhk
						. . dij
						. . ehj
						. . fgj
						. . ghi
at			1			
ds	3	2	1			
es		3	2	1		
fs				3	2	1
hr		2	1			
ir			2	1		
jo	1					
jp			1			
ko		1				
kp		3	2	1		
lo				1		
lp		5	4	3	2	1

TABLE No. 99 (*continued*).

Deg.	13			
Ord.	1	3	5	7
	g^2j	bgo	ag^3	agr
	jq	bt	agq	ajo
	v	g^2k	au	.. b^3o
		gs	b^2gj	.. b^2gk
		.. jr	bdg^2	b^2s
		kq	bdq	.. bdr
		.. mo	bjm	beg^2
			.. dgm	beq
			.. dj^2	bgp
			ghj	.. bho
			.. kr	.. bjn
			.. no	.. bkm
				cgo
				ct
				.. d^2o
				.. dgn
				.. djk
				.. egm
				.. ej^2
				g^2l
				.. ghk
				.. gij
				.. hs
				.. lq
				.. mp
bt	1			
ct			1	
hs	3	2	1	
is		3	2	1
jr	1			
kr		1		
lr				1
mo	1			
mp		2	1	
no		1		
np	4	3	2	1

19 derivs.

TABLE No. 99 (*continued*).

Deg.	14				
Ord.	0	2	4	6	8
		bg^3	bgr	.. ag^2j	$abgo$
		bgq	bjo	ajq	abt
		bu	.. dgo	av	ag^2k
		g^2m	.. dt	b^3g^2	ags
		gj^2	g^2n	b^3q	.. ajr
		mq	gjk	b^2gm	akq
		.. o^2	js	b^2j^2	.. amo
			.. mr	$bdgj$	b^3r
			nq	bg^2h	.. b^2do
				bhq	.. b^2gn
				.. bko	b^2jk
				.. bm^2	.. $bdgk$
				cg^3	.. bds
				cgq	.. $begj$
				cu	bg^2i
				.. d^2g^2	.. bhr
				.. d^2q	biq
				.. djm	.. bjp
				.. ego	.. bmn
				.. et	cgr
				.. ghm	cjo
				.. gk^2	.. d^2r
				.. hj^2	.. deg^2
				.. ks	.. deq
				.. nr	.. dgp
				.. op	.. dho
					.. djn
					.. dkm
					.. ejm
					.. ghn
					.. gim
					.. gjl
					.. hjk
					.. ij^2
dt			1		
et				1	
ft					1
js		1			
ks	3	2	1		
ls			3	2	1
mr	2	1			
nr		2	1		
op			1		
pp		4		2	

17 derivs.

Deg.	15		
Ord.	1	3	5
	g^2o	bg^2j	.. b^2go
	gt	bjq	b^2t
	oq	bv	bg^2k
		dg^3	bgs
		dgq	.. bjr
		.. du	bkq
		gjm	.. bmo
		.. j^3	.. dgr
		.. or	.. djo
			eg^3
			egq
			.. eu
			g^2p
			.. gho
			.. gjn
			.. gkm
			.. ht
			.. j^2k
			.. ms
			pq
bv	1		
cv			1
ht		1	
it			1
ms	2	1	
ns	3	2	1
or	1		
pr		2	1

12 derivs.

TABLE No. 99 (*continued*).

Deg.	16					
Ord.	0	2	4	6	8	Col. 8 concl.
	g^4 g^2q gu q^2	g^2r gjo jt qr	b^2g^3 b^2gq b^2u bg^2m bgj^2 bmq $..bo^2$ dg^2j djq $..dv$ g^3h ghq $..gko$ $..gm^2$ $..hu$ $..j^2m$ $..kt$ $..os$ $..r^2$	ag^2o agt aoq $..b^2gr$ b^2jo $..bdgo$ $..bdt$ bg^2n $bgjk$ bjs $..bmr$ $..bnq$ $..dg^2k$ $..dgs$ $..djr$ $..dkq$ $..dmo$ $..eg^2j$ $..egj$ $..ev$ g^3i $..ghr$ giq $..gjp$ $..gmn$ $..hjo$ $..iu$ $..j^2n$ $..jkm$	$..abg^2j$ $abjq$ abv $..adg^3$ $adgq$ $..adu$ $agjm$ $..aj^3$ $..aor$ b^4g^2 b^4q b^3gm b^3j^2 b^2dgj b^2g^2h b^2hq $..b^2ko$ $..b^2m^2$ bcg^3 $bcgq$ bcu $..bd^2g^2$ $..bd^2q$ $..bdjm$ $..bego$ $..bet$ $..bghm$ $..bgk^2$ $..bhj^2$ $..bks$ $..bnr$ $..bop$ cg^2m cgj^2 cmq $..co^2$ $..d^2gm$ $..d^2j^2$ $..dghj$ $..dkr$ $..dno$ $..eg^2k$ $..egs$ $..ejr$	$..ekq$ $..emo$ $..g^2h^2$ $..gir$ $..gkp$ $..glo$ $..gn^2$ $..h^2q$ $..hko$ $..hm^2$ $..ijo$ $..jkn$ $..k^2m$ $..lt$ $..ps$
dv		1				
ev			1			
fv					1	
jt	1					
kt		1				
lt				1		
op			1			
os		1				

TABLE No. 99 (*continued*).

Deg.	17			18				19	
Ord.	1	3	5	0	2	4	6	1	3
	g^3j	bg^2o	ag^4	w	bg^4	bg^2r	. . ag^3j	g^3o	bg^3j
	gjq	bgt	ag^2q		bg^2q	. . $bgjo$	$agjq$	g^2t	$bgjq$
	gv	boq	agu		bgu	bjt	agv	gqo	bgv
	ju	g^3k	aq^2		bq^2	bqr	aju	ou	bju
		g^2s	b^2g^2j		g^3m	. . dg^2o	b^3g^3	qt	dg^4
		. . gjr	b^2jq		g^2j^2	. . dgt	b^3gq		dg^2q
		gkq	b^2v		gmq	. . doq	b^3u		. . djt
		. . gmo	bdg^3		. . go^2	g^3n	b^2g^2m		dq^2
		. . j^2o	$bdgq$		j^2q	g^2jk	b^2gj^2		g^2jm
		ku	bdu		. . jv	gjs	b^2mq		. . gj^3
		. . mt	$bgjm$		mu	. . gmr	. . b^2o^2		. . gor
		qs	. . bj^3		. . ot	gnq	bdg^2j		jmq
			. . bor			. . j^2r	$bdjq$		. . jo^2
			. . dg^2m			jkq	. . bdv		. . mv
			. . dgj^2			. . jmo	bg^3h		. . rt
			. . dmq			. . kv	$bghq$		
			. . do^2			nu	. . $bgko$		
			g^2hj				. . bgm^2		
			. . gkr				bhu		
			. . gno				. . bj^2m		
			hjq				. . bkt		
			. . hv				. . bos		
			. . jko				. . br^2		
			. . jm^2				cg^4		
			. . nt				cg^2q		
			. . rs				cgu		
							cq^2		
							. . d^2g^3		
							. . d^2gq		
							. . d^2u		
							. . $dgjm$		
							. . dj^3		
							. . dor		
							. . eg^2o		
							. . egt		
							. . eoq		
							. . g^2hm		
							. . g^2k^2		
							. . ghj^2		
							. . gks		
							. . gnr		
							. . gop		
							. . hmq		
							. . ho^2		
							. . jkr		
							. . jno		
							. . k^2q		
							. . kmo		
							. . m^3		
							. . pt		
							. . s^2		

Deg. 17:

	1	3	5
hv		1	
iv			1
mt	1		
nt		1	
rs	2	1	

Deg. 18:

	0	2	4	6
jv	1			
kv		1		
lv				1

Deg. 19:

	1	3
mv	1	
nv		1
rt	1	

TABLE No. 99 (*concluded*).

Deg.	20			21	22		24
Ord.	0	2	4	1	0	2	0
	g^5	bw	b^2g^4	g^4j	gw	bg^5	g^6
	g^3q	g^3r	b^2g^2q	g^2jq		bg^3q	g^4q
	g^2u	g^2jo	b^2gu	g^2v		bg^2u	g^3u
	gq^2	. . gjt	b^2q^2	gju		bgq^2	g^2q^2
	qu	gqr	bg^3m	jq^2		bqu	gqu
		. . joq	bg^2j^2	qv		g^4m	q^3
		. . ov	$bgmq$			g^3j^2	u^2
		ru	. . bgo^2			g^2mq	
			bj^2q			. . g^2o^2	
			. . bjv			gj^2q	
			bmu			. . gjv	
			. . bot			gmu	
			dg^3q			. . got	
			$dgjq$			j^2u	
			. . dgv			mq^2	
			dju			. . o^2q	
			g^4h			. . t^2	
			g^2hq				
			. . g^2ko				
			. . g^2m^2				
			ghu				
			. . gj^2m				
			. . gkt				
			. . gos				
			. . gr^2				
			hq^2				
			. . j^4				
			. . jor				
			. . koq				
			. . m^2q				
			. . mo^2				
			. . st				
ov	1						
pv			1				
sv		1					
ro				1			
sv						1	
tv							1
	3 derivs.			1 deriv.	1 deriv.		1 deriv.

384. The Canonical form (using the divided expressions, Table No. 98) is peculiarly convenient for the calculation of the derivatives. Some attention is required in regard to the numerical determination: it will be observed that A is given in the standard form $(A_0, A_1, A_2, A_3, A_4, A_5 \between x, y)^5$, while the other covariants are given in the denumerate forms $B = (B_0, B_1, B_2 \between x, y)^2$ &c.: these must be converted into the other form $B = (B_0, \tfrac{1}{2}B_1, B_2 \between x, y)^2$, $C = (C_0, \tfrac{1}{6}C_1, \tfrac{1}{15}C_2, \tfrac{1}{20}C_3, \tfrac{1}{15}C_4, \tfrac{1}{6}C_5, C_6 \between x, y)^6$, &c., the numerical coefficients being of course the reciprocals of the binomial coefficients. We thus have, for instance, the leading coefficients,

$$l.c. \text{ of } AC2 = A_0 \,.\, \tfrac{1}{15}C_2 - 2 \,.\; A_1 \,.\, \tfrac{1}{6}C_1 + A_2 \,.\, C_0,$$

but

$$\text{,, ,, } BC2 = B_0 \,.\, \tfrac{1}{15}C_2 - 2 \,.\, \tfrac{1}{2}B_1 \,.\, \tfrac{1}{6}C_1 + B_2 \,.\, C_0.$$

Moreover, as regards the covariants $AA2$, $AA4$, &c., we take what are properly the half-values,

$$l.c. \text{ of } AA2 = A_0A_2 - \;A_1^2 \qquad \text{(instead of } A_0A_2 - 2A_1A_1 + A_2A_0),$$

$$\text{,, ,, } AA4 = A_0A_4 - 4A_1A_3 + 3A_2^2 \text{ (instead of } A_0A_4 - 4A_1A_3 + 6A_2A_2 - 4A_3A_1 - A_4A_0),$$

&c.,

and similarly

$$l.c. \text{ of } BB2 = B_0B_2 - (\tfrac{1}{2}B_1)^2,$$

$$\text{,, ,, } CC2 = C_0 \,.\, \tfrac{1}{15}C_2 - (\tfrac{1}{6}C_1)^2,$$

&c.

Any one of these leading coefficients, for instance $l.c.$ of $AC2$, is equal to the corresponding covariant derivative, multiplied, it may be, by a power of a: the index of this power being at once found by comparing the deg-orders, these in fact differing by a multiple of 1.5 the deg-order of a. Thus

$aa2$, $A_0A_2 - \;A_1^2$, deg-orders are 2.6, 2.6: or $aa2 = A_0A_2 - A_1^2$,

$aa4$, $A_0A_4 - 4A_1A_3 + 3A_2^2$, deg-orders are 2.2, 4.12: or $aa4 = \dfrac{1}{a^2}(A_0A_4 - 4A_1A_3 + 3A_2^2)$;

we have in fact

$$A_0A_2 - \;A_1^2 = 1 \,.\, c - 0^2 = c \qquad\qquad : \text{ and } aa2 = c,$$

$$A_0A_4 - 4A_1A_3 + 3A_2^2 = 1 \,.\, (a^2b - 3c^2) - 4 \,.\, 0 \,.\, f + 3 \,.\, c^2, \; = a^2b: \text{ and } aa4 = b.$$

As another instance, and for the purpose of showing how the calculation is actually effected, consider the derivative $ch2$, which is to be calculated from the leading coefficient of $CH2$, $= C_0 \,.\, \tfrac{1}{6}H_2 - 2 \,.\, \tfrac{1}{6}C_1 \,.\, \tfrac{1}{4}H_1 + \tfrac{1}{15}C_2 \,.\, H_0$: this is

$$\begin{aligned} = \;& c\,(\tfrac{1}{6}a^2g - 2abd - ch) \\ & - 2 \,.\, \tfrac{1}{2}f\,(\tfrac{1}{2}be - l) \\ & + (\tfrac{1}{5}a^2b - c^2)\,h \end{aligned}$$

= column next written down; but this column contains congregate terms which have to be replaced by their segregate values (see Table No. 96, deg-order 8.16); and we thus obtain

	a^3j	a^2b^3	a^2bh	a^2cg	$abcd$	b^2c^2	c^2h
$\frac{1}{5}a^2bh$			$+\frac{1}{5}$				
$+\frac{1}{6}a^2cg$				$+\frac{1}{6}$			
$-2abcd$					-2		
$-\frac{1}{2}bef$			$-\frac{1}{2}$		$+3$	$+2$	
$-2c^2h$							-2
$+fl$	$\frac{1}{3}$	$-\frac{1}{3}$	$+\frac{2}{3}$	$-\frac{1}{3}$	-1	-2	$+2$
$=$	$\frac{1}{3}$	$-\frac{1}{3}$	$+\frac{11}{30}$	$-\frac{1}{6}$	0	0	0

viz. the terms other than those divisible by a^2 all disappear: we may either abbreviate the calculation by omitting them *ab initio*, or retain them for the sake of the verification afforded by their disappearance. The factor a^2 divides out, and the final result is

$$ch2 = \tfrac{1}{3}aj - \tfrac{1}{3}b^3 + \tfrac{11}{30}bh - \tfrac{1}{6}cg,$$

which is the proper segregate expression of the derivative $ch2$: of course, we have deg-order $CH2 = 8.16$, deg-order $ch2 = 6.6$, and the difference is 2.10, the double of 1.5, so that the factor a^2 is as it ought to be.

TABLE No. 100 (The Derivatives up to the Sixth Order).

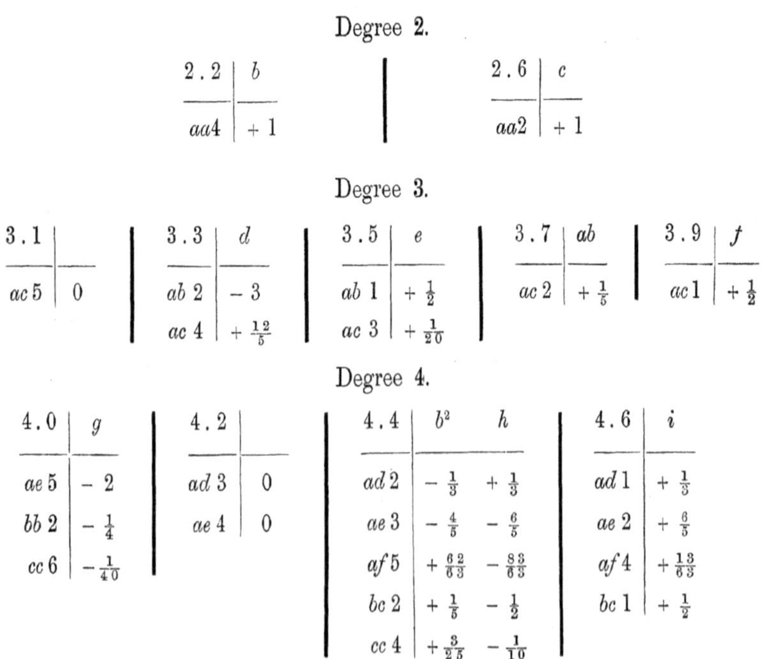

Degree 2.

2.2	b
$aa4$	$+1$

2.6	c
$aa2$	$+1$

Degree 3.

3.1	
$ac\,5$	0

3.3	d
$ab\,2$	-3
$ac\,4$	$+\frac{12}{5}$

3.5	e
$ab\,1$	$+\frac{1}{2}$
$ac\,3$	$+\frac{1}{20}$

3.7	ab
$ac\,2$	$+\frac{1}{5}$

3.9	f
$ac\,1$	$+\frac{1}{2}$

Degree 4.

4.0	g
$ae\,5$	-2
$bb\,2$	$-\frac{1}{4}$
$cc\,6$	$-\frac{1}{40}$

4.2	
$ad\,3$	0
$ae\,4$	0

4.4	b^2	h
$ad\,2$	$-\frac{1}{3}$	$+\frac{1}{3}$
$ae\,3$	$-\frac{4}{5}$	$-\frac{6}{5}$
$af\,5$	$+\frac{62}{63}$	$-\frac{83}{63}$
$bc\,2$	$+\frac{1}{5}$	$-\frac{1}{2}$
$cc\,4$	$+\frac{3}{25}$	$-\frac{1}{10}$

4.6	i
$ad\,1$	$+\frac{1}{3}$
$ae\,2$	$+\frac{6}{5}$
$af\,4$	$+\frac{13}{63}$
$bc\,1$	$+\frac{1}{2}$

TABLE No. 100 (*continued*).

4 . 8	ad	bc
ae 1	$-\frac{6}{5}$	-2
af 3	$+\frac{59}{42}$	$-\frac{5}{6}$
cc 2	$+\frac{1}{4}$	$-\frac{1}{20}$

4 . 10	ae
af 2	$+1$

4 . 12	a^2b	c^2
af 1	$+\frac{2}{9}$	-2

Degree 5.

5 . 1	j
ah 4	$+2$
ai 5	$-\frac{4}{3}$
bd 2	$-\frac{1}{3}$
ce 5	$-\frac{8}{5}$

5 . 3	k
ah 3	$+\frac{1}{2}$
ai 4	$+\frac{1}{3}$
bd 1	$-\frac{1}{6}$
be 2	$-\frac{3}{5}$
cd 3	$+\frac{3}{20}$
ce 4	$+\frac{2}{25}$
cf 6	$-\frac{19}{42}$

5 . 5	ag	bd
ah 2	$+\frac{1}{6}$	-2
ai 3	$+0$	-2
be 1	$-\frac{1}{2}$	$+\frac{24}{5}$
cd 2	0	$-\frac{2}{15}$
ce 3	$-\frac{1}{20}$	$-\frac{48}{5}$
df 5	$-\frac{1}{72}$	$+\frac{8}{63}$

5 . 7	be	l
ah 1	-2	-4
ai 2	0	$+\frac{1}{3}$
bf 2	$-\frac{7}{36}$	$+1$
cd 1	0	$+\frac{1}{6}$
ce 2	$\frac{1}{5}$	$-\frac{2}{5}$
df 4	$-\frac{1}{90}$	$+\frac{43}{315}$

5 . 9	ab^2	ah	cd
ai 1	$-\frac{1}{3}$	$+\frac{1}{3}$	$+3$
bf 1	$+\frac{2}{9}$	$-\frac{1}{2}$	$+3$
ce 1	0	$-\frac{1}{2}$	$+\frac{9}{5}$
df 3	$+\frac{2}{15}$	$-\frac{11}{120}$	$-\frac{439}{420}$

5 . 11	ai	ce
df 2	$+\frac{1}{45}$	$+\frac{1}{180}$

5 . 13	a^2d	abc
df 1	$+\frac{1}{2}$	$-\frac{5}{8}$

Degree 6.

6 . 0	
ci 6	0

6 . 2	bg	m
ak 3	0	-4
al 5	0	$+\frac{20}{7}$
bh 2	$-\frac{1}{3}$	-2
ch 4	$\frac{1}{10}$	$-\frac{2}{5}$
ci 5	0	$+\frac{2}{3}$
dd 2	0	$+\frac{1}{9}$
de 3	0	$-\frac{4}{5}$
ee 4	-1	$-\frac{48}{25}$
ff 8	$-\frac{5}{324}$	$-\frac{68}{567}$

6 . 4	n
aj 1	-1
ak 2	$+\frac{1}{3}$
al 4	$-\frac{1}{35}$
bh 1	$+\frac{1}{2}$
bi 2	$+\frac{1}{3}$
ch 3	$+\frac{3}{10}$
ci 4	$+\frac{1}{15}$
de 2	$-\frac{1}{3}$
ef 5	$-\frac{64}{63}$

6 . 6	aj	b^3	bh	cg
a 1	$-\frac{2}{3}$	$+2$	-3	$+1$
al 3	$-\frac{16}{35}$	$+\frac{2}{7}$	$-\frac{5}{7}$	$+\frac{3}{7}$
bi 1	$-\frac{1}{3}$	$+\frac{5}{6}$	$-\frac{1}{2}$	0
ch 2	$+\frac{1}{3}$	$-\frac{1}{3}$	$+\frac{11}{36}$	$-\frac{1}{6}$
ci 3	$-\frac{1}{6}$	$-\frac{1}{30}$	$+\frac{11}{3}$	$+\frac{7}{60}$
de 1	$-\frac{2}{15}$	$+\frac{2}{15}$	$+\frac{1}{15}$	$-\frac{1}{5}$
df 3	$+\frac{59}{378}$	$-\frac{143}{378}$	$+\frac{425}{756}$	$-\frac{139}{756}$
ee 2	$-\frac{4}{25}$	$+\frac{4}{25}$	$-\frac{38}{25}$	$+\frac{9}{25}$
ef 4	$+\frac{236}{315}$	$-\frac{4}{105}$	$+\frac{10}{63}$	$-\frac{71}{105}$
ff 6	$-\frac{1529}{7938}$	$+\frac{2873}{7938}$	$+\frac{3533}{15876}$	$-\frac{5591}{31752}$

TABLE No. 100 (*concluded*).

6 . 8	ak	bi
al 2	$-\frac{4}{21}$	$+\frac{1}{21}$
ch 1	$+\frac{1}{6}$	$+\frac{1}{3}$
ci 2	$+\frac{1}{9}$	$+\frac{4}{45}$
df 2	$-\frac{11}{108}$	$-\frac{1}{54}$
ef 3	$+\frac{32}{315}$	$+\frac{64}{315}$

6 . 10	a^2g	abd	b^2c	ch
al 1	0	$-\frac{3}{7}$	-1	$+1$
ci 1	0	$-\frac{1}{2}$	$+\frac{1}{6}$	$-\frac{1}{6}$
df 1	0	$-\frac{1}{9}$	$+\frac{1}{3}$	$-\frac{1}{3}$
ef 2	$+\frac{1}{36}$	$+\frac{7}{5}$	$-\frac{19}{45}$	$-\frac{4}{5}$
ff 4	$+\frac{1}{432}$	$-\frac{53}{756}$	$+\frac{89}{756}$	$-\frac{8}{63}$

6 . 12	abe	al	ci
ef 1	$\frac{2}{9}$	$-\frac{2}{5}$	$-\frac{6}{5}$

6 . 14	a^2b^2	a^2h	acd	bc^2
ff 2	$-\frac{4}{81}$	$+\frac{1}{36}$	$+\frac{5}{6}$	$-\frac{2}{9}$

which is complete to the sixth degree. I had calculated the derivatives up to the tenth degree, but the results were not in the segregate form.

On the form of the Numerical Generating Functions: the N.G.F. of a Sextic. Art. Nos. 385, 386.

385. It is to be remarked that the R.G.F. is derived not from the fraction in its least terms, which is algebraically the most simple form of the N.G.F., but from a form which contains common factors in the numerator and denominator: thus for the quadric, the cubic, and the quartic, writing down the two forms (identical in the case of the quadric) these are—

Quadric

$$\text{N.G.F.} = \frac{1}{1-ax^2 . 1-a^2}.$$

Cubic

$$\text{N.G.F.} = \frac{1-ax+a^2x^2}{1-a^4 . 1-ax^3 . 1-ax} \qquad = \frac{1-a^6x^6}{1-a^4 . 1-ax^3 . 1-a^2x^2 . 1-a^3x^3}.$$

Quartic

$$\text{N.G.F.} = \frac{1-ax^2+a^2x^4}{1-a^2 . 1-a^3 . 1-ax^4 . 1-ax^2} \qquad = \frac{1-a^6x^{12}}{1-a^2 . 1-a^3 . 1-ax^4 . 1-a^2x^4 . 1-a^3x^6}.$$

For the quintic the two forms are, N.G.F. =

$($ 1			$-\ a^6$			$+\ a^{12})\ x^0$
$+\ (-\ 1$		$+\ a^4$	$+\ 2a^6$			$-\ a^{12})\ ax^1$
$+\ ($	$+\ a^2$			$-\ a^8$		$+\ a^{10})\ x^2$
$+\ (-\ 1$		$+\ a^4$	$+\ a^6$	$+\ a^8$	$-\ a^{10}$	$-\ a^{12})\ ax^3$
$+\ (+\ 1$	$+\ a^2$	$-\ a^4$	$-\ a^6$	$-\ a^8$		$+\ a^{12})\ a^2x^4$
$+\ ($	$-\ a^2$	$+\ a^4$			$-\ a^{10}$	$)\ a^3x^5$
$+\ (+\ 1$			$-\ 2a^6$	$-\ a^8$		$+\ a^{12})\ a^2x^6$
$+\ (-\ 1$			$+\ a^6$			$-\ a^{12})\ a^3x^7$

divided by

$$1 - a^4 . 1 - a^6 . 1 - a^8 . 1 - ax^5 . 1 - ax^3 . 1 - ax\ ;$$

and

$($ 1									$+\ a^{18})\ x^0$
$($		a^4	$+\ a^6$		$+\ a^{10}$	$+\ a^{12}$			$)\ ax$
$($		a^4	$+\ a^6$	$+\ a^8$	$+\ a^{10}$		$+\ a^{14}$		$-\ a^{18})\ a^2x^2$
$($ 1	$+\ a^2$	$+\ a^4$		$+\ a^8$					$)\ a^3x^3$
$($ 1	$+\ a^2$	$+\ a^4$	$+\ a^6$		$+\ a^{10}$		$-\ a^{14}$		$)\ a^4x^4$
$($ 1		$+\ a^4$	$+\ a^6$					$-\ a^{16}$	$)\ a^3x^5$
$($	a^2					$-\ a^{12}$	$-\ a^{14}$		$)\ a^2x^6$
$($		a^4		$-\ a^8$		$-\ a^{12}$	$-\ a^{14}$	$-\ a^{16}$	$-\ a^{18})\ ax^7$
$($					$-\ a^{10}$	$-\ a^{12}$	$-\ a^{14}$	$-\ a^{16}$	$-\ a^{18})\ a^2x^8$
$($		$-\ a^4$		$-\ a^8$	$-\ a^{10}$	$-\ a^{12}$	$-\ a^{14}$		$)\ a^3x^9$
$($			$-\ a^6$	$-\ a^8$		$-\ a^{12}$	$-\ a^{14}$		$)\ a^4x^{10}$
$(-1$									$-\ a^{18})\ a^5x^{11}$

divided by

$$1 - a^4 . 1 - a^8 . 1 - a^{12} . 1 - ax^5 . 1 - a^2x^2 . 1 - a^2x^6 :$$

this last being in fact equivalent to that used for the determination of the R.G.F.

386. For the sextic the forms are, N.G.F. =

(1	$+a$		$-a^3$	$-a^4$	$-a^5$		$+a^7$	$+a^8)\,x^0$
(−1	$-a$	$+a^2$	$+2a^3$	$+2a^4$	$+a^5$		$-a^7$	$-a^8)\,ax^2$
(−1		$+a^2$	$+a^3$	$+a^4$	$+a^5$		$-a^7$	$-a^8)\,ax^4$
(1	$+a$		$-a^3$	$-a^4$	$-a^5$	$-a^6$		$+a^8)\,a^2x^6$
(1	$+a$		$-a^3$	$-2a^4$	$-2a^5$	$-a^6$	$+a^7$	$+a^8)\,a^2x^8$
(−1	$-a$		$+a^3$	$+a^4$	$+a^5$		$-a^7$	$-a^8)\,a^3x^{10}$

divided by

$$1+a\,.\,1-a^2\,.\,1-a^3\,.\,1-a^4\,.\,1-a^5\,.\,1-ax^6\,.\,1-ax^4\,.\,1-ax^2\,;$$

and

(1															$+a^{15}\ )\,x^0$
+(1		$+a^2$		$+a^4$	$+a^5$		$+a^7$		$+a^9$						$)\,a^3x^2$
+(		$+a^2$	$+a^3$	$+a^4$	$+a^5$	$+a^6$	$+a^7$	$+a^8$	$+a^9$		$+a^{11}$				$)\,a^2x^4$
+(1	$+a$		$+2a^3$		$+a^5$	$+a^6$		$+a^8$					$-a^{13}$		$)\,a^3x^6$
+($+a^{.5}$		$+a^{2.5}$		$+a^{4.5}$						$-a^{10.5}$		$-a^{12.5}$		$-a^{14.5}$	$)\,a^{2.5}x^8$
+(		$+a^2$					$-a^7$		$-a^9$	$-a^{10}$		$-2a^{12}$		$-a^{14}$	$-a^{15}\ )\,a^2x^{10}$
+(				$-a^4$		$-a^6$	$-a^7$	$-a^8$	$-a^9$	$-a^{10}$	$-a^{11}$	$-a^{12}$	$-a^{13}$		$)\,a^3x^{12}$
+(						$-a^6$		$-a^8$		$-a^{10}$	$-a^{11}$		$-a^{13}$		$-a^{15}\ )\,a^2x^{14}$
+(−1															$-a^{15}\ )\,a^5x^{16}$

divided by

$$1-a^2\,.\,1-a^4\,.\,1-a^6\,.\,1-a^{10}\,.\,1-ax^6\,.\,1-a^2x^4\,.\,1-a^2x^8,$$

where observe that in the middle term, although for symmetry $a^{\cdot 5}\,(=\sqrt{a})$ has been introduced into the expression, the coefficient is really rational, viz. the term is

$$(a^3+a^5+a^7-a^{13}-a^{15}-a^{17})\,x^8.$$

The second form or one equivalent to it is due to Sylvester: I do not know whether he divided out the common factors so as to obtain the first form. I assume that it would be possible from this second form to obtain a R.G.F., and thence to establish for the 26 covariants of the sextic a theory such as has been given for the 23 covariants of the quintic: but I have not entered upon this question.

TABLE No. 93 *bis* (The covariant S, adopted form $= -(D, M)$).

In this Table, a, b, c, d, e, f denote, as in the tables of former memoirs, the coefficients of the quintic form $(a, b, c, d, e, f \between x, y)^5$.

$S = ($

$a^3b^0c^3f^3$	$-\ 2$	$a^3b^0c^2df^3$	$-\ 3$	$a^3b^0cd^2f^3$	$+\ 3$	$a^3b^0d^3f^3$	$+\ 2$
c^2def^2	$+\ 15$	$c^2e^2f^2$	$+\ 3$	cde^2f^2	$-\ 6$	$d^2e^2f^2$	$-\ 6$
c^2e^3f	$-\ 9$	cd^2ef^2	$+\ 24$	ce^4f	$+\ 3$	de^4f	$+\ 6$
cd^3f^2	$-\ 9$	cde^3f	$-\ 42$	d^3ef^2	$-\ 3$	e^6	$-\ 2$
cd^2e^2f	$-\ 6$	ce^5	$+\ 18$	d^2e^3f	$+\ 6$	$a^2b\ cd^2f^3$	$-\ 15$
cde^4	$+\ 9$	d^4f^2	$-\ 18$	de^5	$-\ 3$	cde^2f^2	$+\ 30$
d^4ef	$+\ 9$	d^3e^2f	$+\ 33$	$a^2b^2d^2f^3$	$-\ 3$	ce^4f	$-\ 15$
d^3e^3	$-\ 7$	d^2e^4	$-\ 15$	de^2f^2	$+\ 6$	d^3ef^2	$+\ 15$
$a^2b^2c^2f^3$	$+\ 6$	$a^2b^2cdf^3$	$+\ 6$	e^4f	$-\ 3$	d^2e^3f	$-\ 30$
$cdef^2$	$-\ 30$	ce^2f^2	$-\ 6$	,, $b\ c^2df^3$	$-\ 24$	de^5	$+\ 15$
ce^3f	$+\ 18$	d^2ef^2	$-\ 24$	$c^2e^2f^2$	$+\ 24$	,, $b^0c^3df^3$	$+\ 9$
d^3f^2	$+\ 9$	de^3f	$+\ 42$	cd^2ef^2	$+\ 78$	$c^3e^2f^2$	$-\ 9$
d^2e^2f	$+\ 6$	e^5	$-\ 18$	cde^3f	$-\ 108$	$c^2d^2ef^2$	$-\ 21$
de^4	$-\ 9$	,, $b\ c^3f^3$	$+\ 3$	ce^5	$+\ 30$	c^2de^3f	$+\ 15$
,, $b\ c^3ef^2$	$-\ 15$	c^2def^2	$-\ 78$	d^4f^2	$-\ 24$	c^2e^5	$+\ 6$
$c^2d^2f^2$	$+\ 21$	c^2e^3f	$+\ 69$	d^3e^2f	$+\ 24$	cd^4f^2	$+\ 3$
c^2de^2f	$-\ 6$	cd^3f^2	$+\ 93$	,, $b^0c^4f^3$	$+\ 18$	cd^3e^2f	$+\ 21$
c^2e^4	$+\ 18$	cd^2e^2f	$-\ 51$	c^3def^2	$-\ 93$	cd^2e^4	$-\ 24$
cd^3ef	$+\ 30$	cde^4	$-\ 33$	c^3e^3f	$+\ 21$	d^5ef	$-\ 9$
cd^2e^3	$-\ 51$	d^4ef	$-\ 57$	$c^2d^3f^2$	$+\ 36$	d^4e^3	$+\ 9$
d^5f	$-\ 36$	d^3e^3	$+\ 54$	$c^2d^2e^2f$	$+\ 123$	$a\ b^3d^2f^3$	$+\ 9$
d^4e^2	$+\ 39$	,, $b^0c^4ef^2$	$+\ 24$	c^2de^4	$-\ 51$	de^2f^2	$-\ 18$
,, $b^0c^4df^2$	$-\ 3$	$c^3d^2f^2$	$-\ 36$	cd^4ef	$-\ 111$	e^4f	$+\ 9$
c^4e^2f	$+\ 45$	c^3de^2f	$-\ 9$	cd^3e^3	$+\ 39$	,, $b^2c^2df^3$	$+\ 6$
c^3d^2ef	$-\ 84$	c^3e^4	$-\ 54$	d^6f	$+\ 27$	$c^2e^2f^2$	$-\ 6$
c^3de^3	$-\ 63$	c^2d^3ef	$+\ 24$	d^5e^2	$-\ 9$	cd^2ef^2	$+\ 6$
c^2d^4f	$+\ 45$	$c^2d^2e^3$	$+\ 129$	$a\ b^3cdf^3$	$+\ 42$	cde^3f	$-\ 24$
$c^2d^3e^2$	$+\ 150$	cd^5f	$+\ 9$	ce^2f^2	$-\ 42$	ce^5	$+\ 18$
cd^5e	$-\ 117$	cd^4e^2	$-\ 114$	d^2ef^2	$-\ 69$	d^4f^2	$-\ 45$
d^7	$+\ 27$	d^6e	$+\ 27$	de^3f	$+\ 96$	d^3e^2f	$+\ 96$
$a^1b^4cf^3$	$-\ 6$	$a^1b^4df^3$	$-\ 3$	e^5	$-\ 27$	d^2e^4	$-\ 51$
def^2	$+\ 15$	e^2f^2	$+\ 3$	,, $b^2c^3f^3$	$-\ 33$	,, $b\ c^4f^3$	$-\ 9$
e^3f	$-\ 9$	,, $b^3c^2f^3$	$-\ 6$	c^2def^2	$+\ 51$	c^3def^2	$-\ 30$
,, $b^3c^2ef^2$	$+\ 30$	$cdef^2$	$+\ 108$	c^2e^3f	$+\ 48$	c^3e^3f	$+\ 66$
cd^2f^2	$-\ 15$	ce^3f	$-\ 96$	cd^3f^2	$+\ 9$	$c^2d^3f^2$	$+\ 84$
cde^2f	$+\ 24$	d^3f^2	$-\ 21$	cd^2e^2f	$-\ 147$	$c^2d^2e^2f$	$-\ 36$
ce^4	$-\ 45$	d^2e^2f	$-\ 48$	cde^4	$+\ 39$	c^2de^4	$-\ 102$
d^3ef	$-\ 66$	de^4	$+\ 63$	d^4ef	$+\ 78$	cd^4ef	$-\ 174$
d^2e^3	$+\ 72$	,, $b^2c^3ef^2$	$-\ 24$	d^3e^3	$-\ 45$	cd^3e^3	$+\ 210$
,, $b^2c^3df^2$	$-\ 21$	c^2d^2f	$-\ 123$	,, $b\ c^4ef^2$	$+\ 57$	d^6f	$+\ 63$
c^3e^2f	$-\ 96$	c^2de^2f	$+\ 147$	$c^3d^2f^2$	$-\ 24$	d^5e^2	$-\ 72$
c^2d^2ef	$+\ 36$	c^2e^4	$+\ 66$	c^3de^2f	$-\ 78$	,, $b^0c^5ef^2$	$+\ 36$
c^2de^3	$+\ 213$	cd^3ef	$+\ 78$	c^3e^4	$-\ 60$	$c^4d^2f^2$	$-\ 45$
cd^4f	$+\ 120$	cd^2e^3	$-\ 186$	c^2d^3ef	$+\ 36$	c^4de^2f	$-\ 120$
cd^3e^2	$-\ 303$	d^5f	$+\ 51$	$c^2d^2e^3$	$+\ 108$	c^4e^4	$-\ 6$
d^5e	$+\ 51$	d^4e^2	$-\ 9$	cd^5f	$-\ 24$	c^3d^3ef	$+\ 204$

$\between x, y)^8$

(*continued on next page.*)

(continued from last page.)

$a^1b\,c^5f^2$	$+\ 9$	$a^1b\,c^4df^2$	$+\ 111$	$a\,b\,cd^4e^2$	$-\ 6$	$a\,b^0c^3d^2e^3$	$+\ 120$
c^4def	$+\ 174$	c^4e^2f	$-\ 78$	d^3e	$-\ 9$	c^2d^5f	$-\ 66$
c^4e^3	$-\ 36$	c^3d^2ef	$-\ 36$	,, $b^0c^5df^2$	$-\ 9$	$c^2d^4e^2$	$-\ 240$
c^3d^3f	$-\ 204$	c^3de^3	$-\ 54$	c^5e^2f	$-\ 51$	cd^6e	$+\ 144$
$c^3d^2e^2$	$-\ 174$	c^2d^4f	$-\ 96$	c^4d^2ef	$+\ 96$	d^8	$-\ 27$
c^2d^4e	$+\ 330$	$c^2d^3e^2$	$+\ 150$	c^4de^3	$+\ 111$	$a^0b^4cdf^3$	$-\ 9$
cd^6	$-\ 99$	cd^5e	$+\ 30$	c^3d^4f	$-\ 27$	ce^2f^2	$+\ 9$
,, b^0c^6ef	$-\ 63$	d^7	$-\ 27$	$c^3d^3e^2$	$-\ 234$	d^2ef^2	$-\ 18$
c^5d^2f	$+\ 66$	,, $b^0c^6f^2$	$-\ 27$	c^2d^5e	$+\ 141$	de^3f	$+\ 45$
c^5de^2	$+\ 99$	c^5def	$+\ 24$	cd^7	$-\ 27$	e^5	$-\ 27$
c^4d^3e	$-\ 147$	c^5e^3	$+\ 54$	$a^0b^5df^3$	$-\ 18$	,, $b^3c^3f^3$	$+\ 7$
c^3d^5	$+\ 45$	c^4d^3f	$+\ 27$	e^2f^2	$+\ 18$	c^2def^2	$+\ 51$
$a^0b^6f^3$	$+\ 2$	$c^4d^2e^2$	$-\ 93$	,, $b^4c^2f^3$	$+\ 15$	c^2e^3f	$-\ 72$
,, b^5cef^2	$-\ 15$	c^3d^4e	$+\ 6$	$cdef^2$	$+\ 33$	cd^3f^2	$+\ 63$
d^2f^2	$-\ 6$	c^2d^6	$+\ 9$	ce^3f	$-\ 63$	cd^2e^2f	$-\ 213$
de^2f	$-\ 18$	$a^0b^5cf^3$	$+\ 3$	d^3f^2	$+\ 54$	cde^4	$+\ 171$
e^4	$+\ 27$	def^2	$-\ 30$	d^2e^2f	$-\ 66$	d^4ef	$+\ 36$
,, $b^4c^2df^2$	$+\ 24$	e^3f	$+\ 27$	de^4	$+\ 27$	d^3e^3	$-\ 43$
c^2e^2f	$+\ 51$	,, $b^4cd^2f^2$	$+\ 51$	,, $b^3c^3ef^2$	$-\ 54$	,, $b^2c^4ef^2$	$-\ 39$
cd^2ef	$+\ 102$	cde^2f	$-\ 39$	$c^2d^2f^2$	$-\ 129$	$c^3d^2f^2$	$-\ 150$
cde^3	$-\ 171$	ce^4	$-\ 27$	c^2de^2f	$+\ 186$	c^3de^2f	$+\ 303$
d^4f	$+\ 6$	d^3ef	$+\ 60$	c^2e^4	$+\ 45$	c^3e^4	$-\ 18$
d^3e^2	$+\ 18$	d^2e^3	$-\ 45$	cd^3ef	$+\ 54$	c^2d^3ef	$+\ 174$
,, $b^3c^4f^2$	$-\ 9$	,, $b^3c^3df^2$	$-\ 39$	cd^2e^3	$-\ 96$	$c^2d^2e^3$	$-\ 345$
c^3def	$-\ 210$	c^3e^2f	$+\ 45$	d^5f	$-\ 54$	cd^5f	$-\ 99$
c^3e^3	$+\ 43$	c^2d^2ef	$-\ 108$	d^4e^2	$+\ 48$	cd^4e^2	$+\ 192$
c^2d^3f	$-\ 120$	c^2de^3	$+\ 96$	,, $b^2c^4df^2$	$+\ 114$	d^6e	$-\ 18$
$c^2d^2e^2$	$+\ 345$	cd^4f	$-\ 111$	c^4e^2f	$+\ 9$	,, $b\,c^5df^2$	$+\ 117$
cd^4e	$-\ 87$	cd^3e^2	$+\ 147$	c^3d^2ef	$-\ 150$	c^5e^2f	$-\ 51$
d^6	$-\ 2$	d^5e	$-\ 30$	c^3de^3	$-\ 147$	c^4d^2ef	$-\ 330$
,, b^2c^5ef	$+\ 72$	,, $b^2c^5f^2$	$+\ 9$	c^2d^4f	$+\ 93$	c^4de^3	$+\ 87$
c^4d^2f	$+\ 240$	c^4def	$+\ 6$	$c^2d^3e^2$	$+\ 150$	c^3d^4f	$+\ 147$
c^4de^2	$-\ 192$	c^4e^3	$-\ 48$	cd^5e	$-\ 87$	$c^3d^3e^2$	$+\ 186$
c^3d^3e	$-\ 186$	c^3d^3f	$+\ 234$	d^7	$+\ 18$	c^2d^5e	$-\ 201$
c^2d^5	$+\ 96$	$c^3d^2e^2$	$-\ 150$	,, $b\,c^6f^2$	$-\ 27$	cd^7	$+\ 45$
,, $b\,c^6df$	$-\ 144$	c^2d^4e	$-\ 108$	c^5def	$-\ 30$	,, $b^0c^7f^2$	$-\ 27$
c^6e^2	$+\ 18$	cd^6	$+\ 57$	c^5e^3	$+\ 30$	c^6def	$+\ 99$
c^5d^2e	$+\ 201$	,, $b\,c^6ef$	$+\ 9$	c^4d^3f	$-\ 6$	c^6e^3	$+\ 2$
c^4d^4	$-\ 87$	c^5d^2f	$-\ 141$	$c^4d^2e^2$	$+\ 108$	c^5d^3f	$-\ 45$
,, b^0c^8f	$+\ 27$	c^5de^2	$+\ 87$	c^3d^4e	$-\ 96$	$c^5d^2e^2$	$-\ 96$
c^7de	$-\ 45$	c^4d^3e	$+\ 96$	c^2d^6	$+\ 21$	c^4d^4e	$+\ 87$
c^6d^3	$+\ 20$	c^3d^5	$-\ 51$	,, b^0c^7ef	$+\ 27$	c^3d^6	$-\ 20$
		,, b^0c^7df	$+\ 27$	c^6d^2f	$-\ 9$		
		c^7e^2	$-\ 18$	c^6de^2	$-\ 57$		
		c^6d^2e	$-\ 21$	c^5d^3e	$+\ 51$		
		c^5d^4	$+\ 12$	c^4d^5	$-\ 12$		

I remark that I calculated the first two coefficients S_0, S_1, and deduced the other two S_2 from S_1, and S_3 from S_0, by reversing the order of the letters (or which is the same thing, interchanging a and f, b and e, c and d) and reversing also the signs of the numerical coefficients. This process for S_2, S_3 is to a very great extent a verification of the values of S_0, S_1. For, as presently mentioned, the terms of S_0 form subdivisions such that in each subdivision the sum of the numerical coefficients is $=0$: in passing by the reversal process to the value of S_3, the terms are distributed into an entirely new set of subdivisions, and then in each of these subdivisions the sum of the numerical coefficients is found to be $=0$; and the like as regards S_1 and S_2.

If in the expressions for S_0, S_1, S_2, S_3 we first write $d=e=f=1$, thus in effect combining the numerical coefficients for the terms which contain the same powers in a, b, c, we find

$$\begin{aligned}S_0 = \ & a^3(-2c^3+6c^2-6c+2)\\ & + a^2\{b^2(6c^2-12c-6)+b(-15c^3+33c^2-21c+3)\\ & \qquad + b^0(42c^4-147c^3+195c^2-117c+27)\}\\ & + a\ \{b^4.0+b^3(30c^2-36c+6)+b^2(-117c^3+249c^2-183c+51)\\ & \qquad + b(9c^5+138c^4-378c^3+330c^2-99c)+b^0(-63c^6+165c^5-147c^4+45c^3)\}\\ & + a^0\{b^6.2+b^5(-15c+3)+b^4(75c^2-69c+24)+b^3(-9c^4-167c^3+225c^2-87c-2)\\ & \qquad + b^2(72c^5+48c^4-186c^3+96c^2)+b(-126c^6+201c^5-87c^4)\\ & \qquad + b^0(27c^8-45c^7+20c^6)\};\end{aligned}$$

which for $c=1$ becomes

$$=2b^6-12b^5+30b^4-40b^3+30b^2-12b+2, \text{ that is, } 2(b-1)^6,$$

and for $b=1$ becomes $=0$.

$$\begin{aligned}S_2 = \ & a^3(0c^2+0c+0)\\ & + a^2\{b^2(0c+0)+b(3c^3-9c^2+9c-3)+b^0(24c^4-99c^3+153c^2-105c+27)\}\\ & + a\ \{b^4.0\ +b^3(-6c^2+12c-6)+b^2(-24c^3+90c^2-108c+42)\\ & \qquad + b(33c^4-90c^3+54c^2+30c-27)+b^0(-27c^6+78c^5-66c^4+6c^3+9c^2)\}\\ & + a^0\{b^5(3c-3)+b^4(-15c+15)+b^3(6c^3-12c^2+36c-30)\\ & \qquad + b^2(9c^5-42c^4+84c^3-108c^2+57c)+b(9c^6-54c^5+96c^4-51c^3)\\ & \qquad + b^0(9c^7-9c^6)\}:\end{aligned}$$

which for $c=1$ becomes $=0$.

$$\begin{aligned}S_3 = \ & a^3(0c+0)\\ & + a^2\{b^2.0+b(0c^2+0c+0)+b^0(18c^4-72c^3+108c^2-72c+18)\}\\ & + a\ \{b^3(0c+0)+b^2(-33c^3+99c^2-99c+33)+b(57c^4-162c^3+144c^2-30c-9)\\ & \qquad + b^0(-60c^5+207c^4-261c^3+141c^2-27c)\}\\ & + a^0\{b^5.0+b^4(15c^2-30c+15)+b^3(-54c^3+102c^2-42c-6)\\ & \qquad + b^2(123c^4-297c^3+243c^2-87c+18)+b(-27c^5+102c^4-96c^3+21c^2)\\ & \qquad + b^0(27c^7-66c^6+51c^5-12c^4)\}:\end{aligned}$$

which for $c=1$ becomes $=0$.

$$\begin{aligned}S_4 = \ & a^3 . 0 \\ & + a^2 \{b\ (0c + 0) + b^0 (0c^3 + 0c^2 + 0c + 0)\} \\ & + a\ \{b^3 . 0\ + b^2 (0c^2 + 0c + 0) + b (-9c^4 + 36c^3 - 54c^2 + 36c - 9) \\ & \qquad + b^0 (36c^5 - 171c^4 + 324c^3 - 306c^2 + 144c - 27)\} \\ & + a^0 \{b^4 (0c + 0) + b^3 (7c^3 - 21c^2 + 21c - 7) + b^2 (-39c^4 + 135c^3 - 171c^2 + 93c - 18) \\ & \qquad + b (66c^5 - 243c^4 + 333c^3 - 201c^2 + 45c) \\ & \qquad + b^0 (-27c^7 + 101c^6 - 141c^5 + 87c^4 - 20c^3)\} :\end{aligned}$$

which for $c = 1$ becomes $= 0$.

It follows that, for $c = d = e = f = 1$, the value of the covariant S is $= 2(b-1)^6 x^3$, which might be easily verified.

694.

DESIDERATA AND SUGGESTIONS.

No. 1. The theory of groups.

[From the *American Journal of Mathematics*, t. I. (1878), pp. 50—52.]

Substitutions, and (in connexion therewith) groups, have been a good deal studied; but only a little has been done towards the solution of the general problem of groups. I give the theory so far as is necessary for the purpose of pointing out what appears to me to be wanting.

Let $\alpha, \beta, \ldots$ be functional symbols, each operating upon one and the same number of letters and producing as its result the same number of functions of these letters; for instance, $\alpha(x, y, z) = (X, Y, Z)$, where the capitals denote each of them a given function of (x, y, z).

Such symbols are susceptible of repetition and of combination;

$$\alpha^2(x, y, z) = \alpha(X, Y, Z), \text{ or } \beta\alpha(x, y, z) = \beta(X, Y, Z),$$

= in each case three given functions of (x, y, z); and similarly for α^3, $\alpha^2\beta$, &c.

The symbols are not in general commutative, $\alpha\beta$ not $=\beta\alpha$; but they are associative, $\alpha\beta.\gamma = \alpha.\beta\gamma$, each $=\alpha\beta\gamma$, which has thus a determinate signification.

The associativeness of such symbols arises from the circumstance that the definitions of $\alpha, \beta, \gamma, \ldots$ determine the meanings of $\alpha\beta$, $\alpha\gamma$, &c.: if $\alpha, \beta, \gamma, \ldots$ were quasi-quantitative symbols such as the quaternion imaginaries i, j, k, then $\alpha\beta$ and $\beta\gamma$ might have by definition values δ and ϵ such that $\alpha\beta.\gamma$ and $\alpha.\beta\gamma$ ($=\delta\gamma$ and $\alpha\epsilon$ respectively) have unequal values.

Unity as a functional symbol denotes that the letters are unaltered, $1(x, y, z) = (x, y, z)$; whence $1\alpha = \alpha 1 = \alpha$.

The functional symbols *may* be substitutions; $\alpha(x, y, z) = (y, z, x)$, the same letters in a different order: substitutions can be represented by the notation $\alpha = \frac{yzx}{xyz}$, the substitution which changes xyz into yzx, or as products of cyclical substitutions, $\alpha = \frac{yzx\ wu}{xyz\ uw}$, $= (xyz)(uw)$, the product of the cyclical interchanges x into y, y into z, and z into x; and u into w, w into u.

A set of symbols $\alpha, \beta, \gamma, \ldots$, such that the product $\alpha\beta$ of each two of them (in each order, $\alpha\beta$ or $\beta\alpha$), is a symbol of the set, is a group. It is easily seen that 1 is a symbol of every group, and we may therefore give the definition in the form that a set of symbols, $1, \alpha, \beta, \gamma, \ldots$ satisfying the foregoing condition is a group. When the number of the symbols (or terms) is $= n$, then the group is of the nth order; and each symbol α is such that $\alpha^n = 1$, so that a group of the order n is, in fact, a group of symbolical nth roots of unity.

A group is defined by means of the laws of combination of its symbols: for the statement of these we may either (by the introduction of powers and products) diminish as much as may be the number of independent functional symbols, or else, using distinct letters for the several terms of the group, employ a square diagram as presently mentioned.

Thus, in the first mode, a group is $1, \beta, \beta^2, \alpha, \alpha\beta, \alpha\beta^2$ $(\alpha^2 = 1,\ \beta^3 = 1,\ \alpha\beta = \beta^2\alpha)$; where observe that these conditions imply also $\alpha\beta^2 = \beta\alpha$.

Or, in the second mode, calling the same group $(1, \alpha, \beta, \gamma, \delta, \epsilon)$, the laws of combination are given by the square diagram

	1	α	β	γ	δ	ϵ
1	1	α	β	γ	δ	ϵ
α	α	1	γ	β	ϵ	δ
β	β	ϵ	δ	α	1	γ
γ	γ	δ	ϵ	1	α	β
δ	δ	γ	1	ϵ	β	α
ϵ	ϵ	β	α	δ	γ	1

for the symbols $(1, \alpha, \beta, \gamma, \delta, \epsilon)$ are in fact $= (1, \alpha, \beta, \alpha\beta, \beta^2, \alpha\beta^2)$.

The general problem is to find all the groups of a given order n; thus if $n = 2$, the only group is $1, \alpha$ $(\alpha^2 = 1)$; if $n = 3$, the only group is $1, \alpha, \alpha^2$ $(\alpha^3 = 1)$; if $n = 4$, the groups are $1, \alpha, \alpha^2, \alpha^3$ $(\alpha^4 = 1)$, and $1, \alpha, \beta, \alpha\beta$ $(\alpha^2 = 1,\ \beta^2 = 1,\ \alpha\beta = \beta\alpha)$; if $n = 6$, there are three groups, a group $1, \alpha, \alpha^2, \alpha^3, \alpha^4, \alpha^5$ $(\alpha^6 = 1)$, and two groups $1, \beta, \beta^2, \alpha, \alpha\beta,$

$\alpha\beta^2$ ($\alpha^2=1$, $\beta^3=1$); viz. in the first of these $\alpha\beta=\beta\alpha$, while in the other of them (that mentioned above) we have $\alpha\beta=\beta^2\alpha$, $\alpha\beta^2=\beta\alpha$.

But although the theory as above stated is a general one, including as a particular case the theory of substitutions, yet the general problem of finding all the groups of a given order n, is really identical with the apparently less general problem of finding all the groups of the same order n, which can be formed with the substitutions upon n letters; in fact, referring to the diagram, it appears that 1, α, β, γ, δ, ϵ may be regarded as substitutions performed upon the six letters 1, α, β, γ, δ, ϵ, viz. 1 is the substitution unity which leaves the order unaltered, α the substitution which changes $1\alpha\beta\gamma\delta\epsilon$ into $\alpha 1\gamma\beta\epsilon\delta$, and so for β, γ, δ, ϵ. This, however, does not in any wise show that the best or the easiest mode of treating the general problem is thus to regard it as a problem of substitutions: and it seems clear that the better course is to consider the general problem in itself, and to deduce from it the theory of groups of substitutions.

Cambridge, 26th November, 1877.

No. 2. The theory of groups; graphical representation.

[From the *American Journal of Mathematics*, t. I. (1878), pp. 174—176.]

In regard to a substitution-group of the order n upon the same number of letters, I omitted to mention the important theorem that every substitution is *regular* (that is, either cyclical or composed of a number of cycles each of them of the same order). Thus, in the group of 6 given in No. 1, writing a, b, c, d, e, f in place of 1, α, β, γ, δ, ϵ, the substitutions of the group are 1, $ace\,.\,bfd$, $aec\,.\,bdf$, $ab\,.\,cd\,.\,ef$, $ad\,.\,be\,.\,cf$, $af\,.\,bc\,.\,de$.

Let the letters be represented by points; a change a into b will be represented by a directed line (line with an arrow) joining the two points; and therefore a cycle abc, that is, a into b, b into c, c into a, by the three sides of the trilateral abc, with the three arrows pointing accordingly, and similarly for the cycles $abcd$, &c.: the cycle ab means a into b, b into a, and we have here the line ab with a two-headed arrow pointing both ways; such a line may be regarded as a bilateral. A substitution is thus represented by a multilateral or system of multilaterals, each side with its arrow; and in the case of a regular substitution the multilaterals (if more than one) have each of them the same number of sides. To represent two or more substitutions we require different colours, the multilaterals belonging to any one substitution being of the same colour.

In order to represent a group we need to represent only independent substitutions thereof; that is, substitutions such that no one of them can be obtained from the others by compounding them together in any manner. I take as an example a group of the order 12 upon 12 letters, where the number of independent substitutions is $=2$. See the diagram, wherein the continuous lines represent black lines, and the dotted lines, red lines.

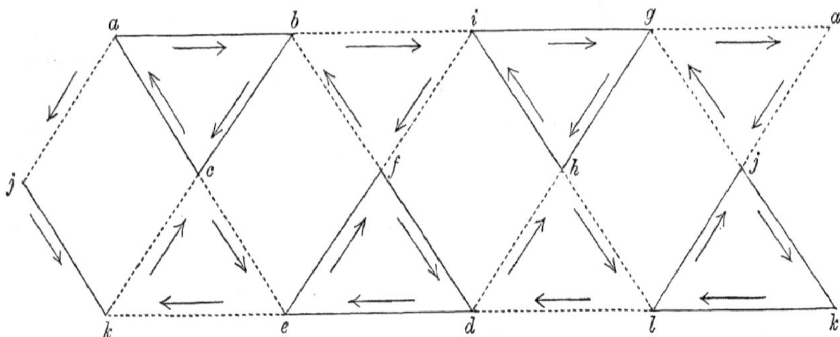

The diagram is drawn, in the first instance, with the arrows but without the letters, which are then affixed *at pleasure;* viz. the *form of the group* is quite independent of the way in which this is done, though the group itself is of course dependent upon it. The diagram shows two substitutions, each of them of the third order: one is represented by the black triangles, and the other by the dotted triangles. It will be observed that there is *from* each point of the diagram (that is, in the direction of the arrow) one and only one black line, and one and only one dotted line; hence a symbol B, "move along a black line," B^2, "move successively along two black lines," BR (read always from right to left), "move first along a dotted line and then along a black line," has in every case a perfectly definite meaning and determines the path when the initial point is given; any such symbol may be spoken of as a "route."

a	*b*	*c*	*d*	*e*	*f*	*g*	*h*	*i*	*j*	*k*	*l*	1
b	*c*	*a*	*e*	*f*	*d*	*h*	*i*	*g*	*k*	*l*	*j*	*abc* . *def* . *ghi* . *jkl* (= B)
c	*a*	*b*	*f*	*d*	*e*	*i*	*g*	*h*	*l*	*j*	*k*	*acb* . *dfe* . *gih* . *jlk*
d	*l*	*h*	*a*	*g*	*j*	*e*	*c*	*k*	*f*	*i*	*b*	*ad* . *bl* . *ch* . *eg* . *fj* . *ik*
e	*j*	*i*	*b*	*h*	*k*	*f*	*a*	*l*	*d*	*g*	*c*	*aeh* . *bjd* . *cil* . *fkg*
f	*k*	*g*	*c*	*i*	*l*	*d*	*b*	*j*	*e*	*h*	*a*	*afl* . *bkh* . *cgd* . *eij*
g	*f*	*k*	*l*	*c*	*i*	*j*	*d*	*b*	*a*	*e*	*h*	*agj* . *bfi* . *cke* . *dlh*
h	*d*	*l*	*j*	*a*	*g*	*k*	*e*	*c*	*b*	*f*	*i*	*ahe* . *bdj* . *cli* . *fgk*
i	*e*	*j*	*k*	*b*	*h*	*l*	*f*	*a*	*c*	*d*	*g*	*ai* . *be* . *cj* . *dk* . *fh* . *gl*
j	*i*	*e*	*h*	*k*	*b*	*a*	*l*	*f*	*g*	*c*	*d*	*ajg* . *bif* . *cek* . *dhl* (= R)
k	*g*	*f*	*i*	*l*	*c*	*b*	*j*	*d*	*h*	*a*	*e*	*ak* . *bg* . *cf* . *di* . *el* . *hj*
l	*h*	*d*	*g*	*j*	*a*	*c*	*k*	*e*	*i*	*b*	*f*	*alf* . *bhk* . *cdg* . *eji*

The diagram has a remarkable property, *in virtue whereof it in fact represents a group.* It may be seen that any route leading from some one point a to itself, leads also from every other point to itself, or say from b to b, from c to $c,\ldots$, and from l to l. We hence see that a route, applied in succession to the whole series of initial points or letters *abcdefghijkl*, gives a new arrangement of these letters, wherein no one of them occupies its original place; a route is thus, in effect, a substitution. Moreover, we may regard as distinct routes, those which lead from a to a, to b, to $c,\ldots$, to l, respectively. We have thus 12 substitutions (the first of them, which leaves the arrangement unaltered, being the substitution unity), and these 12 substitutions form a group. I omit the details of the proof; it will be sufficient to give the square obtained by means of the several routes, or substitutions, performed upon the primitive arrangement *abcdefghijkl*, and the cyclical expressions of the substitutions themselves: it will be observed that the substitutions are unity, 3 substitutions of the order (or index) 2, and 8 substitutions of the order (or index) 3.

It may be remarked that the group of 12 is really the group of the 12 positive substitutions upon 4 letters *abcd*, viz. these are 1, *abc*, *acb*, *abd*, *adb*, *acd*, *adc*, *bcd*, *bdc*, *ab.cd*, *ac.bd*, *ad.bc*.

Cambridge, 16*th May*, 1878.

No. 3. The Newton-Fourier imaginary problem.

[From the *American Journal of Mathematics*, t. II. (1879), p. 97.]

The Newtonian method as completed by Fourier, or say the Newton-Fourier method, for the solution of a numerical equation by successive approximations, relates to an equation $f(x)=0$, with real coefficients, and to the determination of a certain real root thereof a by means of an assumed approximate real value ξ satisfying prescribed conditions: we then, from ξ, derive a nearer approximate value ξ_1 by the formula $\xi_1=\xi-\dfrac{f(\xi)}{f'(\xi)}$; and thence, in like manner, ξ_1, ξ_2, $\xi_3,\ldots$ approximating more and more nearly to the required root a.

In connexion herewith, throwing aside the restrictions as to reality, we have what I call the Newton-Fourier Imaginary Problem, as follows.

Take $f(u)$, a given rational and integral function of u, with real or imaginary coefficients; ξ, a given real or imaginary value, and from this derive ξ_1 by the formula $\xi_1=\xi-\dfrac{f(\xi)}{f'(\xi)}$, and thence ξ_1, ξ_2, $\xi_3,\ldots$ each from the preceding one by the like formula.

A given imaginary quantity $x+iy$ may be represented by a point the coordinates of which are (x, y): the roots of the equation are thus represented by given points

A, B, C,..., and the values ξ, ξ_1, ξ_2,... by points P, P_1, P_2,... the first of which is assumed at pleasure, and the others each from the preceding one by the like given geometrical construction. The problem is to determine the regions of the plane such that, P being taken at pleasure anywhere within one region, we arrive ultimately at the point A; anywhere within another region at the point B; and so for the several points representing the roots of the equation.

The solution is easy and elegant in the case of a quadric equation: but the next succeeding case of the cubic equation appears to present considerable difficulty.

Cambridge, March 3rd, 1879.

No. 4. THE MECHANICAL CONSTRUCTION OF CONFORMABLE FIGURES.

[From the *American Journal of Mathematics*, t. II. (1879), p. 186.]

Is it possible to devise an apparatus for the mechanical construction of conformable figures; that is, figures which are similar as regards corresponding infinitesimal areas? The problem is to connect mechanically two points P_1 and P_2 in such wise that P_1 (1) shall have two degrees of freedom (or be capable of moving over a plane area) its position always determining that of P_2: (2) that if P_1, P_2 describe the infinitesimal lengths P_1Q_1, P_2Q_2, then the ratio of these lengths, and their mutual inclination, shall depend upon the position of P_1, but be independent of the direction of P_1Q_1: or what is the same thing, that if P_1 describe uniformly an indefinitely small circle, then P_2 shall also describe uniformly an indefinitely small circle, the ratio of the radii, and the relative position of the starting points in the two circles respectively, depending on the position of P_1.

Of course a pentagraph is a solution, but the two figures are in this case similar; and this is not what is wanted. Any unadjustable apparatus would give one solution only: the complete solution would be by an apparatus containing, suppose, a flexible lamina, so that P_1 moving in a given right line, the path of P_2 could be made to be any given curve whatever.

Cambridge, July 9th, 1879.

695.

A LINK-WORK FOR x^2: EXTRACT FROM A LETTER TO MR. SYLVESTER.

[From the *American Journal of Mathematics*, t. I. (1878), p. 386.]

I SUPPOSE the following is substantially your link-work for x^2. I use a slot to make D move in the line OA; but this could be replaced by proper link-work. Supposing O and A fixed; the line OB is movable, and I wanted to have the

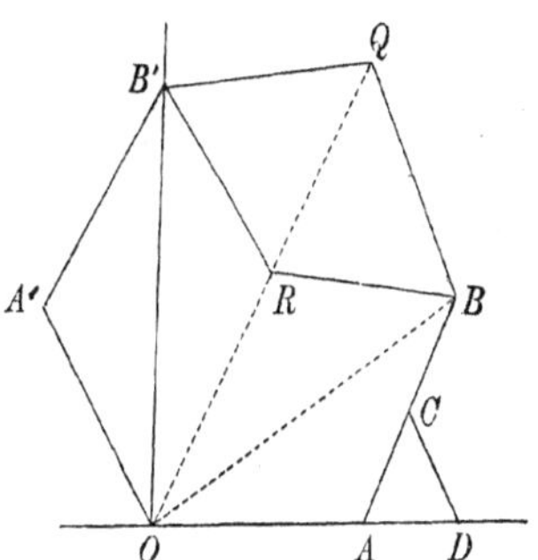

distance OB measured in a fixed direction. This can be done by a hexagon $OABQB'A'$ with equal sides, and two other equal links $B'R$, BR: then of course, if O, R, Q are in lineâ, the hexagon will be symmetrical as to OQ, and OB' will be equal to OB, and B' may be made to move in the fixed line OB'. If

$$BOA = \tfrac{1}{2}\theta, \quad OA = AB = a, \quad AC = CD = \tfrac{1}{2}a,$$

then

$$OB = 2a \cos \tfrac{1}{2}\theta, \quad OD = a(1 + \cos\theta) = 2a\cos^2 \tfrac{1}{2}\theta,$$

or

$$2a \,.\, OD = (OB)^2.$$

November 30, 1877.

696.

CALCULATION OF THE MINIMUM N.G.F. OF THE BINARY SEVENTHIC.

[From the *American Journal of Mathematics*, t. II. (1879), pp. 71—84.]

FOR the binary seventhic $(a, \ldots \between x, y)^7$ the number of the asyzygetic covariants $(a, \ldots)^\theta (x, y)^\mu$, or say of the deg-order $(\theta . \mu)$, is given as the coefficient of $a^\theta x^\mu$ in the function

$$\frac{1-x^{-2}}{1-ax^7 . 1-ax^5 . 1-ax^3 . 1-ax . 1-ax^{-1} . 1-ax^{-3} . 1-ax^{-5} . 1-ax^{-7}}$$

developed in ascending powers of a. See my "Ninth Memoir on Quantics," *Phil. Trans.*, t. CLXI. (1871), pp. 17—50, [462].

This function is in fact

$$= A(x) - \frac{1}{x^2} A\left(\frac{1}{x}\right),$$

where, developing in ascending powers of a, the second term $-\frac{1}{x^2} A\left(\frac{1}{x}\right)$ contains only negative powers of x, and it may consequently be disregarded: the number of asyzygetic covariants of the deg-order $(\theta . \mu)$ is thus equal to the coefficient of $a^\theta x^\mu$ in the function $A(x)$, which function is for this reason called the Numerical Generating Function (N.G.F.) of the binary seventhic; and the function $A(x)$ expressed as a fraction in its least terms is said to be the minimum N.G.F.

According to a theorem of Professor Sylvester's (*Proc. Royal Soc.*, t. XXVIII. (1878), pp. 11—13), this minimum N.G.F. is of the form

$$\frac{Z_0 + aZ_1 + a^2 Z_2 + \ldots + a^{36} Z_{36}}{1-ax . 1-ax^3 . 1-ax^5 . 1-ax^7 . 1-a^4 . 1-a^6 . 1-a^8 . 1-a^{10} . 1-a^{12}}$$

where $Z_0, Z_1, \ldots, Z_{36}$ are rational and integral functions of x of degrees not exceeding 14; and where, as will presently be seen, there is a symmetry in regard to the terms Z_0, Z_{36}; Z_1, Z_{35}; &c., equidistant from the middle term Z_{18}, such that the terms $Z_0, \ldots, Z_{18}$ being known, the remaining terms $Z_{19}, \ldots, Z_{36}$ can be at once written down.

Using only the foregoing properties, I obtained for the N.G.F. an expression which I communicated to Professor Sylvester, and which is published, *Comptes Rendus*, t. LXXXVII. (1878), p. 505, but with an erroneous value for the coefficient of a^7 and for that of the corresponding term a^{29}.* The correct value is

Numerator of Minimum N.G.F. is =

$$\begin{aligned}
&1\\
&+a\ (-x-x^3-x^5)\\
&+a^2\ (x^2+x^4+2x^6+x^8+x^{10})\\
&+a^3\ (-x^7-x^9-x^{11}-x^{13})\\
&+a^4\ (2x^4+x^8+x^{14})\\
&+a^5\ (x+2x^3-x^9-x^{11})\\
&+a^6\ (-1+2x^2-x^4-x^8-x^{10}+x^{12})\\
&+a^7\ (4x+x^3+3x^5-x^9+x^{11})\\
&+a^8\ (2-x^2-3x^6-3x^8-x^{10}-x^{12})\\
&+a^9\ (x+3x^3+x^5-x^7+2x^9+2x^{13})\\
&+a^{10}(-1+4x^2-x^6-2x^8-2x^{10}-x^{14})\\
&+a^{11}(5x+3x^3+2x^5-x^7-2x^9-x^{11}+x^{13})\\
&+a^{12}(5+x^2-4x^6-6x^8-4x^{10}-x^{12}+2x^{14})\\
&+a^{13}(x-4x^5-4x^7-x^9+x^{11}+4x^{13})\\
&+a^{14}(2+5x^2+x^4+x^6-2x^8+3x^{12}-x^{14})\\
&+a^{15}(3x-x^3-x^5-7x^7-5x^9-x^{11}-x^{13})\\
&+a^{16}(6+3x^2+3x^4-4x^6-3x^8-x^{12}+5x^{14})\\
&+a^{17}(-x-2x^3-9x^5-8x^7-4x^9-3x^{11}+4x^{13})\\
&+a^{18}(2+6x^2+x^4+2x^6+2x^8+x^{10}+6x^{12}+2x^{14})\\
&+a^{19}(4x-3x^3-4x^5-8x^7-9x^9-2x^{11}-x^{13})\\
&+a^{20}(5-x^2-3x^6-4x^8+3x^{10}+3x^{12}+6x^{14})\\
&+a^{21}(-x-x^3-5x^5-7x^7-x^9-x^{11}+3x^{13})\\
&+a^{22}(-1+3x^2-2x^4+x^8+x^{10}+5x^{12}+2x^{14})\\
&+a^{23}(4x+x^3-x^5-4x^7-4x^9+x^{13})
\end{aligned}$$

* The existence of these errors was pointed out to me by Professor Sylvester in a letter dated 13th November, 1878.

$$+a^{24}(2-x^2-4x^4-6x^6-4x^8+x^{10}+5x^{14})$$
$$+a^{25}(x-x^3-2x^5-x^7+2x^9+3x^{11}+5x^{13})$$
$$+a^{26}(-1-2x^4-2x^6-x^8+4x^{10}-x^{14})$$
$$+a^{27}(2x+2x^5-x^7+x^9+3x^{11}+x^{13})$$
$$+a^{28}(-x^2-x^4-3x^6-3x^8-x^{12}+2x^{14})$$
$$+a^{29}(x^3-x^5+3x^9+x^{11}+4x^{13})$$
$$+a^{30}(x^2-x^4-x^6-x^{10}+2x^{12}-x^{14})$$
$$+a^{31}(-x^3-x^5+2x^{11}+x^{13})$$
$$+a^{32}(1+x^6+2x^{10})$$
$$+a^{33}(-x-x^3-x^5-x^7)$$
$$+a^{34}(x^4+x^6+2x^8+x^{10}+x^{12})$$
$$+a^{35}(-x^9-x^{11}-x^{13})$$
$$+a^{36}\,.\,x^{14}.$$

Denominator (as mentioned before) is

$$=1-ax\,.\,1-ax^3\,.\,1-ax^5\,.\,1-ax^7\,.\,1-a^4\,.\,1-a^6\,.\,1-a^8\,.\,1-a^{10}\,.\,1-a^{12}.$$

The method of calculation is as follows: write for a moment

$$\phi(a,\,x)=\frac{1-x^{-2}}{1-ax^7\,.\,1-ax^5\,.\,1-ax^3\,.\,1-ax\,.\,1-ax^{-1}\,.\,1-ax^{-3}\,.\,1-ax^{-5}\,.\,1-ax^{-7}};$$

then $\phi(a,\,x)$, developed in ascending powers of a, and rejecting from the result all negative powers of x, is

$$=\frac{Z_0+aZ_1+\ldots+a^{36}Z_{36}}{1-ax\,.\,1-ax^3\,.\,1-ax^5\,.\,1-ax^7\,.\,1-a^4\,.\,1-a^6\,.\,1-a^8\,.\,1-a^{10}\,.\,1-a^{12}},$$

developed in like manner in ascending powers of a; for the determination of the Z's up to Z_{18} we require only the development of $\phi(a,\,x)$ up to a^{18}; and, assuming that each Z is at most of the degree 14 in x, we require the coefficients of the different powers of a in $\phi(a,\,x)$ only up to x^{14}. Assuming then that $\phi(a,\,x)$ developed in ascending powers of a, up to a^{18}, rejecting all negative powers of x, and all positive powers greater than x^{14}, is

$$=X_0+aX_1+\ldots+a^{18}X_{18},$$

we have

$$X_0+aX_1+\ldots+a^{18}X_{18}=\frac{Z_0+aZ_1+\ldots+a^{18}Z_{18}}{1-ax\,.\,1-ax^3\,.\,1-ax^5\,.\,1-ax^7\,.\,1-a^4\,.\,1-a^6\,.\,1-a^8\,.\,1-a^{10}\,.\,1-a^{12}},$$

or say

$$Z_0+aZ_1+\ldots+a^{18}Z_{18}=1-a^4\,.\,1-a^6\,.\,1-a^8\,.\,1-a^{10}\,.\,1-a^{12}\,.$$
$$1-ax\,.\,1-ax^3\,.\,1-ax^5\,.\,1-ax^7\,.\,(X_0+aX_1+\ldots+a^{18}X_{18});$$

viz. developing here the right-hand side as far as a^{18}, but in each term rejecting the powers of x above x^{14}, the coefficients of the several powers a^0, $a^1,\ldots,$ a^{18} give the required values $Z_0, Z_1, \ldots, Z_{18}$. We require, therefore, only to know the values of these functions $X_0, X_1, \ldots, X_{18}$.

To make a break in the calculation, it is convenient to write

$$1-ax.1-ax^3.1-ax^5.1-ax^7(X_0+aX_1+\ldots+a^{18}X_{18})=Y_0+aY_1+\ldots+a^{18}Y_{18};$$

putting then

$$1-ax.1-ax^3.1-ax^5.1-ax^7=1-ap+a^2q-a^3r,$$

where (up to x^{14})

$$p=x+x^3+x^5+x^7,$$
$$q=x^4+x^6+2x^8+x^{10}+x^{12},$$
$$r=x^9+x^{11}+x^{13},$$

we have

$$Y_0+aY_1+a^2Y_2+\ldots+a^{18}Y_{18}=(1-ap+a^2q-a^3r)(X_0+aX_1+a^2X_2+\ldots+a^{18}X_{18}).$$

The values of $Y_0, Y_1, \ldots, Y_{18}$ then are

	Y_0	Y_1	Y_2	$Y_3 \ldots\ldots$	Y_{18}
$=$	X_0	X_1	X_2	X_3	X_{18}
		$-pX_0$	$-pX_1$	$-pX_2$	$-pX_{17}$
			$+qX_0$	$+qX_1$	$+qX_{16}$
				$-rX_0$	$-rX_{15}$

the values being taken to x^{14} only; and we then have

$$Z_0+aZ_1+a^2Z_2+\ldots+a^{18}Z_{18}=1-a^4.1-a^6.1-a^8.1-a^{10}.1-a^{12}(Y_0+aY_1+\ldots+a^{18}Y_{18});$$

viz. the values of $Z_0, Z_1, \ldots, Z_{18}$ are

	Z_0	Z_1	Z_2	Z_3	Z_4	Z_5	Z_6	Z_7	Z_8	Z_9
$=$	Y_0	Y_1	Y_2	Y_3	Y_4	Y_5	Y_6	Y_7	Y_8	Y_9
					$-Y_0$	$-Y_1$	$-Y_2$	$-Y_3$	$-Y_4$	$-Y_5$
							$-Y_0$	$-Y_1$	$-Y_2$	$-Y_3$
									$-Y_0$	$-Y_1$

	Z_{10}	Z_{11}	Z_{12}	Z_{13}	Z_{14}	Z_{15}	Z_{16}	Z_{17}	Z_{18}
$=$	Y_{10}	Y_{11}	Y_{12}	Y_{13}	Y_{14}	Y_{15}	Y_{16}	Y_{17}	Y_{18}
	$-Y_6$	$-Y_7$	$-Y_8$	$-Y_9$	$-Y_{10}$	$-Y_{11}$	$-Y_{12}$	$-Y_{13}$	$-Y_{14}$
	$-Y_4$	$-Y_5$	$-Y_6$	$-Y_7$	$-Y_8$	$-Y_9$	$-Y_{10}$	$-Y_{11}$	$-Y_{12}$
	$-Y_2$	$-Y_3$	$-Y_4$	$-Y_5$	$-Y_6$	$-Y_7$	$-Y_8$	$-Y_9$	$-Y_{10}$
					$+2Y_0$	$+2Y_1$	$+2Y_2$	$+2Y_3$	$+2Y_4$
							$+2Y_0$	$+2Y_1$	$+2Y_2$
									$+\ Y_0.$

The rule of symmetry, before referred to, is that the coefficient Z_{36-p} of a^{36-p} is obtained from the coefficient Z_p of a^p by changing each power x^q into x^{14-q}, the coefficients being unaltered; in particular Z_{18}, the coefficient of a^{18}, must remain unaltered when each power x^q is changed into x^{14-q}; and the verification thus obtained of the value

$$Z_{18}=2+6x^2+x^4+2x^6+2x^8+x^{10}+6x^{12}+2x^{14}$$

is in fact almost a complete verification of the whole work. Some other verifications, which present themselves in the course of the work, will be referred to further on.

We have, therefore, to calculate the coefficients $X_0, X_1, \ldots, X_{18}$; the function $\phi(a, x)$ regarded as a function of a is at once decomposed into simple fractions; viz. we have

$$\phi(a, x)=\frac{1-x^{-2}}{1-ax^7\,.\,1-ax^5\,.\,1-ax^3\,.\,1-ax\,.\,1-ax^{-1}\,.\,1-ax^{-3}\,.\,1-ax^{-5}\,.\,1-ax^{-7}}$$

$$=\frac{x^{54}}{1-x^4\,.\,1-x^6\,.\,1-x^8\,.\,1-x^{10}\,.\,1-x^{12}\,.\,1-x^{14}}\,\frac{1}{1-ax^7}$$

$$-\frac{x^{40}}{1-x^2\,.\,1-x^4\,.\,1-x^6\,.\,1-x^8\,.\,1-x^{10}\,.\,1-x^{12}}\,\frac{1}{1-ax^5}$$

$$+\frac{x^{28}}{1-x^2\,.\,(1-x^4)^2\,.\,1-x^6\,.\,1-x^8\,.\,1-x^{10}}\,\frac{1}{1-ax^3}$$

$$-\frac{x^{18}}{1-x^2\,.\,(1-x^4)^2\,.\,(1-x^6)^2\,.\,1-x^8}\,\frac{1}{1-ax}$$

$$+\frac{x^{10}}{1-x^2\,.\,(1-x^4)^2\,.\,(1-x^6)^2\,.\,1-x^8}\,\frac{1}{1-ax^{-1}}$$

$$-\frac{x^4}{1-x^2\,.\,(1-x^4)^2\,.\,1-x^6\,.\,1-x^8\,.\,1-x^{10}}\,\frac{1}{1-ax^{-3}}$$

$$+\frac{1}{1-x^2\,.\,1-x^4\,.\,1-x^6\,.\,1-x^8\,.\,1-x^{10}\,.\,1-x^{12}}\,\frac{1}{1-ax^{-5}}$$

$$-\frac{x^{-2}}{1-x^4\,.\,1-x^6\,.\,1-x^8\,.\,1-x^{10}\,.\,1-x^{12}\,.\,1-x^{14}}\,\frac{1}{1-ax^{-7}}.$$

In order to obtain the expansion of $\phi(a, x)$ in the assumed form of an expansion in ascending powers of a, we must of course expand the simple fractions $\frac{1}{1-ax^7}$, &c., in ascending powers of a, but it requires a little consideration to see that we must also expand the x-coefficients of these simple fractions *in ascending powers of x.* For instance, as regards the term independent of a, here developing the several coefficients as far as x^{18}, the last five terms give (see *post*)

$$\begin{array}{rcccccccccc}
 & & & & & & & & & & -\ x^{18}\\
 & & & & & & +\ x^{10}+ & x^{12}+ & 3x^{14}+ & 5x^{16}+ & 9x^{18}\\
 & & & -\ x^4- & x^6- & 3x^8- & 4x^{10}- & 8x^{12}- & 11x^{14}- & 18x^{16}- & 24x^{18}\\
 & 1+ & x^2+ & 2x^4+ & 3x^6+ & 5x^8+ & 7x^{10}+ & 11x^{12}+ & 14x^{14}+ & 20x^{16}+ & 26x^{18}\\
-x^{-2} & . & -x^2- & x^4- & 2x^6- & 2x^8- & 4x^{10}- & 4x^{12}- & 6x^{14}- & 7x^{16}- & 10x^{18}\\
\hline
=-x^{-2} & +1 & 0 & 0 & 0 & 0 & 0 & 0 & 0 & 0 & 0
\end{array}$$

viz. the sum is $=1-x^{-2}$ as it should be*.

* To give the last degree of perfection to the beautiful method of Professor Cayley it would seem desirable that a proof should be given of the principle illustrated by the example in the text, and the nature of the mischief resulting from its neglect clearly pointed out.—EDS. of the A. J. M.

The expansion is required only as far as x^{14}: the first four terms are therefore to be disregarded, and, writing for shortness

$$E = \frac{1}{1-x^2.(1-x^4)^2(1-x^6)^2.1-x^8},$$

$$F = \frac{1}{1-x^2.(1-x^4)^2.1-x^6.1-x^8.1-x^{10}},$$

$$G = \frac{1}{1-x^2.1-x^4.1-x^6.1-x^8.1-x^{10}.1-x^{12}},$$

$$H = \frac{1}{1-x^4.1-x^6.1-x^8.1-x^{10}.1-x^{12}.1-x^{14}},$$

we have

$$\phi(a, x) = \frac{x^{10}E}{1-ax^{-1}} - \frac{x^4F}{1-ax^{-3}} + \frac{G}{1-ax^{-5}} - \frac{x^{-2}H}{1-ax^{-7}},$$

which is

$$\begin{aligned} = \quad & x^{10}E\,(1+ax^{-1}+a^2x^{-2}+\dots) \\ - \; & x^4\,F\,(1+ax^{-3}+a^2x^{-6}+\dots) \\ + \; & G\,(1+ax^{-5}+a^2x^{-10}+\dots) \\ - \; & x^{-2}H\,(1+ax^{-7}+a^2x^{-14}+\dots), \end{aligned}$$

where the several series are to be continued up to a^{18}, and, after substituting for E, F, G, H their expansions in ascending powers of x, we are to reject negative powers of x, and also powers higher than x^{14}. The functions E, F, G, H contain each of them only even powers of x, and it is easy to see that we require the expansions up to x^{22}, x^{64}, x^{104} and x^{142} respectively. For the sake of a verification, I in fact calculated E, F up to x^{64} and G, H up to x^{142}: viz. we have

$$(1-x^6)\,E = (1-x^{10})\,F,$$

from the coefficients of E we have those of $(1-x^6)\,E$, and in the process of calculating F we have at the last step but one the coefficients of $(1-x^{10})\,F$, the agreement of the two sets being the verification; similarly,

$$(1-x^2)\,G = (1-x^{14})\,H$$

gives a verification. The process for the calculation of E,

$$= \frac{1}{1-x^2.(1-x^4)^2(1-x^6)^2.1-x^8},$$

is as follows:

	Ind. x											
	0	2	4	6	8	10	12	14	16	18	20	22
$(1-x^2)^{-1}$	1	1	1	1	1	1	1	1	1	1	1	1
			1	1	2	2	3	3	4	4	5	5
$(1-x^4)^{-1}$	1	1	2	2	3	3	4	4	5	5	6	6
			1	1	3	3	6	6	10	10	15	15
$(1-x^4)^{-1}$	1	1	3	3	6	6	10	10	15	15	21	21
				1	1	3	4	7	9	14	17	24
$(1-x^6)^{-1}$	1	1	3	4	7	9	14	17	24	29	38	45
				1	1	3	5	8	12	19	25	36
$(1-x^6)^{-1}$	1	1	3	5	8	12	19	25	36	48	63	81
					1	1	3	5	9	13	22	30
$E=(1-x^8)^{-1}$	1	1	3	5	9	13	22	30	45	61	85	111

the alternate lines giving the developments of the functions

$$(1-x^2)^{-1},\ (1-x^2)^{-1}(1-x^4)^{-1},\ (1-x^2)^{-1}(1-x^4)^{-2},\ \ldots,$$

which are the products of the x-functions down to any particular line. And in like manner we have the expansions of the other functions F, G, H respectively. I give first the expansions of E, F, G, H; next the calculation of the X's; then the calculation of the Y's: and from these the Z's up to Z_{18}, or coefficients of the powers $a^0, a^1, \ldots, a^{18}$ in the numerator of the N.G.F. are at once found; and the coefficients of the remaining powers $a^{19}, \ldots, a^{36}$ are then deduced, as already mentioned.

Writing in the formula $x=0$, we have, for the numerator of the N.G.F. of the *invariants*, the expression

$$1-a^6+2a^8-a^{10}+5a^{12}+2a^{14}+6a^{16}+2a^{18}+5a^{20}-a^{22}+2a^{24}-a^{26}+a^{32},$$

agreeing with a result in my "Second Memoir on Quantics," *Phil. Trans.*, t. CXLVI. (1856), [Number 141, vol. II. in this Collection, p. 266]; this, then, was a known result, and it affords a verification, not only of the terms in x^0, but also of those in x^{14}. Thus, in calculating the foregoing expression of the numerator, we obtain $Z_4=(2x^4+x^8+x^{14})$, viz. the term is

$$a^4(2x^4+x^8+x^{14}),$$

and we thence have the corresponding term $a^{32}(1+x^6+2x^{10})$, which, when $x=0$, becomes $=a^{32}$, a term of the numerator for the invariants: and the term $1x^{14}$ of Z_4

is thus verified, viz. so soon as, in the calculation, we arrive at this term, we know that it is right, and the calculation up to this point is, to a considerable extent, verified. And similarly, in continuing the calculation, we arrive at other terms which are verified in the like manner.

Expansions of the Functions E, F, G, H.

Ind. x	E	F	G	H	Ind. x	E	F	G	H
0	1	1	1	1	16	45	36	20	6
2	1	1	1	0	18	61	47	26	7
4	3	3	2	1	20	85	66	35	10
6	5	4	3	1	22	111	84	44	11
8	9	8	5	2	24		113	58	16
10	13	11	7	2	26		141	71	17
12	22	18	11	4	28		183	90	23
14	30	24	14	4	30		225	110	26

Ind. x	F	G	H	Ind. x	G	H	Ind. x	H
32	284	136	33	70	2172	419	108	2265
34	344	163	37	72	2432	472	110	2426
36	425	199	47	74	2702	515	112	2623
38	508	235	52	76	3009	576	114	2807
40	617	282	64	78	3331	629	116	3026
42	729	331	72	80	3692	699	118	3232
44	872	391	86	82	4070	760	120	3479
46	1020	454	96	84	4494	843	122	3708
48	1205	532	115	86	4935	913	124	3981
50	1397	612	127	88	5427	1007	126	4240
52	1632	709	149	90	5942	1091	128	4541
54	1877	811	166	92	6510	1197	130	4828
56	2172	931	192	94	7104	1293	132	5164
58	2480	1057	212	96	7760	1416	134	5481
60	2846	1206	245	98	8442	1525	136	5850
62	3228	1360	269	100	9192	1663	138	6204
64	3677	1540	307	102	9975	1790	140	6609
66		1729	338	104	10829	1945	142	6998
68		1945	382	106		2088		

Calculation of the X's.

Ind. x even or odd according as suffix X is even or odd.

	0_1	2_3	4_5	6_7	8_9	10_{11}	12_{13}	14
						1	1	3
			−1	−1	−3	−4	−8	−11
	1	1	2	3	5	7	11	14
		−1	−1	−2	−2	−4	−4	−6
$X_0=$	1	0	0	0	0	0	0	0
					1	1	3	
	−1	−1	−3	−4	−8	−11	−18	
	3	5	7	11	14	20	26	
	−2	−4	−4	−6	−7	−10	−11	
$X_1=$	0	0	0	+1	0	0	0	
					1	1	3	5
	−1	−3	−4	−8	−11	−18	−24	−36
	7	11	14	20	26	35	44	58
	−6	−7	−10	−11	−16	−17	−23	−26
$X_2=$	0	+1	0	+1	0	+1	0	+1
				1	1	3	5	
	−4	−8	−11	−18	−24	−36	−47	
	20	26	35	44	58	71	90	
	−16	−17	−23	−26	−33	−37	−47	
$X_3=$	0	+1	+1	+1	+2	+1	+1	
				1	1	3	5	9
	−8	−11	−18	−24	−36	−47	−66	−84
	35	44	58	71	90	110	136	163
	−26	−33	−37	−47	−52	−64	−72	−86
$X_4=$	1	0	+3	+1	+3	+2	+3	+2
			1	1	3	5	9	
	−18	−24	−36	−47	−66	−84	−113	
	71	90	110	136	163	199	235	
	−52	−64	−72	−86	−96	−115	−127	
$X_5=$	1	+2	+3	+4	+4	+5	+4	

	0_1	2_3	4_5	6_7	8_9	10_{11}	12_{13}	14
			1	1	3	5	9	13
	− 24	− 36	− 47	− 66	− 84	− 113	− 141	− 183
	110	136	163	199	235	282	331	391
	− 86	− 96	− 115	− 127	− 149	− 166	− 191	− 212
$X_6 =$	0	+ 4	+ 2	+ 7	+ 5	+ 8	+ 8	+ 9
		1	1	3	5	9	13	
	− 47	− 66	− 84	− 113	− 141	− 183	− 225	
	199	235	282	331	391	454	532	
	− 149	− 166	− 192	− 212	− 245	− 269	− 307	
$X_7 =$	3	+ 4	+ 7	+ 9	+ 10	+ 11	+ 13	
		1	1	3	5	9	13	22
	− 66	− 84	− 113	− 141	− 183	− 225	− 284	− 344
	282	331	391	454	532	612	709	811
	− 212	− 245	− 269	− 307	− 338	− 382	− 419	− 472
$X_8 =$	4	+ 3	+ 10	+ 9	+ 16	+ 14	+ 19	+ 17
	1	1	3	5	9	13	22	
	− 113	− 141	− 183	− 225	− 284	− 344	− 425	
	454	532	612	709	811	931	1057	
	− 338	− 382	− 419	− 472	− 515	− 576	− 629	
$X_9 =$	4	+ 10	+ 13	+ 17	+ 21	+ 24	+ 25	
	1	1	3	5	9	13	22	30
	− 141	− 183	− 225	− 284	− 344	− 425	− 508	− 617
	612	709	811	931	1057	1206	1360	1540
	− 472	− 515	− 576	− 629	− 699	− 760	− 843	− 913
$X_{10} =$	0	+ 12	+ 13	+ 23	+ 23	+ 34	+ 31	+ 40
	1	3	5	9	13	22	30	
	− 225	− 284	− 344	− 425	− 508	− 617	− 729	
	931	1057	1206	1360	1540	1729	1945	
	− 699	− 760	− 843	− 913	− 1007	− 1091	− 1197	
$X_{11} =$	8	+ 16	+ 24	+ 31	+ 38	+ 43	+ 49	
	1	3	5	9	13	22	30	45
	− 284	− 344	− 425	− 508	− 617	− 729	− 872	− 1020
	1206	1360	1540	1729	1945	2172	2432	2702
	− 913	− 1007	− 1091	− 1197	− 1293	− 1416	− 1525	− 1663
$X_{12} =$	10	+ 12	+ 29	+ 33	+ 48	+ 49	+ 65	+ 64

	0_1	2_3	4_5	6_7	8_9	10_{11}	12_{13}	14
	3	5	9	13	22	30	45	
	− 425	− 508	− 617	− 729	− 872	− 1020	− 1205	
	1729	1945	2172	2432	2702	3009	3331	
	− 1293	− 1416	− 1525	− 1663	− 1790	− 1945	− 2088	
$X_{13}=$	14	+ 26	+ 39	+ 53	+ 62	+ 74	+ 83	
	3	5	9	13	22	30	45	61
	− 508	− 617	− 729	− 872	− 1020	− 1205	− 1397	− 1632
	2172	2432	2702	3009	3331	3692	4070	4494
	− 1663	− 1790	− 1945	− 2088	− 2265	− 2426	− 2623	− 2807
$X_{14}=$	4	+ 30	+ 37	+ 62	+ 68	+ 91	+ 95	+ 116
	5	9	13	22	30	45	61	
	− 729	− 872	− 1020	− 1205	− 1397	− 1632	− 1877	
	3009	3331	3692	4070	4494	4935	5427	
	− 2265	− 2426	− 2623	− 2807	− 3026	− 3232	− 3479	
$X_{15}=$	20	+ 42	+ 62	+ 80	+ 101	+ 116	+ 132	
	5	9	13	22	30	45	61	85
	− 872	− 1020	− 1205	− 1397	− 1632	− 1877	− 2172	− 2480
	3692	4070	4494	4935	5427	5942	6510	7104
	− 2807	− 3026	− 3232	− 3479	− 3708	− 3981	− 4240	− 4541
$X_{16}=$	18	+ 33	+ 70	+ 81	+ 117	+ 129	+ 159	+ 168
	9	13	22	30	45	61	85	
	− 1205	− 1397	− 1632	− 1877	− 2172	− 2480	− 2846	
	4935	5427	5942	6510	7104	7760	8442	
	− 3708	− 3981	− 4240	− 4541	− 4828	− 5164	− 5481	
$X_{17}=$	31	+ 62	+ 92	+ 122	+ 149	+ 177	+ 200	
	9	13	22	30	45	61	85	111
	− 1397	− 1632	− 1877	− 2172	− 2480	− 2846	− 3228	− 3677
	5942	6510	7104	7760	8442	9192	9975	10829
	− 4541	− 4828	− 5164	− 5481	− 5850	− 6204	− 6609	− 6998
$X_{18}=$	13	+ 63	+ 85	+ 137	+ 157	+ 203	+ 223	+ 265

Calculation of the Y's.

Ind. x even or odd as suffix X is even or odd.

	0_1	2_3	4_5	6_7	8_9	10_{11}	12_{13}	14
	1							
$Y_0 =$	1							
				1				
	−1	−1	−1	−1				
$Y_1 =$	−1	−1	−1	0	0	0	0	
	0	1	0	1	0	1	0	1
					−1	−1	−1	−1
			1	1	2	1	1	
$Y_2 =$	0	1	1	2	1	1	0	0
		1	1	1	2	1	1	
		−1	−1	−2	−2	−2	−2	
						1	1	
					−1	−1	−1	
$Y_3 =$	0	0	0	−1	−1	−1	−1	
	1	0	3	1	3	2	3	2
			−1	−2	−3	−5	−5	−5
				1	1	3	2	4
$Y_4 =$	1	0	+2	0	+1	0	0	+1
	1	2	3	4	4	5	4	
	−1	−1	−4	−5	−7	−9	−9	
				1	2	4	6	
						−1	−1	
$Y_5 =$	0	+1	−1	0	−1	−1	0	
		4	2	7	5	8	7	9
		−1	−3	−6	−10	−13	−16	−17
			1	1	5	5	11	10
							−1	−2
$Y_6 =$	0	+3	0	+2	0	0	+1	0

	0_1	2_3	4_5	6_7	8_9	10_{11}	12_{13}	14
	3	4	7	9	10	11	13	
		− 4	− 6	− 13	− 18	− 22	− 27	
			1	3	7	12	17	
					− 1	− 1	− 4	
$Y_7 =$	3	0	+ 2	− 1	− 2	0	− 1	
	4	3	10	9	16	14	19	17
		− 3	− 7	− 14	− 23	− 30	− 37	− 43
				4	6	17	20	33
						− 1	− 3	− 6
$Y_8 =$	4	0	+ 3	− 1	− 1	0	− 1	+ 1
	4	10	13	17	21	24	25	
	− 4	− 7	− 17	− 26	− 38	− 49	− 58	
			3	7	17	27	40	
						− 4	− 6	
$Y_9 =$	0	+ 3	− 1	− 2	0	− 2	+ 1	
		12	13	23	23	34	31	40
		− 4	− 14	− 27	− 44	− 61	− 75	− 87
			4	7	21	29	52	61
						− 3	− 7	− 14
$Y_{10} =$	0	+ 8	+ 3	+ 3	0	− 1	+ 1	0
	8	16	24	31	38	43	49	
		− 12	− 25	− 48	− 71	− 93	− 111	
			4	14	31	54	78	
					− 4	− 7	− 17	
$Y_{11} =$	8	+ 4	+ 3	− 3	− 6	− 3	− 1	
	10	12	29	33	48	49	65	64
		− 8	− 24	− 48	− 79	− 109	− 136	− 161
				12	25	60	84	128
						− 4	− 14	− 27
$Y_{12} =$	10	+ 4	+ 5	− 3	− 6	− 4	− 1	+ 4

	0_1	2_3	4_5	6_7	8_9	10_{11}	12_{13}	14
	14	26	39	53	62	74	83	
	- 10	- 22	- 51	- 84	- 122	- 159	- 195	
			8	24	56	95	141	
						- 12	- 25	
$Y_{13}=$	4	+ 4	- 4	- 7	- 4	- 2	+ 4	
	4	30	37	62	68	91	95	116
		- 14	- 40	- 79	- 132	- 180	- 228	- 272
			10	22	61	96	161	204
						- 8	- 24	- 48
$Y_{14}=$	4	+ 16	+ 7	+ 5	- 3	- 1	+ 4	0
	20	42	62	80	101	116	132	
	- 4	- 34	- 71	- 133	- 197	- 258	- 316	
			14	40	93	158	233	
					- 10	- 22	- 51	
$Y_{15}=$	16	+ 8	+ 5	- 13	- 13	- 6	- 2	
	18	33	70	81	117	129	159	168
		- 20	- 62	- 124	- 204	- 285	- 359	- 429
			4	34	75	163	238	350
						- 14	- 40	- 79
$Y_{16}=$	18	+ 13	+ 12	- 9	- 12	- 7	- 2	+ 10
	31	62	92	122	149	177	200	
	- 18	- 51	- 121	- 202	- 301	- 397	- 486	
			20	62	144	246	367	
					- 4	- 34	- 71	
$Y_{17}=$	13	+ 11	- 9	- 18	- 12	- 8	+ 10	
	13	63	85	137	157	203	223	265
		- 31	- 93	- 185	- 307	- 425	- 540	- 648
			18	51	139	235	389	511
						- 20	- 62	- 124
$Y_{18}=$	13	+ 32	+ 10	+ 3	- 11	- 7	+ 10	+ 4

Cambridge, December 7th, 1878.

697.

ON THE DOUBLE ϑ-FUNCTIONS.

[From the *Journal für die reine und angewandte Mathematik* (Crelle), t. LXXXVII. (1878), pp. 74—81.]

I HAVE sought to obtain, in forms which may be useful in regard to the theory of the double ϑ-functions, the integral of the elliptic differential equation

$$\frac{dx}{\sqrt{a-x\,.\,b-x\,.\,c-x\,.\,d-x}}+\frac{dy}{\sqrt{a-y\,.\,b-y\,.\,c-y\,.\,d-y}}=0:$$

the present paper has immediate reference only to this differential equation; but, on account of the design of the investigation, I have entitled it as above.

We may for the general integral of the above equation take a particular integral of the equation

$$\frac{dx}{\sqrt{a-x\,.\,b-x\,.\,c-x\,.\,d-x}}+\frac{dy}{\sqrt{a-y\,.\,b-y\,.\,c-y\,.\,d-y}}\pm\frac{dz}{\sqrt{a-z\,.\,b-z\,.\,c-z\,.\,d-z}}=0;$$

viz. this particular integral, regarding therein z as an arbitrary constant, will be the general integral of the first mentioned equation. And we may further assume that z is the value of y corresponding to the value a of x.

I write for shortness

$$a-x,\ b-x,\ c-x,\ d-x=\mathrm{a}\,,\ \mathrm{b}\,,\ \mathrm{c}\,,\ \mathrm{d}\,,$$
$$a-y,\ b-y,\ c-y,\ d-y=\mathrm{a}_1,\ \mathrm{b}_1,\ \mathrm{c}_1,\ \mathrm{d}_1;$$

and I write also $(xy,\ bc,\ ad)$, or more shortly $(bc,\ ad)$ to denote the determinant

$$\begin{vmatrix} 1, & x+y, & xy \\ 1, & b+c, & bc \\ 1, & a+d, & ad \end{vmatrix};$$

we have of course $(ad,\ bc)=-(bc,\ ad)$, and there are thus the three distinct determinants $(ad,\ bc)$, $(bd,\ ac)$ and $(cd,\ ab)$.

We have then for each of the functions

$$\sqrt{\frac{a-z}{d-z}},\ \sqrt{\frac{b-z}{d-z}},\ \sqrt{\frac{c-z}{d-z}}$$

a set of four equivalent expressions, the whole system being

$$\sqrt{\frac{a-z}{d-z}} = \frac{\sqrt{a-b\,.\,a-c}\,\{\sqrt{\mathrm{adb_1c_1}}+\sqrt{\mathrm{a_1d_1bc}}\}}{(bc,\ ad)} = \frac{\sqrt{a-b\,.\,a-c}\,(x-y)}{\sqrt{\mathrm{adb_1c_1}}-\sqrt{\mathrm{a_1d_1bc}}}$$

$$= \frac{\sqrt{a-b\,.\,a-c}\,\{\sqrt{\mathrm{abc_1d_1}}+\sqrt{\mathrm{a_1b_1cd}}\}}{(a-c)\sqrt{\mathrm{bdb_1d_1}}-(b-d)\sqrt{\mathrm{aca_1c_1}}} = \frac{\sqrt{a-b\,.\,a-c}\,\{\sqrt{\mathrm{acb_1d_1}}+\sqrt{\mathrm{a_1c_1bd}}\}}{(a-b)\sqrt{\mathrm{cdc_1d_1}}-(c-d)\sqrt{\mathrm{aba_1b_1}}};$$

$$\sqrt{\frac{b-z}{d-z}} = \frac{\sqrt{\frac{a-b}{a-d}}\{(a-c)\sqrt{\mathrm{bdb_1d_1}}+(b-d)\sqrt{\mathrm{aca_1c_1}}\}}{(bc,\ ad)} = \frac{\sqrt{\frac{a-b}{a-d}}\{\sqrt{\mathrm{abc_1d_1}}-\sqrt{\mathrm{a_1b_1cd}}\}}{\sqrt{\mathrm{adb_1c_1}}-\sqrt{\mathrm{a_1d_1bc}}}$$

$$= \frac{\sqrt{\frac{a-b}{a-d}}(cd,\ ab)}{(a-c)\sqrt{\mathrm{bdb_1d_1}}-(b-d)\sqrt{\mathrm{aca_1c_1}}} = \frac{\sqrt{\frac{a-b}{a-d}}\{(a-d)\sqrt{\mathrm{bcb_1c_1}}+(b-c)\sqrt{\mathrm{ada_1d_1}}\}}{(a-b)\sqrt{\mathrm{cdc_1d_1}}-(c-d)\sqrt{\mathrm{aba_1b_1}}};$$

$$\sqrt{\frac{c-z}{d-z}} = \frac{\sqrt{\frac{a-c}{a-d}}\{(a-b)\sqrt{\mathrm{cdc_1d_1}}+(c-d)\sqrt{\mathrm{aba_1b_1}}\}}{(bc,\ ad)} = \frac{\sqrt{\frac{a-c}{a-d}}\{\sqrt{\mathrm{acb_1d_1}}-\sqrt{\mathrm{a_1c_1bd}}\}}{\sqrt{\mathrm{adb_1c_1}}-\sqrt{\mathrm{a_1d_1bc}}}$$

$$= \frac{\sqrt{\frac{a-c}{a-d}}\{(a-d)\sqrt{\mathrm{bcb_1c_1}}-(b-c)\sqrt{\mathrm{ada_1d_1}}\}}{(a-c)\sqrt{\mathrm{bdb_1d_1}}-(b-d)\sqrt{\mathrm{aca_1c_1}}} = \frac{\sqrt{\frac{a-c}{a-d}}(bd,\ ac)}{(a-b)\sqrt{\mathrm{cdc_1d_1}}-(c-d)\sqrt{\mathrm{aba_1b_1}}}.$$

The expressions in the like fourfold form for the functions $\operatorname{sn}(u+v)$, $\operatorname{cn}(u+v)$, $\operatorname{dn}(u+v)$ are given p. 63 of my *Treatise on Elliptic Functions.*

It is easy to verify first that the four expressions for the same function of z are identical, and next that the expressions for the three several functions

$$\sqrt{\frac{a-z}{d-z}},\ \sqrt{\frac{b-z}{d-z}},\ \sqrt{\frac{c-z}{d-z}},$$

are consistent with each other. For instance, comparing the first and second expressions of $\sqrt{\frac{a-z}{d-z}}$, the equation to be verified is

$$\mathrm{adb_1c_1}-\mathrm{a_1d_1bc}=(x-y)(bc,\ ad),$$

which is at once shown to be true. Again comparing the first and second expressions for $\sqrt{\frac{b-z}{d-z}}$, we ought to have

$$\{(a-c)\sqrt{\mathrm{bdb_1d_1}}+(b-d)\sqrt{\mathrm{aca_1c_1}}\}\{\sqrt{\mathrm{adb_1c_1}}-\sqrt{\mathrm{a_1d_1bc}}\}=(bc,\ ad)\{\sqrt{\mathrm{abc_1d_1}}-\sqrt{\mathrm{a_1b_1cd}}\}.$$

Here the product on the left-hand side is

$$= (a-c)\{\mathrm{b_1 d}\sqrt{\mathrm{abc_1d_1}} - \mathrm{bd_1}\sqrt{\mathrm{a_1b_1cd}}\} + (b-d)\{-\mathrm{a_1c}\sqrt{\mathrm{abc_1d_1}} + \mathrm{ac_1}\sqrt{\mathrm{a_1b_1cd}}\},$$

viz. this is

$$= \sqrt{\mathrm{abc_1d_1}}\,\{(a-c)\,\mathrm{b_1d} - (b-d)\,\mathrm{a_1c}\} - \sqrt{\mathrm{a_1b_1cd}}\,\{(a-c)\,\mathrm{bd_1} - (b-d)\,\mathrm{ac_1}\},$$

and in this last expression the two terms in { } are at once shown to be each $=(bc,\ ad)$; whence the identity in question.

Comparing in like manner the first expressions for $\sqrt{\dfrac{a-z}{d-z}}$ and $\sqrt{\dfrac{b-z}{d-z}}$ respectively, we have

$$(b-d)(bc,\ ad)^2\frac{a-z}{d-z} = (a-b)(a-c)(b-d)\{\mathrm{adb_1c_1} + \mathrm{a_1d_1bc} + 2\sqrt{\mathrm{abcda_1b_1c_1d_1}}\},$$

$$(d-a)(bc,\ ad)^2\frac{b-z}{d-z} =$$

$$-(a-b)\{(a-c)^2\,\mathrm{bdb_1d_1} + (b-d)^2\,\mathrm{aca_1c_1} + 2(a-c)(b-d)\sqrt{\mathrm{abcda_1b_1c_1d_1}}\},$$

whence, adding, the radical on the right-hand side disappears; the whole equation divides by $-(a-b)$, and omitting this factor, the relation to be verified is

$$(bc,\ ad)^2 = (a-c)^2\,\mathrm{bdb_1d_1} + (b-d)^2\,\mathrm{aca_1c_1} - (a-c)(b-d)(\mathrm{adb_1c_1} + \mathrm{a_1d_1bc});$$

the right-hand side is here

$$= \{(a-c)\,\mathrm{b_1d} - (b-d)\,\mathrm{a_1c}\}\,\{(a-c)\,\mathrm{bd_1} - (b-d)\,\mathrm{ac_1}\},$$

and each of the two factors being $=(bc,\ ad)$, the identity is verified. It thus appears that the twelve equations are in fact equivalent to a single equation in x, y, z.

Writing in the several formulæ $x=a$, b, c, d successively, they become

$x=a,$	$x=b,$	$x=c,$	$x=d,$
$\dfrac{a-z}{d-z}=\dfrac{\mathrm{a_1}}{\mathrm{d_1}},$	$-\dfrac{c-a}{d-b}\cdot\dfrac{\mathrm{b_1}}{\mathrm{c_1}},$	$-\dfrac{b-a}{d-c}\cdot\dfrac{\mathrm{c_1}}{\mathrm{b_1}},$	$\dfrac{a-b\,.\,a-c}{d-b\,.\,d-c}\cdot\dfrac{\mathrm{d_1}}{\mathrm{a_1}},$
$\dfrac{b-z}{d-z}=\dfrac{\mathrm{b_1}}{\mathrm{d_1}},$	$-\dfrac{c-b}{d-a}\cdot\dfrac{\mathrm{a_1}}{\mathrm{c_1}},$	$\dfrac{b-a\,.\,b-c}{d-a\,.\,d-c}\cdot\dfrac{\mathrm{d_1}}{\mathrm{b_1}},$	$-\dfrac{a-b}{d-c}\cdot\dfrac{\mathrm{c_1}}{\mathrm{a_1}},$
$\dfrac{c-z}{d-z}=\dfrac{\mathrm{c_1}}{\mathrm{d_1}},$	$\dfrac{c-a\,.\,c-b}{d-a\,.\,d-b}\cdot\dfrac{\mathrm{d_1}}{\mathrm{c_1}},$	$-\dfrac{b-c}{d-a}\cdot\dfrac{\mathrm{a_1}}{\mathrm{b_1}},$	$-\dfrac{a-c}{d-b}\cdot\dfrac{\mathrm{b_1}}{\mathrm{a_1}},$

viz. for $x=a$, the relation is $z=y$, but in the other three cases respectively the relation is a linear one, $z=\dfrac{\alpha y+\beta}{\gamma y+\delta}$.

Rationalising the first equation for $\sqrt{\dfrac{a-z}{d-z}}$, we have

$$(bc,\ ad)^2(a-z) = (a-b)(a-c)(d-z)\{\mathrm{adb_1c_1} + \mathrm{a_1d_1bc} + 2\sqrt{\mathrm{abcda_1b_1c_1d_1}}\},$$

and thence

$$\{(bc,\ ad)^2(a-z) - (a-b)(a-c)(d-z)(\mathrm{adb_1c_1} + \mathrm{a_1d_1bc})\}^2$$
$$= (a-b)^2(a-c)^2(d-z)^2\,.\,4\mathrm{abcda_1b_1c_1d_1}.$$

Expanding, and observing that

$$(\mathrm{adb_1c_1} + \mathrm{a_1d_1bc})^2 = (\mathrm{adb_1c_1} - \mathrm{a_1d_1bc})^2 + 4\mathrm{abcda_1b_1c_1d_1} = (bc,\ ad)^2(x-y)^2 + 4\mathrm{abcda_1b_1c_1d_1},$$

the whole equation becomes divisible by $(bc,\ ad)^2$, and omitting this factor, the equation is

$$(bc,\ ad)^2(a-z)^2 - 2(a-b)(a-c)(a-z)(d-z)(\mathrm{adb_1c_1} + \mathrm{a_1d_1bc}) \\ + (a-b)^2(a-c)^2(d-z)^2(x-y)^2 = 0,$$

or, as this may also be written,

$$\begin{aligned} &z^2\{(bc,\ ad)^2 - 2(a-b)(a-c)(\mathrm{adb_1c_1} + \mathrm{a_1d_1bc}) + (a-b)^2(a-c)^2(x-y)^2\} \\ -\,&2z\{(bc,\ ad)\,a - (a-b)(a-c)(\mathrm{adb_1c_1} + \mathrm{a_1d_1bc})(a+d) + (a-b)^2(a-c)^2(x-y)^2 d\} \\ +\,&\{(bc,\ ad)a^2 - 2(a-b)(a-c)(\mathrm{adb_1c_1} + \mathrm{a_1d_1bc})\,ad + (a-b)^2(a-c)^2(x-y)^2 d^2\} = 0. \end{aligned}$$

This is really a symmetrical equation in x, y, z of the form

$$\begin{aligned} &A \\ &+ 2B(x+y+z) \\ &+ C(x^2+y^2+z^2) \\ &+ 2D(yz+zx+xy) \\ &+ 2E(y^2z+yz^2+z^2x+zx^2+x^2y+xy^2) \\ &+ 4Fxyz \\ &+ 2G(x^2yz+xy^2z+xyz^2) \\ &+ H(y^2z^2+z^2x^2+x^2y^2) \\ &+ 2I(xy^2z^2+x^2yz^2+x^2y^2z) \\ &+ Jx^2y^2z^2 = 0; \end{aligned}$$

the several coefficients being symmetrical as regards b, c, d, but the a entering unsymmetrically: the actual values are

$$\begin{aligned} A &= a^4\{b^2c^2+b^2d^2+c^2d^2-2bcd(b+c+d)\} + 2a^3bcd(bc+bd+cd) - 3a^2b^2c^2d^2, \\ B &= 2a^4bcd - a^3(b^2c^2+b^2d^2+c^2d^2) + ab^2c^2d^2, \\ C &= -4a^3bcd + a^2(bc+bd+cd)^2 - 2abcd(bc+bd+cd) + b^2c^2d^2, \\ D &= -a^4(bc+bd+cd) + a^3(b^2c+bc^2+b^2d+bd^2+c^2d+cd^2-2bcd) \\ &\qquad + a^2\{b^2c^2+b^2d^2+c^2d^2-bcd(b+c+d)\} - b^2c^2d^2, \\ E &= a^3(bc+bd+cd) - a^2(b^2c+bc^2+b^2d+bd^2+c^2d+cd^2) + abcd(b+c+d), \\ F &= a^4(b+c+d) - a^3(b^2+c^2+d^2+bc+bd+cd) + 6a^2bcd \\ &\qquad - a\{b^2c^2+b^2d^2+c^2d^2+bcd(b+c+d)\} + bcd(bc+bd+cd), \\ G &= -a^4 + a^2(b^2+c^2+d^2-bc-bd-cd) + a(b^2c+bc^2+b^2d+bd^2+c^2d+cd^2-2bcd) \\ &\qquad - bcd(b+c+d), \\ H &= a^4 - 2a^3(b+c+d) + a^2(b+c+d)^2 - 4abcd, \\ I &= a^3 - a(b^2+c^2+d^2) + 2bcd, \\ J &= -3a^2 + 2a(b+c+d) + b^2+c^2+d^2 - 2(bc+bd+cd). \end{aligned}$$

It may be remarked by way of verification that the equation remains unaltered on substituting for x, y, z, a, b, c, d their reciprocals and multiplying the whole by $a^4b^2c^2d^2x^2y^2z^2$.

I further remark that, writing $a=0$, we have

$$A=0,\quad B=0,\quad C=b^2c^2d^2,\quad D=-b^2c^2d^2,\quad E=0,\quad F=bcd\,(bc+bd+cd),$$
$$G=-bcd\,(b+c+d),\quad H=0,\quad I=2bcd,\quad J=b^2+c^2+d^2-2\,(bc+bd+cd);$$

and writing also

$$\epsilon=1,\quad -\delta=(b+c+d),\quad \gamma=bc+bd+cd,\quad -\beta=bcd,$$

(whence

$$a-x\,.\,b-x\,.\,c-x\,.\,d-x=\beta x+\gamma x^2+\delta x^3+\epsilon x^4),$$

we have the formula

$$\begin{aligned}&\beta^2\,(x^2+y^2+z^2-2yz-2zx-2xy)\\ &-4\beta\gamma\,xyz\\ &-2\beta\delta\,xyz\,(x+y+z)\\ &-4\beta\epsilon\,xyz\,(yz+zx+xy)\\ &+\ (\delta^2-4\gamma\epsilon)\,x^2y^2z^2=0,\end{aligned}$$

given p. 348 of my *Elliptic Functions* as a particular integral of the differential equation when the radical is $\sqrt{\beta x+\gamma x^2+\delta x^3+\epsilon x^4}$.

Let the equation in (x, y, z) be called $u=0$; u has been given in the form $u=\mathfrak{C}z^2-2\mathfrak{B}z+\mathfrak{A}$, and we thence have $\tfrac{1}{2}\dfrac{du}{dz}=\mathfrak{C}z-\mathfrak{B}$ which, in virtue of the equation $u=0$ itself, becomes $\tfrac{1}{2}\dfrac{du}{dz}=\sqrt{\mathfrak{B}^2-\mathfrak{A}\mathfrak{C}}$; we find easily

$$\mathfrak{B}^2-\mathfrak{A}\mathfrak{C}=(a-b)^2(a-c)^2(a-d)^2\,\{(\mathrm{adb_1c_1+a_1d_1bc})^2-(bc,\ ad)^2\,(x-y)^2\},$$

or, attending to the relation

$$\begin{aligned}(\mathrm{adb_1c_1+a_1d_1bc})^2&=(\mathrm{adb_1c_1-a_1d_1bc})^2+4\mathrm{abcda_1b_1c_1d_1}\\ &=(bc,\ ad)^2\,(x-y)^2+4\mathrm{abcda_1b_1c_1d_1},\end{aligned}$$

this is

$$\mathfrak{B}^2-\mathfrak{A}\mathfrak{C}=4\,(a-b)^2\,(a-c)^2\,(a-d)^2\,\mathrm{abcda_1b_1c_1d_1},$$

or we have

$$\tfrac{1}{4}\frac{du}{dz}=(a-b)\,(a-c)\,(a-d)\,\sqrt{\mathrm{abcd}}\,\sqrt{\mathrm{a_1b_1c_1d_1}}.$$

Writing

$$a-z,\ b-z,\ c-z,\ d-z=\mathrm{a_2,\ b_2,\ c_2,\ d_2},$$

we have of course the like formulæ

$$\tfrac{1}{4}\frac{du}{dx}=(a-b)\,(a-c)\,(a-d)\,\sqrt{\mathrm{a_1b_1c_1d_1}}\,\sqrt{\mathrm{a_2b_2c_2d_2}},$$

$$\tfrac{1}{4}\frac{du}{dy}=(a-b)\,(a-c)\,(a-d)\,\sqrt{\mathrm{abcd}}\,\sqrt{\mathrm{a_2b_2c_2d_2}};$$

and the equation $du = 0$ then gives

$$\frac{dx}{\sqrt{\text{abcd}}} + \frac{dy}{\sqrt{\text{a}_1\text{b}_1\text{c}_1\text{d}_1}} + \frac{dz}{\sqrt{\text{a}_2\text{b}_2\text{c}_2\text{d}_2}} = 0,$$

as it should do. The differential equation might also have been verified directly from any one of the expressions for

$$\sqrt{\frac{a-z}{d-z}}, \quad \sqrt{\frac{b-z}{d-z}} \text{ or } \sqrt{\frac{c-z}{d-z}}.$$

Writing for shortness

$$X = a - x \,.\, b - x \,.\, c - x \,.\, d - x, \text{ etc.},$$

then the general integral of the differential equation

$$\frac{dx}{\sqrt{X}} + \frac{dy}{\sqrt{Y}} + \frac{dz}{\sqrt{Z}} = 0$$

by Abel's theorem is

$$\begin{vmatrix} x^2, & x, & 1, & \sqrt{X} \\ y^2, & y, & 1, & \sqrt{Y} \\ z^2, & z, & 1, & \sqrt{Z} \\ w^2, & w, & 1, & \sqrt{W} \end{vmatrix} = 0,$$

where w is the constant of integration: and it is to be shown that the value of w which corresponds to the integral given in the present paper is $w = a$. Observe that writing in the determinant $w = a$, the determinant on putting therein $x = a$, would vanish whether z were or were not $= y$; but this is on account of an extraneous factor $a - w$, so that we do not thus prove the required theorem that (w being $= a$) we have $y = z$ when $x = a$.

An equivalent form of Abel's integral is that there exist values A, B, C such that

$$Ax^2 + Bx + C = \sqrt{X},$$
$$Ay^2 + By + C = \sqrt{Y},$$
$$Az^2 + Bz + C = \sqrt{Z},$$
$$Aw^2 + Bw + C = \sqrt{W},$$

or, what is the same thing, that we have identically

$$(A\theta^2 + B\theta + C)^2 - \Theta = (A^2 - 1) \,.\, \theta - x \,.\, \theta - y \,.\, \theta - z \,.\, \theta - w.$$

We have therefore

$$C^2 - abcd = (A^2 - 1)\, xyzw,$$

or say

$$xyzw = \frac{C^2 - abcd}{A^2 - 1};$$

which equation, regarding therein A, B, C as determined by the three equations

$$Ax^2 + Bx + C = \sqrt{X},$$
$$Ay^2 + By + C = \sqrt{Y},$$
$$Aw^2 + Bw + C = \sqrt{W},$$

is a form of Abel's integral, giving z rationally in terms of x, y, w.

Supposing that, when $x = a$, $z = y$: then the last-mentioned integral gives

$$ay^2w = \frac{C^2 - abcd}{A^2 - 1},$$

where A, C are now determined by the equations

$$Aa^2 + Ba + C = 0,$$
$$Ay^2 + By + C = \sqrt{Y},$$
$$Aw^2 + Bw + C = \sqrt{W},$$

and, imagining these values actually substituted, it is to be shown that the equation

$$ay^2w = \frac{C^2 - abcd}{A^2 - 1}$$

is satisfied by the value $w = a$.

We have

$$A\,.\,a-y\,.\,a-w\,.\,w-y = \qquad (a-w)\sqrt{Y} - \qquad (a-y)\sqrt{W},$$
$$B\,.\,a-y\,.\,a-w\,.\,w-y = (a-w)(a+w)\sqrt{Y} - (a-y)(a+y)\sqrt{W},$$
$$C\,.\,a-y\,.\,a-w\,.\,w-y = (a-w)\,aw \quad \sqrt{Y} - (a-y)\,ay \quad \sqrt{W},$$

or writing as before

$$a-y,\ b-y,\ c-y,\ d-y = \mathfrak{a}_1,\ \mathfrak{b}_1,\ \mathfrak{c}_1,\ \mathfrak{d}_1,$$

and also

$$a-w,\ b-w,\ c-w,\ d-w = \mathfrak{a}_3,\ \mathfrak{b}_3,\ \mathfrak{c}_3,\ \mathfrak{d}_3,$$

then $Y = \mathfrak{a}_1\mathfrak{b}_1\mathfrak{c}_1\mathfrak{d}_1$, $W = \mathfrak{a}_3\mathfrak{b}_3\mathfrak{c}_3\mathfrak{d}_3$, and the formulæ become

$$A = \frac{1}{(w-y)\sqrt{\mathfrak{a}_1\mathfrak{a}_3}}\{\sqrt{\mathfrak{a}_3\mathfrak{b}_1\mathfrak{c}_1\mathfrak{d}_1} - \sqrt{\mathfrak{a}_1\mathfrak{b}_3\mathfrak{c}_3\mathfrak{d}_3}\},$$

$$B = \frac{1}{(w-y)\sqrt{\mathfrak{a}_1\mathfrak{a}_3}}\{-(a+w)\sqrt{\mathfrak{a}_3\mathfrak{b}_1\mathfrak{c}_1\mathfrak{d}_1} + (a+y)\sqrt{\mathfrak{a}_1\mathfrak{b}_3\mathfrak{c}_3\mathfrak{d}_3}\},$$

$$C = \frac{1}{(w-y)\sqrt{\mathfrak{a}_1\mathfrak{a}_3}}\{aw\sqrt{\mathfrak{a}_3\mathfrak{b}_1\mathfrak{c}_1\mathfrak{d}_1} - ay\sqrt{\mathfrak{a}_1\mathfrak{b}_3\mathfrak{c}_3\mathfrak{d}_3}\}.$$

If in these formulæ w is indefinitely nearly $=a$, then a_3 is indefinitely small, so that $\sqrt{\mathrm{a}_3\mathrm{b}_1\mathrm{c}_1\mathrm{d}_1}$ may be neglected in comparison with $\sqrt{\mathrm{a}_1\mathrm{b}_3\mathrm{c}_3\mathrm{d}_3}$: also $w-y$ may be put $=\mathrm{a}_1$; the formulæ thus become

$$A=-\frac{\sqrt{\mathrm{b}_3\mathrm{c}_3\mathrm{d}_3}}{\mathrm{a}_1\sqrt{\mathrm{a}_3}},\quad B=(a+y)\frac{\sqrt{\mathrm{b}_3\mathrm{c}_3\mathrm{d}_3}}{\mathrm{a}_1\sqrt{\mathrm{a}_3}},\quad C=-ay\frac{\sqrt{\mathrm{b}_3\mathrm{c}_3\mathrm{d}_3}}{\mathrm{a}_1\sqrt{\mathrm{a}_3}},$$

where the values of A, B, C are each of them indefinitely large on account of the factor $\sqrt{\mathrm{a}_3}$ in the denominator; the value of C is $C=ayA$, and substituting this value in the equation

$$ay^2w=\frac{C^2-abcd}{A^2-1},$$

and then considering A as indefinitely large, the equation becomes $ay^2w=a^2y^2$, that is, $w=a$; so that $w=a$ is a value of w satisfying this equation.

Cambridge, 3 July, 1878.

698.

ON A THEOREM RELATING TO COVARIANTS.

[From the *Journal für die reine und angewandte Mathematik* (Crelle), t. LXXXVII. (1878), pp. 82, 83.]

THE theorem given by Prof. Sylvester, *Crelle*, vol. LXXXV., p. 109, may be stated as follows: If for a binary quantic of the order i in the variables, we consider the whole system of covariants of the degree j in the coefficients, then

$$\Sigma\theta(k+1) = \frac{\Pi(i+j)}{\Pi(i)\,\Pi(j)},$$

where θ denotes the number of asyzygetic covariants of the order θ in the variables, the values of θ being ij, $ij-2$, $ij-4, \ldots, 1$ or 0, according as ij is odd or even.

In the case of the binary quintic $(a, \ldots \between x, y)^5$, $(i=5)$, we have a series of verifications in the Table 88 of my "Ninth Memoir on Quantics," *Phil. Trans.* vol. CLXI. (1871), [462]: viz. writing the small letters $a, b, c, \ldots, u, v, w$ (instead of the capitals A, B, etc.) to denote the covariants of the quintic, a, the quintic itself, degree 1, order 5, or as I express it, deg-order 1.5: b, the covariant deg-order 2.2, etc., the whole series of deg-orders being

a,	b,	c,	d,	e,	f,	g,	h,	i,	j,	k,	l,
1.5,	2.2,	2.6,	3.3,	3.5,	3.9,	4.0,	4.4,	4.6,	5.1,	5.3,	5.7,

m,	n,	o,	p,	q,	r,	s,	t,	u,	v,	w,
6.2,	6.4,	7.1,	7.5,	8.0,	8.2,	9.3,	11.1,	12.0,	13.1,	18.0,

then the table shows for each deg-order, the several covariants of that deg-order, and the number of them which are asyzygetic; for instance, $i=5$ as above, $j=6$, an extract from the table is

j	k	θ		$(k+1)\,\theta$
6	30	1	a^6	31
	28	0		0
	26	1	a^4c	27
	24	1	a^3f	25
	22	2	$a^4b,\ a^2c^2$	46
	20	2	$a^3e,\ acf$	42
	18	3	$a^3d,\ a^2bc,\ c^3,\ f^2$	57
	16	2	$a^2i,\ abf,\ ace$	34
	14	4	$a^2b^2,\ a^2h,\ acd,\ bc^2,\ ef$	60
	12	3	$abe,\ al,\ ce,\ df$	39
	10	4	$a^2g,\ abd,\ b^2c,\ ch,\ e^2$	44
	8	2	$ak,\ bi,\ de$	18
	6	4	$aj,\ b^3,\ bh,\ cg,\ d^2$	28
	4	1	n	5
	2	2	$bg,\ m$	6
	0	0		0

$$462=\frac{\Pi(11)}{\Pi(5)\,\Pi(6)},$$

where, for instance deg-order 6.14, the covariants are a^2b^2, a^2h, acd, bc^2, ef, but the number against these in the third column being (not 5 but) 4, the meaning is that there exists between these five terms one syzygy, making the number of asyzygetic covariants of the deg-order 6.14 to be 4. The second column thus in fact contains the several values of k, and the third column the corresponding values of θ; whence, forming the several products $(k+1)$ as shown, the sum of these is as it should be $=462$.

Cambridge, 13 *July*, 1878.

699.

ON THE TRIPLE ϑ-FUNCTIONS.

[From the *Journal für die reine und angewandte Mathematik* (Crelle), t. LXXXVII. (1878), pp. 134—138.]

THERE should be in all 64 functions proportional to irrational algebraical functions of three independent variables x, y, z; there is no difficulty in obtaining the expression of these 64 functions in the case of the system of differential equations connected with the integral

$$\int dx : \sqrt{a-x \,.\, b-x \,.\, c-x \,.\, d-x \,.\, e-x \,.\, f-x \,.\, g-x \,.\, h-x};$$

but this is *not the general form* of the system for the deficiency (Geschlecht) $p=3$; and I do not know how to deal with the general form: the present note relates therefore exclusively to the above-mentioned hyper-elliptic form.

I.

If in the Memoir, Weierstrass, "Theorie der Abel'schen Functionen," *Crelle*, t. LII. (1856), pp. 285—380, we take $\rho=3$, and write x, y, z; u, v, w; a, b, c, d, e, f, g instead of x_1, x_2, x_3; u_1, u_2, u_3; a_1, a_2, a_3, a_4, a_5, a_6, a_7; then, neglecting throughout mere constant factors, we have

$$X = a-x \,.\, b-x \,.\, c-x \,.\, d-x \,.\, e-x \,.\, f-x \,.\, g-x,$$

with the like values for Y and Z: the differential equations are

$$du = \frac{b-x \,.\, c-x \,.\, dx}{\sqrt{X}} + \frac{b-y \,.\, c-y \,.\, dy}{\sqrt{Y}} + \frac{b-z \,.\, c-z \,.\, dz}{\sqrt{Z}},$$

$$dv = \frac{c-x \,.\, a-x \,.\, dx}{\sqrt{X}} + \frac{c-y \,.\, a-y \,.\, dy}{\sqrt{Y}} + \frac{c-z \,.\, a-z \,.\, dz}{\sqrt{Z}},$$

$$dw = \frac{a-x \,.\, b-x \,.\, dx}{\sqrt{X}} + \frac{a-y \,.\, b-y \,.\, dy}{\sqrt{Y}} + \frac{a-z \,.\, b-z \,.\, dz}{\sqrt{Z}},$$

and if we write the single letters A, B, C, D, E, F, G for $\text{al}(u, v, w)_1$, $\text{al}(u, v, w)_2$, $\text{al}(u, v, w)_3$, $\text{al}(u, v, w)_4$, $\text{al}(u, v, w)_5$, $\text{al}(u, v, w)_6$, $\text{al}(u, v, w)_7$ respectively, each of the capital letters thus denoting a function of (u, v, w), the expressions of these functions in terms of (x, y, z) are

$$A = \sqrt{a-x\,.\,b-x\,.\,c-x}, \quad \text{(seven equations).}$$
$$\vdots \qquad \vdots$$

Similarly, instead of the 21 functions $\text{al}(u, v, w)_{12}, \ldots, \text{al}(u, v, w)_{67}$ writing $AB, \ldots, FG$, each of these binary symbols denoting in like manner a function of (u, v, w), the definition of AB is

$$AB = A\nabla B - B\nabla A,$$

where

$$\nabla = \frac{d}{du} + \frac{d}{dv} + \frac{d}{dw};$$

we have

$$b-c\,.\,c-a\,.\,a-b\,.\frac{dx}{\sqrt{X}} = \frac{a-y\,.\,a-z}{x-y\,.\,x-z}(b-c)\,du + \frac{b-y\,.\,b-z}{x-y\,.\,x-z}(c-a)\,dv + \frac{c-y\,.\,c-z}{x-y\,.\,x-z}(a-b)\,dw,$$

$$\text{,,} \qquad \frac{dy}{\sqrt{Y}} = \frac{a-z\,.\,a-x}{y-z\,.\,y-x}(b-c)\,du + \frac{b-z\,.\,b-x}{y-z\,.\,y-x}(c-a)\,dv + \frac{c-z\,.\,c-x}{y-z\,.\,y-x}(a-b)\,dw,$$

$$\text{,,} \qquad \frac{dz}{\sqrt{Z}} = \frac{a-x\,.\,a-y}{z-x\,.\,z-y}(b-c)\,du + \frac{b-x\,.\,b-y}{z-x\,.\,z-y}(c-a)\,dv + \frac{c-x\,.\,c-y}{z-x\,.\,z-y}(a-b)\,dw;$$

hence

$$\frac{b-c\,.\,c-a\,.\,a-b}{\sqrt{X}}\,\nabla x = \frac{a-y\,.\,a-z}{x-y\,.\,x-z}(b-c) + \frac{b-y\,.\,b-z}{x-y\,.\,x-z}(c-a) + \frac{c-y\,.\,c-z}{x-y\,.\,x-z}(a-b),$$

$$= -\frac{b-c\,.\,c-a\,.\,a-b}{x-y\,.\,x-z},$$

that is,

$$\nabla x = \frac{-\sqrt{X}}{x-y\,.\,x-z};$$

and similarly

$$\nabla y = \frac{-\sqrt{Y}}{y-x\,.\,y-z}, \quad \nabla z = \frac{-\sqrt{Z}}{z-x\,.\,z-y}.$$

Hence from the equation

$$A = \sqrt{a-x\,.\,a-y\,.\,a-z}$$

we have

$$\nabla A = -\tfrac{1}{2}A\left(\frac{1}{a-x}\,\nabla x + \frac{1}{a-y}\,\nabla y + \frac{1}{a-z}\,\nabla z\right),$$

that is,

$$\nabla A = \frac{\tfrac{1}{2}A}{y-z\,.\,z-x\,.\,x-y}\left\{\frac{y-z}{a-x}\sqrt{X} + \frac{z-x}{a-y}\sqrt{Y} + \frac{x-y}{a-z}\sqrt{Z}\right\}:$$

and similarly

$$\nabla B = \frac{\tfrac{1}{2}B}{y-z\,.\,z-x\,.\,x-y}\left\{\frac{y-z}{b-x}\sqrt{X} + \frac{z-x}{b-y}\sqrt{Y} + \frac{x-y}{b-z}\sqrt{Z}\right\};$$

consequently

$$AB = \frac{\tfrac{1}{2}(a-b)\,AB}{y-z\,.\,z-x\,.\,x-y}\left\{\frac{(y-z)\sqrt{X}}{a-x\,.\,b-x} + \frac{(z-x)\sqrt{Y}}{a-y\,.\,b-y} + \frac{(x-y)\sqrt{Z}}{a-z\,.\,b-z}\right\},$$

or substituting for A and B their values, and disregarding the constant factor $\frac{1}{2}(a-b)$, this is

$$AB = \frac{1}{y-z\,.\,z-x\,.\,x-y}\{(y-z)\sqrt{a-y\,.\,b-y\,.\,a-z\,.\,b-z\,.\,c-x\,.\,d-x\,.\,e-x\,.f-x\,.\,g-x}$$
$$+(z-x)\sqrt{a-z\,.\,b-z\,.\,a-x\,.\,b-x\,.\,c-y\,.\,d-y\,.\,e-y\,.f-y\,.\,g-y}$$
$$+(x-y)\sqrt{a-x\,.\,b-x\,.\,a-y\,.\,b-y\,.\,c-z\,.\,d-z\,.\,e-z\,.f-z\,.\,g-z}\}.$$

We have thus in all 21 equations, which exhibit the form of the Weierstrassian functions $\text{al}(u, v, w)_{12}, \ldots, \text{al}(u, v, w)_{67}$.

To complete the system, there should it is clear be 35 new functions $\text{al}(u, v, w)_{123}$, ..., $\text{al}(u, v, w)_{567}$, represented by $ABC, \ldots, EFG$, viz. the whole number of functions would then be

$$7 + \frac{7\,.\,6}{1\,.\,2} + \frac{7\,.\,6\,.\,5}{1\,.\,2\,.\,3}(=7+21+35) = 63, \;= 64-1,$$

since the functions represent ratios of the ϑ-functions.

II.

Starting now with the radical

$$\sqrt{a-x\,.\,b-x\,.\,c-x\,.\,d-x\,.\,e-x\,.f-x\,.\,g-x\,.\,h-x}$$

composed of eight linear factors, and writing, as in my "Memoir on the double ϑ-functions," t. LXXXV. (1878), pp. 214—245, [665]; a, b, c, d, e, f, g, h to denote these factors, and similarly $a_1, b_1, c_1, d_1, e_1, f_1, g_1, h_1$ and $a_2, b_2, c_2, d_2, e_2, f_2, g_2, h_2$ to denote $a-y$, $b-y$, etc., and $a-z$, $b-z$, etc., so that $X = \text{abcdefgh}$, $Y = a_1b_1c_1d_1e_1f_1g_1h_1$, $Z = a_2b_2c_2d_2e_2f_2g_2h_2$; then, instead of the Weierstrassian form, the differential equations may be taken to be

$$du = \frac{dx}{\sqrt{X}} + \frac{dy}{\sqrt{Y}} + \frac{dz}{\sqrt{Z}},$$

$$dv = \frac{x\,dx}{\sqrt{X}} + \frac{y\,dy}{\sqrt{Y}} + \frac{z\,dz}{\sqrt{Z}},$$

$$dw = \frac{x^2\,dx}{\sqrt{X}} + \frac{y^2\,dy}{\sqrt{Y}} + \frac{z^2\,dz}{\sqrt{Z}}.$$

We then have 64 ϑ-functions and an ω-function, viz. writing

$$\theta = y-z\,.\,z-x\,.\,x-y,$$

and then

$$\sqrt{a} = \sqrt{aa_1a_2} \quad \text{(8 equations)}$$
$$\vdots \qquad \vdots$$
$$\sqrt{abc} = \frac{1}{\theta}\{(y-z)\sqrt{a_1b_1c_1a_2b_2c_2\text{defgh}} + (z-x)\sqrt{a_2b_2c_2\text{abc}d_1e_1f_1g_1h_1} + (x-y)\sqrt{\text{abc}a_1b_1c_1d_2e_2f_2g_2h_2}\}$$
$$\vdots \qquad \vdots \qquad\qquad \text{(56 equations)}$$

the equations, which define the ϑ-functions $A, B, \ldots, H, ABC, \ldots, FGH$, and the ω-function Ω, are

$$A = \Omega \sqrt{a} \qquad \text{(8 equations)}$$
$$\vdots \quad \vdots$$
$$ABC = \Omega \sqrt{abc} \quad \text{(56 equations)}$$
$$\vdots \quad \vdots$$

and one other relation which I have not as yet investigated.

As regards the algebraical relations between the 64 ϑ-functions, it is to be remarked that, selecting in a proper manner 8 of the functions, the square of any one of the other functions can be expressed as a linear function of the squares of the 8 selected functions. To explain this somewhat further, observe that, taking any 5 squares such as $(ABC)^2$, we can with these 5 squares form a linear combination which is rational in x, y, z. We have for instance, writing down the irrational part only,

$$(ABC)^2 = \frac{2}{\theta^2}\{\mathrm{abc}\,(z-x)(x-y)\sqrt{YZ} + \mathrm{a_1b_1c_1}\,(x-y)(y-z)\sqrt{ZX} + \mathrm{a_2b_2c_2}\,(y-z)(z-x)\sqrt{XY}\},$$

and forming in all five such equations, then inasmuch as the coefficients abc,... of $(z-x)(x-y)\sqrt{YZ}$ are each of them a cubic function containing terms in x^0, x^1, x^2, x^3, we have a determinate set of constant factors such that the resulting term in $(z-x)(x-y)\sqrt{YZ}$ will be $=0$; but the coefficients $\mathrm{a_1b_1c_1}, \ldots$ of $(x-y)(y-z)\sqrt{ZX}$ only differ from the first set of coefficients by containing y instead of x, and the same set of constant factors will thus make the resulting term in $(x-y)(y-z)\sqrt{ZX}$ to be $=0$; and similarly the same set of constant factors will make the resulting term in $(y-z)(z-x)\sqrt{XY}$ to be $=0$; viz. we have thus a set of constant factors, such that the whole irrational part will disappear. *It seems to be in general true that the same set of constant factors will make the rational part integral;* viz. the rational part is a function of the form $\frac{1}{\theta^2}$ multiplied by a rational and integral function of x, y, z, and if this rational and integral function divide by θ^2, then the final result will be a rational and integral function, which, being symmetrical in x, y, z, is at once seen to be a linear function of the symmetrical combinations 1, $x+y+z$, $yz+zx+xy$, xyz. Such a function is obviously a linear function of any four squares A^2, B^2, C^2, D^2; or the form is, linear function of five squares $(ABC)^2 =$ linear function of four squares A^2, that is, any one of the five squares is a linear function of 8 squares.

As an instance, consider the *three* squares $(ABC)^2$, $(ABD)^2$, $(ABE)^2$, which are such that we have a linear combination which is rational: in fact, we have here in each function the pair of factors ab, which unites itself with $(z-x)(x-y)\sqrt{XY}$, viz. it is only the coefficient of $\mathrm{ab}\,(z-x)(x-y)\sqrt{XY}$ which has to be made $=0$; the required combination is obviously

$$(d-e)(ABC)^2 + (e-c)(ABD)^2 + (c-d)(ABE)^2.$$

Here the irrational part vanishes and the rational part is found to be

$$=\frac{1}{\theta^2}\left[\mathrm{a_1b_1a_2b_2fgh}\,(y-z)^2\left\{\begin{array}{l}(d-e)\,\mathrm{c_1c_2de}\\+(e-c)\,\mathrm{d_1d_2ce}\\+(c-d)\,\mathrm{e_1e_2dc}\end{array}\right\}\right.$$

$$+\mathrm{a_2b_2abf_1g_1h_1}\,(z-x)^2\left\{\begin{array}{l}(d-e)\,\mathrm{c_2cd_1e_1}\\+(e-c)\,\mathrm{d_2dc_1e_1}\\+(c-d)\,\mathrm{e_2ed_1c_1}\end{array}\right\}$$

$$\left.+\mathrm{aba_1b_1f_2g_2h_2}\,(x-y)^2\left\{\begin{array}{l}(d-e)\,\mathrm{cc_1d_2e_2}\\+(e-c)\,\mathrm{dd_1c_2e_2}\\+(c-d)\,\mathrm{ee_1d_2c_2}\end{array}\right\}\right].$$

The three terms in { } are here $=-(c-d)(d-e)(e-c)$ multiplied by $(z-x)(x-y)$, $(x-y)(y-z)$, $(y-z)(z-x)$ respectively; hence the term in [ ] divides by θ and the result is

$$=-\frac{(c-d)(d-e)(e-c)}{\theta}\left[\mathrm{a_1b_1a_2b_2f\,g\,h}\;(y-z)\right.$$

$$+\mathrm{a_2b_2a\,b\,f_1g_1h_1}\,(z-x)$$

$$\left.+\mathrm{a\,b\,a_1b_1f_2g_2h_2}\,(x-y)\right],$$

or finally this is

$$=-(c-d)(d-e)(e-c)$$

multiplied by

$$\{(a^2+ab+b^2)fgh-(a^2b+ab^2)(fg+fh+gh)+a^2b^2(f+g+h)\}$$

$$+\quad(x+y+z)\{\;-(a+b)fgh+\qquad ab(fg+fh+gh)-a^2b^2\qquad\}$$

$$+(yz+zx+xy)\{\qquad fgh-\qquad ab(f+g+h)\quad+a^2b+ab^2\qquad\}$$

$$+\qquad xyz\{-(fg+fh+gh)+\qquad(a+b)(f+g+h)-(a^2+ab+b^2)\;\},$$

that is, we have $(d-e)(ABC)^2+(e-c)(ABD)^2+(c-d)(ABE)^2=$ a sum of four squares, viz. we have here a linear relation between 7 squares.

I have not as yet investigated the forms of the relations between the products of pairs of ϑ-functions.

Cambridge, 30 *September*, 1878.

700.

ON THE TETRAHEDROID AS A PARTICULAR CASE OF THE 16-NODAL QUARTIC SURFACE.

[From the *Journal für die reine und angewandte Mathematik* (Crelle), t. LXXXVII. (1878), pp. 161—164.]

IN the paper "Sur un cas particulier de la surface du quatrième ordre avec seize points singuliers," *Crelle*, t. LXV. (1866), pp. 284—290, [356], I showed how the surface called the Tetrahedroid could be identified as a special form of Kummer's 16-nodal quartic surface; but I was not then in possession of the simplified form of the equation of the 16-nodal surface given in my paper "Note sur la surface du quatrième ordre douée de seize points singuliers et de seize plans singuliers," *Crelle*, t. LXXIII. (1871), pp. 292, 293, [442]; see also my paper, "A third memoir on Quartic surfaces," *Proc. Lond. Math. Soc.* t. III. (1871), p. 250, [454, this Collection, t. VII., p. 281]. Using the equation last referred to, I resume therefore the consideration of the question.

Taking the constants $\alpha, \beta, \gamma, \alpha', \beta', \gamma', \alpha'', \beta'', \gamma''$, such that

$$\alpha+\beta+\gamma=0, \quad \alpha'+\beta'+\gamma'=0, \quad \alpha''+\beta''+\gamma''=0,$$

and writing also

$$\begin{aligned} M &= \alpha'\alpha''(\beta-\gamma)+\beta'\beta''(\gamma-\alpha)+\gamma'\gamma''(\alpha-\beta) \\ &= \alpha''\alpha\,(\beta'-\gamma')+\beta''\beta\,(\gamma'-\alpha')+\gamma''\gamma\,(\alpha'-\beta') \\ &= \alpha\,\alpha'(\beta''-\gamma'')+\beta\,\beta'(\gamma''-\alpha'')+\gamma\,\gamma'(\alpha''-\beta'') \\ &= -\tfrac{1}{3}\{(\beta-\gamma)(\beta'-\gamma')(\beta''-\gamma'')+(\gamma-\alpha)(\gamma'-\alpha')(\gamma''-\alpha'')+(\alpha-\beta)(\alpha'-\beta')(\alpha''-\beta'')\}, \end{aligned}$$

(the equivalence of which different expressions for M is verified without difficulty): writing also X, Y, Z, W as current coordinates, the equation of the 16-nodal surface is

$$0=\left\{\begin{aligned} &W^2(X^2+Y^2+Z^2-2YZ-2ZX-2XY) \\ &+2W\{\alpha\alpha'\alpha''(Y^2Z-YZ^2)+\beta\beta'\beta''(Z^2X-ZX^2)+\gamma\gamma'\gamma''(X^2Y-XY^2)+MXYZ\} \\ &+(\alpha\alpha'\alpha''YZ+\beta\beta'\beta''ZX+\gamma\gamma'\gamma''XY)^2, \end{aligned}\right.$$

where, $\alpha, \beta, \gamma, \alpha', \beta', \gamma', \alpha'', \beta'', \gamma''$ being connected as above, the number of constants is $=6$.

The equations of the 16 singular planes are

$$
\begin{array}{ll}
X=0, & Y=0, \\
\alpha\ (\gamma'\gamma''Y-\beta'\beta''Z)-W=0, & \beta\ (\alpha'\alpha''Z-\gamma'\gamma''X)-W=0, \\
\alpha'\ (\gamma''\gamma\, Y-\beta''\beta\, Z)-W=0, & \beta'\ (\alpha''\alpha\, Z-\gamma''\gamma\, X)-W=0, \\
\alpha''\ (\gamma\gamma'\ Y-\beta\beta'\ Z)-W=0, & \beta''\ (\alpha\alpha'\ Z-\gamma\gamma'\ X)-W=0, \\
Z=0, & W=0, \\
\gamma\ (\beta'\beta''X-\alpha'\alpha''\ Y)-W=0, & \beta\gamma X+\ \gamma\alpha Y+\ \alpha\beta Z=0, \\
\gamma'\ (\beta''\beta\, X-\alpha''\alpha\ Y)-W=0, & \beta'\gamma' X+\ \gamma'\alpha' Y+\ \alpha'\beta' Z=0, \\
\gamma''\ (\beta\beta'\ X-\alpha\alpha'\ Y)-W=0, & \beta''\gamma'' X+\gamma''\alpha'' Y+\alpha''\beta'' Z=0.
\end{array}
$$

Writing x, y, z, w as current coordinates, the equation of the Tetrahedroid is

$$
\begin{aligned}
&m^2n^2f^2x^4+n^2l^2g^2y^4+l^2m^2h^2z^4+f^2g^2h^2w^4 \\
&+(l^2f^2-m^2g^2-n^2h^2)(l^2y^2z^2+f^2x^2w^2)+(-l^2f^2+m^2g^2-n^2h^2)(m^2z^2x^2+g^2y^2w^2) \\
&\qquad +(-l^2f^2-m^2g^2+n^2h^2)(n^2x^2y^2+h^2z^2w^2)=0,
\end{aligned}
$$

where, inasmuch as f, g, h, l, m, n enter homogeneously, the number of constants is $=5$.

The equations of the 16 singular planes, written in an order corresponding to that used for the 16-nodal surface, are

$*\quad ny-mz+fw=0$	$-nx\quad *\quad +lz+gw=0$	$mx-ly\quad *\quad +hw=0$	$-fx-gy-hz\quad *\quad =0$
$fx-gy-hz\quad *\quad =0$	$mx+ly\quad *\quad +hw=0$	$-nx\quad *\quad -lz+gw=0$	$*\quad ny-mz-fw=0$
$-mx-ly\quad *\quad +hw=0$	$-fx+gy-hz\quad *\quad =0$	$*\quad ny+mz+fw=0$	$-nx\quad *\quad +lz-gw=0$
$nx\quad *\quad +lz+gw=0$	$*\quad -ny-mz+fw=0$	$-fx-gy+hz\quad *\quad =0$	$mx-ly\quad *\quad -hw=0.$

These equations can be made to agree each to each with those of the 16 singular planes of the 16-nodal surface, provided that we have

$$\frac{m}{\gamma}=\frac{n}{\beta},\quad \frac{n}{\alpha'}=\frac{l}{\gamma'},\quad \frac{l}{\beta''}=\frac{m}{\alpha''};\quad f=-l\alpha\alpha'\alpha'',\quad g=-m\beta\beta'\beta'',\quad h=-n\gamma\gamma'\gamma'',$$

where observe that the first three equations give $\alpha'\beta''\gamma=\alpha''\beta\gamma'$, which is the relation between the constants when the 16-nodal surface reduces itself to a tetrahedroid in the above manner. And if we then assume

$$X=ny-mz+fw,\quad Y=-nx+lz+gw,\quad Z=mx-ly+hw,\quad W=-fx-gy-hz,$$

the 16 linear functions of X, Y, Z, W will become mere constant multiples of the corresponding 16 linear functions of x, y, z, w; the constants, by which the several functions of x, y, z, w have to be multiplied in order to reduce them each to the corresponding linear function of X, Y, Z, W, being given by the table

$$\begin{array}{cccc}
1, & 1, & 1, & 1, \\
\frac{1}{m\beta}(l\alpha - m\beta), & -\frac{\alpha'\alpha''}{m}(l\alpha - m\beta), & \frac{\alpha'\alpha''}{n}(l\alpha - m\beta), & \frac{\beta\gamma}{\alpha'\alpha''}(\beta'\gamma'' - \beta''\gamma'), \\
\frac{\beta''\beta}{l}(m\beta' - n\gamma'), & \frac{1}{n\gamma'}(m\beta' - n\gamma'), & -\frac{\beta''\beta}{n}(m\beta' - n\gamma'), & \frac{\gamma'\alpha'}{\beta''\beta}(\gamma''\alpha - \gamma\alpha''), \\
-\frac{\gamma\gamma'}{l}(n\gamma'' - l\alpha''), & \frac{\gamma\gamma'}{m}(n\gamma'' - l\alpha''), & \frac{1}{l\alpha''}(n\gamma'' - l\alpha''), & \frac{\alpha''\beta''}{\gamma\gamma'}(\alpha\beta' - \alpha'\beta).
\end{array}$$

For instance, we have

$$\alpha(\gamma'\gamma''Y - \beta'\beta''Z) - W = \frac{1}{m\beta}(l\alpha - m\beta)(fx - gy - hz),$$

viz. substituting for Y, Z, W their values, the relation is

$$\left.\begin{array}{r}
m\alpha\beta\,.\,\gamma'\gamma''\,(-nx \quad * \quad + lz + gw) \\
-\,m\alpha\beta\,.\,\beta'\beta''\,(\ mx - ly \quad * \quad + hw) \\
-\quad m\beta\,(-fx \ - gy - hz \quad * \quad)
\end{array}\right\} = (l\alpha - m\beta)(fx - gy - hz).$$

As regards the terms in y, z, and w, the identity is at once verified. As regards the term in x, we should have

$$m\alpha\beta(-n\gamma'\gamma'' - m\beta'\beta'') - (l\alpha - 2m\beta)f = 0,$$

viz. substituting for f its value, $-l\alpha\alpha'\alpha'' = -m\alpha\alpha'\beta''$, the equation divides by $m\alpha$ and we then have

$$\beta(-n\gamma'\gamma'' - m\beta'\beta'') + \alpha'\beta''(l\alpha - 2m\beta) = 0,$$

that is,

$$l\alpha\alpha'\beta'' - m\beta\beta''(\beta' + 2\alpha') - n\beta\gamma'\gamma'' = 0,$$

or writing herein $m\beta'' = l\alpha''$, $n\gamma' = l\alpha'$, and $\beta' + 2\alpha' = \alpha' - \gamma'$, the equation becomes $\alpha'\alpha\beta'' - \alpha''\beta(\alpha' - \gamma') - \alpha'\beta\gamma'' = 0$, that is, $\alpha'(\alpha\beta'' - \alpha''\beta) = \alpha'\beta\gamma'' - \alpha''\beta\gamma'$; or writing herein $\alpha''\beta\gamma' = \alpha'\beta''\gamma$, the equation divided by α' becomes $\alpha\beta'' - \alpha''\beta = \beta\gamma'' - \beta''\gamma$, which is true in virtue of $\alpha + \beta + \gamma = 0$ and $\alpha'' + \beta'' + \gamma'' = 0$. And in like manner the several other identities may be verified.

The equation $\alpha'\beta''\gamma = \alpha''\beta\gamma'$ might have been obtained as the condition of the intersection, in a common point, of four of the singular planes of the 16-nodal surface; and when this equation is satisfied, there are in fact four systems each of four planes, such that the four planes of a system meet in a common point: viz. we have

Planes

$$\begin{array}{llll}
X = 0, & \beta\gamma X + \gamma\alpha Y + \alpha\beta Z = 0, & \gamma'(\beta''\beta X - \alpha''\alpha Y) - W = 0, & \beta''(\alpha\alpha' Z - \gamma\gamma' X) - W = 0, \\
Y = 0, & \gamma(\beta'\beta'' X - \alpha'\alpha'' Y) - W = 0, & \beta'\gamma' X + \gamma'\alpha' Y + \alpha'\beta' Z = 0, & \alpha''(\gamma\gamma' Y - \beta\beta' Z) - W = 0, \\
Z = 0, & \beta(\alpha'\alpha'' Z - \gamma'\gamma'' X) - W = 0, & \alpha'(\gamma''\gamma Y - \beta''\beta Z) - W = 0, & \beta''\gamma'' X + \gamma''\alpha'' Y + \alpha''\beta'' Z = 0, \\
W = 0, & \alpha(\gamma'\gamma'' Y - \beta'\beta'' Z) - W = 0, & \beta'(\alpha''\alpha Z - \gamma''\gamma X) - W = 0, & \gamma''(\beta\beta' X - \alpha\alpha' Y) - W = 0,
\end{array}$$

meeting in points

$$\begin{array}{cccc} 0, & -\beta, & \gamma, & \alpha''\beta\gamma' . \alpha\ , \\ \alpha', & 0, & -\gamma', & \alpha''\beta\gamma' . \beta', \\ -\alpha'', & \beta'', & 0, & \alpha''\beta\gamma' . \gamma'', \\ \beta'' . \alpha\alpha'\alpha'', & \alpha'' . \beta\beta'\beta'', & \gamma'' . \alpha''\beta\gamma', & 0, \end{array}$$

the four points being in fact the vertices of the tetrahedron formed by the four planes of the tetrahedroid. Observe that, if the singular planes of the 16-nodal surface in their original order are

$$\begin{array}{cccc} 1, & 2, & 3, & 4, \\ 5, & 6, & 7, & 8, \\ 9, & 10, & 11, & 12, \\ 13, & 14, & 15, & 16, \end{array}$$

then the planes forming the last-mentioned four systems of planes are

$$\begin{array}{l} (1,\ 8,\ 11,\ 14), \\ (2,\ 7,\ 12,\ 13), \\ (3,\ 6,\ 9,\ 16), \\ (4,\ 5,\ 10,\ 15), \end{array}$$

viz. they correspond each of them to a term which in the determinant formed with the 16 symbols would have the sign +.

The equation $\alpha'\beta''\gamma = \alpha''\beta\gamma'$ is evidently not unique. The triads (α, β, γ), $(\alpha', \beta', \gamma')$, $(\alpha'', \beta'', \gamma'')$ enter symmetrically into the equation of the 16-nodal surface; by taking the singular planes of one of the surfaces in a different order, the equation would present itself under one or other of the different forms

$$\alpha'\beta''\gamma = \alpha''\beta\gamma', \quad \alpha''\beta\gamma' = \alpha\beta'\gamma'', \quad \alpha\beta'\gamma'' = \alpha'\beta''\gamma,$$
$$\alpha'\beta\gamma'' = \alpha''\beta'\gamma, \quad \alpha''\beta'\gamma = \alpha\beta''\gamma', \quad \alpha\beta''\gamma' = \alpha'\beta\gamma''.$$

Cambridge, 9 December, 1878.

701.

ALGORITHM FOR THE CHARACTERISTICS OF THE TRIPLE ϑ-FUNCTIONS.

[From the *Journal für die reine und angewandte Mathematik* (Crelle), t. LXXXVII. (1878), pp. 165—169.]

THE characteristics of the triple ϑ-functions may be represented, the 28 odd characteristics by the binary symbols or duads, 12, ..., 78, and the even ones $\left(\text{other than } \begin{matrix}000\\000\end{matrix},\ = 0\right)$, say the 35 even characteristics, by the ternary symbols or triads 123, ..., 567: which triads may be regarded as abbreviations for the double tetrads 1238 . 4567, ..., 5678 . 1234, the 8 being always attached to the expressed triad. The correspondence of the symbols is given by the diagram:

upper line of characteristic

lower line of characteristic	000	100	010	110	001	101	011	111
000	0	236	345	137	467	156	124	257
100	237	67	136	12	157	48	256	35
010	245	127	23	68	134	357	15	47
110	126	13	78	145	356	25	46	234
001	567	146	125	247	45	17	38	26
101	147	58	246	34	16	123	27	367
011	135	347	14	57	28	36	167	456
111	346	24	56	235	37	267	457	18

Or, what is the same thing, it is

upper line of characteristic

	000	100	010	110	001	101	011	111
12				100				
13		110						
14			011					
15							010	
16					101			
17						001		
18								111
23			010					
24		111						
25						110		
26								001
27							101	
28					011			
34				101				
35								100
36						011		
37					111			
38							001	
45					001			
46							110	
47								010
48						100		
56			111					
57				011				
58		101						
67		100						
68				010				
78			110					

lower line of characteristic

	000	100	010	110	001	101	011	111
123						101		
124							000	
125			001					
126	110							
127		010						
134					010			
135	011							
136			100					
137				000				
145				110				
146		001						
147	101							
156						000		
157					100			
167							011	
234								110
235				111				
236		000						
237	100							
245	010							
246			101					
247				001				
256							100	
257								000
267						111		
345			000					
346	111							
347		011						
356					110			
357						010		
367								101
456								011
457							111	
467					000			
567	001							

lower line of characteristic

by means of which the two-line-characteristic is at once found when the duad or triad is given.

The new algorithm renders unnecessary the Table I. of Weber's memoir "Theorie der Abel'schen Functionen vom Geschlecht 3" (Berlin, 1876). In fact, the system of six pairs corresponding to an odd characteristic such as 12 is

$$13.23,\quad 14.24,\quad 15.25,\quad 16.26,\quad 17.27,\quad 18.28,$$

and that corresponding to an even characteristic such as 123 (=1238.4567) is

$$12.38,\quad 13.28,\quad 18.23,\quad 45.67,\quad 46.57,\quad 47.56:$$

so that all the (28 + 35 =) 63 systems can be at once formed.

The odd characteristics correspond to the bitangents of a quartic curve, and as regards these bitangents the notation is, in fact, the notation arising out of Hesse's investigations and explained Salmon's *Higher Plane Curves* (2nd Ed. 1873), pp. 222—225. It may be noticed that the geometrical symbols corresponding to the before-mentioned two systems are:

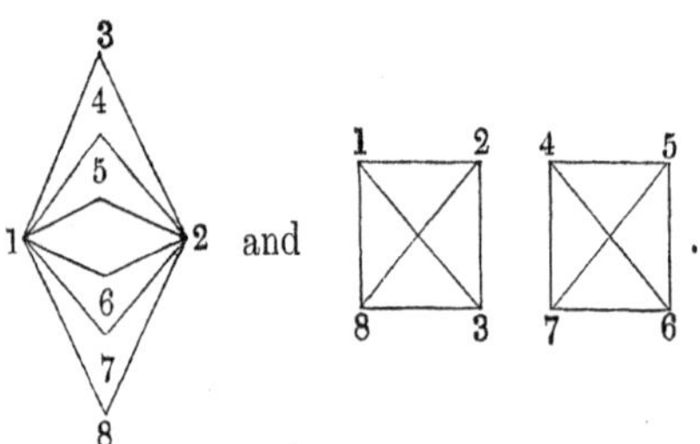

Hence, selecting out of the first system any two pairs, we have a symbol □: but selecting out of the second system any two pairs, we have a symbol which is either □ or ||||; so that in each case (Salmon, p. 224) the four bitangents are such that the eight points of contact lie on a conic.

The 28 bitangents of the general quartic curve

$$\sqrt{x_1\xi_1} + \sqrt{x_2\xi_2} + \sqrt{x_3\xi_3} = 0,$$

represented by the equations given by Weber, *l. c.*, pp. 100, 101, and taken in the order in which they are there written down, have for their duad-characteristics

18, 28, 38, 23, 13, 12, 48, 14, 58, 15, 68, 16, 78, 17, 24, 34, 25, 35, 26, 36, 27, 37, 67, 57, 56, 45, 46, 47

respectively. Taking out of any one of the 63 systems three pairs of bitangents at pleasure, these give rise to an equation of the curve of a form such as

$$\sqrt{x_1\xi_1} + \sqrt{x_2\xi_2} + \sqrt{x_3\xi_3} = 0,$$

and the whole number of the forms of equation is thus $= 1260$. The triads of pairs which enter into the same equation may be

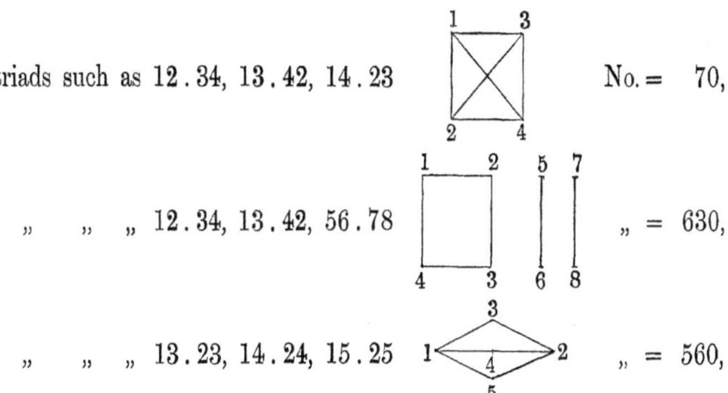

triads such as 12 . 34, 13 . 42, 14 . 23 No. = 70,

" " " 12 . 34, 13 . 42, 56 . 78 " = 630,

" " " 13 . 23, 14 . 24, 15 . 25 " = 560,

making the whole number $= 1260$, as already mentioned.

Cambridge, 7 *December*, 1878.

702.

ON THE TRIPLE ϑ-FUNCTIONS.

[From the *Journal für die reine und angewandte Mathematik* (Crelle), t. LXXXVII. (1878), pp. 190—198.]

A QUARTIC curve has the deficiency 3, and depends therefore on the triple ϑ-functions: and these, as functions of 3 arguments, should be connected with functions of 3 points on the curve; but it is easy to understand that it is possible, and may be convenient, to introduce a fourth point, and so regard them as functions of 4 points on the curve: thus in the circle, the functions $\cos u$, $\sin u$ may be regarded as functions of one point $\cos u = x$, $\sin u = y$, or as functions of two points,

$$\cos u = xx_1 + yy_1, \quad \sin u = xy_1 - x_1y.$$

And accordingly in Weber's memoir "Theorie der Abel'schen Functionen vom Geschlecht 3," (1876), see p. 156, the triple ϑ-functions are regarded as functions of 4 points on the curve: viz. it is in effect shown that (disregarding constant factors) each of the 64 functions is proportional to a determinant, the four lines of which are algebraical functions of the coordinates of the four points respectively: the form of this determinant being different according as the characteristic of the ϑ-function is odd or even, or say according as the ϑ-function is odd or even. But the geometrical signification of these formulæ requires to be developed.

A quartic curve may be touched in six points by a cubic curve: but (Hesse, 1855*) there are two kinds of such tangent cubics, according as the six points of contact are on a conic, or are not on a conic; say we have a conic hexad of points on the quartic, and a cubic hexad of points on the quartic. In either case, three points of the hexad may be assumed at pleasure; we can then in 28 different ways determine the remaining three points of the conic hexad, and in 36 different

* See the two memoirs "Ueber Determinanten und ihre Anwendung in der Geometrie" and "Ueber die Doppeltangenten der Curven vierter Ordnung," *Crelle*, t. XLIX. (1855).

ways the remaining three points of the cubic hexad: or what is the same thing, there are 28 systems of cubics touching in a conic hexad, and 36 systems of cubics touching in a cubic hexad. The condition in order that four points of the quartic curve may belong to a hexad (conic or cubic) is given by an equation $\Omega = 0$, where Ω is a determinant the four lines of which are algebraical functions of the coordinates of the four points respectively: but the form of such determinant is different according as the condition belongs to a conic hexad, or to a cubic hexad: we have thus 28 conic determinants and 36 cubic determinants, Ω; and the 64 ϑ-functions are proportional to constant multiples of these determinants; viz. the odd functions correspond to the conic determinants, and the even functions to the cubic determinants.

First, as to the conic hexads: the points of a conic hexad lie in a conic with the two points of contact of some one of the bitangents of the quartic curve: so that, given any three points of the hexad, these together with the two points of contact of the bitangent determine a conic which meets the quartic in the remaining three points of the hexad. Suppose that a, b, c, f, g, h are linear functions of the coordinates such that the equation of the quartic curve is

$$\sqrt{af} + \sqrt{bg} + \sqrt{ch} = 0;$$

then $a = 0$, $b = 0$, $c = 0$, $f = 0$, $g = 0$, $h = 0$ are six of the bitangents of the curve, and the bitangent $a = 0$ touches the curve at the two points of intersection of this line with the conic $bg - ch = 0$. The general equation of a conic through these two points $a = 0$, $bg - ch = 0$, may be written

$$bg - ch + a(Ax + By + Cz) = 0,$$

where for x, y, z we may if we please substitute any three of the six linear functions a, b, c, f, g, h, or any other linear functions of the coordinates (x, y, z): and the equation may also be written

$$af \pm (bg - ch) + a(Ax + By + Cz) = 0.$$

Adopting this latter form, and considering the intersections of the conic with the quartic, that is, considering the relation

$$\sqrt{af} + \sqrt{bg} + \sqrt{ch} = 0$$

as holding good, we have

$$af + bg - ch = -2\sqrt{afbg},$$

$$af - bg + ch = -2\sqrt{afch},$$

and we thus have at pleasure one or other of the two equations

$$-2\sqrt{afbg} + a(Ax + By + Cz) = 0,$$

$$-2\sqrt{afch} + a(Ax + By + Cz) = 0,$$

that is,

$$-2\sqrt{fbg} + \sqrt{a}(Ax + By + Cz) = 0,$$

$$-2\sqrt{fch} + \sqrt{a}(Ax + By + Cz) = 0.$$

Hence the condition in order that the four points (x_1, y_1, z_1), (x_2, y_2, z_2), (x_3, y_3, z_3), (x_4, y_4, z_4), assumed to be points of the quartic, may belong to the conic hexad, may be written

$$\begin{vmatrix} \sqrt{f_1b_1g_1}, & x_1\sqrt{a_1}, & y_1\sqrt{a_1}, & z_1\sqrt{a_1} \\ \sqrt{f_2b_2g_2}, & x_2\sqrt{a_2}, & y_2\sqrt{a_2}, & z_2\sqrt{a_2} \\ \sqrt{f_3b_3g_3}, & x_3\sqrt{a_3}, & y_3\sqrt{a_3}, & z_3\sqrt{a_3} \\ \sqrt{f_4b_4g_4}, & x_4\sqrt{a_4}, & y_4\sqrt{a_4}, & z_4\sqrt{a_4} \end{vmatrix} = 0, \text{ or } \begin{vmatrix} \sqrt{f_1c_1h_1}, & x_1\sqrt{a_1}, & y_1\sqrt{a_1}, & z_1\sqrt{a_1} \\ \sqrt{f_2c_2h_2}, & x_2\sqrt{a_2}, & y_2\sqrt{a_2}, & z_2\sqrt{a_2} \\ \sqrt{f_3c_3h_3}, & x_3\sqrt{a_3}, & y_3\sqrt{a_3}, & z_3\sqrt{a_3} \\ \sqrt{f_4c_4h_4}, & x_4\sqrt{a_4}, & y_4\sqrt{a_4}, & z_4\sqrt{a_4} \end{vmatrix} = 0,$$

where, as before, the x, y, z may be replaced by any three of the letters a, b, c, f, g, h, or by any other linear functions of (x, y, z): and, moreover, although in obtaining the condition we have used for the quartic the equation

$$\sqrt{af} + \sqrt{bg} + \sqrt{ch} = 0,$$

depending upon six bitangents, yet from the process itself it is clear that the condition can only depend upon the particular bitangent $a=0$: calling the condition $\Omega=0$, all the forms of condition which belong to the same bitangent $a=0$, will be essentially identical, that is, the several determinants Ω will differ only by constant factors; or disregarding these constant factors, we have for the bitangent $a=0$, a single determinant Ω, which may be taken to be any one of the determinants in question. And we have thus 28 determinants Ω, corresponding to the 28 bitangents respectively.

Coming now to the cubic hexads, Hesse showed that the equation of a quartic curve could be (and that in 36 different ways) expressed in the form, symmetrical determinant $=0$, or say

$$\begin{vmatrix} a, & h, & g, & l \\ h, & b, & f, & m \\ g, & f, & c, & n \\ l, & m, & n, & d \end{vmatrix} = 0,$$

where $(a, b, c, d, f, g, h, l, m, n)$ are linear functions of the coordinates; and from each of these forms he obtains the equation of a cubic

$$\begin{vmatrix} a, & h, & g, & l, & \alpha \\ h, & b, & f, & m, & \beta \\ g, & f, & c, & n, & \gamma \\ l, & m, & n, & d, & \delta \\ \alpha, & \beta, & \gamma, & \delta & \end{vmatrix} = 0,$$

containing the four constants α, β, γ, δ, or say the 3 ratios of these constants, touching the quartic in a cubic hexad of points: that the cubic does touch the quartic in six points appears, in fact, from Hesse's identity

$$\begin{vmatrix} a, & h, & g, & l, & \alpha \\ h, & b, & f, & m, & \beta \\ g, & f, & c, & n, & \gamma \\ l, & m, & n, & d, & \delta \\ \alpha, & \beta, & \gamma, & \delta & \end{vmatrix} \begin{vmatrix} a, & h, & g, & l, & \alpha' \\ h, & b, & f, & m, & \beta' \\ g, & f, & c, & n, & \gamma' \\ l, & m, & n, & d, & \delta' \\ \alpha', & \beta', & \gamma', & \delta' & \end{vmatrix} - \begin{vmatrix} a, & h, & g, & l, & \alpha \\ h, & b, & f, & m, & \beta \\ g, & f, & c, & n, & \gamma \\ l, & m, & n, & d, & \delta \\ \alpha', & \beta', & \gamma', & \delta' & \end{vmatrix}^2 = \begin{vmatrix} a, & h, & g, & l \\ h, & b, & f, & m \\ g, & f, & c, & n \\ l, & m, & n, & d \end{vmatrix} U,$$

where U is an easily calculated function of the second order in $a, b, c, d, f, g, h, l, m, n$, and also of the second order in the determinants $\alpha\beta' - \alpha'\beta$, etc.

We can obtain such a form of the equation of the quartic, from the before-mentioned equation

$$\sqrt{af} + \sqrt{bg} + \sqrt{ch} = 0,$$

viz. this equation gives

$$\begin{vmatrix} *, & h, & g, & a \\ h, & *, & f, & b \\ g, & f, & *, & c \\ a, & b, & c, & * \end{vmatrix} = 0,$$

which is of the required form, symmetrical determinant $= 0$; the equation is, in fact,

$$a^2f^2 + b^2g^2 + c^2h^2 - 2bcgh - 2cahf - 2abfg = 0,$$

which is the rationalised form of

$$\sqrt{af} + \sqrt{bg} + \sqrt{ch} = 0,$$

and we hence have the cubic

$$\begin{vmatrix} *, & h, & g, & a, & \alpha \\ h, & *, & f, & b, & \beta \\ g, & f, & *, & c, & \gamma \\ a, & b, & c, & *, & \delta \\ \alpha, & \beta, & \gamma, & \delta, & * \end{vmatrix} = 0,$$

the developed form of which is

$$\begin{aligned} &\alpha^2 bcf + \beta^2 cag + \gamma^2 abh + \delta^2 fgh \\ &-(a\beta\gamma + f\alpha\delta)(-af + bg + ch) \\ &-(b\gamma\alpha \ + g\beta\delta)(\ \ af - bg + ch) \\ &-(c\alpha\beta \ + h\gamma\delta)(\ \ af + bg - ch) = 0. \end{aligned}$$

Considering the intersections with the quartic

$$\sqrt{af} + \sqrt{bg} + \sqrt{ch} = 0,$$

we have

$$-af + bg + ch,\ af - bg + ch,\ af + bg - ch = -2\sqrt{bcgh},\ -2\sqrt{cahf},\ -2\sqrt{abfg},$$

and the equation thus becomes

$$(\alpha\sqrt{bcf} + \beta\sqrt{cag} + \gamma\sqrt{abh} + \delta\sqrt{fgh})^2 = 0\,;$$

viz. for the points of the cubic hexad we have

$$\alpha\sqrt{bcf}+\beta\sqrt{cag}+\gamma\sqrt{abh}+\delta\sqrt{fgh}=0,$$

and hence the condition in order that the four points (x_1, y_1, z_1), (x_2, y_2, z_2), (x_3, y_3, z_3), (x_4, y_4, z_4) may belong to the cubic hexad is

$$\begin{vmatrix} \sqrt{b_1c_1f_1}, & \sqrt{c_1a_1g_1}, & \sqrt{a_1b_1h_1}, & \sqrt{f_1g_1h_1} \\ \sqrt{b_2c_2f_2}, & \sqrt{c_2a_2g_2}, & \sqrt{a_2b_2h_2}, & \sqrt{f_2g_2h_2} \\ \sqrt{b_3c_3f_3}, & \sqrt{c_3a_3g_3}, & \sqrt{a_3b_3h_3}, & \sqrt{f_3g_3h_3} \\ \sqrt{b_4c_4f_4}, & \sqrt{c_4a_4g_4}, & \sqrt{a_4b_4h_4}, & \sqrt{f_4g_4h_4} \end{vmatrix}=0,$$

viz. we have thus the form of the determinant Ω which belongs to a cubic hexad.

It is to be observed that the equation

$$\sqrt{af}+\sqrt{bg}+\sqrt{ch}=0$$

remains unaltered by any of the interchanges a and f, b and g, c and h; but we thus obtain only two cubic hexads; those answering to the equations

$$\alpha\sqrt{bcf}+\beta\sqrt{cag}+\gamma\sqrt{abh}+\delta\sqrt{fgh}=0,$$

and

$$\alpha\sqrt{agh}+\beta\sqrt{bhf}+\gamma\sqrt{cfg}+\delta\sqrt{abc}=0,$$

which give distinct hexads. The whole number of ways in which the equation of the quartic can be expressed in a form such as

$$\sqrt{af}+\sqrt{bg}+\sqrt{ch}=0,$$

attending only to the pairs of bitangents, and disregarding the interchanges of the two bitangents of a pair, is $=1260$, and hence the number of forms for the determinant Ω of a cubic hexad is the double of this, $=2520$, which is $=36\times 70$: but the number of distinct hexads is $=36$, and thus there must be for each hexad, 70 equivalent forms.

To explain this, observe that every even characteristic except $\begin{matrix}000\\000\end{matrix}$, and every odd characteristic, can be (and that in 6 ways) expressed as a sum of two different odd characteristics; we have thus (see Weber's Table I.) a system of $(35+28=)$ 63 hexpairs; and selecting at pleasure any three pairs out of the same hexpair, we have a system of $(63\times 20=)$ 1260 tripairs; giving the 1260 representations of the quartic in a form such as

$$\sqrt{af}+\sqrt{bg}+\sqrt{ch}=0.$$

Each even characteristic $\left(\text{not excluding } \begin{matrix}000\\000\end{matrix}\right)$ can be in 56 different ways (Weber, p. 23) expressed as a sum of three different odd characteristics, and these are such that no two of them belong to the same pair, in any tripair; or we may say that each even characteristic gives rise to 56 hemi-tripairs. But a hemi-tripair can be in 5 different ways completed into a tripair; and we have thus, belonging to the same even characteristic ($56 \times 5 =$) 280 tripairs, which are however 70 tripairs each taken 4 times. A tripair contains in all ($2^3 =$) 8 hemi-tripairs, but these divide themselves into two sets each of 4 hemi-tripairs such that for each hemi-tripair of the first set the three characteristics have a given sum, and for each hemi-tripair of the second set the three characteristics have a different given sum. Hence considering the 70 tripairs corresponding as above to a given even characteristic, in any one of the 70 tripairs, there is a set of 4 hemi-tripairs such that in each of them the sum of the three characteristics is equal to the given even characteristic; and taking the bitangents f, g, h to correspond to any one of these hemi-tripairs, the bitangents which correspond to the other three hemi-tripairs will be b, c, f; c, a, g and a, b, h respectively; and we thus obtain from any one of these one and the same representation

$$\alpha\sqrt{bcf} + \beta\sqrt{cag} + \gamma\sqrt{abh} + \delta\sqrt{fgh} = 0$$

of the cubic hexad. And the 70 tripairs give thus the 70 representations of the same cubic hexad.

The whole number of hemi-tripairs is $36 \times 56 = 2016$: it may be remarked that there exists a system of 288 heptads, each of 7 odd characteristics such that selecting at pleasure any 3 characteristics out of the heptad, we obtain always a hemi-tripair: we have thus in all $288 \times 35 = 10080$ hemi-tripairs: this is $= 2016 \times 5$, or we have the 2016 hemi-tripairs each taken 5 times. Weber's Table II. exhibits 36 out of the 288 heptads.

I recall that in the algorithm derived from Hesse's theory the bitangents are represented by the duads 12, 13, ..., 78 formed with the eight figures 1, 2, 3, 4, 5, 6, 7, 8; these duads correspond to the odd characteristics as shown in the Table, and the table shows also triads corresponding to all the even characteristics except $\begin{matrix}000\\000\end{matrix}$.

Top line of characteristic.

Bottom line of characteristic.	000	100	010	110	001	101	011	111
000		236	345	137	467	156	124	257
100	237	67	136	12	157	48	256	35
010	245	127	23	68	134	357	15	47
110	126	13	78	145	356	25	46	234
001	567	146	125	247	45	17	38	26
101	147	58	246	34	16	123	27	367
011	135	347	14	57	17	36	167	456
111	346	24	56	235	37	267	457	18

See my "Algorithm of the triple ϑ-functions," *Crelle*, t. LXXXVII. p. 165, [701]. The $(35+28=)$ 63 hexpairs then are

35 hexpairs such as , say this is 1234.5678 or for shortness 567 (the 8 going always with the expressed triad): that is, 567 denotes the hexpair

$$12.34;\ 13.24;\ 14.23;\ 56.78;\ 57.68;\ 58.67:$$

and

28 hexpairs such as 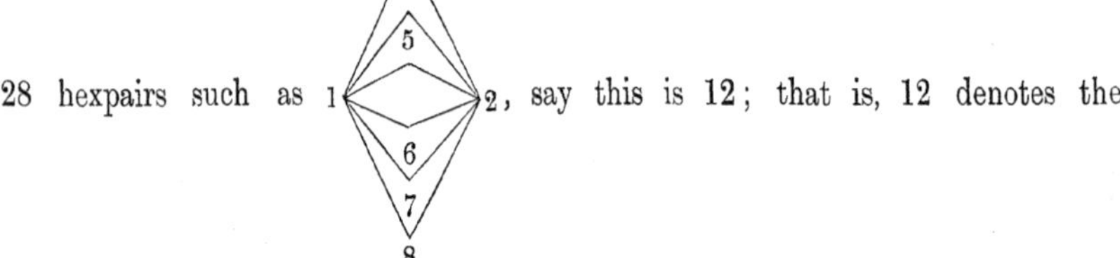

, say this is 12; that is, 12 denotes the hexpair

$$13.32;\ 14.42;\ 15.52;\ 16.62;\ 17.72;\ 18.82.$$

It is to be noticed that the odd characteristics, as represented by their duad symbols, can be added by the formulæ

$$12+23=13, \text{ etc.},$$

or, what is the same thing,

$$12+13+23=0,\ =\frac{000}{000}, \text{ etc.},$$

and

$$12+34=13+24=14+23=56+78=57+68=58+67=567, \text{ etc.}$$

Thus, referring to the table,

$$12+23=\ 13 \text{ means } \frac{110}{100}+\frac{010}{010}=\frac{100}{110},$$

and

$$12+34=567 \text{ means } \frac{110}{100}+\frac{110}{101}=\frac{000}{001},$$

which are right.

The 288 heptads are

8 heptads such as , say this is the heptad 1, denoting the seven duads 12, 13, 14, 15, 16, 17, 18:

and

280 heptads such as 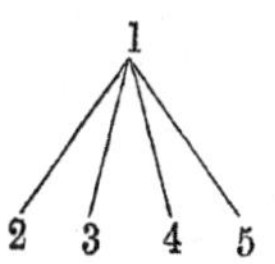

, say this is the heptad 1.678,

denoting the seven duads 12, 13, 14, 15, 67, 68, 78.

We hence see that the 2016 hemi-tripairs are:

280 hemi-tripairs 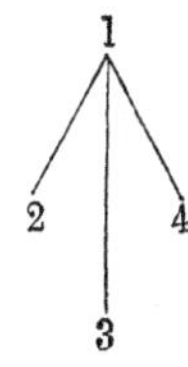

(I.), say this is 1.234, denoting the three duads 12, 13, 14:

1680 hemi-tripairs 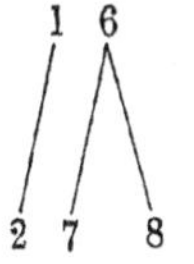

(II.), say this is 12 (6.78), denoting the three duads 12, 67, 68:

56 hemi-tripairs 

(III.), say this is 123, denoting the three duads 12, 13, 23:

2016.

We further see how each hemi-tripair may be completed into a tripair in 5 different ways: thus (I.) gives the 5 tripairs 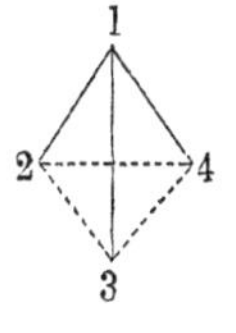

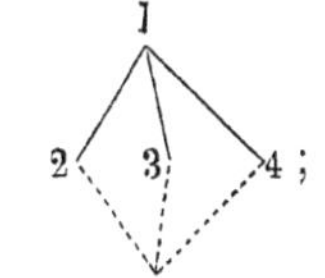

; (III.) gives the 5 tripairs 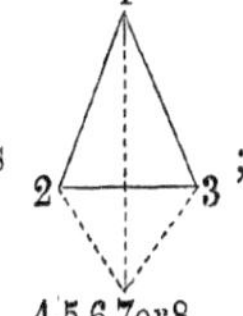

; while (II.) gives the 3 tripairs

1	34
	35
	or
2	45

6, 7, 8, 5, 4 or 3 and the 2 tripairs 1, 6, 2, 7, 8 and 2, 6, 1, 7, 8

To each even characteristic there belongs a system of 56 hemi-tripairs; thus for the characteristic $\begin{matrix}000\\000\end{matrix}$, the 56 hemi-tripairs are 123, that is, 12, 13, 23, etc.: whence the 70 tripairs are 1234, that is, 12.34; 13.24; 14.23, etc.; and in any such tripair, say in 1234, we have the set of four hemi-tripairs 123, 124, 134, 234, for each of which the sum of the three characteristics is

$$=\begin{matrix}000\\000\end{matrix}\left(12+23+13=\begin{matrix}000\\000\end{matrix},\text{ etc.}\right):$$

and the other set 1.234, 2.134, 3.124, 4.123, for each of which the sum of the three characteristics is

$$=567\left(12+13+14,\ =23+14,\ =567,\ =\begin{matrix}000\\001\end{matrix}\right).$$

To find the hemi-tripairs that belong to any other even characteristic; for instance, $\begin{matrix}000\\001\end{matrix}$, corresponds to 567: we have 4 such as 1.234; 24 such as (5.12)34; 4 such as 5.678; and 24 such as (1.56)78; in all 4+24+4+24, =56. The tripairs are the 2, 1234, 5678; 16 such as 54(123); 16 such as 15(678); 36 such as (5162)34.78; in all 2+16+16+36, =70; and in each of these it is easy to select the hemi-tripairs for which the sum of the 3 duads is =567.

Cambridge, 27 *December*, 1878.

703.

ON THE ADDITION OF THE DOUBLE ϑ-FUNCTIONS.

[From the *Journal für die reine und angewandte Mathematik* (Crelle), t. LXXXVIII. (1879), pp. 74—81.]

I ASSUME in general

$$\Theta = a-\theta \,.\, b-\theta \,.\, c-\theta \,.\, d-\theta \,.\, e-\theta \,.\, f-\theta,$$

and I consider the variables x, y, z, w, p, q, connected by the equations

$$\begin{vmatrix} 1, & 1, & 1, & 1, & 1, & 1 \\ x, & y, & z, & w, & p, & q \\ x^2, & y^2, & z^2, & w^2, & p^2, & q^2 \\ x^3, & y^3, & z^3, & w^3, & p^3, & q^3 \\ \sqrt{X}, & \sqrt{Y}, & \sqrt{Z}, & \sqrt{W}, & \sqrt{P}, & \sqrt{Q} \end{vmatrix} = 0,$$

equivalent to two independent equations, which in fact serve to determine p, q, or say the symmetrical functions $p+q$ and pq, in terms of x, y, z, w.

These equations, it is well known, constitute a particular integral of the differential equations

$$\frac{dx}{\sqrt{X}} + \frac{dy}{\sqrt{Y}} + \frac{dz}{\sqrt{Z}} + \frac{dw}{\sqrt{W}} + \frac{dp}{\sqrt{P}} + \frac{dq}{\sqrt{Q}} = 0,$$

$$\frac{x\,dx}{\sqrt{X}} + \frac{y\,dy}{\sqrt{Y}} + \frac{z\,dz}{\sqrt{Z}} + \frac{w\,dw}{\sqrt{W}} + \frac{p\,dp}{\sqrt{P}} + \frac{q\,dq}{\sqrt{Q}} = 0,$$

or what is the same thing, regarding p, q as arbitrary constants, they constitute the general integral of the differential equations

$$\frac{dx}{\sqrt{X}} + \frac{dy}{\sqrt{Y}} + \frac{dz}{\sqrt{Z}} + \frac{dw}{\sqrt{W}} = 0,$$

$$\frac{x\,dx}{\sqrt{X}} + \frac{y\,dy}{\sqrt{Y}} + \frac{z\,dz}{\sqrt{Z}} + \frac{w\,dw}{\sqrt{W}} = 0.$$

I attach the numbers 1, 2, 3, 4, 5, 6 to the variables x, y, z, w, p, q, respectively: and write

$$A_{12} = \sqrt{a-x\,.\,a-y}\,;\quad A_{34} = \sqrt{a-z\,.\,a-w}\,;\quad A_{56} = \sqrt{a-p\,.\,a-q}\,;$$

$\vdots$

(six equations),

$$AB_{12} = \frac{1}{x-y}\{\sqrt{a-x\,.\,b-x\,.\,f-x\,.\,c-y\,.\,d-y\,.\,e-y} - \sqrt{a-y\,.\,b-y\,.\,f-y\,.\,c-x\,.\,d-x\,.\,e-x}\}; \text{ etc.}$$

$\vdots$

(ten equations),

where it is to be borne in mind that AB is an abbreviation for $ABF\,.\,CDE$, and so in other cases, the letter F belonging always to the expressed duad: there are thus in all the sixteen functions A, B, C, D, E, F, AB, AC, AD, AE, BC, BD, BE, CD, CE, DE, these being functions of x and y, of z and w, and of p and q, according as the suffix is 12, 34, or 56.

It is to be shown that the 16 functions A_{56}, AB_{56} of p and q can be by means of the given equations expressed as proportional to rational and integral functions of the 16 functions A_{12}, AB_{12}, A_{34}, AB_{34} of x and y, and of z and w respectively: and it is clear that in so expressing them we have in effect the solution of the problem of the addition of the double ϑ-functions.

I use when convenient the abbreviated notations

$$a-x = \mathrm{a}_1,\quad a-y = \mathrm{a}_2,\quad \text{etc.},$$

$$b-x = \mathrm{b}_1,\quad \text{etc.},$$

$$\theta_{12} = x-y,\quad \theta_{34} = z-w,\quad \theta_{56} = p-q\,;$$

we have of course

$$X = \mathrm{a}_1\mathrm{b}_1\mathrm{c}_1\mathrm{d}_1\mathrm{e}_1\mathrm{f}_1,$$

$$A_{12} = \sqrt{\mathrm{a}_1\mathrm{a}_2},$$

$$AB_{12} = \frac{1}{\theta_{12}}\{\sqrt{\mathrm{a}_1\mathrm{b}_1\mathrm{f}_1\mathrm{c}_2\mathrm{d}_2\mathrm{e}_2} - \sqrt{\mathrm{a}_2\mathrm{b}_2\mathrm{f}_2\mathrm{c}_1\mathrm{d}_1\mathrm{e}_1}\},\ \text{etc.}$$

Proceeding to the investigation, the equations between the variables are obviously those obtained by the elimination of the arbitrary multipliers α, β, γ, δ, ϵ from the six equations obtained from

$$\alpha\theta^3 + \beta\theta^2 + \gamma\theta + \delta = \epsilon\sqrt{\Theta},$$

by writing therein for θ the values x, y, z, w, p, q successively; we may consider the four equations

$$\alpha x^3 + \beta x^2 + \gamma x + \delta = \epsilon\sqrt{X},$$

$$\alpha y^3 + \beta y^2 + \gamma y + \delta = \epsilon\sqrt{Y},$$

$$\alpha z^3 + \beta z^2 + \gamma z + \delta = \epsilon\sqrt{Z},$$

$$\alpha w^3 + \beta w^2 + \gamma w + \delta = \epsilon\sqrt{W},$$

as serving to determine the ratios $\alpha : \beta : \gamma : \delta : \epsilon$ in terms of x, y, z, w; and we have then for the determination of p, q the remaining two equations

$$\alpha p^3 + \beta p^2 + \gamma p + \delta = \epsilon \sqrt{P},$$

$$\alpha q^3 + \beta q^2 + \gamma q + \delta = \epsilon \sqrt{Q},$$

which two equations may be replaced by the identity

$$(\alpha\theta^3 + \beta\theta^2 + \gamma\theta + \delta)^2 - \epsilon^2\Theta = \alpha^2 - \epsilon^2 \,.\, \theta - x \,.\, \theta - y \,.\, \theta - z \,.\, \theta - w \,.\, \theta - p \,.\, \theta - q.$$

Writing herein $\theta =$ any one of the values a, b, c, d, e, f, for instance $\theta = a$, and taking the square root of each side, we have

$$\alpha a^3 + \beta a^2 + \gamma a + \delta = \sqrt{\alpha^2 - \epsilon^2}\, \sqrt{a - x \,.\, a - y}\, \sqrt{a - z \,.\, a - w}\, \sqrt{a - p \,.\, a - q},$$

or as this may be written

$$\alpha a^3 + \beta a^2 + \gamma a + \delta = \sqrt{\alpha^2 - \epsilon^2}\, A_{12} \,.\, A_{34} \,.\, A_{56},$$

which equation when reduced gives the proportional value of A_{56}.

For the reduction we require the value of $\alpha a^3 + \beta a^2 + \gamma a + \delta$: calling this for the moment Ω, we join to the four equations a fifth equation

$$\alpha a^3 + \beta a^2 + \gamma a + \delta = \Omega.$$

Eliminating α, β, γ, δ, we find

$$\begin{vmatrix} x^3, & x^2, & x, & 1, & \epsilon\sqrt{X} \\ y^3, & y^2, & y, & 1, & \epsilon\sqrt{Y} \\ z^3, & z^2, & z, & 1, & \epsilon\sqrt{Z} \\ w^3, & w^2, & w, & 1, & \epsilon\sqrt{W} \\ a^3, & a^2, & a, & 1, & \Omega \end{vmatrix} = 0,$$

or, what is the same thing,

$$\Omega \begin{vmatrix} x^3, & x^2, & x, & 1 \\ y^3, & y^2, & y, & 1 \\ z^3, & z^2, & z, & 1 \\ w^3, & w^2, & w, & 1 \end{vmatrix} + \epsilon \begin{vmatrix} \sqrt{X}, & x^3, & x^2, & x, & 1 \\ \sqrt{Y}, & y^3, & y^2, & y, & 1 \\ \sqrt{Z}, & z^3, & z^2, & z, & 1 \\ \sqrt{W}, & w^3, & w^2, & w, & 1 \\ & a^3, & a^2, & a, & 1 \end{vmatrix} = 0;$$

viz. this is

$$\begin{aligned} \Omega \,.\, x-y \,.\, x-z \,.\, x-w \,.\, y-z \,.\, y-w \,.\, z-w = -\epsilon\, \{ & \sqrt{X} \,.\, y-z \,.\, y-w \,.\, y-a \,.\, z-w \,.\, z-a \,.\, w-a \\ & + \sqrt{Y} \,.\, z-w \,.\, z-a \,.\, z-x \,.\, w-a \,.\, w-x \,.\, a-x \\ & + \sqrt{Z} \,.\, w-a \,.\, w-x \,.\, w-y \,.\, a-x \,.\, a-y \,.\, x-y \\ & + \sqrt{W} \,.\, a-x \,.\, a-y \,.\, a-z \,.\, x-y \,.\, x-z \,.\, y-z \}, \end{aligned}$$

or as it may be written

$$\Omega\,.\,x-z\,.\,x-w\,.\,y-z\,.\,y-w=\frac{\epsilon\,.\,a-z\,.\,a-w}{x-y}\{y-z\,.\,y-w\,.\,a-y\,.\,\sqrt{X}-x-z\,.\,x-w\,.\,a-x\,.\sqrt{Y}\}$$

$$+\frac{\epsilon\,.\,a-x\,.\,a-y}{z-w}\{w-x\,.\,w-y\,.\,a-w\,.\,\sqrt{Z}-z-x\,.\,z-y\,.\,a-z\,.\,\sqrt{W}\},$$

an equation for the determination of Ω.

Consider first the expression which multiplies $\epsilon\,.\,a-z\,.\,a-w$; this is

$$=\frac{1}{\theta_{12}}\{y-z\,.\,y-w\,.\,\mathrm{a}_2\sqrt{X}-x-z\,.\,x-w\,.\,\mathrm{a}_1\sqrt{Y}\};$$

we have

$$BE_{12}=\frac{1}{\theta_{12}}\{\sqrt{\mathrm{b_1e_1f_1a_2c_2d_2}}-\sqrt{\mathrm{b_2e_2f_2a_1c_1d_1}}\},$$

and multiplying this by

$$A_{12}\,.\,C_{12}\,.\,D_{12},\ =\sqrt{\mathrm{a_1c_1d_1a_2c_2d_2}},$$

we derive

$$BE_{12}\,.\,C_{12}\,.\,D_{12}\,.\,A_{12}=\frac{1}{\theta_{12}}\{\mathrm{c_2d_2a_2}\sqrt{X}-\mathrm{c_1d_1a_1}\sqrt{Y}\},$$

and similarly two other equations; the system may be written

$$BE\,.\,C\,.\,D\,.\,A=\frac{1}{\theta_{12}}\{\mathrm{c_2d_2a_2}\sqrt{X}-\mathrm{c_1d_1a_1}\sqrt{Y}\},$$

$$CE\,.\,D\,.\,B\,.\,A=\ ,,\ \{\mathrm{d_2b_2}\,,,\ ,,\ -\mathrm{d_1b_1}\,,,\ ,,\ \},$$

$$DE\,.\,B\,.\,C\,.\,A=\ ,,\ \{\mathrm{b_2c_2}\,,,\ ,,\ -\mathrm{b_1c_1}\,,,\ ,,\ \},$$

the suffixes on the left-hand side being always 12. The letters b, c, d which enter cyclically into these equations are any three of the five letters other than a; the remaining two letters e and f enter symmetrically, for BE is a mere abbreviation for the double triad $BEF\,.\,ACD$; and the like for CE, and DE.

Multiplying these equations by

$$\frac{b-z\,.\,b-w}{b-c\,.\,b-d},\quad \frac{c-z\,.\,c-w}{c-d\,.\,c-b},\quad \frac{d-z\,.\,d-w}{d-b\,.\,d-c},$$

respectively, and then adding, the right-hand side becomes

$$=\frac{1}{\theta_{12}}\{y-z\,.\,y-w\,.\,\mathrm{a}_2\sqrt{X}-x-z\,.\,x-w\,.\,\mathrm{a}_1\sqrt{Y}\}.$$

Writing

$$\frac{b-z\,.\,b-w}{b-c\,.\,b-d}=\frac{-1}{c-d\,.\,d-b\,.\,b-c}\,.\,c-d\,.\,B_{34}^{\,2},\ \text{etc.},$$

the left-hand side becomes

$$=\frac{-A_{12}}{c-d.d-b.b-c}\{c-d.B_{34}{}^2.BE_{12}.C_{12}.D_{12}+d-b.C_{34}{}^2.CE_{12}.D_{12}.B_{12}+b-c.D_{34}{}^2.DE_{12}.B_{12}.C_{12}\},$$

which for shortness may be written

$$=\frac{-A_{12}}{c-d.d-b.b-c}\Sigma\{c-d.B_{34}{}^2.BE_{12}.C_{12}.D_{12}\},$$

the summation referring to the three terms obtained by the cyclical interchange of the letters b, c, d. The result thus is

$$\frac{1}{\theta_{12}}\{y-z.y-w.\mathrm{a}_2\sqrt{X}-x-z.x-w.\mathrm{a}_1\sqrt{Y}\}$$

$$=\frac{-A_{12}}{c-d.d-b.b-c}\Sigma\{c-d.B_{34}{}^2.BE_{12}.C_{12}.D_{12}\}.$$

Interchanging x, y with z, w respectively, we have of course to interchange the suffixes 1, 2 and 3, 4; we thus find

$$\frac{1}{\theta_{34}}\{w-x.w-y.\mathrm{a}_4\sqrt{Z}-z-x.z-y.\mathrm{a}_3\sqrt{W}\}$$

$$=\frac{-A_{34}}{c-d.d-b.b-c}\Sigma\{c-d.B_{12}{}^2.BE_{34}.C_{34}.D_{34}\},$$

and we hence find the value of $\Omega.x-z.x-w.y-z.y-w$. But Ω, $=\alpha a^3+\beta a^2+\gamma a+\delta$, is $=\sqrt{\alpha^2-\epsilon^2}.A_{12}.A_{34}.A_{56}$: the resulting equation divides by $A_{12}.A_{34}$: throwing out this factor, we have

$$-\frac{\sqrt{\alpha^2-\epsilon^2}}{\epsilon}(x-z.x-w.y-z.y-w)(c-d.d-b.b-c)A_{56}$$

$$=A_{34}\Sigma\{c-d.B_{34}{}^2.BE_{12}.C_{12}.D_{12}\}+A_{12}\Sigma\{c-d.B_{12}{}^2.BE_{34}.C_{34}.D_{34}\},$$

where, as before, the summations refer to the three terms obtained by the cyclical interchange of the letters b, c, d; these being any three of the five letters other than a; and the remaining two letters e, f enter into the formula symmetrically. The formula gives thus for A_{56} ten values which are of course equal to each other.

Writing for a each letter in succession, we obtain formulæ for each of the six single-letter functions A_{56} of p and q; and the factor

$$-\frac{\sqrt{\alpha^2-\epsilon^2}}{\epsilon}(x-z.x-w.y-z.y-w)$$

is the same in all the formulæ.

We require further the expressions for the double-letter functions of p, q. Considering for example the function DE_{56}, which is

$$=\frac{1}{\theta_{56}}\{\sqrt{\mathrm{d}_5\mathrm{e}_5\mathrm{f}_5\mathrm{a}_6\mathrm{b}_6\mathrm{c}_6}-\sqrt{\mathrm{d}_6\mathrm{e}_6\mathrm{f}_6\mathrm{a}_5\mathrm{b}_5\mathrm{c}_5}\},$$

then multiplying by

$$A_{56}\,.\,B_{56}\,.\,C_{56},\ =\sqrt{\mathfrak{a}_5\mathfrak{b}_5\mathfrak{c}_5\mathfrak{a}_6\mathfrak{b}_6\mathfrak{c}_6},$$

we have

$$DE_{56}\,.\,A_{56}\,.\,B_{56}\,.\,C_{56}=\frac{1}{\theta_{56}}\{\mathfrak{a}_6\mathfrak{b}_6\mathfrak{c}_6\sqrt{P}-\mathfrak{a}_5\mathfrak{b}_5\mathfrak{c}_5\sqrt{Q}\},$$

$$=\frac{1}{p-q}\{a-q\,.\,b-q\,.\,c-q\,.\sqrt{P}-a-p\,.\,b-p\,.\,c-p\,.\sqrt{Q}\},$$

or recollecting that $\epsilon\sqrt{P}$, $\epsilon\sqrt{Q}$ are $=\alpha p^3+\beta p^2+\gamma p+\delta$ and $\alpha q^3+\beta q^2+\gamma q+\delta$ respectively, this is

$$\epsilon\,.\,DE_{56}\,.\,A_{56}\,.\,B_{56}\,.\,C_{56}$$

$$=\frac{1}{p-q}\{a-q\,.\,b-q\,.\,c-q\,.\,(\alpha p^3+\beta p^2+\gamma p+\delta)-a-p\,.\,b-p\,.\,c-p\,.\,(\alpha q^3+\beta q^2+\gamma q+\delta)\}.$$

Using the well-known identity

$$\begin{aligned}\alpha p^3+\beta p^2+\gamma p+\delta=&\ \alpha a^3+\beta a^2+\gamma a+\delta\,.\,\frac{b-p\,.\,c-p\,.\,d-p}{b-a\,.\,c-a\,.\,d-a}\\&+\alpha b^3+\beta b^2+\gamma b+\delta\,.\,\frac{c-p\,.\,d-p\,.\,a-p}{c-b\,.\,d-b\,.\,a-b}\\&+\alpha c^3+\beta c^2+\gamma c+\delta\,.\,\frac{d-p\,.\,a-p\,.\,b-p}{d-c\,.\,a-c\,.\,b-c}\\&+\alpha d^3+\beta d^2+\gamma d+\delta\,.\,\frac{a-p\,.\,b-p\,.\,c-p}{a-d\,.\,b-d\,.\,c-d},\end{aligned}$$

and the like expression for $\alpha q^3+\beta q^2+\gamma q+\delta$, there will be on the right-hand side terms involving

$$\alpha a^3+\beta a^2+\gamma a+\delta,\quad \alpha b^3+\beta b^2+\gamma b+\delta,\quad \alpha c^3+\beta c^2+\gamma c+\delta:$$

but the term in $\alpha d^3+\beta d^2+\gamma d+\delta$ will disappear of itself.

The term in $\alpha a^3+\beta a^2+\gamma a+\delta$ is

$$\frac{1}{p-q}\,\frac{\alpha a^3+\beta a^2+\gamma a+\delta}{b-a\,.\,c-a\,.\,d-a}\,.\,b-q\,.\,c-q\,.\,b-p\,.\,c-p\,.\,(a-q\,.\,d-p-a-p\,.\,d-q),$$

where the expression in ( ) is $=d-a\,.\,p-q$: hence the term is

$$=\frac{\alpha a^3+\beta a^2+\gamma a+\delta}{b-a\,.\,c-a}\,.\,b-q\,.\,c-q\,.\,b-p\,.\,c-p,$$

which is

$$=\frac{\alpha a^3+\beta a^2+\gamma a+\delta}{b-a\,.\,c-a}\,B_{56}{}^2\,.\,C_{56}{}^2.$$

Forming the two other like terms, the equation is

$$\begin{aligned}\epsilon\,.\,DE_{56}\,.\,A_{56}\,.\,B_{56}\,.\,C_{56}=&\ \frac{\alpha a^3+\beta a^2+\gamma a+\delta}{b-a\,.\,c-a}\,B_{56}{}^2\,.\,C_{56}{}^2\\&+\frac{\alpha b^3+\beta b^2+\gamma b+\delta}{c-b\,.\,a-b}\,C_{56}{}^2\,.\,A_{56}{}^2\\&+\frac{\alpha c^3+\beta c^2+\gamma c+\delta}{a-c\,.\,b-c}\,A_{56}{}^2\,.\,B_{56}{}^2.\end{aligned}$$

But the expressions

$$\alpha a^3+\beta a^2+\gamma a+\delta,\quad \alpha b^3+\beta b^2+\gamma b+\delta,\quad \alpha c^3+\beta c^2+\gamma c+\delta,$$

are

$$=\sqrt{\alpha^2-\epsilon^2}\,A_{12}\,.\,A_{34}\,.\,A_{56},\quad \sqrt{\alpha^2-\epsilon^2}\,B_{12}\,.\,B_{34}\,.\,B_{56},\quad \sqrt{\alpha^2-\epsilon^2}\,C_{12}\,.\,C_{34}\,.\,C_{56},$$

respectively: the whole equation thus divides by $A_{56}\,.\,B_{56}\,.\,C_{56}$; throwing out this factor, and then multiplying each side by $-\dfrac{\sqrt{\alpha^2-\epsilon^2}}{\epsilon}$, we find

$$-\frac{\sqrt{\alpha^2-\epsilon^2}}{\epsilon}DE_{56}=\frac{1}{b-c\,.\,c-a\,.\,a-b}\left(-\frac{\sqrt{\alpha^2-\epsilon^2}}{\epsilon}\right)^2\{\;b-c\,.\,A_{12}\,.\,A_{34}\,.\,B_{56}\,.\,C_{56}$$
$$+\,c-a\,.\,B_{12}\,.\,B_{34}\,.\,C_{56}\,.\,A_{56}$$
$$+\,a-b\,.\,C_{12}\,.\,C_{34}\,.\,A_{56}\,.\,B_{56}\},$$

in which formula if we imagine

$$-\frac{\sqrt{\alpha^2-\epsilon^2}}{\epsilon}A_{56},\quad -\frac{\sqrt{\alpha^2-\epsilon^2}}{\epsilon}B_{56},\quad -\frac{\sqrt{\alpha^2-\epsilon^2}}{\epsilon}C_{56}$$

each replaced by its value in terms of the xy- and zw-functions, we have an equation of the form

$$-\frac{\sqrt{\alpha^2-\epsilon^2}}{\epsilon}(x-z\,.\,x-w\,.\,y-z\,.\,y-w)\,DE_{56}=\frac{1}{x-z\,.\,x-w\,.\,y-z\,.\,y-w}M,$$

where M is a given rational and integral function of the 16 and 16 functions A_{12}, AB_{12} and A_{34}, AB_{34} of x and y and of z and w respectively. The factor

$$-\frac{\sqrt{\alpha^2-\epsilon^2}}{\epsilon}(x-z\,.\,x-w\,.\,y-z\,.\,y-w)$$

is retained on the left-hand side as being the same factor which enters into the equations for A_{56}, etc.: but on the right-hand side $x-z\,.\,x-w\,.\,y-z\,.\,y-w$ should be expressed in terms of the xy- and zw-functions. This can be done by means of the identity

$$x-z\,.\,x-w\,.\,y-z\,.\,y-w=\Sigma\frac{\begin{vmatrix}1, & x+y, & xy\\ 1, & z+w, & zw\\ 1, & a+b, & ab\end{vmatrix}\begin{vmatrix}1, & x+y, & xy\\ 1, & z+w, & zw\\ 1, & a+c, & ac\end{vmatrix}}{a-b\,.\,a-c},$$

where the summation refers to the three terms obtained by the cyclical interchange of the letters a, b, c. The first determinant, multiplied by $a-b$, is in fact

$$=\begin{vmatrix}a-z\,.\,a-w, & a-x\,.\,a-y\\ b-z\,.\,b-w, & b-x\,.\,b-y\end{vmatrix},$$

and the second determinant, multiplied by $a-c$, is

$$=\begin{vmatrix}a-z\,.\,a-w, & a-x\,.\,a-y\\ c-z\,.\,c-w, & c-x\,.\,c-y\end{vmatrix},$$

so that the formula may also be written

$$x-z\,.\,x-w\,.\,y-z\,.\,y-w=\Sigma\frac{\begin{vmatrix} a-z\,.\,a-w, & a-x\,.\,a-y \\ b-z\,.\,b-w, & b-x\,.\,b-y \end{vmatrix}\,.\,\begin{vmatrix} a-z\,.\,a-w, & a-x\,.\,a-y \\ c-z\,.\,c-w, & c-x\,.\,c-y \end{vmatrix}}{(a-b)^2\,(a-c)^2};$$

or, what is the same thing, it is

$$x-z\,.\,x-w\,.\,y-z\,.\,y-w=\Sigma\frac{(A_{34}{}^2B_{12}{}^2-A_{12}{}^2B_{34}{}^2)\,(A_{34}{}^2C_{12}{}^2-A_{12}{}^2C_{34}{}^2)}{(a-b)^2\,(a-c)^2},$$

which is the required expression for $x-z\,.\,x-w\,.\,y-z\,.\,y-w$; the letters a, b, c, which enter into the formula, are any three of the six letters.

As regards the verification of the identity, observe that it may be written

$$x-z\,.\,x-w\,.\,y-z\,.\,y-w=\Sigma\frac{\{L+M(a+b)+Nab\}\,\{L+M(a+c)+Nac\}}{a-b\,.\,a-c},$$

where L, M, N are

$$=(x+y)\,zw-(z+w)\,xy,\quad xy-zw,\quad\text{and}\quad z+w-x-y:$$

this is readily reduced to

$$x-z\,.\,x-w\,.\,y-z\,.\,y-w=M^2-NL,$$

which can be at once verified.

Cambridge, 12th March, 1879.

I take the opportunity of remarking that, in the double-letter formulæ, the sign of the second term is, not as I have in general written it −, but is +,

$$AB=\frac{1}{x-y}\{\sqrt{\mathrm{abfc_1d_1e_1}}+\sqrt{\mathrm{a_1b_1f_1cde}}\},\text{ etc.}$$

In fact, introducing a factor ω which is a function of x and y, the odd and even ϑ-functions are $=\omega\sqrt{\mathrm{aa_1}}$, etc., and

$$\frac{\omega}{x-y}\{\sqrt{\mathrm{abfc_1d_1e_1}}+\sqrt{\mathrm{a_1b_1f_1cde}}\},\text{ etc.},$$

respectively; ω is a function which on the interchange of x, y changes only its sign; and this being so, then when x and y are interchanged, each single-letter function changes its sign, and each double-letter function remains unaltered.

Cambridge, 29th July, 1879.

704.

A MEMOIR ON THE SINGLE AND DOUBLE THETA-FUNCTIONS.

[From the *Philosophical Transactions of the Royal Society of London*, vol. 171, Part III., (1880), pp. 897—1002. Received November 14,—Read November 28, 1879.]

THE Theta-Functions, although arising historically from the Elliptic Functions, may be considered as in order of simplicity preceding these, and connecting themselves directly with the exponential function (e^x or) exp. x; viz. they may be defined each of them as a sum of a series of exponentials, singly infinite in the case of the single functions, doubly infinite in the case of the double functions; and so on. The number of the single functions is $=4$; and the quotients of these, or say three of them each divided by the fourth, are the elliptic functions sn, cn, dn; the number of the double functions is ($4^2=$) 16; and the quotients of these, or say fifteen of them each divided by the sixteenth, are the hyper-elliptic functions of two arguments depending on the square root of a sextic function. Generally, the number of the p-tuple theta-functions is $=4^p$; and the quotients of these, or say all but one of them each divided by the remaining function, are the Abelian functions of p arguments depending on the irrational function y defined by the equation $F(x, y)=0$ of a curve of deficiency p. If, instead of connecting the ratios of the functions with a plane curve, we consider the functions themselves as coordinates of a point in a space of (4^p-1) dimensions, then we have the single functions as the four coordinates of a point on a quadri-quadric curve (one-fold locus) in ordinary space; and the double functions as the sixteen coordinates of a point on a quadri-quadric two-fold locus in 15-dimensional space, the deficiency of this two-fold locus being of course $=2$.

The investigations contained in the First Part of the present Memoir, although for simplicity of notation exhibited only in regard to the double functions are, in fact, applicable to the general case of the p-tuple functions; but in the main the

Memoir relates only to the single and double functions, and the title has been given to it accordingly. The investigations just referred to extend to the single functions; and there is, it seems to me, an advantage in carrying on the two theories simultaneously up to and inclusive of the establishment of what I call the Product-theorem: this is a natural point of separation for the theories of the single and the double functions respectively. The ulterior developments of the two theories are indeed closely analogous to each other; but on the one hand the course of the single theory would be only with difficulty perceptible in the greater complexity of the double theory; and on the other hand we need the single theory as a guide for the course of the double theory.

I accordingly stop to point out in a general manner the course of the single theory, and, in connexion with it but more briefly, that of the double theory; and I then, in the Second and the Third Parts respectively, consider in detail the two theories separately; first, that of the single functions, and then that of the double functions. The paragraphs of the Memoir are numbered consecutively.

The definition adopted for the theta-functions differs somewhat from that which is ordinarily used.

The earlier memoirs on the double theta-functions are the well-known ones:—

Rosenhain, "Mémoire sur les fonctions de deux variables et à quatre périodes, qui sont les inverses des intégrales ultra-elliptiques de la première classe." [1846.] Paris: *Mém. Savans Étrang.*, t. XI. (1851), pp. 361—468.

Göpel, "Theoriæ transcendentium Abelianarum primi ordinis adumbratio levis," *Crelle*, t. XXXV. (1847), pp. 277—312.

My first paper—Cayley, "On the Double θ-Functions in connexion with a 16-nodal Surface," *Crelle-Borchardt*, t. LXXXIII. (1877), pp. 210—219, [662]—was founded directly upon these, and was immediately followed by Dr Borchardt's paper,

Borchardt, "Ueber die Darstellung der *Kummer*sche Fläche vierter Ordnung mit sechzehn Knotenpunkten durch die *Göpel*sche biquadratische Relation zwischen vier Thetafunctionen mit zwei Variabeln," *Ditto*, pp. 234—244.

My other later papers, [663, 664, 665, 697, 703], are contained in the same Journal.

FIRST PART.—INTRODUCTORY.

Definition of the theta-functions.

1. The p-tuple functions depend upon $\frac{1}{2}p(p-1)$ parameters which are the coefficients of a quadric function of p ultimately disappearing integers, upon p arguments, and upon $2p$ characters, each $=0$ or 1, which form the characteristic of the 4^p functions; but it will be sufficient to write down the formulæ in the case $p=2$.

As already mentioned, the adopted definition differs somewhat from that which is ordinarily used. I use, as will be seen, a quadric function $\frac{1}{4}(a,\ h,\ b\between m,\ n)^2$ with *even* integer values of m, n, instead of $(a, h, b \between m, n)^2$ with even or odd values; and I write the other term $\frac{1}{2}\pi i\,(mu+nv)$, instead of $mu+nv$; this comes to affecting the arguments u, v with a factor πi, so that the quarter-periods (instead of being πi) are made to be $=1$.

2. We write

$$\begin{pmatrix} m, & n \\ u, & v \end{pmatrix} = \tfrac{1}{4}(a, h, b \between m, n)^2 + \tfrac{1}{2}\pi i\,(mu+nv),$$

and in like manner

$$\begin{pmatrix} m+\alpha, & n+\beta \\ u+\gamma, & v+\delta \end{pmatrix} = \tfrac{1}{4}(a, h, b \between m+\alpha, n+\beta)^2 + \tfrac{1}{2}\pi i\,\{(m+\alpha)(u+\gamma)+(n+\beta)(v+\delta)\},$$

and prefixing to either of these the functional symbol exp. we have the exponential of the function in question, that is, e with the function as an exponent.

We then write, as the definition of the double theta-functions,

$$\vartheta\begin{pmatrix} \alpha, & \beta \\ \gamma, & \delta \end{pmatrix}(u, v) = \Sigma \text{ exp.} \begin{pmatrix} m+\alpha, & n+\beta \\ u+\gamma, & v+\delta \end{pmatrix},$$

where the summation extends to all positive and negative even integer values (zero included) of m and n respectively: α, β, γ, δ might denote any quantities whatever, but for the theta-functions they are regarded as denoting positive or negative integers; this being so, it will appear that the only effect of altering each or any of them by an even integer is to reverse (it may be) the sign of the function; and the distinct functions are consequently the $(4^2=)$ 16 functions obtained by giving to each of the quantities α, β, γ, δ the two values 0 and 1 successively.

3. We thus have the double theta-functions, depending on the parameters (a, h, b) which determine the quadric function $(a, h, b \between m, n)^2$ of the disappearing even integers (m, n), and on the two arguments (u, v): in the symbol $\begin{pmatrix} \alpha, & \beta \\ \gamma, & \delta \end{pmatrix}$, which is called the characteristic, the characters α, β, γ, δ are each of them $=0$ or 1; and we thus have the 16 functions.

The parameters (a, h, b) may be real or imaginary, but they must be such that reducing each of them to its real part the resulting function $(\ast \between m, n)^2$ is invariable in its sign, and negative for all real values of m and n: this is, in fact, the condition for the convergency of the series which give the values of the theta-functions.

4. The characteristic $\begin{pmatrix} \alpha, & \beta \\ \gamma, & \delta \end{pmatrix}$ is said to be even or odd according as the sum $\alpha\gamma+\beta\delta$ is even or odd.

Allied functions.

5. As already remarked, the definition of

$$\vartheta\begin{pmatrix} \alpha, & \beta \\ \gamma, & \delta \end{pmatrix}(u, v)$$

is not restricted to the case where the α, β, γ, δ represent integers, and there is actually occasion to consider functions of this form where they are not integers: in particular, α, β may be either or each of them of the form, integer $+\frac{1}{2}$. But the functions thus obtained *are not regarded as theta-functions*, and the expression theta-function will consequently not extend to include them.

Properties of the Theta-Functions: Various sub-headings.

Even-integer alteration of characters.

6. If x, y be integers, then m, n having the several even integer values from $-\infty$ to $+\infty$ respectively, it is obvious that $m+\alpha+2x$, $n+\beta+2y$ will have the same series of values with $m+\alpha$, $n+\beta$ respectively; and it thence follows that

$$\vartheta\begin{pmatrix}\alpha+2x, & \beta+2y\\ \gamma\quad, & \delta\end{pmatrix}(u,\ v)=\vartheta\begin{pmatrix}\alpha, & \beta\\ \gamma, & \delta\end{pmatrix}(u,\ v).$$

Similarly if z, w are integers, then in the function

$$\vartheta\begin{pmatrix}\alpha\quad, & \beta\\ \gamma+2z, & \delta+2w\end{pmatrix}(u,\ v)$$

the argument of the exponential function contains the term

$$\tfrac{1}{2}\pi i\{m+\alpha\,.\,u+\gamma+2z+n+\beta\,.\,v+\delta+2w\};$$

this differs from its original value by

$$\tfrac{1}{2}\pi i\,(m+\alpha\,.\,2z+n+\beta\,.\,2w),$$

$$=\pi i\,(mz+nw)+\pi i\,(\alpha z+\beta w),$$

and then, m and n being even integers, $mz+nw$ is also an even integer, and the term $\pi i\,(mz+nw)$ does not affect the value of the exponential: we thus introduce into each term of the series the factor $\exp.\,\pi i\,(\alpha z+\beta w)$, which is, in fact, $=(-)^{\alpha z+\beta w}$; and we consequently have

$$\vartheta\begin{pmatrix}\alpha\quad, & \beta\\ \gamma+2z, & \delta+2w\end{pmatrix}(u,\ v)=(-)^{\alpha z+\beta w}\,\vartheta\begin{pmatrix}\alpha, & \beta\\ \gamma, & \delta\end{pmatrix}(u,\ v);$$

or, uniting the two results,

$$\vartheta\begin{pmatrix}\alpha+2x, & \beta+2y\\ \gamma+2z, & \delta+2w\end{pmatrix}(u,\ v)=(-)^{\alpha z+\beta w}\,\vartheta\begin{pmatrix}\alpha, & \beta\\ \gamma, & \delta\end{pmatrix}(u,\ v).$$

This sustains the before-mentioned conclusion that the only distinct functions are the 16 functions obtained by giving to the characters α, β, γ, δ the values 0 and 1 respectively.

Odd-integer alteration of characters.

7. The effect is obviously to interchange the different functions.

Even and odd functions.

8. It is clear that $-m-\alpha$, $-n-\beta$ have precisely the same series of values with $m+\alpha$, $n+\beta$ respectively: hence considering the function

$$\vartheta\begin{pmatrix}\alpha, & \beta\\ \gamma, & \delta\end{pmatrix}(-u,\ -v)$$

the linear term in the argument of the exponential may be taken to be

$$\tfrac{1}{2}\pi i\,\{-m-\alpha\,.-u+\gamma+-n-\beta\,.-v+\delta\},$$

which is

$$=\tfrac{1}{2}\pi i\,\{m+\alpha\,.\,u+\gamma+n+\beta\,.\,v+\delta\}-\pi i\,\{m+\alpha\,.\,\gamma+n+\beta\,.\,\delta\};$$

the second term is here

$$=-\pi i\,(m\gamma+n\delta)-\pi i\,(\alpha\gamma+\beta\delta),$$

where, $m\gamma+n\delta$ being an even integer, the part $-\pi i\,(m\gamma+n\delta)$ does not alter the value of the exponential: the effect of the remaining part $-\pi i\,(\alpha\gamma+\beta\delta)$ is to affect each term of the series with the factor exp. $-\pi i\,(\alpha\gamma+\beta\delta)$, or what is the same thing, exp. $\pi i\,(\alpha\gamma+\beta\delta)$, each of these being, in fact, $=(-)^{\alpha\gamma+\beta\delta}$.

We have thus

$$\vartheta\begin{pmatrix}\alpha, & \beta\\ \gamma, & \delta\end{pmatrix}(-u,\ -v)=(-)^{\alpha\gamma+\beta\delta}\,\vartheta\begin{pmatrix}\alpha, & \beta\\ \gamma, & \delta\end{pmatrix}(u,\ v),$$

viz. $\vartheta\begin{pmatrix}\alpha, & \beta\\ \gamma, & \delta\end{pmatrix}(u,\ v)$ is an even or odd function of the two arguments $(u,\ v)$ conjointly, according as the characteristic $\begin{pmatrix}\alpha, & \beta\\ \gamma, & \delta\end{pmatrix}$ is even or odd.

The quarter-periods unity.

9. Taking z and w integers, we have from the definition

$$\vartheta\begin{pmatrix}\alpha, & \beta\\ \gamma, & \delta\end{pmatrix}(u+z,\ v+w)=\vartheta\begin{pmatrix}\alpha & ,\ \beta\\ \gamma+z, & \delta+w\end{pmatrix}(u,\ v),$$

viz. the effect of altering the arguments u, v into $u+z$, $v+w$ is simply to interchange the functions as shown by this formula.

If z and w are each of them even, then replacing them by $2z$, $2w$ respectively, we have

$$\vartheta\begin{pmatrix}\alpha, & \beta\\ \gamma, & \delta\end{pmatrix}(u+2z,\ v+2w)=\vartheta\begin{pmatrix}\alpha & ,\ \beta\\ \gamma+2z, & \delta+2w\end{pmatrix}(u,\ v),$$

which by a preceding formula is

$$=(-)^{\alpha z+\beta w}\,\vartheta\begin{pmatrix}\alpha, & \beta\\ \gamma, & \delta\end{pmatrix}(u,\ v),$$

or the function is altered at most in its sign. And again writing $2z$, $2w$ for z, w, we have

$$\vartheta\begin{pmatrix}\alpha, & \beta\\ \gamma, & \delta\end{pmatrix}(u+4z,\ v+4w)=\vartheta\begin{pmatrix}\alpha, & \beta\\ \gamma, & \delta\end{pmatrix}(u,\ v).$$

In reference to the foregoing results we say that the theta-functions have the quarter-periods (1, 1), the half-periods (2, 2), and the whole periods (4, 4).

The conjoint quarter quasi-periods.

10. Taking x, y integers, we consider the effect of the change of u, v into

$$u+\frac{1}{\pi i}(ax+hy),\ \ v+\frac{1}{\pi i}(hx+by).$$

It is convenient to start from the function

$$\vartheta\begin{pmatrix}\alpha-x, & \beta-y\\ \gamma\quad, & \delta\end{pmatrix}\left(u+\frac{1}{\pi i}(ax+hy),\ \ v+\frac{1}{\pi i}(hx+by)\right);$$

the argument of the exponential is here

$$\begin{aligned}&\tfrac{1}{4}(a,\ h,\ b\between m+\alpha-x,\ n+\beta-y)^2\\ &\quad+\tfrac{1}{2}\pi i\left\{m+\alpha-x\,.\,u+\gamma+\frac{1}{\pi i}(ax+hy)+n+\beta-y\,.\,v+\delta+\frac{1}{\pi i}(hx+by)\right\},\end{aligned}$$

which is

$$=\tfrac{1}{4}(a,\ h,\ b\between m+\alpha,\ n+\beta)^2+\tfrac{1}{2}\pi i\,(m+\alpha\,.\,u+\gamma+n+\beta\,.\,v+\delta)$$

+ other terms which are as follows: viz. they are

$$\begin{array}{ll}-\tfrac{1}{2}(a,\ h,\ b\between m+\alpha,\ n+\beta\between x,\ y) & +\tfrac{1}{2}(m+\alpha\,.\,ax+hy+n+\beta\,.\,hx+by)\\ +\tfrac{1}{4}(a,\ h,\ b\between x,\ y)^2 & -\tfrac{1}{2}\pi i\,(x\,.\,u+\gamma+y\,.\,v+\delta)\\ & -\tfrac{1}{2}(x\,.\,ax+hy+y\,.\,hx+by),\end{array}$$

where the terms of the right-hand column are, in fact,

$$\begin{aligned}&=+\tfrac{1}{2}(a,\ h,\ b\between m+\alpha,\ n+\beta\between x,\ y)\\ &\ .\ -\tfrac{1}{2}\pi i\,(x\,.\,u+\gamma+y\,.\,v+\delta)\\ &\quad-\tfrac{1}{2}(a,\ h,\ b\between x,\ y)^2,\end{aligned}$$

and the other terms in question thus reduce themselves to

$$-\tfrac{1}{4}(a,\ h,\ b\between x,\ y)^2\qquad-\tfrac{1}{2}\pi i\,(x\,.\,u+\gamma+y\,.\,v+\delta),$$

which are independent of m, n, and they thus affect each term of the series with the same exponential factor. The result is

$$\begin{aligned}&\vartheta\begin{pmatrix}\alpha-x, & \beta-y\\ \gamma\quad, & \delta\end{pmatrix}\left(u+\frac{1}{\pi i}(ax+hy),\ \ v+\frac{1}{\pi i}(hx+by)\right)\\ &=\exp.\{-\tfrac{1}{4}(a,\ h,\ b\between x,\ y)^2-\tfrac{1}{2}\pi i\,(x\,.\,u+\gamma+y\,.\,v+\delta)\}\,\vartheta\begin{pmatrix}\alpha, & \beta\\ \gamma, & \delta\end{pmatrix}(u,\ v);\end{aligned}$$

or (what is the same thing) for α, β, writing $\alpha+x$, $\beta+y$ respectively, we have

$$\vartheta\begin{pmatrix}\alpha, & \beta\\ \gamma, & \delta\end{pmatrix}\left(u+\frac{1}{\pi i}(ax+hy),\ v+\frac{1}{\pi i}(hx+by)\right)$$

$$=\exp.\{-\tfrac{1}{4}(a,\ h,\ b\between x,\ y)^2-\tfrac{1}{2}\pi i(x.u+\gamma+y.v+\delta)\}\,\vartheta\begin{pmatrix}\alpha+x, & \beta+y\\ \gamma\quad, & \delta\end{pmatrix}(u,\ v).$$

Taking x, y even, or writing $2x$, $2y$ for x, y, then on the right-hand side we have

$$\vartheta\begin{pmatrix}\alpha+2x, & \beta+2y\\ \gamma\quad, & \delta\end{pmatrix}(u,\ v),$$

which is

$$=\vartheta\begin{pmatrix}\alpha, & \beta\\ \gamma, & \delta\end{pmatrix}(u,\ v):$$

but there is still the exponential factor.

11. The formulæ show that the effect of the change u, v into $u+\frac{1}{\pi i}(ax+hy)$, $v+\frac{1}{\pi i}(hx+by)$, where x, y are integers, is to interchange the functions, affecting them however with an exponential factor; and we hence say that $\frac{1}{\pi i}(a,\ h)$, $\frac{1}{\pi i}(h,\ b)$ are conjoint quarter quasi-periods.

The product-theorem.

12. We multiply two theta-functions

$$\vartheta\begin{pmatrix}\alpha, & \beta\\ \gamma, & \delta\end{pmatrix}(u+u',\ v+v'),\quad \vartheta\begin{pmatrix}\alpha', & \beta'\\ \gamma', & \delta'\end{pmatrix}(u-u',\ v-v');$$

it is found that the result is a sum of four products

$$\Theta\begin{pmatrix}\tfrac{1}{2}(\alpha+\alpha')+p, & \tfrac{1}{2}(\beta+\beta')+q\\ \gamma+\gamma'\quad, & \delta+\delta'\end{pmatrix}(2u,\ 2v)\,.\,\Theta\begin{pmatrix}\tfrac{1}{2}(\alpha-\alpha')+p, & \tfrac{1}{2}(\beta-\beta')+q\\ \gamma-\gamma'\quad, & \delta-\delta'\end{pmatrix}(2u',\ 2v'),$$

where p, q have in the four products respectively the values (0, 0), (1, 0), (0, 1), and (1, 1); Θ is written in place of ϑ to denote that the parameters $(a,\ h,\ b)$ are to be changed into $(2a,\ 2h,\ 2b)$. It is to be noticed that, if α, α' are both even or both odd, then $\tfrac{1}{2}(\alpha+\alpha')$, $\tfrac{1}{2}(\alpha-\alpha')$ are integers; and so, if β, β' are both even or both odd, then $\tfrac{1}{2}(\beta+\beta')$, $\tfrac{1}{2}(\beta-\beta')$ are integers; and these conditions being satisfied (and in particular they are so if $\alpha=\alpha'$, $\beta=\beta'$) then the functions on the right-hand side of the equation are theta-functions (with new parameters as already mentioned); but if the conditions are not satisfied, then the functions on the right-hand side are only allied functions. In the applications of the theorem the functions on the right-hand side are eliminated between the different equations, as will appear.

13. The proof is immediate: in the first of the theta-functions, the argument of the exponential is

$$\begin{pmatrix}m+\alpha\qquad, & n+\beta\\ u+u'+\gamma, & v+v'+\delta\end{pmatrix},$$

and in the second, writing m', n' instead of m, n, the argument is

$$\begin{pmatrix} m'+\alpha' & , n'+\beta' \\ u-u'+\gamma', & v-v'+\delta' \end{pmatrix};$$

hence in the product, the argument of the exponential is the sum of these two functions, viz.

$$\begin{aligned} &= \tfrac{1}{4}(a,\ h,\ b \between m+\alpha,\ n+\beta)^2 + \tfrac{1}{2}\pi i\,(m+\alpha\,.\,u+u'+\gamma+n+\beta\,.\,v+v'+\delta) \\ &+ \tfrac{1}{4}(a,\ h,\ b \between m'+\alpha',\ n'+\beta')^2 + \tfrac{1}{2}\pi i\,(m'+\alpha'\,.\,u-u'+\gamma'+n'+\beta'\,.\,v-v'+\delta'). \end{aligned}$$

Comparing herewith the sum of the two functions

$$\begin{pmatrix} \mu+\tfrac{1}{2}(\alpha+\alpha'), & \nu+\tfrac{1}{2}(\beta+\beta') \\ 2u+\gamma+\gamma' & , 2v+\delta+\delta' \end{pmatrix}, \quad \begin{pmatrix} \mu'+\tfrac{1}{2}(\alpha-\alpha'), & \nu'+\tfrac{1}{2}(\beta-\beta') \\ 2u'+\gamma-\gamma' & , 2v'+\delta-\delta' \end{pmatrix},$$

$$\begin{aligned} &= \tfrac{1}{4}(2a,\ 2h,\ 2b \between \mu+\tfrac{1}{2}(\alpha+\alpha'),\ \nu+\tfrac{1}{2}(\beta+\beta'))^2 \\ &\qquad + \tfrac{1}{2}\pi i\,\{\mu+\tfrac{1}{2}(\alpha+\alpha')\,.\,2u+\gamma+\gamma'+\nu+\tfrac{1}{2}(\beta+\beta')\,.\,2v+\delta+\delta'\} \\ &+ \tfrac{1}{4}(2a,\ 2h,\ 2b \between \mu'+\tfrac{1}{2}(\alpha-\alpha'),\ \nu'+\tfrac{1}{2}(\beta-\beta'))^2 \\ &\qquad + \tfrac{1}{2}\pi i\,\{\mu'+\tfrac{1}{2}(\alpha-\alpha')\,.\,2u'+\gamma-\gamma'+\nu'+\tfrac{1}{2}(\beta-\beta')\,.\,2v'+\delta-\delta'\}, \end{aligned}$$

the two sums are identical if only

$$m+m'=2\mu,\quad n+n'=2\nu,$$
$$m-m'=2\mu',\quad n-n'=2\nu',$$

as may easily be verified by comparing the quadric and the linear terms separately. The product of the two theta-functions is thus

$$= \Sigma\,\exp.\begin{pmatrix} \mu+\tfrac{1}{2}(\alpha+\alpha'), & \nu+\tfrac{1}{2}(\beta+\beta') \\ 2u+\gamma+\gamma' & , 2v+\delta+\delta' \end{pmatrix}.\ \Sigma\,\exp.\begin{pmatrix} \mu'+\tfrac{1}{2}(\alpha-\alpha'), & \nu'+\tfrac{1}{2}(\beta-\beta') \\ 2u'+\gamma-\gamma' & , 2v'+\delta-\delta' \end{pmatrix},$$

with the proper conditions as to the values of μ, ν and of μ', ν' in the two sums respectively. As to this, observe that m, m' are even integers; say for a moment that they are similar when they are both $\equiv 0$ or both $\equiv 2 \pmod 4$, but dissimilar when they are one of them $\equiv 0$ and the other of them $\equiv 2 \pmod 4$; and the like as regards n, n'. Hence if m, m' are similar, μ, μ' are both of them even; but if m, m' are dissimilar, then μ, μ' are both of them odd. And so if n, n' are similar, ν, ν' are both of them even; but if n, n' are dissimilar, then ν, ν' are both odd.

14. There are four cases:

m, m' similar, n, n' similar,

m, m' dissimilar, n, n' similar,

m, m' similar, n, n' dissimilar,

m, m' dissimilar, n, n' dissimilar.

In the first of these, μ, ν, μ', ν' are all of them even, and the product is

$$= \Theta\begin{pmatrix} \tfrac{1}{2}(\alpha+\alpha'), & \tfrac{1}{2}(\beta+\beta') \\ \gamma+\gamma', & \delta+\delta' \end{pmatrix}(2u,\ 2v)\,.\,\Theta\begin{pmatrix} \tfrac{1}{2}(\alpha-\alpha'), & \tfrac{1}{2}(\beta-\beta') \\ \gamma-\gamma', & \delta-\delta' \end{pmatrix}(2u',\ 2v').$$

In the second case, writing $\mu+1$, $\mu'+1$ for μ, μ', the new values of μ, μ' will be both even, and we have the like expression with only the characters $\tfrac{1}{2}(\alpha+\alpha')$, $\tfrac{1}{2}(\alpha-\alpha')$ each increased by 1; so in the third case we obtain the like expression with only the characters $\tfrac{1}{2}(\beta+\beta')$, $\tfrac{1}{2}(\beta-\beta')$ each increased by 1; and in the fourth case the like expression with the four upper characters each increased by 1. The product of the two theta-functions is thus equal to the sum of the four products, according to the theorem.

Résumé of the ulterior theory of the single functions.

15. For the single theta-functions the Product-theorem comprises 16 equations, and for the double theta-functions, 256 equations: these systems will be given in full in the sequel. But attending at present to the single functions, I write down here the first four of the 16 equations, viz. these are

$$\begin{array}{llll} 0.0 & \vartheta\binom{0}{0}(u+u').\vartheta\binom{0}{0}(u-u') & = & XX'+YY', \\ 1.0 & \vartheta\,\begin{matrix}1\\0\end{matrix} \quad ,, \quad \vartheta\,\begin{matrix}1\\0\end{matrix} \quad ,, & = & YX'+XY', \\ 0.1 & \vartheta\,\begin{matrix}0\\1\end{matrix} \quad ,, \quad \vartheta\,\begin{matrix}0\\1\end{matrix} \quad ,, & = & XX'-YY', \\ 1.1 & \vartheta\,\begin{matrix}1\\1\end{matrix} \quad ,, \quad \vartheta\,\begin{matrix}1\\1\end{matrix} \quad ,, & = & -YX'+XY'; \end{array}$$

where X, Y denote $\Theta\binom{0}{0}(2u)$, $\Theta\binom{1}{0}(2u)$ respectively, and X', Y' the same functions of $2u'$ respectively. In the other equations we have on the left-hand the product of *different* theta-functions of $u+u'$, $u-u'$ respectively, and on the right-hand expressions involving other functions, X_1, Y_1, X_1', Y_1', &c., of $2u$ and $2u'$ respectively.

16. By writing $u'=0$, we have on the left-hand, squares or products of theta-functions of u, and on the right-hand expressions containing functions of $2u$: in particular, the above equations show that the squares of the four theta-functions are equal to linear functions of X, Y; that is, there exist between the squared functions two linear relations: or again, introducing a variable argument x, the four squared functions may be taken to be proportional to linear functions

$$\mathfrak{A}(a-x),\ \mathfrak{B}(b-x),\ \mathfrak{C}(c-x),\ \mathfrak{D}(d-x),$$

where $\mathfrak{A}$, $\mathfrak{B}$, $\mathfrak{C}$, $\mathfrak{D}$, a, b, c, d, are constants. This suggests a new notation for the four functions, viz. we write

$$\begin{array}{cccc} \vartheta\binom{0}{0}(u), & \vartheta\binom{1}{0}(u), & \vartheta\binom{0}{1}(u), & \vartheta\binom{1}{1}(u) \\ =Au, & Bu, & Cu, & Du; \end{array}$$

and the result just mentioned then is

$$\begin{array}{ccccccc} A^2u & : & B^2u & : & C^2u & : & D^2u \\ =\mathfrak{A}(a-x) & : & \mathfrak{B}(b-x) & : & \mathfrak{C}(c-x) & : & \mathfrak{D}(d-x), \end{array}$$

which expresses that the four functions are the coordinates of a point on a quadri-quadric curve in ordinary space.

17. The remaining 12 of the 16 equations then contain on the left-hand products such as

$$A\,(u+u').\,B\,(u-u');$$

and by suitably combining them we obtain equations such as

$$\frac{\overset{u+u'}{B}\,.\,\overset{u-u'}{A}-\overset{u+u'}{A}\,.\,\overset{u-u'}{B}}{C\,.\,D+D\,.\,C}=\text{function }(u'),$$

where for brevity the arguments are written above; viz. the numerator of the fraction is

$$B\,(u+u')\,A\,(u-u')-A\,(u+u')\,B\,(u-u'),$$

and its denominator is

$$C\,(u+u')\,D\,(u-u')+D\,(u+u')\,C\,(u-u').$$

Admitting the form of the equation, the value of the function of u' is at once found by writing in the equation $u=0$; it is, as it ought to be, a function vanishing for $u'=0$.

18. Take in this equation u' indefinitely small; each side divides by u', and the resulting equation is

$$\frac{AuB'u-BuA'u}{CuDu}=\text{const.},$$

where $A'u$, $B'u$ are the derived functions, or differential coefficients in regard to u. It thus appears that the combination $AuB'u-BuA'u$ is a constant multiple of $CuDu$: or, what is the same thing, that the differential coefficient of the quotient-function $\dfrac{Bu}{Au}$ is a constant multiple of the product of the two quotient-functions $\dfrac{Cu}{Au}$ and $\dfrac{Du}{Au}$.

19. And then substituting for the several quotient-functions their values in terms of x, we obtain a differential relation between x, u; viz. the form hereof is

$$du=\frac{Mdx}{\sqrt{a-x\,.\,b-x\,.\,c-x\,.\,d-x}},$$

and it thus appears that the quotient-functions are in fact elliptic-functions: the actual values as obtained in the sequel are

$$\text{sn}\,Ku=-\frac{1}{\sqrt{k}}\,Du\div Cu,$$

$$\text{cn}\,Ku=\sqrt{\frac{k'}{k}}\,Bu\div Cu,$$

$$\text{dn}\,Ku=\ \sqrt{k'}\,Au\div Cu;$$

and we thus of course identify the functions Au, Bu, Cu, Du with the H and the Θ functions of Jacobi.

20. If in the above-mentioned four equations we write first $u=0$, and then $u'=0$, and by means of the results eliminate from the original equations the quantities X, Y, X', Y' which occur therein, we obtain expressions for the four products such as $A(u+u')A(u-u')$. One of these equations is

$$C^2 0 \,.\, C(u+u')\,C(u-u') = C^2u\,C^2u' - D^2u\,D^2u'.$$

Taking herein u' indefinitely small, we obtain

$$\frac{CuC''u-(C'u)^2}{C^2u} = \frac{C''0}{C0} - \left(\frac{D'0}{C0}\right)^2 \frac{D^2u}{C^2u},$$

where the left-hand side is in fact $\dfrac{d^2}{du^2}\log Cu$, or this second derived function of the theta-function Cu is given in terms of the quotient-function $\dfrac{Du}{Cu}$: hence, integrating twice and taking the exponential of each side, we obtain Cu as an exponential the argument of which contains the double integral $\displaystyle\iint \frac{D^2u}{C^2u}(du)^2$, of a squared quotient-function. This, in fact, corresponds to Jacobi's equation

$$\Theta u = \sqrt{\frac{2Kk'}{\pi}}\, e^{\frac{1}{2}u^2\left(1-\frac{E}{K}\right) - k^2\int_0 du \int_0 du\, \mathrm{sn}^2 u}.$$

21. From the same equation

$$C^2 0 \,.\, C(u+u')\,C(u-u') = C^2u\,C^2u' - D^2u\,D^2u',$$

differentiating logarithmically in regard to u' and integrating in regard to u, we obtain an equation containing on the left-hand side a term $\log\dfrac{C(u-u')}{C(u+u')}$, and on the right-hand an integral in regard to u; this, in fact, corresponds to Jacobi's equation

$$\begin{aligned} u\frac{\Theta' a}{\Theta a} + \tfrac{1}{2}\log\frac{\Theta(u-a)}{\Theta(u+a)} &= \Pi(u, a) \\ &= \int_0 \frac{k^2\,\mathrm{sn}\,a\,\mathrm{cn}\,a\,\mathrm{dn}\,a\,\mathrm{sn}^2 u\,du}{1-k^2\,\mathrm{sn}^2 a\,\mathrm{sn}^2 u}. \end{aligned}$$

22. It may further be noticed that if, in the equation in question and in the three other equations of the system, we introduce into the integral the variable x in place of u, and the corresponding quantity ξ in place of u', then the integral is that of an expression such as

$$\frac{dx}{T\sqrt{a-x\,.\,b-x\,.\,c-x\,.\,d-x}},$$

where T is $=x-\xi$, or is $=$ any one of three forms such as

$$\begin{vmatrix} 1, & x+\xi, & x\xi \\ 1, & a+b, & ab \\ 1, & c+d, & cd \end{vmatrix}.$$

Résumé of the ulterior theory of the double functions.

23. The ulterior theory of the double functions is intended to be carried out on the like plan. As regards these, it is to be observed here that we have not only the 16 equations leading to linear relations between the squared functions, but that the remaining 240 equations lead also to linear relations between binary products of different functions. We have thus between the 16 functions a system of quadric relations, which in fact determine the ratios of the 16 functions in terms of two variable parameters x, y. (The 16 functions are thus the coordinates of a point on a quadri-quadric two-fold locus in 15-dimensional space.) The forms depend upon six constants, a, b, c, d, e, f: writing for shortness

$$\sqrt{a} = \qquad \sqrt{a-x \,.\, a-y},$$
$$\vdots$$
$$\sqrt{ab} = \frac{1}{x-y}\{\sqrt{a-x \,.\, b-x \,.\, f-x \,.\, c-y \,.\, d-y \,.\, e-y} + \sqrt{a-y \,.\, b-y \,.\, f-y \,.\, c-x \,.\, d-x \,.\, e-x}\},$$
$$\vdots$$

(observe that in the symbols $\sqrt{ab}$ it is always f that accompanies the two expressed letters a, b—or, what is the same thing, the duad ab is really an abbreviation for the double triad $abf.cde$): then the 16 functions are proportional to properly determined constant multiples of

$$\sqrt{a},\ \sqrt{b},\ \sqrt{c},\ \sqrt{d},\ \sqrt{e},\ \sqrt{f},\ \sqrt{ab},\ \sqrt{ac},\ \sqrt{ad},\ \sqrt{ae},\ \sqrt{bc},\ \sqrt{bd},\ \sqrt{be},\ \sqrt{cd},\ \sqrt{ce},\ \sqrt{de}:$$

and this suggests that the functions should be represented by the single and double letter notation $A(u, v), \ldots, AB(u, v), \ldots$; viz. if for shortness the arguments are omitted, then we have

$$A,\ B,\ C,\ D,\ E,\ F,\ AB,\ AC,\ AD,\ AE,\ BC,\ BD,\ BE,\ CD,\ CE,\ DE,$$

proportional to determinate constant multiples of the before-mentioned functions $\sqrt{a}, \ldots, \sqrt{ab}, \ldots$, of x and y.

24. It is interesting to notice why in the expressions for $\sqrt{ab}$, &c., the sign connecting the two radicals is $+$; the effect of the interchange of x, y is, in fact, to change (u, v) into $(-u, -v)$; consequently to change the sign of the odd functions, and to leave unaltered those of the even functions: the interchange does in fact leave $\sqrt{a}$, &c., unaltered, while it changes $\sqrt{ab}$, &c., into $-\sqrt{ab}$, &c.; and thus, since only the ratios are attended to, there is a change of sign as there should be.

25. The equations of the product-theorem lead to expressions for

$$\overset{u+u'}{A} \,.\, \overset{u-u'}{B} - \overset{u+u'}{B} \,.\, \overset{u-u'}{A},$$

where the arguments, written above, are used to denote the *two* arguments, viz. $u+u'$ to denote $(u+u', v+v')$ and $u-u'$ to denote $(u-u', v-v')$; and where the letters A, B denote each or either of them a single or double letter. These expressions are found in terms of the functions of (u, v) and of (u', v'): in any such expression taking u', v' each of them indefinitely small, but with their ratio arbitrary, we obtain the value of

$$\overset{u}{A}\,.\,\partial\overset{u}{B} - \overset{u}{B}\,.\,\partial\overset{u}{A},$$

(viz. u here stands for the two arguments (u, v), and ∂ denotes total differentiation

$$\partial A = du\frac{d}{du}A(u, v) + dv\frac{d}{dv}A(u, v)\Big),$$

as a quadric function of the functions of (u, v): or dividing by A^2, the form is $\partial\,\dfrac{B}{A}$ equal to a function of the quotient-functions $\dfrac{B}{A}$, &c., that is, we have the differentials of the quotient-functions in terms of the quotient-functions themselves. Substituting for the quotient-functions their values in terms of x, y, we should obtain the differential relations between dx, dy, du, dv, viz. putting for shortness

$$X = a-x\,.\,b-x\,.\,c-x\,.\,d-x\,.\,e-x\,.\,f-x,$$

and

$$Y = a-y\,.\,b-y\,.\,c-y\,.\,d-y\,.\,e-y\,.\,f-y,$$

these are of the form

$$\frac{dx}{\sqrt{X}} - \frac{dy}{\sqrt{Y}},\quad \frac{x\,dx}{\sqrt{X}} - \frac{y\,dy}{\sqrt{Y}},$$

each of them equal to a linear function of du and dv: so that the quotient-functions are in fact the 15 hyperelliptic functions belonging to the integrals $\displaystyle\int\frac{dx}{\sqrt{X}}$, $\displaystyle\int\frac{x\,dx}{\sqrt{X}}$; and there is thus an addition-theorem for them, in accordance with the theory of these integrals.

26. The first 16 equations of the product-theorem, putting therein first $u=0$, $v=0$, and then $u'=0$, $v'=0$, and using the results to eliminate the functions on the right-hand side, give expressions for

$$\overset{u+u'}{A}\,.\,\overset{u-u'}{B},\ \text{\&c.},$$

that is, they give $A(u+u',\ v+v').B(u-u',\ v-v')$, &c., in terms of the functions of (u, v) and (u', v'): and we have thus an addition-with-subtraction theorem for the double theta-functions. And we have thence also consequences analogous to those which present themselves in the theory of the single functions.

Remark as to notation.

27. I remark, as regards the single theta-functions, that the characteristics

$$\binom{0}{0},\ \binom{1}{0},\ \binom{0}{1},\ \binom{1}{1},$$

might for shortness be represented by a series of current numbers

$$0,\quad 1,\quad 2,\quad 3:$$

and the functions be accordingly called $\vartheta_0 u$, $\vartheta_1 u$, $\vartheta_2 u$, $\vartheta_3 u$; but that, instead of this, I prefer to use throughout the before-mentioned functional symbols

$$A, \quad B, \quad C, \quad D.$$

As regards the double functions, I do, however, denote the characteristics

00	10	01	11	00	10	01	11	00	10	01	11	00	10	01	11
00'	00'	00'	00	10'	10'	10'	10	01'	01'	01'	01	11'	11'	11'	11

by a series of current numbers

0, 1, 2, 3, 4, 5, 6, 7, 8, 9, 10, 11, 12, 13, 14, 15,

and write the functions as $\vartheta_0, \vartheta_1, \ldots, \vartheta_{15}$ accordingly; and I use also, as and when it is convenient, the foregoing single and double letter notation A, $AB, \ldots$, which correspond to them in the order

BD, CE, CD, BE, AC, C, AB, B, BC, DE, F, A, AD, D, E, AE.

Moreover, I write down for the most part a single argument only: thus, $A(u+u')$ stands for $A(u+u', v+v')$, $A(0)$ for $A(0, 0)$: and so in other cases.

SECOND PART.—THE SINGLE THETA-FUNCTIONS.

Notation, &c.

28. Writing exp. $a = q$, and converting the exponentials into circular functions, we have, directly from the definition,

$$\vartheta\begin{matrix}0\\0\end{matrix}(u) = \vartheta u \ = Au = 1 + 2q\cos\pi u + 2q^4\cos 2\pi u + 2q^9\cos 3\pi u + \ldots,$$

$$\vartheta\begin{matrix}1\\0\end{matrix}(u) = \vartheta_1 u = Bu = \quad 2q^{\frac{1}{4}}\cos\tfrac{1}{2}\pi u + 2q^{\frac{9}{4}}\cos\tfrac{3}{2}\pi u + 2q^{\frac{25}{4}}\cos\tfrac{5}{2}\pi u + \ldots,$$

$$\vartheta\begin{matrix}0\\1\end{matrix}(u) = \vartheta_2 u = Cu = 1 - 2q\cos\pi u + 2q^4\cos 2\pi u - 2q^9\cos 3\pi u + \ldots (=\Theta(Ku), \text{ Jacobi}),$$

$$\vartheta\begin{matrix}1\\1\end{matrix}(u) = \vartheta_3 u = Du = \ -2q^{\frac{1}{4}}\sin\tfrac{1}{2}\pi u + 2q^{\frac{9}{4}}\sin\tfrac{3}{2}\pi u - 2q^{\frac{25}{4}}\cos\tfrac{5}{2}\pi u + \ldots (=-\mathrm{H}(Ku), \text{ Jacobi}),$$

where a is of the form $a = -\alpha + \beta i$, α being non-evanescent and positive: hence $q = \exp.(-\alpha + \beta i) = e^{-\alpha}(\cos\beta + i\sin\beta)$, where $e^{-\alpha}$, the modulus of q, is positive and less than 1; $\cos\beta$ may be either positive or negative, and $q^{\frac{1}{4}}$ is written to denote exp. $\frac{1}{4}(-\alpha+\beta i)$, viz. this is $= e^{-\frac{1}{4}\alpha}\{\cos\frac{1}{4}\beta + i\sin\frac{1}{4}\beta\}$. But usually $\beta = 0$, viz. q is a real positive quantity less than 1, and $q^{\frac{1}{4}}$ denotes the real fourth root of q.

I have given above the three notations but, as already mentioned, I propose to employ for the four functions the notation Au, Bu, Cu, Du: it will be observed that Du is an odd function, but that Au, Bu, Cu are even functions, of u.

The constants of the theory.

29. We have

$$\begin{aligned} A0 &= 1 + 2q + 2q^4 + 2q^9 + \dots, \\ B0 &= \quad 2q^{\frac{1}{4}} + 2q^{\frac{9}{4}} + 2q^{\frac{25}{4}} + \dots, \\ C0 &= 1 - 2q + 2q^4 - 2q^9 + \dots, \\ D0 &= 0, \\ D'0 &= -\pi\{q^{\frac{1}{4}} - 3q^{\frac{9}{4}} + 5q^{\frac{25}{4}} - \dots\}. \end{aligned}$$

If, as definitions of k, k', K, we assume

$$k = \frac{B^2 0}{A^2 0},\quad k' = \frac{C^2 0}{A^2 0},\quad K = -\frac{A0}{B0}\cdot\frac{D'0}{C0},$$

then we have

$$\begin{aligned} k &= 4\sqrt{q}\left\{\frac{1+q^2+q^6+\dots}{1+2q+2q^4+\dots}\right\}^2, = 4\sqrt{q}\,(1-4q+14q^2+\dots), \\ k' &= \left\{\frac{1-2q+2q^4-\dots}{1+2q+2q^4+\dots}\right\}^2, = 1-8q+32q^2-96q^3+\dots, \\ K &= \frac{\pi(1+2q+2q^4+\dots)(1-3q^2+5q^6-\dots)}{2(1-2q+2q^4-\dots)(1+q^2+q^6+\dots)}, = \tfrac{1}{2}\pi(1+4q+4q^2+0q^3+\dots), \end{aligned}$$

where I have added the first few terms of the expansions of these quantities. We have identically

$$k^2 + k'^2 = 1.$$

It will be convenient to write also, as the definition of E,

$$K(K-E) = \frac{C''0}{C0}:$$

we have then

$$E = K - \frac{1}{K}\frac{C''0}{C0}, = \frac{1}{A0\,.\,B0\,.\,C0\,.\,D'0}\{-A^2 0\,(D'0)^2 + B^2 0\,.\,C0\,.\,C''0\};$$

moreover,

$$1 - \frac{E}{K} = \frac{1}{K^2}\cdot\frac{C''0}{C0}, = \frac{2\pi^2}{K^2}\cdot\frac{q-4q^4+9q^9-\dots}{1-2q+2q^4+\dots},$$

giving

$$\frac{E}{K} = 1 - 8q + 48q^2 - 224q^3 + \dots,$$

and thence

$$E = \tfrac{1}{2}\pi\{1 - 4q + 20q^2 - 64q^3 + \dots\}.$$

30. Other formulæ are

$$\begin{aligned} k &= 4\sqrt{q}\left\{\frac{1+q^2\,.\,1+q^4\dots}{1+q\,.\,1+q^3\dots}\right\}^4, \\ k' &= \left\{\frac{1-q\,.\,1-q^3\dots}{1+q\,.\,1+q^3\dots}\right\}^4, \\ K &= \tfrac{1}{2}\pi\left\{\frac{1+q\,.\,1+q^3\dots 1-q^2\,.\,1-q^4\dots}{1-q\,.\,1-q^3\dots 1+q^2\,.\,1+q^4\dots}\right\}^2. \end{aligned}$$

31. Jacobi's definition of q is from a different point of view altogether, viz. we have $q = \exp. -\dfrac{\pi K'}{K}$, where

$$K = \int_0^1 \frac{d\phi}{\sqrt{1-k^2\sin^2\phi}};$$

and K' is the like function of k'; the equation gives $\log q = -\dfrac{\pi K'}{K}$, viz. we have

$$K' = -\frac{K}{\pi}\log q,$$

and, regarding herein K as a given function of q, this equation gives K' as a function of q.

The product-theorem.

32. The product-theorem is

$$\vartheta\begin{pmatrix}\alpha\\ \gamma\end{pmatrix}(u+u')\,.\,\vartheta\begin{pmatrix}\alpha'\\ \gamma'\end{pmatrix}(u-u')$$

$$=\Theta\begin{pmatrix}\tfrac{1}{2}(\alpha+\alpha')\\ \gamma+\gamma'\end{pmatrix}2u\,.\,\Theta\begin{pmatrix}\tfrac{1}{2}(\alpha-\alpha')\\ \gamma-\gamma'\end{pmatrix}2u'+\Theta\begin{pmatrix}\tfrac{1}{2}(\alpha+\alpha')+1\\ \gamma+\gamma'\end{pmatrix}2u\,.\,\Theta\begin{pmatrix}\tfrac{1}{2}(\alpha-\alpha')+1\\ \gamma-\gamma'\end{pmatrix}2u'.$$

Here giving to $\begin{matrix}\alpha\\ \gamma\end{matrix}, \begin{matrix}\alpha'\\ \gamma'\end{matrix}$ their different values, and introducing unaccented and accented capitals to denote the functions of $2u$ and $2u'$ respectively, the 16 equations are

$$\begin{array}{llcll}
A\,.\,A & \vartheta^{0}_{0}\,u+u'\,\vartheta^{0}_{0}\,u-u' & = & XX'+\ YY', & \text{(square-set)}\\
B\,.\,B & \vartheta^{1}_{0}\quad ,, \quad \vartheta^{1}_{0}\quad ,, & = & YX'+\ XY', & \\
C\,.\,C & \vartheta^{0}_{1}\quad ,, \quad \vartheta^{0}_{1}\quad ,, & = & XX'-\ YY', & \\
D\,.\,D & \vartheta^{1}_{1}\quad ,, \quad \vartheta^{1}_{1}\quad ,, & =- & YX'+\ XY'; & \\
\\
C\,.\,A & \vartheta^{0}_{1}\,u+u'\,\vartheta^{0}_{0}\,u-u' & = & X_{\prime}X'_{\prime}+Y_{\prime}Y'_{\prime}, & \text{(first product-set)}\\
A\,.\,C & \vartheta^{0}_{0}\quad ,, \quad \vartheta^{0}_{1}\quad ,, & = & X_{\prime}X'_{\prime}-Y_{\prime}Y'_{\prime}, & \\
D\,.\,B & \vartheta^{1}_{1}\quad ,, \quad \vartheta^{1}_{0}\quad ,, & = & Y_{\prime}X'_{\prime}+X_{\prime}Y'_{\prime}, & \\
B\,.\,D & \vartheta^{1}_{0}\quad ,, \quad \vartheta^{1}_{1}\quad ,, & = & Y_{\prime}X'_{\prime}-X_{\prime}Y'_{\prime}; &
\end{array}$$

$$\begin{array}{llll}
B.A & \vartheta^{1}_{0}u+u'\,\vartheta^{0}_{0}u-u' = & PP'+\ QQ', & \text{(second product-set)} \\
A.B & \vartheta^{0}_{0}\quad ,, \quad \vartheta^{1}_{0}\quad ,, \quad = & PQ'+\ QP', & \\
D.C & \vartheta^{1}_{1}\quad ,, \quad \vartheta^{0}_{1}\quad ,, \quad = & iPP'-\ iQQ', & \\
C.D & \vartheta^{0}_{1}\quad ,, \quad \vartheta^{1}_{1}\quad ,, \quad = & iPQ'-\ iQP'; & \\
 & & & \\
D.A & \vartheta^{1}_{1}u+u'\,\vartheta^{0}_{0}u-u' = & P_{,}P'_{,}+\ Q_{,}Q'_{,}, & \text{(third product-set)} \\
A.D & \vartheta^{0}_{0}\quad ,, \quad \vartheta^{1}_{1}\quad ,, \quad = & iP_{,}Q'_{,}-iQ_{,}P'_{,}, & \\
B.C & \vartheta^{1}_{0}\quad ,, \quad \vartheta^{0}_{1}\quad ,, \quad = & -iP_{,}P'_{,}+iQ_{,}Q'_{,}, & \\
C.B & \vartheta^{0}_{1}\quad ,, \quad \vartheta^{1}_{0}\quad ,, \quad = & P_{,}Q'_{,}+\ Q_{,}P'_{,}. &
\end{array}$$

33. Here, and subsequently, we have

$$\begin{array}{l|l}
\Theta^{0}_{0},\ \Theta^{1}_{0},\ \Theta^{0}_{1},\ \Theta^{1}_{1}(2u)=X,\ Y,\ X_{,},\ Y_{,} & \Theta^{\frac{1}{2}}_{0},\ \Theta^{\frac{3}{2}}_{0},\ \Theta^{\frac{1}{2}}_{1},\ \Theta^{\frac{3}{2}}_{1}(2u)=P,\ Q,\ P_{,},\ Q_{,}, \\
\quad ,, \quad ,, \quad ,, \quad ,, (2u')=X',\ Y',\ X'_{,},\ Y'_{,} & \quad ,, \quad ,, \quad ,, \quad ,, (2u')=P',\ Q',\ P'_{,},\ Q'_{,}, \\
\quad ,, \quad ,, \quad ,, \quad ,, (0)=\alpha,\ \beta,\ \alpha_{,},\ \beta_{,} & \quad ,, \quad ,, \quad ,, \quad ,, (0)=\mathfrak{p},\ \mathfrak{q},\ \mathfrak{p}_{,},\ \mathfrak{q}_{,};
\end{array}$$

viz. we use also α, β, $\alpha_{,}$, $\beta_{,}$ and $\mathfrak{p}$, $\mathfrak{q}$, $\mathfrak{p}_{,}$, $\mathfrak{q}_{,}$ to denote the zero-functions; $\beta_{,}$ is $=0$, but we use $\beta'_{,}$ to denote the zero-value of $\frac{d}{du}Y_{,}$.

34. In order to obtain the foregoing relations, it is necessary to observe that

$$\Theta^{\alpha+2}_{\gamma}=\Theta^{\alpha}_{\gamma};$$

by which the upper character is always reduced to 0, 1, ½ or $\frac{3}{2}$; and that, for reducing the lower character, we have

$$\Theta^{0}_{\gamma+2}=\Theta^{0}_{\gamma};\ \Theta^{1}_{\gamma+2}=-\Theta^{1}_{\gamma};$$

$$\Theta^{\frac{1}{2}}_{\gamma+2}=i\Theta^{\frac{1}{2}}_{\gamma},\ \Theta^{\frac{1}{2}}_{\gamma-2}=-i\Theta^{\frac{1}{2}}_{\gamma};\ \Theta^{\frac{3}{2}}_{\gamma+2}=-i\Theta^{\frac{3}{2}}_{\gamma},\ \Theta^{\frac{3}{2}}_{\gamma-2}=i\Theta^{\frac{3}{2}}_{\gamma};$$

by means of which the lower character is always reduced to 0 or 1: in all these formulæ the argument is arbitrary, and it is thus $=2u$, or $2u'$ as the case requires. The formulæ are obtained without difficulty directly from the definition of the functions Θ.

35. As an instance, taking $\genfrac{}{}{0pt}{}{\alpha}{\gamma}, \genfrac{}{}{0pt}{}{\alpha'}{\gamma'} = \genfrac{}{}{0pt}{}{1}{1}, \genfrac{}{}{0pt}{}{0}{1}$, the product-equation is

$$\begin{aligned}\vartheta\genfrac{}{}{0pt}{}{1}{1}(u+u')\,.\,\vartheta\genfrac{}{}{0pt}{}{0}{1}(u-u') &= \Theta\genfrac{}{}{0pt}{}{\frac{1}{2}}{2}(2u)\,.\,\Theta\genfrac{}{}{0pt}{}{\frac{1}{2}}{0}(2u') + \Theta\genfrac{}{}{0pt}{}{\frac{3}{2}}{2}(2u)\,.\,\Theta\genfrac{}{}{0pt}{}{\frac{3}{2}}{0}(2u'),\\ &= i\Theta\genfrac{}{}{0pt}{}{\frac{1}{2}}{0}(2u)\,.\,\Theta\genfrac{}{}{0pt}{}{\frac{1}{2}}{0}(2u') - i\Theta\genfrac{}{}{0pt}{}{\frac{3}{2}}{0}(2u)\,.\,\Theta\genfrac{}{}{0pt}{}{\frac{3}{2}}{0}(2u'),\\ &= iP\,.\,P' \qquad\qquad -iQ\,.\,Q',\end{aligned}$$

which agrees with the before-given value.

36. The following values are not actually required: but I give them to fix the ideas and to show the meaning of the quantities with which we work.

	$u=0$
$X = \Theta\genfrac{}{}{0pt}{}{0}{0}(2u) = 1 + 2q^2\cos 2\pi u + 2q^8\cos 4\pi u + \ldots,$	$\alpha = 1 + 2q^2 + 2q^8 + \ldots,$
$Y = \Theta\genfrac{}{}{0pt}{}{1}{0}(2u) = 2q^{\frac{1}{2}}\cos \pi u + 2q^{\frac{9}{2}}\cos 3\pi u + \ldots,$	$\beta = 2q^{\frac{1}{2}} + 2q^{\frac{9}{2}} + \ldots,$
$X_{\prime} = \Theta\genfrac{}{}{0pt}{}{0}{1}(2u) = 1 - 2q^2\cos 2\pi u + 2q^8\cos 4\pi u - \ldots,$	$\alpha_{\prime} = 1 - 2q^2 + 2q^8 - \ldots,$
$Y_{\prime} = \Theta\genfrac{}{}{0pt}{}{1}{1}(2u) = -2q^{\frac{1}{2}}\sin \pi u + 2q^{\frac{9}{2}}\sin 3\pi u - \ldots,$	$\beta_{\prime}' = 2\pi(-q^{\frac{1}{2}} + 3q^{\frac{9}{2}} - \ldots)$
	$= \frac{d}{du}Y_{\prime}$ for $u=0$.

$$\begin{aligned}P &= \Theta\genfrac{}{}{0pt}{}{\frac{1}{2}}{0}(2u) = q^{\frac{1}{8}}(\cos\tfrac{1}{2}\pi u + i\sin\tfrac{1}{2}\pi u) + q^{\frac{9}{8}}(\cos\tfrac{3}{2}\pi u - i\sin\tfrac{3}{2}\pi u)\\ &\qquad + q^{\frac{25}{8}}(\cos\tfrac{5}{2}\pi u + i\sin\tfrac{5}{2}\pi u) + \ldots,\\ Q &= \Theta\genfrac{}{}{0pt}{}{\frac{3}{2}}{0}(2u) = q^{\frac{1}{8}}(\cos\tfrac{1}{2}\pi u - i\sin\tfrac{1}{2}\pi u) + q^{\frac{9}{8}}(\cos\tfrac{3}{2}\pi u + i\sin\tfrac{3}{2}\pi u)\\ &\qquad + q^{\frac{25}{8}}(\cos\tfrac{5}{2}\pi u - i\sin\tfrac{5}{2}\pi u) + \ldots,\\ P_{\prime} &= \Theta\genfrac{}{}{0pt}{}{\frac{1}{2}}{1}(2u) = \frac{1+i}{\sqrt{2}}\Big\{q^{\frac{1}{8}}(\cos\tfrac{1}{2}\pi u + i\sin\tfrac{1}{2}\pi u) - q^{\frac{9}{8}}(\cos\tfrac{3}{2}\pi u - i\sin\tfrac{3}{2}\pi u)\\ &\qquad - q^{\frac{25}{8}}(\cos\tfrac{5}{2}\pi u + i\sin\tfrac{5}{2}\pi u) + \ldots\Big\},\\ Q_{\prime} &= \Theta\genfrac{}{}{0pt}{}{\frac{3}{2}}{1}(2u) = \frac{1-i}{\sqrt{2}}\Big\{q^{\frac{1}{8}}(\cos\tfrac{1}{2}\pi u - i\sin\tfrac{1}{2}\pi u) - q^{\frac{9}{8}}(\cos\tfrac{3}{2}\pi u + i\sin\tfrac{3}{2}\pi u)\\ &\qquad - q^{\frac{25}{8}}(\cos\tfrac{5}{2}\pi u - i\sin\tfrac{5}{2}\pi u) + \ldots\Big\};\end{aligned}$$

and therefore also

$$\mathfrak{p} = \mathfrak{q} = q^{\frac{1}{8}} + q^{\frac{9}{8}} + q^{\frac{25}{8}} + \ldots\ldots,$$

$$\mathfrak{p}_{\prime} = \frac{1+i}{\sqrt{2}}\Big\{q^{\frac{1}{8}} - q^{\frac{9}{8}} - q^{\frac{25}{8}} + q^{\frac{49}{8}} + q^{\frac{81}{8}} - - \ldots\Big\},\quad \mathfrak{q}_{\prime} = \frac{1-i}{\sqrt{2}}\Big\{\text{Do.}\Big\};\ \mathfrak{p}_{\prime} = i\mathfrak{q}_{\prime}.$$

The square set, $u'=0$; and x-formulæ.

37. We use the square-set, in the first instance by writing therein $u'=0$; the equations become

$$\begin{aligned}A^2u &= \alpha X + \beta Y, = \omega^2\mathfrak{A}(a-x),\\ B^2u &= \beta X + \alpha Y, = \omega^2\mathfrak{B}(b-x),\\ C^2u &= \alpha X - \beta Y, = \omega^2\mathfrak{C}(c-x),\\ D^2u &= \beta X - \alpha Y, = \omega^2\mathfrak{D}(d-x),\end{aligned}$$

viz. the equations without their last members show that there exist functions ω^2 and $x\omega^2$, linear functions of X and Y, such that $\mathfrak{A}$, $\mathfrak{B}$, $\mathfrak{C}$, $\mathfrak{D}$, $\mathfrak{A}a$, $\mathfrak{B}b$, $\mathfrak{C}c$, $\mathfrak{D}d$, being constants, the squared functions may be assumed equal to $\mathfrak{A}a\,.\,\omega^2 - \mathfrak{A}\,.\,\omega^2 x$, &c., that is, $\omega^2\mathfrak{A}(a-x)$, &c., respectively: the squared functions are then *proportional* to the values $\mathfrak{A}(a-x)$, &c.

To show the meaning of the factor ω^2, observe that, from any two of the equations, for instance from the first and second, we have an equation without ω,

$$A^2u \div B^2u = \mathfrak{A}(a-x) \div \mathfrak{B}(b-x);$$

and using this to determine x, and then substituting in $\omega^2 = A^2u \div \mathfrak{A}(a-x)$, we find

$$\omega^2 = \frac{\mathfrak{B}A^2u - \mathfrak{A}B^2u}{(a-b)\,\mathfrak{A}\mathfrak{B}},$$

where the numerator is a function not in anywise more important than any other linear function of A^2u and B^2u.

38. The function Du vanishes for $u=0$, and we may assume that the corresponding value of x is $=d$. Writing in the other equations $u=0$, they become

$$\begin{aligned}A^2 0 &= (\alpha^2+\beta^2) = \omega_0{}^2\mathfrak{A}(a-d),\\ B^2 0 &= \quad 2\alpha\beta \quad = \omega_0{}^2\mathfrak{B}(b-d),\\ C^2 0 &= \alpha^2-\beta^2 \;= \omega_0{}^2\mathfrak{C}(c-d),\end{aligned}$$

where $\omega_0{}^2$ is what ω^2 becomes on writing therein $x=d$. It is convenient to omit altogether these factors ω^2 and $\omega_0{}^2$; it being understood that, without them, the equations denote not absolute equalities but only equalities of ratios: thus, without the $\omega_0{}^2$, the last-mentioned equations would denote

$$A^2 0 : B^2 0 : C^2 0 = \alpha^2+\beta^2 : 2\alpha\beta : \alpha^2-\beta^2, = \mathfrak{A}(a-d) : \mathfrak{B}(b-d) : \mathfrak{C}(c-d).$$

The quantities $\mathfrak{A}$, $\mathfrak{B}$, $\mathfrak{C}$, $\mathfrak{D}$ only present themselves in the products $\mathfrak{A}\omega^2$, &c., and their absolute magnitudes are therefore essentially indeterminate: but regarding ω^2 as containing a constant factor of properly determined value, the absolute values of $\mathfrak{A}$, $\mathfrak{B}$, $\mathfrak{C}$, $\mathfrak{D}$ may be regarded as determinate, and this is accordingly done in the formulæ $\mathfrak{A}^2 = -\text{agh}$, &c., which follow.

Relations between the constants.

39. The formulæ contain the differences of the quantities a, b, c, d; denoting these differences

$$b-c,\quad c-a,\quad a-b,\quad a-d,\quad b-d,\quad c-d,$$

in the usual manner by

$$\mathrm{a},\quad \mathrm{b},\quad \mathrm{c},\quad \mathrm{f},\quad \mathrm{g},\quad \mathrm{h},$$

so that

$$\begin{array}{cccccc} . & -\mathrm{h} & +\mathrm{g} & -\mathrm{a} & = 0, \\ \mathrm{h} & . & -\mathrm{f} & -\mathrm{b} & = 0, \\ -\mathrm{g} & +\mathrm{f} & . & -\mathrm{c} & = 0, \\ \mathrm{a} & +\mathrm{b} & +\mathrm{c} & . & = 0, \end{array}$$

and also

$$\mathrm{af}+\mathrm{bg}+\mathrm{ch} = 0,$$

and then assuming the absolute value of one of the quantities $\mathfrak{A}$, $\mathfrak{B}$, $\mathfrak{C}$, $\mathfrak{D}$, we have the system of relations

$$\begin{array}{llll} \mathfrak{A}^2 = -\mathrm{agh}, & \mathfrak{B}\mathfrak{C}\mathrm{a} = \mathfrak{A}\mathfrak{D}\mathrm{f}, & \mathfrak{A}\mathrm{bcf} = -\mathfrak{B}\mathfrak{C}\mathfrak{D}, & \mathfrak{A}\mathfrak{B}\mathfrak{C}\mathfrak{D} = \mathrm{abcfgh}, \\ \mathfrak{B}^2 = \mathrm{bhf}, & \mathfrak{C}\mathfrak{A}\mathrm{b} = -\mathfrak{B}\mathfrak{D}\mathrm{g}, & \mathfrak{B}\mathrm{cag} = \mathfrak{C}\mathfrak{A}\mathfrak{D}, & \\ \mathfrak{C}^2 = \mathrm{cfg}, & \mathfrak{A}\mathfrak{B}\mathrm{c} = -\mathfrak{C}\mathfrak{D}\mathrm{h}, & \mathfrak{C}\mathrm{abh} = \mathfrak{A}\mathfrak{B}\mathfrak{D}, & \\ \mathfrak{D}^2 = -\mathrm{abc}, & & \mathfrak{D}\mathrm{fgh} = -\mathfrak{A}\mathfrak{B}\mathfrak{C}, & \end{array}$$

$$\begin{array}{lllll} & \mathrm{c}^2\mathfrak{B}^2 & +\mathrm{b}^2\mathfrak{C}^2 & -\mathrm{f}^2\mathfrak{D}^2 & = \mathrm{bcf}\,(\mathrm{af}+\mathrm{bg}+\mathrm{ch}),\ = 0, \\ -\mathrm{c}^2\mathfrak{A}^2 & . & +\mathrm{a}^2\mathfrak{C}^2 & -\mathrm{g}^2\mathfrak{D}^2 & = \mathrm{cag}\,(\quad\text{,,}\quad),\ = 0, \\ -\mathrm{b}^2\mathfrak{A}^2 & +\mathrm{a}^2\mathfrak{B}^2 & . & -\mathrm{h}^2\mathfrak{D}^2 & = \mathrm{abh}\,(\quad\text{,,}\quad),\ = 0, \\ -\mathrm{f}^2\mathfrak{A}^2 & +\mathrm{g}^2\mathfrak{B}^2 & +\mathrm{h}^2\mathfrak{C}^2 & . & = \mathrm{fgh}\,(\quad\text{,,}\quad),\ = 0. \end{array}$$

It is to be remarked that, taking c, a, b, d in the order of decreasing magnitude, we have $-\mathrm{a}$, b, c, f, g, h all positive; hence $\mathfrak{A}^2$, $\mathfrak{B}^2$, $\mathfrak{C}^2$, $\mathfrak{D}^2$ all real; and taking as we may do, $\mathfrak{D}$ negative, then $\mathfrak{A}$, $\mathfrak{B}$, $\mathfrak{C}$ may be taken positive; that is, we have $-\mathrm{a}$, b, c, f, g, h, $\mathfrak{A}$, $\mathfrak{B}$, $\mathfrak{C}$, $-\mathfrak{D}$ all of them positive.

40. We have

$$A^2 0 = \alpha^2+\beta^2 = \mathfrak{A}\mathrm{f},$$

$$B^2 0 = \quad 2\alpha\beta \quad = \mathfrak{B}\mathrm{g},$$

$$C^2 0 = \alpha^2-\beta^2 = \mathfrak{C}\mathrm{h}.$$

The foregoing equations

$$k=\frac{B^2 0}{A^2 0},\qquad k'=\frac{C^2 0}{A^2 0},$$

give

$$k=\frac{\mathfrak{B}\mathrm{g}}{\mathfrak{A}\mathrm{f}},\qquad k'=\frac{\mathfrak{C}\mathrm{h}}{\mathfrak{A}\mathrm{f}},$$

and we thence have

$$k^2=\frac{\mathrm{bg}}{-\mathrm{af}},\quad k'^2=\frac{\mathrm{ch}}{-\mathrm{af}},$$

satisfying

$$k^2+k'^2=1.$$

41. Observe further that, substituting for a, b, c, f, g, h their values, we have

$$\begin{aligned}
\mathfrak{A}^2 &= c-b.b-d.c-d, &&= c-d.d-b.b-c,\\
\mathfrak{B}^2 &= c-a.c-d.a-d, &&= d-a.a-c.c-d,\\
\mathfrak{C}^2 &= a-b.a-d.b-d, &&= -a-b.b-d.d-a,\\
\mathfrak{D}^2 &= c-b.c-a.a-b, &&= -b-c.c-a.a-b,
\end{aligned}$$

where in the first set of values all the differences are positive, but in the second set of values, we take the triads of *abcd*, in the cyclical order *bcd*, *cda*, *dab*, *abc*. There is in this last form an apparent want of symmetry as to the signs (viz. the order which might have been expected is $+-+-$), but taking the order of the letters to be $\mathfrak{C}$, $\mathfrak{A}$, $\mathfrak{B}$, $\mathfrak{D}$ and *c*, *a*, *b*, *d*, then the cyclical arrangement is

$$\begin{aligned}
\mathfrak{C}^2 &= -b-d.d-a.a-b,\\
\mathfrak{A}^2 &= -d-c.c-b.b-d,\\
\mathfrak{B}^2 &= -c-a.a-d.d-c,\\
\mathfrak{D}^2 &= -a-b.b-c.c-a,
\end{aligned}$$

where the four outside signs are all $-$. Observe that the triads of *abcd*, and *abdc*, are

bcd, *cda*, *dab*, *abc*,

and

bdc, *dca*, *cab*, *abd*,

where in the first and second columns the terms of the same column correspond to each other with a reversal of sign, whereas in the third and fourth columns the lower term of either column corresponds to the upper term of the other column, but without a reversal of sign.

The product-sets, $u\pm u'$: and u' indefinitely small, differential formulæ.

42. Coming now to the product-sets, these may be written

$$\begin{aligned}
&\tfrac12\{\overset{u+u'}{C}.\overset{u-u'}{A}+\overset{u+u'}{A}.\overset{u-u'}{C}\}=X_{,}X'_{,},\\
&\;\;,,\;\{D.B+B.D\}=Y_{,}X'_{,},\\
&\tfrac12\{B.A+A.B\}=(P+Q)(P'+Q'),\\
&\;\;,,\;\{D.C+C.D\}=i(P-Q)(P'+Q'),\\
&\tfrac12\{D.A+A.D\}=(P_{,}-iQ_{,})(P'_{,}+iQ'_{,}),\\
&\;\;,,\;\{B.C+C.B\}=-i(P_{,}-iQ_{,})(P'_{,}+iQ'_{,}),
\end{aligned}$$

$$\begin{aligned}
&\tfrac12\{\overset{u+u'}{C}.\overset{u-u'}{A}-\overset{u+u'}{A}.\overset{u-u'}{C}\}=Y_{,}Y'_{,},\\
&\;\;,,\;\{D.B-B.D\}=X_{,}Y'_{,},\\
&\tfrac12\{B.A-A.B\}=(P-Q)(P'-Q'),\\
&\;\;,,\;\{D.C-C.D\}=i(P+Q)(P'-Q'),\\
&\tfrac12\{D.A-A.D\}=(P_{,}+iQ_{,})(P'_{,}-iQ'_{,}),\\
&\;\;,,\;\{B.C-C.B\}=-i(P_{,}-iQ_{,})(P'_{,}-iQ'_{,}).
\end{aligned}$$

43. We can from each set form two fractions (each of them a function of $u+u'$ and $u-u'$), which are equal to one and the same function of u' only: for instance, from the first set we have two fractions, each $\frac{Y'}{X'}$: putting in such equation $u=0$, we obtain a new expression for the function of u' involving only the theta-functions Au', &c., which new expression we may then substitute in the equations first obtained: we thus arrive at the six equations

$$\begin{array}{c} u+u'\ u-u'\ u+u'\ u-u'\ u+u'\ u-u'\ u+u'\ u-u' \end{array}$$

$$\frac{C.A-A.C}{D.B+B.D}=\frac{D.B-B.D}{C.A+A.C}=\frac{Du'.Bu'}{Cu'.Au'},$$

$$-\frac{B.A-A.B}{D.C+C.D}=\frac{D.C-C.D}{B.A+A.B}=\frac{Du'.Cu'}{Bu'.Au'},$$

$$-\frac{B.C-C.B}{D.A+A.D}=\frac{D.A-A.D}{B.C+C.B}=\frac{Du'.Au'}{Bu'.Cu'},$$

where observe that the expressions all vanish for $u'=0$.

44. Taking herein u' indefinitely small, we obtain

$$\frac{Au.C'u-Cu.A'u}{Bu.Du}=\frac{Bu.D'u-Du.B'u}{Cu.Au}=\frac{D'0.B0}{C0.A0}=-K\frac{B^20}{A^20},$$

$$-\frac{Au.B'u-Bu.A'u}{Cu.Du}=\frac{Cu.D'u-Du.C'u}{Au.Bu}=\frac{D'0.C0}{A0.B0}=-K\frac{C^20}{A^20},$$

$$-\frac{Cu.B'u-Bu.C'u}{Au.Du}=\frac{Au.D'u-Du.A'u}{Bu.Cu}=\frac{D'0.A0}{B0.C0}=-K,$$

where the last column is added in order to introduce K in place of $D'0$.

45. These formulæ in effect give the derivatives of the quotient-functions in terms of quotient-functions: for instance, one of the equations is

$$\frac{d}{du}\frac{Du}{Au}=-K\frac{Bu}{Au}\cdot\frac{Cu}{Au};$$

substituting herein for the quotient-fractions their values in terms of x, this becomes

$$\frac{d}{du}\sqrt{\frac{d-x}{a-x}}=-K\sqrt{\frac{\mathfrak{BC}}{\mathfrak{AD}}}\,\frac{\sqrt{b-x\,.\,c-x}}{a-x},\quad =-K\sqrt{\frac{\mathrm{f}}{\mathrm{a}}}\,\frac{\sqrt{b-x\,.\,c-x}}{a-x},$$

or the left-hand being

$$=\frac{-\frac{1}{2}\mathrm{f}}{(a-x)^{\frac{3}{2}}\sqrt{d-x}}\frac{dx}{du},$$

this is

$$Kdu=\frac{\frac{1}{2}\sqrt{\mathrm{af}}\,.\,dx}{\sqrt{a-x\,.\,b-x\,.\,c-x\,.\,d-x}},$$

where on the right-hand side it would be better to write $\sqrt{-\mathrm{af}}$ in the numerator and $x-d$ in place of $d-x$ in the denominator.

Comparison with Jacobi's formulæ.

46. The comparison of the formulæ with Jacobi's formulæ gives

$$\text{sn}\, Ku = -\frac{1}{\sqrt{k}} Du \div Cu, \quad = \sqrt{\frac{\text{a}}{\text{g}}}\sqrt{\frac{d-x}{c-x}}\left(\text{or better } \sqrt{\frac{-\text{a}}{\text{g}}}\sqrt{\frac{x-d}{c-x}}\right),$$

$$\text{cn}\, Ku = \sqrt{\frac{k'}{k}}\, Bu \div Cu, \quad = \sqrt{\frac{\text{h}}{\text{g}}}\sqrt{\frac{b-x}{c-x}},$$

$$\text{dn}\, Ku = \sqrt{k'}\, Au \div Cu, \quad = \sqrt{\frac{\text{h}}{\text{f}}}\sqrt{\frac{a-x}{c-x}},$$

where it will be recollected that

$$k^2 = \frac{\text{bg}}{-\text{af}}, \quad k'^2 = \frac{\text{ch}}{-\text{af}}.$$

It may be remarked that we seek to determine everything in terms of a, b, c, d. The formula just written down, $k^2 = \text{bg} \div - \text{af}$, gives k in terms of these quantities; and k, K being each given in terms of q, we have virtually K as a function of k, that is, of a, b, c, d: but it would not be easy from the expressions of k, K, each in terms of q, to deduce the actual expression

$$K = \int_0^{\frac{\pi}{2}} \frac{d\phi}{\sqrt{1-k^2\sin^2\phi}},$$

of K as a function of k.

The square-set, $u \pm u'$.

47. Reverting to the square-set

$$\begin{aligned}
A(u+u')\,A(u-u') &= XX'+YY', \\
B(u+u')\,B(u-u') &= YX'+XY', \\
C(u+u')\,C(u-u') &= XX'-YY', \\
D(u+u')\,D(u-u') &= -YX'+XY',
\end{aligned}$$

if we first write herein $u'=0$, and then $u=0$, using the results to determine the values of X, Y, X', Y' we find

$$\begin{aligned}
\alpha C^2u - \beta D^2u &= (\alpha^2-\beta^2)\,X, \qquad & \alpha C^2u' - \beta D^2u' &= (\alpha^2-\beta^2)\,X', \\
\beta C^2u - \alpha D^2u &= \quad ,, \qquad Y, & \beta C^2u' - \alpha D^2u' &= \quad ,, \qquad Y',
\end{aligned}$$

and thence

$$\begin{aligned}
(\alpha^2-\beta^2)^2 XX' &= \alpha^2 . C^2uC^2u' + \beta^2 . D^2uD^2u' - \alpha\beta\,(C^2uD^2u' + D^2uC^2u'), \\
,, \quad YY' &= \beta^2 \quad ,, \quad + \alpha^2 \quad ,, \quad - \alpha\beta \quad ,, \quad ,
\end{aligned}$$

whence

$$\begin{aligned}
(\alpha^2-\beta^2)^2(XX'+YY') &= (\alpha^2+\beta^2)(C^2uC^2u' + D^2uD^2u') - 2\alpha\beta\,(C^2uD^2u' + D^2uC^2u'), \\
(\alpha^2-\beta^2)\;(XX'-YY') &= \qquad (C^2uC^2u' - D^2uD^2u'),
\end{aligned}$$

where observe that, in taking the difference, the right-hand side becomes divisible by $\alpha^2-\beta^2$, and therefore in the final result we have on the left-hand side the simple factor $\alpha^2-\beta^2$ instead of $(\alpha^2-\beta^2)^2$.

Similarly

$$(\alpha^2-\beta^2)\,YX' = \alpha\beta\,(C^2uC^2u'+D^2uD^2u') - \alpha^2D^2uC^2u' - \beta^2C^2uD^2u',$$

$$\text{,,}\qquad XY' = \alpha\beta \qquad \text{,,} \qquad -\beta^2 \text{ ,,} \quad -\alpha^2 \text{ ,,}\quad ,$$

and thence

$$(\alpha^2-\beta^2)^2(\;\;YX'+XY') = 2\alpha\beta\,(C^2uC^2u'+D^2uD^2u') - (\alpha^2+\beta^2)(C^2uD^2u'+D^2uC^2u'),$$

$$(\alpha^2-\beta^2)\,(-YX'+XY') = \qquad D^2uC^2u' - C^2uD^2u'.$$

48. Hence recollecting that

$$A^20 = \alpha^2+\beta^2,$$

$$B^20 = 2\alpha\beta,$$

$$C^20 = \alpha^2-\beta^2,$$

the original equations become

$$C^40\,.\,A\,(u+u')\,A\,(u-u') = A^20\,(C^2uC^2u'+D^2uD^2u') - B^20\,(C^2uD^2u'+D^2uC^2u'),$$

$$C^40\,.\,B\,(u+u')\,B\,(u-u') = B^20\,(C^2uC^2u'+D^2uD^2u') - A^20\,(C^2uD^2u'+D^2uC^2u'),$$

$$C^20\,.\,C\,(u+u')\,C\,(u-u') = \qquad C^2uC^2u' - D^2uD^2u',$$

$$C^20\,.\,D\,(u+u')\,D\,(u-u') = \qquad D^2uC^2u' - C^2uD^2u'.$$

49. It will be observed that the four products $A\,(u+u')\,A\,(u-u')$, &c., are each of them expressed in terms of C^2u, D^2u, C^2u', D^2u'. Since each of the squared functions A^2u, B^2u, C^2u, D^2u is a linear function of any two of them, and the like as regards A^2u', B^2u', C^2u', D^2u', it is clear that in each equation we can on the right-hand side introduce any two at pleasure of the squared functions of u, and any two at pleasure of the squared functions of u'. But all the forms so obtained are of course identical if, taking x' the same function of u' that x is of u, we introduce on the right-hand side x, x' instead of u, u'; and the values of $A\,(u+u')\,.\,A\,(u-u')$, &c., are found to be equal to multiples of ∇, ∇_1, ∇_2, ∇_3, where

$$\nabla = x-x',\quad \nabla_1 = \begin{vmatrix} 1, & x+x', & xx' \\ 1, & a+d, & ad \\ 1, & b+c, & bc \end{vmatrix},\quad \nabla_2 = \begin{vmatrix} 1, & x+x', & xx' \\ 1, & b+d, & bd \\ 1, & c+a, & ca \end{vmatrix},\quad \nabla_3 = \begin{vmatrix} 1, & x+x', & xx' \\ 1, & c+d, & cd \\ 1, & a+b, & ab \end{vmatrix}.$$

50. In fact, from the equations

$$A^2u = \mathfrak{A}\,(a-x),\quad A^2u' = \mathfrak{A}\,(a-x'),$$

we have

$$\nabla = \frac{1}{\mathrm{a}\mathfrak{B}\mathfrak{C}}(B^2uC^2u'-C^2uB^2u'),\quad = \frac{1}{\mathrm{b}\mathfrak{C}\mathfrak{A}}(C^2uA^2u'-A^2uC^2u'),\quad = \frac{1}{\mathrm{c}\mathfrak{A}\mathfrak{B}}(A^2uB^2u'-B^2uA^2u'),$$

$$= \frac{1}{\mathrm{f}\mathfrak{A}\mathfrak{D}}(A^2uD^2u'-D^2uA^2u'),\quad = \frac{1}{\mathrm{g}\mathfrak{B}\mathfrak{D}}(B^2uD^2u'-D^2uB^2u'),\quad = \frac{1}{\mathrm{h}\mathfrak{C}\mathfrak{D}}(C^2uD^2u'-D^2uC^2u'),$$

where it will be recollected that

$$\mathrm{f}\mathfrak{A}\mathfrak{D}=\mathrm{a}\mathfrak{B}\mathfrak{C},\quad -\mathrm{g}\mathfrak{B}\mathfrak{D}=\mathrm{b}\mathfrak{C}\mathfrak{A},\quad -\mathrm{h}\mathfrak{C}\mathfrak{D}=\mathrm{c}\mathfrak{A}\mathfrak{B}.$$

Moreover

$$(b-c)\nabla_1=-\begin{vmatrix} b-x.b-x', & c-x.c-x' \\ b-a.b-d, & c-a.c-d \end{vmatrix},\quad (a-d)\nabla_1=\begin{vmatrix} a-x.a-x', & d-x.d-x' \\ a-b.a-c, & d-b.d-c \end{vmatrix},$$

$$(c-a)\nabla_2=-\begin{vmatrix} c-x.c-x', & a-x.a-x' \\ c-b.c-d, & a-b.a-d \end{vmatrix},\quad (b-d)\nabla_2=\begin{vmatrix} b-x.b-x', & d-x.d-x' \\ b-c.b-a, & d-c.d-a \end{vmatrix},$$

$$(a-b)\nabla_3=-\begin{vmatrix} a-x.a-x', & b-x.b-x' \\ a-c.a-d, & b-c.b-d \end{vmatrix},\quad (c-d)\nabla_3=\begin{vmatrix} c-x.c-x', & d-x.d-x' \\ c-a.c-b, & d-a.d-b \end{vmatrix},$$

or as these may be written

$$\nabla_1=-\frac{1}{\mathrm{a}}\{\mathrm{bh}.b-x.b-x'.+\mathrm{cg}.c-x.c-x'\},\quad =\frac{1}{\mathrm{f}}\{\mathrm{gh}.a-x.a-x'.+\mathrm{bc}.d-x.d-x'\},$$

$$\nabla_2=-\frac{1}{\mathrm{b}}\{\mathrm{cf}.c-x.c-x'.+\mathrm{ah}.a-x.a-x'\},\quad =\frac{1}{\mathrm{g}}\{\mathrm{hf}.b-x.b-x'.+\mathrm{ca}.d-x.d-x'\},$$

$$\nabla_3=-\frac{1}{\mathrm{c}}\{\mathrm{ag}.a-x.a-x'.+\mathrm{bf}.b-x.b-x'\},\quad =\frac{1}{\mathrm{h}}\{\mathrm{fg}.c-x.c-x'.+\mathrm{ab}.d-x.d-x'\},$$

that is,

$$\nabla_1=-\frac{1}{\mathrm{a}}\left\{\frac{\mathrm{bh}}{\mathfrak{B}^2}B^2uB^2u'+\frac{\mathrm{cg}}{\mathfrak{C}^2}C^2uC^2u'\right\},\quad =\frac{1}{\mathrm{f}}\left\{\frac{\mathrm{gh}}{\mathfrak{A}^2}A^2uA^2u'+\frac{\mathrm{bc}}{\mathfrak{D}^2}D^2uD^2u'\right\},$$

$$\nabla_2=-\frac{1}{\mathrm{b}}\left\{\frac{\mathrm{cf}}{\mathfrak{C}^2}C^2uC^2u'+\frac{\mathrm{ah}}{\mathfrak{A}^2}A^2uA^2u'\right\},\quad =\frac{1}{\mathrm{g}}\left\{\frac{\mathrm{hf}}{\mathfrak{B}^2}B^2uB^2u'+\frac{\mathrm{ca}}{\mathfrak{D}^2}D^2uD^2u'\right\},$$

$$\nabla_3=-\frac{1}{\mathrm{c}}\left\{\frac{\mathrm{ag}}{\mathfrak{A}^2}A^2uA^2u'+\frac{\mathrm{bf}}{\mathfrak{B}^2}B^2uB^2u'\right\},\quad =\frac{1}{\mathrm{h}}\left\{\frac{\mathrm{fg}}{\mathfrak{C}^2}C^2uC^2u'+\frac{\mathrm{ab}}{\mathfrak{D}^2}D^2uD^2u'\right\},$$

or finally

$$\nabla_1=-\frac{1}{\mathrm{af}}(\ B^2uB^2u'+C^2uC^2u'),\quad =-\frac{1}{\mathrm{af}}(A^2uA^2u'+D^2uD^2u'),$$

$$\nabla_2=-\frac{1}{\mathrm{bg}}(\ C^2uC^2u'-A^2uA^2u'),\quad =\ \frac{1}{\mathrm{bg}}(B^2uB^2u'-D^2uD^2u'),$$

$$\nabla_3=-\frac{1}{\mathrm{ch}}(-A^2uA^2u'+B^2uB^2u'),\quad =\ \frac{1}{\mathrm{ch}}(C^2uC^2u'-D^2uD^2u').$$

51. Hence ∇, ∇_1, ∇_2, ∇_3 denoting these functions of x, x' or of u, u', we have

$$A(u+u')A(u-u')=\frac{\mathfrak{A}}{\mathrm{gh}}\nabla_1,$$

$$B(u+u')B(u-u')=\frac{\mathfrak{B}}{\mathrm{hf}}\nabla_2,$$

$$C(u+u')C(u-u')=\frac{\mathfrak{C}}{\mathrm{fg}}\nabla_3,$$

$$D(u+u')D(u-u')=\mathfrak{D}\nabla.$$

The square-set $u \pm u'$, u' indefinitely small: differential formulæ of the second order.

52. I consider the original form

$$C^2 0\, C(u+u')\, C(u-u') = C^2 u C^2 u' - D^2 u D^2 u',$$

which is of course included in the last-mentioned equations.

Writing this in the form

$$C^2 0 \frac{C(u+u')\, C(u-u')}{C^2 u} = C^2 u' - \frac{D^2 u D^2 u'}{C^2 u},$$

and taking u' indefinitely small, whence

$$C(u+u') = Cu + u'C'u + \tfrac{1}{2}u'^2 C''u, \quad Cu' = C0,$$

$$C(u-u') = Cu - u'C'u + \tfrac{1}{2}u'^2 C''u, \quad Du' = u'D'0,$$

$$C(u+u')\, C(u-u') = C^2 u + u'^2 \{CuC''u - (C'u)^2\},$$

the equation becomes

$$C^2 0 \left(1 + u'^2 \left\{\frac{C''u}{Cu} - \left(\frac{C'u}{Cu}\right)^2\right\}\right) = C^2 0 + u'^2 \left\{C0C''0 - (D'0)^2 \frac{D^2 u}{C^2 u}\right\},$$

that is,

$$\frac{C''u}{Cu} - \left(\frac{C'u}{Cu}\right)^2 = \frac{C''0}{C0} - \left(\frac{D'0}{C0}\right)^2 \frac{D^2 u}{C^2 u},$$

viz. we have $\left(\frac{d}{du}\right)^2 \log Cu$ expressed in terms of the quotient-function $\frac{D^2 u}{C^2 u}$, and consequently Cu given as an exponential, the argument of which depends on the double integral $\int du \int du \frac{D^2 u}{C^2 u}$.

53. To complete the result, I write the equation in the form

$$\frac{d^2}{du^2} \log Cu = \frac{C''0}{C0} - \frac{1}{k}\left(\frac{D'0}{C0}\right)^2 + \frac{1}{k}\left(\frac{D'0}{C0}\right)^2 \left(1 - k\frac{D^2 u}{C^2 u}\right);$$

$\frac{D'0}{C0}$ is $= -\sqrt{k}K$, and $\frac{C''0}{C0}$ is $= K(K-E)$; hence the equation is

$$\frac{d^2}{du^2} \log Cu = K^2 \left(1 - \frac{E}{K} - k\frac{D^2 u}{C^2 u}\right), \quad = K^2 \left(1 - \frac{E}{K} - k^2 \operatorname{sn}^2 Ku\right),$$

or integrating twice, and observing that $\frac{d}{du}\log Cu$ and $\log Cu$, for $u=0$, become $=0$ and $\log C0$ respectively, we have

$$\log Cu = \log C0 + \tfrac{1}{2}\left(1 - \frac{E}{K}\right) K^2 u^2 - k^2 \int_0 du \int_0 du\, K^2 \operatorname{sn}^2 Ku,$$

which is in fact

$$\log \Theta(Ku) = \log C0 + \tfrac{1}{2}\left(1 - \frac{E}{K}\right) K^2 u^2 - k^2 \int_0 du \int_0 du\, K^2 \operatorname{sn}^2 Ku,$$

agreeing with Jacobi's formula

$$\log \Theta u = \log \Theta 0 + \tfrac{1}{2}\left(1 - \frac{E}{K}\right) u^2 - k^2 \int_0 du \int_0 du \,\mathrm{sn}^2 u.$$

Elliptic integrals of the third kind.

54. We may write

$$\frac{A(u+u')\,A(u-u')}{A^2u A^2u'} = \frac{1}{\mathfrak{A}\mathrm{gh}}\,\frac{\nabla_1}{a-x\,.\,a-x'},$$

$$\frac{B(u+u')\,B(u-u')}{B^2u B^2u'} = \frac{1}{\mathfrak{B}\mathrm{hf}}\,\frac{\nabla_2}{b-x\,.\,b-x'},$$

$$\frac{C(u+u')\,C(u-u')}{C^2u C^2u'} = \frac{1}{\mathfrak{C}\mathrm{fg}}\,\frac{\nabla_3}{c-x\,.\,c-x'},$$

$$\frac{D(u+u')\,D(u-u')}{D^2u D^2u'} = \frac{1}{\mathfrak{D}}\,\frac{x-x'}{d-x\,.\,d-x'}.$$

We differentiate logarithmically in regard to u'. Observing that

$$K du' = \frac{\tfrac{1}{2}\sqrt{\mathrm{af}}\,dx'}{\sqrt{a-x'\,.\,b-x'\,.\,c-x'\,.\,d-x'}},\quad = \frac{\tfrac{1}{2}\sqrt{\mathrm{af}}}{\sqrt{X'}}\,dx',$$

suppose, the first equation gives

$$\frac{A'u}{Au} + \tfrac{1}{2}\frac{A'(u-u')}{A(u-u')} - \tfrac{1}{2}\frac{A'(u+u')}{A(u+u')} = -\frac{K\sqrt{X'}}{\sqrt{\mathrm{af}}}\,\frac{d}{dx'}\log\frac{\nabla_1}{a-x'},$$

and if for a moment

$$\nabla_1, = \begin{vmatrix} 1, & x+x', & xx' \\ 1, & a+d, & ad \\ 1, & b+c, & bc \end{vmatrix},$$

is put

$$= P(a-x') + Q(d-x'),$$

then

$$\frac{d}{dx'}\log\frac{\nabla_1}{a-x'},\ = \frac{d}{dx'}\log\left(P + Q\,\frac{d-x'}{a-x'}\right)\ \text{is}\ = \frac{Q(d-a)}{(a-x')\nabla_1},\ = -\frac{Q\mathrm{f}}{(a-x')\nabla_1}.$$

But, writing $x' = a$, we have

$$Q(d-a),\ = -Q\mathrm{f} = \begin{vmatrix} 1, & a+x, & ax \\ 1, & a+d, & ad \end{vmatrix},\ = (a-b)(a-c)(d-x),\ = -\mathrm{bc}\,(d-x),$$

that is,

$$Q\mathrm{f} = -\mathrm{bc}\,(d-x),$$

or

$$\frac{d}{dx'}\log\frac{\nabla_1}{a-x'} = \frac{\mathrm{bc}\,(d-x)}{(a-x')\nabla_1}.$$

Hence the equation is

$$2\frac{A'(u')}{A(u')} + \frac{A'(u-u')}{A(u-u')} - \frac{A'(u+u')}{A(u+u')} = \frac{2K\mathrm{bc}}{\sqrt{\mathrm{af}}}\sqrt{X'}\,\frac{d-x}{(a-x')\nabla_1};$$

and similarly

$$2\frac{B'(u')}{B(u')}+\frac{B'(u-u')}{B(u-u')}-\frac{B'(u+u')}{B(u+u')}=\frac{2K\text{ca}}{\sqrt{\text{af}}}\sqrt{X'}\,\frac{d-x}{(b-x')\nabla_2},$$

$$2\frac{C'(u')}{C(u')}+\frac{C'(u-u')}{C(u-u')}-\frac{C'(u+u')}{C(u+u')}=\frac{2K\text{ab}}{\sqrt{\text{af}}}\sqrt{X'}\,\frac{d-x}{(c-x')\nabla_3},$$

$$2\frac{D'(u')}{D(u')}+\frac{D'(u-u')}{D(u-u')}-\frac{D'(u+u')}{D(u+u')}=\frac{2K}{\sqrt{\text{af}}}\sqrt{X'}\,\frac{d-x}{(d-x')(x-x')}.$$

55. Multiply each of these equations by du, $=\frac{1}{2}\frac{\sqrt{\text{af}}}{K}\frac{dx}{\sqrt{X}}$, and integrate. We have equations such as

$$2u\frac{A'(u')}{A(u')}+\log\frac{A(u-u')}{A(u+u')}=\text{const.}+\frac{\text{bc}\sqrt{X'}}{\sqrt{\text{af}}(a-x')}\int\frac{(d-x)\,dx}{\nabla_1\sqrt{X}},$$

showing how the integrals of the third kind

$$\int\frac{(d-x)\,dx}{\nabla_1\sqrt{X}},\quad \int\frac{(d-x)\,dx}{\nabla_2\sqrt{X}},\quad \int\frac{(d-x)\,dx}{\nabla_3\sqrt{X}},\quad \int\frac{(d-x)\,dx}{(x-x')\sqrt{X}}$$

depend on the theta-functions.

If, instead, we work with the original equation

$$C^2 0\,\frac{C(u+u')\,C(u-u')}{C^2u\,.\,C^2u'}=1-\frac{D^2u}{C^2u}\frac{D^2u'}{C^2u'},$$

we find in the same way

$$2\frac{C'(u')}{C(u')}+\frac{C'(u-u')}{C(u-u')}-\frac{C'(u+u')}{C(u+u')}=-\frac{d}{du'}\log\left(1-\frac{D^2uD^2u'}{C^2uC^2u'}\right),$$

$$=-\frac{d}{du'}\log(1-k^2\,\text{sn}^2\,Ku\,\text{sn}^2\,Ku'),$$

$$=\frac{2k^2K\,\text{sn}\,Ku'\,\text{cn}\,Ku'\,\text{dn}\,Ku'\,\text{sn}^2\,Ku}{1-k^2\,\text{sn}^2\,Ku'\,\text{sn}^2\,Ku};$$

or, multiplying by $\frac{1}{2}du$ and integrating, we have

$$u\frac{C'(u')}{C(u')}+\tfrac{1}{2}\log\frac{C(u-u')}{C(u+u')}=\int\frac{k^2\,\text{sn}\,Ku'\,\text{cn}\,Ku'\,\text{dn}\,Ku'\,\text{sn}^2\,Ku\,.\,Kdu}{1-k^2\,\text{sn}^2\,Ku'\,\text{sn}^2\,Ku},$$

which is in fact Jacobi's equation

$$u\frac{\Theta' a}{\Theta a}+\tfrac{1}{2}\log\frac{\Theta(u-a)}{\Theta(u+a)}=\int\frac{\text{sn}\,a\,\text{cn}\,a\,\text{dn}\,a\,\text{sn}^2\,u\,du}{1-k^2\,\text{sn}^2\,a\,\text{sn}^2\,u},\ =\Pi(u,\,a).$$

I do not effect the operation but consider the forms first obtained,

$$A(u+u')\,A(u-u')=\frac{\mathfrak{A}}{\text{gh}}\nabla_1,\ \text{\&c.},$$

as the equivalent of Jacobi's last-mentioned equation.

Addition-formulæ.

56. The addition-theorem for the quotient-functions is of course given by means of the theorem for the elliptic functions: but more elegantly by the formulæ relating to the form $dx \div \sqrt{a-x \,.\, b-x \,.\, c-x \,.\, d-x}$ obtained in my paper "On the Double ϑ-Functions" (*Crelle*, t. LXXXVII. (1879), pp. 74—81, [697]); viz. for the differential equation

$$\frac{dx}{\sqrt{X}} + \frac{dy}{\sqrt{Y}} - \frac{dz}{\sqrt{Z}} = 0,$$

to obtain the particular integral which for $y = d$ reduces itself to $z = x$, we must, in the formulæ of the paper just referred to, interchange a and d: and writing for shortness a, b, c, $\mathrm{d} = a - x,\ b - x,\ c - x,\ d - x$, and similarly $\mathrm{a}_{\prime}, \mathrm{b}_{\prime}, \mathrm{c}_{\prime}, \mathrm{d}_{\prime} = a - y$, $b - y,\ c - y,\ d - y$, then when the interchange is made, the formulæ become

$$\sqrt{\frac{d-z}{a-z}}$$

$$= \frac{\sqrt{d-b \,.\, d-c}\,\{\sqrt{\mathrm{adb_{\prime}c_{\prime}}} + \sqrt{\mathrm{a_{\prime}d_{\prime}bc}}\}}{(bc,\ ad)},$$

$$= \frac{\sqrt{d-b \,.\, d-c \,.\, x-y}}{\sqrt{\mathrm{adb_{\prime}c_{\prime}}} - \sqrt{\mathrm{a_{\prime}d_{\prime}bc}}},$$

$$= \frac{\sqrt{d-b \,.\, d-c}\,\{\sqrt{\mathrm{bdc_{\prime}a_{\prime}}} + \sqrt{\mathrm{b_{\prime}d_{\prime}ca}}\}}{(d-c)\sqrt{\mathrm{aba_{\prime}b_{\prime}}} - (b-a)\sqrt{\mathrm{cdc_{\prime}d_{\prime}}}},$$

$$= \frac{\sqrt{d-b \,.\, d-c}\,\{\sqrt{\mathrm{cda_{\prime}b_{\prime}}} + \sqrt{\mathrm{abc_{\prime}d_{\prime}}}\}}{(d-b)\sqrt{\mathrm{aca_{\prime}c_{\prime}}} - (c-a)\sqrt{\mathrm{bdb_{\prime}d_{\prime}}}},$$

$$\sqrt{\frac{b-z}{a-z}}$$

$$= \frac{\sqrt{\dfrac{d-b}{d-a}}\,\{(d-c)\sqrt{\mathrm{aba_{\prime}b_{\prime}}} + (b-a)\sqrt{\mathrm{cdc_{\prime}d_{\prime}}}\}}{(bc,\ ad)},$$

$$= \frac{\sqrt{\dfrac{d-b}{d-a}}\,\{\sqrt{\mathrm{bda_{\prime}c_{\prime}}} - \sqrt{\mathrm{b_{\prime}d_{\prime}ac}}\}}{\sqrt{\mathrm{adb_{\prime}c_{\prime}}} - \sqrt{\mathrm{a_{\prime}d_{\prime}bc}}},$$

$$= \frac{\sqrt{\dfrac{d-b}{d-a}}\,(ac,\ bd)}{(d-c)\sqrt{\mathrm{aba_{\prime}b_{\prime}}} - (b-a)\sqrt{\mathrm{cdc_{\prime}d_{\prime}}}},$$

$$= \frac{\sqrt{\dfrac{d-b}{d-a}}\,\{(d-a)\sqrt{\mathrm{bcb_{\prime}c_{\prime}}} + (b-c)\sqrt{\mathrm{aba_{\prime}b_{\prime}}}\}}{(d-b)\sqrt{\mathrm{aca_{\prime}c_{\prime}}} - (c-a)\sqrt{\mathrm{bdb_{\prime}d_{\prime}}}},$$

$$\sqrt{\frac{c-z}{a-z}}$$

$$= \frac{\sqrt{\dfrac{d-c}{d-a}}\,\{(d-b)\sqrt{\mathrm{cac_{\prime}a_{\prime}}} + (c-a)\sqrt{\mathrm{bdb_{\prime}d_{\prime}}}\}}{(bc,\ ad)},$$

$$= \frac{\sqrt{\dfrac{d-c}{d-a}}\,\{\sqrt{\mathrm{cda_{\prime}b_{\prime}}} - \sqrt{\mathrm{abc_{\prime}d_{\prime}}}\}}{\sqrt{\mathrm{adb_{\prime}c_{\prime}}} - \sqrt{\mathrm{a_{\prime}d_{\prime}bc}}},$$

$$= \frac{\sqrt{\dfrac{d-c}{d-a}}\,\{(d-a)\sqrt{\mathrm{bcb_{\prime}c_{\prime}}} - (b-c)\sqrt{\mathrm{ada_{\prime}d_{\prime}}}\}}{(d-c)\sqrt{\mathrm{aba_{\prime}b_{\prime}}} - (b-a)\sqrt{\mathrm{cdc_{\prime}d_{\prime}}}},$$

$$= \frac{\sqrt{\dfrac{d-c}{d-a}}\,(ab,\ cd)}{(d-b)\sqrt{\mathrm{aca_{\prime}c_{\prime}}} - (c-a)\sqrt{\mathrm{bdb_{\prime}d_{\prime}}}}.$$

57. In the foregoing formulæ, $(bc,\ ad)$, $(ac,\ bd)$, and $(ab,\ cd)$ denote respectively

$$\begin{vmatrix} 1, & x+y, & xy \\ 1, & b+c, & bc \\ 1, & a+d, & ad \end{vmatrix}, \quad \begin{vmatrix} 1, & x+y, & xy \\ 1, & c+a, & ca \\ 1, & b+d, & bd \end{vmatrix}, \quad \begin{vmatrix} 1, & x+y, & xy \\ 1, & a+b, & ab \\ 1, & c+d, & cd \end{vmatrix};$$

and substituting for $\mathfrak{A}$, $\mathfrak{B}$, $\mathfrak{C}$, $\mathfrak{D}$ their values, and for a, b, &c., writing again $a-x$, $b-x$, &c., we have moreover

$$\begin{array}{ll|ll}
A^2u = \sqrt{c-b.b-d.c-d} & (a-x), & A^2v = \sqrt{\qquad ,, \qquad} & (a-y), \\
B^2u = \sqrt{c-a.c-d.a-d} & (b-x), & B^2v = \sqrt{\qquad ,, \qquad} & (b-y), \\
C^2u = \sqrt{a-b.a-d.b-d} & (c-x), & C^2v = \sqrt{\qquad ,, \qquad} & (c-y), \\
D^2u = \sqrt{c-b.c-a.a-b} & (d-x), & D^2v = \sqrt{\qquad ,, \qquad} & (d-y),
\end{array}$$

$$\begin{aligned}
A^2(u+v) &= \sqrt{\qquad ,, \qquad} \quad (a-z), \\
B^2(u+v) &= \sqrt{\qquad ,, \qquad} \quad (b-z), \\
C^2(u+v) &= \sqrt{\qquad ,, \qquad} \quad (c-z), \\
D^2(u+v) &= \sqrt{\qquad ,, \qquad} \quad (d-z),
\end{aligned}$$

the constant multipliers being of course the same in the three columns respectively. According to what precedes, the radical of the fourth line should be taken with the sign $-$. The functions $(bc,\ ad)$, &c., contained in the formulæ, require a transformation such as

$$(b-c)(bc,\ ad) = \begin{vmatrix} b-x.b-y, & c-x.c-y \\ b-a.b-d, & c-a.c-d \end{vmatrix},$$

in order to make them separately homogeneous in the differences $a-x$, $b-x$, $c-x$, $d-x$, and $a-y$, $b-y$, $c-y$, $d-y$, and therefore to make them expressible as linear functions of the squared functions A^2u, &c., and A^2v, &c., respectively: the formulæ then give the quotient-functions $A(u+v) \div D(u+v)$, &c., in terms of the quotient-functions of u and v respectively.

Doubly infinite product-forms.

58. The functions Au, Bu, Cu, Du may be expressed each as a doubly infinite product. Writing for shortness

$$\begin{aligned}
m \quad + \quad n.\frac{a}{\pi i} &= (m,\ n), \\
m+1+ \quad n.\frac{a}{\pi i} &= (\overline{m},\ n), \\
m \quad +(n+1)\frac{a}{\pi i} &= (m,\ \overline{n}), \\
m+1+(n+1)\frac{a}{\pi i} &= (\overline{m},\ \overline{n}),
\end{aligned}$$

then the formulæ are

$$Au = A0 \,.\, \Pi\Pi\left\{1+\frac{u}{(\bar{m},\ \bar{n})}\right\},$$

$$Bu = B0 \,.\, \Pi\Pi\left\{1+\frac{u}{(\bar{m},\ n)}\right\},$$

$$Cu = C0 \,.\, \Pi\Pi\left\{1+\frac{u}{(m,\ \bar{n})}\right\},$$

$$Du = D'0 \,.\, u\Pi\Pi\left\{1+\frac{u}{(m,\ n)}\right\},$$

where in all the formulæ, m and n denote even integers having all values whatever, zero included, from $-\infty$ to $+\infty$; only in the formula for Du, the term for which m and n are simultaneously $=0$, is to be omitted.

59. But a further definition in regard to the limits is required: first, we assume that m has the corresponding positive and negative values, and similarly that n has corresponding positive and negative values*; or say, in the four formulæ respectively, we consider m, n as extending

$$\begin{array}{llllll} m \text{ from } -\mu \text{ to } \mu+2, & n \text{ from } -\nu \text{ to } \nu+2, \\ \ \ ,, \quad ,, \quad -\mu \ ,, \ \mu+2, & \ \ ,, \quad ,, \quad -\nu \ ,, \ \nu, \\ \ \ ,, \quad ,, \quad -\mu \ ,, \ \mu\ , & \ \ ,, \quad ,, \quad -\nu \ ,, \ \nu+2, \\ \ \ ,, \quad ,, \quad -\mu \ ,, \ \mu\ , & \ \ ,, \quad ,, \quad -\nu \ ,, \ \nu, \end{array}$$

where μ and ν are each of them ultimately infinite. But, secondly, it is necessary that μ should be indefinitely larger than ν, or say that ultimately $\dfrac{\nu}{\mu}=0$.

60. In fact, transforming the q-series into products as in the *Fundamenta Nova*, and neglecting for the moment mere constant factors, we have

$$\begin{aligned} Au &= \phantom{\cos\tfrac{1}{2}\pi u\,} (1+2q\ \cos\pi u+q^2)(1+2q^3\cos\pi u+q^6)\ldots, \\ Bu &= \cos\tfrac{1}{2}\pi u\,(1+2q^2\cos\pi u+q^4)(1+2q^4\cos\pi u+q^8)\ldots, \\ Cu &= \phantom{\cos\tfrac{1}{2}\pi u\,} (1-2q\ \cos\pi u+q^2)(1-2q^3\cos\pi u+q^6)\ldots, \\ Du &= \sin\tfrac{1}{2}\pi u\,(1-2q^2\cos\pi u+q^4)(1-2q^4\cos\pi u+q^8)\ldots, \end{aligned}$$

and writing for a moment $\alpha=\dfrac{a}{\pi i}$ and therefore $q^{\frac{1}{2}}+q^{-\frac{1}{2}},\ =e^{\frac{1}{2}\pi\alpha i}+e^{-\frac{1}{2}\pi\alpha i},\ =2\cos\tfrac{1}{2}\pi\alpha$, &c., each of these expressions is readily converted into a singly infinite product of sines or cosines

$$\begin{aligned} Au &= \Pi\,.\cos\tfrac{1}{2}\pi\,(u+\bar{n}\alpha), \\ Bu &= \Pi\,.\cos\tfrac{1}{2}\pi\,(u+n\alpha), \\ Cu &= \Pi\,.\sin\tfrac{1}{2}\pi\,(u+\bar{n}\alpha), \\ Du &= \Pi\,.\sin\tfrac{1}{2}\pi\,(u+n\alpha), \end{aligned}$$

* This is more than is necessary; it would be enough if the ultimate values of m were $-\mu$, μ', μ and μ' being in a ratio of equality; and the like as regards n. But it is convenient that the numbers should be absolutely equal.

where $\bar{n}$ is written to denote $n+1$, and n has all positive or negative even values (zero included) from $-\infty$ to $+\infty$, or more accurately from $-\nu$ to ν, if ν is ultimately infinite.

61. The sines and cosines are converted into infinite products by the ordinary formulæ, which neglecting constant factors may conveniently be written

$$\sin \tfrac{1}{2}\pi u = \Pi\,(u+m), \quad \cos \tfrac{1}{2}\pi u = \Pi\,(u+\bar{m}),$$

where $\bar{m}$ is written to denote $m+1$, and m has all positive or negative even values (zero included) from $-\infty$ to $+\infty$, or more accurately from $-\mu$ to μ, if μ be ultimately infinite. But in applying these formulæ to the case of a function such as

$$\sin \tfrac{1}{2}\pi\,(u+n\alpha),$$

it is assumed that the limiting values μ, $-\mu$ of m are indefinitely large in regard to $u+n\alpha$; and therefore, since n may approach to its limiting value $\pm\nu$, it is necessary that μ should be indefinitely large in comparison with ν, or that $\frac{\nu}{\mu}=0$.

62. It is on account of this unsymmetry as to the limits $\frac{\nu}{\mu}=0$, $\frac{\mu}{\nu}=\infty$, that we have 1 as a quarter-period, $\frac{a}{\pi i}$ only as a quarter-quasi-period of the single theta-functions.

The transformation q to r, $\log q \log r = \pi^2$.

63. In general, if we consider the ratio of two such infinite products, where for the first the limits are $(\pm\mu, \pm\nu)$, and for the second they are $(\pm\mu', \pm\nu')$, and where for convenience we take $\mu>\mu'$, $\nu>\nu'$, then the quotient, say $[\mu, \nu]\div[\mu', \nu']$ is $=\exp.\,(Mu^2)$, where

$$M=-\tfrac{1}{8}\iint\frac{dm\,dn}{(m,\ n)^2}$$

taken over the area included between the two rectangles. We have

$$(m,\ n)=m+\frac{a}{\pi i}n, \ =m+i\theta n$$

suppose, where (a being negative) $\theta=-\frac{a}{\pi}$ is positive: the integral is

$$\iint\frac{dm\,dn}{(m+i\theta n)^2}, \ =\int dm\,.\,-\frac{1}{i\theta}\left(\frac{1}{m+i\theta n}\right)_{-\nu}^{\nu},$$

$$=\frac{1}{i\theta}\int dm\left(\frac{1}{m-i\theta\nu}-\frac{1}{m+i\theta\nu}\right),$$

$$=\frac{1}{i\theta}\log\frac{m-i\theta\nu}{m+i\theta\nu};$$

or finally between the proper limits the value is

$$= \frac{2}{i\theta}\left\{\log\left(\frac{\mu - i\theta\nu}{\mu + i\theta\nu}\right) - \log\left(\frac{\mu' - i\theta\nu'}{\mu' + i\theta\nu'}\right)\right\},$$

where the logarithms are

$$\log(\mu - i\theta\nu) = \log\sqrt{\mu^2 + \nu^2} - i\tan^{-1}\frac{\theta\nu}{\mu}, \text{ \&c.,}$$

and the $\tan^{-1}$ denotes always an arc between the limits $-\tfrac{1}{2}\pi$, $+\tfrac{1}{2}\pi$. Hence, if $\frac{\mu}{\nu} = \infty$, $\frac{\mu'}{\nu'} = 0$, the value is

$$\frac{2}{i\theta}(-0i - 0i + \tfrac{1}{2}\pi i + \tfrac{1}{2}\pi i) = \frac{2\pi}{\theta}, \ = -\frac{2\pi^2}{a}; \text{ or } K = \tfrac{1}{4}\frac{\pi^2}{a}.$$

Hence finally

$$[\mu \div \nu, \ = \infty] \div [\mu \div \nu, \ = 0] = \exp.\left(\tfrac{1}{4}\frac{\pi^2}{a}u^2\right).$$

64. We have a, $= \log q$, negative; hence taking r such that $\log q \log r = \pi^2$, we have $a' = \log r$, also negative; and r, like q, is positive and less than 1. We consider the theta-functions which depend on r in the same manner that the original functions did on q, say these are

$$A(u,\ r) = A\ (0,\ r)\ \ \Pi\Pi\left\{1 + \frac{u}{\bar{m} + \bar{n}\frac{a'}{\pi i}}\right\},$$

$$B(u,\ r) = B\ (0,\ r)\ \ \Pi\Pi\left\{1 + \frac{u}{\bar{m} + n\frac{a'}{\pi i}}\right\},$$

$$C(u,\ r) = C\ (0,\ r)\ \ \Pi\Pi\left\{1 + \frac{u}{m + \bar{n}\frac{a'}{\pi i}}\right\},$$

$$D(u,\ r) = D'(0,\ r)\,u\,\Pi\Pi\left\{1 + \frac{u}{m + n\frac{a'}{\pi i}}\right\},$$

limits as before, and in particular $\frac{\mu}{\nu} = \infty$; it is at once seen that if in the original functions, which I now call $A(u,\ q)$, $B(u,\ q)$, $C(u,\ q)$, $D(u,\ q)$, we write $\frac{au}{\pi i}$ for u, we obtain the same infinite products which present themselves in the expressions of the new functions $A(u,\ r)$, &c., only with a different condition as to the limits; we have for instance

$$\Pi\Pi\left(1 + \frac{\frac{au}{\pi i}}{m + n\frac{a}{\pi i}}\right) = \Pi\Pi\left(1 + \frac{u}{n - m\frac{a'}{\pi i}}\right), \ = \Pi\Pi\left(1 + \frac{u}{n + m\frac{a'}{\pi i}}\right),$$

which, interchanging m, n, and of course also μ, ν, is

$$=\Pi\Pi\left(1+\frac{u}{m+n\dfrac{a'}{\pi i}}\right),$$

with the condition $\dfrac{\mu}{\nu}=0$ instead of $\dfrac{\mu}{\nu}=\infty$. Hence disregarding for the moment constant factors, and observing that a is replaced by a', we have

$$D(u,\ r)\div D\left(\frac{au}{\pi i},\ q\right)=[\mu\div\nu,\ =\infty]\div[\mu\div\nu,\ =0]$$

$$=\exp.\left(\tfrac{1}{4}\frac{\pi^2}{a'}u^2\right),\ =\exp.\left(\tfrac{1}{4}u^2\log q\right).$$

65. We have equations of this form for the four functions, but with a proper constant multiplier in each equation: the equations, in fact, are

$$A(u,\ r)=\{A\ (0,\ r)\div A\ (0,\ q)\}\exp.\left(\tfrac{1}{4}u^2\log q\right)A\left(\frac{au}{\pi i},\ q\right),$$

$$B(u,\ r)=\{B\ (0,\ r)\div B\ (0,\ q)\}\quad\text{,,}\quad B\left(\frac{au}{\pi i},\ q\right),$$

$$C(u,\ r)=\{C\ (0,\ r)\div C\ (0,\ q)\}\quad\text{,,}\quad C\left(\frac{au}{\pi i},\ q\right),$$

$$D(u,\ r)=\{D'(0,\ r)\div D'(0,\ q)\}\frac{\pi i}{a}\quad\text{,,}\quad D\left(\frac{au}{\pi i},\ q\right).$$

It is to be observed that r is the same function of k' that q is of k. This would at once follow from Jacobi's equation $\log q=-\dfrac{\pi K'}{K}$, for then $\log q\log r=\pi^2$ and therefore $\log r=-\dfrac{\pi K'}{K}$ $\Big($only we are not at liberty to use the relation in question $\log q=-\dfrac{\pi K'}{K}\Big)$: assuming it to be true, we have

$$k=\frac{B^2(0,\ q)}{A^2(0,\ q)},\quad k'=\frac{C^2(0,\ q)}{A^2(0,\ q)},\quad K=-\frac{A(0,\ q)D'(0,\ q)}{B(0,\ q)C(0,\ q)},$$

$$k=\frac{C^2(0,\ r)}{A^2(0,\ r)},\quad k'=\frac{B^2(0,\ r)}{A^2(0,\ r)},\quad K'=-\frac{A(0,\ r)D'(0,\ r)}{B(0,\ r)C(0,\ r)},$$

$$\log q=-\frac{\pi K'}{K},\quad \log r=-\frac{\pi K}{K'},$$

where, if the identity of the two values of k or of the two values of k' were proved independently (as might doubtless be done), the required theorem, viz. that r is the same function of k' that q is of k, would follow conversely: and thence the other equations of the system.

66. We have, in the *Fundamenta Nova*, p. 175, [Jacobi's *Ges. Werke*, t. I., p. 227], the equation

$$\frac{H(iu, k)}{\Theta(0, k)} = i\sqrt{\frac{k}{k'}}\, e^{\frac{\pi u^2}{4KK'}}\, \frac{H(u, k')}{\Theta(0, k')};$$

writing here $K'u$ instead of u the equation becomes

$$\frac{H(iK'u, k)}{\Theta(0, k)} = i\sqrt{\frac{k}{k'}}\, \exp.\left(\tfrac{1}{4}\frac{\pi K'}{K}u^2\right).\frac{H(K'u, k')}{\Theta(0, k')},$$

or, what is the same thing,

$$\frac{D\left(\dfrac{au}{\pi i}, q\right)}{C(0, q)} = i\sqrt{\frac{k}{k'}}\, \exp.\left(-\tfrac{1}{4}u^2 \log q\right).\frac{D(u, r)}{C(0, r)}$$

which can be readily identified with the foregoing equation between $D\left(\dfrac{au}{\pi i}, q\right)$ and $D(u, r)$. But the real meaning of the transformation is best seen by means of the double-product formulæ.

THIRD PART.—THE DOUBLE THETA-FUNCTIONS.

Notations, &c.

67. We have here 16 functions $\vartheta\begin{pmatrix}\alpha\beta\\ \gamma\delta\end{pmatrix}(u, v)$: this notation by characteristics, containing each of them four numbers, is too cumbrous for ordinary use, and I therefore replace it by the current-number notation, in which the characteristics are denoted by the series of numbers 0, 1, 2,..., 15: we cannot in place of this introduce the single-and-double-letter notation A, B,..., AB, &c., for there is not here any correspondence of the two notations, nor is there anything in the definition of the functions which in anywise suggests the single-and-double-letter notation: this first presents itself in connexion with the relations between the functions given by the product-theorem: and as the product-theorem is based upon the notation by characteristics, it is proper to present the theorem in this notation, or in the equivalent current-number notation: and then to show how by the relations thus obtained between the functions we are led to the single-and-double-letter notation.

68. There are some other notations which may be referred to: and for showing the correspondence between them I annex the following table:—

THE DOUBLE THETA-FUNCTIONS.

Asterisk denotes the odd functions.	1. Current number.	2. Character.	3. Single-and-double-letter, Cayley.	4. Göpel.	5. Göpel-Cayley.	6. Rosenhain.	7. Weierstrass.	8. Kummer.
	ϑ_0	$\vartheta\begin{matrix}00\\00\end{matrix}$	BD	P'''	P_3	ϑ_{22}	ϑ_5	12
	1	$\begin{matrix}10\\00\end{matrix}$	CE	R'''	R_3	32	4	8
	2	$\begin{matrix}01\\00\end{matrix}$	CD	Q'''	Q_3	23	01	10
	3	$\begin{matrix}11\\00\end{matrix}$	BE	S'''	S_3	33	23	6
	4	$\begin{matrix}00\\10\end{matrix}$	AC	P'	P_1	02	34	4
*	5	$\begin{matrix}10\\10\end{matrix}$	C	iR'	R_1	12	3	16
	6	$\begin{matrix}01\\10\end{matrix}$	AB	Q'	Q_1	03	2	2
*	7	$\begin{matrix}11\\10\end{matrix}$	B	iS'	S_1	13	24	14
	8	$\begin{matrix}00\\01\end{matrix}$	BC	P''	P_2	20	12	9
	9	$\begin{matrix}10\\01\end{matrix}$	DE	R''	R_2	30	03	5
*	10	$\begin{matrix}01\\01\end{matrix}$	F	iQ''	Q_2	21	02	11
*	11	$\begin{matrix}11\\01\end{matrix}$	A	iS''	S_2	31	13	7
	12	$\begin{matrix}00\\11\end{matrix}$	AD	P	P	00	0	1
*	13	$\begin{matrix}10\\11\end{matrix}$	D	iR	R	10	04	13
*	14	$\begin{matrix}01\\11\end{matrix}$	E	iQ	Q	01	1	3
	15	$\begin{matrix}11\\11\end{matrix}$	AE	S	S	11	14	15

69. These are the notations:—

1. By current-numbers. It may be remarked that the series was taken 0, 1, ..., 15, instead of 1, 2, ..., 16, in order that 0 might correspond to the characteristic $\begin{matrix}00\\00\end{matrix}$;

2. By characteristics;

3. By single-and-double letters;

4. Göpel's, in his paper above referred to, and

5. The same as used in my paper above referred to;

6. Rosenhain's, in his paper above referred to;

7. Weierstrass', as quoted by Königsberger in his paper "Ueber die Transformation der *Abel*schen Functionen erster Ordnung," *Crelle-Borchardt*, t. LXIV. (1865), p. 17, and by Borchardt in his paper above referred to;

8. Not a theta-notation, but the series of current-numbers used in Kummer's Memoir "Ueber die algebraischen Strahlen-systeme," *Berl. Abh.* 1866, for the nodes of his 16-nodal quartic surface, and connected with the double theta-functions in my paper above referred to.

But in the present memoir only the first three columns of the table need be attended to.

70. It will be convenient to introduce here some other remarks as to notation, &c.

The letter c is used in connexion with the zero values $u=0$, $v=0$ of the arguments, viz.:—

$$\vartheta_0,\ \vartheta_1,\ \vartheta_2,\ \vartheta_3,\ \vartheta_4,\ \vartheta_6,\ \vartheta_8,\ \vartheta_9,\ \vartheta_{12},\ \vartheta_{15}$$

are even functions, and the corresponding zero-functions are denoted by

$$c_0,\ c_1,\ c_2,\ c_3,\ c_4,\ c_6,\ c_8,\ c_9,\ c_{12},\ c_{15}:$$

there are thus 10 constants c.

When (u, v) are indefinitely small each of these functions is of course equal to its zero-value *plus* a quadric term in (u, v), and we may write in general

$$\vartheta = c + \tfrac{1}{2}(c''', c^{\text{iv}}, c^{\text{v}} \between u, v)^2:$$

there are thus 30 constants c''', c^{iv}, c^{v}.

The remaining functions

$$\vartheta_5,\ \vartheta_7,\ \vartheta_{10},\ \vartheta_{11},\ \vartheta_{13},\ \vartheta_{14}$$

are odd functions vanishing for $u=0$, $v=0$; when these arguments are indefinitely small, we may write in general

$$\vartheta = (c', c'' \between u, v):$$

there are thus 12 constants c', c''.

71. All these constants are in the first instance given as transcendental functions of the parameters, or say rather of exp. a, exp. h, exp. b, which exponentials correspond to the q of the single theory: viz., in a notation which will be readily understood, the constants c, c''', c^{iv}, c^{v} of the even functions are

$$\Sigma \text{ exp.} \begin{pmatrix} m+\alpha, & n+\beta \\ \gamma & \delta \end{pmatrix};$$

$$-\tfrac{1}{2}\pi^2\Sigma\,(m+\alpha)^2,\ 2\,(m+\alpha)\,(n+\beta),\ (n+\beta)^2,\ \text{exp.} \begin{pmatrix} m+\alpha, & n+\beta \\ \gamma & \delta \end{pmatrix};$$

and the constants c', c'' of the odd functions are

$$\tfrac{1}{2}\pi i\Sigma\,(m+\alpha),\ (n+\beta),\ \text{exp.} \begin{pmatrix} m+\alpha, & n+\beta \\ \gamma & \delta \end{pmatrix}.$$

72. It is convenient for the verification of results and otherwise, to have the values of the functions, belonging to small values of exp. $(-a)$, exp. $(-b)$; if to avoid fractional exponents we regard these and exp. $(-h)$ as fourth powers, and write

$$\text{exp.}\,(-a) = Q^4,\ \text{exp.}\,(-h) = R^4,\ \text{exp.}\,(-b) = S^4,$$

also

$$QR^2S = \Lambda,\ QR^{-2}S = \Lambda',\ \text{and therefore } \Lambda\Lambda' = Q^2S^2,$$

then attending only to the lowest powers of Q, S we find without difficulty

$\vartheta_0\,(u) = 1,$	and therefore also	$c_0 = 1,$
$\vartheta_1 = 2Q\cos\tfrac{1}{2}\pi u,$		$c_1 = 2Q,$
$\vartheta_2 = 2S\cos\tfrac{1}{2}\pi v,$		$c_2 = 2S,$
$\vartheta_3 = 2\Lambda\cos\tfrac{1}{2}\pi\,(u+v) + 2\Lambda'\cos\tfrac{1}{2}\pi\,(u-v),$		$c_3 = 2\Lambda + 2\Lambda',$
$\vartheta_4 = 1,$		$c_4 = 1,$
$\vartheta_5 = -2Q\sin\tfrac{1}{2}\pi u,$		
$\vartheta_6 = 2S\cos\tfrac{1}{2}\pi v,$		$c_6 = 2S,$
$\vartheta_7 = -2\Lambda\sin\tfrac{1}{2}\pi\,(u+v) - 2\Lambda'\sin\tfrac{1}{2}\pi\,(u-v),$		
$\vartheta_8 = 1,$		$c_8 = 1,$
$\vartheta_9 = 2Q\cos\tfrac{1}{2}\pi u,$		$c_9 = 2Q,$
$\vartheta_{10} = -2S\sin\tfrac{1}{2}\pi v,$		
$\vartheta_{11} = -2\Lambda\sin\tfrac{1}{2}\pi\,(u+v) + 2\Lambda'\sin\tfrac{1}{2}\pi\,(u-v),$		
$\vartheta_{12} = 1,$		$c_{12} = 1,$
$\vartheta_{13} = -2Q\sin\tfrac{1}{2}\pi u,$		
$\vartheta_{14} = -2S\sin\tfrac{1}{2}\pi v,$		
$\vartheta_{15} = -2\Lambda\cos\tfrac{1}{2}\pi\,(u+v) + 2\Lambda'\cos\tfrac{1}{2}\pi\,(u-v),$		$c_{15} = -2\Lambda + 2\Lambda'.$

73. The expansions might be carried further; we have for instance

$$\vartheta_0\ (u) = 1 + 2Q^4 \cos \pi u + 2S^4 \cos \pi v, \qquad c_0 = 1 + 2Q^4 + 2S^4,$$

$$\vartheta_4 \quad = 1 - 2Q^4 \quad ,, \quad + 2S^4 \quad ,, \quad , \qquad c_4 = 1 - 2Q^4 + 2S^4,$$

$$\vartheta_8 \quad = 1 + 2Q^4 \quad ,, \quad - 2S^4 \quad ,, \quad , \qquad c_8 = 1 + 2Q^4 - 2S^4,$$

$$\vartheta_{12} \quad = 1 - 2Q^4 \quad ,, \quad - 2S^4 \quad ,, \quad , \qquad c_{12} = 1 - 2Q^4 - 2S^4,$$

$$\vartheta_1 \quad = \quad 2Q \cos \tfrac{1}{2}\pi u + 2Q^9 \cos \tfrac{3}{2}\pi u + 2A \cos \tfrac{1}{2}\pi (u + 2v) + 2A' \cos \tfrac{1}{2}\pi (u - 2v),$$

$$c_1 = 2Q + 2Q^9 + 2A + 2A',$$

$$\vartheta_5 \quad = -2Q \sin \tfrac{1}{2}\pi u + 2Q^9 \sin \tfrac{3}{2}\pi u - 2A \sin \tfrac{1}{2}\pi (u + 2v) - 2A' \sin \tfrac{1}{2}\pi (u - 2v),$$

$$\vartheta_9 \quad = \quad 2Q \cos \tfrac{1}{2}\pi u + 2Q^9 \cos \tfrac{3}{2}\pi u - 2A \cos \tfrac{1}{2}\pi (u + 2v) - 2A' \cos \tfrac{1}{2}\pi (u - 2v),$$

$$c_9 = 2Q + 2Q^9 - 2A - 2A',$$

$$\vartheta_{13} \quad = -2Q \sin \tfrac{1}{2}\pi u + 2Q^9 \sin \tfrac{3}{2}\pi u + 2A \sin \tfrac{1}{2}\pi (u + 2v) + 2A' \sin \tfrac{1}{2}\pi (u - 2v),$$

in which last formulæ

$$A = QR^4S^4, = \frac{\Lambda^2 S^2}{Q};\ A' = QR^{-4}S^4, = \frac{\Lambda'^2 S^2}{Q}.$$

74. In the single-and-double-letter notation we have six letters A, B, C, D, E, F; and the duads AB, AC, ..., DE are used as abbreviations for the double triads ABF, CDE, &c., the letter F always accompanying the expressed duad; there are thus in all six single-letter symbols, and 10 double-letter symbols, in all 16 symbols, corresponding to the double-theta functions, as already mentioned in the order

ϑ 0	1	2	3	4	5	6	7	8	9	10	11	12	13	14	15
BD,	CE,	CD,	BE,	AC,	C,	AB,	B,	BC,	DE,	F,	A,	AD,	D,	E,	AE,

where observe that the single letters C, B, F, A, D, E correspond to the odd functions 5, 7, 10, 11, 13, 14 respectively.

75. The ground of the notation is as follows:—

The algebraical relations between the double theta-functions are such that, introducing six constant quantities a, b, c, d, e, f and two variable quantities x, y, it is allowable to express the 16 functions as proportional to given functions of these quantities $(a, b, c, d, e, f;\ x, y)$; viz. there are six functions the notations of which depend on the single letters a, b, c, d, e, f respectively, and 10 functions the notations of which depend on the pairs ab, ac, ad, ae, bc, bd, be, cd, ce, de respectively: each of the 16 functions is, in fact, proportional to the product of two factors, viz. a constant factor depending only on (a, b, c, d, e, f), and a variable factor containing also x and y. Attending in the first instance to the variable factors, and writing for shortness

$$a - x,\ b - x,\ c - x,\ d - x,\ e - x,\ f - x = \mathrm{a},\ \mathrm{b},\ \mathrm{c},\ \mathrm{d},\ \mathrm{e},\ \mathrm{f};\ x - y = \theta;$$

$$a - y,\ b - y,\ c - y,\ d - y,\ e - y,\ f - y = \mathrm{a}_{,},\ \mathrm{b}_{,},\ \mathrm{c}_{,},\ \mathrm{d}_{,},\ \mathrm{e}_{,},\ \mathrm{f}_{,};$$

these are of the forms

$$\sqrt{a} = \sqrt{\text{aa}_{,}}, \quad \sqrt{ab} = \frac{1}{\theta}\{\sqrt{\text{abfc}_{,}\text{d}_{,}\text{e}_{,}} + \sqrt{\text{a}_{,}\text{b}_{,}\text{f}_{,}\text{cde}}\}.$$

I remark that to avoid ambiguity the squares of these expressions are throughout written as $(\sqrt{a})^2$ and $(\sqrt{ab})^2$ respectively.

76. There is, for the constant factors, a like single-and-double-letter notation which will be mentioned presently; but in the first instance I use for the even functions the before-mentioned 10 letters c, and for the odd ones six letters k. I assume that the values $x, y = \infty, \infty$ (ratio not determined) correspond to the values $u=0$, $v=0$ of the arguments; and that ω is a function of (x, y) which, when (x, y) are interchanged, changes only its sign, and which for indefinitely large values of (x, y) becomes $=\dfrac{x-y}{(xy)^{\frac{3}{4}}}$. This being so, we write (adding a second column which will be afterwards explained)

$$\begin{array}{ll}
\vartheta & \\
0 = BD = \omega c_0 \sqrt{bd}, & c_0 = \lambda \sqrt[4]{bd}, \\
1 = CE = \text{,,}\, c_1 \sqrt{ce}, & c_1 = \text{,,} \sqrt[4]{ce}, \\
2 = CD = \text{,,}\, c_2 \sqrt{cd}, & c_2 = \text{,,} \sqrt[4]{cd}, \\
3 = BE = \text{,,}\, c_3 \sqrt{be}, & c_3 = \text{,,} \sqrt[4]{be}, \\
4 = AC = \text{,,}\, c_4 \sqrt{ac}, & c_4 = \text{,,} \sqrt[4]{ac}, \\
5 = \ C\ = \text{,,}\, k_5 \sqrt{c}, & k_5 = \text{,,} \sqrt[4]{c}, \\
6 = AB = \text{,,}\, c_6 \sqrt{ab}, & c_6 = \text{,,} \sqrt[4]{ab}, \\
7 = \ B\ = \text{,,}\, k_7 \sqrt{b}, & k_7 = \text{,,} \sqrt[4]{b}, \\
8 = BC = \text{,,}\, c_8 \sqrt{bc}, & c_8 = \text{,,} \sqrt[4]{bc}, \\
9 = DE = \text{,,}\, c_9 \sqrt{de}, & c_9 = \text{,,} \sqrt[4]{de}, \\
10 = \ F\ = \text{,,}\, k_{10} \sqrt{f}, & k_{10} = \text{,,} \sqrt[4]{f}, \\
11 = \ A\ = \text{,,}\, k_{11} \sqrt{a}, & k_{11} = \text{,,} \sqrt[4]{a}, \\
12 = AD = \text{,,}\, c_{12} \sqrt{ad}, & c_{12} = \text{,,} \sqrt[4]{ad}, \\
13 = \ D\ = \text{,,}\, k_{13} \sqrt{d}, & k_{13} = \text{,,} \sqrt[4]{d}, \\
14 = \ E\ = \text{,,}\, k_{14} \sqrt{e}, & k_{14} = \text{,,} \sqrt[4]{e}, \\
15 = AE = \text{,,}\, c_{15} \sqrt{ae}, & c_{15} = \text{,,} \sqrt[4]{ae};
\end{array}$$

viz. here, on writing $x, y = \infty, \infty$, each of the functions $\sqrt{bd}$, &c. becomes $= 2\dfrac{(xy)^{\frac{3}{4}}}{x-y}$; and each of the functions $\sqrt{a}$, &c., becomes $=\sqrt{xy}$; hence by reason of the assumed form of ω, the odd functions each vanish (their evanescent values being proportional to k_5, k_7, k_{10}, k_{11}, k_{13}, k_{14} respectively), while the even functions become equal to c_0, c_1, c_2, c_3, c_4, c_6, c_8, c_9, c_{12}, c_{15} respectively.

Observe further that on interchanging x, y, the even functions remain unaltered, while the odd functions change their sign; that is, the interchange of x, y corresponds to the change u, v into $-u$, $-v$.

77. As to the values of the 10 c's and the six k's in terms of a, b, c, d, e, f, these are proportional to fourth roots, $\sqrt[4]{a}$, &c., $\sqrt[4]{ab}$, &c.; in $\sqrt[4]{a}$, a is in the first instance regarded as standing for the pentad $bcdef$, and then this is used to denote a product of differences $bc.bd.be.bf.cd.ce.cf.de.df.ef$; similarly ab is in the first instance regarded as standing for the double triad $abf.cde$, and then each of these triads is used to denote a product of differences, $ab.af.bf$ and $cd.ce.de$ respectively. The order of the letters is always the alphabetical one, viz. the single letters and duads denote pentads and double triads, thus:

$$\begin{array}{ll} a = bcdef, & ab = abf.cde, \\ b = acdef, & ac = acf.bde, \\ c = abdef, & ad = adf.bce, \\ d = abcef, & ae = aef.bcd, \\ e = abcdf, & bc = bcf.ade, \\ f = abcde, & bd = bdf.ace, \\ & be = bef.acd, \\ & cd = cdf.abe, \\ & ce = cef.abd, \\ & de = def.abc. \end{array}$$

There is no fear of ambiguity in writing (and we accordingly write) the squares of $\sqrt[4]{a}$ and $\sqrt[4]{ab}$ as $\sqrt{a}$ and $\sqrt{ab}$ respectively; the fourth powers are written $(\sqrt{a})^2$ and $(\sqrt{ab})^2$; the double stroke of the radical symbol $\sqrt{}$ is in every case perfectly distinctive.

This being so we have as above $c_0 = \lambda\sqrt[4]{bd}$, &c., $k_5 = \lambda\sqrt[4]{c}$, &c.: it is, however, important to notice that the fourth roots in question do not denote positive values, but they are fourth roots each taken with its proper sign ($+$, $-$, $+i$, $-i$, as the case may be) so as to satisfy the identical relations which exist between the sixteen constants; and it is not easy to determine the signs.

The variables x, y are connected with u, v by the differential relations

$$\sigma\, du + \tau\, dv = -\tfrac{1}{2}\left\{\frac{dx}{\sqrt{X}} - \frac{dy}{\sqrt{Y}}\right\},$$

$$\varpi\, du + \rho\, dv = -\tfrac{1}{2}\left\{\frac{x\,dx}{\sqrt{X}} - \frac{y\,dy}{\sqrt{Y}}\right\},$$

where $X = \text{abcdef}$, $Y = \text{a}_{,}\text{b}_{,}\text{c}_{,}\text{d}_{,}\text{e}_{,}\text{f}_{,}$; which equations contain the constants ϖ, ρ, σ, τ, the values of which will be afterwards connected with the other constants.

78. The c's are expressed as functions of four quantities α, β, γ, δ, and connected with each other, and with the constants a, b, c, d, e, f, by the formulæ

$$\begin{aligned}
c^2 \qquad & \\
0 &= \alpha^2+\beta^2+\gamma^2+\delta^2 = \omega_0{}^2\sqrt{bd},\\
1 &= 2(\alpha\beta+\gamma\delta) \quad = \;,, \;\sqrt{ce},\\
2 &= 2(\alpha\gamma+\beta\delta) \quad = \;,, \;\sqrt{cd},\\
3 &= 2(\alpha\delta+\beta\gamma) \quad = \;,, \;\sqrt{be},\\
4 &= \alpha^2-\beta^2+\gamma^2-\delta^2 = \;,, \;\sqrt{ac},\\
6 &= 2(\alpha\gamma-\beta\delta) \quad = \;,, \;\sqrt{ab},\\
8 &= \alpha^2+\beta^2-\gamma^2-\delta^2 = \;,, \;\sqrt{bc},\\
9 &= 2(\alpha\beta-\gamma\delta) \quad = \;,, \;\sqrt{de},\\
12 &= \alpha^2-\beta^2-\gamma^2-\delta^2 = \;,, \;\sqrt{ad},\\
15 &= 2(\alpha\delta-\beta\gamma) \quad = \;,, \;\sqrt{ae}.
\end{aligned}$$

It hence appears that we can form an arrangement

$$\left|\begin{matrix} c_{12}^2, & c_1^2, & c_6^2 \\ c_9^2, & -c_4^2, & c_3^2 \\ c_2^2, & -c_{15}^2, & -c_8^2 \end{matrix}\right| \div c_0^2 \qquad = \left|\begin{matrix} a\,, & b\,, & c \\ a', & b', & c' \\ a'', & b'', & c'' \end{matrix}\right|,$$

a system of coefficients in the transformation between two sets of rectangular coordinates.

We have, between the squares of these coefficients of transformation, a system of $6+9$ equations

$$\begin{aligned}
a^2 + b^2 + c^2 &= 1,\\
a'^2 + b'^2 + c'^2 &= 1,\\
a''^2 + b''^2 + c''^2 &= 1,\\
a^2 + a'^2 + a''^2 &= 1,\\
b^2 + b'^2 + b''^2 &= 1,\\
c^2 + c'^2 + c''^2 &= 1,
\end{aligned}$$

$$\begin{gathered}
b^2+c^2 = a'^2+a''^2, \quad b'^2+c'^2 = a''^2+a^2, \quad b''^2+c''^2 = a^2+a'^2,\\
c^2+a^2 = b'^2+b''^2, \quad c'^2+a'^2 = b''^2+b^2, \quad c''^2+a''^2 = b^2+b'^2,\\
a^2+b^2 = c'^2+c''^2, \quad a'^2+b'^2 = c''^2+c^2, \quad a''^2+b''^2 = c^2+c'^2:
\end{gathered}$$

that is,

c^4	c^4	c^4	c^4
12	+ 1	+ 6	− 0
9	+ 4	+ 3	− 0
2	+ 15	+ 8	− 0
12	+ 9	+ 2	− 0
1	+ 4	+ 15	− 0
6	+ 3	+ 8	− 0
1	+ 6	− 9	− 2
6	+ 12	− 4	− 15
12	+ 1	− 3	− 8
4	+ 3	− 2	− 12
3	+ 9	− 15	− 1
9	+ 4	− 8	− 6
15	+ 8	− 12	− 9
8	+ 2	− 1	− 4
2	+ 15	− 6	− 3

$= 0$;

and we have, between their products, a system of $6+9$ equations

$$a'a'' + b'b'' + c'c'' = 0,$$

$$a''a + b''b + c''c = 0,$$

$$aa' + bb' + cc' = 0,$$

$$bc + b'c' + b''c'' = 0,$$

$$ca + c'a' + c''a'' = 0,$$

$$ab + a'b' + a''b'' = 0,$$

$$\begin{array}{lll} a\,,\ \ b\,,\ \ c & = b'c''-b''c', & c'a''-c''a', & a'b''-a''b', \\ a',\ \ b',\ \ c' & b''c-bc'', & c''a-ca'', & ab''-a''b, \\ a'',\ \ b'',\ \ c'' & bc'-b'c\,, & ca'-c'a\,, & ab'-a'b\,: \end{array}$$

that is,

c^2	c^2	c^2	c^2	c^2	c^2
9	2	+ 4	15	− 3	8
2	12	− 15	1	− 8	6
12	9	− 1	4	+ 6	3
1	6	− 4	3	+ 15	8
6	12	+ 3	9	− 8	2
12	1	− 9	4	− 2	15
− 0	12	+ 4	8	+ 3	15
− 0	1	+ 3	2	+ 8	9
− 0	6	− 9	15	+ 2	4
− 0	9	− 15	6	+ 8	1
+ 0	4	− 8	12	− 6	2
− 0	3	− 12	15	− 1	2
− 0	2	+ 1	3	+ 4	6
+ 0	15	+ 6	9	− 3	12
+ 0	8	− 12	4	− 9	1

$= 0$;

each of the first set of 15 giving a homogeneous linear relation between four terms c^4; and each of the second set giving a homogeneous linear relation between three terms $c^2 . c^2$, formed with the 10 constants c. Thus the first equation is

$$c_{12}{}^4 + c_1{}^4 + c_6{}^4 - c_0{}^4 = 0;$$

and so for the other lines of the two diagrams.

79. I form in the two notations the following tables:—

TABLE OF THE 16 KUMMER HEXADS.

A	A	A	A	A	B	B	B	B	C	C	C	D	D	E	A
B	C	D	E	F	C	D	E	F	D	E	F	E	F	F	B
AB	AC	AD	AE	AB	BC	BD	BE	AB	CD	CE	AC	DE	AD	AE	C
CD	BD	BC	BC	AC	AD	AC	AC	BC	AB	AB	BC	AB	BD	BE	D
CE	BE	BE	BD	AD	AE	AE	AD	BD	AE	AD	CD	AC	CD	CE	E
DE	DE	CE	CD	AE	DE	CE	CD	BE	BE	BD	CE	BC	DE	DE	F

=

11	11	11	11	11	7	7	7	7	5	5	5	13	13	14	11
7	5	13	14	10	5	13	14	10	13	14	10	14	10	10	7
6	4	14	12	6	8	0	3	6	2	1	4	9	12	15	5
2	0	8	8	4	12	4	4	8	6	6	8	6	0	3	13
1	3	3	3	12	15	15	12	0	15	12	2	4	2	1	14
9	9	1	2	15	9	1	2	3	3	0	1	8	9	9	10

80. TABLE OF THE 120 PAIRS.

A.B	A.C	A.D	A.E	A.F	B.C	B.D	B.E	B.F	C.D	C.E	C.F	D.E	D.F	E.F
AC.BC	AB.BC	AB.BD	AB.BE	BC.DE	AB.AC	AB.AD	AB.AE	AC.DE	AC.AD	AC.AE	AB.DE	AD.AE	AB.CE	AB.CD
AD.BD	AD.CD	AC.CD	AC.CE	BD.CE	BD.CD	BC.CD	BC.CE	AD.CE	BC.BD	BC.BE	AD.BE	BD.BE	AC.BE	AC.BD
AE.BE	AE.CE	AE.DE	AD.DE	BE.CD	BE.CE	BE.DE	BD.DE	AE.CD	CE.DE	CD.DE	AE.BD	CD.CE	AE.BC	AD.BC
F.AB	F.AC	F.AD	F.AE	B.AB	F.BC	F.BD	F.BE	A.AB	F.CD	F.CE	A.AC	F.DE	A.AD	A.AE
C.DE	B.DE	B.CE	B.CD	C.AC	A.DE	A.CE	A.CD	C.BC	A.BE	A.BD	B.BC	A.BC	B.BD	B.BE
D.CE	D.BE	C.BE	C.BD	D.AD	D.AE	C.AE	C.AD	D.BD	B.AE	B.AD	D.CD	B.AC	C.CD	C.CE
E.CD	E.BD	E.BC	D.BC	E.AE	E.AD	E.AC	D.AC	E.BE	E.AB	D.AB	E.CE	C.AB	E.DE	D.DE

11.7	11.5	11.13	11.14	11.10	7.5	7.13	7.14	7.10	5.13	5.14	5.10	13.14	13.10	14.10
4.8	6.8	6.0	6.3	8.9	6.4	6.12	6.15	4.9	4.12	4.15	6.9	12.15	6.1	6.2
12.0	12.2	4.2	4.1	0.1	0.2	8.2	8.1	12.1	8.0	8.3	12.3	0.3	4.3	4.0
15.3	15.1	15.9	12.9	3.2	3.1	3.9	0.9	15.2	1.9	2.9	15.0	2.1	15.8	12.8
10.6	10.4	10.12	10.15	7.6	10.8	10.0	10.3	11.6	10.2	10.3	11.4	10.9	11.12	11.15
5.9	7.9	7.1	7.2	5.4	11.9	11.1	11.2	5.8	11.3	11.0	7.8	11.8	7.0	7.3
13.1	13.3	5.3	5.0	13.12	13.15	5.15	5.12	13.0	7.15	7.12	13.2	7.4	5.2	5.1
14.2	14.0	14.8	13.8	14.15	14.12	14.4	13.4	14.3	14.6	13.6	14.1	5.6	14.9	13.9

81. TABLE OF THE 60 GÖPEL TETRADS.

$A.B.AE.BE$ $A.B.AD.BD$ $A.B.AC.BC$	$C.D.CE.DE$ $C.E.CD.DE$ $C.F.AB.DE$	$E.F.AB.CD$ $D.F.AB.CE$ $D.E.CD.CE$	$AC.BD.AD.BC$ $AC.BE.AE.BC$ $AD.BE.AE.BD$
$A.C.AE.CE$ $A.C.AD.CD$ $A.C.AB.BC$	$B.D.BE.DE$ $B.E.BD.DE$ $B.F.AC.DE$	$E.F.AC.BD$ $D.F.AC.BE$ $D.E.BD.BE$	$AB.CD.AD.BC$ $AB.CE.AE.BC$ $AD.CE.AE.CD$
$A.D.AE.DE$ $A.D.AC.CD$ $A.D.AB.BD$	$B.C.BE.CE$ $B.E.BC.CE$ $B.F.AD.CE$	$E.F.AD.BC$ $C.F.AD.BE$ $C.E.CD.DE$	$AB.CD.AC.BD$ $AB.DE.AE.BD$ $AC.DE.AE.CD$
$A.E.AD.DE$ $A.E.AC.CE$ $A.E.AB.BE$	$B.C.BD.CD$ $B.D.BC.CD$ $B.F.AE.CD$	$D.F.AE.BC$ $C.F.AE.BD$ $C.D.BC.BD$	$AB.CE.AC.BE$ $AB.DE.AD.BE$ $AC.DE.AD.CE$
$A.F.BC.DE$ $A.F.BD.CE$ $A.F.BE.CD$	$B.C.AB.AC$ $B.D.AB.AD$ $B.E.AB.AE$	$D.E.AD.AE$ $C.E.AC.AE$ $C.D.AC.AD$	$BD.CE.BE.CD$ $BC.DE.BE.CD$ $BC.DE.BD.CE$

=

11 7 15 3 11 7 12 0 11 7 4 8	5 13 1 9 5 14 2 9 5 10 6 9	14 10 6 2 13 10 6 1 13 14 2 1	4 0 12 8 4 3 15 8 12 3 15 0
11 5 15 1 11 5 12 2 11 5 6 8	7 13 3 9 7 14 0 9 7 10 4 9	14 10 4 0 13 10 4 3 13 14 0 3	6 2 12 8 6 1 15 8 12 1 15 2
11 13 15 9 11 13 4 2 11 13 6 0	7 5 3 1 7 14 8 1 7 10 12 1	14 10 12 8 5 10 12 3 5 14 2 9	6 2 4 0 6 9 15 0 4 9 15 2
11 14 12 9 11 14 4 1 11 14 6 3	7 5 0 2 7 13 8 2 7 10 15 2	13 10 15 8 5 10 15 0 5 13 8 0	6 1 4 3 6 9 12 3 4 9 12 1
11 10 8 9 11 10 0 1 11 10 3 2	7 5 6 4 7 13 6 12 7 14 6 15	13 14 12 15 5 14 4 15 5 13 4 12	0 1 3 2 8 9 3 2 8 9 0 1

The product-theorem, and its results.

82. The product-theorem was

$$\vartheta\begin{pmatrix}\alpha, & \beta \\ \gamma, & \delta\end{pmatrix}(u+u').\vartheta\begin{pmatrix}\alpha', & \beta' \\ \gamma', & \delta'\end{pmatrix}(u-u')$$

$$=\Sigma\Theta\begin{matrix}\tfrac{1}{2}(\alpha+\alpha')+p, & \tfrac{1}{2}(\beta+\beta')+q \\ \gamma+\gamma' \quad , & \delta+\delta\end{matrix}(2u).\Theta\begin{matrix}\tfrac{1}{2}(\alpha-\alpha')+p, & \tfrac{1}{2}(\beta-\beta')+q \\ \gamma-\gamma' \quad , & \delta-\delta'\end{matrix}(2u'),$$

where only one argument is exhibited, viz. $u+u'$, $u-u'$, $2u$, $2u'$ are written in place of $(u+u', v+v')$, $(u-u', v-v')$, $(2u, 2v)$, $(2u', 2v')$ respectively. The expression on the right-hand side is always a sum of four terms, corresponding to the values $(0, 0)$, $(1, 0)$, $(0, 1)$, and $(1, 1)$ of (p, q). For the development of the results it was found convenient to use the following auxiliary diagram.

Upper half of characteristic.

α	β	α'	β'	$\frac{1}{2}(\alpha+\alpha')$	$\frac{1}{2}(\beta+\beta')$	$\frac{1}{2}(\alpha-\alpha')$	$\frac{1}{2}(\beta-\beta')$	$\frac{1}{2}(\alpha+\alpha')+1$	$\frac{1}{2}(\beta+\beta')$	$\frac{1}{2}(\alpha-\alpha')+1$	$\frac{1}{2}(\beta-\beta')$	$\frac{1}{2}(\alpha+\alpha')$	$\frac{1}{2}(\beta+\beta')+1$	$\frac{1}{2}(\alpha-\alpha')$	$\frac{1}{2}(\beta-\beta')+1$	$\frac{1}{2}(\alpha+\alpha')+1$	$\frac{1}{2}(\beta+\beta')+1$	$\frac{1}{2}(\alpha-\alpha')+1$	$\frac{1}{2}(\beta-\beta')+1$
0	0	0	0	0	0	0	0	1	0	1	0	0	1	0	1	1	1	1	1
1	0	0	0	$\frac{1}{2}$	0	$\frac{1}{2}$	0	$\frac{3}{2}$	0	$\frac{3}{2}$	0	$\frac{1}{2}$	1	$\frac{1}{2}$	1	$\frac{3}{2}$	1	$\frac{3}{2}$	1
0	1	0	0	0	$\frac{1}{2}$	0	$\frac{1}{2}$	1	$\frac{1}{2}$	1	$\frac{1}{2}$	0	$\frac{3}{2}$	0	$\frac{3}{2}$	1	$\frac{3}{2}$	1	$\frac{3}{2}$
1	1	0	0	$\frac{1}{2}$	$\frac{1}{2}$	$\frac{1}{2}$	$\frac{1}{2}$	$\frac{3}{2}$	$\frac{1}{2}$	$\frac{3}{2}$	$\frac{1}{2}$	$\frac{1}{2}$	$\frac{3}{2}$	$\frac{1}{2}$	$\frac{3}{2}$	$\frac{3}{2}$	$\frac{3}{2}$	$\frac{3}{2}$	$\frac{3}{2}$
0	0	1	0	$\frac{1}{2}$	0	$\frac{3}{2}$	0	$\frac{1}{2}$	0	$\frac{1}{2}$	0	$\frac{1}{2}$	1	$\frac{3}{2}$	1	$\frac{3}{2}$	1	$\frac{1}{2}$	1
1	0	1	0	1	0	0	0	0	0	1	0	1	1	0	1	0	1	1	1
0	1	1	0	$\frac{1}{2}$	$\frac{1}{2}$	$\frac{3}{2}$	$\frac{1}{2}$	$\frac{3}{2}$	$\frac{1}{2}$	$\frac{1}{2}$	$\frac{1}{2}$	$\frac{1}{2}$	$\frac{3}{2}$	$\frac{3}{2}$	$\frac{3}{2}$	$\frac{3}{2}$	$\frac{3}{2}$	$\frac{1}{2}$	$\frac{3}{2}$
1	1	1	0	1	$\frac{1}{2}$	0	$\frac{1}{2}$	0	$\frac{1}{2}$	1	$\frac{1}{2}$	1	$\frac{3}{2}$	0	$\frac{3}{2}$	0	$\frac{3}{2}$	1	$\frac{3}{2}$
0	0	0	1	0	$\frac{1}{2}$	0	$\frac{3}{2}$	1	$\frac{1}{2}$	1	$\frac{3}{2}$	0	$\frac{3}{2}$	0	$\frac{1}{2}$	1	$\frac{3}{2}$	1	$\frac{1}{2}$
1	0	0	1	$\frac{1}{2}$	$\frac{1}{2}$	$\frac{1}{2}$	$\frac{3}{2}$	$\frac{3}{2}$	$\frac{1}{2}$	$\frac{3}{2}$	$\frac{3}{2}$	$\frac{1}{2}$	$\frac{3}{2}$	$\frac{1}{2}$	$\frac{1}{2}$	$\frac{3}{2}$	$\frac{3}{2}$	$\frac{3}{2}$	$\frac{1}{2}$
0	1	0	1	0	1	0	0	1	1	1	0	0	0	0	1	1	0	1	1
1	1	0	1	$\frac{1}{2}$	1	$\frac{1}{2}$	0	$\frac{3}{2}$	1	$\frac{3}{2}$	0	$\frac{1}{2}$	0	$\frac{1}{2}$	1	$\frac{3}{2}$	0	$\frac{3}{2}$	1
0	0	1	1	$\frac{1}{2}$	$\frac{1}{2}$	$\frac{3}{2}$	$\frac{3}{2}$	$\frac{3}{2}$	$\frac{1}{2}$	$\frac{1}{2}$	$\frac{3}{2}$	$\frac{1}{2}$	$\frac{3}{2}$	$\frac{3}{2}$	$\frac{1}{2}$	$\frac{3}{2}$	$\frac{3}{2}$	$\frac{1}{2}$	$\frac{1}{2}$
1	0	1	1	1	$\frac{1}{2}$	0	$\frac{3}{2}$	0	$\frac{1}{2}$	1	$\frac{3}{2}$	1	$\frac{3}{2}$	0	$\frac{1}{2}$	0	$\frac{3}{2}$	1	$\frac{1}{2}$
0	1	1	1	$\frac{1}{2}$	1	$\frac{3}{2}$	0	$\frac{3}{2}$	1	$\frac{1}{2}$	0	$\frac{1}{2}$	0	$\frac{3}{2}$	1	$\frac{3}{2}$	0	$\frac{1}{2}$	1
1	1	1	1	1	1	0	0	0	1	1	0	1	0	0	1	0	0	1	1

Lower half of characteristic.

γ	δ	γ'	δ'	$\gamma+\gamma'$	$\delta+\delta'$	$\gamma-\gamma'$	$\delta-\delta'$	$\gamma+\gamma'$	$\delta+\delta'$	$\gamma-\gamma'$	$\delta-\delta'$	$\gamma+\gamma'$	$\delta+\delta'$	$\gamma-\gamma'$	$\delta-\delta'$	$\gamma+\gamma'$	$\delta+\delta'$	$\gamma-\gamma'$	$\delta-\delta'$
0	0	0	0	0	0	0	0	0	0	0	0	0	0	0	0	0	0	0	0
1	0	0	0	1	0	1	0	1	0	1	0	1	0	1	0	1	0	1	0
0	1	0	0	0	1	0	1	0	1	0	1	0	1	0	1	0	1	0	1
1	1	0	0	1	1	1	1	1	1	1	1	1	1	1	1	1	1	1	1
0	0	1	0	1	0	−1	0	1	0	−1	0	1	0	−1	0	1	0	−1	0
1	0	1	0	2	0	0	0	2	0	0	0	2	0	0	0	2	0	0	0
0	1	1	0	1	1	−1	1	1	1	−1	1	1	1	−1	1	1	1	−1	1
1	1	1	0	2	1	0	1	2	1	0	1	2	1	0	1	2	1	0	1
0	0	0	1	0	1	0	−1	0	1	0	−1	0	1	0	−1	0	1	0	−1
1	0	0	1	1	1	1	−1	1	1	1	−1	1	1	1	−1	1	1	1	−1
0	1	0	1	0	2	0	0	0	2	0	0	0	2	0	0	0	2	0	0
1	1	0	1	1	2	1	0	1	2	1	0	1	2	1	0	1	2	1	0
0	0	1	1	1	1	−1	−1	1	1	−1	−1	1	1	−1	−1	1	1	−1	−1
1	0	1	1	2	1	0	−1	2	1	0	−1	2	1	0	−1	2	1	0	−1
0	1	1	1	1	2	−1	0	0	2	−1	0	0	2	−1	0	1	2	−1	0
1	1	1	1	2	2	0	0	2	2	0	0	2	2	0	0	2	2	0	0

83. The upper characters of the Θ's have thus the values 0, 1, $\frac{1}{2}$, $\frac{3}{2}$; the lower characters are originally 2, 1, 0, or −1, and these have when necessary to be, by the addition or subtraction of 2, reduced to 0 or 1; the effect of this change is either to leave the Θ unaltered, or to multiply it by −1 or $\pm i$, as follows:

$$\Theta^{0}_{\gamma\pm2}=\Theta^{0}_{\gamma},\qquad \Theta^{\frac12}_{\gamma+2}=i\Theta^{\frac12}_{\gamma},\quad \Theta^{\frac12}_{\gamma-2}=-i\Theta^{\frac12}_{\gamma},$$

$$\Theta^{1}_{\gamma\pm2}=-\Theta^{1}_{\gamma},\qquad \Theta^{\frac32}_{\gamma+2}=-i\Theta^{\frac32}_{\gamma},\quad \Theta^{\frac32}_{\gamma+2}=i\Theta^{\frac32}_{\gamma},$$

where only the first column of characters is shown, but the same rule applies to the second column; and where we must of course combine the multipliers corresponding to the first and second columns respectively: for instance

$$\Theta^{\frac32\ \frac12}_{\gamma+2\ \delta+2}=(-i\,.\,-i\Theta^{\frac32\ \frac12}_{\gamma\ \delta}=)-\Theta^{\frac32\ \frac12}_{\gamma\ \delta}.$$

Thus taking the tenth line of the upper half, and the fifth line of the lower half, we have

10	01	$\frac{1}{2}$ $\frac{1}{2}$ $\frac{1}{2}$ $\frac{3}{2}$	$\frac{3}{2}$ $\frac{1}{2}$ $\frac{3}{2}$ $\frac{3}{2}$	$\frac{1}{2}$ $\frac{3}{2}$ $\frac{1}{2}$ $\frac{1}{2}$	$\frac{3}{2}$ $\frac{3}{2}$ $\frac{3}{2}$ $\frac{1}{2}$
00	10	1 0 −1 0	1 0 −1 0	1 0 −1 0	1 0 −1 0

giving the value of

$$\vartheta\begin{matrix}1&0\\0&0\end{matrix}(u+u').\vartheta\begin{matrix}0&1\\1&0\end{matrix}(u-u'):$$

viz. this is

$$\begin{aligned}
=\ &\Theta\begin{matrix}\frac{1}{2}&\frac{1}{2}\\1&0\end{matrix}(2u).\Theta\begin{matrix}\frac{1}{2}&\frac{3}{2}\\-1&0\end{matrix}(2u') = &&i\Theta\begin{matrix}\frac{1}{2}&\frac{1}{2}\\1&0\end{matrix}(2u).\Theta\begin{matrix}\frac{1}{2}&\frac{3}{2}\\1&0\end{matrix}(2u')\\
+&\Theta\begin{matrix}\frac{3}{2}&\frac{1}{2}\\1&0\end{matrix}(\,,,\,).\Theta\begin{matrix}\frac{3}{2}&\frac{3}{2}\\-1&0\end{matrix}(\,,,\,) &&-i\Theta\begin{matrix}\frac{3}{2}&\frac{1}{2}\\1&0\end{matrix}(\,,,\,).\Theta\begin{matrix}\frac{3}{2}&\frac{3}{2}\\1&0\end{matrix}(\,,,\,)\\
+&\Theta\begin{matrix}\frac{1}{2}&\frac{3}{2}\\1&0\end{matrix}(\,,,\,).\Theta\begin{matrix}\frac{1}{2}&\frac{1}{2}\\-1&0\end{matrix}(\,,,\,) &&+i\Theta\begin{matrix}\frac{1}{2}&\frac{3}{2}\\1&0\end{matrix}(\,,,\,).\Theta\begin{matrix}\frac{1}{2}&\frac{1}{2}\\1&0\end{matrix}(\,,,\,)\\
+&\Theta\begin{matrix}\frac{3}{2}&\frac{3}{2}\\1&0\end{matrix}(\,,,\,).\Theta\begin{matrix}\frac{3}{2}&\frac{1}{2}\\-1&0\end{matrix}(\,,,\,) &&-i\Theta\begin{matrix}\frac{3}{2}&\frac{3}{2}\\1&0\end{matrix}(\,,,\,).\Theta\begin{matrix}\frac{1}{2}&\frac{1}{2}\\1&0\end{matrix}(\,,,\,),
\end{aligned}$$

where the first column is the value given directly by the diagram: it is then reduced to that given by the second column.

84. But instead of the Θ's, we introduce single letters (X, Y, Z, W), (E, F, G, H), (I, J, K, L), (M, N, P, Q), with the suffixes $(0, 1, 2, 3)$, in all 64 symbols, thus

Θ	00	10	01	11 $(2u)=$		X	Y	Z	W
00					0				
10					1				
01					2				
11					3				

that is,

$$\Theta\begin{matrix}00\\00\end{matrix}(2u)=X,\quad \Theta\begin{matrix}10\\00\end{matrix}=Y,\ \&c.,$$

$$\Theta\begin{matrix}00\\10\end{matrix}(2u)=X_1,\ldots$$

$$\vdots$$

Θ	$\frac{1}{2}0$	$\frac{1}{2}1$	$\frac{3}{2}0$	$\frac{3}{2}1$ $(2u)=$		E	F	G	H
00					0				
10					1				
01					2				
11					3				

that is,

$$\Theta\begin{matrix}\frac{1}{2}0\\00\end{matrix}(2u)=E,\ \&c.,$$

$\Theta\ 0\frac{1}{2}\ \ 1\frac{1}{2}\ \ 0\frac{3}{2}\ \ 1\frac{3}{2}\ (2u) =$		I	J	K	L
00	0				
10	1				
01	2				
11	3				

The functions of $(2u')$ are denoted in like manner by accented letters

$$\Theta\begin{matrix}00\\00\end{matrix}(2u') = X', \text{ \&c.,}$$

$\Theta\ \frac{1}{2}\frac{1}{2}\ \ \frac{3}{2}\frac{1}{2}\ \ \frac{1}{2}\frac{3}{2}\ \ \frac{3}{2}\frac{3}{2}\ (2u) =$		M	N	P	Q
00	0				
10	1				
01	2				
11	3				

85. To simplify the expression of the results, instead of in each case writing down the suffixes, I have indicated them by means of the column headed "Suff."

Thus

$$|\,8-0\,| \qquad \vartheta\begin{matrix}01\\01\end{matrix} u+u' \,.\, \vartheta\begin{matrix}00\\00\end{matrix} u-u' = XX' + YY' + ZZ' + WW' \quad \overset{\text{Suff.}}{|\,2\,|}$$

means that the equation is to be read

$$= X_2X_2' + Y_2Y_2' + Z_2Z_2' + W_2W_2'.$$

It is hardly necessary to mention that the $|\,8-0\,|$ of the left-hand column shows the current numbers of the theta-functions; viz. the left-hand side of the equation is

$$\vartheta_8(u+u') \,.\, \vartheta_0(u-u').$$

And by a preceding remark the single arguments $u+u'$ and $u-u'$ are written in place of $(u+u',\ v+v')$ and $(u-u',\ v-v')$ respectively.

The 256 equations now are

86. First set, 64 equations.

				Suffixes.
0 – 0	$\vartheta^{00}_{00}u+u'$.	$\vartheta^{00}_{00}u-u'$	$= XX' + YY' + ZZ' + WW'$	0
4 – 0	$^{00}_{10}$	$^{00}_{00}$	$= XX' + YY' + ZZ' + WW'$	1
8 – 0	$^{00}_{01}$	$^{00}_{00}$	$= XX' + YY' + ZZ' + WW'$	2
12 – 0	$^{00}_{11}$	$^{00}_{00}$	$= XX' + YY' + ZZ' + WW'$	3
0 – 4	$\vartheta^{00}_{00}u+u'$.	$\vartheta^{00}_{10}u-u'$	$= XX' - YY' + ZZ' - WW'$	1
4 – 4	$^{00}_{10}$	$^{00}_{10}$	$= XX' - YY' + ZZ' - WW'$	0
8 – 4	$^{00}_{01}$	$^{00}_{10}$	$= XX' - YY' + ZZ' - WW'$	3
12 – 4	$^{00}_{11}$	$^{00}_{10}$	$= XX' - YY' + ZZ' - WW'$	2
0 – 8	$\vartheta^{00}_{00}u+u'$.	$\vartheta^{00}_{01}u-u'$	$= XX' + YY' - ZZ' - WW'$	2
4 – 8	$^{00}_{10}$	$^{00}_{01}$	$= XX' + YY' - ZZ' - WW'$	3
8 – 8	$^{00}_{01}$	$^{00}_{01}$	$= XX' + YY' - ZZ' - WW'$	0
12 – 8	$^{00}_{11}$	$^{00}_{01}$	$= XX' + YY' - ZZ' - WW'$	1
0 – 12	$\vartheta^{00}_{00}u+u'$.	$\vartheta^{00}_{11}u-u'$	$= XX' - YY' - ZZ' + WW'$	3
4 – 12	$^{00}_{10}$	$^{00}_{11}$	$= XX' - YY' - ZZ' + WW'$	2
8 – 12	$^{00}_{01}$	$^{00}_{11}$	$= XX' - YY' - ZZ' + WW'$	1
12 – 12	$^{00}_{11}$	$^{00}_{11}$	$= XX' - YY' - ZZ' + WW'$	0

First set, 64 equations (*continued*).

				Suffixes.
1 − 1	$\vartheta\begin{smallmatrix}10\\00\end{smallmatrix}u+u'$.	$\vartheta\begin{smallmatrix}10\\00\end{smallmatrix}u-u'$	$= YX' + XY' + WZ' + ZW'$	0
5 − 1	$\begin{smallmatrix}10\\10\end{smallmatrix}$	$\begin{smallmatrix}10\\00\end{smallmatrix}$	$= YX' + XY' + WZ' + ZW'$	1
9 − 1	$\begin{smallmatrix}10\\01\end{smallmatrix}$	$\begin{smallmatrix}10\\00\end{smallmatrix}$	$= YX' + XY' + WZ' + ZW'$	2
13 − 1	$\begin{smallmatrix}10\\11\end{smallmatrix}$	$\begin{smallmatrix}10\\00\end{smallmatrix}$	$= YX' + XY' + WZ' + ZW'$	3
1 − 5	$\vartheta\begin{smallmatrix}10\\00\end{smallmatrix}u+u'$.	$\vartheta\begin{smallmatrix}10\\10\end{smallmatrix}u-u'$	$= YX' - XY' + WZ' - ZW'$	1
5 − 5	$\begin{smallmatrix}10\\10\end{smallmatrix}$	$\begin{smallmatrix}10\\10\end{smallmatrix}$	$= - YX' + XY' - WZ' + ZW'$	0
9 − 5	$\begin{smallmatrix}10\\01\end{smallmatrix}$	$\begin{smallmatrix}10\\10\end{smallmatrix}$	$= YX' - XY' + WZ' - ZW'$	3
13 − 5	$\begin{smallmatrix}10\\11\end{smallmatrix}$	$\begin{smallmatrix}10\\10\end{smallmatrix}$	$= - YX' + XY' - WZ' + ZW'$	2
1 − 9	$\vartheta\begin{smallmatrix}10\\00\end{smallmatrix}u+u'$.	$\vartheta\begin{smallmatrix}10\\01\end{smallmatrix}u-u'$	$= YX' + XY' - WZ' - ZW'$	2
5 − 9	$\begin{smallmatrix}10\\10\end{smallmatrix}$	$\begin{smallmatrix}10\\01\end{smallmatrix}$	$= YX' + XY' - WZ' - ZW'$	3
9 − 9	$\begin{smallmatrix}10\\01\end{smallmatrix}$	$\begin{smallmatrix}10\\01\end{smallmatrix}$	$= YX' + XY' - WZ' - ZW'$	0
13 − 9	$\begin{smallmatrix}10\\11\end{smallmatrix}$	$\begin{smallmatrix}10\\01\end{smallmatrix}$	$= YX' + XY' - WZ' - ZW'$	1
1 − 13	$\vartheta\begin{smallmatrix}10\\00\end{smallmatrix}u+u'$.	$\vartheta\begin{smallmatrix}10\\11\end{smallmatrix}u-u'$	$= YX' - XY' - WZ' + ZW'$	3
5 − 13	$\begin{smallmatrix}10\\10\end{smallmatrix}$	$\begin{smallmatrix}10\\11\end{smallmatrix}$	$= - YX' + XY' + WZ' - ZW'$	2
9 − 13	$\begin{smallmatrix}10\\01\end{smallmatrix}$	$\begin{smallmatrix}10\\11\end{smallmatrix}$	$= YX' - XY' - WZ' + ZW'$	1
13 − 13	$\begin{smallmatrix}10\\11\end{smallmatrix}$	$\begin{smallmatrix}10\\11\end{smallmatrix}$	$= - YX' + XY' + WZ' - ZW'$	0

First set, 64 equations (*continued*).

				Suffixes.
$2-2$	$\vartheta\begin{matrix}01\\00\end{matrix}u+u'$.	$\vartheta\begin{matrix}01\\00\end{matrix}u-u'$	$=\ ZX'+WY'+XZ'+YW'$	0
$6-2$	$\begin{matrix}01\\10\end{matrix}$	$\begin{matrix}01\\00\end{matrix}$	$=\ ZX'+WY'+XZ'+YW'$	1
$10-2$	$\begin{matrix}01\\01\end{matrix}$	$\begin{matrix}01\\00\end{matrix}$	$=\ ZX'+WY'+XZ'+YW'$	2
$14-2$	$\begin{matrix}01\\11\end{matrix}$	$\begin{matrix}01\\00\end{matrix}$	$=\ ZX'+WY'+XZ'+YW'$	3
$2-6$	$\vartheta\begin{matrix}01\\00\end{matrix}u+u'$.	$\vartheta\begin{matrix}01\\10\end{matrix}u-u'$	$=\ ZX'-WY'+XZ'-YW'$	1
$6-6$	$\begin{matrix}01\\10\end{matrix}$	$\begin{matrix}01\\10\end{matrix}$	$=\ ZX'-WY'+XZ'-YW'$	0
$10-6$	$\begin{matrix}01\\01\end{matrix}$	$\begin{matrix}01\\10\end{matrix}$	$=\ ZX'-WY'+XZ'-YW'$	3
$14-6$	$\begin{matrix}01\\11\end{matrix}$	$\begin{matrix}01\\10\end{matrix}$	$=\ ZX'-WY'+XZ'-YW'$	2
$2-10$	$\vartheta\begin{matrix}01\\00\end{matrix}u+u'$.	$\vartheta\begin{matrix}01\\01\end{matrix}u-u'$	$=\ ZX'+WY'-XZ'-YW'$	2
$6-10$	$\begin{matrix}01\\10\end{matrix}$	$\begin{matrix}01\\01\end{matrix}$	$=\ ZX'+WY'-XZ'-YW'$	3
$10-10$	$\begin{matrix}01\\01\end{matrix}$	$\begin{matrix}01\\01\end{matrix}$	$=-ZX'-WY'+XZ'+YW'$	0
$14-10$	$\begin{matrix}01\\11\end{matrix}$	$\begin{matrix}01\\01\end{matrix}$	$=-ZX'-WY'+XZ'+YW'$	1
$2-14$	$\vartheta\begin{matrix}01\\00\end{matrix}u+u'$.	$\vartheta\begin{matrix}01\\11\end{matrix}u-u'$	$=\ ZX'-WY'-XZ'+YW'$	3
$6-14$	$\begin{matrix}01\\10\end{matrix}$	$\begin{matrix}01\\11\end{matrix}$	$=\ ZX'-WY'-XZ'+YW'$	2
$10-14$	$\begin{matrix}01\\01\end{matrix}$	$\begin{matrix}01\\11\end{matrix}$	$=-ZX'+WY'+XZ'-YW'$	1
$14-14$	$\begin{matrix}01\\11\end{matrix}$	$\begin{matrix}01\\11\end{matrix}$	$=-ZX'+WY'+XZ'-YW'$	0

First set, 64 equations (*concluded*).

			Suffixes.
3 – 3	$\vartheta\begin{matrix}11\\00\end{matrix}u+u'\,.\,\vartheta\begin{matrix}11\\00\end{matrix}u-u'$	$= WX' + ZY' + YZ' + XW'$	0
7 – 3	$\begin{matrix}11\\10\end{matrix}\quad\begin{matrix}11\\00\end{matrix}$	$= WX' + ZY' + YZ' + XW'$	1
11 – 3	$\begin{matrix}11\\01\end{matrix}\quad\begin{matrix}11\\00\end{matrix}$	$= WX' + ZY' + YZ' + XW'$	2
15 – 3	$\begin{matrix}11\\11\end{matrix}\quad\begin{matrix}11\\00\end{matrix}$	$= WX' + ZY' + YZ' + XW'$	3
3 – 7	$\vartheta\begin{matrix}11\\00\end{matrix}u+u'\,.\,\vartheta\begin{matrix}11\\10\end{matrix}u-u'$	$= WX' - ZY' + YZ' - XW'$	1
7 – 7	$\begin{matrix}11\\10\end{matrix}\quad\begin{matrix}11\\10\end{matrix}$	$=- WX' + ZY' - YZ' + XW'$	0
11 – 7	$\begin{matrix}11\\01\end{matrix}\quad\begin{matrix}11\\10\end{matrix}$	$= WX' - ZY' + YZ' - XW'$	3
15 – 7	$\begin{matrix}11\\11\end{matrix}\quad\begin{matrix}11\\10\end{matrix}$	$=- WX' + ZY' - YZ' + XW'$	2
3 – 11	$\vartheta\begin{matrix}11\\00\end{matrix}u+u'\,.\,\vartheta\begin{matrix}11\\01\end{matrix}u-u'$	$= WX' + ZY' - YZ' - XW'$	2
7 – 11	$\begin{matrix}11\\10\end{matrix}\quad\begin{matrix}11\\01\end{matrix}$	$= WX' + ZY' - YZ' - XW'$	3
11 – 11	$\begin{matrix}11\\01\end{matrix}\quad\begin{matrix}11\\01\end{matrix}$	$=- WX' - ZY' + YZ' + XW'$	0
15 – 11	$\begin{matrix}11\\11\end{matrix}\quad\begin{matrix}11\\01\end{matrix}$	$=- WX' - ZY' + YZ' + XW'$	1
3 – 15	$\vartheta\begin{matrix}11\\00\end{matrix}u+u'\,.\,\vartheta\begin{matrix}11\\11\end{matrix}u-u'$	$= WX' - ZY' - YZ' + XW'$	3
7 – 15	$\begin{matrix}11\\10\end{matrix}\quad\begin{matrix}11\\11\end{matrix}$	$=- WX' + ZY' + YZ' - XW'$	2
11 – 15	$\begin{matrix}11\\01\end{matrix}\quad\begin{matrix}11\\11\end{matrix}$	$=- WX' + ZY' + YZ' - XW'$	1
15 – 15	$\begin{matrix}11\\11\end{matrix}\quad\begin{matrix}11\\11\end{matrix}$	$= WX' - ZY' - YZ' + XW'$	0

87. Second set, 64 equations.

			Suffixes.
1 − 0	$\vartheta\,{}^{10}_{00}\,u+u'\,.\,\vartheta\,{}^{00}_{00}\,u-u'$	$= \quad EE' + GG' + FF' + HH'$	0
5 − 0	${}^{10}_{10} \qquad {}^{00}_{00}$	$= \quad EE' + GG' + FF' + HH'$	1
9 − 0	${}^{10}_{01} \qquad {}^{00}_{00}$	$= \quad EE' + GG' + FF' + HH'$	2
13 − 0	${}^{10}_{11} \qquad {}^{00}_{00}$	$= \quad EE' + GG' + FF' + HH'$	3
1 − 4	$\vartheta\,{}^{10}_{00}\,u+u'\,.\,\vartheta\,{}^{00}_{10}\,u-u'$	$= -iEE' + iGG' - iFF' + iHH'$	1
5 − 4	${}^{10}_{10} \qquad {}^{00}_{10}$	$= \quad iEE' - iGG' + iFF' - iHH'$	0
9 − 4	${}^{10}_{01} \qquad {}^{00}_{10}$	$= -iEE' + iGG' - iFF' + iHH'$	3
13 − 4	${}^{10}_{11} \qquad {}^{00}_{10}$	$= \quad iEE' - iGG' + iFF' - iHH'$	2
1 − 8	$\vartheta\,{}^{10}_{00}\,u+u'\,.\,\vartheta\,{}^{00}_{01}\,u-u'$	$= \quad EE' + GG' - FF' - HH'$	2
5 − 8	${}^{10}_{10} \qquad {}^{00}_{01}$	$= \quad EE' + GG' - FF' - HH'$	3
9 − 8	${}^{10}_{01} \qquad {}^{00}_{01}$	$= \quad EE' + GG' - FF' - HH'$	0
13 − 8	${}^{10}_{11} \qquad {}^{00}_{01}$	$= \quad EE' + GG' - FF' - HH'$	1
1 − 12	$\vartheta\,{}^{10}_{00}\,u+u'\,.\,\vartheta\,{}^{00}_{11}\,u-u'$	$= -iEE' + iGG' + iFF' - iHH'$	3
5 − 12	${}^{10}_{10} \qquad {}^{00}_{11}$	$= \quad iEE' - iGG' - iFF' + iHH'$	2
9 − 12	${}^{10}_{01} \qquad {}^{00}_{11}$	$= -iEE' + iGG' + iFF' - iHH'$	1
13 − 12	${}^{10}_{11} \qquad {}^{00}_{11}$	$= \quad iEE' - iGG' - iFF' + iHH'$	0

Second set, 64 equations (*continued*).

				Suffixes.
0 − 1	$\vartheta\begin{matrix}00\\00\end{matrix}u+u'\,.\,\vartheta\begin{matrix}10\\00\end{matrix}u-u'$	=	$EG'+GE'+FH'+HF'$	0
4 − 1	$\begin{matrix}00\\10\end{matrix}\qquad\begin{matrix}10\\00\end{matrix}$	=	$EG'+GE'+FH'+HF'$	1
8 − 1	$\begin{matrix}00\\01\end{matrix}\qquad\begin{matrix}10\\00\end{matrix}$	=	$EG'+GE'+FH'+HF'$	2
12 − 1	$\begin{matrix}00\\11\end{matrix}\qquad\begin{matrix}10\\00\end{matrix}$	=	$EG'+GE'+FH'+HF'$	3
0 − 5	$\vartheta\begin{matrix}00\\00\end{matrix}u+u'\,.\,\vartheta\begin{matrix}10\\10\end{matrix}u-u'$	=	$iEG'-iGE'+iFH'-iHF'$	1
4 − 5	$\begin{matrix}00\\10\end{matrix}\qquad\begin{matrix}10\\10\end{matrix}$	=	$iEG'-iGE'+iFH'-iHF'$	0
8 − 5	$\begin{matrix}00\\01\end{matrix}\qquad\begin{matrix}10\\10\end{matrix}$	=	$iEG'-iGE'+iFH'-iHF'$	3
12 − 5	$\begin{matrix}00\\11\end{matrix}\qquad\begin{matrix}10\\10\end{matrix}$	=	$iEG'-iGE'+iFH'-iHF'$	2
0 − 9	$\vartheta\begin{matrix}00\\00\end{matrix}u+u'\,.\,\vartheta\begin{matrix}10\\01\end{matrix}u-u'$	=	$EG'+GE'-FH'-HF'$	2
4 − 9	$\begin{matrix}00\\10\end{matrix}\qquad\begin{matrix}10\\01\end{matrix}$	=	$EG'+GE'-FH'-HF'$	3
8 − 9	$\begin{matrix}00\\01\end{matrix}\qquad\begin{matrix}10\\01\end{matrix}$	=	$EG'+GE'-FH'-HF'$	0
12 − 9	$\begin{matrix}00\\11\end{matrix}\qquad\begin{matrix}10\\01\end{matrix}$	=	$EG'+GE'-FH'-HF'$	1
0 − 13	$\vartheta\begin{matrix}00\\00\end{matrix}u+u'\,.\,\vartheta\begin{matrix}10\\11\end{matrix}u-u'$	=	$iEG'-iGE'-iFH'+iHF'$	3
4 − 13	$\begin{matrix}00\\10\end{matrix}\qquad\begin{matrix}10\\11\end{matrix}$	=	$iEG'-iGE'-iFH'+iHF'$	2
8 − 13	$\begin{matrix}00\\01\end{matrix}\qquad\begin{matrix}10\\11\end{matrix}$	=	$iEG'-iGE'-iFH'+iHF'$	1
12 − 13	$\begin{matrix}00\\11\end{matrix}\qquad\begin{matrix}10\\11\end{matrix}$	=	$iEG'-iGE'-iFH'+iHF'$	0

Second set, 64 equations (*continued*).

				Suffixes.
3 − 2	$\vartheta\begin{smallmatrix}11\\00\end{smallmatrix}u+u'\,.$	$\vartheta\begin{smallmatrix}01\\00\end{smallmatrix}u-u'$	$= FE' + HG' + EF' + GH'$	0
7 − 2	$\begin{smallmatrix}11\\10\end{smallmatrix}$	$\begin{smallmatrix}01\\00\end{smallmatrix}$	$= FE' + HG' + EF' + GH'$	1
11 − 2	$\begin{smallmatrix}11\\01\end{smallmatrix}$	$\begin{smallmatrix}01\\00\end{smallmatrix}$	$= FE' + HG' + EF' + GH'$	2
15 − 2	$\begin{smallmatrix}11\\11\end{smallmatrix}$	$\begin{smallmatrix}01\\00\end{smallmatrix}$	$= FE' + HG' + EF' + GH'$	3
3 − 6	$\vartheta\begin{smallmatrix}11\\00\end{smallmatrix}u+u'\,.$	$\vartheta\begin{smallmatrix}01\\10\end{smallmatrix}u-u'$	$= -iFE' + iHG' - iEF' + iGH'$	1
7 − 6	$\begin{smallmatrix}11\\10\end{smallmatrix}$	$\begin{smallmatrix}01\\10\end{smallmatrix}$	$= iFE' - iHG' + iEF' - iGH'$	0
11 − 6	$\begin{smallmatrix}11\\01\end{smallmatrix}$	$\begin{smallmatrix}01\\10\end{smallmatrix}$	$= -iFE' + iHG' - iEF' + iGH'$	3
15 − 6	$\begin{smallmatrix}11\\11\end{smallmatrix}$	$\begin{smallmatrix}01\\10\end{smallmatrix}$	$= iFE' - iHG' + iEF' - iGH'$	2
3 − 10	$\vartheta\begin{smallmatrix}11\\00\end{smallmatrix}u+u'\,.$	$\vartheta\begin{smallmatrix}01\\01\end{smallmatrix}u-u'$	$= FE' + HG' - EF' - GH'$	2
7 − 10	$\begin{smallmatrix}11\\10\end{smallmatrix}$	$\begin{smallmatrix}01\\01\end{smallmatrix}$	$= FE' + HG' - EF' - GH'$	3
11 − 10	$\begin{smallmatrix}11\\01\end{smallmatrix}$	$\begin{smallmatrix}01\\01\end{smallmatrix}$	$= - FE' - HG' + EF' + GH'$	0
15 − 10	$\begin{smallmatrix}11\\11\end{smallmatrix}$	$\begin{smallmatrix}01\\01\end{smallmatrix}$	$= - FE' - HG' + EF' + GH'$	1
3 − 14	$\vartheta\begin{smallmatrix}11\\00\end{smallmatrix}u+u'\,.$	$\vartheta\begin{smallmatrix}01\\11\end{smallmatrix}u-u'$	$= -iFE' + iHG' + iEF' - iGH'$	3
7 − 14	$\begin{smallmatrix}11\\10\end{smallmatrix}$	$\begin{smallmatrix}01\\11\end{smallmatrix}$	$= iFE' - iHG' - iEF' + iGH'$	2
11 − 14	$\begin{smallmatrix}11\\01\end{smallmatrix}$	$\begin{smallmatrix}01\\11\end{smallmatrix}$	$= iFE' - iHG' - iEF' + iGH'$	1
15 − 14	$\begin{smallmatrix}11\\11\end{smallmatrix}$	$\begin{smallmatrix}01\\11\end{smallmatrix}$	$= -iFE' + iHG' + iEF' - iGH'$	0

Second set, 64 equations (*concluded*).

				Suffixes.
2 − 3	$\vartheta\,{}^{01}_{00}\,u+u'$.	$\vartheta\,{}^{11}_{00}\,u-u'$	$= FG' + HE' + EH' + GF'$	0
6 − 3	${}^{01}_{10}$	${}^{11}_{00}$	$= FG' + HE' + EH' + GF'$	1
10 − 3	${}^{01}_{01}$	${}^{11}_{00}$	$= FG' + HE' + EH' + GF'$	2
14 − 3	${}^{01}_{11}$	${}^{11}_{00}$	$= FG' + HE' + EH' + GF'$	3
2 − 7	$\vartheta\,{}^{01}_{00}\,u+u'$.	$\vartheta\,{}^{11}_{10}\,u-u'$	$= iFG' - iHE' + iEH' - iGF'$	1
6 − 7	${}^{01}_{10}$	${}^{11}_{10}$	$= iFG' - iHE' + iEH' - iGF'$	0
10 − 7	${}^{01}_{01}$	${}^{11}_{10}$	$= iFG' - iHE' + iEH' - iGF'$	3
14 − 7	${}^{01}_{11}$	${}^{11}_{10}$	$= iFG' - iHE' + iEH' - iGF'$	2
2 − 11	$\vartheta\,{}^{01}_{00}\,u+u'$.	$\vartheta\,{}^{11}_{01}\,u-u'$	$= FG' + HE' - EH' - GF'$	2
6 − 11	${}^{01}_{10}$	${}^{11}_{01}$	$= FG' + HE' - EH' - GF'$	3
10 − 11	${}^{01}_{01}$	${}^{11}_{01}$	$= - FG' - HE' + EH' + GF'$	0
14 − 11	${}^{01}_{11}$	${}^{11}_{01}$	$= - FG' - HE' + EH' + GF'$	1
2 − 15	$\vartheta\,{}^{01}_{00}\,u+u'$.	$\vartheta\,{}^{11}_{11}\,u-u'$	$= iFG' - iHE' - iEH' + iGF'$	3
6 − 15	${}^{01}_{10}$	${}^{11}_{11}$	$= iFG' - iHE' - iEH' + iGF'$	2
10 − 15	${}^{01}_{01}$	${}^{11}_{11}$	$= -iFG' + iHE' + iEH' - iGF'$	1
14 − 15	${}^{01}_{11}$	${}^{11}_{11}$	$= -iFG' + iHE' + iEH' - iGF'$	0

88. Third set, 64 equations.

			Suffixes.
2 − 0	$\vartheta\begin{smallmatrix}01\\00\end{smallmatrix}u+u'\,.\,\vartheta\begin{smallmatrix}00\\00\end{smallmatrix}u-u'$	$= II' + JJ' + KK' + LL'$	0
6 − 0	$\begin{smallmatrix}01\\10\end{smallmatrix}\quad\begin{smallmatrix}00\\00\end{smallmatrix}$	$= II' + JJ' + KK' + LL'$	1
10 − 0	$\begin{smallmatrix}01\\01\end{smallmatrix}\quad\begin{smallmatrix}00\\00\end{smallmatrix}$	$= II' + JJ' + KK' + LL'$	2
14 − 0	$\begin{smallmatrix}01\\11\end{smallmatrix}\quad\begin{smallmatrix}00\\00\end{smallmatrix}$	$= II' + JJ' + KK' + LL'$	3
2 − 4	$\vartheta\begin{smallmatrix}01\\00\end{smallmatrix}u+u'\,.\,\vartheta\begin{smallmatrix}00\\10\end{smallmatrix}u-u'$	$= II' - JJ' + KK' - LL'$	1
6 − 4	$\begin{smallmatrix}01\\10\end{smallmatrix}\quad\begin{smallmatrix}00\\10\end{smallmatrix}$	$= II' - JJ' + KK' - LL'$	0
10 − 4	$\begin{smallmatrix}01\\01\end{smallmatrix}\quad\begin{smallmatrix}00\\10\end{smallmatrix}$	$= II' - JJ' + KK' - LL'$	3
14 − 4	$\begin{smallmatrix}01\\11\end{smallmatrix}\quad\begin{smallmatrix}00\\10\end{smallmatrix}$	$= II' - JJ' + KK' - LL'$	2
2 − 8	$\vartheta\begin{smallmatrix}01\\00\end{smallmatrix}u+u'\,.\,\vartheta\begin{smallmatrix}00\\01\end{smallmatrix}u-u'$	$=-iII' - iJJ' + iKK' + iLL'$	2
6 − 8	$\begin{smallmatrix}01\\10\end{smallmatrix}\quad\begin{smallmatrix}00\\01\end{smallmatrix}$	$=-iII' - iJJ' + iKK' + iLL'$	3
10 − 8	$\begin{smallmatrix}01\\01\end{smallmatrix}\quad\begin{smallmatrix}00\\01\end{smallmatrix}$	$= iII' + iJJ' - iKK' - iLL'$	0
14 − 8	$\begin{smallmatrix}01\\11\end{smallmatrix}\quad\begin{smallmatrix}00\\01\end{smallmatrix}$	$= iII' + iJJ' - iKK' - iLL'$	1
2 − 12	$\vartheta\begin{smallmatrix}01\\00\end{smallmatrix}u+u'\,.\,\vartheta\begin{smallmatrix}00\\11\end{smallmatrix}u-u'$	$=-iII' + iJJ' + iKK' - iLL'$	3
6 − 12	$\begin{smallmatrix}01\\10\end{smallmatrix}\quad\begin{smallmatrix}00\\11\end{smallmatrix}$	$=-iII' + iJJ' + iKK' - iLL'$	2
10 − 12	$\begin{smallmatrix}01\\01\end{smallmatrix}\quad\begin{smallmatrix}00\\11\end{smallmatrix}$	$= iII' - iJJ' - iKK' + iLL'$	1
14 − 12	$\begin{smallmatrix}01\\11\end{smallmatrix}\quad\begin{smallmatrix}00\\11\end{smallmatrix}$	$= iII' - iJJ' - iKK' + iLL'$	0

Third set, 64 equations (*continued*).

			Suffixes.
3 − 1	$\vartheta\begin{matrix}11\\00\end{matrix}u+u'\,.\,\vartheta\begin{matrix}10\\00\end{matrix}u-u'$	$= JI' + IJ' + LK' + KL'$	0
7 − 1	$\begin{matrix}11\\10\end{matrix}\quad\begin{matrix}10\\00\end{matrix}$	$= JI' + IJ' + LK' + KL'$	1
11 − 1	$\begin{matrix}11\\01\end{matrix}\quad\begin{matrix}10\\00\end{matrix}$	$= JI' + IJ' + LK' + KL'$	2
15 − 1	$\begin{matrix}11\\11\end{matrix}\quad\begin{matrix}10\\00\end{matrix}$	$= JI' + IJ' + LK' + KL'$	3
3 − 5	$\vartheta\begin{matrix}11\\00\end{matrix}u+u'\,.\,\vartheta\begin{matrix}10\\10\end{matrix}u-u'$	$= JI' - IJ' + LK' - KL'$	1
7 − 5	$\begin{matrix}11\\10\end{matrix}\quad\begin{matrix}10\\10\end{matrix}$	$= - JI' + IJ' - LK' + KL'$	0
11 − 5	$\begin{matrix}11\\01\end{matrix}\quad\begin{matrix}10\\10\end{matrix}$	$= JI' - IJ' + LK' - KL'$	3
15 − 5	$\begin{matrix}11\\11\end{matrix}\quad\begin{matrix}10\\10\end{matrix}$	$= - JI' + IJ' - LK' + KL'$	2
3 − 9	$\vartheta\begin{matrix}11\\00\end{matrix}u+u'\,.\,\vartheta\begin{matrix}10\\01\end{matrix}u-u'$	$= - iJI' - iIJ' + iLK' + iKL'$	2
7 − 9	$\begin{matrix}11\\10\end{matrix}\quad\begin{matrix}10\\01\end{matrix}$	$= - iJI' - iIJ' + iLK' + iKL'$	3
11 − 9	$\begin{matrix}11\\01\end{matrix}\quad\begin{matrix}10\\01\end{matrix}$	$= iJI' + iIJ' - iLK' - iKL'$	0
15 − 9	$\begin{matrix}11\\11\end{matrix}\quad\begin{matrix}10\\01\end{matrix}$	$= iJI' + iIJ' - iLK' - iKL'$	1
3 − 13	$\vartheta\begin{matrix}11\\00\end{matrix}u+u'\,.\,\vartheta\begin{matrix}10\\11\end{matrix}u-u'$	$= - iJI' + iIJ' + iLK' - iKL'$	3
7 − 13	$\begin{matrix}11\\10\end{matrix}\quad\begin{matrix}10\\11\end{matrix}$	$= iJI' - iIJ' - iLK' + iKL'$	2
11 − 13	$\begin{matrix}11\\01\end{matrix}\quad\begin{matrix}10\\11\end{matrix}$	$= iJI' - iIJ' - iLK' + iKL'$	1
15 − 13	$\begin{matrix}11\\11\end{matrix}\quad\begin{matrix}10\\11\end{matrix}$	$= - iJI' + iIJ' + iLK' - iKL'$	0

Third set, 64 equations (*continued*).

				Suffixes.
0 – 2	$\vartheta\begin{matrix}00\\00\end{matrix}u+u' \,.\, \vartheta\begin{matrix}01\\00\end{matrix}u-u'$	=	$IK' + JL' + KI' + LJ'$	0
4 – 2	$\begin{matrix}00\\10\end{matrix} \qquad \begin{matrix}01\\00\end{matrix}$	=	$IK' + JL' + KI' + LJ'$	1
8 – 2	$\begin{matrix}00\\01\end{matrix} \qquad \begin{matrix}01\\00\end{matrix}$	=	$IK' + JL' + KI' + LJ'$	2
12 – 2	$\begin{matrix}00\\11\end{matrix} \qquad \begin{matrix}01\\00\end{matrix}$	=	$IK' + JL' + KI' + LJ'$	3
0 – 6	$\vartheta\begin{matrix}00\\00\end{matrix}u+u' \,.\, \vartheta\begin{matrix}01\\10\end{matrix}u-u'$	=	$IK' - JL' + KI' - LJ'$	1
4 – 6	$\begin{matrix}00\\10\end{matrix} \qquad \begin{matrix}01\\10\end{matrix}$	=	$IK' - JL' + KI' - LJ'$	0
8 – 6	$\begin{matrix}00\\01\end{matrix} \qquad \begin{matrix}01\\10\end{matrix}$	=	$IK' - JL' + KI' - LJ'$	3
12 – 6	$\begin{matrix}00\\11\end{matrix} \qquad \begin{matrix}01\\10\end{matrix}$	=	$IK' - JL' + KI' - LJ'$	2
0 – 10	$\vartheta\begin{matrix}00\\00\end{matrix}u+u' \,.\, \vartheta\begin{matrix}01\\01\end{matrix}u-u'$	=	$iIK' + iJL' - iKI' - iLJ'$	2
4 – 10	$\begin{matrix}00\\10\end{matrix} \qquad \begin{matrix}01\\01\end{matrix}$	=	$iIK' + iJL' - iKI' - iLJ'$	3
8 – 10	$\begin{matrix}00\\01\end{matrix} \qquad \begin{matrix}01\\01\end{matrix}$	=	$iIK' + iJL' - iKI' - iLJ'$	0
12 – 10	$\begin{matrix}00\\11\end{matrix} \qquad \begin{matrix}01\\01\end{matrix}$	=	$iIK' + iJL' - iKI' - iLJ'$	1
0 – 14	$\vartheta\begin{matrix}00\\00\end{matrix}u+u' \,.\, \vartheta\begin{matrix}01\\11\end{matrix}u-u'$	=	$iIK' - iJL' - iKI' + iLJ'$	3
4 – 14	$\begin{matrix}00\\10\end{matrix} \qquad \begin{matrix}01\\11\end{matrix}$	=	$iIK' - iJL' - iKI' + iLJ'$	2
8 – 14	$\begin{matrix}00\\01\end{matrix} \qquad \begin{matrix}01\\11\end{matrix}$	=	$iIK' - iJL' - iKI' + iLJ'$	1
12 – 14	$\begin{matrix}00\\11\end{matrix} \qquad \begin{matrix}01\\11\end{matrix}$	=	$iIK' - iJL' - iKI' + iLJ'$	0

Third set, 64 equations (*concluded*).

			Suffixes.
1 − 3	$\vartheta\begin{matrix}10\\00\end{matrix}u+u' . \vartheta\begin{matrix}11\\00\end{matrix}u-u'$	$= JK' + IL' + LI' + KJ'$	0
5 − 3	$\begin{matrix}10\\10\end{matrix} \quad \begin{matrix}11\\00\end{matrix}$	$= JK' + IL' + LI' + KJ'$	1
9 − 3	$\begin{matrix}10\\01\end{matrix} \quad \begin{matrix}11\\00\end{matrix}$	$= JK' + IL' + LI' + KJ'$	2
13 − 3	$\begin{matrix}10\\11\end{matrix} \quad \begin{matrix}11\\00\end{matrix}$	$= JK' + IL' + LI' + KJ'$	3
1 − 7	$\vartheta\begin{matrix}10\\00\end{matrix}u+u' . \vartheta\begin{matrix}11\\10\end{matrix}u-u'$	$= JK' - IL' + LI' - KJ'$	1
5 − 7	$\begin{matrix}10\\10\end{matrix} \quad \begin{matrix}11\\10\end{matrix}$	$=- JK' + IL' - LI' + KJ'$	0
9 − 7	$\begin{matrix}10\\01\end{matrix} \quad \begin{matrix}11\\10\end{matrix}$	$= JK' - IL' + LI' - KJ'$	3
13 − 7	$\begin{matrix}10\\11\end{matrix} \quad \begin{matrix}11\\10\end{matrix}$	$=- JK' + IL' - LI' + KJ'$	2
1 − 11	$\vartheta\begin{matrix}10\\00\end{matrix}u+u' . \vartheta\begin{matrix}11\\01\end{matrix}u-u'$	$= iJK' + iIL' - iLI' - iKJ'$	2
5 − 11	$\begin{matrix}10\\10\end{matrix} \quad \begin{matrix}11\\01\end{matrix}$	$= iJK' + iIL' - iLI' - iKJ'$	3
9 − 11	$\begin{matrix}10\\01\end{matrix} \quad \begin{matrix}11\\01\end{matrix}$	$= iJK' + iIL' - iLI' - iKJ'$	0
13 − 11	$\begin{matrix}10\\11\end{matrix} \quad \begin{matrix}11\\01\end{matrix}$	$= iJK' + iIL' - iLI' - iKJ'$	1
1 − 15	$\vartheta\begin{matrix}10\\00\end{matrix}u+u' . \vartheta\begin{matrix}11\\11\end{matrix}u-u'$	$= iJK' - iIL' - iLI' + iKJ'$	3
5 − 15	$\begin{matrix}10\\10\end{matrix} \quad \begin{matrix}11\\11\end{matrix}$	$=-iJK' + iIL' + iLI' - iKJ'$	2
9 − 15	$\begin{matrix}10\\01\end{matrix} \quad \begin{matrix}11\\11\end{matrix}$	$= iJK' - iIL' - iLI' + iKJ'$	1
13 − 15	$\begin{matrix}10\\11\end{matrix} \quad \begin{matrix}11\\11\end{matrix}$	$=-iJK' + iIL' + iLI' - iKJ'$	0

89. Fourth set, 64 equations.

			Suffixes.
3 − 0	$\vartheta\begin{matrix}11\\00\end{matrix}u+u'\,.\,\vartheta\begin{matrix}00\\00\end{matrix}u-u'$	$= MM' + NN' + PP' + QQ'$	0
7 − 0	$\begin{matrix}11\\10\end{matrix}\quad\begin{matrix}00\\00\end{matrix}$	$= MM' + NN' + PP' + QQ'$	1
11 − 0	$\begin{matrix}11\\01\end{matrix}\quad\begin{matrix}00\\00\end{matrix}$	$= MM' + NN' + PP' + QQ'$	2
15 − 0	$\begin{matrix}11\\11\end{matrix}\quad\begin{matrix}00\\00\end{matrix}$	$= MM' + NN' + PP' + QQ'$	3
3 − 4	$\vartheta\begin{matrix}11\\00\end{matrix}u+u'\,.\,\vartheta\begin{matrix}00\\10\end{matrix}u-u'$	$=-iMM' + iNN' - iPP' + iQQ'$	1
7 − 4	$\begin{matrix}11\\10\end{matrix}\quad\begin{matrix}00\\10\end{matrix}$	$=+iMM' - iNN' + iPP' - iQQ'$	0
11 − 4	$\begin{matrix}11\\01\end{matrix}\quad\begin{matrix}00\\10\end{matrix}$	$=-iMM' + iNN' - iPP' + iQQ'$	3
15 − 4	$\begin{matrix}11\\11\end{matrix}\quad\begin{matrix}00\\10\end{matrix}$	$=+iMM' - iNN' + iPP' - iQQ'$	2
3 − 8	$\vartheta\begin{matrix}11\\00\end{matrix}u+u'\,.\,\vartheta\begin{matrix}00\\01\end{matrix}u-u'$	$=-iMM' - iNN' + iPP' + iQQ'$	2
7 − 8	$\begin{matrix}11\\10\end{matrix}\quad\begin{matrix}00\\01\end{matrix}$	$=-iMM' - iNN' + iPP' + iQQ'$	3
11 − 8	$\begin{matrix}11\\01\end{matrix}\quad\begin{matrix}00\\01\end{matrix}$	$= iMM' + iNN' - iPP' - iQQ'$	0
15 − 8	$\begin{matrix}11\\11\end{matrix}\quad\begin{matrix}00\\01\end{matrix}$	$= iMM' + iNN' - iPP' - iQQ'$	1
3 − 12	$\vartheta\begin{matrix}11\\00\end{matrix}u+u'\,.\,\vartheta\begin{matrix}00\\11\end{matrix}u-u'$	$=- MM' + NN' + PP' - QQ'$	3
7 − 12	$\begin{matrix}11\\10\end{matrix}\quad\begin{matrix}00\\11\end{matrix}$	$=+ MM' - NN' - PP' + QQ'$	2
11 − 12	$\begin{matrix}11\\01\end{matrix}\quad\begin{matrix}00\\11\end{matrix}$	$=+ MM' - NN' - PP' + QQ'$	1
15 − 12	$\begin{matrix}11\\11\end{matrix}\quad\begin{matrix}00\\11\end{matrix}$	$=- MM' + NN' + PP' - QQ'$	0

Fourth set, 64 equations (*continued*).

			Suffixes.
$2-1$	$\vartheta\genfrac{}{}{0pt}{}{01}{00}u+u'\,.\,\vartheta\genfrac{}{}{0pt}{}{10}{00}u-u'$	$= MN' + NM' + PQ' + QP'$	0
$6-1$	$\genfrac{}{}{0pt}{}{01}{10}\qquad\genfrac{}{}{0pt}{}{10}{00}$	$= MN' + NM' + PQ' + QP'$	1
$10-1$	$\genfrac{}{}{0pt}{}{01}{01}\qquad\genfrac{}{}{0pt}{}{10}{00}$	$= MN' + NM' + PQ' + QP'$	2
$14-1$	$\genfrac{}{}{0pt}{}{01}{11}\qquad\genfrac{}{}{0pt}{}{10}{00}$	$= MN' + NM' + PQ' + QP'$	3
$2-5$	$\vartheta\genfrac{}{}{0pt}{}{01}{00}u+u'\,.\,\vartheta\genfrac{}{}{0pt}{}{10}{10}u-u'$	$= iMN' - iNM' + iPQ' - iQP'$	1
$6-5$	$\genfrac{}{}{0pt}{}{01}{10}\qquad\genfrac{}{}{0pt}{}{10}{10}$	$= iMN' - iNM' + iPQ' - iQP'$	0
$10-5$	$\genfrac{}{}{0pt}{}{01}{01}\qquad\genfrac{}{}{0pt}{}{10}{10}$	$= iMN' - iNM' + iPQ' - iQP'$	3
$14-5$	$\genfrac{}{}{0pt}{}{01}{11}\qquad\genfrac{}{}{0pt}{}{10}{10}$	$= iMN' - iNM' + iPQ' - iQP'$	2
$2-9$	$\vartheta\genfrac{}{}{0pt}{}{01}{00}u+u'\,.\,\vartheta\genfrac{}{}{0pt}{}{10}{01}u-u'$	$= -iMN' - iNM' + iPQ' + iQP'$	2
$6-9$	$\genfrac{}{}{0pt}{}{01}{10}\qquad\genfrac{}{}{0pt}{}{10}{01}$	$= -iMN' - iNM' + iPQ' + iQP'$	3
$10-9$	$\genfrac{}{}{0pt}{}{01}{01}\qquad\genfrac{}{}{0pt}{}{10}{01}$	$= iMN' + iNM' - iPQ' - iQP'$	0
$14-9$	$\genfrac{}{}{0pt}{}{01}{11}\qquad\genfrac{}{}{0pt}{}{10}{01}$	$= iMN' + iNM' - iPQ' - iQP'$	1
$2-13$	$\vartheta\genfrac{}{}{0pt}{}{01}{00}u+u'\,.\,\vartheta\genfrac{}{}{0pt}{}{10}{11}u-u'$	$= MN' - NM' - PQ' + QP'$	3
$6-13$	$\genfrac{}{}{0pt}{}{01}{10}\qquad\genfrac{}{}{0pt}{}{10}{11}$	$= MN' - NM' - PQ' + QP'$	2
$10-13$	$\genfrac{}{}{0pt}{}{01}{01}\qquad\genfrac{}{}{0pt}{}{10}{11}$	$= -MN' + NM' + PQ' - QP'$	1
$14-13$	$\genfrac{}{}{0pt}{}{01}{11}\qquad\genfrac{}{}{0pt}{}{10}{11}$	$= -MN' + NM' + PQ' - QP'$	0

Fourth set, 64 equations (*continued*).

			Suffixes.
1 − 2	$\vartheta\begin{matrix}10\\00\end{matrix}u+u' \,.\, \vartheta\begin{matrix}01\\00\end{matrix}u-u'$	$= MP' + NQ' + PM' + QN'$	0
5 − 2	$\begin{matrix}10\\10\end{matrix} \quad \begin{matrix}01\\00\end{matrix}$	$= MP' + NQ' + PM' + QN'$	1
9 − 2	$\begin{matrix}10\\01\end{matrix} \quad \begin{matrix}01\\00\end{matrix}$	$= MP' + NQ' + PM' + QN'$	2
13 − 2	$\begin{matrix}10\\11\end{matrix} \quad \begin{matrix}01\\00\end{matrix}$	$= MP' + NQ' + PM' + QN'$	3
1 − 6	$\vartheta\begin{matrix}10\\00\end{matrix}u+u' \,.\, \vartheta\begin{matrix}01\\10\end{matrix}u-u'$	$= -iMP' + iNQ' - iPM' + iQN'$	1
5 − 6	$\begin{matrix}10\\10\end{matrix} \quad \begin{matrix}01\\10\end{matrix}$	$= iMP' - iNQ' + iPM' - iQN'$	0
9 − 6	$\begin{matrix}10\\01\end{matrix} \quad \begin{matrix}01\\10\end{matrix}$	$= -iMP' + iNQ' - iPM' + iQN'$	3
13 − 6	$\begin{matrix}10\\11\end{matrix} \quad \begin{matrix}01\\10\end{matrix}$	$= iMP' - iNQ' + iPM' - iQN'$	2
1 − 10	$\vartheta\begin{matrix}10\\00\end{matrix}u+u' \,.\, \vartheta\begin{matrix}01\\01\end{matrix}u-u'$	$= iMP' + iNQ' - iPM' - iQN'$	2
5 − 10	$\begin{matrix}10\\10\end{matrix} \quad \begin{matrix}01\\01\end{matrix}$	$= iMP' + iNQ' - iPM' - iQN'$	3
9 − 10	$\begin{matrix}10\\01\end{matrix} \quad \begin{matrix}01\\01\end{matrix}$	$= iMP' + iNQ' - iPM' - iQN'$	0
13 − 10	$\begin{matrix}10\\11\end{matrix} \quad \begin{matrix}01\\01\end{matrix}$	$= iMP' + iNQ' - iPM' - iQN'$	1
1 − 14	$\vartheta\begin{matrix}10\\00\end{matrix}u+u' \,.\, \vartheta\begin{matrix}01\\11\end{matrix}u-u'$	$= MP' - NQ' - PM' + QN'$	3
5 − 14	$\begin{matrix}10\\10\end{matrix} \quad \begin{matrix}01\\11\end{matrix}$	$= -MP' + NQ' + PM' - QN'$	2
9 − 14	$\begin{matrix}10\\01\end{matrix} \quad \begin{matrix}01\\11\end{matrix}$	$= MP' - NQ' - PM' + QN'$	1
13 − 14	$\begin{matrix}10\\11\end{matrix} \quad \begin{matrix}01\\11\end{matrix}$	$= -MP' + NQ' + PM' - QN'$	0

Fourth set, 64 equations (*concluded*).

				Suffixes.
0 – 3	$\vartheta\begin{matrix}00\\00\end{matrix}u+u'\,.$	$\vartheta\begin{matrix}11\\00\end{matrix}u-u'$	$= MQ' + NP' + PN' + QM'$	0
4 – 3	$\begin{matrix}00\\10\end{matrix}$	$\begin{matrix}11\\00\end{matrix}$	$= MQ' + NP' + PN' + QM'$	1
8 – 3	$\begin{matrix}00\\01\end{matrix}$	$\begin{matrix}11\\00\end{matrix}$	$= MQ' + NP' + PN' + QM'$	2
12 – 3	$\begin{matrix}00\\11\end{matrix}$	$\begin{matrix}11\\00\end{matrix}$	$= MQ' + NP' + PN' + QM'$	3
0 – 7	$\vartheta\begin{matrix}00\\00\end{matrix}u+u'\,.$	$\vartheta\begin{matrix}11\\10\end{matrix}u-u'$	$= iMQ' - iNP' + iPN' - iQM'$	1
4 – 7	$\begin{matrix}00\\10\end{matrix}$	$\begin{matrix}11\\10\end{matrix}$	$= iMQ' - iNP' + iPN' - iQM'$	0
8 – 7	$\begin{matrix}00\\01\end{matrix}$	$\begin{matrix}11\\10\end{matrix}$	$= iMQ' - iNP' + iPN' - iQM'$	3
12 – 7	$\begin{matrix}00\\11\end{matrix}$	$\begin{matrix}11\\10\end{matrix}$	$= iMQ' - iNP' + iPN' - iQM'$	2
0 – 11	$\vartheta\begin{matrix}00\\00\end{matrix}u+u'\,.$	$\vartheta\begin{matrix}11\\01\end{matrix}u-u'$	$= iMQ' + iNP' - iPN' - iQM'$	2
4 – 11	$\begin{matrix}00\\10\end{matrix}$	$\begin{matrix}11\\01\end{matrix}$	$= iMQ' + iNP' - iPN' - iQM'$	3
8 – 11	$\begin{matrix}00\\01\end{matrix}$	$\begin{matrix}11\\01\end{matrix}$	$= iMQ' + iNP' - iPN' - iQM'$	0
12 – 11	$\begin{matrix}00\\11\end{matrix}$	$\begin{matrix}11\\01\end{matrix}$	$= iMQ' + iNP' - iPN' - iQM'$	1
0 – 15	$\vartheta\begin{matrix}00\\00\end{matrix}u+u'\,.$	$\vartheta\begin{matrix}11\\11\end{matrix}u-u'$	$= - MQ' + NP' + PN' - QM'$	3
4 – 15	$\begin{matrix}00\\10\end{matrix}$	$\begin{matrix}11\\11\end{matrix}$	$= - MQ' + NP' + PN' - QM'$	2
8 – 15	$\begin{matrix}00\\01\end{matrix}$	$\begin{matrix}11\\11\end{matrix}$	$= - MQ' + NP' + PN' - QM'$	1
12 – 15	$\begin{matrix}00\\11\end{matrix}$	$\begin{matrix}11\\11\end{matrix}$	$= - MQ' + NP' + PN' - QM'$	0

90. I re-arrange these in sets of 16 equations, the equations of the first or square-set of 16 being taken as they stand, but those of the other sets being combined in pairs by addition and subtraction as will be seen. And I now drop altogether the characteristics, retaining only the current numbers: thus, in the set of equations next written down, the first equation is

$$\vartheta_0(u+u')\vartheta_0(u-u') = XX' + YY' + ZZ' + WW':$$

in the second set, the first equation is

$$\tfrac{1}{2}\{\vartheta_4(u+u')\vartheta_0(u-u') + \vartheta_0(u+u')\vartheta_4(u-u')\} = X_1X_1' + Z_1Z_1',$$

and so in other cases.

First or square-set of 16.

$\overset{u+u'}{\vartheta}$	$\overset{u-u'}{\vartheta}$	(Suffixes 0.)				
0	0	=	XX'	$+YY'$	$+ZZ'$	$+WW'$
4	4	=	XX'	$-YY'$	$+ZZ'$	$-WW'$
8	8	=	XX'	$+YY'$	$-ZZ'$	$-WW'$
12	12	=	XX'	$-YY'$	$-ZZ'$	$+WW'$
1	1	=	YX'	$+XY'$	$+WZ'$	$+ZW'$
5	5	=	$-YX'$	$+XY'$	$-WZ'$	$+ZW'$
9	9	=	YX'	$+XY'$	$-WZ'$	$-ZW'$
13	13	=	$-YX'$	$+XY'$	$+WZ'$	$-ZW'$
2	2	=	ZX'	$+WY'$	$+XZ'$	$+YW'$
6	6	=	ZX'	$-WY'$	$+XZ'$	$-YW'$
10	10	=	$-ZX'$	$-WY'$	$+XZ'$	$+YW'$
14	14	=	$-ZX'$	$+WY'$	$+XZ'$	$-YW'$
3	3	=	WX'	$+ZY'$	$+YZ'$	$+XW'$
7	7	=	$-WX'$	$+ZY'$	$-YZ'$	$+XW'$
11	11	=	$-WX'$	$-ZY'$	$+YZ'$	$+XW'$
15	15	=	WX'	$-ZY'$	$-YZ'$	$+XW'$

91. Second set of 16.

$\tfrac{1}{2}\{\overset{u+u'}{\vartheta}$	. $\overset{u-u'}{\vartheta}$	$+\ \overset{u+u'}{\vartheta}$	. $\overset{u-u'}{\vartheta}\}$	(Suffixes 1.)	
4	0	0	4 =	XX'	$+ZZ'$
12	8	8	12	XX'	$-ZZ'$
5	1	1	5	YX'	$+WZ'$
13	9	9	13	YX'	$-WZ'$
6	2	2	6	ZX'	$+XZ'$
14	10	10	14	$-ZX'$	$+XZ'$
7	3	3	7	WX'	$+YZ'$
15	11	11	15	$-WX'$	$+YZ'$

$\tfrac{1}{2}\{\overset{u+u'}{\vartheta}$	. $\overset{u-u'}{\vartheta}$	$-\ \overset{u+u'}{\vartheta}$	. $\overset{u-u'}{\vartheta}\}$	(Suffixes 1.)	
4	0	0	4 =	YY'	$+WW'$
12	8	8	12	YY'	$-WW'$
5	1	1	5	XY'	$+ZW'$
13	9	9	13	XY'	$-ZW'$
6	2	2	6	WY'	$+YW'$
14	10	10	14	$-WY'$	$+YW'$
7	3	3	7	ZY'	$+XW'$
15	11	11	15	$-ZY'$	$+XW'$

92. Third set of 16.

$\frac{1}{2}\{\overset{u+u'}{\vartheta} \,.\, \overset{u-u'}{\vartheta} + \overset{u+u'}{\vartheta} \,.\, \overset{u-u'}{\vartheta}\}$				(Suffixes 2.)
$u+u'$	$u-u'$	$u+u'$	$u-u'$	
8	0	0	8 =	$XX' + YY'$
12	4	4	12	$XX' - YY'$
9	1	1	9	$YX' + XY'$
13	5	5	13	$-YX' + XY'$
10	2	2	10	$ZX' + WY'$
14	6	6	14	$ZX' - WY'$
11	3	3	11	$WX' + ZY'$
15	7	7	15	$-WX' + ZY'$

$\frac{1}{2}\{\overset{u+u'}{\vartheta} \,.\, \overset{u-u'}{\vartheta} - \overset{u+u'}{\vartheta} \,.\, \overset{u-u'}{\vartheta}\}$				(Suffixes 2.)
$u+u'$	$u-u'$	$u+u'$	$u-u'$	
8	0	0	8 =	$ZZ' + WW'$
12	4	4	12	$ZZ' - WW'$
9	1	1	9	$WZ' + ZW'$
13	5	5	13	$-WZ' + ZW'$
10	2	2	10	$XZ' + YW'$
14	6	6	14	$XZ' - YW'$
11	3	3	11	$YZ' + XW'$
15	7	7	15	$-YZ' + XW'$

93. Fourth set of 16.

$\frac{1}{2}\{\overset{u+u'}{\vartheta} \,.\, \overset{u-u'}{\vartheta} + \overset{u+u'}{\vartheta} \,.\, \overset{u-u'}{\vartheta}\}$				(Suffixes 3.)
$u+u'$	$u-u'$	$u+u'$	$u-u'$	
12	0	0	12 =	$XX' + WW'$
8	4	4	8	$XX' - WW'$
13	1	1	13	$YX' + ZW'$
9	5	5	9	$YX' - ZW'$
14	2	2	14	$ZX' + YW'$
10	6	6	10	$ZX' - YW'$
15	3	3	15	$WX' + XW'$
11	7	7	11	$WX' - XW'$

$\frac{1}{2}\{\overset{u+u'}{\vartheta} \,.\, \overset{u-u'}{\vartheta} - \overset{u+u'}{\vartheta} \,.\, \overset{u-u'}{\vartheta}\}$				(Suffixes 3.)
$u+u'$	$u-u'$	$u+u'$	$u-u'$	
12	0	0	12 =	$YY' + ZZ'$
8	4	4	8	$-YY' + ZZ'$
13	1	1	13	$XY' + WZ'$
9	5	5	9	$-XY' + WZ'$
14	2	2	14	$WY' + XZ'$
10	6	6	10	$-WY' + XZ'$
15	3	3	15	$ZY' + YZ'$
11	7	7	11	$-ZY' + YZ'$

94. Fifth set of 16.

$\tfrac{1}{2}\{\overset{u+u'}{\vartheta}\,.\,\overset{u-u'}{\vartheta}+\overset{u+u'}{\vartheta}\,.\,\overset{u-u'}{\vartheta}\}$ (Suffixes 0.)

1	0	0	1	=	$E+G\,.\,E'+G'$	+	$F+H\,.\,F'+H'$	
5	4	4	5		$i\,.\,E-G$ „	+	$i\,.\,F-H$	„
9	8	8	9		$E+G$ „	−	$.\,F+H$	„
13	12	12	13		$i\,.\,E-G$ „	−	$i\,.\,F-H$	„
3	2	2	3		$F+H$ „	+	$E+G$	„
7	6	6	7		$i\,.\,F-H$ „	+	$i\,.\,E-G$	„
11	10	10	11		$-\ .\,F+H$ „	+	$E+G$	„
15	14	14	15		$-i\,.\,F-H$ „	+	$i\,.\,E-G$	„

$\tfrac{1}{2}\{\overset{u+u'}{\vartheta}\,.\,\overset{u-u'}{\vartheta}-\overset{u+u'}{\vartheta}\,.\,\overset{u-u'}{\vartheta}\}$ (Suffixes 0.)

1	0	0	1	=	$E-G\,.\,E'-G'$	+	$F-H\,.\,F'-H'$	
5	4	4	5		$i\,.\,E+G$ „	+	$i\,.\,F+H$	„
9	8	8	9		$E-G$ „	−	$.\,F-H$	„
13	12	12	13		$i\,.\,E+G$ „	−	$i\,.\,F+H$	„
3	2	2	3		$F-H$ „	+	$E-G$	„
7	6	6	7		$i\,.\,F+H$ „	+	$i\,.\,E+G$	„
11	10	10	11		$-\ .\,F-H$ „	+	$E-G$	„
15	14	14	15		$-i\,.\,F+H$ „	+	$i\,.\,E+G$	„

95. Sixth set of 16.

$\tfrac{1}{2}\{\overset{u+u'}{\vartheta}\,.\,\overset{u-u'}{\vartheta}+\overset{u+u'}{\vartheta}\,.\,\overset{u-u'}{\vartheta}\}$ (Suffixes 1.)

5	0	0	5	=	$E-iG\,.\,E'+iG'$	+	$F-iH\,.\,F'+iH'$	
1	4	4	1		$-i\,.\,E+iG$ „	−	$i\,.\,F+iH$	„
8	13	13	8		$E-iG$ „	−	$.\,F-iH$	„
9	12	12	9		$-i\,.\,E+iG$ „	+	$i\,.\,F+iH$	„
7	2	2	7		$F-iH$ „	+	$E-iG$	„
3	6	6	3		$-i\,.\,F+iH$ „	−	$i\,.\,E+iG$	„
15	10	10	15		$-\ .\,F-iH$ „	+	$E-iG$	„
11	14	14	11		$i\,.\,F+iH$ „	−	$i\,.\,E+iG$	„

$\tfrac{1}{2}\{\overset{u+u'}{\vartheta}\,.\,\overset{u-u'}{\vartheta}-\overset{u+u'}{\vartheta}\,.\,\overset{u-u'}{\vartheta}\}$ (Suffixes 1.)

5	0	0	5	=	$E+iG\,.\,E'-iG'$	+	$F+iH\,.\,F'-iH'$	
1	4	4	1		$-i\,.\,E-iG$ „	−	$i\,.\,F-iH$	„
13	8	8	13		$E+iG$ „	−	$.\,F+iH$	„
9	12	12	9		$-i\,.\,E-iG$ „	+	$i\,.\,F-iH$	„
7	2	2	7		$F+iH$ „	+	$E+iG$	„
3	6	6	3		$-i\,.\,F-iH$ „	−	$i\,.\,E-iG$	„
15	10	10	15		$-\ .\,F+iH$ „	+	$.\,E+iG$	„
11	14	14	11		$+i\,.\,F-iH$ „	−	$i\,.\,E-iG$	„

96. Seventh set of 16.

$\tfrac{1}{2}\{\overset{u+u'}{\vartheta}\;.\;\overset{u-u'}{\vartheta}+\overset{u+u'}{\vartheta}\;.\;\overset{u-u'}{\vartheta}\}$ (Suffixes 2.)

$\overset{u+u'}{\vartheta}$	$\overset{u-u'}{\vartheta}$	$\overset{u+u'}{\vartheta}$	$\overset{u-u'}{\vartheta}$						
9	0	0	9	=	$E+G\,.\,E'+G'$		+	$F-H\,.\,F'-H'$	
13	4	4	13		$i\,.\,E-G$	„	+	$i\,.\,F+H$	„
1	8	8	1		$E+G$	„	−	$.\,F-H$	„
5	12	12	5		$i\,.\,E-G$	„	−	$i\,.\,F+H$	„
11	2	2	11		$F+H$	„	+	$E-G$	„
15	6	6	15		$i\,.\,F-H$	„	+	$i\,.\,E+G$	„
3	10	10	3		$F+H$	„	−	$.\,E-G$	„
7	14	14	7		$i\,.\,F-H$	„	−	$i\,.\,E+G$	„

$\tfrac{1}{2}\{\overset{u+u'}{\vartheta}\;.\;\overset{u-u'}{\vartheta}-\overset{u+u'}{\vartheta}\;.\;\overset{u-u'}{\vartheta}\}$ (Suffixes 2.)

$\overset{u+u'}{\vartheta}$	$\overset{u-u'}{\vartheta}$	$\overset{u+u'}{\vartheta}$	$\overset{u-u'}{\vartheta}$						
9	0	0	9	=	$E-G\,.\,E'-G'$		+	$F+H\,.\,F'+H'$	
13	4	4	13		$i\,.\,E+G$	„	+	$i\,.\,F-H$	„
1	8	8	1		$E-G$	„	−	$.\,F+H$	„
5	12	12	5		$i\,.\,E+G$	„	−	$i\,.\,F-H$	„
11	2	2	11		$F-H$	„	+	$E+G$	„
15	6	6	15		$i\,.\,F+H$	„	+	$i\,.\,E-G$	„
3	10	10	3		$F-H$	„	−	$.\,E+G$	„
7	14	14	7		$i\,.\,F+H$	„	−	$i\,.\,E-G$	„

97. Eighth set of 16.

$\tfrac{1}{2}\{\overset{u+u'}{\vartheta}\;.\;\overset{u-u'}{\vartheta}+\overset{u+u'}{\vartheta}\;.\;\overset{u-u'}{\vartheta}\}$ (Suffixes 3.)

$\overset{u+u'}{\vartheta}$	$\overset{u-u'}{\vartheta}$	$\overset{u+u'}{\vartheta}$	$\overset{u-u'}{\vartheta}$						
13	0	0	13	=	$E-iG\,.\,E'+iG'$		+	$F+iH\,.\,F'-iH'$	
9	4	4	9		$-i\,.\,E+iG$	„	−	$i\,.\,F-iH$	„
5	8	8	5		$E-iG$	„	−	$.\,F+iH$	„
1	12	12	1		$-i\,.\,E+iG$	„	+	$i\,.\,F-iH$	„
15	2	2	15		$F-iH$	„	+	$E+iG$	„
11	6	6	11		$-i\,.\,F+iH$	„	−	$i\,.\,E-iG$	„
7	10	10	7		$F-iH$	„	−	$.\,E+iG$	„
3	14	14	3		$-i\,.\,F+iH$	„	+	$i\,.\,E-iG$	„

$\tfrac{1}{2}\{\overset{u+u'}{\vartheta}\;.\;\overset{u-u'}{\vartheta}-\overset{u+u'}{\vartheta}\;.\;\overset{u-u'}{\vartheta}\}$ (Suffixes 3.)

$\overset{u+u'}{\vartheta}$	$\overset{u-u'}{\vartheta}$	$\overset{u+u'}{\vartheta}$	$\overset{u-u'}{\vartheta}$						
13	0	0	13	=	$E+iG\,.\,E'-iG'$		+	$F-iH\,.\,F'+iH'$	
9	4	4	9		$-i\,.\,E-iG$	„	−	$i\,.\,F+iH$	„
5	8	8	5		$E+iG$	„	−	$.\,F-iH$	„
1	12	12	1		$-i\,.\,E-iG$	„	+	$i\,.\,F+iH$	„
15	2	2	15		$F+iH$	„	+	$E-iG$	„
11	6	6	11		$-i\,.\,F-iH$	„	−	$i\,.\,E+iG$	„
7	10	10	7		$F+iH$	„	−	$.\,E-iG$	„
3	14	14	3		$-i\,.\,F-iH$	„	+	$i\,.\,E+iG$	„

98. Ninth set of 16.

$$\tfrac{1}{2}\{\overset{u+u'}{\vartheta}\,.\,\overset{u-u'}{\vartheta}+\overset{u+u'}{\vartheta}\,.\,\overset{u-u'}{\vartheta}\}$$ (Suffixes 0.)

2	0	0	2	=	$I+K\,.\,I'+K'$		+	$J+L\,.\,J'+L'$
6	4	4	6		$I+K$	,,	−	$.J+L$,,
10	8	8	10		$i.I-K$	,,	+	$i.J-L$,,
14	12	12	14		$i.I-K$	,,	−	$i.J-L$,,
3	1	1	3		$J+L$	,,	+	$I+K$,,
7	5	5	7	−	$.J+L$	,,	−	$.I+K$,,
11	9	9	11		$i.J-L$	,,	+	$i.I-K$,,
15	13	13	15	−	$i.J-L$	,,	−	$i.I-K$,,

$$\tfrac{1}{2}\{\overset{u+u'}{\vartheta}\,.\,\overset{u-u'}{\vartheta}-\overset{u+u'}{\vartheta}\,.\,\overset{u-u'}{\vartheta}\}$$ (Suffixes 0.)

2	0	0	2	=	$I-K\,.\,I'-K'$		+	$J-L\,.\,J'-L'$
6	4	4	6		$I-K$	,,	−	$.J-L$,,
10	8	8	10		$i.I+K$	,,	+	$i.J+L$,,
14	12	12	14		$i.I+K$	,,	−	$i.J+L$,,
3	1	1	3		$J-L$	,,	+	$I-K$,,
7	5	5	7	−	$.J-L$	,,	+	$I-K$,,
11	9	9	11		$i.J+L$	,,	+	$i.I+K$,,
15	13	13	15	−	$i.J+L$	,,	+	$i.I+K$,,

99. Tenth set of 16.

$$\tfrac{1}{2}\{\overset{u+u'}{\vartheta}\,.\,\overset{u-u'}{\vartheta}+\overset{u+u'}{\vartheta}\,.\,\overset{u-u'}{\vartheta}\}$$ (Suffixes 1.)

6	0	0	6	=	$I+K\,.\,I'+K'$		+	$J-L\,.\,J'-L'$
2	4	4	2		$I+K$	,,	−	$.J-L$,,
14	8	8	14		$i.I-K$	,,	+	$i.J+L$,,
10	12	12	10		$i.I-K$	,,	−	$i.J+L$,,
7	1	1	7		$J+L$	,,	+	$I-K$,,
3	5	5	3		$J+L$	,,	−	$.I-K$,,
15	9	9	15		$i.J-L$	,,	+	$i.I+K$,,
11	13	13	11		$i.J-L$	,,	−	$i.I+K$,,

$$\tfrac{1}{2}\{\overset{u+u'}{\vartheta}\,.\,\overset{u-u'}{\vartheta}-\overset{u+u'}{\vartheta}\,.\,\overset{u-u'}{\vartheta}\}$$ (Suffixes 1.)

6	0	0	6	=	$I-K\,.\,I'-K'$		+	$J+L\,.\,J'+L'$
2	4	4	2		$I-K$	,,	−	$.J+L$,,
14	8	8	14		$i.I+K$	,,	+	$i.J-L$,,
10	12	12	10		$i.I+K$	,,	−	$i.J-L$,,
7	1	1	7		$J-L$	,,	+	$I+K$,,
3	5	5	3		$J-L$	,,	−	$.I+K$,,
15	9	9	15		$i.J+L$	,,	+	$i.I-K$,,
11	13	13	11		$i.J+L$	,,	−	$i.I-K$,,

100. Eleventh set of 16.

$\tfrac{1}{2}\{\overset{u+u'}{\vartheta}\;.\;\overset{u-u'}{\vartheta}\;+\;\overset{u+u'}{\vartheta}\;.\;\overset{u-u'}{\vartheta}\}$ (Suffixes 2.)

10	0	0	10	=	$I-iK\,.\,I'+iK'$		+	$.\,J-iL\,.\,J'+iL'$	
14	4	4	14		$I-iK$	,,	−	$.\,J-iL$	,,
2	8	8	2		$-i\,.\,I+iK$	,,	−	$i\,.\,J+iL$	,,
6	12	12	6		$-i\,.\,I+iK$	,,	+	$i\,.\,J+iL$	,,
11	1	1	11		$J-iL$	,,	+	$I-iK$	,,
15	5	5	15		$-\;.\,J-iL$	,,	+	$I-iK$	,,
3	9	9	3		$-i\,.\,J+iL$	,,	−	$i\,.\,I+iK$	,,
7	13	13	7		$+i\,.\,J+iL$	,,	−	$i\,.\,I+iK$	,,

$\tfrac{1}{2}\{\overset{u+u'}{\vartheta}\;.\;\overset{u-u'}{\vartheta}\;-\;\overset{u+u'}{\vartheta}\;.\;\overset{u-u'}{\vartheta}\}$ (Suffixes 2.)

10	0	0	10	=	$I+iK\,.\,I'-iK'$		+	$J+iL\,.\,J'-iL'$	
14	4	4	14		$I+iK$	,,	−	$J+iL$	,,
2	8	8	2		$-i\,.\,I-iK$	,,	−	$i\,.\,J-iL$	,,
6	12	12	6		$-i\,.\,I-iK$	,,	+	$i\,.\,J-iL$	,,
11	1	1	11		$J+iL$	,,	+	$I+iK$	,,
15	5	5	15		$-\;.\,J+iL$	,,	+	$I+iK$	,,
3	9	9	3		$-i\,.\,J-iL$	,,	−	$i\,.\,I-iK$	,,
7	13	13	7		$+i\,.\,J-iL$	,,	−	$i\,.\,I-iK$	,,

101. Twelfth set of 16.

$\tfrac{1}{2}\{\overset{u+u'}{\vartheta}\;.\;\overset{u-u'}{\vartheta}\;+\;\overset{u+u'}{\vartheta}\;.\;\overset{u-u'}{\vartheta}\}$ (Suffixes 3.)

14	0	0	14	=	$I-iK\,.\,I'+iK'$		+	$J+iL\,.\,J'-iL'$	
10	4	4	10		$I-iK$	,,	−	$.\,J+iL$	,,
6	8	8	6		$-i\,.\,I+iK$	,,	−	$i\,.\,J-iL$	,,
2	12	12	2		$-i\,.\,I+iK$	,,	+	$i\,.\,J-iL$	,,
15	1	1	15		$J-iL$	,,	+	$I+iK$	,,
11	5	5	11		$J-iL$	,,	−	$.\,I+iK$	,,
7	9	9	7		$-i\,.\,J+iL$	,,	−	$i\,.\,I-iK$	,,
3	13	13	3		$-i\,.\,J+iL$	,,	+	$i\,.\,I-iK$	,,

$\tfrac{1}{2}\{\overset{u+u'}{\vartheta}\;.\;\overset{u-u'}{\vartheta}\;-\;\overset{u+u'}{\vartheta}\;.\;\overset{u-u'}{\vartheta}\}$ (Suffixes 3.)

14	0	0	14	=	$I+iK\,.\,I'-iK'$		+	$J-iL\,.\,J'+iL'$	
10	4	4	10		$I+iK$	,,	−	$.\,J-iL$	,,
6	8	8	6		$-i\,.\,I-iK$	,,	−	$i\,.\,J+iL$	,,
2	12	12	2		$-i\,.\,I-iK$	,,	+	$i\,.\,J+iL$	,,
15	1	1	15		$J+iL$	,,	+	$I-iK$	,,
11	5	5	11		$J+iL$	,,	−	$.\,I-iK$	,,
7	9	9	7		$-i\,.\,J-iL$	,,	−	$i\,.\,I+iK$	,,
3	13	13	3		$-i\,.\,J-iL$	,,	+	$i\,.\,I+iK$	,,

102. Thirteenth set of 16.

$\frac{1}{2}\{ \overset{u+u'}{\vartheta} \,.\, \overset{u-u'}{\vartheta} + \overset{u+u'}{\vartheta} \,.\, \overset{u-u'}{\vartheta} \}$ (Suffixes 0.)

3	0	0	3	=	$M+Q\,.\,M'+Q'$	+	$N+P\,.\,N'+P'$
7	4	4	7		$i\,.\,M-Q$,,	−	$i\,.\,N-P$,,
11	8	8	11		$i\,.\,M-Q$,,	+	$i\,.\,N-P$,,
15	12	12	15	−	$.\,M+Q$,,	+	$N+P$,,
2	1	1	2		$N+P$,,	+	$M+Q$,,
6	5	5	6	−	$i\,.\,N-P$,,	+	$i\,.\,M-Q$,,
10	9	9	10		$i\,.\,N-P$,,	+	$i\,.\,M-Q$,,
14	13	13	14		$N+P$,,	−	$.\,M+Q$,,

$\frac{1}{2}\{ \overset{u+u'}{\vartheta} \,.\, \overset{u-u'}{\vartheta} - \overset{u+u'}{\vartheta} \,.\, \overset{u-u'}{\vartheta} \}$ (Suffixes 0.)

3	0	0	3	=	$M-Q\,.\,M'-Q'$	+	$N-P\,.\,N'-P'$
7	4	4	7		$i\,.\,M+Q$,,	−	$i\,.\,N+P$,,
11	8	8	11		$i\,.\,M+Q$,,	+	$i\,.\,N+P$,,
15	12	12	15	−	$.\,M-Q$,,	+	$N-P$,,
2	1	1	2		$N-P$,,	+	$M-Q$,,
6	5	5	6	−	$i\,.\,N+P$,,	+	$i\,.\,M+Q$,,
10	9	9	10		$i\,.\,N+P$,,	+	$i\,.\,M+Q$,,
14	13	13	14		$N-P$,,	−	$.\,M-Q$,,

103. Fourteenth set of 16.

$\frac{1}{2}\{ \overset{u+u'}{\vartheta} \,.\, \overset{u-u'}{\vartheta} + \overset{u+u'}{\vartheta} \,.\, \overset{u-u'}{\vartheta} \}$ (Suffixes 1.)

7	0	0	7	=	$M-iQ\,.\,M'+iQ'$	+	$N+iP\,.\,N'-iP'$
3	4	4	3	−	$i\,.\,M+iQ$,,	+	$i\,.\,N-iP$,,
15	8	8	15		$i\,.\,M+iQ$,,	+	$i\,.\,N-iP$,,
11	12	12	11		$M-iQ$,,	−	$.\,N+iP$,,
6	1	1	6		$N-iP$,,	+	$M+iQ$,,
2	5	5	2	−	$i\,.\,N-iP$,,	+	$i\,.\,M-iQ$,,
14	9	9	14	+	$i\,.\,N+iP$,,	+	$i\,.\,M-iQ$,,
10	13	13	10		$N-iP$,,	−	$.\,M+iQ$,,

$\frac{1}{2}\{ \overset{u+u'}{\vartheta} \,.\, \overset{u-u'}{\vartheta} - \overset{u+u'}{\vartheta} \,.\, \overset{u-u'}{\vartheta} \}$ (Suffixes 1.)

7	0	0	7	=	$M+iQ\,.\,M'-iQ'$	+	$N-iP\,.\,N'+iP$
3	4	4	3	−	$i\,.\,M-iQ$,,	+	$i\,.\,N+iP$,,
15	8	8	15	−	$i\,.\,M-iQ$,,	+	$i\,.\,N+iP$,,
11	12	12	11		$M+iQ$,,	−	$.\,N+iP$,,
6	1	1	6		$N+iP$,,	+	$M-iQ$,,
2	5	5	2	−	$i\,.\,N-iP$,,	+	$i\,.\,M+iQ$,,
14	9	9	14	+	$i\,.\,N-iP$,,	+	$i\,.\,M+iQ$,,
10	13	13	10		$N+iP$,,	−	$.\,M-iQ$,,

104. Fifteenth set of 16.

$\frac{1}{2}\{ \overset{u+u'}{\vartheta} . \overset{u-u'}{\vartheta} + \overset{u+u'}{\vartheta} . \overset{u-u'}{\vartheta} \}$ (Suffixes 2.)

11	0	0	11 =	$M-iQ$.	$M'+iQ'$	+	$N-iP$.	$N'+iP'$
15	4	4	15	$i.M+iQ$	,,	−	$i.N+iP$	,,
3	8	8	3	$-i.M+iQ$	,,	−	$i.N+iP$	,,
7	12	12	7	$M-iQ$	,,	−	$.N-iP$	,,
10	1	1	10	$N-iP$	,,	+	$M-iQ$	,,
14	5	5	14	$-i.N+iP$	,,	+	$i.M+iQ$	,,
2	9	9	2	$-i.N+iP$	,,	−	$i.M+iQ$	,,
6	13	13	6	$-\ .N-iP$	,,	+	$M-iQ$	,,

$\frac{1}{2}\{ \overset{u+u'}{\vartheta} . \overset{u-u'}{\vartheta} - \overset{u+u'}{\vartheta} . \overset{u-u'}{\vartheta} \}$ (Suffixes 2.)

11	0	0	11 =	$M+iQ$.	$M'-iQ'$	+	$N+iP$.	$N'-iP'$
15	4	4	15	$i.M-iQ$	,,	−	$i.N-iP$	,,
3	8	8	3	$-i.M-iQ$	,,	−	$i.N-iP$	,,
7	12	12	7	$M+iQ$	,,	−	$.N+iP$	,,
10	1	1	10	$N+iP$	,,	+	$M+iQ$	,,
14	5	5	14	$-i.N-iP$	,,	+	$i.M-iQ$	,,
2	9	9	2	$-i.N-iP$	,,	−	$i.M-iQ$	,,
6	13	13	6	$-\ .N+iP$	,,	+	$M+iQ$	,,

105. Sixteenth set of 16.

$\frac{1}{2}\{ \overset{u+u'}{\vartheta} . \overset{u-u'}{\vartheta} + \overset{u+u'}{\vartheta} . \overset{u-u'}{\vartheta} \}$ (Suffixes 3.)

15	0	0	15 =	$M-Q$.	$M'-Q'$	+	$N+P$.	$N'+P'$
11	4	4	11	$-i.M+Q$	,,	+	$i.N-P$	,,
7	8	8	7	$-i.M+Q$	,,	−	$i.N-P$	,,
3	12	12	3	$-\ .M-Q$	,,	+	$N+P$	,,
14	1	1	14	$N-P$	,,	+	$M+Q$	,,
10	5	5	10	$-i.N+P$	,,	+	$i.M-Q$	,,
6	9	9	6	$-i.N+P$	,,	−	$i.M-Q$	,,
2	13	13	2	$-\ .N-P$	,,	+	$M+Q$	,,

$\frac{1}{2}\{ \overset{u+u'}{\vartheta} . \overset{u-u'}{\vartheta} - \overset{u+u'}{\vartheta} . \overset{u-u'}{\vartheta} \}$ (Suffixes 3.)

15	0	0	15 =	$M+Q$.	$M'+Q'$	+	$N-P$.	$N'-P'$
11	4	4	11	$-i.M-Q$	,,	+	$i.N+P$	,,
7	8	8	7	$-i.M-Q$	,,	−	$i.N+P$	,,
3	12	12	3	$-\ .M+Q$	,,	+	$N-P$	,,
14	1	1	14	$N+P$	,,	+	$M-Q$	,,
10	5	5	10	$-i.N-P$	,,	+	$i.M+Q$	,,
6	9	9	6	$-i.N-P$	,,	−	$i.M+Q$	,,
2	13	13	2	$-\ .N+P$	,,	+	$M-Q$	,,

106. In the square set, writing $u'=v'=0$, and α, β, γ, δ for X', Y', Z', W'; also slightly altering the arrangement,

the system becomes : and further writing herein $u=0$, $v=0$, it becomes

u			X	Y	Z	W	0					c^2
ϑ^2	0	=	α	β	γ	δ	ϑ^2	0	=	$\alpha^2+\beta^2+\gamma^2+\delta^2$	=	0
	4	=	α	$-\beta$	γ	$-\delta$		4	=	$\alpha^2-\beta^2+\gamma^2-\delta^2$	=	4
	8	=	α	β	$-\gamma$	$-\delta$		8	=	$\alpha^2-\beta^2-\gamma^2-\delta^2$	=	8
	12	=	α	$-\beta$	$-\gamma$	δ		12	=	$\alpha^2-\beta^2-\gamma^2+\delta^2$	=	12
	1	=	β	α	δ	γ		1	=	$2(\alpha\beta+\gamma\delta)$	=	1
	5	=	β	$-\alpha$	δ	$-\gamma$		5		0		
	9	=	β	α	$-\delta$	$-\gamma$		9	=	$2(\alpha\beta-\gamma\delta)$	=	9
	13	=	β	$-\alpha$	$-\delta$	γ		13		0		
	2	=	γ	δ	α	β		2	=	$2(\alpha\gamma+\beta\delta)$	=	2
	6	=	γ	$-\delta$	α	$-\beta$		6	=	$2(\alpha\gamma-\beta\delta)$	=	6
	10	=	γ	δ	$-\alpha$	$-\beta$		10		0		
	14	=	γ	$-\delta$	$-\alpha$	β		14		0		
	3	=	δ	γ	β	α		3	=	$2(\alpha\delta+\beta\gamma)$	=	3
	7	=	δ	$-\gamma$	β	$-\alpha$		7		0		
	11	=	δ	γ	$-\beta$	$-\alpha$		11		0		
	15	=	δ	$-\gamma$	$-\beta$	α		15	=	$2(\alpha\delta-\beta\gamma)$	=	15

viz. this last is the before-mentioned system of equations giving the values of the 10 zero-functions c in terms of the four constants α, β, γ, δ.

107. The system first obtained is a system of 16 equations

$$\vartheta_0^2(u,\ v)=\alpha X+\beta Y+\gamma Z+\delta W, \text{ \&c.},$$

showing that the squares of the theta-functions are each of them a linear function of the four quantities X, Y, Z, W. If the functions on the right-hand side were independent (asyzygetic) linear functions of X, Y, Z, W, it would follow that any four (selected at pleasure) of the squared theta-functions are linearly independent, and that we could in terms of these four express linearly each of the remaining 12 squared functions. But this is not so; the form of the linear functions of (X, Y, Z, W) is such that we can (and that in 16 different ways) select out of the 16 linear functions six functions, such that any four of them are connected by a linear equation; and there are consequently 16 hexads of squared theta-functions, such that any four out of the same hexad are connected by a linear relation. The hexads are shown by the foregoing "Table of the 16 Kummer hexads."

108. The *à posteriori* verification is immediately effected; taking for instance the first column, the equations are

	$\vartheta^2 u$		X	Y	Z	W
A	11	=	δ	γ	$-\beta$	$-\alpha,$
B	7		δ	$-\gamma$	β	$-\alpha,$
AB	6		γ	$-\delta$	α	$-\beta,$
CD	2		γ	δ	α	$\beta,$
CE	1		β	α	δ	$\gamma,$
DE	9		β	α	δ	$-\gamma;$

viz. it should thence follow that there is a linear relation between any four of the six squared functions 11, 7, 6, 2, 1, 9: and it is accordingly seen that this is so. It further appears that, in the several linear relations, the coefficients (obtained in the first instance as functions of α, β, γ, δ) are in fact the 10 constants c: the 15 relations connecting the several systems of four out of the six squared functions are given in the following table.

109.

ϑ^2	11	7	6	2	1	9	= 0.
c^2			6	− 2	1	− 9	
	6			+ 15	− 12	+ 4	
	− 2		− 15		+ 8	− 0	
	1		+ 12	− 8		+ 3	
	− 9		− 4	+ 0	− 3		
		6		3	− 0	+ 8	
		− 2	− 3		+ 4	− 12	
		1	+ 0	− 4		− 15	
		− 9	− 8	+ 12	+ 15		
	− 15	+ 3	+ 2	− 6			
	− 12	+ 0	+ 1		− 6		
	− 4	+ 8	+ 9			− 6	
	− 3	+ 15			+ 9	− 1	
	− 0	+ 12		+ 9		− 2	
	− 8	+ 4		+ 1	− 2		

Read

$$\cdot \quad c_6^2\vartheta_6^2 - c_2^2\,\vartheta_2^2 + c_1^2\,\vartheta_1^2 - c_9^2\vartheta_9^2 = 0,$$

$$c_6^2\vartheta_{11}^2 \quad \cdot \quad + c_{15}^2\vartheta_2^2 - c_{12}^2\vartheta_1^2 + c_4^2\vartheta_9^2 = 0, \text{ \&c.}$$

110. The first set of 16 equations is the square-set, which has been already considered. If in each of the other sets of 16 equations we write in like manner $u'=0$, each set in fact reduces itself to eight equations; sets 2, 3, 4 give thus $8+8+8$, $=24$ equations; sets 5 to 8, 9 to 12, and 13 to 15, give each $8+8+8+8$, $=32$ equations; or we have sets of 24, 32, 32, 32, together 120 equations, the number being of course one half of $256-16$, the number of equations after deducting the 16 equations of the square set.

111. The first set, 24 equations.

This is derived from the second, third, and fourth sets, each of 16 equations, by writing therein $u'=0$. Taking α_1, β_1, γ_1, δ_1 for the zero-functions corresponding to X_1, Y_1, Z_1, W_1, then on writing $u'=0$, X_1', Y_1', Z_1', W_1' become α_1, β_1, γ_1, δ_1. In the second set of 16 equations, the first equations thus are

$$\vartheta_4 u\,.\,\vartheta_0 u = \alpha_1 X_1 + \gamma_1 Z_1, \qquad 0 = \beta_1 Y_1 + \delta_1 W_1,$$
$$\vartheta_{12} u\,.\,\vartheta_8 u = \alpha_1 X_1 - \gamma_1 Z_1, \qquad 0 = \beta_1 Y_1 - \delta_1 W_1,$$
$$\vdots \qquad\qquad\qquad \vdots$$

viz. the equations of the column require that, and are all satisfied if, $\beta_1=0$, $\delta_1=0$: hence the zero-functions are α_1, 0, γ_1, 0; and this being so we have only the equations of the first column. And similarly as regards the third and fourth sets; the zero values corresponding to

	X_1,	Y_1,	Z_1,	W_1	X_2,	Y_2,	Z_2,	W_2	X_3,	Y_3,	Z_3,	W_3
are	α_1	0	γ_1	0	α_2	β_2	0	0	α_3	0	0	δ_3;

and we have in all $8+8+8$, $=24$ equations. These are

ϑu	. ϑu		(Suffixes 1.) X	Z	ϑu	. ϑu		(Suffixes 2.) X	Y	ϑu	. ϑu		(Suffixes 3.) X	W
4	0	=	α	γ	8	0	=	α	β	12	0	=	α	δ
12	8	=	α	$-\gamma$	12	4	=	α	$-\beta$	8	4	=	α	$-\delta$
6	2	=	γ	α	9	1	=	β	α	15	3	=	δ	α
14	10	=	γ	$-\alpha$	13	5	=	β	$-\alpha$	11	7	=	$-\delta$	α
			Y	W				Z	W				Y	Z
5	1	=	α	γ	10	2	=	α	β	13	1	=	α	δ
13	9	=	α	$-\gamma$	14	6	=	α	$-\beta$	9	5	=	α	$-\delta$
7	3	=	γ	α	11	3	=	β	α	14	2	=	δ	α
15	11	=	γ	$-\alpha$	15	7	=	β	$-\alpha$	10	6	=	$-\delta$	α
$\vartheta 0$	. $\vartheta 0$				$\vartheta 0$	. $\vartheta 0$				$\vartheta 0$	. $\vartheta 0$			
4	0	=	$\alpha^2+\gamma^2$		8	0	=	$\alpha^2+\beta^2$		12	0	=	$\alpha^2+\delta^2$	
12	8	=	$\alpha^2-\gamma^2$		12	4	=	$\alpha^2-\beta^2$		8	4	=	$\alpha^2-\delta^2$	
6	2	=	$2\alpha\gamma$		9	1	=	$2\alpha\beta$		15	3	=	$2\alpha\delta$.	

112. The second set, 32 equations.

To exhibit these in a convenient form, I alter the notation, viz. I write

$$\begin{array}{cccc|cccc} E+G, & i(E-G), & (F+H), & i(F-H) & E_1+iG_1, & E_1-iG_1, & F_1+iH_1, & F_1-iH_1 \\ = X, & Y, & Z, & W & X_1, & Y_1, & Z_1, & W_1 \end{array}$$

$$\begin{array}{cccc|cccc} (E_2+G_2), & i(E_2-G_2), & (F_2+H_2), & i(F_2-H_2) & E_3+iG_3, & E_3-iG_3, & F_3+iH_3, & F_3-iH_3 \\ = X_2, & Y_2, & Z_2, & W_2 & X_3, & Y_3, & Z_3, & W_3, \end{array}$$

so that as regards the present set of equations, X, Y, Z, W, signify as just mentioned. And, this being so, the corresponding zero-values are

$$\alpha,\ 0,\ \gamma,\ 0 \ \big|\ \alpha_1,\ 0,\ \gamma_1,\ 0 \ \big|\ \alpha_2,\ 0,\ 0,\ \delta_2 \ \big|\ \alpha_3,\ 0,\ 0,\ \delta_3.$$

The equations then are

(Suffixes 0.)

ϑu .	ϑu	X	Z
1	0 =	α	γ
9	8 =	α	$-\gamma$
3	2 =	γ	α
11	10 =	γ	$-\alpha$
		Y	W
5	4 =	α	γ
13	12 =	α	$-\gamma$
7	6 =	γ	α
15	14 =	γ	$-\alpha$
$\vartheta 0$.	$\vartheta 0$		
1	0 =	$\alpha^2+\gamma^2$	
9	8 =	$\alpha^2-\gamma^2$	
3	2 =	$2\alpha\gamma$	

(Suffixes 1.)

ϑu .	ϑu	X	Z
1	4 =	$-i\alpha$	$-i\gamma$
9	12 =	$-i\alpha$	$+i\gamma$
3	6 =	$-i\gamma$	$-i\alpha$
11	14 =	$-i\gamma$	$+i\alpha$
		Y	W
5	0 =	α	γ
13	8 =	α	$-\gamma$
7	2 =	γ	α
15	10 =	γ	$-\alpha$
$\vartheta 0$.	$\vartheta 0$		
1	4 =	$-i(\alpha^2+\gamma^2)$	
9	12 =	$-i(\alpha^2-\gamma^2)$	
3	6 =	$-2i\alpha\gamma$	

(Suffixes 2.)

ϑu .	ϑu	X	W
9	0 =	α	$-\delta$
1	8 =	α	δ
15	6 =	δ	α
7	14 =	$-\delta$	α
		Y	Z
13	4 =	α	δ
5	12 =	α	$-\delta$
11	2 =	$-\delta$	α
3	10 =	δ	α
$\vartheta 0$.	$\vartheta 0$		
9	0 =	$\alpha^2-\delta^2$	
1	8 =	$\alpha^2+\delta^2$	
15	6 =	$2\alpha\delta$	

(Suffixes 3.)

ϑu .	ϑu	X	W
9	4 =	$-i\alpha$	$-i\delta$
1	12 =	$-i\alpha$	$+i\delta$
15	2 =	δ	α
7	10 =	$-\delta$	α
		Y	Z
13	0 =	α	δ
5	8 =	α	$-\delta$
11	6 =	$-i\delta$	$-i\alpha$
3	14 =	$i\delta$	$-i\alpha$
$\vartheta 0$.	$\vartheta 0$		
9	4 =	$-i(\alpha^2+\delta^2)$	
1	12 =	$-i(\alpha^2-\delta^2)$	
15	2 =	$2\alpha\delta$.	

113. Third set, 32 equations.

We again change the notation, writing

$$\begin{array}{cccc|cccc} I+K, & i(I-K), & J+L, & i(J-L) & I_1+K_1, & i(I_1-K_1), & (J_1+L_1), & i(J_1-L_1) \\ = X, & Y, & Z, & W & X_1, & Y_1, & Z_1, & W_1 \end{array}$$

$$\begin{array}{cccc|cccc} I_2+iK_2, & I_2-iK_2, & J_2+iL_2, & J_2-iL_2 & I_3+iK_3, & I_3-iK_3, & J_3+iL_3, & J_3-iL_3 \\ = X_2, & Y_2, & Z_2, & W_2 & X_3, & Y_3, & Z_3, & W_3, \end{array}$$

the zero values being

$$\alpha,\ 0,\ \gamma,\ 0 \mid \alpha_1,\ 0,\ 0,\ \delta_1 \mid \alpha_2,\ 0,\ \gamma_2,\ 0 \mid \alpha_3,\ 0,\ 0,\ \delta_3.$$

Then equations are

(Suffixes 0.)

$\vartheta u \,.\, \vartheta u$		X	Z
2	0 =	α	γ
6	4 =	α	$-\gamma$
3	1 =	γ	α
7	5 =	γ	$-\alpha$
		Y	W
10	8 =	α	γ
14	12 =	α	$-\gamma$
11	9 =	γ	α
15	13 =	γ	$-\alpha$

$\vartheta 0 \,.\, \vartheta 0$		
2	0 =	$\alpha^2+\gamma^2$
6	4 =	$\alpha^2-\gamma^2$
3	1 =	$2\alpha\gamma$

(Suffixes 1.)

$\vartheta u \,.\, \vartheta u$		X	W
6	0 =	α	$-\delta$
2	4 =	α	δ
15	9 =	δ	α
11	13 =	$-\delta$	α
		Y	Z
14	8 =	α	δ
10	12 =	α	$-\delta$
7	1 =	$-\delta$	α
3	5 =	δ	α

$\vartheta 0 \,.\, \vartheta 0$		
6	0 =	$\alpha^2-\delta^2$
2	4 =	$\alpha^2+\delta^2$
15	9 =	$2\alpha\delta$

(Suffixes 2.)

$\vartheta u \,.\, \vartheta u$		X	Z
2	8 =	$-i\alpha$	$-i\gamma$
6	12 =	$-i\alpha$	$+i\gamma$
3	9 =	$-i\gamma$	$-i\alpha$
7	13 =	$-i\gamma$	$+i\alpha$
		Y	W
10	0 =	α	γ
14	4 =	α	$-\gamma$
11	1 =	γ	α
15	5 =	γ	$-\alpha$

$\vartheta 0 \,.\, \vartheta 0$		
2	8 =	$-i(\alpha^2+\gamma^2)$
6	12 =	$-i(\alpha^2-\gamma^2)$
3	9 =	$-\ 2i\alpha\gamma$

(Suffixes 3.)

$\vartheta u \,.\, \vartheta u$		X	W
6	8 =	$-i\alpha$	$-i\delta$
2	12 =	$-i\alpha$	$i\delta$
15	1 =	δ	α
11	5 =	$-\delta$	α
		Y	Z
14	0 =	α	δ
10	4 =	α	$-\delta$
7	9 =	$-i\delta$	$-i\alpha$
3	13 =	$i\delta$	$-i\alpha$

$\vartheta 0 \,.\, \vartheta 0$		
6	8 =	$-i(\alpha^2+\delta^2)$
2	12 =	$-i(\alpha^2-\delta^2)$
15	1 =	$2\alpha\delta.$

114. Fourth set, 32 equations.

Again changing the notation, we write

$$\begin{array}{lllll|llll} & M_2+iQ_2, & M_2-iQ_2, & N_2+iP_2, & N_2-iP_2 & M_3+Q_3, & i(M_3-Q_3), & N_3+P_3, & i(N_3-P_3) \\ = & X, & Y, & Z, & W & X_1, & Y_1, & Z_1, & W_1, \end{array}$$

$$\begin{array}{lllll|llll} & M+Q, & i(M-Q), & N+P, & i(N-P) & M_1+iQ_1, & M_1-iQ_1, & N_1+iP_1, & N_1-iP_1 \\ = & X_2, & Y_2, & Z_2, & W_2 & X_3, & Y_3, & Z_3, & W_3, \end{array}$$

the zero values being

$$\alpha,\ 0,\ \gamma,\ 0 \ \big|\ \alpha_1,\ 0,\ 0,\ \delta_1 \ \big|\ \alpha_2,\ 0,\ \gamma_2,\ 0 \ \big|\ 0,\ \beta_3,\ \gamma_3,\ 0.$$

The equations then are

(Suffixes 0.)

$\vartheta u\,.\,\vartheta u$		X	Z
0	3 =	α	γ
15	12 =	$-\alpha$	γ
2	1 =	γ	α
14	13 =	$-\gamma$	α
		Y	W
4	7 =	α	$-\gamma$
8	11 =	α	γ
6	5 =	γ	$-\alpha$
10	9 =	γ	α

$\vartheta 0\,.\,\vartheta 0$		
0	3 =	$\alpha^2+\gamma^2$
15	12 =	$-(\alpha^2-\gamma^2)$
2	1 =	$2\alpha\gamma$

(Suffixes 1.)

$\vartheta u\,.\,\vartheta u$		X	W
3	4 =	$-i\alpha$	$i\delta$
15	8 =	$i\alpha$	$i\delta$
6	1 =	δ	α
10	13 =	$-\delta$	α
		Y	Z
7	0 =	α	δ
11	12 =	α	$-\delta$
2	5 =	$i\delta$	$-i\alpha$
14	9 =	$i\delta$	$i\alpha$

$\vartheta 0\,.\,\vartheta 0$		
3	4 =	$-i(\alpha^2-\delta^2)$
15	8 =	$i(\alpha^2+\delta^2)$
6	1 =	$2\alpha\delta$

(Suffixes 2.)

$\vartheta u\,.\,\vartheta u$		X	Z
15	4 =	$i\alpha$	$-i\gamma$
3	8 =	$-i\alpha$	$-i\gamma$
14	5 =	$i\gamma$	$-i\alpha$
2	9 =	$-i\gamma$	$-i\alpha$
		Y	W
11	0 =	α	γ
7	12 =	α	$-\gamma$
10	1 =	γ	α
6	13 =	γ	$-\alpha$

$\vartheta 0\,.\,\vartheta 0$		
15	4 =	$i(\alpha^2-\gamma^2)$
3	8 =	$-i(\alpha^2+\gamma^2)$
2	9 =	$-2i\alpha\gamma$

(Suffixes 3.)

$\vartheta u\,.\,\vartheta u$		Y	Z
15	0 =	$-\beta$	γ
3	12 =	β	γ
10	5 =	γ	$-\beta$
6	9 =	$-\gamma$	$-\beta$
		X	W
11	4 =	$-\beta$	γ
7	8 =	$-\beta$	$-\gamma$
14	1 =	γ	$-\beta$
2	13 =	γ	β

$\vartheta 0\,.\,\vartheta 0$		
15	0 =	$-(\beta^2-\gamma^2)$
3	12 =	$\beta^2+\gamma^2$
6	9 =	$-2\beta\gamma$.

115. It will be noticed that the pairs of theta-functions which present themselves in these equations are the same as in the foregoing "Table of the 120 pairs." And the equations show that the four products, each of a pair of theta-functions, belonging to the upper half or to the lower half of any column of the table, are such that any three of the four products are connected by a linear equation. The coefficients of these linear relations are, in fact, functions such as the $\alpha^2+\delta^2$, $\alpha^2-\delta^2$, $2\alpha\delta$ written down at the foot of the several systems of eight equations, and they are consequently products each of two zero-functions c.

Thus (see "The first set, 24 equations") we have

ϑu . ϑu	(Suffixes 3.) X	W	ϑu . ϑu	(Suffixes 3.) Y	Z	$\vartheta 0$. $\vartheta 0$	(Suffixes 3.)
4 8 =	α	$-\delta$	5 9 =	α	$-\delta$	4 8	$= \alpha^2-\delta^2$
0 12 =	α	δ	1 13 =	α	δ	0 12	$= \alpha^2+\delta^2$
3 15 =	δ	α	2 14 =	δ	α	15 3	$= 2\alpha\delta$.
7 11 =	$-\delta$	α	6 10 =	$-\delta$	α		

116. In the left-hand four of these, omitting successively the first, second, third, and fourth equation, and from the remaining three eliminating the X_3 and W_3, we write down, almost mechanically,

ϑu . ϑu				
4 8	.	$+2\alpha\delta$,	$-\delta^2-\alpha^2$,	$\alpha^2-\delta^2$
0 12	$-2\alpha\delta$,	.	$-\delta^2+\alpha^2$,	$\alpha^2+\delta^2$
3 15	$\alpha^2+\delta^2$,	$\delta^2-\alpha^2$,	.	$2\alpha\delta$
7 11	$-\alpha^2+\delta^2$,	$\delta^2+\alpha^2$,	$-2\alpha\delta$	

and thence derive the first of the next following system of equations; read

$$\begin{aligned}
c_3c_{15}\vartheta_0\vartheta_{12} - c_0c_{12}\vartheta_3\vartheta_{15} + c_4c_8\,\vartheta_7\vartheta_{11} &= 0,\\
-\,c_3c_{15}\vartheta_4\vartheta_8 \qquad\qquad + c_4c_8\,\vartheta_3\vartheta_{15} - c_0c_{12}\vartheta_7\vartheta_{11} &= 0,\\
c_0c_{12}\vartheta_4\vartheta_8 - c_4c_8\,\vartheta_0\vartheta_{12} \qquad\qquad + c_3c_{15}\vartheta_7\vartheta_{11} &= 0,\\
-\,c_4c_8\,\vartheta_4\vartheta_8 + c_0c_{12}\vartheta_0\vartheta_{12} - c_3c_{15}\vartheta_3\vartheta_{15} \qquad\qquad &= 0,
\end{aligned}$$

where the theta-functions have the arguments u, v.

Observe that, on writing herein $u=0$, $v=0$, the first three equations become each of them identically $0=0$; the fourth equation becomes

$$-\,c_4^2c_8^2 + c_0^2c_{12}^2 - c_3^2c_{15}^2 = 0,$$

which is one of the relations between the c's and serves as a verification.

But in the right-hand system, on writing $u=v=0$, each of the four equations becomes identically $0=0$.

117. The equations are

ϑ	4.8	0.12	3.15	7.11	= 0,
c		3.15	− 0.12	4.8	
	− 3.15		− 4.8	− 0.12	
	0.12	− 4.8		3.15	
	− 4.8	0.12	− 3.15		

ϑ	5.9	1.13	2.14	6.10	= 0,
c		3.15	− 0.12	4.8	
	− 3.15		4.8	− 0.12	
	0.12	− 4.8		3.15	
	− 4.8	− 0.12	− 3.15		

ϑ	6.8	2.12	1.15	5.11	= 0,
c		1.15	− 2.12	6.8	
	− 1.15		6.8	− 2.11	
	2.12	− 6.8		1.15	
	− 6.8	2.11	− 1.15		

ϑ	7.9	3.13	0.14	4.10	= 0,
c		1.15	− 2.12	6.8	
	− 1.15		6.8	− 2.12	
	2.12	− 6.8		1.15	
	− 6.8	2.12	− 1.15		

ϑ	0.6	2.4	9.15	11.13	= 0,
c		9.15	− 2.4	0.6	
	− 9.15		0.6	− 2.4	
	2.4	− 0.6		9.15	
	− 0.6	2.4	− 9.15		

ϑ	1.7	3.5	8.14	10.12	= 0,
c		9.15	− 2.4	0.6	
	− 9.15		0.6	− 2.4	
	2.4	− 0.6		9.15	
	− 0.6	2.4	− 9.15		

ϑ	3.6	1.4	9.12	14.11	= 0,
c		9.12	− 1.4	3.6	
	− 9.12		3.6	− 1.4	
	1.4	− 3.6		9.12	
	− 3.6	1.4	− 9.12		

ϑ	2.7	0.5	8.13	10.15	= 0,
c		9.12	− 1.4	3.6	
	− 9.12		3.6	− 1.4	
	1.4	− 3.6		9.12	
	− 3.6	1.4	− 9.12		

ϑ	8.9	0.1	2.3	10.11	= 0,
c		− 2.3	0.1	8.9	
	2.3		− 8.9	− 0.1	
	− 0.1	8.9		2.3	
	− 8.9	0.1	− 2.3		

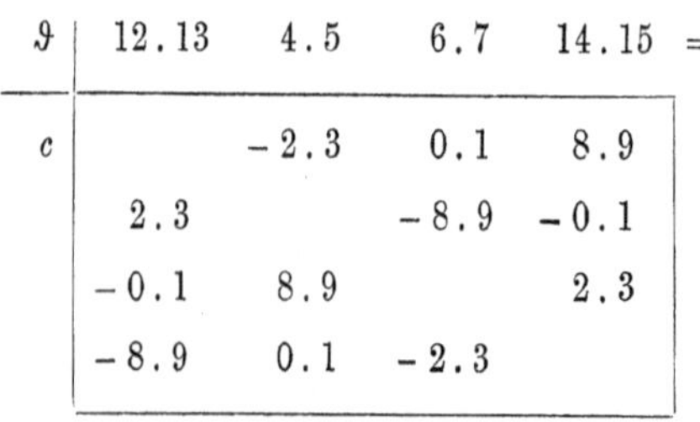

ϑ	12.13	4.5	6.7	14.15	= 0,
c		− 2.3	0.1	8.9	
	2.3		− 8.9	− 0.1	
	− 0.1	8.9		2.3	
	− 8.9	0.1	− 2.3		

ϑ	4.6	0.2	1.3	5.7	=0,
c		−1.3	0.2	4.6	
	1.3		−4.6	−0.2	
	−0.2	4.6		1.3	
	−4.6	0.2	−1.3		

ϑ	9.11	13.15	12.14	8.10	=0,
c		−1.3	0.2	4.6	
	1.3		−4.6	−0.2	
	−0.2	−4.6		1.3	
	4.6	0.2	−1.3		

ϑ	6.12	2.8	3.9	7.13	=0,
c		3.9	−2.8	−6.12	
	−3.9		6.12	2.8	
	2.8	−6.12		−3.9	
	6.12	−2.8	3.9		

ϑ	1.11	5.15	4.14	0.10	=0,
c		3.9	−2.8	6.12	
	−3.9		−6.12	2.8	
	2.8	6.12		−3.9	
	−6.12	−2.8	3.9		

ϑ	6.15	1.8	0.9	7.14	=0,
c		0.9	−1.8	−6.15	
	−0.9		6.15	1.8	
	1.8	−6.15		−0.9	
	6.15	−1.8	0.9		

ϑ	2.11	5.12	4.13	3.10	=0,
c		0.9	−1.8	6.15	
	−0.9		−6.15	1.8	
	1.8	6.15		−0.9	
	−6.15	−1.8	0.9		

ϑ	4.9	1.12	2.15	7.10	=0,
c		2.15	−1.12	4.9	
	−2.15		4.9	−1.12	
	1.12	−4.9		2.15	
	−4.9	1.12	−2.15		

ϑ	0.13	5.8	6.11	3.14	=0,
c		2.15	1.12	−4.9	
	−2.15		−4.9	1.12	
	−1.12	4.9		2.15	
	4.9	−1.12	−2.15		

ϑ	4.12	0.8	1.9	5.13	=0,
c		−1.9	0.8	4.12	
	1.9		−4.12	−0.8	
	−0.8	4.12		1.9	
	−4.12	0.8	−1.9		

ϑ	3.11	7.15	6.14	2.10	=0,
c		1.9	−0.8	4.12	
	−1.9		−4.12	0.8	
	0.8	4.12		−1.9	
	−4.12	−0.8	1.9		

ϑ	4.15	3.8	2.9	5.14	= 0,
c		−2.9	3.8	4.15	
	2.9		−4.15	−3.8	
	−3.8	4.15		2.9	
	−4.15	3.8	−2.9		

ϑ	0.11	7.12	6.13	1.10	= 0,
c		−2.9	3.8	−4.15	
	2.9		4.15	−3.8	
	−3.8	−4.15		2.9	
	4.15	3.8	−2.9		

ϑ	6.9	3.12	0.15	5.10	= 0,
c		−0.15	3.12	−6.9	
	0.15		−6.9	3.12	
	−3.12	6.9		−0.15	
	6.9	−3.12	0.15		

ϑ	2.13	7.8	4.11	1.14	= 0,
c		0.15	3.12	−6.9	
	−0.15		−6.9	3.12	
	−3.12	6.9		0.15	
	6.9	−3.12	−0.15		

ϑ	12.15	0.3	1.2	13.14	= 0,
c		1.2	−0.3	−12.15	
	−1.2		12.15	0.3	
	0.3	−12.15		−1.2	
	12.15	−0.3	1.2		

ϑ	8.11	4.7	5.6	9.10	= 0,
c		1.2	−0.3	12.15	
	−1.2		−12.15	0.3	
	0.3	12.15		−1.2	
	−12.15	−0.3	1.2		

ϑ	1.6	3.4	8.15	10.13	= 0,
c		8.15	−3.4	1.6	
	−8.15		1.6	−3.4	
	3.4	−1.6		8.15	
	−1.6	3.4	−8.15		

ϑ	2.5	0.7	11.12	9.14	= 0,
c		−8.15	−3.4	1.6	
	8.15		1.6	−3.4	
	3.4	−1.6		−8.15	
	−1.6	3.4	8.15		

ϑ	2.6	0.4	8.12	10.14	= 0,
c		−8.12	0.4	−2.6	
	8.12		−2.6	0.4	
	−0.4	2.6		−8.12	
	2.6	−0.4	8.12		

ϑ	1.5	3.7	11.15	9.13	= 0.
c		−8.12	−0.4	2.6	
	8.12		2.6	−0.4	
	0.4	−2.6		−8.12	
	−2.6	0.4	8.12		

118. The foregoing equations may be verified, and it is interesting to verify them, by means of the approximate values of the functions: thus, for one of the equations, we have

$$\begin{array}{l|l}
c_3c_{15}\vartheta_0\vartheta_{12}, \text{ i.e.,} & (2\Lambda+2\Lambda')(-2\Lambda+2\Lambda') \quad . \quad 1 \quad . \quad 1 \\
-c_0c_{12}\vartheta_3\vartheta_{15} & - \quad 1 \quad . \quad 1 \quad . \quad 2\Lambda\cos\tfrac{1}{2}\pi(u+v)+2\Lambda'\cos\tfrac{1}{2}\pi(u-v). \\
 & \qquad\qquad\qquad -2\Lambda\cos\tfrac{1}{2}\pi(u+v)+2\Lambda'\cos\tfrac{1}{2}\pi(u-v) \\
+c_4c_8\vartheta_7\vartheta_{11} & + \quad 1 \quad . \quad 1 \quad . -2\Lambda\sin\tfrac{1}{2}\pi(u+v)-2\Lambda'\sin\tfrac{1}{2}\pi(u-v). \\
 & \qquad\qquad\qquad -2\Lambda\sin\tfrac{1}{2}\pi(u+v)+2\Lambda'\sin\tfrac{1}{2}\pi(u-v); \\
=0, & =0,
\end{array}$$

viz. the equation to be verified is here

$$\begin{aligned}
&-4\Lambda^2 \qquad\qquad\qquad +4\Lambda'^2 \\
&+4\Lambda^2\cos^2\tfrac{1}{2}\pi(u+v)-4\Lambda'^2\cos^2\tfrac{1}{2}\pi(u-v) \\
&+4\Lambda^2\sin^2\tfrac{1}{2}\pi(u+v)-4\Lambda'^2\sin^2\tfrac{1}{2}\pi(u-v) \\
&=0,
\end{aligned}$$

which is right.

119. In the equation

$$\begin{array}{l|l}
c_9c_{12}\vartheta_1\vartheta_4, \text{ i.e.,} & 2Q\,.\,1\,.\,2Q\cos\tfrac{1}{2}\pi u\,.\,1 \\
-c_1c_4\vartheta_9\vartheta_{12} & -2Q\,.\,1\,.\,2Q\cos\tfrac{1}{2}\pi u\,.\,1 \\
+c_3c_6\vartheta_{14}\vartheta_{11} & \\
=0, & =0;
\end{array}$$

this is right, but there is no verification as to the term $c_3c_6\vartheta_{14}\vartheta_{11}$; taking the more approximate values, the term in question taken negatively, that is, $-c_3c_6\vartheta_{14}\vartheta_{11}$ is

$$=-(2\Lambda+2\Lambda').\quad 2S.\quad -2S\sin\tfrac{1}{2}\pi v.\quad -2\Lambda\sin\tfrac{1}{2}\pi(u+v)+2\Lambda'\sin\tfrac{1}{2}\pi(u-v),$$

which is

$$=-8S^2(\Lambda+\Lambda')^2\cos\tfrac{1}{2}\pi u+8S^2(\Lambda+\Lambda')\Lambda\cos\tfrac{1}{2}\pi(u+2v)+8S^2(\Lambda+\Lambda')\Lambda'\cos\tfrac{1}{2}\pi(u-2v),$$

and this ought therefore to be the value of the first two terms, that is, of

$$\begin{aligned}
&(2Q+2Q^9-2A-2A')(1-2Q^4-2S^4)\{2Q\cos\tfrac{1}{2}\pi u+2Q^9\cos\tfrac{3}{2}\pi u \\
&\qquad\qquad +2A\cos\tfrac{1}{2}\pi(u+2v)+2A'\cos\tfrac{1}{2}\pi(u-2v)\}(1-2Q^4\cos\pi u+2S^4\cos\pi v) \\
&-(2Q+2Q^9+2A+2A')(1-2Q^4+2S^4)\{2Q\cos\tfrac{1}{2}\pi u+2Q^9\cos\tfrac{3}{2}\pi u \\
&\qquad\qquad -2A\cos\tfrac{1}{2}\pi(u+2v)-2A'\cos\tfrac{1}{2}\pi(u-2v)\}(1-2Q^4\cos\pi u-2S^4\cos\pi v),
\end{aligned}$$

which to the proper degree of approximation is

$$\begin{aligned}
&=(2Q-4Q^5-4QS^4+2Q^9-2A-2A')\{2Q\cos\tfrac{1}{2}\pi u-4Q^5\cos\tfrac{1}{2}\pi u\cos\pi u \\
&\qquad +4QS^4\cos\tfrac{1}{2}\pi u\cos\pi v+2Q^9\cos\tfrac{3}{2}\pi u+2A\cos\tfrac{1}{2}\pi(u+2v)+2A'\cos\tfrac{1}{2}\pi(u-2v)\} \\
&-(2Q-4Q^5+4QS^4+2Q^9+2A+2A')\{2Q\cos\tfrac{1}{2}\pi u-4Q^5\cos\tfrac{1}{2}\pi u\cos\pi u \\
&\qquad -4QS^4\cos\tfrac{1}{2}\pi u\cos\pi v+2Q^9\cos\tfrac{3}{2}\pi u-2A\cos\tfrac{1}{2}\pi(u+2v)-2A'\cos\tfrac{1}{2}\pi(u-2v)\}.
\end{aligned}$$

This is

$$\begin{aligned}&(2M_0 - 2\Omega_0)(2M + 2\Omega)\\ &-(2M_0 + 2\Omega_0)(2M - 2\Omega), \quad = 8(M_0\Omega - M\Omega_0),\end{aligned}$$

if for a moment

$$M = Q\cos\tfrac{1}{2}\pi u - 2Q^5\cos\tfrac{1}{2}\pi u\cos\pi u + Q^9\cos\tfrac{3}{2}\pi u, \qquad M_0 = Q - 2Q^5 + Q^9,$$

$$\Omega = 2QS^4\cos\pi u\cos\pi v + A\cos\tfrac{1}{2}\pi(u+2v) + A'\cos\tfrac{1}{2}\pi(u-2v), \quad \Omega_0 = 2QS^4 \quad + A + A',$$

or substituting and reducing, the value of $8(M_0\Omega - M\Omega_0)$ to the proper degree of approximation is found to be

$$\begin{aligned}&= -8Q(2QS^4 + A + A')\cos\tfrac{1}{2}\pi u\\ &\qquad + 8(Q^2S^4 + 8QA)\cos\tfrac{1}{2}\pi(u+2v) + 8(Q^2S^4 + 8QA')\cos\tfrac{1}{2}\pi(u-2v),\end{aligned}$$

which in virtue of the relations $QA = \Lambda^2S^2$, $QA' = \Lambda'^2S^2$, $Q^2S^2 = \Lambda\Lambda'$, is equal to the foregoing value of $c_3c_6\vartheta_{14}\vartheta_{11}$. I have thought it worth while to give this somewhat elaborate verification.

Résumé of the foregoing results.

120. In what precedes we have all the quadric relations between the 16 double theta-functions: or say we have the linear relations between squares (squared functions) and the linear relations between pairs (products of two functions): the number of the asyzygetic linear relations between squares is obviously $=12$; and that of the asyzygetic linear relations between pairs is $=60$ (since each of the 30 tetrads of pairs gives two asyzygetic relations): there are thus in all $12+60$, $=72$, asyzygetic linear relations. But these constitute only a 13-fold relation between the functions, viz. they are such as to give for the ratios of the 16 functions expressions depending upon two arbitrary parameters, x, y. Or taking the 16 functions as the coordinates of a point in 15-dimensional space, these coordinates are connected by a 13-fold relation (expressed by means of the foregoing system of 72 quadric equations), and the locus is thus a 13-fold, or two-dimensional, locus in 15-dimensional space.

Hence, taking any four of the functions, these are connected by a single equation; that is, regarding the four functions as the coordinates of a point in ordinary space, the locus of the point is a surface.

In particular, the four functions may be any four functions belonging to a hexad: by what precedes there is then a linear relation between the squares of the four functions: or the locus is a quadric surface. Each hexad gives 15 such surfaces, or the number of quadric surfaces is $(16 \times 15 =)$ 240.

The 16-nodal quartic surfaces.

121. If the four functions are those contained in any two pairs out of a tetrad of pairs (see the foregoing "Table of the 120 pairs"), then the locus is a quartic

surface, which is, in fact, a Kummer's 16-nodal quartic surface. For if for a moment $x.y$ and $z.w$ are two pairs out of a tetrad, and $r.s$ be either of the remaining pairs of the tetrad; then we have rs a linear function of xy and zw: squaring, r^2s^2 is a linear function of x^2y^2, $xyzw$, z^2w^2; but we then have r^2 and s^2, each of them a linear function of x^2, y^2, z^2, w^2; or substituting we have an equation of the fourth order, containing terms of the second order in (x^2, y^2, z^2, w^2), and also a term in $xyzw$. It is clear that, if instead of $r.s$ we had taken the remaining pair of the tetrad, we should have obtained the same quartic equation in (x, y, z, w). And moreover it appears by inspection that, if xy and zw are pairs in a tetrad, then xz and yw are pairs in a second tetrad, and xw and yz are pairs in a third tetrad: we obtain in each case the same quartic equation. We have from each tetrad of pairs six sets of four functions (x, y, z, w): and the number of such sets is thus $(\frac{1}{3}6.30=)\,60$: these are shown in the foregoing "Table of the 60 Göpel tetrads," viz. taking as coordinates of a point the four functions in any tetrad of this table, the locus is a 16-nodal quartic surface.

122. To exhibit the process I take a tetrad 4, 7, 8, 11 containing two odd functions; and representing these for convenience by x, y, z, w, viz. writing

$$\vartheta_4,\ \vartheta_7,\ \vartheta_8,\ \vartheta_{11}(u)=x,\ y,\ z,\ w,$$

we have then X, Y, Z, W linear functions of the four squares, viz. it is easy to obtain

$$\begin{aligned}
\alpha(x^2+z^2)-\delta(y^2+w^2)&=2(\alpha^2-\delta^2)X,\\
\delta(\ \ ,,\ \ )-\alpha(\ \ ,,\ \ )&=2(\ \ ,,\ \ )W,\\
-\beta(x^2-z^2)+\gamma(y^2-w^2)&=2(\beta^2-\gamma^2)Y,\\
-\gamma(\ \ ,,\ \ )+\beta(\ \ ,,\ \ )&=2(\ \ ,,\ \ )Z.
\end{aligned}$$

Also considering two other functions $\vartheta_0(u)$ and $\vartheta_{12}(u)$, or as for shortness I write them, ϑ_0 and ϑ_{12}, we have

$$\begin{aligned}
\vartheta_0^2&=\alpha X+\beta Y+\gamma Z+\delta W,\\
\vartheta_{12}^2&=\alpha X-\beta Y-\gamma Z+\delta W,
\end{aligned}$$

and substituting the foregoing values of X, Y, Z, W, we find

$$\begin{aligned}
M\vartheta_0^2&=Ax^2+By^2+Cz^2+Dw^2,\\
M\vartheta_{12}^2&=Cx^2+Dy^2+Az^2+Bw^2,
\end{aligned}$$

where, writing down the values first in terms of α, β, γ, δ and then in terms of the c's, we have

$$\begin{aligned}
M&=\ (\alpha^2-\delta^2)(\beta^2-\gamma^2) &&=\tfrac{1}{4}\ .\ c_8^4-c_4^4,\\
A&=\ \beta^2\delta^2-\alpha^2\gamma^2 &&=\ ,,\ \ -c_2^2c_6^2,\\
B&=-\alpha\delta(\beta^2-\gamma^2)+\beta\gamma(\alpha^2-\delta^2) &&=\ ,,\ \ c_3^2c_4^2-c_{15}^2c_8^2,\\
C&=\ \alpha^2\beta^2-\gamma^2\delta^2 &&=\ ,,\ \ c_1^2c_9^2,\\
D&=-\alpha\delta(\beta^2-\gamma^2)-\beta\gamma(\alpha^2-\delta^2) &&=\ ,,\ \ c_{15}^2c_4^2-c_3^2c_8^2;
\end{aligned}$$

and we then have further

$$c_4c_8\vartheta_0\vartheta_{12} = c_0c_{12}\vartheta_4\vartheta_8 + c_3c_{15}\vartheta_7\vartheta_{11},$$

that is,

$$c_4c_8\vartheta_0\vartheta_{12} = c_0c_{12}xz \quad + c_3c_{15}yw;$$

whence equating the two values of $\vartheta_0^2\vartheta_{12}^2$ we have the required quartic equation in x, y, z, w.

123. But the reduction is effected more simply if instead of the c's we introduce the rectangular coefficients a, b, c, &c. We then have

$$M = (c''^2 - b'^2), \quad A = -a''c, \quad C = a'b,$$
$$B = -b'c' - b''c'', = bc, \quad D = b'b'' + c'c'', = a'a'';$$

and the equations become

$$(c''^2 - b'^2)\vartheta_0^2 = -a''cx^2 + \quad bcy^2 + a'bz^2 - a'a''w^2,$$
$$(c''^2 - b'^2)\vartheta_{12}^2 = \quad a'bx^2 - a'a''y^2 - a''cz^2 + \quad bcw^2,$$
$$\sqrt{b'c''}\,\vartheta_0\vartheta_{12} = \sqrt{a}\,xz + \sqrt{-b''c'}\,yw,$$

so that the elimination gives

$$b'c''(-a''cx^2 + bcy^2 + a'bz^2 - a'a''w^2)(a'bx^2 - a'a''y^2 - a''cz^2 + bcw^2)$$
$$= (c''^2 - b'^2)^2\{ax^2z^2 - b''c'y^2w^2 + 2\sqrt{-ab''c'}\,xyzw\},$$

viz. this is

$$\begin{aligned}
&- a'a''bb'cc''(x^4 + y^4 + z^4 + w^4)\\
&+ a'b'cc''(a''^2 + b^2)(x^2y^2 + z^2w^2)\\
&+ \{b'c''(a'^2b^2 + a''^2c^2) - \quad a(b'^2 - c''^2)^2\}\,x^2z^2\\
&+ \{b'c''(a'^2a''^2 + b^2c^2) + b''c'(b'^2 - c''^2)^2\}\,y^2w^2\\
&- a''bb'c''(a'^2 + c^2)(x^2w^2 + y^2z^2)\\
&- 2(b'^2 - c''^2)^2\sqrt{-ab''c'}\,xyzw = 0.
\end{aligned}$$

124. In this equation the coefficients of x^2z^2 and y^2w^2 are each $= a'a''bc(b'^2 + c''^2)$, as at once appears from the identities

$$\begin{cases} a'b\,.\,b' \ - c''\,.\,a''c = a(b'^2 - c''^2),\\ a'b\,.\,c'' - b'\,.\,a''c = \quad (b'^2 - c''^2),\end{cases}$$
$$\begin{cases} a'a''\,.\,b' - c''\,.\,bc = -b''(b'^2 - c''^2),\\ a'a''\,.\,c \ - \ b'\,.\,bc = \quad c'(b'^2 - c''^2),\end{cases}$$

by multiplying together in each pair the left-hand and the right-hand sides respectively. Substituting and dividing by $-a'a''bb'cc''$, we have

$$\begin{aligned}
&x^4 + y^4 + z^4 + w^4\\
&- \frac{a''^2 + b^2}{a''b}(x^2y^2 + z^2w^2) - \frac{b'^2 + c''^2}{b'c''}(x^2z^2 + y^2w^2) + \frac{a'^2 + c^2}{a'c}(x^2w^2 + y^2z^2)\\
&+ \frac{2(b'^2 - c''^2)^2\sqrt{-ab''c'}}{a'a''bb'cc''}\,xyzw = 0;
\end{aligned}$$

or, if we herein restore the c's in place of the rectangular coefficients, this is

$$\begin{aligned}&x^4+y^4+z^4+w^4\\ &-\frac{c_1^4+c_2^4}{c_1^2c_2^2}(x^2y^2+z^2w^2)-\frac{c_4^4+c_8^4}{c_4^2c_8^2}(x^2z^2+y^2w^2)+\frac{c_6^4+c_9^4}{c_6^2c_9^2}(x^2w^2+y^2z^2)\\ &+2\frac{c_0c_3c_{12}c_{15}(c_4^4-c_8^4)^2}{c_1^2c_2^2c_4^2c_6^2c_8^2c_9^2}xyzw=0,\end{aligned}$$

which is the equation of the 16-nodal quartic surface.

Substituting for x, y, z, w their values ϑ_4, ϑ_7, ϑ_8, $\vartheta_{11}(u)$, we have the equation connecting the four theta-functions 4, 7, 8, 11 of a Göpel tetrad. And there is an equation of the like form between the four functions of any other Göpel tetrad: for obtaining the actual equations some further investigation would be necessary.

The xy-expressions of the theta-functions.

125. The various quadric relations between the theta-functions, admitting that they constitute a 13-fold relation, show that the theta-functions may be expressed as proportional to functions of two arbitrary parameters x, y; and two of these functions being assumed at pleasure the others of them would be determinate; we have of course (though it would not be easy to arrive at it in this manner) such a system in the foregoing expressions of the 16 functions in terms of x, y; and conversely these expressions must satisfy identically the quadric relations between the theta-functions.

126. To show that this is so as to the general form of the equations, consider first the xy-factors $\sqrt{a}$, $\sqrt{ab}$, &c. As regards the squared functions $(\sqrt{ab})^2$, we have for instance

$$(\sqrt{ab})^2=\frac{1}{\theta^2}\{\text{abfc}_{\prime}\text{d}_{\prime}\text{e}_{\prime}+\text{a}_{\prime}\text{b}_{\prime}\text{f}_{\prime}\text{cde}+2\sqrt{XY}\},$$

$$(\sqrt{cd})^2=\frac{1}{\theta^2}\{\text{cdfa}_{\prime}\text{b}_{\prime}\text{e}_{\prime}+\text{c}_{\prime}\text{d}_{\prime}\text{f}_{\prime}\text{abe}+2\sqrt{XY}\};$$

each of these contains the same irrational part $\frac{2}{\theta^2}\sqrt{XY}$, and the difference is therefore rational: and it is moreover integral, for we have

$$(\sqrt{ab})^2-(\sqrt{cd})^2=\frac{1}{\theta^2}(\text{abc}_{\prime}\text{d}_{\prime}-\text{a}_{\prime}\text{b}_{\prime}\text{cd})(\text{fe}_{\prime}-\text{f}_{\prime}\text{e}),$$

where each factor divides by θ, and consequently the product by θ^2; the value is in fact

$$=(e-f)\begin{vmatrix}1, & x+y, & xy\\ 1, & a+b, & ab\\ 1, & c+d, & cd\end{vmatrix},$$

a linear function of 1, $x+y$, xy. This is the case as regards the difference of any two of the squares $(\sqrt{ab})^2$, $(\sqrt{ac})^2$, &c.; hence selecting any one of these squares, for instance $(\sqrt{de})^2$, any other of the squares is of the form

$$\lambda + \mu (x+y) + \nu xy + \rho (\sqrt{de})^2, \quad (\rho = 1);$$

and obviously, the other squares $(\sqrt{a})^2$, &c., are of the like form, the last coefficient ρ being $=0$. We hence have the theorem that each square can be expressed as a linear function of any four (properly selected) squares.

127. But we have also the theorem of the 16 Kummer hexads.

Obviously the six squares

$$(\sqrt{a})^2, \quad (\sqrt{b})^2, \quad (\sqrt{c})^2, \quad (\sqrt{d})^2, \quad (\sqrt{e})^2, \quad (\sqrt{f})^2$$

are a hexad, viz. each of these is a linear function of 1, $x+y$, xy: and therefore selecting any three of them, each of the remaining three can be expressed as a linear function of these.

But further the squares $(\sqrt{a})^2$, $(\sqrt{b})^2$, $(\sqrt{ab})^2$, $(\sqrt{cd})^2$, $(\sqrt{ce})^2$, $(\sqrt{de})^2$ form a hexad. For reverting to the expression obtained for $(\sqrt{ab})^2 - (\sqrt{cd})^2$, the determinant contained therein is a linear function of $\mathrm{aa}_{,}$ and $\mathrm{bb}_{,}$, that is, of $(\sqrt{a})^2$ and $(\sqrt{b})^2$; we, in fact, have

$$(a-b)\begin{vmatrix} 1, & x+y, & xy \\ 1, & a+b, & ab \\ 1, & c+d, & cd \end{vmatrix} = (b-c)(b-d)(a-x)(a-y) - (a-c)(a-d)(b-x)(b-y).$$

Hence $(\sqrt{ab})^2 - (\sqrt{cd})^2$ is a linear function of $(\sqrt{a})^2$, $(\sqrt{b})^2$; and by a mere interchange of letters $(\sqrt{ab})^2 - (\sqrt{ce})^2$, $(\sqrt{ab})^2 - (\sqrt{de})^2$, are each of them also a linear function of $(\sqrt{a})^2$ and $(\sqrt{b})^2$; whence the theorem. And we have thus all the remaining 15 hexads.

128. We have a like theory as regards the products of pairs of functions. A tetrad of pairs is of one of the two forms

$$\sqrt{a}\sqrt{b},\ \sqrt{ac}\sqrt{bc},\ \sqrt{ad}\sqrt{bd},\ \sqrt{ae}\sqrt{be}, \text{ and } \sqrt{f}\sqrt{ab},\ \sqrt{c}\sqrt{de},\ \sqrt{d}\sqrt{ce},\ \sqrt{e}\sqrt{cd};$$

in the first case the terms are

$$\sqrt{\mathrm{aa_{,}bb_{,}}},$$

$$\frac{1}{\theta^2}\{(\mathrm{ab_{,}+a_{,}b})\sqrt{\mathrm{cdefc_{,}d_{,}e_{,}f_{,}}} + (\mathrm{cfd_{,}e_{,}+c_{,}f_{,}de})\sqrt{\mathrm{aa_{,}bb_{,}}}\},$$

$$\frac{1}{\theta^2}\{\quad '' \qquad '' \quad + (\mathrm{dfc_{,}e_{,}+d_{,}f_{,}ce}) \quad '' \quad \},$$

$$\frac{1}{\theta^2}\{\quad '' \qquad '' \quad + (\mathrm{efc_{,}d_{,}+e_{,}f_{,}cd}) \quad '' \quad \},$$

and as regards the last three terms the difference of any two of them is a mere constant multiple of $\sqrt{\mathrm{aa_{,}bb_{,}}}$; for instance, the second term $-$ the third term is

$$= \frac{1}{\theta^2}(\mathrm{cd_{,}-c_{,}d})(\mathrm{fe_{,}-f_{,}e})\sqrt{\mathrm{aa_{,}bb_{,}}}, = (c-d)(f-e)\sqrt{\mathrm{aa_{,}bb_{,}}};$$

we have thus a tetrad such that, selecting any two terms, each of the remaining terms is a linear function of these.

In the second case, the terms are

$$\frac{1}{\theta}\{\mathrm{f}\sqrt{\mathrm{abc}_{,}\mathrm{d}_{,}\mathrm{e}_{,}\mathrm{f}_{,}}+\mathrm{f}_{,}\sqrt{\mathrm{a}_{,}\mathrm{b}_{,}\mathrm{cdef}}\},$$

$$\frac{1}{\theta}\{\mathrm{c} \quad ,, \quad +\mathrm{c}_{,} \quad ,, \quad \},$$

$$\frac{1}{\theta}\{\mathrm{d} \quad ,, \quad +\mathrm{d}_{,} \quad ,, \quad \},$$

$$\frac{1}{\theta}\{\mathrm{e} \quad ,, \quad +\mathrm{e}_{,} \quad ,, \quad \},$$

whence clearly the four terms are a tetrad as above. And it may be added that any linear function of the four terms is of the form

$$\frac{1}{\theta}\{(\lambda+\mu x)\sqrt{\mathrm{abc}_{,}\mathrm{d}_{,}\mathrm{e}_{,}\mathrm{f}_{,}}+(\lambda+\mu y)\sqrt{\mathrm{a}_{,}\mathrm{b}_{,}\mathrm{cdef}}\}.$$

129. Considering next the actual equations between the squared theta-functions, take as a specimen

$$c_6^2\vartheta_6^2-c_2^2\vartheta_2^2+c_1^2\vartheta_1^2-c_9^2\vartheta_9^2=0,$$

that is,

$$c_6^4(\sqrt{ab})^2-c_2^4(\sqrt{cd})^2+c_1^4(\sqrt{ce})^2-c_9^4(\sqrt{de})^2=0,$$

where $c_6,\ c_2,\ c_1,\ c_9=\sqrt[4]{\overline{ab}},\ \sqrt[4]{\overline{cd}},\ \sqrt[4]{\overline{ce}},\ \sqrt[4]{\overline{de}}$ respectively. Since the functions $(\sqrt{ab})^2$, &c., contain the same irrational term $\frac{2}{\theta^2}\sqrt{XY}$, it is clear that the equation can only be true if

$$c_6^4-c_2^4+c_1^4-c_9^4=0:$$

and, this being so, it will be true if

$$c_2^4\{(\sqrt{ab})^2-(\sqrt{cd})^2\}-c_1^4\{(\sqrt{ab})^2-(\sqrt{ce})^2\}+c_9^4\{(\sqrt{ab})^2-(\sqrt{de})^2\}=0,$$

where, by what precedes, each of the terms in { } is a linear function of $(\sqrt{a})^2$ and $(\sqrt{b})^2$. Attending first to the term in $(\sqrt{a})^2$, the coefficient hereof is

$$ef.bc.bd.c_2^4-df.bc.be.c_1^4+cf.bd.be.c_9^4,$$

where for shortness bc, bd, &c., are written to denote the differences $b-c$, $b-d$, &c.: substituting for c_2^4 its value $(\sqrt{\overline{cd}})^4$, $=cd.cf.df.ab.ae.be$, and similarly for c_1^4 and c_9^4 their values, $=ce.cf.ef.ab.ad.bd$, and $de.df.ef.ab.ac.bc$ respectively, the whole expression contains the factor $ab.bc.bd.be.cf.df.ef$, and throwing this out, the equation to be verified becomes

$$cd.ae-ce.ad+de.ac=0,$$

which is true identically. The verification is thus made to depend upon that of $c_6^4 - c_2^4 + c_1^4 - c_9^4 = 0$; and similarly for the other relations between the squared functions, the verification depends upon relations containing the fourth powers, or the products of squares, of the constants c and k.

130. Among these are included the before-mentioned system of equations involving the fourth powers or the products of squares of only the constants c; and it is interesting to show how these are satisfied identically by the values $c_0 = \sqrt[4]{bd}$, &c.

Thus one of these equations is $c_{12}^4 + c_1^4 + c_6^4 = c_0^4$; substituting the values, there is a factor ce which divides out, and the resulting equation is

$$ad\,.\,af\,.\,df\,.\,bc\,.\,be + cf\,.\,ef\,.\,ab\,.\,ad\,.\,bd + ab\,.\,af\,.\,bf\,.\,cd\,.\,de - ac\,.\,ae\,.\,bd\,.\,bf\,.\,df = 0.$$

There are here terms in a^2, a, a^0 which should separately vanish; for the terms in a^2, the equation becomes

$$df\,.\,bc\,.\,be + bd\,.\,cf\,.\,ef + bf\,.\,cd\,.\,de - bd\,.\,bf\,.\,df = 0,$$

which is easily verified; and the equations in a and a^0 may also be verified.

An equation involving products of the squares is $c_{12}^2 c_9^2 - c_1^2 c_4^2 + c_3^2 c_6^2 = 0$. The term $c_{12}^2 c_9^2$ is here $\sqrt{adf\,.\,bce}\,\sqrt{def\,.\,abc}$ which is $= \sqrt{(bc)^2 (df)^2\,.\,ab\,.\,ac\,.\,ad\,.\,af\,.\,be\,.\,ce\,.\,de\,.\,ef}$, which is taken $= bc\,.\,df \sqrt{ab\,.\,ac\,.\,ad\,.\,af\,.\,be\,.\,ce\,.\,de\,.\,ef}$; similarly the values of $c_1^2 c_4^2$ and $c_3^2 c_6^2$ are $= bd\,.\,cf$ and $bf\,.\,cd$ each multiplied by the same radical, and the equation to be verified is

$$bc\,.\,df - bd\,.\,cf + bf\,.\,cd = 0,$$

which is right: the other equations may be verified in a similar manner.

131. Coming next to the equations connecting the pairs of theta-functions, for instance

$$c_3 c_{15} \vartheta_0 \vartheta_{12} - c_0 c_{12} \vartheta_3 \vartheta_{15} + c_4 c_8 \vartheta_7 \vartheta_{11} = 0,$$

this is

$$c_3 c_{15} c_0 c_{12} \{\sqrt{bd}\,\sqrt{ad} - \sqrt{be}\,\sqrt{ae}\} + c_4 c_8 k_7 k_{11}\,.\,\sqrt{b}\,\sqrt{a} = 0,$$

the products $\sqrt{bd}\,\sqrt{ad}$ and $\sqrt{be}\,\sqrt{ae}$ contain besides a common term the terms

$$\frac{1}{\theta^2}(\mathrm{dfc_, e_,} + \mathrm{d_, f_, ce}) \sqrt{\mathrm{aa_, bb_,}}, \text{ and } \frac{1}{\theta^2}(\mathrm{efc_, d_,} + \mathrm{e_, f_, cd}) \sqrt{\mathrm{aa_, bb_,}},$$

hence their difference contains $\frac{1}{\theta^2}(\mathrm{de_,} - \mathrm{d_, e})(\mathrm{fc_,} - \mathrm{f_, c}) \sqrt{\mathrm{aa_, bb_,}}$ which is $= de\,.\,fc \sqrt{\mathrm{aa_, bb_,}}$, that is, $de\,.\,fc \sqrt{a}\,\sqrt{b}$: hence the equation to be verified is

$$de\,.\,fc\,.\,c_3 c_{15} c_0 c_{12} + c_4 c_8 k_7 k_{11} = 0;$$

$c_3 c_{15} c_0 c_{12}$ is $= \sqrt[4]{bef\,.\,acd}\,\sqrt[4]{aef\,.\,bcd}\,\sqrt[4]{bdf\,.\,ace}\,\sqrt[4]{adf\,.\,bce}$, where under the fourth root we have 24 factors, which are, in fact, 12 factors twice repeated; and if we write Π, $= ab\,.\,ac\,.\,ad\,.\,ae\,.\,af\,.\,bc\,.\,bd\,.\,be\,.\,bf\,.\,cd\,.\,ce\,.\,cf\,.\,de\,.\,df\,.\,ef$, for the product of all the 15 factors, then the 12 factors are in fact all those of Π, except ab, cf, de; viz. we have

$$c_3 c_{15} c_0 c_{12} = \sqrt[4]{\Pi^2 \div (ab)^2 (cf)^2 (de)^2}.$$

Again, $c_4c_8k_7k_{11}, = \sqrt[4]{acf.bde}\sqrt[4]{bcf.ade}\sqrt[4]{acdef}\sqrt[4]{bcdef}$, is a fourth root of a product of 32 factors, which are in fact 16 factors twice repeated, and in the 16 factors, ab does not occur, cf and de occur each twice, and the other 12 factors each once: we thus have

$$c_4c_8k_7k_{11} = \sqrt[4]{\Pi^2 (cf)^2 (de)^2 \div (ab)^2},$$

and the relation to be verified assumes the form

$$fc.de\sqrt[4]{1 \div (cf)^2 (de)^2} + \sqrt[4]{(cf)^2 (de)^2} = 0,$$

which, taking $fc.de = -\sqrt[4]{(cf)^4 (de)^4}$, is right. And so for the other equations. It will be observed that, in the equation $de.fc.c_3c_{15}c_0c_{12} + c_4c_8k_7k_{11} = 0$, and in the other equations upon which the verifications depend, there is no ambiguity of sign: the signs of the radicals have to be determined consistently with all the equations which connect the c's and the k's: that this is possible appears evident *à priori*, but the actual verification presents some difficulty. I do not here enter further into the question.

Further results of the product-theorem, the $u \pm u'$ formulæ.

132. Recurring now to the equations in $u+u'$, $u-u'$, by putting therein $u'=0$, we can express X, Y, Z, W in terms of four of the squared functions of u, and by putting $u=0$ we can express X', Y', Z', W' in terms of four of the squared functions of u'; and, substituting in the original equations, we have the products

$$\vartheta(\ )u+u'.\vartheta(\ )u-u'$$

in terms of the squared functions of u and u'.

Selecting as in a former investigation the functions 4, 7, 8, 11, which were called x, y, z, w, it is more convenient to use single letters to represent the squared functions. I write

$\vartheta(u+u').\vartheta(u-u')$		$\vartheta^2 u$	$\vartheta^2 u'$	$\vartheta^2 0$
4	$4 = P$,	$4 = p$,	$4 = p'$,	$4 = p_0 (= c_4^2)$,
7	$7 = Q$,	$7 = q$,	$7 = q'$,	$7 = 0$,
8	$8 = R$,	$8 = r$,	$8 = r'$,	$8 = r_0 (= c_8^2)$,
11	$11 = S$,	$11 = s$,	$11 = s'$,	$11 = 0$.

Then

$X \quad Y \quad Z \quad W$	$X \quad Y \quad Z \quad W$	$X' \quad Y' \quad Z' \quad W'$
$P = X' - Y' + Z' - W'$,	$p = \alpha - \beta + \gamma - \delta$,	$p' = \alpha - \beta + \gamma - \delta$,
$Q = W' - Z' + Y' - X'$,	$q = \delta - \gamma + \beta - \alpha$,	$q' = \delta - \gamma + \beta - \alpha$,
$R = X' + Y' - Z' - W'$,	$r = \alpha + \beta - \gamma - \delta$,	$r' = \alpha + \beta - \gamma - \delta$,
$S = W' + Z' - Y' - X'$,	$s = \delta + \gamma - \beta - \alpha$,	$s' = \delta + \gamma - \beta - \alpha$.

Hence

$$\alpha(p+r) - \delta(q+s) = 2(\alpha^2 - \delta^2) X, \qquad \alpha(p'+r') - \delta(q'+s') = 2(\alpha^2 - \delta^2) X',$$

$$\delta \quad " \quad -\alpha \quad " \quad = 2 \quad " \quad W, \qquad \delta \quad " \quad -\alpha \quad " \quad = 2 \quad " \quad W',$$

$$-\beta(p-r) + \gamma(q-s) = 2(\beta^2 - \gamma^2) Y, \qquad -\beta(p'-r') + \gamma(q'-s') = 2(\beta^2 - \gamma^2) Y',$$

$$-\gamma \quad " \quad +\beta \quad " \quad = 2 \quad " \quad Z, \qquad -\gamma \quad " \quad +\beta \quad " \quad = 2 \quad " \quad Z'.$$

By means of these values, we have

$$4(\alpha^2-\delta^2)^2 X'X = \alpha^2(p+r)(p'+r')+\delta^2(q+s)(q'+s')-\alpha\delta\,[(p+r)(q'+s')+(p'+r')(q+s)],$$

$$4 \quad ,, \quad W'W = \delta^2 \quad ,, \quad +\alpha^2 \quad ,, \quad -\alpha\delta\,[\quad ,, \quad ,, \quad],$$

$$4(\beta^2-\gamma^2)^2 Y'Y = \beta^2(p-r)(p'-r')+\gamma^2(q-s)(q'-s')-\beta\gamma\,[(p-r)(q'-s')+(p'-r')(q-s)],$$

$$4 \quad ,, \quad Z'Z = \gamma^2 \quad ,, \quad +\beta^2 \quad ,, \quad -\beta\gamma\,[\quad ,, \quad ,, \quad].$$

Hence

$$4(\alpha^2-\delta^2)(X'X-W'W)=(p+r)(p'+r')-(q+s)(q'+s'),$$

$$4(\beta^2-\gamma^2)(Y'Y-Z'Z)=(p-r)(p'-r')-(q-s)(q'-s'),$$

and substituting in the expressions for P and R,

$$4(\alpha^2-\delta^2)(\beta^2-\gamma^2)P=$$

$$(\beta^2-\gamma^2)[(p+r)(p'+r')-(q+s)(q'+s')]-(\alpha^2-\delta^2)[(p-r)(p'-r')-(q-s)(q'-s')],$$

$$4 \quad ,, \quad R=$$

$$,, \;[\quad ,, \quad ,, \quad]+ \;,, \;[\quad ,, \quad ,, \quad].$$

Similarly

$$4(\alpha^2-\delta^2)^2 W'X=$$

$$\alpha\delta\,[(p+r)(p'+r')+(q+s)(q'+s')]-\alpha^2(p+r)(q'+s')-\delta^2(q+s)(p'+r'),$$

$$4 \quad ,, \quad X'W=$$

$$,, \;[\quad ,, \quad ,, \quad]-\delta^2 \quad ,, \quad -\alpha^2 \quad ,, \quad ,$$

$$4(\beta^2-\gamma^2)^2 Z'Y=$$

$$\beta\gamma\,[(p-r)(p'-r')+(q-s)(q'-s')]-\beta^2(p-r)(q'-s')-\gamma^2(q-s)(p'-r'),$$

$$4 \quad ,, \quad Y'Z=$$

$$,, \;[\quad ,, \quad ,, \quad]-\gamma^2 \quad ,, \quad -\beta^2 \quad ,, \quad ;$$

whence

$$4(\alpha^2-\delta^2)(W'X-X'W)=-[(p+r)(q'+s')-(p'+r')(q+s)],$$

$$4(\beta^2-\gamma^2)(Z'Y-Y'Z)=-[(p-r)(q'-s')-(p'-r')(q-s)],$$

and substituting in the expressions for Q and S

$$4(\alpha^2-\delta^2)(\beta^2-\gamma^2)Q=$$

$$-(\beta^2-\gamma^2)[(p+r)(q'+s')-(p'+r')(q+s)]+(\alpha^2-\delta^2)[(p-r)(q'-s')-(p'-r')(q-s)],$$

$$4 \quad ,, \quad S=$$

$$- \;,, \;[\quad ,, \quad ,, \quad]- \;,, \;[\quad ,, \quad ,, \quad].$$

133. Hence, collecting and reducing,

$$4(\alpha^2-\delta^2)(\beta^2-\gamma^2)P=$$
$$-(\alpha^2-\beta^2+\gamma^2-\delta^2)(pp'-qq'+rr'-ss')+(\alpha^2+\beta^2-\gamma^2-\delta^2)(pr'+p'r-qs'-q's),$$

$$4 \quad ,, \quad R=$$
$$(\alpha^2+\beta^2-\gamma^2-\delta^2)(\quad ,, \quad)-(\alpha^2-\beta^2+\gamma^2-\delta^2)(\quad ,, \quad),$$

$$4 \quad ,, \quad Q=$$
$$(\alpha^2-\beta^2+\gamma^2-\delta^2)(\quad ,, \quad)-(\alpha^2+\beta^2-\gamma^2-\delta^2)(\quad ,, \quad),$$

$$4 \quad ,, \quad S=$$
$$-(\alpha^2+\beta^2-\gamma^2-\delta^2)(\quad ,, \quad)+(\alpha^2-\beta^2+\gamma^2-\delta^2)(\quad ,, \quad);$$

we have

$$p_0(=c_4^2)=\alpha^2-\beta^2+\gamma^2-\delta^2,\quad r_0(=c_8^2)=\alpha^2+\beta^2-\gamma^2-\delta^2,$$

and thence

$$r_0^2-p_0^2=4(\alpha^2-\delta^2)(\beta^2-\gamma^2);$$

the equations hence become

$$(r_0^2-p_0^2)P=-p_0(pp'-qq'+rr'-ss')+r_0(pr'+p'r-qs'-q's),$$
$$,, \quad R=\ r_0(\quad ,, \quad)-p_0(\quad ,, \quad),$$
$$,, \quad Q=\ p_0(pq'-p'q+rs'-r's)-r_0(\quad ,, \quad),$$
$$,, \quad S=-r_0(\quad ,, \quad)+p_0(\quad ,, \quad).$$

On writing in the equations $u'=0$, then $P, Q, R, S, p', q', r', s'$ become $=p, q, r, s, p_0, 0, r_0, 0$; and the equations are (as they should be) true identically. The equations may be written

$(c^4\ \ c^4)$	$\overset{u+u'}{\vartheta}$.	$\overset{u-u'}{\vartheta}$		$c^2(\overset{u}{\vartheta}{}^2.\overset{u'}{\vartheta}{}^2\ \ \overset{u}{\vartheta}{}^2.\overset{u'}{\vartheta}{}^2\ \ \overset{u}{\vartheta}{}^2.\overset{u'}{\vartheta}{}^2\ \ \overset{u}{\vartheta}{}^2.\overset{u'}{\vartheta}{}^2)$	$c^2(\overset{u}{\vartheta}{}^2.\overset{u'}{\vartheta}{}^2\ \ \overset{u}{\vartheta}{}^2.\overset{u'}{\vartheta}{}^2\ \ \overset{u}{\vartheta}{}^2.\overset{u'}{\vartheta}{}^2\ \ \overset{u}{\vartheta}{}^2.\overset{u'}{\vartheta}{}^2)$
(8 − 4)	4	4	=	−4(4.4 − 7.7 + 8.8 − 11.11)	+8(4.8 + 8.4 − 7.11 − 11.7),
(,,)	8	8	=	+8(,,)	−4(,,),
(,,)	7	7	=	+4(4.7 − 7.4 + 8.11 − 11.8)	−8(4.11 − 11.4 + 8.7 − 7.8),
(,,)	11	11	=	−8(,,)	+4(,,).

There is of course such a system for each of the 60 Göpel tetrads.

Differential relations connecting the theta-functions with the quotient-functions.

134. Imagine p, q, r, s, &c., changed into x^2, y^2, z^2, w^2, &c.; that is, let x, y, z, w represent the theta-functions 4, 7, 8, 11 of u, v; and similarly x', y', z', w' those of u', v', and $x_0, 0, z_0, 0$ those of 0, 0. Let u', v' be each of them indefinitely small; and take $\partial, =u'\dfrac{d}{du}+v'\dfrac{d}{dv}$, as the symbol of total differentiation in regard to u, v, the infinitesimals u' and v' being arbitrary: then, as far as the second order, we have in general

$$\vartheta(u+u',\ v+v')=\vartheta(u,\ v)+\partial\vartheta(u,\ v)+\tfrac{1}{2}\partial^2\vartheta(u,\ v),$$

and hence

$$P = (x + \partial x + \tfrac{1}{2}\partial^2 x)(x - \partial x + \tfrac{1}{2}\partial^2 x), \ = x^2 + \{x\partial^2 x - (\partial x)^2\},$$

and similarly for Q, R, S. Moreover, observing that x' and z' are even functions, y' and w' are odd functions, of u', v', we have

$$x', \ y', \ z', \ w' = x_0 + \tfrac{1}{2}\partial^2 x_0, \ \partial y_0, \ z_0 + \tfrac{1}{2}\partial^2 z_0, \ \partial w_0,$$

where $\partial^2 x_0$, ∂y_0, &c., are what $\partial^2 x$, ∂y, &c., become on writing therein $u = 0$, $v = 0$; ∂y_0, ∂w_0 are of course linear functions, $\partial^2 x_0$, $\partial^2 z_0$ quadric functions of u' and v'. The values of x'^2, y'^2, z'^2, w'^2 are thus $x_0^2 + x_0\partial^2 x_0$, $(\partial y_0)^2$, $z_0^2 + z_0\partial^2 z_0$, $(\partial w_0)^2$; and we have

$$\begin{array}{llllllllll}
 & & & & & & x_0\partial^2 x_0 & (\partial y_0)^2 & z_0\partial^2 z_0 & (\partial w_0)^2 \\ \hline
x^2x'^2 & -y^2y'^2 & +z^2z'^2 & -w^2w'^2 = & x^2x_0^2 & +z^2z_0^2 & +x^2 & -y^2 & +z^2 & -w^2, \\
x^2y'^2 & -y^2x'^2 & +z^2w'^2 & -w^2z'^2 = & -y^2x_0^2 & -w^2z_0^2 & -y^2 & +x^2 & -w^2 & +z^2, \\
x^2z'^2 & -y^2w'^2 & +z^2x'^2 & -w^2y'^2 = & z^2x_0^2 & +x^2z_0^2 & +z^2 & -w^2 & +x^2 & -y^2, \\
x^2w'^2 & -y^2z'^2 & +z^2y'^2 & -w^2x'^2 = & -w^2x_0^2 & -y^2z_0^2 & -w^2 & +z^2 & -y^2 & +x^2.
\end{array}$$

135. On substituting these values, the constant terms (or terms independent of u', v') disappear of themselves; and the equations, transposing the second and third of them, become

$$\begin{array}{rlllll}
 & & x_0\partial^2 x_0 & (\partial y_0)^2 & z_0\partial^2 z_0 & (\partial w_0)^2 \\ \hline
(z_0^4 - x_0^4)\{x\partial^2 x - (\partial x)^2\} = & & (-x_0^2x^2 + z_0^2z^2) & +(x_0^2y^2 - z_0^2w^2) & +(-x_0^2z^2 + z_0^2x^2) & +(x_0^2w^2 - z_0^2y^2), \\
\text{,,} \quad \{y\partial^2 y - (\partial y)^2\} = & & -(x_0^2y^2 - z_0^2w^2) & -(-x_0^2x^2 + z_0^2z^2) & -(x_0^2w^2 - z_0^2y^2) & -(-x_0^2z^2 + z_0^2x^2), \\
\text{,,} \quad \{z\partial^2 z - (\partial z)^2\} = & & (-x_0^2z^2 + z_0^2x^2) & +(x_0^2w^2 - z_0^2y^2) & +(-x_0^2x^2 + z_0^2z^2) & +(x_0^2y^2 - z_0^2w^2), \\
\text{,,} \quad \{w\partial^2 w - (\partial w)^2\} = & & -(x_0^2w^2 - z_0^2y^2) & -(-x_0^2z^2 + z_0^2x^2) & -(x_0^2y^2 - z_0^2w^2) & -(-x_0^2x^2 + z_0^2z^2),
\end{array}$$

where it will be recollected that x, y, z, w mean ϑ_4, ϑ_7, ϑ_8, $\vartheta_{11}(u)$; x_0 is $\vartheta_4(0)$, that is, c_4, and z_0 is $\vartheta_8(0)$, that is, c_8. But the formulæ contain also

$$\partial^2 x_0 = (c_4''', \ c_4^{\text{iv}}, \ c_4^{\text{v}} \between u', \ v')^2, \quad \partial y_0 = (c_7', \ c_7'' \between u', \ v'),$$

$$\partial^2 z_0 = (c_8''', \ c_8^{\text{iv}}, \ c_8^{\text{v}} \between u', \ v')^2, \quad \partial w_0 = (c_{11}', \ c_{11}'' \between u', \ v').$$

The formulæ may be written

$$\begin{array}{rl|l|l|l|l}
 & & c_4\partial^2 c_4 & (\partial c_7)^2 & c_8\partial^2 c_8 & (\partial c_{11})^2 \\
 & \{\overset{u}{\vartheta}\,.\,\partial^2\overset{u}{\vartheta} - (\partial\overset{u}{\vartheta})^2\} & c^2.\vartheta^2 \quad c^2.\vartheta^2 & c^2.\vartheta^2 \quad c^2.\vartheta^2 & c^2.\vartheta^2 \quad c^2.\vartheta^2 & c^2.\vartheta^2 \quad c^2.\vartheta^2 \\ \hline
(c_8^4 - c_4^4) & \{\ 4 \quad 4 \qquad 4\ \} = & (-4 \quad 4 + 8 \quad 8) & +(\ 4 \quad 7 - 8 \ 11) & +(-4 \quad 8 + 8 \quad 4) & +(\ 4 \ 11 - 8 \quad 7), \\
\text{,,} & \{\ 7 \quad 7 \qquad 7\ \} = & -(\ 4 \quad 7 - 8 \ 11) & -(-4 \quad 4 + 8 \quad 8) & -(\ 4 \ 11 - 8 \quad 7) & -(-4 \quad 8 + 8 \quad 4), \\
\text{,,} & \{\ 8 \quad 8 \qquad 8\ \} = & (-4 \quad 8 + 8 \quad 4) & +(\ 4 \ 11 - 8 \quad 7) & +(-4 \quad 4 + 8 \quad 8) & +(\ 4 \quad 7 - 8 \ 11), \\
\text{,,} & \{11 \ 11 \qquad 11\ \} = & -(\ 4 \ 11 - 8 \quad 7) & -(-4 \quad 8 + 8 \quad 4) & -(\ 4 \quad 7 - 8 \ 11) & -(-4 \quad 4 + 8 \quad 8),
\end{array}$$

where $\partial^2 c_4$, $\partial^2 c_8$, ∂c_7, ∂c_{11} are written in place of $\partial^2 x_0$, $\partial^2 z_0$, ∂y_0, ∂w_0. There is of course a like system of equations for each of the Göpel tetrads.

136. Observe that, dividing the first equation by $\vartheta_4^2(u)$, or say by ϑ_4^2, the left-hand side is a mere constant multiple of $\partial^2 \log \vartheta_4$; and the right-hand side depends only on the quotient-functions $\vartheta_7 \div \vartheta_4$, $\vartheta_8 \div \vartheta_4$, $\vartheta_{11} \div \vartheta_4$; each side is a quadric function of u', v'. Equating the terms in u'^2, $u'v'$, v'^2 respectively, we have

$$\frac{d^2}{du^2} \log \vartheta_4, \quad \frac{d^2}{du\,dv} \log \vartheta_4, \quad \frac{d^2}{dv^2} \log \vartheta_4,$$

each of them expressed as a linear function of the squares of the quotient-functions $\vartheta_7 \div \vartheta_4$, $\vartheta_8 \div \vartheta_4$, $\vartheta_{11} \div \vartheta_4$. The formula is thus a second-derivative formula serving for the expression of a double theta-function by means of three quotient-functions.

Differential relations of the theta-functions.

137. In "The second set of 16," selecting the eight equations which contain Y_1 and W_1, these are

$$\begin{array}{rrcrrcl}
 & \scriptstyle u+u' & \scriptstyle u-u' & \scriptstyle u+u' & \scriptstyle u-u' & & \text{(Suffixes 1.)} \\
 & \vartheta \,. & \vartheta & \vartheta \,. & \vartheta & & Y \quad W \\
\tfrac{1}{2}\{ & 4 & 0\ - & 0 & 4\} & = & Y' + W', \\
 & 12 & 8\ - & 8 & 12 & = & Y' - W', \\
 & 6 & 2\ - & 2 & 6 & = & W' + Y', \\
 & 14 & 10\ - & 10 & 14 & = & W' - Y', \\
\hline
\tfrac{1}{2}\{ & 5 & 1\ + & 1 & 5\} & = & X' + Z', \\
 & 13 & 9\ + & 9 & 13 & = & X' - Z', \\
 & 7 & 3\ + & 3 & 7 & = & Z' + X', \\
 & 15 & 11\ + & 11 & 15 & = & Z' - X'.
\end{array}$$

Then, considering any line in the upper half and any two lines in the lower half, we can from the three equations eliminate Y_1 and W_1, thus obtaining an equation such as

$$\begin{vmatrix} \vartheta_4\vartheta_0 - \vartheta_0\vartheta_4, & Y', & W' \\ \vartheta_5\vartheta_1 + \vartheta_1\vartheta_5, & X', & Z' \\ \vartheta_{13}\vartheta_9 + \vartheta_9\vartheta_{13}, & X', & -Z' \end{vmatrix} = 0,$$

viz. this is

$$\begin{aligned} &-2X'Z' && (\vartheta_4\vartheta_0 - \vartheta_0\vartheta_4) \\ &+(\ \ X'W' + Y'Z')&&(\vartheta_5\vartheta_1 + \vartheta_1\vartheta_5) \\ &+(-X'W' + Y'Z')&&(\vartheta_{13}\vartheta_9 + \vartheta_9\vartheta_{13}) = 0, \end{aligned}$$

where the arguments of the theta-functions are as above, $u+u'$, $u-u'$, $u+u'$, $u-u'$; and the suffixes of the X', Y', Z', W' are all $=1$.

138. Suppose that in this equation u' becomes indefinitely small. If u' were $=0$, the values of X', Y', Z', W' would be α, 0, γ, 0: hence u' being indefinitely small, we take them to be α, $\partial\beta$, γ, $\partial\delta$, where

$$\partial\beta, =\left(u'\frac{d}{du}+v'\frac{d}{dv}\right)Y, \text{ and } \partial\delta, =\left(u'\frac{d}{du}+v'\frac{d}{dv}\right)W, \qquad (u=v=0),$$

are, in fact, linear functions of u' and v'.

We have $\vartheta_4\vartheta_0-\vartheta_0\vartheta_4$ standing for

$$\vartheta_4(u+u')\,\vartheta_0(u-u')-\vartheta_0(u+u')\,\vartheta_4(u-u'),$$

and here

$$\vartheta_4(u\pm u')=\vartheta_4\pm\partial\vartheta_4,\quad \vartheta_0(u\pm u')=\vartheta_0\pm\partial\vartheta_0;$$

the function in question is thus

$$(\vartheta_4+\partial\vartheta_4)(\vartheta_0-\partial\vartheta_0)-(\vartheta_4-\partial\vartheta_4)(\vartheta_0+\partial\vartheta_0)=2\{\vartheta_0\partial\vartheta_4-\vartheta_4\partial\vartheta_0\},$$

where the arguments are u, v, and the ∂ denotes $u'\dfrac{d}{du}+v'\dfrac{d}{dv}$.

Also $\vartheta_5\vartheta_1+\vartheta_1\vartheta_5$, that is, $\vartheta_5(u+u')\,\vartheta_1(u-u')+\vartheta_1(u+u')\,\vartheta_5(u-u')$, becomes simply $=2\vartheta_5\vartheta_1$, and similarly $\vartheta_{13}\vartheta_9+\vartheta_9\vartheta_{13}$ becomes $=2\vartheta_{13}\vartheta_9$; and the equation thus is

$$-2\alpha_1\gamma_1(\vartheta_0\partial\vartheta_4-\vartheta_4\partial\vartheta_0)+(\alpha_1\partial\delta_1+\gamma_1\partial\beta_1)\,\vartheta_5\vartheta_1+(-\alpha_1\partial\delta_1+\gamma_1\partial\beta_1)\,\vartheta_{13}\vartheta_9=0,$$

where the proper suffix 1 is restored to the α, $\partial\beta$, γ, and $\partial\delta$.

139. The equation shows that the differential combination $\vartheta_0\partial\vartheta_4-\vartheta_4\partial\vartheta_0$ is a linear function of $\vartheta_5\vartheta_1$ and $\vartheta_{13}\vartheta_9$, the coefficients of these products being of course linear functions of u' and v'. Writing the equation

$$\vartheta_0\partial\vartheta_4-\vartheta_4\partial\vartheta_0=A\vartheta_5\vartheta_1+B\vartheta_{13}\vartheta_9,$$

we can if we please determine the coefficients in terms of the constants c', c'', c''', c^{iv}, c^{v}; viz. taking u, v indefinitely small, we have

$$\begin{aligned}
\vartheta_0&=c_0, & \partial\vartheta_4&=u'(c_4'''u+c_4^{\text{iv}}v)+v'(c_4^{\text{iv}}u+c_4^{\text{v}}v),\\
\vartheta_4&=c_4, & \partial\vartheta_0&=u'(c_0'''u+c_0^{\text{iv}}v)+v'(c_0^{\text{iv}}u+c_0^{\text{v}}v),\\
\vartheta_1&=c_1, & \vartheta_5&=c_5'u+c_5''v,\\
\vartheta_9&=c_9, & \vartheta_{13}&=c_{13}'u+c_{13}''v,
\end{aligned}$$

or substituting, and equating the coefficients of u and v respectively, we have

$$\begin{aligned}
c_0(c_4'''u'+c_4^{\text{iv}}v')-c_4(c_0'''u'+c_0^{\text{iv}}v')&=Ac_1c_5'+Bc_9c_{13}',\\
c_0(c_4^{\text{iv}}u'+c_4^{\text{v}}v')-c_4(c_0^{\text{iv}}u'+c_0^{\text{v}}v')&=Ac_1c_5''+Bc_9c_{13}'',
\end{aligned}$$

which equations give the values of A, B.

140. Disregarding the values of the coefficients, and attending only to the form of the equation

$$\vartheta_0\partial\vartheta_4-\vartheta_4\partial\vartheta_0=A\vartheta_5\vartheta_1+B\vartheta_{13}\vartheta_9,$$

this is one of a system of 120 equations; viz. referring to the foregoing table of the 120 pairs, it in fact appears that taking any pair such as $\vartheta_0\vartheta_4$ out of the upper compartment or the lower compartment of any column of the table, the corresponding differential combination $\vartheta_0\partial\vartheta_4 - \vartheta_4\partial\vartheta_0$ is a linear function of any two of the four pairs in the other compartment of the same column.

Differential relation of u, v and x, y.

141. We have, as before, in the two notations, the pairs

$A\,.\,B$	11 . 7
$C\,.\,DE$	5 . 9
$D\,.\,CE$	13 . 1
$E\,.\,CD$	14 . 2
$F\,.\,AB$	10 . 6.

From the expressions given above for the four pairs below the line, it is clear that any linear function of these four pairs may be represented by

$$(a-b)\frac{1}{\theta}\{(\lambda+\mu y)\sqrt{\mathrm{cdefa_,b_,}}+(\lambda+\mu x)\sqrt{\mathrm{c_,d_,e_,f_,ab}}\},$$

where λ, μ are constant coefficients: the factor $(a-b)$ has been introduced for convenience, as will appear.

We have consequently a relation

$$\sqrt{\mathrm{aa_,}}\,\partial\sqrt{\mathrm{bb_,}}-\sqrt{\mathrm{bb_,}}\,\partial\sqrt{\mathrm{aa_,}}=\frac{a-b}{\theta}\{(\lambda+\mu y)\sqrt{\mathrm{cdefa_,b_,}}+(\lambda+\mu x)\sqrt{\mathrm{c_,d_,e_,f_,ab}}\},$$

where, as before, ∂ is used to denote $u'\dfrac{d}{du}+v'\dfrac{d}{dv}$, u' and v' being arbitrary multipliers; considering u, v as functions of x, y, we have

$$\frac{d}{du}=\frac{dx}{du}\frac{d}{dx}+\frac{dy}{du}\frac{d}{dy},$$

$$\frac{d}{dv}=\frac{dx}{dv}\frac{d}{dx}+\frac{dy}{dv}\frac{d}{dy},$$

and thence $\partial=P\dfrac{d}{dx}+Q\dfrac{d}{dy}$, if for shortness P and Q are written to denote $u'\dfrac{dx}{du}+v'\dfrac{dx}{dv}$ and $u'\dfrac{dy}{du}+v'\dfrac{dy}{dv}$ respectively.

142. The left-hand side then is

$$=P\left(\sqrt{\mathrm{aa_,}}\frac{d}{dx}\sqrt{\mathrm{bb_,}}-\sqrt{\mathrm{bb_,}}\frac{d}{dx}\sqrt{\mathrm{aa_,}}\right)+Q\left(\sqrt{\mathrm{aa_,}}\frac{d}{dy}\sqrt{\mathrm{bb_,}}-\sqrt{\mathrm{bb_,}}\frac{d}{dy}\sqrt{\mathrm{aa_,}}\right);$$

the coefficients of P and Q are at once found to be

$$= -\tfrac{1}{2}\frac{(\mathrm{a}-\mathrm{b})\sqrt{\mathrm{a}_{\prime}\mathrm{b}_{\prime}}}{\sqrt{\mathrm{ab}}},\quad -\tfrac{1}{2}\frac{(\mathrm{a}_{\prime}-\mathrm{b}_{\prime})\sqrt{\mathrm{ab}}}{\sqrt{\mathrm{a}_{\prime}\mathrm{b}_{\prime}}},$$

respectively, or observing that $\mathrm{a}-\mathrm{b}, =\mathrm{a}_{\prime}-\mathrm{b}_{\prime}, = a-b$, the equation becomes

$$P\frac{\sqrt{\mathrm{a}_{\prime}\mathrm{b}_{\prime}}}{\sqrt{\mathrm{ab}}}+Q\frac{\sqrt{\mathrm{ab}}}{\sqrt{\mathrm{a}_{\prime}\mathrm{b}_{\prime}}} = -\frac{2}{\theta}\{(\lambda+\mu y)\sqrt{\mathrm{cdefa}_{\prime}\mathrm{b}_{\prime}}+(\lambda+\mu x)\sqrt{\mathrm{c}_{\prime}\mathrm{d}_{\prime}\mathrm{e}_{\prime}\mathrm{f}_{\prime}\mathrm{ab}}\};$$

or multiplying by $\sqrt{\mathrm{aba}_{\prime}\mathrm{b}_{\prime}}$ and writing for shortness $\mathrm{abcdef}=X$, $\mathrm{a}_{\prime}\mathrm{b}_{\prime}\mathrm{c}_{\prime}\mathrm{d}_{\prime}\mathrm{e}_{\prime}\mathrm{f}_{\prime}=Y$, this becomes

$$\mathrm{a}_{\prime}\mathrm{b}_{\prime}\{P+\frac{2}{\theta}(\lambda+\mu y)\sqrt{X}\}+\mathrm{ab}\{Q+\frac{2}{\theta}(\lambda+\mu x)\sqrt{Y}\}=0.$$

143. There are, it is clear, the like equations

$$\mathrm{b}_{\prime}\mathrm{c}_{\prime}\{P+\frac{2}{\theta}(\lambda'+\mu' y)\sqrt{X}\}+\mathrm{bc}\{Q+\frac{2}{\theta}(\lambda'+\mu' x)\sqrt{Y}\}=0,$$

$$\mathrm{c}_{\prime}\mathrm{a}_{\prime}\{P+\frac{2}{\theta}(\lambda''+\mu'' y)\sqrt{X}\}+\mathrm{ca}\{Q+\frac{2}{\theta}(\lambda''+\mu'' x)\sqrt{Y}\}=0,$$

and it is to be shown that $\lambda=\lambda'=\lambda''$ and $\mu=\mu'=\mu''$. For this purpose, recurring to the forms

$$\sqrt{\mathrm{aa}_{\prime}}\,\partial\sqrt{\mathrm{bb}_{\prime}}-\sqrt{\mathrm{bb}_{\prime}}\,\partial\sqrt{\mathrm{aa}_{\prime}}=\frac{a-b}{\theta}\{(\lambda+\mu y)\sqrt{\mathrm{cdefa}_{\prime}\mathrm{b}_{\prime}}+(\lambda+\mu x)\sqrt{\mathrm{c}_{\prime}\mathrm{d}_{\prime}\mathrm{e}_{\prime}\mathrm{f}_{\prime}\mathrm{ab}}\},$$

$$\sqrt{\mathrm{bb}_{\prime}}\,\partial\sqrt{\mathrm{cc}_{\prime}}-\sqrt{\mathrm{cc}_{\prime}}\,\partial\sqrt{\mathrm{bb}_{\prime}}=\frac{b-c}{\theta}\{(\lambda'+\mu' y)\sqrt{\mathrm{adefb}_{\prime}\mathrm{c}_{\prime}}+(\lambda'+\mu' x)\sqrt{\mathrm{a}_{\prime}\mathrm{d}_{\prime}\mathrm{e}_{\prime}\mathrm{f}_{\prime}\mathrm{bc}}\},$$

$$\sqrt{\mathrm{cc}_{\prime}}\,\partial\sqrt{\mathrm{aa}_{\prime}}-\sqrt{\mathrm{aa}_{\prime}}\,\partial\sqrt{\mathrm{cc}_{\prime}}=\frac{c-a}{\theta}\{(\lambda''+\mu'' y)\sqrt{\mathrm{bdefc}_{\prime}\mathrm{a}_{\prime}}+(\lambda''+\mu'' x)\sqrt{\mathrm{b}_{\prime}\mathrm{d}_{\prime}\mathrm{e}_{\prime}\mathrm{f}_{\prime}\mathrm{ca}}\},$$

multiply the first equation by $\sqrt{\mathrm{cc}_{\prime}}$, the second by $\sqrt{\mathrm{aa}_{\prime}}$, and the third by $\sqrt{\mathrm{bb}_{\prime}}$, and add: the left-hand side vanishes, and therefore the right-hand side must also vanish identically.

144. But on the right-hand side we have the term $\frac{1}{\theta}\sqrt{\mathrm{defa}_{\prime}\mathrm{b}_{\prime}\mathrm{c}_{\prime}}$ multiplied by

$$(a-b)\,\mathrm{c}\,(\lambda+\mu y)+(b-c)\,\mathrm{a}\,(\lambda'+\mu' y)+(c-a)\,\mathrm{b}\,(\lambda''+\mu'' y),$$

and the term $-\frac{1}{\theta}\sqrt{\mathrm{d}_{\prime}\mathrm{e}_{\prime}\mathrm{f}_{\prime}\mathrm{abc}}$ multiplied by

$$(a-b)\,\mathrm{c}_{\prime}(\lambda+\mu x)+(b-c)\,\mathrm{a}_{\prime}(\lambda'+\mu' x)+(c-a)\,\mathrm{b}_{\prime}(\lambda''+\mu'' x),$$

and it is clear that the whole can vanish only if these two coefficients separately vanish. This will be the case if we have for λ, λ', λ'' the equations

$$(a-b)\,\lambda+(b-c)\,\lambda'+(c-a)\,\lambda''=0,$$

$$c\quad ,,\quad +a\quad ,,\quad +b\quad ,,\quad =0,$$

and the like equations for μ, μ', μ''. The equations written down give

$$(a-b)\lambda : (b-c)\lambda' : (c-a)\lambda'' = a-b : b-c : c-a,$$

that is, $\lambda = \lambda' = \lambda''$: and similarly $\mu = \mu' = \mu''$.

145. But this being so, the three equations in P, Q give

$$P + \frac{2}{\theta}(\lambda + \mu y)\sqrt{X} = 0, \quad Q + \frac{2}{\theta}(\lambda + \mu x)\sqrt{Y} = 0,$$

that is,

$$u'\frac{dx}{du} + v'\frac{dx}{dv} = -\frac{2}{x-y}(\lambda + \mu y)\sqrt{X},$$

$$u'\frac{dy}{du} + v'\frac{dy}{dv} = -\frac{2}{x-y}(\lambda + \mu x)\sqrt{Y}.$$

In these equations u' and v' are arbitrary; hence λ and μ must be linear functions of u' and v'; say their values are $= \varpi u' + \rho v'$, $\sigma u' + \tau v'$ respectively. We have therefore

$$\frac{dx}{du} = -\frac{2}{\theta}(\varpi + \sigma y)\sqrt{X}, \quad \frac{dx}{dv} = -\frac{2}{\theta}(\rho + \tau y)\sqrt{X},$$

$$\frac{dy}{du} = -\frac{2}{\theta}(\varpi + \sigma x)\sqrt{Y}, \quad \frac{dy}{dv} = -\frac{2}{\theta}(\rho + \tau x)\sqrt{Y},$$

or, what is the same thing,

$$-\tfrac{1}{2}\theta\frac{dx}{\sqrt{X}} = (\varpi + \sigma y)\,du + (\rho + \tau y)\,dv,$$

$$-\tfrac{1}{2}\theta\frac{dy}{\sqrt{Y}} = (\varpi + \sigma x)\,du + (\rho + \tau x)\,dv,$$

whence also

$$\sigma du + \tau dv = \tfrac{1}{2}\left(\frac{dx}{\sqrt{X}} - \frac{dy}{\sqrt{Y}}\right),$$

$$\varpi du + \rho dv = -\tfrac{1}{2}\left(\frac{xdx}{\sqrt{X}} - \frac{ydy}{\sqrt{Y}}\right),$$

which are the required relations, depending on the square roots of the sextic functions $X = \text{abcdef}$, and $Y = \text{a}_{\prime}\text{b}_{\prime}\text{c}_{\prime}\text{d}_{\prime}\text{e}_{\prime}\text{f}_{\prime}$ of x and y respectively; but containing the constants ϖ, ρ, σ, τ, the values of which are not as yet ascertained.

146. I commence the integration of these equations on the assumption that the values $u=0$, $v=0$ correspond to indefinitely large values of x and y. We have

$$X = x^6\left(1 - \frac{S}{x} + \ldots\right), \quad Y = y^6\left(1 - \frac{S}{y} + \ldots\right),$$

where $S = a+b+c+d+e+f$; and thence the equations are

$$\sigma du + \tau dv = \tfrac{1}{2}\frac{dx}{x^3}\left(1 + \frac{\frac{1}{2}S}{x} + \ldots\right) - \tfrac{1}{2}\frac{dy}{y^3}\left(1 + \frac{\frac{1}{2}S}{y} + \ldots\right),$$

$$\varpi du + \rho dv = -\tfrac{1}{2}\frac{dx}{x^2}\left(1 + \frac{\frac{1}{2}S}{x} + \ldots\right) + \tfrac{1}{2}\frac{dy}{y^2}\left(1 + \frac{\frac{1}{2}S}{y} + \ldots\right).$$

Hence integrating, we have

$$\sigma u + \tau v = -\tfrac{1}{2}\left(\frac{1}{x^2} - \frac{1}{y^2}\right) + \dots,$$

$$\varpi u + \rho v = \tfrac{1}{2}\left(\frac{1}{x} - \frac{1}{y}\right) + \tfrac{1}{8}S\left(\frac{1}{x^2} - \frac{1}{y^2}\right) + \dots,$$

and thence

$$\varpi u + \rho v + \tfrac{1}{4}S(\sigma u + \tau v) = \tfrac{1}{2}\left(\frac{1}{x} - \frac{1}{y}\right) + \dots,$$

where the omitted terms depend on $\frac{1}{x^3}, \frac{1}{y^3}$ &c.

Hence, neglecting these terms, we have

$$\frac{\sigma u + \tau v}{\varpi u + \rho v + \tfrac{1}{4}S(\sigma u + \tau v)} = -\left(\frac{1}{x} + \frac{1}{y}\right),$$

an equation connecting the indefinitely small values of u, v, with the indefinitely large values of x, y.

147. From the equations $A = k_{11}\varpi\sqrt{a}$, $B = k_7\varpi\sqrt{b}$, taking (u, v) indefinitely small and therefore (x, y) indefinitely large, we deduce

$$\frac{c_{11}'u + c_{11}''v}{c_7'u + c_7''v} = \frac{k_{11}}{k_7}\,\frac{1 - \tfrac{1}{2}a\left(\frac{1}{x} + \frac{1}{y}\right)}{1 - \tfrac{1}{2}b\left(\frac{1}{x} + \frac{1}{y}\right)};$$

hence substituting for $\frac{1}{x} + \frac{1}{y}$ the foregoing value, and introducing an indeterminate multiplier M, we obtain

$$c_{11}'u + c_{11}''v = Mk_{11}\{\varpi u + \rho v + \tfrac{1}{4}S(\sigma u + \tau v) + \tfrac{1}{2}a(\sigma u + \tau v)\},$$

which breaks up into the two equations

$$c_{11}' = Mk_{11}\{\varpi + (\tfrac{1}{4}S + \tfrac{1}{2}a)\sigma\},\quad c_{11}'' = Mk_{11}\{\rho + (\tfrac{1}{4}S + \tfrac{1}{2}a)\tau\}.$$

Similarly

$$\begin{aligned}
c_7' &= Mk_7\{\quad ,, \quad b\ \}, & c_7'' &= Mk_7\{\quad ,, \quad b\ \},\\
c_5' &= Mk_5\{\quad ,, \quad c\ \}, & c_5'' &= Mk_5\{\quad ,, \quad c\ \},\\
c_{13}' &= Mk_{13}\{\quad ,, \quad d\ \}, & c_{13}'' &= Mk_{13}\{\quad ,, \quad d\ \},\\
c_{14}' &= Mk_{14}\{\quad ,, \quad e\ \}, & c_{14}'' &= Mk_{14}\{\quad ,, \quad e\ \},\\
c_{10}' &= Mk_{10}\{\quad ,, \quad f\ \}, & c_{10}'' &= Mk_{10}\{\quad ,, \quad f\ \},
\end{aligned}$$

which twelve equations determine the coefficients ϖ, σ, ρ, τ in terms of the c', c'' of the odd functions 5, 7, 10, 11, 13, 14; and moreover give rise to relations connecting these c', c'' with each other and with the constants a, b, c, d, e, f.

148. It is observed that if, as before,

$$\partial = u'\frac{d}{du} + v'\frac{d}{dv},\quad = P\frac{d}{dx} + Q\frac{d}{dy},$$

then, substituting for P and Q their values, we have

$$\partial = -\frac{2}{\theta}(\varpi u' + \rho v')\left(\sqrt{X}\frac{d}{dx} + \sqrt{Y}\frac{d}{dy}\right) - \frac{2}{\theta}(\sigma u' + \tau v')\left(y\sqrt{X}\frac{d}{dx} + x\sqrt{Y}\frac{d}{dy}\right),$$

$$= (\varpi u' + \rho v')\,\partial_1 + (\sigma u' + \tau v')\,\partial_2,$$

if for shortness

$$\partial_1 = -\frac{2}{\theta}\left(\sqrt{X}\frac{d}{dx} + \sqrt{Y}\frac{d}{dy}\right),\quad \partial_2 = -\frac{2}{\theta}\left(y\sqrt{X}\frac{d}{dx} + x\sqrt{Y}\frac{d}{dy}\right);$$

then operating with ∂ on the equations $A = \varpi k_{11}\sqrt{ab}$, &c., we have for instance

$$\begin{aligned} A\partial B - B\partial A = \varpi^2 k_{11} k_7 \{ & \quad (\varpi u' + \rho v')(\sqrt{a}\,\partial_1\sqrt{b} - \sqrt{b}\,\partial_1\sqrt{a}) \\ & + (\sigma u' + \tau v')(\sqrt{a}\,\partial_2\sqrt{b} - \sqrt{b}\,\partial_2\sqrt{a})\}, \end{aligned}$$

which is one of a system of 120 equations, the A, B being in fact any two of the 16 functions.

These are in fact nothing else than the foregoing system of 120 equations giving the values of the differential combinations $\vartheta_0\partial\vartheta_1 - \vartheta_1\partial\vartheta_0$, &c., each as a sum of products of pairs of functions, only on the right-hand sides we have expressions such as $\sqrt{a}\,\partial_1\sqrt{b} - \sqrt{b}\,\partial_1\sqrt{a}$, &c., which present themselves as perfectly determinate functions of x, y: so that, regarding $\varpi u' + \rho v'$, $\sigma u' + \tau v'$ as given linear functions of the arbitrary quantities u', v', there is no longer anything indeterminate in the form of the equations.

705.

PROBLEMS AND SOLUTIONS.

[From the *Mathematical Questions with their Solutions from the Educational Times*, vols. XIV. to LXI. (1871—1894).]

[Vol. XIV., July to December, 1870, pp. 17—19.]

3002. (PROPOSED by MATTHEW COLLINS, B.A.)—If every two of five circles A, B, C, D, E touch each other, except D and E, prove that the common tangent of D and E is just twice as long as it would be if D and E touched each other.

Solution by PROFESSOR CAYLEY.

Consider the ellipse $\frac{x^2}{a^2}+\frac{y^2}{b^2}=1$, foci R, S; the coordinates of a point U on the ellipse may be taken to be $(a\cos u,\ b\sin u)$, and then the distances of this point from the foci will be

$$r=a(1-e\cos u),\quad s=a(1+e\cos u).$$

Taking k arbitrarily, with centre R describe a circle radius $a-k$, with centre S a circle radius $a+k$, and with centre U a circle radius $k-ae\cos u$: say these are the circles R, S, U respectively; the circle U will touch each of the circles R, S (viz. assuming $ae<k<a$, so that the foregoing radii are all positive, it will touch the circle R externally and the circle S internally).

Considering next a point V, coordinates $(a\cos v,\ b\sin v)$, and the circle described about this point with the radius $k-ae\cos v$, say the circle V; this will touch in like manner the circles R, S respectively. And the circles U, V may be made to touch each other externally; viz. this will be the case if squared sum of radii = squared distance of centres, or what is the same thing, squared difference of radii + 4 times the product of radii = squared distance of centres; that is,

$$a^2e^2(\cos u - \cos v)^2 + 4(k - ae\cos u)(k - ae\cos v) = a^2(\cos u - \cos v)^2 + b^2(\sin u - \sin v)^2,$$

or

$$2(k - ae\cos u)(k - ae\cos v) = b^2\{1 - \cos(u - v)\}.$$

If for a moment we write $\tan\frac{1}{2}u = x$, $\tan\frac{1}{2}v = y$, and therefore

$$\cos u = \frac{1-x^2}{1+x^2},\quad \cos v = \frac{1-y^2}{1+y^2},\quad \sin u = \frac{2x}{1+x^2},\quad \sin v = \frac{2y}{1+y^2},$$

$$\cos(u-v) = \frac{(1-x^2)(1-y^2)+4xy}{(1+x^2)(1+y^2)},\quad 1-\cos(u-v) = \frac{2(x-y)^2}{(1+x^2)(1+y^2)},$$

we have

$$\left\{k - \frac{ae(1-x^2)}{1+x^2}\right\}\left\{k - \frac{ae(1-y^2)}{1+y^2}\right\} = \frac{b^2(x-y)^2}{(1+x^2)(1+y^2)},$$

or

$$\{k - ae + (k+ae)x^2\}\{k - ae + (k+ae)y^2\} = b^2(x-y)^2,$$

which is readily identified with the circular relation

$$\tan^{-1} y\left(\frac{k+ae}{k-ae}\right)^{\frac{1}{2}} - \tan^{-1} x\left(\frac{k+ae}{k-ae}\right)^{\frac{1}{2}} = \tan^{-1}\left(\frac{k^2-a^2e^2}{a^2-k^2}\right)^{\frac{1}{2}}:$$

or, what is the same thing, in order that the circles U, V may touch, the relation between the parameters u, v must be

$$\tan^{-1}\left\{\left(\frac{k+ae}{k-ae}\right)^{\frac{1}{2}}\tan\tfrac{1}{2}v\right\} - \tan^{-1}\left\{\left(\frac{k+ae}{k-ae}\right)^{\frac{1}{2}}\tan\tfrac{1}{2}u\right\} = \tan^{-1}\left(\frac{k^2-a^2e^2}{a^2-k^2}\right)^{\frac{1}{2}}.$$

Considering in like manner a circle, centre the point W, coordinates $(a\cos w, b\sin w)$, and radius $k - ae\cos w$, say the circle W; this will, as before, touch the circles R, S; and we may make W touch each of the circles U, V; viz. we must have

$$\tan^{-1}\left\{\left(\frac{k+ae}{k-ae}\right)^{\frac{1}{2}}\tan\tfrac{1}{2}w\right\} - \tan^{-1}\left\{\left(\frac{k+ae}{k-ae}\right)^{\frac{1}{2}}\tan^{-1}\tfrac{1}{2}v\right\} = \tan^{-1}\left(\frac{k^2-a^2e^2}{a^2-k^2}\right)^{\frac{1}{2}},$$

$$\tan^{-1}\left\{\left(\frac{k+ae}{k-ae}\right)^{\frac{1}{2}}\tan\tfrac{1}{2}u\right\} - \tan^{-1}\left\{\left(\frac{k+ae}{k-ae}\right)^{\frac{1}{2}}\tan^{-1}\tfrac{1}{2}w\right\} = \tan^{-1}\left(\frac{k^2-a^2e^2}{a^2-k^2}\right)^{\frac{1}{2}},$$

where, in the last equation, $\tan^{-1}\left\{\left(\frac{k+ae}{k-ae}\right)^{\frac{1}{2}}\tan\frac{1}{2}u\right\}$ must be considered as denoting its value in the first equation increased by π. Hence, adding the three equations, we have

$$\pi = 3\tan^{-1}\left(\frac{k^2-a^2e^2}{a^2-k^2}\right)^{\frac{1}{2}},$$

that is,

$$\left(\frac{k^2-a^2e^2}{a^2-k^2}\right)^{\frac{1}{2}} = \tan\tfrac{1}{3}\pi = \sqrt{3};$$

or

$$k^2 - a^2e^2 = 3(a^2 - k^2),$$

that is,

$$3a^2 - 4k^2 + a^2e^2 = 0;$$

viz. this is the condition for the existence of the three circles U, V, W, each touching the two others, and also the circles R, S.

The circle R lies inside the circle S, and the tangential distance is thus imaginary; but defining it by the equation

squared tangential dist. = squared dist. of centres − squared sum of radii,

the squared tangential distance is

$$= 4a^2e^2 - 4a^2.$$

But if the circles were brought into contact, the distance of the centres would be $2k$, and the value of the squared tangential distance $= 4k^2 - 4a^2$; hence, if this be = one-fourth of the former value, we have

$$4(k^2 - a^2) = a^2e^2 - a^2,$$

that is,

$$3a^2 - 4k^2 + a^2e^2 = 0,$$

the same condition as above. The solution might easily be varied in such wise that the circles R, S should be external to each other, and therefore the tangential distance real; but the case here considered, where the locus of the centres of the circles U, V, W is an ellipse, is the more convenient, and may be regarded as the standard case.

[Vol. XIV., p. 19.]

3144. (Proposed by Professor CAYLEY.)—If the extremities A, A' of a given line AA' describe given lines respectively, show that there is a point rigidly connected with AA' which describes a circle.

[Vol. XIV., pp. 67, 68.]

3120. (Proposed by Professor CAYLEY.)—To find the equation of the Jacobian of the quadric surfaces through the six points

$$(1, 0, 0, 0), (0, 1, 0, 0), (0, 0, 1, 0), (0, 0, 0, 1), (1, 1, 1, 1), (\alpha, \beta, \gamma, \delta).$$

Solution by the PROPOSER.

Writing for shortness

$$a = \beta - \gamma,\quad b = \gamma - \alpha,\quad c = \alpha - \beta,\quad f = \alpha - \delta,\quad g = \beta - \delta,\quad h = \gamma - \delta,$$

(so that $a+h-g=0$, &c., $a+b+c=0$, $af+bg+ch=0$), the six points lie in each of the plane-pairs

$$x(hy-gz+aw)=0,\quad y(-hx+fz+bw)=0,$$
$$z(gx-fy+cw)=0,\quad w(-ax-by-cz)=0.$$

We cannot take these as the four quadrics, on account of the identical equation $0=0$, which is obtained by adding the four equations; but we may take the first three of them for three of the quadrics, and for the fourth quadric the cone, vertex (0, 0, 0, 1), which passes through the other five points; viz. this is

$$a\alpha yz+b\beta zx+c\gamma xy=0.$$

We write therefore

$$P=x(hy-gz+aw),\quad Q=y(-hx+fz+bw),$$
$$R=z(gx-fy+cw),\quad S=a\alpha yz+b\beta zx+c\gamma xy;$$

and we equate to zero the determinant formed with the derived functions of P, Q, R, S in regard to the coordinates (x, y, z, w) respectively. If, for a moment, we write A, B, C to denote $bg-ch$, $ch-af$, $af-bg$ respectively, it is easily found that the term containing d_xS is

$$(b\beta z+c\gamma y)\,x\,(-agh,\ bhf,\ cfg,\ abc,\ -af^2,\ -gB,\ hC,\ aA,\ b^2g,\ -c^2h\between x,\ y,\ z,\ w)^2:$$

the terms containing d_yS and d_zS are derived from this by a mere cyclical interchange of the letters (x, y, z), (A, B, C), (a, b, c), and (f, g, h). Collecting and reducing, it is found that the whole equation divides by $2abc$; and that, omitting this factor, the result is

$$\left.\begin{array}{r} ayz(\alpha w^2-\delta x^2)+fxw(\beta z^2-\gamma y^2) \\ +bzx(\beta w^2-\delta y^2)+gyw(\gamma x^2-\alpha z^2) \\ +cxy(\gamma w^2-\delta z^2)+hzw(\alpha y^2-\beta x^2) \end{array}\right\}=0,$$

which, substituting for a, b, c, f, g, h their values, is the required form.

If, in the equation, we write for instance $x=0$, the equation becomes

$$\alpha yzw(hy-gz+aw)=0;$$

or, the section by the plane is made up of four lines. Calling the given points 1, 2, 3, 4, 5, 6, it thus appears that the surface contains the fifteen lines 12, 13, ..., 56, and also the ten lines 123.456, &c.; in all twenty-five lines. Moreover, since the surface contains the lines 12, 13, 14, 15, 16, it is clear that the point 1 is a node (conical point) on the surface; and the like as to the points 2, 3, 4, 5, 6.

[Vol. XIV., pp. 104, 105.]

3249. (Proposed by Professor CAYLEY.)—Given on a given conic two quadrangles $PQRS$ and $pqrs$, having the same centres, and such that P, p; Q, q; R, r; S, s are the corresponding vertices (that is, the four lines PQ, RS, pq, rs all pass through the same point; and similarly the lines PR, QS, pr, qs, and the lines PS, QR, ps, qr): it is required to show that a conic may be drawn, passing through the points p, q, r, s and touched at these points by the lines pP, qQ, rR, sS, respectively.

Solution by the PROPOSER.

Taking the centres for the vertices of the fundamental triangle, the equation of the given conic may be taken to be $x^2+y^2+z^2=0$; and then the coordinates of P, Q, R, S to be (A, B, C), $(A, -B, C)$, $(A, B, -C)$, $(A, -B, -C)$ respectively, where $A^2+B^2+C^2=0$; and those of p, q, r, s to be (α, β, γ), $(\alpha, -\beta, \gamma)$, $(\alpha, \beta, -\gamma)$, $(\alpha, -\beta, -\gamma)$ respectively, where $\alpha^2+\beta^2+\gamma^2=0$. The required conic, assuming it to exist, will be given by an equation of the form $lx^2+my^2+nz^2=0$. This must pass through the point (α, β, γ), and the tangent at this point must be

$$x(B\gamma-C\beta)+y(C\alpha-A\gamma)+z(A\beta-B\alpha)=0;$$

that is, we must have $l\alpha^2+m\beta^2+n\gamma^2=0$, and

$$l\alpha : m\beta : n\gamma = B\gamma-C\beta : C\alpha-A\gamma : A\beta-B\alpha.$$

The first condition is obviously included in the second; and the second condition remains unaltered if we reverse the signs of B, β, or of C, γ, or of B, β and C, γ. Hence the conic passing through p, and touched at this point by pP, will also pass through the points q, r, s, and be touched at these points by the lines qQ, rR, sS, respectively; that is, the equation of the required conic is

$$\frac{B\gamma-C\beta}{\alpha}x^2+\frac{C\alpha-A\gamma}{\beta}y^2+\frac{A\beta-B\alpha}{\gamma}z^2=0;$$

or, what is the same thing,

$$\begin{vmatrix} \beta\gamma x^2, & \gamma\alpha y^2, & \alpha\beta z^2 \\ A\;, & B\;, & C \\ \alpha\;, & \beta\;, & \gamma \end{vmatrix}=0.$$

[Vol. XV., January to June, 1871, pp. 17—20.]

3206. (Proposed by Professor CAYLEY.)—In how many geometrically distinct ways can nine points lie in nine lines, each through three points?

3278. (Proposed by Professor CAYLEY.)—It is required, with nine numbers each taken three times, to form nine triads containing twenty-seven distinct duads (or, what is the same thing, no duad twice), and to find in how many essentially distinct ways this can be done.

Solution by the PROPOSER.

Let the numbers be 1, 2, 3, 4, 5, 6, 7, 8, 9. Any number, say 1, enters into three triads, no two of which have any number in common. We may take these triads to be 123, 145, 167. There remain the two numbers 8, 9; and these are, or are not, a duad of the system.

First Case.—8 and 9 a duad. In the triad which contains 89, the remaining number cannot be 1; it must therefore be one of the numbers 2, 3; 4, 5; 6, 7; and it is quite immaterial which; the triad may therefore be taken to be 289. There is one other triad containing 2, the remaining two numbers thereof being taken from the numbers 4, 5; 6, 7. They cannot be 4, 5 or 6, 7; and it is indifferent whether they are taken to be 4, 6; 4, 7; 5, 6, or 5, 7: the triad is taken to be 247. We have thus the triads

123, 145, 167, 289, 247;

and we require two triads containing 8 and two triads containing 9. These must be made up with the numbers 3, 4, 5, 6, 7: but as no one of them can contain 47, it follows that, of the two pairs which contain 8 and 9 respectively, one pair must be made up with 3, 5, 6, 7, and the other pair with 3, 5, 6, 4; say, the pairs which contain 8 are made up with 3, 5, 6, 7, and those which contain 9 are made up with 3, 5, 6, 4 (since obviously no distinct case would arise by the interchange of the numbers 8, 9). The triads which contain 8 must contain each of the numbers 3, 5, 6, 7, and they cannot be 835, 867, since we have 67 in the triad 167; similarly the triads which contain 9 must contain each of the numbers 3, 5, 6, 4, and they cannot be 845, 836, since we have 45 in 145. Hence the triads can only be

836, 857	934, 956,
837, 856	935, 946;

and clearly the top row of 8 must combine with the top row of 9, and the bottom row of 8 with the bottom row of 9; that is, the system of the nine triads is

123, 145, 167, 289, 247,

in combination with

836, 857, 934, 956,

or else in combination with

837, 856, 935, 946.

These are really systems of the same form, that is, each of them is of the form

$BC\alpha$	$\beta\gamma a$	bcC
$CA\beta$	γab	caA
$AB\gamma$	$a\beta c$	abB;

viz. in the first and second systems respectively we have

A	B	C	α	β	γ	a	b	c	
6	1	3	2	8	7	5	4	9	(First system),
5	1	3	2	9	4	6	7	8	(Second system),

as one out of many ways of effecting the identification. Observe that there is not in the system any triad of triads containing all the numbers. It thus appears that 8, 9, a duad, gives only a single form of the system.

Cor.—It is possible to find in a plane nine points such that the points belonging to the same triad lie *in lineâ*. The nine points are, in fact, on a cubic curve; and the figure is that belonging to a theorem of Prof. Sylvester's, according to which it is possible to find on a cubic curve a system of points 1, 2, 4, 5, 7, 8, &c., (a series of

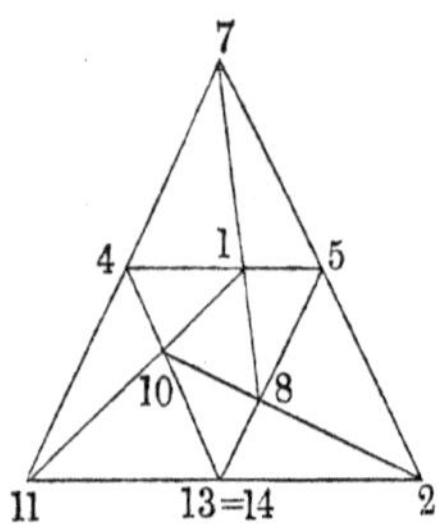

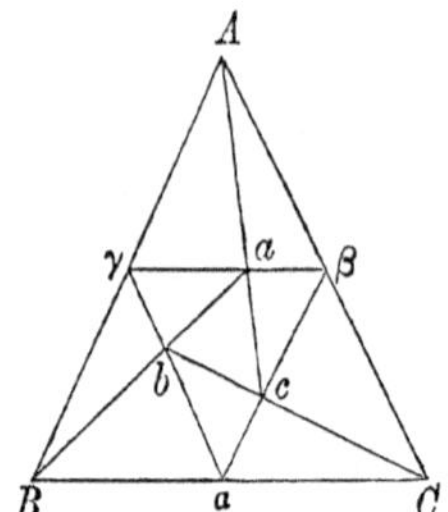

numbers not divisible by 3), such that for any triad (such as 145) where the sum of the numbers, one taken negatively, $=0$, the three points are *in lineâ*; and so also that, if two of the points become identical, in the figure $13=14$, then there is not any new point, but the preceding points are indefinitely repeated; thus, 2, 14, 16 being *in lineâ*, and 14 being $=13$, 16 must be $=11$, and so on.

Second and Third Cases.—8 and 9 do not form a duad. There are thus three triads composed of 8 with (2, 3; 4, 5; 6, 7), and three triads composed of 9 with (2, 3; 4, 5; 6, 7). If with these numbers (2, 3; 4, 5; 6, 7) we form all the arrangements of three duads other than those which contain all or any of the duads 23, 45, 67, there are the eight arrangements

$$\begin{aligned} A &= 24,\ 37,\ 56, & E &= 26,\ 35,\ 47, \\ B &= 24,\ 36,\ 57, & F &= 26,\ 34,\ 57, \\ C &= 25,\ 36,\ 47, & G &= 27,\ 34,\ 56, \\ D &= 25,\ 37,\ 46, & H &= 27,\ 35,\ 46, \end{aligned}$$

where A has a duad in common with B, with D, and with G: but it has no duad in common with C, E, F, or H. We have thus the sixteen pairs

$$\begin{array}{llll} AC, & AE, & AF, & AH, \\ BD, & BE, & BG, & BH, \\ CF, & CG, & CH, & \\ DE, & DF, & DG, & \\ EG, & FH, & & \end{array}$$

where each pair contains six different duads.

Combining AC with 8, 9, we have the triads 8 (24, 37, 56) and 9 (24, 36, 57), that is, the triads

$$824,\ 837,\ 856:\ 924,\ 936,\ 957:$$

which, with the original three triads 123, 145, 167, form a system of nine triads; 8 and 9 might, of course, be interchanged, but no essentially distinct system would arise thereby. Hence we have a system of nine triads by combining the original three triads 123, 145, 167, with any one of the sixteen pairs AC, AE, &c. But it is sufficient to consider the combinations of the three triads with each of the pairs AC, AE, AF, AH; in fact, these are the only systems which contain the triad 824; and since there is no distinction between the two pairs 4, 5 and 6, 7, or between the two numbers of the same pair, it is allowable to take 824 as a triad of the system. Hence—

Second Case.—The system consists of the three triads combined with AE; viz. it is

$$123,\ 145,\ 167:\ 824,\ 837,\ 856:\ 926,\ 935,\ 947:$$

which, it is to be observed, consists of three triads of triads, each triad of triads containing all the nine numbers; viz. the system is

$$123,\ 479,\ 568:\ 145,\ 269,\ 378:\ 167,\ 248,\ 359.$$

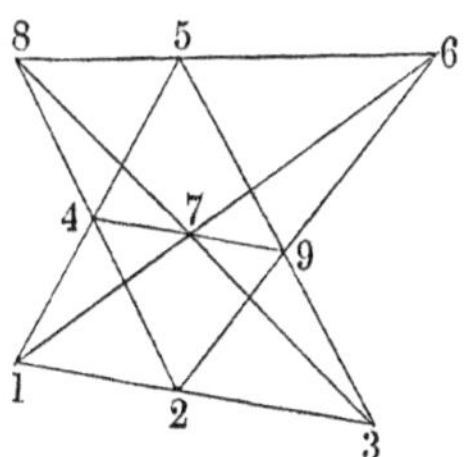

Cor.—We may have nine points such that the points belonging to the same triad lie *in lineâ*, viz. the figure is that of Pascal's hexagon when the conic is a line-pair.

Third Case.—Combining the three triads with AC, AF, or AH, it is readily seen that we obtain in each case a system of the form

$$\begin{array}{lll} A\alpha\alpha', & A\beta\gamma, & A\beta'\gamma', \\ B\beta\beta', & B\gamma\alpha, & B\gamma'\alpha', \\ C\gamma\gamma', & C\alpha\beta, & C\alpha'\beta', \end{array}$$

viz. in the case where the pair is AC; that is, the system is

$$123,\ 145,\ 167:\ 824,\ 837,\ 856:\ 925,\ 936,\ 947;$$

and in the cases where the pair is AF or AH, the identifications may be taken to be

A	B	C	α	β	γ	α'	β'	γ'	
9,	8,	1;	4,	5,	2;	7,	6,	3	...... (AC),
9,	8,	1;	2,	3,	4;	6,	7,	5	...... (AF),
9,	8,	1;	5,	4,	6;	3,	2,	7	...... (AH).

Observe that there is in the system a single triad of triads $A\alpha\alpha'$, $B\beta\beta'$, $C\gamma\gamma'$, containing all the numbers; viz. for the system with AC, this is 123, 856, 947; for the system with AF, it is 145, 837, 926; and for the system with AH, it is 167, 824, 935.

Cor.—It is possible to find a system of nine points such that the points belonging to the same triad lie *in lined*. Such a figure is this:—

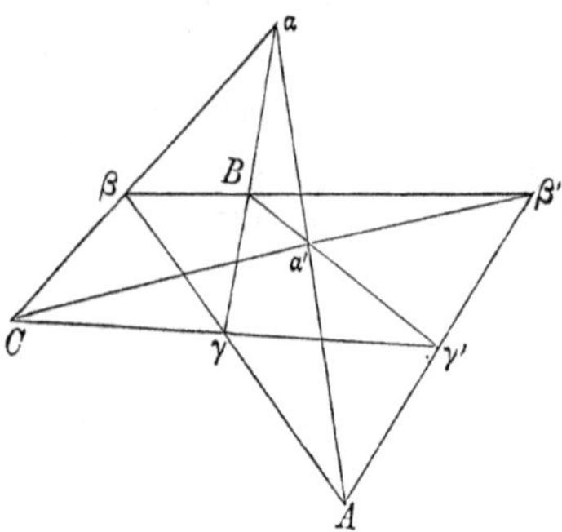

The solution shows that these are the only systems of nine points satisfying the prescribed conditions.

[Vol. xv., pp. 66, 67.]

3329. (Proposed by Professor CAYLEY.)—It is required to show that every permutation of 12345 can be produced by means of the cyclical substitution (12345), and the interchange (12).

Solution by the PROPOSER.

It is sufficient to show that the interchanges (13), (14), (15) can be so produced; for then, with the interchanges (12), (13), (14), (15), we can, by at most two such interchanges, bring any number into any place.

Writing $P = (12345)$, $\alpha = (12)$, we have

$$(12) = \alpha,$$
$$(13) = \alpha P\alpha P^4\alpha,$$
$$(14) = \alpha P\alpha P^4\alpha P^2\alpha P^3\alpha P\alpha P^4\alpha,$$
$$(15) = P^4\alpha P,$$

as can be at once verified; and the theorem is thus proved.

I remark that, starting with any two or more substitutions, and combining them in every possible manner (each of them being repeatable an indefinite number of times), we obtain a "group"; viz. this is either (as in the problem proposed) the system of all the substitutions (or say the entire group), or else it is a system the number of whose terms is a submultiple of the whole number of substitutions. The interesting question is, to determine those two or more substitutions, which, by their combination as above, do *not* give the entire group; for in this way we should arrive at all the forms of a submultiple group.

[Vol. XV., p. 80.]

3356. (Proposed by Professor CAYLEY.)—If the roots $(\alpha, \beta, \gamma, \delta)$ of the equation $(a, b, c, d, e)(u, 1)^4 = 0$ are no two of them equal; and if there exist unequal magnitudes θ and ϕ, such that

$$(\theta+\alpha)^4 : (\theta+\beta)^4 : (\theta+\gamma)^4 : (\theta+\delta)^4 = (\phi+\alpha)^4 : (\phi+\beta)^4 : (\phi+\gamma)^4 : (\phi+\delta)^4;$$

show that the cubinvariant

$$ace - ad^2 - b^2e - c^3 + 2bcd = 0;$$

and find the values of θ, ϕ.

[Vol. XVI., June to December, 1871, p. 65.]

3507. (Proposed by Professor CAYLEY.)—Show that, for the quadric cones which pass through six given points, the locus of the vertices is a quartic surface having upon it twenty-five right lines; and, thence or otherwise, that for the quadric cones passing through seven given points the locus of the vertices is a sextic curve.

[Vol. XVI., p. 90.]

3536. (Proposed by Professor CAYLEY.)—A particle describes an ellipse under the simultaneous action of given central forces, each varying as $(\text{distance})^{-2}$, at the two foci respectively: find the differential relation between the time and the excentric anomaly.

[Vol. XVII., January to June, 1872, p. 35.]

3591. (Proposed by Professor CAYLEY.)—If in a plane A, B, C, D are fixed points and P a variable point, find the linear relation

$$\alpha . PAB + \beta . PBC + \gamma . PCD + \delta . PDA = 0,$$

which connects the areas of the triangles PAB, &c.

[Vol. XVII., p. 49.]

2652. (Proposed by Professor CAYLEY.)—Find the differential equation of the parallel surfaces of an ellipsoid.

[Vol. XVII., p. 60.]

3677. (Proposed by Professor CAYLEY.)—Find at any point of a plane curve the angle between the normal and the line drawn from the point to the centre of the chord parallel and indefinitely near to the tangent at the point; and examine whether a like question applies to a point on a surface and the indicatrix section at such point.

[Vol. XVII., p. 72.]

3564. (Proposed by Professor CAYLEY.)—To determine the least circle enclosing three given points.

[Vol. XVIII., July to December, 1872, p. 68.]

3875. (Proposed by Professor CAYLEY.)—Given the constant a and the variables x, y, to construct mechanically $\dfrac{a^2-x^2}{y}$; or what is the same thing, given the fixed points A, B, and the moving point P, to mechanically connect therewith a point P' such that PP' shall be always at right angles to AB, and the point P' in the circle APB.

[Vol. XX., July to December, 1873, pp. 106, 107.]

3430. (Proposed by W. J. C. MILLER.)—Find the equation of the first negative focal pedal of (1) an ellipsoid, and (2) an ellipse.

Solution by PROFESSOR CAYLEY.

1. It is easily seen that if a sphere be drawn, passing through the centre of the given quadric and touching it at any point (x', y', z'), then the point (x, y, z) on the required surface, which corresponds to (x', y', z'), is the extremity of the diameter of this sphere which passes through the centre of the quadric. We thus easily find the expressions

$$x = x'\left(2-\frac{t}{a^2}\right),\quad y = y'\left(2-\frac{t}{b^2}\right),\quad z = z'\left(2-\frac{t}{c^2}\right);$$

where

$$t = x'^2 + y'^2 + z'^2.$$

Solving these equations for x', y', z', and substituting in the two equations

$$xx' + yy' + zz' = x'^2 + y'^2 + z'^2,\quad \frac{x'^2}{a^2}+\frac{y'^2}{b^2}+\frac{z'^2}{c^2}=1,$$

we get

$$\frac{x^2}{\left(2-\dfrac{t}{a^2}\right)}+\frac{y^2}{\left(2-\dfrac{t}{b^2}\right)}+\frac{z^2}{\left(2-\dfrac{t}{c^2}\right)}=t \quad \text{......................(1)},$$

$$\frac{x^2}{a^2\left(2-\dfrac{t}{a^2}\right)^2}+\frac{y^2}{b^2\left(2-\dfrac{t}{b^2}\right)^2}+\frac{z^2}{c^2\left(2-\dfrac{t}{c^2}\right)^2}=1 \quad \text{......................(2)}.$$

Since (2) is the differential with respect to t of (1), the result of eliminating t between these two equations is the discriminant of (1). Hence the equation of the required surface is the discriminant of (1) with respect to t. Since (1) is only of the fourth degree, this discriminant is easily formed. If (1) be written in the form

$$At^4+4Bt^3+6Ct^2+4Dt+E=0,$$

it will be found that A and B do not contain x, y, z, while C, D, E contain them, each in the second degree. Now the discriminant is of the sixth degree in the coefficients, and of the form $A\phi+B^2\psi$ (see Salmon's *Higher Algebra*, § 107); hence it contains x, y, z only in the tenth degree. This is therefore the degree of the required surface.

The section of this derived surface by the principal plane z consists of the discriminant of

$$\frac{x^2}{2-\dfrac{t}{a^2}}+\frac{y^2}{2-\dfrac{t}{b^2}}=t \quad \text{.................................. (3)},$$

$\left(\text{which is of the sixth degree, and is the first negative pedal of } \dfrac{x^2}{a^2}+\dfrac{y^2}{b^2}=1\right)$, together with the conic (taken twice), which is obtained by putting $t=2c^2$ in (3).

This conic, which is a double curve on the surface, touches the curve of the sixth degree in four points.

2. The formulæ for the conic are quite analogous to those for the ellipsoid, viz. we have

$$x=X\left\{2-\frac{1}{a^2}(X^2+Y^2)\right\},\quad y=Y\left\{2-\frac{1}{b^2}(X^2+Y^2)\right\},$$

leading to the equations

$$\theta=\frac{x^2}{2-\dfrac{\theta}{a^2}}+\frac{y^2}{2-\dfrac{\theta}{b^2}},$$

and its derived equation, from which to eliminate θ. The first is the cubic equation $(A, B, C, D)(\theta, 1)^3=0$, where

$$A=1,\quad B=-\tfrac{2}{3}(a^2+b^2),\quad C=\tfrac{1}{3}(a^2x^2+b^2y^2+4a^2b^2),\quad D=-2a^2b^2(x^2+y^2).$$

Equating the discriminant to zero, this is

$$0 = A^2\nabla = 4(AC - B^2)^3 - (3ABC - A^2D - 2B^3)^2.$$

Or finally

$$(3a^2x^2 + 3b^2y^2 - 4a^4 + 4a^2b^2 - 4b^4)^3$$
$$+ \{9(a^2 - 2b^2)a^2x^2 + 9(b^2 - 2a^2)b^2y^2 - 8a^6 + 12a^4b^2 + 12a^2b^4 - b^6\}^2 = 0,$$

which is the required equation.

[Vol. XXI., January to June, 1874, pp. 29, 30.]

4298. (Proposed by J. W. L. GLAISHER, B.A.)—With four given straight lines to form a quadrilateral inscribable in a circle.

Solution by PROFESSOR CAYLEY.

Let the sides of the quadrilateral taken in order be a, b, c, d; and let its diagonals be x, y; viz. x the diagonal joining the intersection of the sides a, b with that of the sides c, d; y the diagonal joining the intersection of the sides a, d with that of the sides b, c; then, the quadrilateral being inscribed in a circle, the opposite angles are supplementary to each other. Suppose for a moment that the angles subtended by the diagonal x are θ, $\pi - \theta$, we have

$$x^2 = b^2 + c^2 + 2bc\cos\theta, \quad x^2 = a^2 + d^2 - 2ad\cos\theta;$$

and thence

$$(ad + bc)x^2 = ad(b^2 + c^2) + bc(a^2 + d^2) = (ac + bd)(ab + cd),$$

that is,

$$x^2 = (ac + bd)\frac{ab + cd}{ad + bc},$$

and similarly

$$y^2 = (ac + bd)\frac{ad + bc}{ab + cd},$$

agreeing as they should do with the known relation $xy = ac + bd$: the quadrilateral is thus determined by means of either of its diagonals. It is however interesting to treat the question in a different manner.

Considering a, b, c, d, x, y as the sides and diagonals of a quadrilateral, we have between these quantities a given relation, say

$$F(a, b, c, d, x, y) = 0,$$

and the quadrilateral being inscribed in a circle, we have also the relation $xy = ac + bd$; which two equations determine x, y; and thus give the solution of the problem.

The expression of the function F is in effect given in my paper, "Note on the value of certain determinants, &c.," *Quarterly Mathematical Journal*, t. III. (1860), pp. 275—277, [286]; viz. a, b, c being the edges of any face, and f, g, h the remaining edges of a tetrahedron, then

$$\begin{aligned} \text{volume} = \tfrac{1}{144}\ \{ & b^2c^2\ (g^2+h^2) + c^2a^2\,(h^2+f^2) + a^2b^2\,(f^2+g^2) \\ & + g^2h^2\ (b^2+c^2) + h^2f^2\,(c^2+a^2) + f^2g^2\,(a^2+b^2) \\ & - a^2f^2\ (a^2+f^2) - b^2g^2\,(b^2+g^2) + c^2h^2\,(c^2+h^2) \\ & - a^2g^2h^2 - b^2h^2f^2 - c^2f^2g^2 - a^2b^2c^2\}, \end{aligned}$$

where, when the tetrahedron becomes a quadrilateral, the volume is $=0$.

In this formula, changing c, b, h, g, f, a into a, h, c, d, x, y, we have the required equation $F=0$; viz. this is found to be

$$\begin{aligned} a^2b^2c^2 + b^2c^2d^2 + c^2d^2a^2 + d^2a^2b^2 - b^2d^2\,(b^2+d^2) - a^2c^2\,(a^2+c^2) + x^2y^2\,(a^2+b^2+c^2+d^2-x^2-y^2) \\ + x^2\,(a^2c^2+b^2d^2-a^2d^2-b^2c^2) + y^2\,(a^2c^2+b^2d^2-a^2b^2-c^2d^2) = 0, \end{aligned}$$

which, with $xy = ac+bd$, determines x, y. Substituting in the foregoing equation for xy its value, the equation becomes

$$(ad+bc)^2\,x^2 + (ab+cd)^2\,y^2 = 2\,\{a^2b^2c^2 + b^2c^2d^2 + c^2d^2a^2 + d^2a^2b^2 + abcd\,(a^2+b^2+c^2+d^2)\},$$

or

$$(ad+bc)^2\,x^2 + (ab+cd)^2\,y^2 = 2\,(ad+bc)\,(ab+cd)\,(ac+bd).$$

To show more clearly how this equation arises, I observe that we have identically

$$\begin{aligned} F - (a^2+b^2+c^2+d^2-x^2-y^2)\,(xy+ac+bd)\,(xy-ac-bd) - 2\,(ad+bc)\,(ab+cd)\,(xy-ac-bd) \\ = \{(ad+bc)\,x - (ab+cd)\,y\}^2. \end{aligned}$$

The resulting equation $(ad+bc)\,x-(ab+cd)\,y=0$, together with $xy=ac+bd$, gives for x, y the foregoing values.

[Vol. XXI., pp. 81, 82.]

4392. (Proposed by S. ROBERTS, M.A.)—If N_p denotes the number of terms in a symmetrical determinant of p rows and columns, show that the successive numbers are given by the equation

$$N_k - N_{k-1} - (k-1)^2\,N_{k-2} + \tfrac{1}{2}\,(k-1)\,(k-2)\,\{N_{k-3} + (k-3)\,N_{k-4}\} = 0,$$

k being positive and N_0 being taken equal to unity.

Solution by PROFESSOR CAYLEY.

It is a curious coincidence that the question of determining the number of distinct terms in a symmetrical determinant has been recently solved by Captain Allan

Cunningham in a paper in the last number of the *Quarterly Journal of Science**; and the question having been proposed to me by Mr Glaisher, I have also solved it in a paper [580] printed in the April Number of the *Monthly Notices of the Royal Astronomical Society*. I there obtain

$$N_k = 1\,.\,2 \ldots k \text{ coeff. } x^k \text{ in } \frac{e^{\frac{1}{2}x+\frac{1}{4}x^2}}{(1-x)^{\frac{1}{2}}},$$

viz. writing

$$u = N_0 + N_1 \frac{x}{1} + N_2 \frac{x^2}{1\,.\,2} + \ldots,$$

I show that u satisfies the differential equation

$$2\frac{du}{dx} = \left\{1 + x + \frac{1}{1-x}\right\} u,$$

giving when the constant is determined

$$u = \frac{e^{\frac{1}{2}x+\frac{1}{4}x^2}}{(1-x)^{\frac{1}{2}}}.$$

Writing the differential equation in the form

$$2(1-x)\frac{du}{dx} = (2 - x^2)\,u,$$

we at once obtain for N_k the equation of differences

$$N_k - kN_{k-1} + \tfrac{1}{2}(k-1)(k-2)\,N_{k-3} = 0,$$

which is in fact a particular first integral of Mr Roberts's equation; viz. from the above equation we have

$$N_{k-1} - (k-1)\,N_{k-2} + \tfrac{1}{2}(k-2)(k-3)\,N_{k-4} = 0,$$

and multiplying this last by $k-1$ and adding, we have

$$N_k - N_{k-1} - (k-1)^2 N_{k-2} + \tfrac{1}{2}(k-1)(k-2)\{N_{k-3} + (k-3)\,N_{k-4}\} = 0,$$

which is the equation obtained by Mr Roberts. It thence appears that the *general* first integral of his equation is

$$N_k - kN_{k-1} + \tfrac{1}{2}(k-1)(k-2)\,N_{k-3} = (-)^k C 1\,.\,2 \ldots (k-1).$$

The equation

$$N_k = kN_{k-1} - \tfrac{1}{2}(k-1)(k-2)\,N_{k-3}$$

gives very readily the numerical values, viz.

$1 = 1\,.\,1 - 0$	$17 = 4\,.\,5 - 3\,.\,1$	$2461 = 7\,.\,388 - 15\,.\,17$
$2 = 2\,.\,1 - 0$	$73 = 5\,.\,17 - 6\,.\,2$	$18155 = 8\,.\,2461 - 21\,.\,73.$
$5 = 3\,.\,2 - 1\,.\,1$	$388 = 6\,.\,73 - 10\,.\,5$	

* I have not the volume at hand to refer to, but he obtains an equation of differences, and gives the numbers 1, 2, 5, 73, 398 (should be 388), ...

[Vol. XXII., July to December, 1874, pp. 20, 21.]

4354. (Proposed by R. TUCKER, M.A.)—Solve the equations

$$-x^2+xy+xz=a=4 \ldots\ldots (1),$$

$$-y^2+xy+yz=b=-20 \ldots\ldots (2),$$

$$-z^2+xz+yz=c=-8 \ldots\ldots (3).$$

Note on Question 4354. *By* PROFESSOR CAYLEY.

A question of simple algebra such as this, becomes more interesting when interpreted geometrically: thus, writing the equations in the form

$$-x^2+xy+xz=aw^2,\quad yx-y^2+yz=bw^2,\quad zx+zy-z^2=cw^2,$$

and then putting for shortness

$$\alpha=-a+b+c,\quad \beta=a-b+c,\quad \gamma=a+b-c,$$

the solutions obtained are

$$x:y:z:w=a\alpha:b\beta:c\gamma:(\alpha\beta\gamma)^{\frac{1}{2}},$$

$$x:y:z:w=a\alpha:b\beta:c\gamma:-(\alpha\beta\gamma)^{\frac{1}{2}};$$

say these are

$$\{a\alpha,\ b\beta,\ c\gamma,\ (\alpha\beta\gamma)^{\frac{1}{2}}\} \text{ and } \{a\alpha,\ b\beta,\ c\gamma,\ -(\alpha\beta\gamma)^{\frac{1}{2}}\}.$$

But the equations are also satisfied by

$$(x=0,\ y=z,\ w=0),\quad (y=0,\ z=x,\ w=0),\quad (z=0,\ x=y,\ w=0),$$

or what is the same thing, (0, 1, 1, 0), (1, 0, 1, 0), (1, 1, 0, 0). The three equations represent quadric surfaces, each two of them intersecting in a proper quadric curve, and the three having in common 8 points; viz. these are made up of the first mentioned two points each once, and the last mentioned three points each twice: $2+3.2,\ =8$.

To verify this, observe that, at each of the three points, the tangent planes of the surfaces have a common line of intersection; this line is the tangent of the curve of intersection of any two of the surfaces, and the curve of intersection therefore touches the third surface; wherefore the point counts for two intersections. In fact, taking (X, Y, Z, W) as current coordinates, the equations of the tangent planes at the point (x, y, z, w) are

$$\begin{array}{llll} X(2x-y-z)-Yx & -Zx & & +2aWw=0,\\ -Xy & +Y(-x+2y-z)-Zy & & +2bWw=0,\\ -Xz & -Yz & +Z(-x-y+2z) & +2cWw=0: \end{array}$$

hence at the point (0, 1, 1, 0) these equations are

$$-2X=0,\quad X+Y-Z=0,\quad -X-Y+Z=0,$$

which three planes meet in the line $X=0,\ Y-Z=0$; and similarly for the other two of the three points.

[Vol. XXII., pp. 60—64.]

4458. (Proposed by Professor CAYLEY.)—Find (1) the intersections of the two quartic curves

$$\lambda(ab-xy)^2 = abx(a-y)(b-y), \quad \mu(ab-xy)^2 = aby(a-x)(b-x);$$

and (2) trace the curves in some particular cases; for instance, when $a=1$, $b=2$, $\lambda=1$, $\mu=-2$.

Solution by the PROPOSER.

1. The 16 intersections are made up as follows: 5 points at infinity on the line $x=0$, 5 at infinity on the line $y=0$, the two points $(x=a,\ y=b)$, $(x=b,\ y=a)$, and 4 other points, $16=5+5+2+4$. As to the points at infinity, observe that, as regards the first curve, the point at infinity on the line $x=0$ is a flecnode having this line for a tangent to the flecnodal branch; and, as regards the second curve, the same point is a cusp, having this line for its tangent; hence the point in question counts as $2+3$, $=5$ intersections; and the like as to the point at infinity on the line $y=0$. It remains to find the coordinates of the 4 points of intersection. Assume $xy=ab\omega$, then the equations become

$$\lambda(1-\omega)^2 = x+\omega y-(a+b)\omega, \quad \mu(1-\omega)^2 = \omega x+y-(a+b)\omega;$$

hence, eliminating successively y and x, the factor $1-\omega$ divides out,—this factor belongs to the points $(x=a,\ y=b)$, $(x=b,\ y=a)$ for which obviously $\omega=1$—, and the equations become

$$(\lambda-\mu\omega)(1-\omega)+(a+b)\omega=(1+\omega)x, \quad (\mu-\lambda\omega)(1-\omega)+(a+b)\omega=(1+\omega)y.$$

Multiplying these two equations together, and substituting for xy its value $ab\omega$, we find

$$\{(\lambda-\mu\omega)(\mu-\lambda\omega)+(a+b)(\lambda+\mu)\omega\}(1-\omega)^2+(a+b)^2\omega^2-(1+\omega)^2\omega ab=0.$$

Write, for shortness, $p=(\lambda+\mu)(a+b)-\lambda^2-\mu^2$, then, dividing by ω^2, and writing $\omega+\dfrac{1}{\omega}=\Omega$, the equation is

$$(\lambda\mu\Omega+p)(\Omega-2)+(a+b)^2-ab(\Omega+2)=0;$$

viz. this is a quadric equation for Ω. But, instead of Ω, it is convenient to introduce the quantity θ, $=\dfrac{\Omega-2}{\Omega+2}$, $=\left(\dfrac{\omega-1}{\omega+1}\right)^2$. The equation thus becomes

$$\left\{2\lambda\mu\,\frac{1+\theta}{1-\theta}+p\right\}\frac{4\theta}{1-\theta}+(a+b)^2-ab\,\frac{4}{1-\theta}=0,$$

or

$$\{2\lambda\mu(1+\theta)+p(1-\theta)\}\,4\theta+(a+b)^2(1-\theta)^2-4ab(1-\theta)=0,$$

or

$$\theta^2\{(a+b)^2-4(p-2\lambda\mu)\}+\theta\{-2a^2-2b^2+4(p+2\lambda\mu)\}+(a-b)^2=0;$$

viz. substituting for p its values, this is

$$\theta^2(a+b-2\lambda-2\mu)^2+2\theta\{-a^2-b^2+2(\lambda+\mu)(a+b)-2(\lambda-\mu)^2\}^2+(a-b)^2=0;$$

or if we write

$$A=a^2-2a(\lambda+\mu)+(\lambda-\mu)^2,\quad B=b^2-2b(\lambda+\mu)+(\lambda-\mu)^2,$$

this is

$$\theta^2(a+b-2\lambda-2\mu)^2-2(A+B)\theta+(a-b)^2=0,$$

whence

$$\{(a-b)^2-(A+B)\theta\}^2=\theta^2\{(A+B)^2-(a-b)^2(a+b-2\lambda-2\mu)^2\}$$
$$=\theta^2\{(A+B)^2-(A-B)^2\}=4AB\theta^2;$$

viz. taking for convenience the sign $-$ on the right-hand side, this is

$$(a-b)^2-(A+B)\theta=-2\theta\sqrt{AB};$$

and we have thus

$$\theta=\frac{(a-b)^2}{(\sqrt{A}-\sqrt{B})^2},$$

that is,

$$\theta,=\frac{\omega-1}{\omega+1}=\frac{a-b}{\sqrt{A}-\sqrt{B}};\quad \omega=\frac{\sqrt{A}-\sqrt{B}+a-b}{\sqrt{A}-\sqrt{B}-a+b}.$$

We may write

$$x=\mu(\omega-1)+\tfrac{1}{2}(a+b)+\tfrac{1}{2}(a+b-2\lambda-2\mu)\frac{\omega-1}{\omega+1},$$

$$y=\lambda(\omega-1)+\tfrac{1}{2}(a+b)+\tfrac{1}{2}(a+b-2\lambda-2\mu)\frac{\omega-1}{\omega+1};$$

whence also $x-y=(\mu-\lambda)(\omega-1)$, as is also seen at once from the original equations; then we have

$$\tfrac{1}{2}(a+b-2\lambda-2\mu)\frac{\omega-1}{\omega+1}=\frac{\tfrac{1}{2}(a-b)(a+b-2\lambda-2\mu)}{\sqrt{A}-\sqrt{B}}$$
$$=\frac{\tfrac{1}{2}(A-B)}{\sqrt{A}-\sqrt{B}},\ =\tfrac{1}{2}(\sqrt{A}+\sqrt{B});$$

and the values are

$$x=\frac{2\mu(a-b)}{\sqrt{A}-\sqrt{B}-a+b}+\tfrac{1}{2}(\sqrt{A}+\sqrt{B}+a+b)$$
$$=\frac{(a-b)(\mu-\lambda)+b\sqrt{A}-a\sqrt{B}}{\sqrt{A}-\sqrt{B}-a+b},$$

$$y=\frac{2\lambda(a-b)}{\sqrt{A}-\sqrt{B}-a+b}+\tfrac{1}{2}(\sqrt{A}+\sqrt{B}+a+b)$$
$$=\frac{(a-b)(\lambda-\mu)+b\sqrt{A}-a\sqrt{B}}{\sqrt{A}-\sqrt{B}-a+b},$$

which may be expressed in the more simple form

$$x=\frac{1}{4\lambda}(a+\lambda-\mu+\sqrt{A})(b+\lambda-\mu+\sqrt{B}),$$

$$y=\frac{1}{4\mu}(a-\lambda+\mu+\sqrt{A})(b-\lambda+\mu+\sqrt{B}),$$

the transformations depending on the identity

$$\frac{8\lambda\mu(a-b)}{\sqrt{A}-\sqrt{B}-a+b}=ab-(\lambda+\mu)(a+b)+(\lambda-\mu)^2+\sqrt{A}(b-\lambda-\mu)+\sqrt{B}(a-\lambda-\mu)+\sqrt{AB},$$

which is easily verified. Of course, since the signs of $\sqrt{A}$, $\sqrt{B}$ are arbitrary, we have 4 systems of values of (x, y), which is right.

In the original equations, for a, b, λ, μ, x, y, write 1, k^{-2}, λ^2, $-\mu^2$, x^2, $-y^2$; then the equations become

$$\lambda^2(1+k^2x^2y^2)^2=x^2(1+y^2)(1+k^2y^2),\quad \mu^2(1+k^2x^2y^2)^2=y^2(1-x^2)(1-k^2x^2),$$

and we thence have

$$\lambda+\mu i=\frac{x\sqrt{(1+y^2)(1+k^2y^2)}+iy\sqrt{(1-x^2)(1-k^2x^2)}}{1+k^2x^2y^2};$$

viz. assuming $x=\text{sn}\,\alpha\,(\text{sinam}\,\alpha)$, $iy=\text{sn}\,i\beta$, this is $\lambda+\mu i=\text{sn}(\alpha+\beta i)$; viz. the problem is (for a given modulus k, assumed as usual to be real, positive, and less than 1) to reduce a given imaginary quantity $\lambda+\mu i$ to the form $\text{sn}(\alpha+\beta i)$. The proper solution is that in which the signs of the radicals are each $-$, viz. it may in this case be shown that the value of x^2 is positive and less than 1, that of y^2 positive. The values thus are

$$x^2=\frac{1}{4\lambda^2}(1+\lambda^2+\mu^2-\sqrt{A})\left(\frac{1}{k^2}+\lambda^2+\mu^2-\sqrt{B}\right),$$

$$y^2=\frac{1}{4\mu^2}(1-\lambda^2-\mu^2-\sqrt{A})\left(\frac{1}{k^2}-\lambda^2-\mu^2-\sqrt{B}\right),$$

where

$$A=1-2\lambda^2+2\mu^2+(\lambda^2+\mu^2),\quad B=\frac{1}{k^2}-\frac{2}{k^2}\lambda^2+\frac{2}{k^2}\mu^2+(\lambda^2+\mu^2)^2.$$

The solution is really equivalent to that given by Richelot (*Crelle*, t. XLV., 1853, p. 225). To verify this partially, observe that, writing σ, τ for Richelot's $\tan\frac{1}{2}\phi$, $\tan\frac{1}{2}\psi$, we have

$$\left(\sigma+\frac{1}{\sigma}\right)\lambda=1+\lambda^2+\mu^2,$$

giving

$$\left(\sigma-\frac{1}{\sigma}\right)\lambda=-\sqrt{A};$$

$$\left(\tau+\frac{1}{\tau}\right)\frac{\lambda}{k}=\frac{1}{k^2}+\lambda^2+\mu^2,$$

giving

$$\left(\tau-\frac{1}{\tau}\right)\frac{\lambda}{k}=-\sqrt{B};$$

whence

$$2\sigma\lambda=1+\lambda^2+\mu^2-\sqrt{A},\quad 2\tau\frac{\lambda}{k}=\frac{1}{k^2}+\lambda^2+\mu^2-\sqrt{B},$$

or the above value of x^2 is $=k^{-1}\sigma\tau$, agreeing with his; the value of y^2 is, however, presented under a somewhat different form.

2. The curves are

$$(2-xy)^2=2x(1-y)(2-y), \quad -(2-xy)^2=y(1-x)(2-x) \ldots\ldots\ldots (1,\ 2),$$

each passing through the points (1, 2) and (2, 1); the four points of intersection found by the foregoing general theory are all real, viz. these are

$x=\frac{1}{2}(2+\sqrt{3})(5+\sqrt{17})$,	$y=-\frac{1}{8}(-1+\sqrt{3})(-1+\sqrt{17})$,	say	$+17{\cdot}00$ and	$-0{\cdot}57$,
$-\sqrt{3},\ +\sqrt{17}$	"	"	$+\ 1{\cdot}65$	$+0{\cdot}94$,
$+\sqrt{3},\ -\sqrt{17}$	"	"	$+\ 1{\cdot}22$	$+2{\cdot}13$,
$-\sqrt{3},\ -\sqrt{17}$	"	"	$-\ 0{\cdot}12$	$-3{\cdot}49$.

The equation of the curve (1) may also be written in the forms

$$y^2(x^2-2x)+2yx-4x+4=0, \quad x^2y^2+x(-2y^2+2y-4)+4=0.$$

The original form shows that, if y is between 1 and 2, x is negative—(but by a further examination it appears that there is not in fact any branch of the curve between these limits of y)—but y being outside these limits, then x is positive; in fact, the whole curve lies on the positive side of the axis of y. And then the inspection of the first quadric equation shows that the lines $x=0$ and $x=2$ are each an asymptote.

The point at infinity on the axis of y is in fact a flecnode, the tangent to the flecnodal branch being $x=0$, and that of the ordinary branch $x=2$.

Similarly, from the second quadric equation, it appears that the line $y=0$ is an asymptote; the point at infinity on the axis of x is in fact a cusp, the axis in question $y=0$ being the cuspidal tangent.

The equation of the curve (2) may also be written in the forms

$$x^2y^2+(x^2-7x+2)y+4=0, \quad (y^2+y)x^2-7yx+2y+4=0.$$

The original form shows that, if x is between 1 and 2, y is positive; but that x being beyond these limits, y is negative; and as regards the first case, x between 1 and 2, we at once establish the existence of an oval, meeting the line $y=1$ in the points $x=2$ and $\frac{3}{2}$, and the line $y=2$ in the points $x=1$ and $\frac{4}{3}$; it is further easy to see that the horizontal tangents of the oval are $y=\frac{1}{16}(25\pm\sqrt{113})$, $=$ say $2{\cdot}2$ and $0{\cdot}9$.

The remainder of the curve lies wholly below the line $y=0$. The first quadric equation shows the asymptote $x=0$; the point at infinity on the axis of y is in fact a cusp, having the axis itself for the cuspidal tangent. The second quadric equation shows the asymptotes $y=0$, $y=-1$; the point at infinity on the axis of x is in fact a flecnode, having the line $y=0$ for the tangent to the flecnodal branch, and $y=-1$ for that of the other branch. It is further seen that there are two vertical tangents $x=\frac{1}{2}(11\pm\sqrt{113})=10{\cdot}8$ or $0{\cdot}2$; the former of these touches a branch lying wholly between the two asymptotes $y=0$, $y=-1$; the latter one of the branches belonging to the cuspidal asymptote $x=0$; this last branch cuts the asymptote $x=0$ at $y=-2$, and then, cutting the asymptote $y=-1$ and $x=-\frac{2}{7}(=-0{\cdot}3)$, goes on to touch at infinity the asymptote $y=0$. It is now easy to trace the curve.

The figure shows the two curves. The curve (1) is shown by a continuous line, the curve (2) by a thick dotted line; the points 1, 2, 3, 4 show the above mentioned

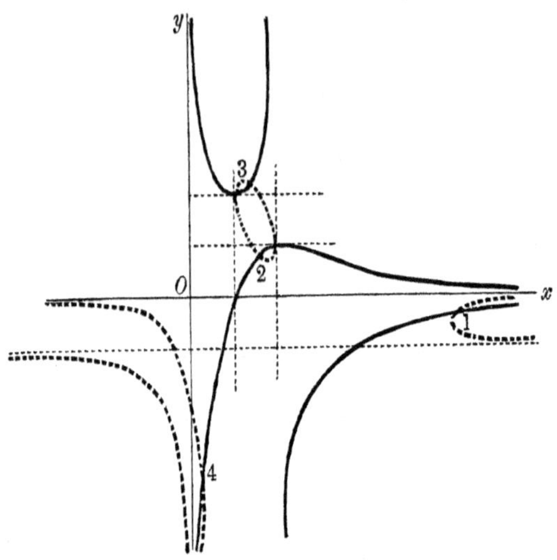

four intersections of the curves; the point 1 and the dotted branch through it are of necessity drawn considerably out of their true positions; viz. as above appearing, the x-coordinate of 1 is $=17{\cdot}00$, and the equation of the vertical tangent to the branch is $x=10{\cdot}8$.

[Vol. XXII., pp. 78, 79.]

4520. (Proposed by A. B. EVANS, M.A.)—Find the least integral values of x and y that will satisfy the equation $x^2-953y^2=-1$.

Solution by PROFESSOR CAYLEY.

The values are given in Degen's Tables, viz.

$$x=2746864744,\quad y=88979677.$$

The work referred to is entitled "Canon Pellianus, sive Tabula simplicissimam æquationis celebratissimæ $y^2=ax^2+1$ solutionem pro singulis numeri dati valoribus ab 1 usque ad 1000 in numeris rationalibus iisdemque integris exhibens. Auctore C. F. Degen, Hafniæ (Copenhagen), 1817."

Table I., pp. 3—106 gives, for all numbers 1 to 1000, the denominators, (?) the quotients of the convergent fraction of $\sqrt{a}$, and also the least values of x, y which will satisfy the equation $x^2 - ay^2 = +1$. Thus

953	30, 1, 6, 1, 2, 1, 3, 8, 1, 1, (4, 4),
	1, 53, 8, 41, 17, 37, 16, 7, 32, 29, (13, 13),
	488830275367615376, 15090531843660371073.

Table II., pp. 109—112, is described as giving for all those values of a between 1 and 1000, for which there exists a solution of the equation $x^2 - ay^2 = -1$, the least values of x and y which satisfy this equation: thus 953, x and y as above. It is, however, to be noticed that the values of $a = \beta^2 + 1$, for which there is the obvious solution $x = \beta$, $y = 1$, are omitted from the table. The reason for this appears, but the heading should have been different.

[Vol. XXIII., January to July, 1875, pp. 18, 19.]

4528. (Proposed by Professor CAYLEY.)—A lottery is arranged as follows:—There are n tickets representing a, b, c pounds respectively. A person draws once; looks at his ticket; and if he pleases, draws again (out of the remaining $n-1$ tickets); looks at his ticket, and if he pleases draws again (out of the remaining $n-2$ tickets); and so on, drawing in all not more than k times; and he receives the value of the last drawn ticket. Supposing that he regulates his drawings in the manner most advantageous to him according to the theory of probabilities, what is the value of his expectation?

Solution by the PROPOSER.

Let the expression "a or α" signify "a or α, whichever of the two is greatest," and let $M_1(a, b, c, \ldots)$ denote the mean of the quantities $(a, b, c, \ldots)$, viz. their sum, divided by the number of them.

To fix the ideas, consider five quantities a, b, c, d, e, and write

$$M_1(a, b, c, d, e) = M_1(a, b, c, d, e),$$

$$M_2(a, b, c, d, e) = M_1\{a \text{ or } M_1(b, c, d, e),\ b \text{ or } M_1(a, c, d, e), \ldots, e \text{ or } M_1(a, b, c, d)\},$$

$$M_3(a, b, c, d, e) = M_1\{a \text{ or } M_2(b, c, d, e),\ b \text{ or } M_2(a, c, d, e), \ldots, e \text{ or } M_2(a, b, c, d)\},$$

and so on. And the like in the case of any number of quantities a, b, c, ...

Then the value of the expectation is $= M_k(a, b, c, \ldots)$.

For, when $k = 1$, the value is obviously $= M_1(a, b, c, \ldots)$.

When $k=2$, if a is drawn, the adventurer will be satisfied or he will draw again, according as a or $M_1(b, c, \ldots)$ is greatest, viz. in this case the value of the expectation is "a or $M_1(b, c, \ldots)$."

So if b is drawn, the adventurer will be satisfied or he will draw again, according as b or $M_1(a, c, \ldots)$ is greatest; viz. in this case the value of the expectation is "b or $M_1(a, c, \ldots)$"; and so on: and the several cases being equally probable, the value of the total expectation is

$$= M_1\{a \text{ or } M_1(b, c, \ldots),\ \ b \text{ or } M_1(a, c, \ldots), \ldots\} = M_2(a, b, c, \ldots):$$

and the like for $k=3$, $k=4$, &c.

For instance, $a, b, c, d = 1, 2, 3, 4$, $M_1(1, 2, 3, 4) = \frac{10}{4}$,

$$M_2(1, 2, 3, 4) = M_1(1 \text{ or } \tfrac{9}{3},\ 2 \text{ or } \tfrac{8}{3},\ 3 \text{ or } \tfrac{7}{3},\ 4 \text{ or } \tfrac{6}{3}) = M_1(\tfrac{9}{3}, \tfrac{8}{3}, \tfrac{9}{3}, \tfrac{12}{3}) = \tfrac{38}{12},$$

$$M_2(2, 3, 4) = M_1(2 \text{ or } \tfrac{7}{2},\ 3 \text{ or } \tfrac{6}{2},\ 4 \text{ or } \tfrac{5}{2}) = M_1(\tfrac{7}{2}, \tfrac{6}{2}, \tfrac{8}{2}) = \tfrac{21}{6},$$

$$M_2(1, 3, 4) = M_1(1 \text{ or } \tfrac{7}{2},\ 3 \text{ or } \tfrac{5}{2},\ 4 \text{ or } \tfrac{4}{2}) = M_1(\tfrac{7}{2}, \tfrac{6}{2}, \tfrac{8}{2}) = \tfrac{21}{6},$$

$$M_2(1, 2, 4) = M_1(1 \text{ or } \tfrac{6}{2},\ 2 \text{ or } \tfrac{5}{2},\ 4 \text{ or } \tfrac{3}{2}) = M_1(\tfrac{6}{2}, \tfrac{5}{2}, \tfrac{8}{2}) = \tfrac{19}{6},$$

$$M_2(1, 2, 3) = M_1(1 \text{ or } \tfrac{5}{2},\ 2 \text{ or } \tfrac{4}{2},\ 3 \text{ or } \tfrac{3}{2}) = M_1(\tfrac{5}{2}, \tfrac{4}{2}, \tfrac{6}{2}) = \tfrac{15}{6},$$

$$M_3(1, 2, 3, 4) = M_1(1 \text{ or } \tfrac{21}{6},\ 2 \text{ or } \tfrac{21}{6},\ 3 \text{ or } \tfrac{19}{6},\ 4 \text{ or } \tfrac{15}{6}) = M_1(\tfrac{21}{6}, \tfrac{21}{6}, \tfrac{19}{6}, \tfrac{24}{6}) = \tfrac{85}{24},$$

$$M_3(1, 2, 3) = 3 \text{ \&c.},$$

$$M_4(1, 2, 3, 4) = M_1(1 \text{ or } 4,\ 2 \text{ or } 4,\ 3 \text{ or } 4,\ 4 \text{ or } 3) = M_1(4, 4, 4, 4) = 4.$$

Or finally

$$M_1,\ M_2,\ M_3,\ M_4 = \tfrac{10}{4},\ \tfrac{38}{12},\ \tfrac{85}{24},\ 4 = \tfrac{60}{24},\ \tfrac{76}{24},\ \tfrac{85}{24},\ \tfrac{96}{24}.$$

COR. If the $a, b, c, \ldots$ denote penalties instead of prizes, then the solution is the same, except that "a or α" must now denote "a or α, whichever of them is least."

[Vol. XXIII., pp. 47, 48.]

4581. (Proposed by the Rev. M. M. U. WILKINSON.)—A witness, whose statement is what he opines once in m times, and whose opinion is correct once in n times, asserted that the number of a note, issued by a bank universally known to have issued notes numbered from B to $B+A-1$ inclusive, was $B+P$, where P is either 0, 1, 2, ..., or $A-1$. Prove (1) that the probability that the note in question was that note is

$$\frac{1}{mn}\left\{1 + \frac{(m-1)(n-1)}{A-1}\right\}.$$

The above witness also said that the note was signed by X, it being universally known that X has signed one note, and Y the remaining $A-1$ notes; find (2) the probability that this last statement was correct.

Remark by PROFESSOR CAYLEY.

There is a serious difficulty in the question, or the answer; I think, in the question. Try the answer in numbers $m=10$, $n=10$. The witness says what he opines once out of 10 times—he is in fact an atrocious liar; and he opines rightly once out of 10 times, that is, wrongly 9 times out of 10; he is therefore a blunderer—but a remarkably ingenious one, in that the chances are so greatly against his blundering upon a right result.

He says that the note was signed by X, and the chance of this being so is found to be $\frac{1}{100}+\frac{81}{100}=\frac{82}{100}$, or more than $\frac{8}{10}$; the larger part $\frac{81}{100}$ of this is obtained as follows:—the witness having said that the note was signed by X, the chances are 9 out of 10 that he thought the reverse; and, thinking the reverse, the chance is 9 out of 10 that he thought wrongly, viz. that the note was signed by X. But can the statement of such a witness create any probability in favour of the event?

The fallacy seems to consist in the assumption that n can have a determinate value irrespective of the nature of the opinion. Suppose there are 500 notes, and that the opinion is that the note was a definite number 99; it is quite conceivable that, in forming a series of such opinions, the witness may be wrong 9 times out of 10. But let the opinion be that the note was *not* 99; no amount of ingenuity of blundering can make him wrong 9 times out of 10 in a series of such opinions. If it could, a friend who knew the true opinion of the witness, would be able 9 times out of 10 to know the number of the note, from the mere fact that the witness opines that the note is not a named number.

[Vol. XXIII., p. 58.]

4638. (Proposed by Professor CAYLEY.)—Find the equation of the surface which is the envelope of the quadric surface $ax^2+by^2+cz^2+dw^2=0$, where a, b, c, d are variable parameters connected by the equation $Abc+Bca+Cab+Fad+Gbd+Hcd=0$; and consider in particular the case in which the constants A, B, C, F, G, H satisfy the condition

$$(AF)^{\frac{1}{3}}+(BG)^{\frac{1}{3}}+(CH)^{\frac{1}{3}}=0.$$

[Vol. XXIV., July to December, 1875, p. 41.]

4694. (Proposed by Professor CAYLEY.)—Taking F, F' a pair of reciprocal points in respect to a circle, centre O; then if F, F' are centres of force, each force varying as (distance)$^{-n}$, prove that (1) the resultant force upon any point P on the circle is in the direction of a fixed point S on the axis OFF'; and if, moreover, the forces at the unit of distance are as $(OF)^{\frac{1}{2}(n-1)}$ to $(OF')^{\frac{1}{2}(n-1)}$, then (2) the resultant force is proportional to

$$(SP)^{-\frac{1}{2}(n-1)}.(PV)^{-\frac{1}{2}(n+1)},$$

where PV is the chord through S.

[Vol. XXIV., pp. 72—74.]

4793. (Proposed by Professor WOLSTENHOLME, M.A.)—If $y = x^n (\log x)^r$, where n and r are integers, prove that

$$x^r \frac{d^{n+r}y}{dx^{n+r}} + \frac{r(r-1)}{2} x^{r-1} \frac{d^{n+r-1}y}{dx^{n+r-1}} + \frac{r(r-1)(r-2)(3r-5)}{24} x^{r-2} \frac{d^{n+r-2}y}{dx^{n+r-2}} + \dots$$

$$\dots + (2^{r-1}-1)\, x^2 \frac{d^{n+2}y}{dx^{n+2}} + x \frac{d^{n+1}y}{dx^{n+1}} = \underline{|r}\, \underline{|n},$$

the coefficients being

$$\frac{\Delta^{r-1}1^{r-1}}{\underline{|r-1}},\quad \frac{\Delta^{r-2}1^{r-1}}{\underline{|r-2}},\quad \frac{\Delta^{r-3}1^{r-1}}{\underline{|r-3}},\ \dots\dots,\ \frac{\Delta 1^{r-1}}{\underline{|1}},\ \text{and } 1;$$

so that the result may be symbolically written

$$\epsilon^{xD\Delta}\left(1^{r-1}\frac{d^{n+1}y}{dx^{n+1}}\right) = \frac{\underline{|r}\,\underline{|n}}{x},$$

where D denotes $\dfrac{d}{dx}$ and operates on $\dfrac{d^n y}{dx^n}$ only, and Δ operates on 1^{r-1} only, the terms after the rth all vanishing since $\Delta^m x^n = 0$, when m is an integer $> n$. The calculations involved prove that, when $x = 1$,

$$\Delta^{n-1}x^n = \frac{\underline{|n+1}}{2},\quad \Delta^{n-2}x^n = \underline{|n+1}\,.\frac{3n-2}{24},$$

$$\Delta^{n-3}x^n = \underline{|n+1}\,.\frac{(n-1)(n-2)}{48}.$$

Solution by PROFESSOR CAYLEY.

Since $y = x^n (\log x)^r$, therefore $(xd_x - n)\, y = rx^n (\log x)^{r-1}$; by repeating the same operation, we have

$$(xd_x - n)^r y = [r]^r x^n;\ \text{whence } d_x^{\,n}(xd_x - n)^r y = [r]^r [n]^n.$$

Now, for any value whatever of the function y, we have

$$d_x^{\,n}(xd_x - n)^r y = Ax^r d_x^{\,r+n} y + Bx^{r-1} d_x^{\,r+n-1} y + Cx^{r-2} d_x^{\,r+n-2} y + \text{\&c.},$$

the coefficients $A, B, C, \dots$ being functions, presumably of r, n, but independent of the form of the function y. It will, however, appear that $A, B, C, \dots$ are, in fact, functions of r only.

To see how this is, observe that $(xd_x - n)^r$ consists of a set of terms

$$(xd_x)^\theta,\quad (\theta = 0 \text{ to } r),$$

where $(xd_x)^\theta$ denotes θ repetitions of the operation xd_x; by a well-known theorem, this is $= [xd_x + \theta - 1]^\theta$, where, after expansion of the factorial, $(xd_x)^s$ is to be replaced by $x^s d_x^{\,s}$, thus

$$(xd_x)^2 = [xd_x + 1]^2 = x^2 d_x^{\,2} + xd_x,\quad (xd_x)^3 = [xd_x + 2]^3 = x^3 d_x^{\,3} + 3x^2 d_x^{\,2} + 2xd_x,\ \text{\&c.};$$

thus $(xd_x - n)^r$ consists of a series of terms $x^\theta d_x^\theta$, $(\theta = 0$ to $r)$, and, operating with d_x^n, this last, $= (d_x + d_x')^n$, consists of a series of terms such as $d_x^\alpha d_x'^{n-\alpha}$, where the unaccented symbol operates on the x^θ, and the accented symbol on the y; the term is thus $x^{\theta-\alpha} d_x^{n+\theta-\alpha}$, or observing that $\theta - \alpha$ is at most $= r$, and putting it $= r - k$, the term is $x^{r-k} d_x^{n+k}$, viz. $d_x^n (xd_x - n)^r$ consists of a series of terms of the form $x^{r-k} d_x^{n+k}$; or, what is the same thing, $d_x^n (xd_x - n)^r y$ is a series of the form in question.

To understand how it can be that the coefficients A, B, C, ... are independent of n, take the particular case $r = 2$; then we have here

$$d_x^n (xd_x - n)^2 y = Ax^2 d_x^{n+2} y + Bxd_x^{n+1} y + Cd_x^n y.$$

The right-hand side is

$$d_x^n \{x^2 d_x^2 - (2n - 1)\, xd_x + n^2\}\, y,$$

which is

$$\begin{aligned} = \quad & \{x^2 d^{n+2} + 2nxd_x^{n+1} + (n^2 - n)\, d_x^n\}\, y \\ - (2n-1)\, & \{\qquad\qquad xd_x^{n+1} + \qquad nd_x^n\}\, y \\ + n^2\, & \{\qquad\qquad\qquad\qquad\qquad d_x^n\}\, y; \end{aligned}$$

hence

$$A = 1, \quad B = 2n - (2n - 1),\ = 1, \quad C = (n^2 - n) - n\,(2n - 1) + n^2,\ = 0;$$

and we thus see also how in this particular case the last coefficient is $= 0$, viz. that we have

$$d_x^n (xd_x - n)^2 y = x^2 d_x^{n+2} y + xd_x^{n+1} y,$$

without any term in $d_x^n y$.

To find the coefficients A, B, C, ... generally, write $y = x^{r+n+\theta}$, then $xd_x - n = r + \theta$, and consequently

$$d_x^n (xd_x - n)^r y, \quad = (r + \theta)^r d_x^n x^{r+n+\theta}, \quad = (r + \theta)^r [r + n + \theta]^n x^{r+\theta};$$

whence

$$(r + \theta)^r [r + n + \theta]^n = A\, [r + \theta + n]^{n+r} + B\, [r + \theta + n]^{n+r-1} + \ldots\ldots;$$

or, what is the same thing,

$$(r + \theta)^r = A\, [r + \theta]^r + B\, [r + \theta]^{r-1} + \ldots\ldots.$$

Since the left-hand side, and every term $[r + \theta]^s$ on the right-hand side, contains the factor $r + \theta$, there is not on the right-hand side any term $[r + \theta]^0$; dividing the equation by $r + \theta$, it then becomes

$$(r + \theta)^{r-1} = A\, [r + \theta - 1]^{r-1} + B\, [r + \theta - 1]^{r-2} + \ldots\ldots,$$

and we thus have

$$A = \frac{\Delta^{r-1} 1^{r-1}}{[r-1]^{r-1}}\, (= 1), \quad B = \frac{\Delta^{r-2} 1^{r-1}}{[r-2]^{r-2}};$$

viz. writing $r + \theta = 1 + x$, $u_x = (1 + x)^{r-1}$, and taking the terms in the reverse order, the series is the well-known one

$$u_x = u_0 + \frac{x}{1} \Delta u_0 + \frac{x \,.\, x - 1}{1 \,.\, 2} \Delta^2 u_0 + \&\text{c}.$$

Hence, in general,

$$d_x^{\,n}(xd_x - n)^r y = \frac{\Delta^{r-1}1^{r-1}}{[r-1]^{r-1}} x^r d_x^{\,r+n} y + \frac{\Delta^{r-2}1^{r-1}}{[r-2]^{r-2}} x^{r-1} d_x^{\,r+n-1} y + \&c.,$$

where observe that the last term is $= xd_x^{\,n+1}y$.

For the function $y = x^n(\log x)^n$, the value of each side is $=[r]^r[n]^n$.

[Vol. XXIV., pp. 89—91.]

4752. (Proposed by Professor CAYLEY.)—Mr Wolstenholme's Question 3067 may evidently be stated as follows:—

If (a, b, c) are the coordinates of a point on the cubic curve

$$a^3 + b^3 + c^3 = (b+c)(c+a)(a+b),$$

and if

$$(b^2+c^2-a^2)x = (c^2+a^2-b^2)y = (a^2+b^2-c^2)z;$$

then (x, y, z) are the coordinates of a point on the same cubic curve.

This being so, it is required to find the geometrical relation of the two points to each other.

Solution by PROFESSOR CAYLEY.

1. On referring to Professor Wolstenholme's Solution of the original Question 3067 (*Reprint*, Vol. XIII., p. 70), it appears that the coordinates (x, y, z) of the point in question may be expressed in the more simple form

$$x : y : z = a(-a+b+c) : b(a-b+c) : c(a+b-c);$$

viz. the given relation between (a, b, c) being equivalent to

$$4abc + (-a+b+c)(a-b+c)(a+b-c) = 0,$$

we have

$$a^2 - (b-c)^2 = \frac{-4abc}{-a+b+c},$$

and thence

$$b^2+c^2-a^2 = 2bc\left(1 + \frac{2a}{-a+b+c}\right) = 2bc\left(\frac{a+b+c}{-a+b+c}\right);$$

and consequently

$$(b^2+c^2-a^2)x = \frac{2abc(a+b+c)x}{a(-a+b+c)},$$

whence the transformation in question.

2. Writing for greater symmetry (x, y, z) in place of (a, b, c), and (x', y', z') in place of (x, y, z), the coordinates (x, y, z) and (x', y', z') of the two points are connected by the relation

$$x' : y' : z' = x(-x+y+z) : y(x-y+z) : z(x+y-z),$$

and we thence at once deduce the converse relation

$$x : y : z = x'(-x'+y'+z') : y'(x'-y'+z') : z'(x'+y'-z').$$

Hence, writing

$$(-x+y+z,\quad x-y+z,\quad x+y-z) = (\xi,\ \eta,\ \zeta),$$

and similarly

$$(-x'+y'+z',\quad x'-y'+z',\quad x'+y'-z') = (\xi',\ \eta',\ \zeta'),$$

we have

$$x' : y' : z' = x\xi : y\eta : z\zeta,\quad x : y : z = x'\xi' : y'\eta' : z'\zeta',$$

and thence also $\xi\xi' = \eta\eta' = \zeta\zeta'$; so that, regarding $(\xi,\ \eta,\ \zeta)$, $(\xi',\ \eta',\ \zeta')$ as the coordinates of the two points, we see that these are inverse points one of the other in regard to the triangle $\xi = 0$, $\eta = 0$, $\zeta = 0$.

To complete the solution, we must introduce these new coordinates into the equation of the cubic curve. Writing this under the form

$$8xyz + 2(-x+y+z)(x-y+z)(x+y-z) = 0,$$

and observing that

$$(2x,\ 2y,\ 2z) = (\eta+\zeta,\ \zeta+\xi,\ \xi+\eta),$$

the equation is

$$(\eta+\zeta)(\zeta+\xi)(\xi+\eta) + 2\xi\eta\zeta = 0;$$

viz. this is a cubic curve inverting into itself. And the two points in question are thus any two inverse points on this cubic curve.

3. In regard to the original form, that the point $(x,\ y,\ z)$ defined by the equations

$$x(-a^2+b^2+c^2) = y(a^2-b^2+c^2) = z(a^2+b^2-c^2),$$

lies on the cubic curve

$$a^3+b^3+c^3-(b+c)(c+a)(a+b) = 0,$$

Professor Sylvester proceeds as follows:—Writing

$$(x,\ y,\ z) = \{a^4-(b^2-c^2)^2,\ b^4-(c^2-a^2)^2,\ c^4-(a^2-b^2)^2\},\ = (A,\ B,\ C),$$

suppose; and

$$F(a,\ b,\ c) = a^3+b^3+c^3-(b+c)(c+a)(a+b),$$

he observes that the truth of the theorem depends on the identity

$$F(A,\ B,\ C) + F(a,\ b,\ c)\,F(a,\ -b,\ c)\,F(a,\ b,\ -c)\,F(a,\ -b,\ -c) = 0,$$

and that, in order to prove the identity generally, it is sufficient to prove it for the three cases $a^2 = 0$, $a^2 = b^2 + c^2$, $a^2 = b^2$, which may be effected without difficulty.

4. But, for a general proof of the identity, write

$$\lambda = b^2 + c^2,\quad \mu = b^2 - c^2,$$

so that

$$A = a^4 - \mu^2,\quad B = (a^2+\mu)(-a^2+\lambda),\quad C = (-a^2+\lambda)(a^2-\mu),$$

whence

$$-F(A, B, C) = -(a^4-\mu^2)^3 + 2(a^2-\lambda)^3(a^6+3a^2\mu^2) - 8a^2b^2c^2(a^4-\mu^2)(a^2-\lambda),$$

$$= a^{12} - 6\lambda a^{10} + (6\lambda^2+9\mu^2-8b^2c^2)a^8 + \lambda(-2\lambda^2-18\mu^2+8b^2c^2)a^6$$

$$+ \mu^2(18\lambda^2-3\mu^2+8b^2c^2)a^4 + \lambda\mu^2(-6\lambda^2-8b^2c^2)a^2+\mu^6.$$

Moreover

$$F(a, b, c) = a\{a^2-(b+c)^2\} - (b+c)\{a^2-(b-c)^2\},$$

therefore

$$F(a, -b, -c) = a\{a^2-(b+c)^2\} + (b+c)\{a^2-(b-c)^2\};$$

whence

$$F(a, b, c)\,F(a, -b, -c) = a^2\{a^2-(b+c)^2\}^2 - (b+c)^2\{a^2-(b-c)^2\}^2$$

$$= a^6 - 3\gamma^2a^4 + \gamma^2(\gamma^2+2\delta^2)a^2 - \gamma^2\delta^2,$$

if $\gamma = b+c$, $\delta = b-c$. By changing the sign of c, we interchange γ and δ, and we thus have

$$F(a, b, -c)\,F(a, -b, c) = a^6 - 3\delta^2a^4 + \delta^2(2\gamma^2+\delta^2)a^2 - \gamma^4\delta^2,$$

and the identity to be verified is thus

$$\{a^6-3\gamma^2a^4+\gamma^2(\gamma^2+2\delta^2)a^2-\gamma^2\delta^4\}\{a^6-3\delta^2a^4+\delta^2(2\gamma^2+\delta^2)a^2-\gamma^4\delta^2\}$$

$$= a^{12} - 6\lambda a^{10} + \ldots\ldots + \mu^6, \text{ ut suprà};$$

the values of λ, μ in terms of γ, δ are $\lambda = \frac{1}{2}(\gamma^2+\delta^2)$, $\mu = \gamma\delta$; substituting these values on the right-hand side, the verification can be completed without difficulty.

[Vol. XXV., January to July, 1876, p. 82.]

4946. (Proposed by Professor CAYLEY.)—Show that the attraction of an indefinitely thin double convex lens on a point at the centre of one of its faces is equal to that of the infinite plate included between the tangent plane at the point and the parallel tangent plane of the other face of the lens.

[Vol. XXVI., July to December, 1876, pp. 41, 42.]

5020. (Proposed by W. S. B. WOOLHOUSE, F.R.A.S.)—Let $1, \delta_1, \delta_2, \delta_3, \ldots, \delta_n$ be the first differences of the coefficients of the expansion of the binomial $(1+x)^{2n}$ taken as far as the central or maximum coefficient; also let

$$\nu = \tfrac{1}{2}(n+1)n, \quad \nu' = \tfrac{1}{2}n(n-1), \quad \nu'' = \tfrac{1}{2}(n-1)(n-2), \ \&c.;$$

then show that the algebraic function

$$x^\nu - \delta_1 x^{\nu'} + \delta_2 x^{\nu''} - \delta_3 x^{\nu'''} + \&c.$$

is divisible by $(x-1)^n$ without a remainder; and that the sum of the numerical coefficients of the quotient is equal to $1.3.5\ldots 2n-1$.

[See Solution to Question 1894, *Reprint*, vol. v., p. 113.]

Solution by PROFESSOR CAYLEY.

Mr Woolhouse's elegant theorem depends ultimately on the property of triangular numbers $\phi(n)$, $=\frac{1}{2}(n^2-n)$; then $\phi(n+1)=\phi(-n)$, so that, writing down the series of triangular numbers backwards and forwards,

$$\begin{array}{l}\ldots,\ 10,\ 6,\ 3,\ 1,\ 0,\ 0,\ 1,\ 3,\ 6,\ 10,\ldots\\ \ldots\quad ,\ a,\ b,\ c,\ d,\ e,\ f,\ g,\ h,\quad \ldots\end{array}$$

we have, in fact, a continuous single series obtained by giving to n the different negative and positive integer values, zero included.

Thus a particular case is

$$(1+x)^6-5(1+x)^3+9(1+x)-5\equiv 0\,(\text{mod. } x^3)=1.3.5x^3+\&\text{c}.\,x^4+\ldots,$$

where, on the left-hand side, the exponents are the triangular numbers $\phi(n+1)$, $n=0$ to 3; and the coefficients, after the first, are the differences of the binomial coefficients of the power $2n$ (in the particular case, $n=3$); viz. the binomial coefficients being

1, 6, 15, 20, 15, 6, 1,

the differences taken as far as they are positive are

5, 9, 5.

Expanding the several terms and writing down only the coefficients, we have a diagram

$$\begin{array}{r|l}
 & 1,\ 6,\ 15,\ 20,\ 15,\ 6,\ 1,\\
-5 & 1,\ 3,\ 3,\ 1,\\
+9 & 1,\ 1,\\
-5 & 1,
\end{array}$$

The theorem in the particular case depends on the identities

$$\begin{aligned}1-5+9-5&=0,\\ 6-15+9&=0,\\ 5-15&=0,\\ 20-5&=1.3.5;\end{aligned}$$

or writing, as above, h, g, f, e, to denote the triangular numbers 6, 3, 1, 0, these may be replaced by

$$\begin{aligned}
&h^0 - 5g^0 + 9f^0 - 5e^0 = 0,\\
&h - 5g + 9f - 5e = 0,\\
&\tfrac{1}{2}h(h-1) - 5.\tfrac{1}{2}g(g-1) + \ldots = 0,\\
&\tfrac{1}{6}h(h-1)(h-2) - 5.\tfrac{1}{6}g(g-1)(g-2) + \ldots = 1.3.5;
\end{aligned}$$

or, reducing each equation by those which precede it, these become

$$h^0 - 5g^0 + 9f^0 - 5e^0 = 0,$$
$$h^1 - 5g^1 + 9f^1 - 5e^1 = 0,$$
$$h^2 - 5g^2 + 9f^2 - 5e^2 = 0,$$
$$h^3 - 5g^3 + 9f^3 - 5e^3 = 1.2.3.1.3.5.$$

Consider any one of these, for instance the third; the function on the left-hand is

$$1h^2 - (6-1)\,g^2 + (15-6)\,f^2 - (20-15)\,e^2,$$

or, introducing the values b, c, d as above,

$$1h^2 - 6g^2 + 15f^2 - 20e^2 + 15d^2 - 6c^2 + 1b^2,$$

which is, in fact, $=0$, if b, c, d, e, f, g, h are *any* successive triangular numbers; viz. this is an immediate consequence of the well-known theorem

$$1\,(\theta+6)^m - 6\,(\theta+5)^m + 15\,(\theta+4)^m - 20\,(\theta+3)^m + 15\,(\theta+2)^m - 6\,(\theta+1)^m + \theta^m$$
$$= \Delta^6\theta^m, \ = 0 \text{ for any value of } m \text{ up to } m=5, \text{ and}$$
$$= 1.2.3.4.5.6 \text{ for } m=6.$$

We have thus all the equations except the last; and as regards the last equation, observe that the equation to be verified is

$$1\,[\tfrac{1}{2}(\theta+6)(\theta+5)]^3 - 6\,[\tfrac{1}{2}(\theta+5)(\theta+4)]^3 + \ldots = 1.2.3.1.3.5,$$

viz. this may be replaced by

$$1\,(\theta+6)^6 - 6\,(\theta+5)^6 + \ldots = 2^3.1.2.3.1.3.5 = 2.4.6.1.3.5 = 1.2.3.4.5.6,$$

which is right.

It is clear that the proof, although worked out on a particular case, is perfectly general; and Mr Woolhouse's theorem is thus proved.

[Vol. XXVI., pp. 77, 78.]

5079. (Proposed by Professor CAYLEY.)—Show that the curve

$$\{(\beta-\gamma i)^2 - \delta^2\}^{\frac{1}{2}}\,\{(x-\beta i)^2 + y^2\}^{\frac{1}{2}} + q\,\{(\beta+\gamma i)^2 - \delta^2\}^{\frac{1}{2}}\,\{(x+\beta i)^2 + y^2\}^{\frac{1}{2}}$$
$$= \left\{(1-q^2)\frac{\beta}{\delta}\right\}^{\frac{1}{2}}\{\beta^2 - (\gamma-\delta i)^2\}^{\frac{1}{2}}\,\{(x-\gamma-\delta i)^2 + y^2\}^{\frac{1}{2}},$$

where $i = (\sqrt{-1})$ as usual, is a real bicircular quartic having the axial foci

$$\beta i,\ -\beta i,\ \gamma+\delta i,\ \gamma-\delta i.$$

Solution by the PROPOSER.

Consider the equation

$$(l+mi)^{\frac{1}{2}}[(x-\beta i)^2+y^2]^{\frac{1}{2}}+q(l-mi)^{\frac{1}{2}}[(x+\beta i)^2+y^2]^{\frac{1}{2}}=(\lambda+\mu i)^{\frac{1}{2}}[(x-\gamma-\delta i)^2+y^2]^{\frac{1}{2}}.$$

This is

$$\begin{aligned}(l+mi)\{x^2+y^2-\beta^2-2\beta xi\}+q^2(l-mi)\{x^2+y^2-\beta^2+2\beta xi\}\\ -(\lambda+\mu i)\{x^2+y^2-\beta^2+\beta^2+\gamma^2-\delta^2-2\gamma x-2(x-\gamma)\delta i\}\\ +2q(l^2+m^2)^{\frac{1}{2}}[(x^2+y^2-\beta^2)^2+4\beta^2x^2]^{\frac{1}{2}}=0,\end{aligned}$$

where, putting the imaginary part equal to zero, we have

$$m(1-q^2)(x^2+y^2-\beta^2)-2l(1-q^2)\beta x-\mu\{x^2+y^2-\beta^2+(\beta^2+\gamma^2-\delta^2)-2\gamma x\}+2\lambda(x-\gamma)\delta=0,$$

which will be true identically if

$$\begin{aligned}m(1-q^2)-\mu&=0,\\ -l(1-q^2)\beta+\mu\gamma+\lambda\delta&=0,\\ -\mu(\beta^2+\gamma^2-\delta^2)-2\lambda\gamma\delta&=0.\end{aligned}$$

The last gives

$$\lambda=\theta(\beta^2+\gamma^2-\delta^2),\quad \mu=-2\theta\gamma\delta,\quad \theta \text{ arbitrary};$$

and then

$$l(1-q^2)\beta=\theta\delta(\beta^2+\gamma^2-\delta^2-2\gamma^2)=\theta\delta(\beta^2-\gamma^2-\delta^2),$$
$$m(1-q^2)=-2\theta\delta\gamma;$$

so that, putting

$$\theta\delta=(1-q^2)\beta,\text{ or }\theta=(1-q^2)\frac{\beta}{\delta},$$

we have

$$l=\beta^2-\gamma^2-\delta^2,\quad m=-2\beta\gamma,$$
$$\lambda=(1-q^2)\frac{\beta}{\delta}(\beta^2+\gamma^2-\delta^2),\quad \mu=(1-q^2)\frac{\beta}{\delta}(-2\gamma\delta).$$

Therefore

$$l\pm mi=(\beta\mp\gamma i)^2-\delta^2,$$
$$\lambda\pm\mu i=(1-q^2)\frac{\beta}{\delta}[\beta^2+(\gamma\mp\delta i)^2];$$

and the equation is

$$\begin{aligned}\{(\beta-\gamma i)^2-\delta^2\}^{\frac{1}{2}}\{(x-\beta i)^2+y^2\}^{\frac{1}{2}}+q\{(\beta+\gamma i)^2-\delta^2\}^{\frac{1}{2}}\{(x+\beta i)^2+y^2\}^{\frac{1}{2}}\\ =\left\{(1-q^2)\frac{\beta}{\delta}\right\}^{\frac{1}{2}}\{\beta^2-(\gamma-\delta i)^2\}^{\frac{1}{2}}\{(x-\gamma-\delta i)^2+y^2\}^{\frac{1}{2}},\end{aligned}$$

which is a real curve having the axial foci $+\beta i$, $-\beta i$; $\gamma+\delta i$; $\gamma-\delta i$; viz. $\gamma+\delta i$ being a focus, and the curve being real, it is clear that $\gamma-\delta i$ is also a focus.

[Vol. XXVII., January to June, 1877, p. 20.]

5130. (Proposed by Professor CAYLEY.)—Show that the envelope of a variable circle, having its centre on a given conic and cutting at right angles a given circle, is a bicircular quartic; which, when the given conic and the circle have double contact, becomes a pair of circles; and, by means of the last-mentioned particular case of the theorem, connect together the porisms arising out of the two problems—

(i) Given two conics, to find a polygon of n sides inscribed in the one and circumscribed about the other.

(ii) Given two circles, to find a closed series of n circles each touching the two circles and the two adjacent circles of the series.

[Vol. XXVII., pp. 81—83.]

5208. (Proposed by Professor SYLVESTER.)—Let the *magnitude* of any ramification signify the number of its branches, and let its partial magnitudes in respect to any node signify the magnitudes of the ramifications which come together at that node. If at any node the largest magnitude exceeds by k the sum of the other magnitudes, let the node be called superior by k, or be said to be of superiority k; but if no magnitude exceeds the sum of the other magnitudes, let the node be called subequal. Then the theorem is, in any ramification, *either* there is one and only one subequal node; *or else* there are two and only two nodes each superior by unity, these two nodes being contiguous.

Solution by PROFESSOR CAYLEY.

The proof consists in showing that (1) there cannot be more than one subequal node; (2) there cannot be more than two nodes each superior by unity: and if there is one such node, then there is, contiguous to it, another such node; (3) starting from a node which is superior by more than unity, there is always a contiguous node which is either of smaller superiority, or else subequal; for, these theorems holding good, we can, by (3), always arrive at a node which is either subequal or else superior by unity; in the former case, by (1), the subequal node thus arrived at is unique; in the latter case, by (2), we have, contiguous to the node arrived at, a second node superior by unity; and we have thus a unique pair of nodes each superior by unity.

I will prove only (3), as it is easy to see that the like process applies to the proof of (1) and (2).

Let the whole magnitude be n; and suppose at a node P which is superior by k, the largest magnitude is α, and that the other magnitudes are, say, β, γ, δ. We have $\alpha=\beta+\gamma+\delta+k$; and since $n=\alpha+\beta+\gamma+\delta$, we have thence $n=2\alpha-k$, or $\alpha=\frac{1}{2}(n+k)$, $\beta+\gamma+\delta=\frac{1}{2}(n-k)$: clearly k is even or odd, according as n is even or odd.

Suppose now that we pass from P, along the branch of magnitude α, to a contiguous node Q; and let the magnitudes for Q be α', β', γ', δ', ϵ', where α' denotes the magnitude for the branch QP. We have $\alpha'=\beta+\gamma+\delta+1$: for the ramification consists of the branch QP and of the ramifications of magnitudes β, γ, δ which meet in P. We have thus

$$\alpha'=\tfrac{1}{2}(n-k)+1=\tfrac{1}{2}n-\tfrac{1}{2}(k-2);$$

and thence

$$\beta'+\gamma'+\delta'+\epsilon'=\tfrac{1}{2}n+\tfrac{1}{2}(k-2).$$

Supposing here that k is greater than 1, viz. that it is $=$ or >2, $k-2$ is 0 or positive; and if α' be the greatest magnitude belonging to the node Q, this is a subequal node. But it may be that α' is not the greatest magnitude; supposing then that the greatest magnitude is β', we have

$$\beta'=\tfrac{1}{2}n+\tfrac{1}{2}(k-2)-\gamma'-\delta'-\epsilon',$$

$$\alpha'+\gamma'+\delta'+\epsilon'=\tfrac{1}{2}n-\tfrac{1}{2}(k-2)+\gamma'+\delta'+\epsilon',$$

and thence

$$\beta'-(\alpha'+\gamma'+\delta'+\epsilon')=k-2-2(\gamma'+\delta'+\epsilon');$$

viz. either the node is subequal, or else, being superior, the superiority is at most $=k-2$; that is, if from the node P, of superiority $=$ or >2, we pass along the branch of greatest magnitude to the contiguous node Q, this is either subequal, or else of superiority less than that of P; which is the foregoing proposition (3).

The subequal node, and the two nodes of superiority 1, in the cases where they respectively exist, may be termed the centre and the bicentre respectively; and the theorem thus is, every ramification has either a centre or else a bicentre. But the centre and the bicentre here considered, due (as remarked by Professor Sylvester) to M. Camille Jordan, and which may for distinction be termed the centre and the bicentre of *magnitude*, are quite distinct from the centre and the bicentre discovered by Professor Sylvester, and considered in my researches upon trees, *British Association Report*, 1875, [610]. These last may for distinction be termed the centre and the bicentre of *distance*: viz. we here consider, not the magnitude, but the length of a ramification, such length being measured by the number of branches to be travelled over in order to reach the most distant terminal node. The ramification has either a centre or else a bicentre of distance: viz. the centre is a node such that, for two or more of the ramifications which proceed from it, the lengths are equal and superior to those of the other ramifications, if any; the bicentre a pair of contiguous nodes such that, disregarding the branch which unites the two nodes, there are from the two nodes respectively (one at least from each of them) two or more ramifications the lengths of which are equal to each other and superior to those of the other ramifications, if any.

It is very noticeable how close the agreement is between the proofs for the existence of the two kinds of centre or bicentre respectively. Say, first as regards distance, if at any node the length of the longest branch exceeds by k the length of the next longest branch or branches, then the node is superior by k, or is of the superiority k; but, if there are two or more longest branches, then the node is subequal. And say next, in regard to magnitude, if at any node the largest magnitude exceeds by k the sum of all the other magnitudes, the node is superior by k, or has a superiority k; but if the largest magnitude does not exceed the sum of the other magnitudes, then the node is subequal. Then, whether we attend to distance or to magnitude, the three propositions hold good: (1) there cannot be more than one subequal node; (2) there cannot be more than two nodes each superior by unity: and if there is one such node, there is contiguous to it another such node; (3) starting from a node which is superior by more than unity, there is always a contiguous node which is of smaller superiority or else subequal; whence, as in the solution just referred to, there is always, as regards distance, a centre or a bicentre; and there is always, as regards magnitude, a centre or a bicentre.

[Vol. XXVII., pp. 89, 90.]

On Mr Artemas Martin's First Question in Probabilities. By Professor Cayley.

The question was, "A says that B says that a certain event took place: required the probability that the event did take place, p_1 and p_2 being A's and B's respective probabilities of speaking the truth."

The solutions, referred to or given on pp. 77—79 of volume XXVII. of the *Reprint*, give the following values for the probability in question:—

Todhunter's *Algebra* .... $p_1p_2 + (1-p_1)(1-p_2)$.

Artemas Martin .... $p_1[p_1p_2 + (1-p_1)(1-p_2)]$.

American Mathematicians and Woolhouse... p_1p_2.

It seems to me that the true answer cannot be expressed in terms of only p_1 and p_2, but that it involves two other constants β and k; and my value is—

Cayley .... $p_1p_2 + \beta(1-p_1)(1-p_2) + k(1-\beta)(1-p_1)$.

In obtaining this I introduce, *but I think of necessity*, elements which Mr Woolhouse calls extraneous and imperfect.

B told A that the event happened, or he did not tell A this; the only evidence is A's statement that B told him that the event happened; and the chances are p_1 and $1-p_1$. But, in the latter case, either B told A that the event did not happen, or he did not tell him at all; the chances (on the supposition of the incorrectness of A's statement) are β and $1-\beta$; and the chances of the three cases are thus p_1, $\beta(1-p_1)$, and $(1-\beta)(1-p_1)$. On the suppositions of the first and second cases respectively, the chances for the event having happened are p_2 and $1-p_2$; on the supposition of the third case (viz. here there is no information as to the event) the chance is k, the antecedent probability; and the whole chance in favour of the event is

$$p_1p_2+\beta(1-p_1)(1-p_2)+k(1-\beta)(1-p_1).$$

If $\beta=1$, we have Todhunter's solution; if $\beta=0$, and also $k=0$, we have the solution preferred by Woolhouse; but we do not (otherwise than by establishing between k and β a relation which is quite arbitrary) obtain Martin's solution. The error in this seems to be as follows:—A says that B told him as to the event, and says further that B told him that the event did happen; the probability of the truth of the compound statement is taken to be $=p_1^2$; whereas, in calling the probability of A's speaking the truth p_1, we mean that if A makes the statement, "B says that the event took place," this is to be regarded as a simple statement, and the probability of the truth of the statement is $=p_1$; viz. I think that Martin introduces into his solution a hypothesis contradictory to the assumptions of the question.

I remark further that in my solution I assume that the event is of such a nature that, when there is *any* testimony in regard to it, the probability is determined by that testimony, irrespectively of the antecedent probability. This is quite consistent with the antecedent probability being, not zero, but as small as we please; so that, if k is (as it may very well be) indefinitely small, the whole probability is the same as if k were $=0$. But there is absolutely no reason for assigning any determinate value to β; so that the solutions $p_1p_2+(1-p_1)(1-p_2)$ and p_1p_2, which assume respectively $\beta=1$ and $\beta=0$, seem to me on this ground erroneous.

[Vol. XXVIII., June to December, 1877, p. 17.]

5306. (Proposed by Professor CAYLEY.)—If α, β, γ, δ; α_1, β_1, γ_1, δ_1, are such that

$$(\alpha_1-\delta_1)(\beta_1-\gamma_1)=(\alpha-\delta)(\beta-\gamma),$$

$$(\beta_1-\delta_1)(\gamma_1-\alpha_1)=(\beta-\delta)(\gamma-\alpha),\quad (\gamma_1-\delta_1)(\alpha_1-\beta_1)=(\gamma-\delta)(\alpha-\beta);$$

show that the three equations

$$\frac{x_1-\alpha_1}{x_1-\delta_1}=\frac{1}{(\beta_1-\delta_1)(\gamma_1-\delta_1)}\{(x-\alpha)^{\frac{1}{2}}(x-\delta)^{\frac{1}{2}}-(x-\beta)^{\frac{1}{2}}(x-\gamma)^{\frac{1}{2}}\}^2,$$

$$\frac{x_1-\beta_1}{x_1-\delta_1}=\frac{1}{(\gamma_1-\delta_1)(\alpha_1-\delta_1)}\{(x-\beta)^{\frac{1}{2}}(x-\delta)^{\frac{1}{2}}-(x-\gamma)^{\frac{1}{2}}(x-\alpha)^{\frac{1}{2}}\}^2,$$

$$\frac{x_1-\gamma_1}{x_1-\delta_1}=\frac{1}{(\alpha_1-\delta_1)(\beta_1-\delta_1)}\{(x-\gamma)^{\frac{1}{2}}(x-\delta)^{\frac{1}{2}}-(x-\alpha)^{\frac{1}{2}}(x-\beta)^{\frac{1}{2}}\}^2,$$

are equivalent to each other; and show also that, consistently with the foregoing relations between the constants, the differences $\alpha_1-\delta_1$, $\beta_1-\delta_1$, $\gamma_1-\delta_1$ may be so determined that the equations in (x, x_1) constitute a particular integral of the differential equation

$$\frac{dx}{\{(x-\alpha)(x-\beta)(x-\gamma)(x-\delta)\}^{\frac{1}{2}}}=\frac{dx_1}{\{(x_1-\alpha_1)(x_1-\beta_1)(x_1-\gamma_1)(x_1-\delta_1)\}^{\frac{1}{2}}}.$$

[Vol. XXIX., January to June, 1878, p. 20.]

4870. (Proposed by Professor CAYLEY.)—Given three conics passing through the same four points; and on the first a point A, on the second a point B, and on the third a point C. It is required to find on the first a point A', on the second a point B', and on the third a point C', such that the intersections of the lines

$A'B'$ and AC, $A'C'$ and AB, lie on the first conic;

$B'C'$ and BA, $B'A'$ and BC, lie on the second conic;

$C'A'$ and CB, $C'B'$ and CA, lie on the third conic.

[Vol. XXIX., pp. 96, 97.]

5625. (Proposed by Professor CAYLEY.)—The equation

$$\{q^2(x+y+z)^2-yz-zx-xy\}^2=4(2q+1)\,xyz\,(x+y+z)$$

represents a trinodal quartic curve having the lines $x=0$, $y=0$, $z=0$, $x+y+z=0$ for its four bitangents; it is required to transform to the coordinates X, Y, Z, where $X=0$, $Y=0$, $Z=0$ represent the sides of the triangle formed by the three nodes.

[Vol. XXXI., January to June, 1879, p. 38.]

5387. (Proposed by Professor CAYLEY.)—Show that a cubic surface has at most 4 conical points, and a quartic surface at most 16 conical points.

[Vol. XXXII., July to December, 1879, p. 35.]

5927. (Proposed by Professor CAYLEY.)—If $\{\alpha+\beta+\gamma+\ldots\}^p$, denote the expansion of $(\alpha+\beta+\gamma+\ldots)^p$, retaining those terms $N\alpha^a\beta^b\gamma^c\delta^d\ldots$ only in which

$$b+c+d+\ldots \not> p-1,\quad c+d+\ldots \not> p-2, \text{ \&c. \&c.};$$

prove that

$$x^n=(x+\alpha)^n-(n)_1\{\alpha\}^1(x+\alpha+\beta)^{n-1}+\frac{n(n-1)}{1.2}\{\alpha+\beta\}^2(x+\alpha+\beta+\gamma)^{n-2}$$

$$-\frac{n(n-1)(n-2)}{1.2.3}\{\alpha+\beta+\gamma\}^3(x+\alpha+\beta+\gamma+\delta)^{n-3}+\text{\&c.}\ldots(1).$$

[Vol. XXXIII., January to July, 1880, p. 17.]

6155. (Proposed by Professor CAYLEY.)—Given, by means of their metrical coordinates, any two lines; it is required to find their inclination, shortest distance, and the coordinates of the line of shortest distance.

N.B.—If λ, μ, ν are the inclinations of a line to three rectangular axes, and α, β, γ the coordinates to the same axes of a point on the line, then the metrical coordinates of the line are

$$\begin{array}{cccccc} a, & b, & c, & f, & g, & h, \\ =\cos\lambda, & \cos\mu, & \cos\nu, & \beta\cos\nu-\gamma\cos\mu, & \gamma\cos\lambda-\alpha\cos\nu, & \alpha\cos\mu-\beta\cos\lambda, \end{array}$$

satisfying identically the relations

$$a^2+b^2+c^2=1, \quad af+bg+ch=0.$$

[Vol. XXXVI., 1881, p. 21.]

6470. (Proposed by Professor CAYLEY.)—It is required, by a real or imaginary linear transformation, to express the equation of a given cubic curve in the form

$$xy-z^2=\{(z^2-x^2)(z^2-k^2x^2)\}^{\frac{1}{2}}.$$

[Vol. XXXVI., p. 64.]

6766. (Proposed by Professor CAYLEY.)—Find the stationary and the double tangents of the curve $x^4+y^4+z^4=0$.

Solution by the PROPOSER.

Take l a fourth root of -1; m and n fourth roots of $+1$; then the 28 double tangents are the lines $x=ly$, $x=lz$, $y=lz$, $(4+4+4=)$ 12 lines; and the lines $x+my+nz=0$, 16 lines; and the first 12 of these, each counted twice, are the 24 stationary tangents. In fact, any one of the 12 lines is an osculating tangent, or line meeting the curve in 4 coincident points; it counts therefore once as a double tangent, and twice as a stationary tangent. There should consequently be 16 other double tangents; and it only needs to be shown that these are the 16 lines $x+my+nz=0$. Consider any one line $x+my+nz=0$; for its intersections with the curve $x^4+y^4+z^4=0$, we have

$$(my+nz)^4+y^4+z^4=0,$$

or, as this may be written,

$$(my+nz)^4+m^4y^4+n^4z^4=0;$$

viz. this is

$$2(1,\ 2,\ 3,\ 2,\ 1\between my,\ nz)^4=0,$$

or, what is the same thing,

$$2[(1,\ 1,\ 1\between my,\ nz)^2]^2=0:$$

so that the line is a double tangent, the two points of contact being given by means of the equation $(1, 1, 1 \between my, nz)^2 = 0$; viz. ω being an imaginary cube root of unity, we have $nz = \omega my$ or $\omega^2 my$: and thence, for the points of contact,

$$x : y : z = 1 : \frac{\omega}{m} : \frac{\omega^2}{n}, \text{ or } = 1 : \frac{\omega^2}{m} : \frac{\omega}{n};$$

values which satisfy, as they should do, the two equations

$$x + my + nz = 0 \text{ and } x^4 + y^4 + z^4 = 0.$$

[Vol. XXXVI., pp. 106, 107.]

6800. (Proposed by W. J. C. MILLER, B.A.)—Prove that, if

$$\frac{ayz}{y^2 + z^2} = \frac{bzx}{z^2 + x^2} = \frac{cxy}{x^2 + y^2} = 1,$$

then

$$a^2 + b^2 + c^2 = abc + 4.$$

Note on Question 6800. *By* PROFESSOR CAYLEY.

The identity given by the solution is a very interesting one. Instead of a, b, c, writing $(a, b, c) \div d$, we have

$$4d^3 - d(a^2 + b^2 + c^2) + abc = 0,$$

satisfied by

$$a : b : c : d = x(y^2 + z^2) : y(z^2 + x^2) : z(x^2 + y^2) : xyz;$$

or, considering (a, b, c, d) as the coordinates of a point in space, and (x, y, z) as the coordinates of a point in a plane, we have thus a correspondence between the points of the cubic surface $4d^3 - d(a^2 + b^2 + c^2) + abc = 0$, and the points of the plane. To a given system of values of (x, y, z) there corresponds, it is clear, a single system of values of (a, b, c, d); and it may be shown without difficulty that, to a given system of values of (a, b, c, d) satisfying the equation of the surface, there correspond two systems of values of (x, y, z); the plane and cubic surface have thus a (1, 2) correspondence with each other.

[Vol. XXXVII., 1882, p. 74.]

5244. (Proposed by Professor CAYLEY.)—Writing for shortness

$$L = \beta^2\gamma^2 - \alpha^2\delta^2, \quad M = \gamma^2\alpha^2 - \beta^2\delta^2, \quad N = \alpha^2\beta^2 - \gamma^2\delta^2,$$

$$F = \alpha^2 + \delta^2 - \beta^2 - \gamma^2, \quad G = \beta^2 + \delta^2 - \gamma^2 - \alpha^2, \quad H = \gamma^2 + \delta^2 - \alpha^2 - \beta^2,$$

$$\Delta = \alpha^2 + \beta^2 + \gamma^2 + \delta^2;$$

show that the equation

$$LMN(x^4+y^4+z^4+w^4)+MN(F\Delta+2L)(y^2z^2+x^2w^2)+NL(G\Delta+2M)(z^2x^2+y^2w^2)$$
$$+LM(H\Delta+2N)(x^2y^2+z^2w^2)-2\alpha\beta\gamma\delta\, FGH\Delta\, xyzw=0$$

belongs to a 16-nodal quartic surface, having the nodes

$x=\alpha$	α	α	α	β	β	β	β	γ	γ	γ	γ	δ	δ	δ	δ
$y=\beta$	$-\beta$	$-\beta$	β	α	$-\alpha$	$-\alpha$	α	δ	$-\delta$	$-\delta$	δ	γ	$-\gamma$	$-\gamma$	γ
$z=\gamma$	$-\gamma$	γ	$-\gamma$	δ	$-\delta$	δ	$-\delta$	α	$-\alpha$	α	$-\alpha$	β	$-\beta$	β	$-\beta$
$w=\delta$	δ	$-\delta$	$-\delta$	γ	γ	$-\gamma$	$-\gamma$	β	β	$-\beta$	$-\beta$	α	α	$-\alpha$	$-\alpha$

and the sixteen singular tangent planes represented by the equations

$$(\alpha, \beta, \gamma, \delta)(x, y, z, w)=0, \text{ \&c.}$$

[Vol. XXXVIII., 1883, pp. 87—89.]

7190. (Proposed by Professor WOLSTENHOLME, M.A.)—If x, y, z be three quantities satisfying the two symmetrical equations

$$yz+zx+xy=0, \quad x^3+y^3+z^3+4xyz=0;$$

prove that (1) they will also satisfy one of the two pairs of semi-symmetrical expressions

$$y^2z+z^2x+x^2y=(y-z)(z-x)(x-y), \ =+xyz,$$
$$yz^2+zx^2+xy^2=(y-z)(z-x)(x-y), \ =-xyz;$$

and (2) one set of the following equations will also be satisfied:—

$$(x^2+yz-y^2=0, \quad y^2+zx-z^2=0, \quad z^2+xy-x^2=0);$$
$$(x^2+yz-z^2=0, \quad z^2+zx-x^2=0, \quad z^2+xy-y^2=0).$$

Solution by PROFESSOR CAYLEY.

The two symmetrical equations represent a conic and a cubic respectively; they intersect therefore in 6 points, and if we denote by α a root of the equation

$$u^3+u^2-2u-1=0,$$

then the other two roots of this equation are

$$\beta, =-1-\frac{1}{\alpha}, \quad \gamma=\frac{-1}{\alpha+1};$$

viz. if $\alpha^3+\alpha^2-2\alpha-1=0$, then we have

$$(u-\alpha)\left(u+1+\frac{1}{\alpha}\right)\left(u+\frac{1}{\alpha+1}\right)=u^3+u^2-2u-1,$$

an identity which is easily verified. It may be remarked that, if

$$\phi\alpha = -1 - \frac{1}{\alpha},$$

then

$$\phi^2\alpha = \frac{-1}{\alpha+1}, \quad \phi^3\alpha = \alpha;$$

the left-hand side of the last mentioned equation thus is $(u-\alpha)(u-\phi\alpha)(u-\phi^2\alpha)$, which remains unaltered when α is changed into $\phi\alpha$ or $\phi^2\alpha$. Then the coordinates of the six points of intersection can be expressed indifferently in terms of any one of the roots (α, β, γ), viz. the coordinates are

$$(\alpha^2-1,\ -\alpha,\ -1),\quad (-1,\ \alpha^2-1,\ -\alpha),\quad (-\alpha,\ -1,\ \alpha^2-1), \ldots (1,\ 2,\ 3),$$
$$(\alpha^2-1,\ -1,\ -\alpha),\quad (-\alpha,\ \alpha^2-1,\ -1),\quad (-1,\ -\alpha,\ \alpha^2-1), \ldots (4,\ 5,\ 6);$$

or they are equal to the like expressions in β and in γ respectively; say these are the coordinates of the points 1, 2, 3, 4, 5, 6 respectively, as shown by the attached numbers. Thus, writing

$$x,\ y,\ z = \alpha^2-1,\ -\alpha,\ -1,$$

we find

$$yz+zx+xy = \alpha-\alpha^2+1-\alpha^3+\alpha = -(\alpha^3+\alpha^2-2\alpha-1) = 0,$$

$$x^3+y^3+z^3+4xyz = (\alpha^2-1)^3-\alpha^3-1+4\alpha(\alpha^2-1)$$
$$= \alpha^6-3\alpha^4+3\alpha^3+3\alpha^2-4\alpha-2 = (\alpha^3+\alpha^2-2\alpha-1)(\alpha^3-\alpha^2+2) = 0,$$

which verifies the formulæ for the six points of intersection. Take, again,

$$x,\ y,\ z = \alpha^2-1,\ -\alpha,\ -1;$$

then we find

$$yz^2+zx^2+xy^2 = -\alpha \ -(\alpha^2-1)^2+\alpha^2(\alpha^2-1) \ = \alpha^2-\alpha-1,$$
$$y^2z+z^2x+x^2y = -\alpha^2+(\alpha^2-1) \ -\alpha\ (\alpha^2-1)^2 = -\alpha^5+2\alpha^3-\alpha-1.$$

Or, since $\alpha^3 = -\alpha^2+2\alpha+1$, and thence

$$\alpha^4 = 3\alpha^2-\alpha-1, \quad \alpha^5 = -4\alpha^2+5\alpha+3,$$

the last equation becomes

$$y^2z+z^2x+x^2y = 2\alpha^2-2\alpha-2.$$

We have also

$$xyz = \alpha^3-\alpha, \ = -\alpha^2+\alpha+1;$$

hence the point in question is situate on each of the cubics

$$yz^2+zx^2+xy^2+xyz = 0, \quad y^2z+z^2x+x^2y+2xyz = 0,$$
$$y^2z+z^2x+x^2y-2(yz^2+zx^2+xy^2) = 0;$$

and this, of course, shows the points 1, 2, 3 are all three of them situate upon each of the three cubics; and in precisely the same manner it appears that the points 4, 5, 6 are all three of them situate on each of the three cubics

$$yz^2+zx^2+xy^2+2xyz = 0, \quad y^2z+z^2x+x^2y+xyz = 0,$$
$$yz^2+zx^2+xy^2-2(y^2z+z^2x+x^2y) = 0.$$

Again, from the values $x, y, z = \alpha^2 - 1, \ -\alpha, \ -1$, we have

$$x^2 + yz - y^2 = 0, \quad y^2 + zx - z^2 = 0, \quad z^2 + xy - x^2 = 0;$$

viz. the point 1 lies on each of these conics; similarly the point 2 lies on each of the same conics; and the point 3 lies on each of the same conics; that is, the conics in question have in common the points 1, 2, 3.

In like manner, the conics

$$x^2 + yz - z^2 = 0, \quad y^2 + zx - x^2 = 0, \quad z^2 + xy - y^2 = 0,$$

have in common the points 4, 5, 6.

The general result is that the given conic and the cubic meet in six points forming two groups of points (1, 2, 3) and (4, 5, 6); through the points (1, 2, 3) we have three cubics and three conics; and through the points (4, 5, 6) we have three cubics and three conics.

If in the equation $x^3 + x^2 - 2x - 1 = 0$, whose roots are α, $\phi(\alpha)$, $\phi^2(\alpha)$, we put $x = 2\cos\theta$, the equation becomes

$$2(3\cos\theta + \cos 3\theta) + 2(1 + \cos 2\theta) - 4\cos\theta - 2 = 0,$$

or

$$2\cos 3\theta + 2\cos 2\theta + 2\cos\theta = 0, \quad \text{or} \quad \frac{\sin\frac{7}{2}\theta}{\sin\frac{1}{2}\theta} = 0;$$

or the three roots are $2\cos\frac{2}{7}\pi$, $2\cos\frac{4}{7}\pi$, $2\cos\frac{6}{7}\pi$. The two equations

$$yz + zx + xy = 0, \quad x^3 + y^3 + z^3 + 3xyz = 0,$$

are satisfied if $x : y : z =$ these three roots in any order, giving the six solutions. The semi-symmetrical systems are satisfied, the one by

$$x : y : z, \ \text{or} \ y : z : x, \ \text{or} \ z : x : y, = \cos\tfrac{2}{7}\pi : \cos\tfrac{4}{7}\pi : \cos\tfrac{6}{7}\pi;$$

and the other by

$$z : y : x, \ \text{or} \ y : x : z, \ \text{or} \ x : z : y, = \cos\tfrac{2}{7}\pi : \cos\tfrac{4}{7}\pi : \cos\tfrac{6}{7}\pi.$$

[Vol. XXXIX., 1883, p. 31.]

5689. (Proposed by Professor CAYLEY.)—Show (1) that the apparent contour of a Steiner's surface $(2xyz + y^2z^2 + z^2x^2 + x^2y^2 = 0)$, as seen from an exterior point on a nodal line (say the axis of z), projected on the plane of the other two nodal lines, is an ellipse passing through the four points $(\pm 1, 0)$ and $(0, \pm 1)$; and (2) find the surface-contour, or curve of contact, of the cone and surface.

[Vol. XXXIX., p. 49.]

4722. (Proposed by Professor CAYLEY.)—1. Show that the conditions in regard to the reality of the roots of the equation

$$(x^2-\alpha)^2+16A(x-m)=0,$$

are, if

$$(4m^2-3\alpha)^3-(8m^3-9m\alpha-27A)^2=-,$$

then the roots are two real, two imaginary; but if

$$(4m^2-3\alpha)^3-(8m^3-9m\alpha-27A)^2=+,$$

then, if simultaneously

$$\alpha=+,\quad A(m\alpha-9A)=+,$$

the roots are all real, but otherwise they are all imaginary.

2. If the roots of the foregoing equation are all imaginary, then for any real value whatever of y, the roots of the equation

$$(x^2+y^2-\alpha)^2+16A(x-m)=0$$

are all imaginary.

[Vol. XXXIX., pp. 69, 70.]

4387. (Proposed by Professor CAYLEY.)—Using the term "Cassinian" to denote a bi-circular quartic having four foci in a right line; show that the equation of a Cassinian having for its four foci the points $x=a$, $x=b$, $x=c$, $x=d$ on the axis of x, may be written in the four equivalent forms

$$\left(\begin{array}{cccc} \cdot\ , & \tau(d-c), & \sigma(b-d), & \rho(c-b) \\ \tau(c-d), & \cdot\ , & \rho(d-a), & \sigma(a-c) \\ \sigma(d-b), & \rho(a-d), & \cdot\ , & \tau(b-a) \\ \rho(b-c), & \sigma(c-a), & \tau(a-b), & \cdot \end{array}\right)(A^{\frac{1}{2}},\ B^{\frac{1}{2}},\ C^{\frac{1}{2}},\ D^{\frac{1}{2}})=0,$$

that is,

$$\cdot\quad \tau(d-c)B^{\frac{1}{2}}+\sigma(b-d)C^{\frac{1}{2}}+\rho(c-b)D^{\frac{1}{2}}=0,$$

$$\tau(c-d)A^{\frac{1}{2}}\quad \cdot\quad +\rho(d-a)C^{\frac{1}{2}}+\sigma(a-c)D^{\frac{1}{2}}=0,$$

&c., &c.,

where $A^{\frac{1}{2}}$, $B^{\frac{1}{2}}$, $C^{\frac{1}{2}}$, $D^{\frac{1}{2}}$ are the distances from the four foci respectively, and the parameters ρ, σ, τ are connected by the equation

$$\rho^2(a-d)(b-c)+\sigma^2(b-d)(c-a)+\tau^2(c-d)(a-b)=0.$$

Show also that the curve has, at right angles to the axis of x, two double tangents, the equation whereof is any one of the three equivalent forms

$$(a+d-2x)(b+c-2x) : (b+d-2x)(c+a-2x) : (c+d-2x)(a+b-2x)=\rho^2 : \sigma^2 : \tau^2.$$

[Vol. XL., 1884, p. 32.]

7376. (Proposed by Professor CAYLEY.)—Show how the construction of a regular heptagon may be made to depend on the trisection of the angle $\cos^{-1}\left(\frac{1}{2\sqrt{7}}\right)$.

[Vol. XL., p. 110.]

7352. (Proposed by Professor CAYLEY.)—Denoting by x, y, z, ξ, η, ζ homogeneous linear functions of four coordinates, such that identically

$$x+y+z+\xi+\eta+\zeta=0,\quad ax+by+cz+f\xi+g\eta+h\zeta=0,$$

where $af=bg=ch=1$; show that

$$\sqrt{(x\xi)}+\sqrt{(y\eta)}+\sqrt{(z\zeta)}=0$$

is the equation of a quartic surface having the sixteen singular tangent planes (each touching it along a conic)

$$x=0,\quad y=0,\quad z=0,\quad \xi=0,\quad \eta=0,\quad \zeta=0,$$

$$x+y+z=0,\quad x+\eta+z=0,\quad ax+by+cz=0,\quad ax+g\eta+cz=0,$$

$$\xi+y+z=0,\quad x+y+\zeta=0,\quad f\xi+by+cz=0,\quad ax+by+h\zeta=0,$$

$$\frac{x}{1-bc}+\frac{y}{1-ca}+\frac{z}{1-ab}=0,\quad \frac{\xi}{1-gh}+\frac{\eta}{1-hf}+\frac{\zeta}{1-fg}=0.$$

[Vol. XLI., 1884, p. 37.]

5421. (Proposed by Professor CAYLEY.)—Suppose

$$S_x=m_1(x-a_1),\ m_2(x-a_2),\ m_3(x-a_3),\ m_4(x-a_4);$$

where, for any given value of x, we write +, −, or 0, according as the linear function is positive, negative, or zero, and where the order of the terms is not attended to. If x is any one of the values a_1, a_2, a_3, a_4, the corresponding S is $0+++$, $0---$, $0++-$, or $0+--$: and if I denote indifferently the first or the second form, and R denote indifferently the third or the fourth form: then it is to be shown that the four S's are R, R, R, R, or else R, R, I, I.

[Vol. XLIV., 1886, p. 109.]

8340. (Proposed by F. MORLEY, B.A.)—Show that (1) on a chess-board the number of squares visible is 204, and the number of rectangles (including squares) visible is 1,296; and (2) on a similar board, with n squares in each side, the number of squares is the sum of the first n square numbers, and the number of rectangles (including squares) is the sum of the first n cube numbers.

Solution by PROFESSOR CAYLEY.

In a board of n^2 squares, the number of pairs of vertical lines at a distance from each other of $n-r+1$ squares is $=r$; and the number of pairs of horizontal

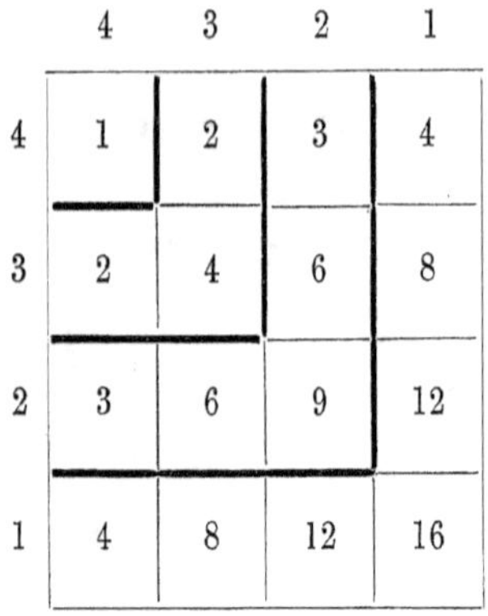

	4	3	2	1
4	1	2	3	4
3	2	4	6	8
2	3	6	9	12
1	4	8	12	16

lines at a distance from each other of $n-s$ squares is $=s$. Hence the number of rectangles, breadth $n-r+1$ and depth $n-s+1$, or say the number of

$$(n-r+1)(n-s+1)$$

rectangles, is $=rs$.

For instance, $n=4$, the number of rectangles 44, 43, 34, &c., is shown in the diagram; hence the whole number of rectangles is $(1+2+3+4)^2 = 1^3+2^3+3^3+4^3$, and so for any value of n.

The same diagram shows that the whole number of squares is $=1^2+2^2+3^2+4^2$; and so for any value of n.

[Vol. XLVI., 1887, pp. 49, 50.]

8636. (Proposed by Professor MAHENDRA NATH RAY, M.A., LL.B.)—Show that the following equations are satisfied by the *same* value of x, and find this value:—

$$ax(x^2-a^2)^{\frac{1}{2}} + bx(x^2-b^2)^{\frac{1}{2}} + cx(x^2-c^2)^{\frac{1}{2}} = 2abc,$$

$$2(x^2-a^2)^{\frac{1}{2}}(x^2-b^2)^{\frac{1}{2}}(x^2-c^2)^{\frac{1}{2}} = x(a^2+b^2+c^2-2x^2).$$

Solution by PROFESSOR CAYLEY.

The second equation rationalised gives

$$4x^6 - 4x^4(a^2+b^2+c^2) + 4x^2(b^2c^2+c^2a^2+a^2b^2) - 4a^2b^2c^2 = 4x^6 - 4x^4(a^2+b^2+c^2) + x^2(a^2+b^2+c^2)^2;$$

that is,

$$\nabla x^2 = 4a^2b^2c^2,$$

if, for shortness,

$$\nabla = -a^4 - b^4 - c^4 + 2b^2c^2 + 2c^2a^2 + 2a^2b^2.$$

We thence find

$$\nabla (x^2 - a^2) = a^2 (-a^2 + b^2 + c^2)^2,$$

$$\nabla (x^2 - b^2) = b^2 (a^2 - b^2 + c^2)^2, \quad \nabla (x^2 - c^2) = c^2 (a^2 + b^2 - c^2)^2,$$

and therefore also

$$\nabla^2 a^2x^2 (x^2 - a^2) = 4a^6b^2c^2 (-a^2 + b^2 + c^2)^2, \text{ \&c.}$$

Or, assuming the sign of the square roots,

$$\nabla ax (x^2 - a^2)^{\frac{1}{2}} = 2abc (-a^4 + a^2b^2 + a^2c^2), \quad \nabla bx (x^2 - b^2)^{\frac{1}{2}} = 2abc (+a^2b^2 - b^4 + b^2c^2),$$

$$\nabla cx (x^2 - c^2)^{\frac{1}{2}} = 2abc (a^2c^2 + b^2c^2 - c^4),$$

whence, adding, the whole divides by ∇ and we have

$$ax (x^2 - a^2)^{\frac{1}{2}} + bx (x^2 - b^2)^{\frac{1}{2}} + cx (x^2 - c^2)^{\frac{1}{2}} = 2abc,$$

the second equation. Observe that the second equation rationalised gives an equation of the form $(x^2, 1)^4 = 0$; the foregoing value $x^2 = 4a^2b^2c^2/\Delta$ is thus one of the four values of x^2.

[Vol. XLVII., 1887, p. 141.]

5271. (Proposed by Professor CAYLEY.)—If ω be an imaginary cube root of unity, show that, if

$$y = \frac{(\omega - \omega^2)\, x + \omega^2 x^3}{1 - \omega^2 (\omega - \omega^2)\, x^2},$$

then

$$\frac{dy}{(1 - y^2)^{\frac{1}{2}} (1 + \omega y^2)^{\frac{1}{2}}} = \frac{(\omega - \omega^2)\, dx}{(1 - x^2)^{\frac{1}{2}} (1 + \omega x^2)^{\frac{1}{2}}};$$

and explain the general theory.

[Vol. L., 1889, p. 189.]

3105. (Proposed by Professor CAYLEY.)—The following singular problem of literal partitions arises out of the geometrical theory given in Professor Cremona's Memoir, "Sulle trasformazioni geometriche delle figure piane," *Mem. di Bologna*, tom. V. (1865). It is best explained by an example:—A number is made up in any manner with the parts 2, 5, 8, 11, &c., viz. the parts are always the positive integers $\equiv 2 \pmod{3}$; for instance, $27 = 1.11 + 8.2$. Forming, then, the product of 27 factors $a^{11} (bcdefghi)^2$, this may be partitioned on the same type $1.11 + 8.2$ as follows,

$$a^3bcdefghi, \; ab, \; ac, \; ad, \; ae, \; af, \; ag, \; ah, \; ai.$$

(Observe that the partitionment is to be symmetrical as regards the letters which have a common index.) But, to take another example,

$$37 = 0.11 + 3.8 + 1.5 + 4.2 = 1.11 + 0.8 + 4.5 + 3.2.$$

The first of these gives the product $(abc)^8 d^5 (efgh)^2$, which cannot be partitioned (symmetrically as above) on its own type, though it may be on the second type; and the second gives the product $a^{11} (bcde)^5 (fgh)^2$, which cannot be partitioned (symmetrically as above) on its own type, though it may be on the first type; viz. the partitions of the two products respectively are:

First product on second type,

$$(abc)^2 defgh,\quad abcde,\quad abcdf,\quad abcdg,\quad abcdh,\quad ab,\quad ac,\quad bc;$$

Second product on first type,

$$a^2bcdefg,\quad a^2bcdefh,\quad a^2bcdegh,\quad abcde,\quad ab,\quad ac,\quad ad,\quad ae;$$

so that in the first example the type is sibi-reciprocal, but in the second example there are two conjugate types. Other examples are:

Parts	48	54	55	56	53		55		No.
2	14	3	1	0	3	6	0	2	Reciprocals.
5	0	2	3	0	6	0	5	0	
8	0	3	2	7	0	1	2	5	
11	0	0	2	0	0	3	0	1	
14	0	1	0	0	0	0	1	0	
17	0	0	0	0	1	0	0	0	
20	1	0	0	0	0	0	0	0	

viz. the first four columns give each of them a sibi-reciprocal type, but the last two double columns give conjugate types. It is required to investigate the general solution.

[Vol. L., p. 191.]

3304. (Proposed by Professor CAYLEY.)—The coordinates x, y, z being proportional to the perpendicular distances from the sides of an equilateral triangle, it is required to trace the curve

$$(y-z)\sqrt{x}+(z-x)\sqrt{y}+(x-y)\sqrt{z}=0.$$

[Prof. Cayley remarks that the curve in question is a particular case of that which presents itself in the following theorem, communicated to him (with a demonstration) several years ago by Mr J. Griffiths:—

The locus of a point (x, y, z) such that its pedal circle (that is, the circle which passes through the feet of the perpendiculars drawn from the point in question upon the sides of the triangle of reference) touches the nine-point circle, is the sextic curve

$$\left\{x\cos A\,(y\cos B - z\cos C)\left(\frac{y}{\cos B} - \frac{z}{\cos C}\right)\right\}^{\frac{1}{3}}$$
$$+\left\{y\cos B\,(z\cos C - x\cos A)\left(\frac{z}{\cos C} - \frac{x}{\cos A}\right)\right\}^{\frac{1}{3}}$$
$$+\left\{z\cos C\,(x\cos A - y\cos B)\left(\frac{x}{\cos A} - \frac{y}{\cos B}\right)\right\}^{\frac{1}{3}} = 0.$$

It would be an interesting problem to trace this more general curve.]

[Vol. L., p. 192.]

3481. (Proposed by Professor CAYLEY.)—Find, in the Hamiltonian form

$$\frac{d\eta}{dt} = \frac{dH}{d\varpi}, \quad \frac{d\varpi}{dt} = -\frac{dH}{d\eta}, \text{ \&c.,}$$

the equations for the motion of a particle acted on by a central force.

[Vol. LV., 1891, p. 27.]

10716. (Proposed by Professor CAYLEY.)—In a hexahedron $ABCDA'B'C'D'$ the plane faces of which are $ABCD$, $A'B'C'D'$, $A'ADD'$, $D'DCC'$, $C'CBB'$, $B'BAA'$, the edges AA', BB', CC', DD' intersect in four points, say AA', DD' in α; BB', CC' in β; CC', DD' in γ; AA', BB' in δ: that is, starting with the duad of lines $\alpha\beta$, $\gamma\delta$, the four edges AA', BB', CC', DD' are the lines $\alpha\delta$, $\beta\delta$, $\beta\gamma$, $\alpha\gamma$ which join the extremities of these duads. Similarly, the four edges AB, CD, $A'B'$, $C'D'$ are the lines joining the extremities of a duad; and the four edges AD, BC, $A'D'$, $B'C'$ are the lines joining the extremities of a duad. The question arises, "Given two duads, is it possible to place them in space so that the two tetrads of joining lines may be eight of the twelve edges of a hexahedron?" The duad $\alpha\beta$, $\gamma\delta$ is considered to be given when there is given the tetrahedron $\alpha\beta\gamma\delta$, which determines the relative position of the two finite lines $\alpha\beta$ and $\gamma\delta$.

[Vol. LXI., 1894, pp. 122, 123.]

3162. (Proposed by Professor CAYLEY.)—By a proper determination of the co-ordinates, the skew cubic through any six given points may be taken to be

$$x : y : z = y : z : w;$$

or, what is the same thing, the coordinates of the six given points may be taken to be

$$(1,\ t_1,\ t_1^2,\ t_1^3),\ \ldots,(1,\ t_6,\ t_6^2,\ t_6^3).$$

Assuming this, it is required to show that if

$$p_1=\Sigma t_1,\ p_2=\Sigma t_1t_2,\ \ldots,\ p_6=t_1t_2t_3t_4t_5t_6,$$

and if

$$\nabla = 6xyzw-4xz^3-4y^3w+3y^2z^2-x^2w^2;$$

then the equation of the Jacobian surface of the six points is

$$\left.\begin{array}{l} 3(\ xp_3 \qquad\quad + \ zp_1-2w\ )\,\delta_x\nabla \\ +\ (\qquad\qquad\qquad 2zp_2-wp_1)\,\delta_y\nabla \\ +\ (\ xp_5-2yp_4 \qquad\qquad\quad)\,\delta_z\nabla \\ +\ (2xp_6-\ yp_5 \qquad\quad -wp_3)\,\delta_w\nabla \end{array}\right\}=0.$$

[Vol. LXI., p. 123.]

3185. (Proposed by Professor CAYLEY.)—An unclosed polygon of $(m+1)$ vertices is constructed as follows: viz. the abscissæ of the several vertices are 0, 1, 2, ..., m, and corresponding to the abscissa k, the ordinate is equal to the chance of $m+k$ heads in $2m$ tosses of a coin; and m then continually increases up to any very large value; what information in regard to the successive polygons, and to the areas of any portions thereof, is afforded by the general results of the Theory of Probabilities?

[Vol. LXI., p. 124.]

3229. (Proposed by Professor CAYLEY.)—It is required to find the value of the elliptic integral $F(c,\ \theta)$ when c is very nearly $=1$ and θ very nearly $=\frac{1}{2}\pi$; that is, the value of

$$\int_0^{\frac{1}{2}\pi-a}\frac{d\theta}{\{1-(1-b^2)\sin^2\theta\}^{\frac{1}{2}}},$$

where a, b are each of them indefinitely small.

N.B.—Observe that, for $a=0$, b small, the value is equal $\log 4/b$, and for $b=0$, a small, the value is $-\log\cot\frac{1}{2}a$.

In the following Contents, the Problems are referred to, each by its number and the proposer's name; and the subject is briefly indicated. An asterisk shews that no solution was given. A line —— shews that there is no number.

END OF VOL. X.

CAMBRIDGE: PRINTED BY J. AND C. F. CLAY, AT THE UNIVERSITY PRESS.

www.ingramcontent.com/pod-product-compliance
Lightning Source LLC
LaVergne TN
LVHW011244110826
845149LV00001B/46